Food Engineering Series

AF773820

For further volumes:
http://www.springer.com/series/5996

Joyce I. Boye · Yves Arcand
Editors

Green Technologies in Food Production and Processing

Springer

Editors
Joyce I. Boye
Food Research and Development Centre
Agriculture and Agri-Food Canada
Saint Hyacinthe, QC, Canada

Yves Arcand
Food Research and Development Centre
Agriculture and Agri-Food Canada
Saint Hyacinthe, QC, Canada

ISBN 978-1-4899-9207-9 ISBN 978-1-4614-1587-9 (eBook)
DOI 10.1007/978-1-4614-1587-9
Springer New York Dordrecht Heidelberg London

Printed on acid-free paper

Springer is part of Springer Science+Business Media (www.springer.com)

Protecting the sustainability of our resources is the only way to ensure access to food for the present as well as for future generations.

Anonymous, 2011

Preface

As we move into the future, decisions on food purchases will increasingly be influenced not only by price and quality, but also by social and environmental factors such as the sustainability of technologies used for food production and processing and their environmental and health impacts. Growing consumer awareness about the impact of processing and production practices on the environment, the high energy consumption of certain processes, health impacts of some of the technologies used in processing, and a heightened social and industrial consciousness to reduce the carbon-footprint are examples of factors influencing food choice. These factors have been made quite evident in the growing "buy-local," "fair-trade," "certified organic" trends. As a result, farmers and food manufacturers will increasingly be interested in identifying and using greener economically viable technologies for food production and processing. Some producers are already responding with the use of organic inputs in processing, use of recyclable and good-for-the-environment packaging, establishing just employer–employee relationships, and reducing animal testing in product development.

There is little doubt as we move into the future that nutrition will continue to be one of most fundamental needs of human beings. The food chain has evolved tremendously from a "do it yourself" mindset before the Industrial Revolution, to a fully developed industrial sector in which more than 90% of the current population do not know where their food comes from or how it is made. Since the Industrial Revolution, worldwide food production has increased significantly but at a slower rate than global population. As an example, in the year 2000, of a worldwide population of nearly 6.5 billion people, 70% were sufficiently fed, 15% were malnourished, and the remaining 15% were severely malnourished. At the same time most land and aquatic resources were overused.

The challenge as we move toward the year 2050 will be how to feed the projected nine billion global population who are likely to have a higher standard of living and are therefore likely to eat more, including eating more meat. To do this successfully will require that we produce 30% more food with much less impact on our environment.

Some of the most serious environmental concerns we face include, land use change and tremendous reduction of biodiversity (>10× the sustainable rate), aquatic eutrophication by nitrogenous substances caused by over-fertilization (3.5×), global warming caused by enteric fermentation and use of fossil fuels (1.5×), aquatic eutrophication by phosphorus substances caused by fertilizer overuse (1.0×), water shortages owing to irrigation (0.7×), ecotoxicity and human effects of pesticides (0.5×), and so on.

But there is hope.

First, we are now aware of the problem. We now know that about 30–50% of food produced is wasted. Tools such as life cycle analysis (LCA) are currently being developed to provide a good global picture of the current situation and the main contributors to the present challenges. Initiatives are emerging from different parts of the world to identify more sustainable approaches to food production, transformation, and distribution. It is quite evident that one of the most important challenges of the next 40 years will be to find a balance between food supply and demand in a manner that is sustainable and that ensures the long-term survival of humankind.

Green Technologies in Food Production and Processing provides a comprehensive review of the current status of the agriculture and agri-food sectors in regard to environmental sustainability and material and energy stewardship and provides strategies that can be used by industries to enhance the use of environmentally friendly technologies for food production and processing.

In the first section, the book provides a global look at today's supplies, demands, and key players of the food supply chain. Key drivers of the evolution in the food supply chain are described based on past, present and future trends. Part II provides an in-depth description of food production and processing using the life cycle assessment (LCA) tool. Chapter 3 starts with a description of LCA as applied to the food sector followed by chapters describing in more details crop production, animal production, and food processing. Part III of the book provides information on approaches to use to improve food production practices (e.g., optimizing the management of nutrient cycles, analyzing benefits of organic farming, and quantifying transportation impacts, etc.). In Part IV, information on more sustainable food processing approaches are provided, such as greener separation technologies (supercritical and subcritical fluid extraction, aqueous extraction), enzyme assisted food processing, electrotechnologies, greener technologies for microbial control, reduced energy unit operations and low environmental impact packaging. Part V further provides an in-depth look at some emerging analytical techniques for research and development that reduce solvent, chemical, and energy use. In the last section of the book, a critical analysis of some of the challenges associated with the use of agricultural resources to grow biofuels and bio-based products are addressed. In addition, technologies to reduce the generation of process-induced toxins are also reviewed. Furthermore, social factors that influence consumer perceptions about some of the current and emerging agri-food technologies, and the need and importance of biodiversity in maintaining sustainable diets of human populations are discussed in detail.

We wish to thank all the contributors to this book for the stellar job in compiling the most comprehensive information on *Green Technologies in Food Production and Processing* to date. We trust that this work will be a significant contribution to the agriculture and agri-food sectors and of use to policy makers, the food industry, and the general public.

Although much effort was made to avoid duplication of the information presented, as often happens with books of this nature, some minor duplication is unavoidable. We have endeavored to reduce this as much as possible. Additionally, we have made every effort to avoid errors. However, there is always a possibility that some oversight may have occurred. We take full responsibility for any errors and would kindly ask that we be informed of any such errors so corrections can be made in any future editions.

Finally, we would like to give special acknowledgment to Danielle Palardy at the Food Research and Development Center of Agriculture and Agri-food Canada, St. Hyacinthe, Quebec, Canada for the monumental secretarial help in the preparation of the book. Special thanks also go to Pierre Di Campo and Hélène Simard-Vermette, librarians at the same location, and to graduate students Zhen Ma and Xin Rui for their wonderful assistance. And, last but not the least, to all our readers, we thank you and hope that you enjoy reading this book.

Saint Hyacinthe, QC, Canada

Joyce I. Boye
Yves Arcand

Contents

Contributors

Humayoun Akhtar Guelph Food Research Centre, Agriculture and Agri-Food Canada, Guelph, ON, Canada

Yves Arcand Food Research and Development Centre, Agriculture and Agri-Food Canada, Saint Hyacinthe, QC, Canada

René Audet Department of Strategy and Social and Environmental Responsibility, Management Sciences School, Quebec University at Montreal, Montreal, QC, Canada

Eric J. Beckman Chemical Engineering Department, University of Pittsburgh, Pittsburgh, PA, USA

Steve Bittner Food Research and Development Centre, Agriculture and Agri-Food Canada, Saint Hyacinthe, QC, Canada

Joyce I. Boye Food Research and Development Centre, Agriculture and Agri-Food Canada, Saint Hyacinthe, QC, Canada

Frank Brentrup Yara International, Research Centre Hanninghof, Dülmen, Germany

Barbara Burlingame Nutrition and Consumer Protection Division, Food and Agriculture Organization of the United Nations, Rome, Italy

U. Ruth Charrondiere Nutrition and Consumer Protection Division, Food and Agriculture Organization of the United Nations, Rome, Italy

Susan Cholette San Francisco State University, San Francisco, CA, USA

Raffaele Coppola Istituto di Scienze dell'Alimentazione, Avellino, Italy

Sandro Dernini Nutrition and Consumer Protection Division, Food and Agriculture Organization of the United Nations, Rome, Italy

Raymond L. Desjardins Agriculture & Agri-Food Canada, Ottawa, Canada

James A. Dyer Agro-environmental Consultant, Cambridge, ON, Canada

Florinda Fratianni Istituto di Scienze dell'Alimentazione, Avellino, Italy

Nadia Gagnon Agriculture and Agri-Food Canada (AAFC), Research and Analysis Directorate, Ottawa, ON, Canada

Andreas Gattinger Research Institute of Organic Agriculture (FiBL), Frick, Switzerland

Elisabeth Gauthier Agriculture and Agri-Food Canada, Food Research and Development Centre, Saint Hyacinthe, QC, Canada

Corinne Gendron Department of Strategy and Social and Environmental Responsibility, Management Sciences School, Quebec University at Montreal, Montreal, QC, Canada

Stefan Grabowski Food Research and Development Centre, Agriculture and Agri-Food Canada, Saint Hyacinthe, QC, Canada

John E. Hermansen Department of Agroecology, Aarhus University, Research Centre Foulum, Tjele, Denmark

Denis Ippersiel Agriculture and Agri-Food Canada, Food Research and Development Centre, Saint-Hyacinthe, QC, Canada

Yueming Jiang South China Botanical Garden, Chinese Academy of Sciences, Guangzhou, China

Lamin Kassama Department of Food and Animal Sciences, Alabama A&M University, Normal, AL, USA

François Lamarche Agriculture and Agri-Food Canada, Food Research and Development Centre, Saint-Hyacinthe, QC, Canada

Mohsin Bin Latheef Department of Bioresource Engineering, McGill University, Montreal, Québec, Canada

Ying Ma College of Food Science and Engineering, Harbin Institute of Technology, Harbin, China

Anne Maltais Food Research and Development Centre, Agriculture and Agri-Food Canada, Saint Hyacinthe, QC, Canada

Dominique Maxime Département de génie chimique, Interuniversity Research Centre for the Life Cycle of Products, Processes and Services, International Chair in Life Cycle Assessment, École Polytechnique de Montréal, Montréal, QC, Canada

Brian G. McConkey Agriculture & Agri-Food Canada, Swift Current, Canada

Martin Mondor Agriculture and Agri-Food Canada, Food Research and Development Centre, Saint-Hyacinthe, QC, Canada

Stefano Mondovì Nutrition and Consumer Protection Division, Food and Agriculture Organization of the United Nations, Rome, Italy

Filomena Nazzaro Istituto di Scienze dell'Alimentazione, Avellino, Italy

Michael O. Ngadi Department of Bioresource Engineering, McGill University, Montreal, Québec, Canada

Thu Lan T. Nguyen Department of Agroecology, Aarhus University, Research Centre Foulum, Tjele, Denmark

Mariana Cristina Rufino Plant Production Systems, Department of Plant Sciences, Wageningen University, Wageningen, The Netherlands

Sustainable Livestock Futures, International Livestock Research Institute, Nairobi, Kenya

Xin Rui Bioresource Engineering Department, McGill University (Macdonald Campus), QC, Canada

Teemu Santonen Laurea University of Applied Sciences, Espoo, Finland

Christian Schader Research Institute of Organic Agriculture (FiBL), Frick, Switzerland

Rita Schenck Institute for Environmental Research and Education, Tacoma, WA, USA

Susan E.M. Selke School of Packaging, Michigan State University, East Lansing, MI, USA

John Shi Guelph Food Research Center, Agriculture and Agri-Food Canada, Guelph, ON, Canada

Benjamin K. Simpson Food Science and Agricultural Chemistry Department, McGill University (Macdonald Campus), QC, Canada

Barbara Stadlmayr Nutrition and Consumer Protection Division, Food and Agriculture Organization of the United Nations, Rome, Italy

Matthias Stolze Research Institute of Organic Agriculture (FiBL), Frick, Switzerland

Jonathan Vayssières Livestock Systems Research Unit, Centre de Coopération Internationale en Recherche Agronomique pour le Développement (CIRAD), La Réunion, France

Kumar Venkat CleanMetrics Corporation, Portland, OR, USA

Xavier P.C. Vergé Agriculture & Agri-Food Canada, Ottawa, Canada

Wayne Wakeland Portland State University, Portland, OR, USA

Devon E. Worth Agriculture & Agri-Food Canada, Ottawa, Canada

Jiang XiuJie Hei Long Jiang Ba Yi Agricultural University College of Food Science, Da Qing, PR, China

Sophia Jun Xue Guelph Food Research Center, Agriculture and Agri-Food Canada, Guelph, ON, Canada

Xingqian Ye School of Biosystems Engineering and Food Science, Zhejiang University, Hangzhou, China

Dianyu Yu College of Food Science and Engineering, Northeast Agricultural University, Harbin, China

Reza Zareifard Food Research and Development Centre, Agriculture and Agri-Food Canada, Saint Hyacinthe, QC, Canada

Part I
The food chain

Chapter 1
Introduction to the global agri-food system

Nadia Gagnon

1.1 Introduction

The global agri-food system has evolved rapidly over the past few decades. Many influencing factors, particularly globalization and trade liberalization, which came about in the early 1990s with the signing of the Uruguay Round trade negotiations and the World Trade Organization's (WTO) multilateral trade agreement have contributed to these changes. The WTO agreement led to a substantial decrease in many tariffs and barriers to trade that had inhibited trade in agricultural products before that time. Since 1993, therefore, trade in agriculture products has grown immensely. World agriculture and agri-food exports alone rose from $US280 billion in 1999 to $US920 billion in 2009. Similar products sourced from various countries around the world can now be found in grocery stores everywhere from China to Brazil to North America.

At the same time, the development of new technologies, new business models, and evolving consumer demands have affected the global agri-food system. Since the early 1990s, such significant innovations as biotechnology, no ill conservation tillage, Geographical Positioning Systems (GPS), and the Internet have transformed agriculture and food production across the globe. Increasingly wealthy consumers around the world are demanding more from their food, including more health benefits, ethical concerns, humane treatment of animals, fair trade, and environmental benefits such as low carbon, water, and energy use. The food industry is responding by developing private standards that can ensure these consumer demands are met by differentiating their products. This requires increased global supply chain integration.

N. Gagnon (✉)
Agriculture and Agri-Food Canada (AAFC), Research and Analysis Directorate,
Ottawa, ON, Canada
e-mail: nadia.gagnon@agr.gc.ca

J.I. Boye and Y. Arcand (eds.), *Green Technologies in Food Production and Processing*,
Food Engineering Series, DOI 10.1007/978-1-4614-1587-9_1,

This chapter presents a broad overview of the global agri-food system or supply chain. It begins with a context in which the global agri-food system is evolving, particularly related to consumer demands and world agricultural production. It then goes on to describe what a "supply chain" is, and explores the structure of each component of the supply chain or system as well as the principal actors within the chain and their contribution to the economy. The various components of the agri-food system or supply chain are described, from input industries to the players involved in selling and delivering food to consumers with regard to their size, importance, interaction, and contribution to the "greening" of the supply chain.

1.2 Global context related to consumer demands

Evolving consumer demands are a vital factor that has led to a significant change in the global agri-food system. Traditionally, food price and consumer income have been widely recognized as playing a key role in influencing the purchasing decisions of food consumers. However, there are increasingly other considerations that influence consumer choice of one food product over another. For instance, consumers are increasingly paying more attention to food safety concerns. In recent years, a number of food-related public health and environmental incidents has raised consumer interest in agricultural production methods. Bovine spongiform encephalopathy (BSE) (mad cow disease), avian influenza, and genetically modified (GM) foods have all led to consumers becoming quite interested in the safety of their food. Several cases of BSE were reported in North America and Europe, leading to trade embargoes, animal destruction, and a complete breakdown in trust in the food safety systems of the United Kingdom (WOAH 2010). The outbreak of avian influenza in poultry across Asia and Eastern Europe in 2003–2004 caused serious social and economic losses, because birds had to be destroyed and several human deaths were attributed to the disease. In Vietnam alone, where poultry production was down by around 15%, economic losses resulting from the avian flu were estimated at about 0.1% of gross domestic product (GDP) (World Bank 2005). Additional losses occurred because of lower egg production and reduced activity in distribution channels.

In addition, the development of biotechnology and the resulting consumer resistance to GM crops, the result of a revolution in technology that recombines genetic material across different species, makes news headlines every now and then with significant negative repercussions for producers and markets. In Europe in particular there is significant consumer resistance to these GM products, and exports from North America, where some GM crops are produced, are routinely scrutinized. Recently the agro-chemical producer BASF, with its headquarters in Germany, has been in the spotlight after seeds from a new strain of GM potato were found in a Swedish field (BASF 2010). Similarly, a genetically modified strain of flax was found in a shipment of flax from Canada in Germany. Such episodes have raised consumer concerns about the characteristics of food to which the agri-food system must respond.

In addition to food safety and genetically modified food products, consumers are increasingly interested in ethical considerations when purchasing food. Ethical consumption is driven by consumer preferences related to broader societal goals that go beyond their immediate need for safe and nutritious food products. For instance, environmental concerns related to ecological footprints (carbon, water etc.), fair and equitable labour market outcomes (fair trade), animal welfare (free-range rearing and humane methods of slaughtering), poverty, and underdevelopment in the Less Developed World are increasingly important and are affecting food purchasing decisions. Growing interest in local food and cooperative distribution networks has also become important. Despite the fact that the largest food retailers in industrialized countries continue to capture the largest market share, local food demand is being influenced by environmental concerns as well as consumer interest in supporting local farmers.

1.3 Characteristics of the global agri-food system

Some of the main characteristics and dynamics of the global agri-food system are explored in the following, with a focus on the various activities performed by different players in the system, including inputs and services suppliers, agricultural production, food processing, retailing, and consumption practices. In addition a variety of technological, environmental, health and socioeconomic issues related to these activities are explored. A typology is presented in Table 1.1, which characterizes agri-food systems or supply chains in different parts of the world. The first type is a traditional food system with traditional, unorganized supply chains and limited market infrastructure. The second type of food system is considered a modern well-structured agricultural system, with traditional actors, subject to more rules and regulations and a more efficient market infrastructure. The third type is considered the industrialized food system found in most developed countries. It is characterized by a high degree of coordination among actors, a highly consolidated processing and

Table 1.1 Main characteristics of the three types of global agri-food systems

Food system characteristics	Traditional agriculture	Modernizing agriculture	Industrialized agriculture
Share of GDP from agriculture	>30%	10–30%	<10%
Share of labor force in agriculture	>50%	15–50%	<15%
Market orientation	Subsistence	National	International
Output mix	Food staples	Food staples and high value crops	Highly differentiated
Scale economies	Not important	May be important	Important
Examples	Kenya, Uganda	Most countries in Asia, Mexico, Eastern Europe	Canada, US, EU, NZ

Source: McCullough et al. (2008)

Table 1.2 Comparison of linkages in agri-food supply chain across agri-food systems typology

Constituents of the food supply chain	Traditional agriculture	Modernizing agriculture	Industrialized agriculture
Inputs intensity	Low input use	High level of use	Enhanced input use efficiency
Primary agriculture	Diversified	Specialization of cropping systems	Specialization and focus on conservation
Processing sector	Very limited	Processed products are seen as value-added and provide employment	Large processing sectors for domestic and export markets
Wholesalers	Traditional wholesalers with retailers bypassing for exports	Traditional and specialized wholesalers	Specialized wholesalers and distribution centers
Retailers	Small market	Spread of supermarkets, less penetration for fruits and vegetables	Widespread supermarkets
Consumers	Rising caloric intake	Diet diversification, switch to processed foods	High value, processed foods
Traceability	No traceability	In some chains with private standards	HACCP programs

Source: McCullough et al. (2008)

retailing sector, and organized retailers. The main driving forces that have shaped this last food system include high income, trade liberalization and foreign direct investment, a high level of technology use in chain logistics such as transportation and food handling, and demographic shifts related to urbanization (McCullough et al. 2008).

Given the importance of labor and other resources engaged in low–value-generating subsistence agricultural production, productivity growth in these countries characterized by traditional agriculture is very low. The scope for enhancing vertical coordination among the different players is very limited given the lack of appropriate market conditions and infrastructure. Countries associated with a "modernizing agriculture and food system" are still characterized by subsistence agriculture but have some sectors that have become export oriented and industrialized. At least a considerable share of producers have adopted better and improved technologies in their operations. However, more improvements in the organizational aspects and infrastructure are needed for such systems to become competitive suppliers of high value agri-food products. For industrialized agriculture, regulations and institutions facilitate investment in the sector (Globerman and Shapiro 2002) and such investments can be recouped through enhanced productivity and economic growth in the sector. However, because resources of such industrialized economies are engaged in more profitable economic activities, such as service provision and manufacturing, the share of agricultural GDP and labor is less than 10% and 15%, respectively. Table 1.2 provides a snapshot of the roles of the main players across the supply chain and their key differences.

One immediately notices that the scope of adopting greener technologies are conditioned by some important differences among the three agri-food systems. For instance, for green technologies to be successfully adopted in the traditional agricultural system, they need to be scale neutral, whereas in industrialized agricultural systems, such technologies may have scale biases. In contrast, if these green technologies are labor intensive, there is a better prospect for their adoption in the traditional food system. Coordination of agri-food chain partners for countries having industrialized agriculture is very important because the volume of production or sales are much more important compared with traditional agriculture. Also, fewer individuals are required to manage the flow of products in industrialized agriculture. Modernized agriculture is considered an emerging economy having many opportunities for gains in efficiency for the agricultural production itself and the flow of food products throughout the supply chain.

1.4 Global agriculture

At the global level, the primary agriculture sector is more important in developing countries as a source of livelihood for a sizable portion of the population. However, as countries develop and the manufacturing and services sectors become more important, the significance of the primary agricultural sector in the economy tends to shrink. The relative significance of the primary agriculture sector has declined in some of the new major economies. The BRIC (Brazil, Russia, India, and China) group of countries is a prime example of how agriculture becomes less important as countries industrialize (Table 1.3).

Between 2000 and 2008, GDP in the BRIC group of countries increased by an average of 7% per annum, well above the rates for most industrialized countries. Sub-Saharan Africa also grew at a comparable, rate and projections clearly show the continuous growth for these regions. Estimates of global GDP growth rates from the Food and Agriculture Organization of the United Nations (FAO) report for 2030–2050 give a similar picture, with constant world GDP growth and an emphasis on the countries in South Asia and East Asia as sources of high GDP growth. Since 1970, total agricultural output has grown at a faster pace than has world population, as shown in Fig. 1.1.

Table 1.3 Annual percent change in Gross Domestic Product (GDP) by region 2000–2014 (constant prices)

	2000	2004	2008	2009[a]	2010[a]	2014[a]
World	4.8	4.9	3.0	−1.1	3.1	4.5
Major advanced economies (G7)	3.8	2.9	0.3	−3.6	1.3	2.1
Newly industrialized Asian economies	7.7	5.9	1.5	−2.4	3.6	4.6
European Union	4.0	2.7	1.0	−4.2	0.5	2.1
Africa: sub-Sahara	3.7	7.1	5.5	1.3	4.1	5.4
Brazil, Russia, China and India	7.1	7.7	6.8	1.4	5.1	6.6

Source: International Monetary Fund (2010)
[a]IMF estimates

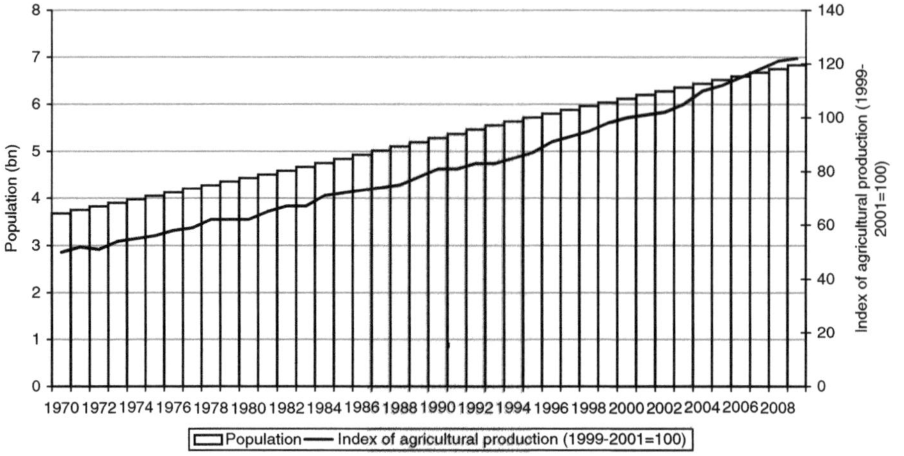

Fig. 1.1 World agricultural production and population, 1970–2009

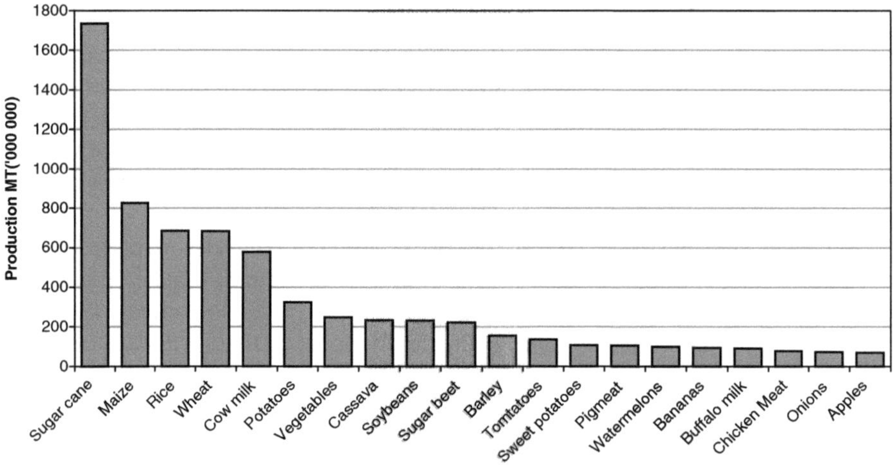

Fig. 1.2 Production of the top 20 world agricultural commodities, 2008 (Source: FAOSTAT 2010)

According to the FAO, total agricultural production has continuously increased since 1970. At the global level, agricultural production will have to continue to increase to feed a growing population which will also be wealthier. Further increases in agricultural productivity for the future may be required to achieve this objective (Cabinet Office 2008). Figure 1.2 shows the production of the top world agricultural commodities in terms of tonnage for 2008.

The main crop produced in 2008 was sugar cane (1,736 gigatons) followed by maize, rice, wheat, and cow milk in 2008. The major source of the world's sugar production is from sugar cane in tropical zones located in the Southern Hemisphere. The leading sugar cane producers are Brazil, India, China, Thailand, and Pakistan.

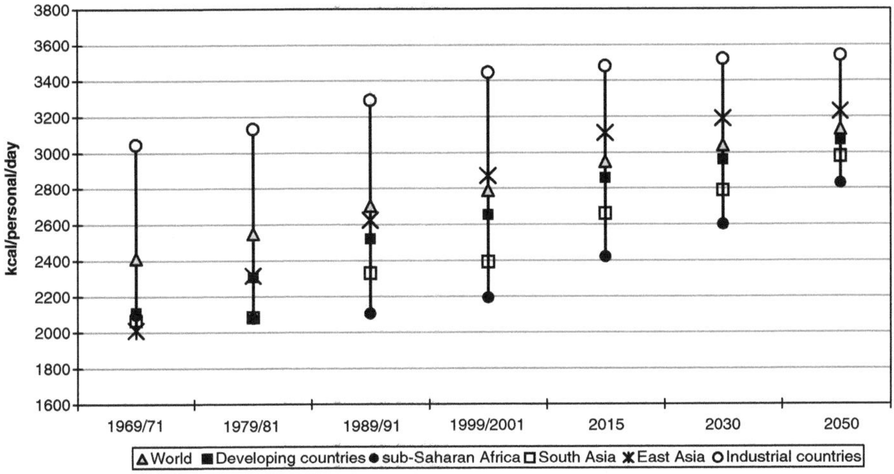

Fig. 1.3 Per capita food consumption, 1969–2050 (Source: FAO, World Agriculture Towards 2030/2050 (2006))

Another source of sugar production is sugar beets grown in temperate zones of the Northern Hemisphere: France, Russia, the United States, Germany, and Turkey. A variety of vegetables, fruits, and meat represent a growing share of the list of top agricultural commodities by production. In the near future, some of the developing countries will be sources of growth in world agricultural production, consumption, and trade. In recent years, there has also been an increase in demand for biofuel feedstock (maize and sugarcane).

Agricultural diversification toward higher-value products is a demand-driven process. Food consumption, in terms of kcal/person/day, is an essential variable used for measuring and evaluating the evolution of the world food situation (commonly referred as the "national average apparent food consumption of availability"). Countries have made significant improvements in raising food consumption per person. As is shown in Fig. 1.3, the food consumption (caloric intake) averaged 2,400 kcal/person/day in 1969, reaching almost 2,800 kcal/person/day in 2000 with projections to reach 3,130 kcal/person/day in 2050. The lowest levels of food consumption will most likely be found in South Asia (2,980) and sub-Saharan countries (2,830) (FAO 2006).

In a growing number of countries, in combination with lifestyle changes that are largely associated with rapid urbanization, growth in per capita food consumption is accompanied by a shift in diets (Table 1.4). Some of these changes in diet experienced by many countries imply changes toward energy-dense diets high in fat, particularly saturated fat, sugar, and salt and low in unrefined carbohydrates. In contrast, other countries with different consumer preferences and food systems may shift away from staple foods to vegetables, fruits, meat, dairy, and fish (FAO 2006). The adoption of a Western fast food industry in Asia, nonexistent just a few decades ago, is an example of a trend toward increasing demand for ready-to-cook and ready-to-eat foods (Pingali 2006).

Table 1.4 Change in food consumption (kg/person/year), ratio 2007/1970

	Milk	Meat	Fish and seafood	Sugar, raw eq.	Vegetable oils	Pulses	Cereals	Vegetables	Fruits
World	1.1	1.5	1.5	1.1	1.9	0.8	1.1	2.0	1.6
Africa	1.2	1.2	1.3	1.3	1.4	1.0	1.1	1.3	1.1
Americas	1.1	1.3	1.4	1.1	1.8	1.0	1.2	1.3	1.2
Asia	2.4	3.5	2.0	1.3	2.8	0.7	1.1	2.9	2.5
Eastern Asia	4.2	5.2	2.6	1.7	4.2	0.3	1.1	4.6	6.0
Southern Asia	2.0	1.3	1.7	0.9	2.1	0.7	1.0	1.6	1.8
Europe	1.1	1.3	1.1	1.1	1.7	0.7	0.9	1.2	1.4

Source: FAOSTAT (2010)

The consumption of milk, cereals, and sugar remained almost unchanged between 1970 and 2007, whereas the increase is noticeable for vegetable oils, vegetables, and fruits. The consumption of vegetable oils and vegetables doubled in contrast to the decrease in pulse consumption over the same time period. The shift in food consumption is particularly noticeable in Eastern and Southern Asia. High population growth, an increase in food consumption per capita, and non-food uses are some of the factors leading to a need to increase agricultural commodity production to meet the sustained demand from emerging countries.

1.5 The agri-food supply chain

The term "supply chain" entered the business lexicon in the 1980s to characterize "a system whose constituent parts include material suppliers, production facilities, distribution services and customers linked together via a feed-forward flow of materials and feedback flow of information" (Stevens 1989). The agri-food supply chain in particular refers to a system of actors linked from "farm to fork" to produce consumer-oriented products in a more effective manner and optimize the flow of agricultural products through the different steps of the chain. Some authors suggest that there are some critical dimensions of food supply chains that make them different from those in other production sectors (Mena and Stevens 2010), including: (1) seasonality of demand and supply, (2) consumer issues of traceability and risk management related to health, nutrition, and safety, and (3) the impact of food production on the environment through extensive resource use, including water and land and from the significant greenhouse gas emissions and waste resulting from agricultural production.

Another common definition of "supply chains" proposed by other authors defines supply chain as "a network of organizations that are involved, through upstream and downstream linkages, in different processes and activities that produce value in the form of products and services delivered to the ultimate producer" (Christopher 1992). In other words, the supply chain consists of multiple firms, both upstream (i.e., supply) and downstream (i.e., distribution) and the final consumer (Mentzer et al. 2001).

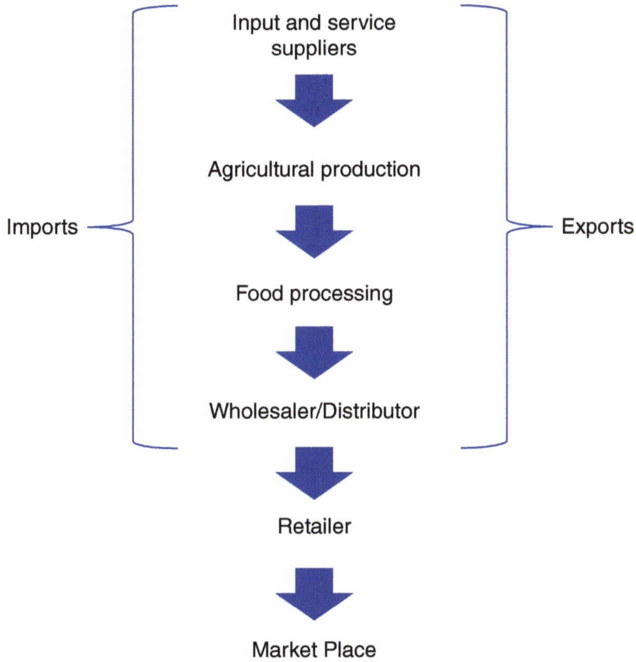

Fig. 1.4 The basic agri-food supply chain (Source: Adapted from Agricultural and Agri-Food Canada 2010)

The early supply chains were local and relatively simple. More recently, with the globalization of agricultural markets and the help of more effective transportation systems and telecommunications, the agri-food supply chain has become increasingly complex. The modern agri-food supply chain links many stakeholders from around the world, including growers, pickers, packers, processors, storage and transport facilitators, marketers, exporters, importers, distributors, wholesalers, and retailers. Food production and processing are no longer restricted to a local or regional supply: Fresh produce, such as fruits and vegetables, can be shipped from halfway across the world at low costs and sold at very competitive prices.

The basic agri-food supply chain is represented in Fig. 1.4.

1.5.1 The canadian agri-food system supply chain

The Canadian agriculture and agri-food system is a chain of industries including agricultural input and service suppliers, primary agriculture producers, food, beverage, and tobacco (FBT) processors, food retailers/wholesalers, and foodservice establishments. It is commonly referred to as the Canadian Agricultural and Agri-Food System (AAFS), which includes all enterprises categorized under the North American Classification System (NAICS) from input suppliers to foodservices.

Table 1.5 Contribution of the agriculture and agri-food system to Canada's GDP

Agriculture and agri-food system GDP	Gross domestic product (Chained billions dollars 2002) 1997	2000	2003	2006	2009	As percent of agriculture and agri-food (subtotal) 2009	As percent of Canada's GDP 2009
Input and service suppliers	6	7	7	7	7	7.1	0.6
Primary agriculture	17	20	17	20	20	20.6	1.7
FBT processing	20	22	23	24	24	24.3	2.0
Food retail/wholesale	19	21	25	28	30	30.2	2.5
Foodservice	14	16	16	17	17	17.8	1.5
Total AAFS GDP	77	86	88	96	98	100	8
Canada, all industries	885	1,026	1,091	1,191	1,195		100.0

Source: Statistics Canada (2006), Table 329-0027, 301-0003, and 301-0006, and Agricultural and Agri-Food Canada special tabulations. 2009 data are preliminary

In 2009, it was estimated that the Canadian agriculture and agri-food system contributed $98 billion (chained 2002 dollars) to Canada's GDP, representing 8.2% of it (AAFC 2011). GDP is a measure of the value of all goods and services produced in a country during a given period. More information on the various components of the agriculture and agri-food system is presented in Table 1.5.

In 2009, primary agriculture GDP represented 21% of the total GDP of the agricultural and agri-food system and 1.7% of the GDP of the total Canadian economy. The FBT processing industry was the largest manufacturing sector in Canada, accounting for 16% of total manufacturing GDP in 2009. The value of food and beverage shipments for 2009 was $78.4 billion, up by $30 billion since the beginning of the 1990s (AAFC 2011). Within the agriculture and agri-food system, the food retail and wholesale industries contribute the most to AAFS GDP (at 30%) followed by the FBT processing sector (at 24%). Both of these sectors contributed more than the primary sector, at 20%. Finally, foodservice s accounted for 18% of AAFS GDP, whereas input and services suppliers accounted for only 7%. Primary agricultural products such as crops and livestock are destined for a variety of end uses. Figures 1.5 and 1.6 show the destinations of primary agricultural products in 2006, based on the Input/Output Accounts.

The food processing sector is the major user of primary agricultural products in Canada, purchasing about 42% of primary agricultural production in 2006. The food processing and beverage industry includes companies engaged in processing and packaging meats, fish, animal feeds, fruit juices, and dairy products along with grain milling, crop grading and packaging, animal slaughtering and packing operations, seafood processing, freezing, canning operations, and beverage production. About 23% of primary agricultural products are exported, whereas about 16% of primary agricultural products are sold to input suppliers, especially as feed grain, which is directly used on the farm itself as livestock feed or sold to other farmers or feed mills. In addition, a small share of primary agricultural production (2%) is sold directly to final consumers, and foodservice accounts only for 2%.

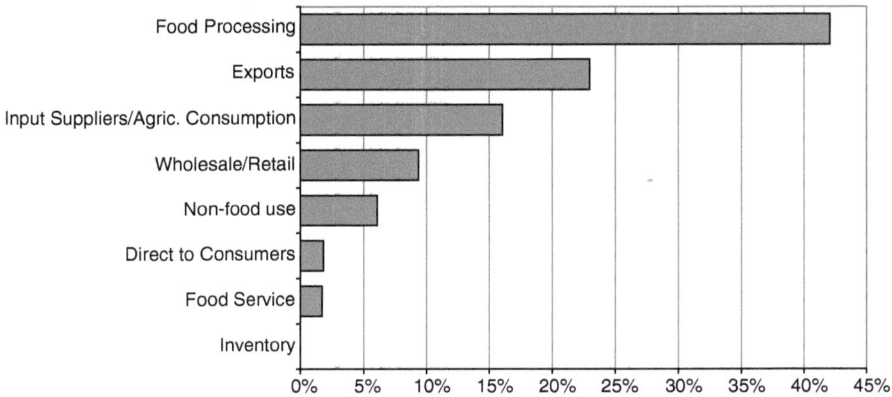

Fig. 1.5 Percentage distribution of on-farm agricultural production by use, Canada, 2006 (Source: Statistics Canada, Input-Output Accounts)

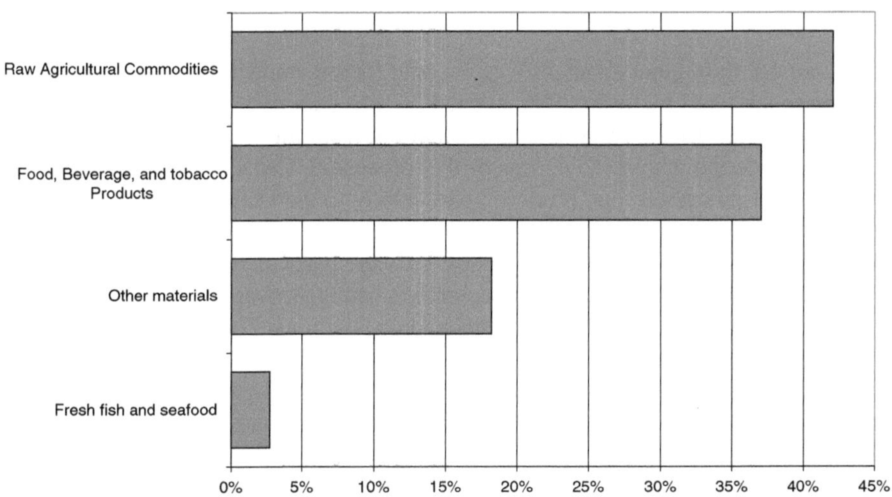

Fig. 1.6 Percent distribution by source of material inputs in the food processing sector, Canada, 2006 (Source: Statistics Canada, Input-Output Accounts)

As already noted, the FBT processing sector is one of the most important players in the agri-food system. Figure 1.6 depicts the origin of the material inputs flowing into the FBT processing sector.

About 42% of material inputs into the FBT processing sector are primary agricultural products. The food processing sector also makes use of processed food products for further processing (e.g., cheese and pepperoni in pizzas), at about 37% of the FBT processing industries inputs. Finally, the input–output tables provide information on the destination of the output of the food processing industries.

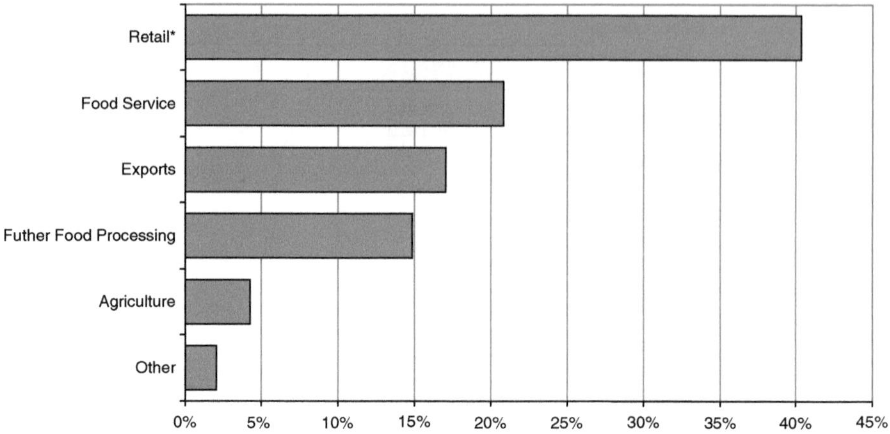

Fig. 1.7 Percent distribution by destination of food processing sector outputs, Canada, 2006 (Source: Statistics Canada, Input-Output Accounts)

Figure 1.7 shows that about 40% of the total food manufacturing output was sold directly to the retail sector in 2006: about 17% was exported, whereas 21% went to foodservice industries (hotels, restaurants, and/or institutions). A sizable proportion of food processing output (15%) was further processed. Canada is highly dependent on exports of processed food products, particularly to the United States, which is its main export destination with 68% (AAFC calculations). The grain milling, oilseed processing, and red meat processing industries were particularly dependent on trade.

The structure of the food retail sector has changed over the years, becoming more consolidated as a result of mergers and acquisitions. There are only two pan-Canadian grocery chains: Loblaws Co. Ltd. and Sobeys Inc., with Metro being the third largest grocery store chain in Canada. About 60% of the retail food sales are from grocery store chains, whereas the balance is from drugstores, general merchandise stores, and others. On the other hand, the dominance of chains as a percentage of sales varies greatly by province. Chains are most important in the Atlantic provinces (76.9%), but much less important in Quebec (36.2%). In 2009, chains saw a slight reduction of their share in eastern Canada, but increased their share slightly in western Canada (AAFC 2011).

The foodservices sector is also a significant component of the Canadian agri-food system. There are about twice as many independent restaurants as there are chain restaurants in Canada, as shown in Fig. 1.8. However, independent restaurants account for only 39% of total sales and 27% of customer traffic. In 2009, chain restaurants sales grew by a modest 0.3%, whereas sales at independent restaurants declined by 3.2%. In 2008, the top 50 foodservice companies accounted for 48% of industry sales and about 27% of establishments (AAFC 2011).

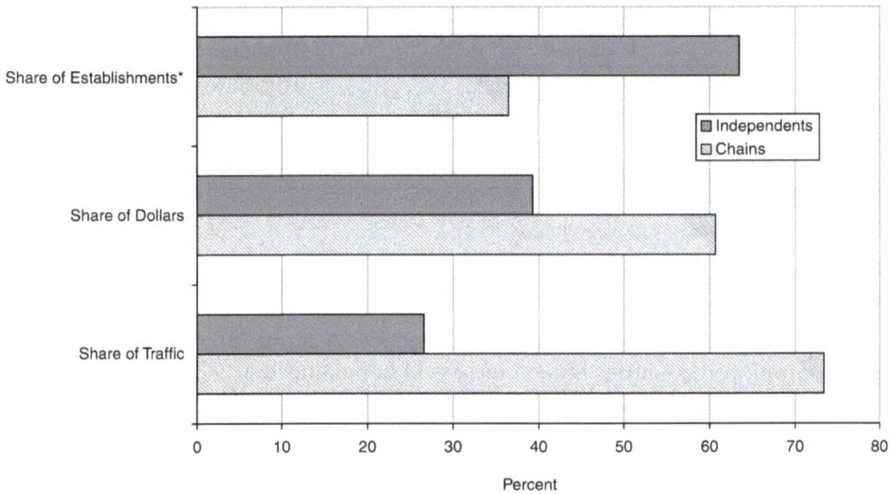

Fig. 1.8 Foodservice market share, chains versus independents, 2009 (Source: Canadian Restaurant and Foodservices Association (2010), Data refers to chain and independent foodservice sourced from restaurants and retail stores)

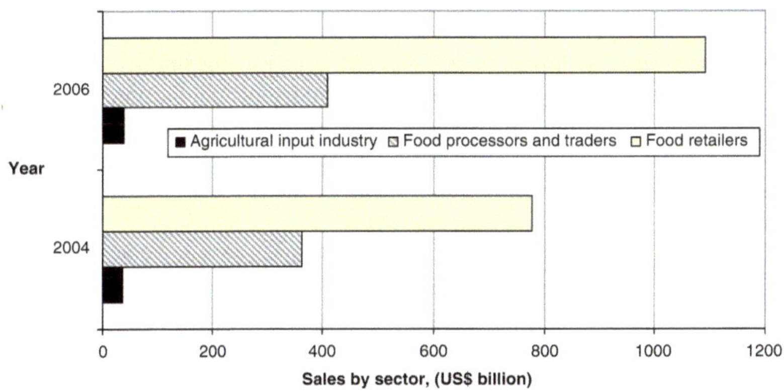

Fig. 1.9 Sales of top ten companies by sector of activity (Source: von Braun (2007)

1.5.2 The global agri-food system: structure and layers

A significant trend in the structure of agri-food supply chains is the growing concentration and consolidation that has been taking place at all levels of the chain (see Table 1.2). As an example, Fig. 1.9 shows the sales of the top ten companies in these sectors between 2004 and 2006. Trends indicate that the top firms have been increasing in size, and this may indicate a greater degree of market power with implications for the supply chain (von Braun 2007).

Table 1.6 Some major players of the agricultural input industry

Seeds industry	Pesticides	Fertilizers
Monsanto (United States)	Bayer (Germany)	PotashCrop (Canada)
DuPont (United States)	Syngenta (United States)	Yara (Norway)
Syngenta (United States)	BASF (Germany)	Mosaic (Cargill) (United States)
Group Limagrain (France)	Dow AgroSciences (United States)	Israel Chemicals (Israel)

Source: Author
Country of headquarters is in parentheses

The notion of "market power" in the supply chain is a concern if it results in collusion, price fixing, and unfair market practices. Competition policy is in place in most developed countries around the world to monitor this behavior. There is no doubt that the activities controlled by such large firms have significant implications for the structure and management of the supply chain (Mena and Stevens 2010). The following section briefly discusses the three main categories shown in Fig. 1.9—agricultural input industry, food processors, and food retailers—with some specific examples.

1.5.3 Input sector

The agricultural input sector is composed of the following industries: machinery, fertilizers, chemical, and seeds producers. Farmers produce some of their own inputs, such as feed for their livestock or seed. However, increasingly, specialized inputs must be purchased from input suppliers. The agricultural input sector is heterogeneous, ranging from multinational firms to small local businesses and cooperatives selling seed, feed, and pesticides (Table 1.6).

The intensity of pesticide use varies across European countries. For instance, greenhouse cultivation and irrigated farming make use of many inputs. On the other hand, small-scale farming is considered a low user of agricultural chemicals (Van der Grijp and den Hond 1999). Farmers in countries at different stages of economic development adopt new technology by buying inputs from agricultural input industries.

The prospect of increasing yields and effectiveness of commercial seed varieties and other desirable properties has led the industry to engage in research in engineered seed and complementary farming techniques (USDA 2001). Significant progress has been made in the seed input industry through plant breeding methods and related changes in intellectual property right regimes. The traditional asexual plant breeding techniques have been overtaken by the progress made in biotechnological research and the development of herbicide- and insecticide-tolerant corn, cotton, soybean, and canola varieties.

New opportunities are also being explored in the biotechnology sector for further enhancement of low input use crop varieties and livestock breeds. Globalization of agricultural input markets as a result of an increase in global market demand and a

Table 1.7 Top five food processing industries in the United States and Canada in 2009

Rank	Company	Food sales, 2009 ($US billions)
1	Nestlé (United States and Canada)	28
2	Tyson Foods Inc. (10/3/09)	25
3	Kraft Foods Inc.	23
4	Pepsico Inc.	22
5	Anheuser-Busch InBev	15

Source: foodprocessing.com

decrease in trade barriers have provided opportunities for private input industries to increase their sales, improve value-added products and research investments in many countries. At the same time, food processors and retailers are responding to consumer demands for reduced chemical inputs by insisting on more sustainable production practices and assurances of those practices in the products they purchase.

1.5.4 Food processors

Primary producers and food processors are under constant pressure from domestic and international markets to improve the efficiency of their operations and reduce costs. Regulatory requirements are crucial for accessing international markets and remaining competitive. Also, trading partners must take into account standards required and realities of the competitiveness of the food processing industries. Table 1.7 represents the top five food processing companies based on their sales for consumer-ready goods that were processed in U.S. and Canadian facilities. Nestlé has the highest food sales, closely followed by Tyson Foods, producing a wide variety of protein-based and prepared food products. Kraft Foods, PepsiCo, and Anheuser-Busch InBen are also some of the top food processing companies in the United States and Canada.

Nestlé is a well-known company that has operations around the globe. Founded in the 1860s, Nestlé companies own many brands, such as Perrier water, cereals, chocolate, and even pet food. Within the company, Nestlé establishes minimum standard requirements for their suppliers and contractors to which they must adhere at all times when conducting business. The company fosters good company practices in their supply chain. Operating practices, farming practices, and agricultural production systems are supported and encouraged by the company. (For more information, please consult the Policy on Environmental Sustainability published in 2008 on the Nestlé website). In a 2003 report, Nestlé mentioned the need for having a reliable supply of local raw material to foster investments in technology among growers. These investments in technologies may have significant and positive effects across their supply chain partners. Nestlé also developed a preference system that provides quality assurance in partnership with agricultural producers, covering the entire food supply chain from raw materials to food bought by consumers. The areas of focus include animal management and environmental sustainability such as the use of nitrogen fertilizers or irrigation practices. By making use of auditors and a quality assurance system, consumers can be guaranteed food safety and quality requirements (Nestlé 2003). Of the top 50 food processors in 2008, about 42 had environmental sustainability commitments such as:

- Energy reduction goals
- Water issues
- Waste reduction
- Packaging reduction goals
- GHG emissions
- Transportation
- Sustainable agriculture initiatives
- Renewable energy
- Footprint

New standards and requirements for production in the food processing sector also include implementation of food safety standards, which can lead to challenges to the industry's ability to manufacture products competitively. Food safety is a minimum standard required by all participants in the agri-food supply chain. In 2008, Maple Leaf Canada suffered from a deadly outbreak of listeriosis in processed meats at one of its plants (Canadian Food Inspection Agency 2008). This incident reminds us of the need to have higher quality standards and enhance coordination across the supply chain. Keeping food safe as it moves through the supply chain is a significant challenge. Today, perishable products may be processed several times before reaching the consumer, and contamination may occur either by accident or mishandling. With the globalization of food chains, new requirements are even more important to ensure the safety of food products.

1.5.5 Retail industry

Environmental issues, consolidation in the industry, changing demographics, and an increasing interest in health, wellness, and food safety by consumers are some of the most important trends that the food retail industry faces. There has been a trend toward consolidation in food retailing in many countries. This trend is likely to continue because one of the essential objectives of food retailers has been to achieve increased economies of scale and access new markets through acquisitions. The food processing sector is facing fewer and larger distributors and retailers as a result, as retailers have increasingly centralized their buying operations. Wal-Mart stores dominate the retail industry, with almost $US405 billion in sales in 2009, followed by Carrefour, which has the largest number of stores in the world, Metro Group, Tesco, and the Schwarz Group (Table 1.8).

Table 1.8 Top five food retailer companies in the world, 2009

Rank	Stores	No. of stores	Sales ($US billions)
1	Wal-Mart stores (United States)	8,416	405
2	Carrefour (France)	14,215	120
3	Metro group (Germany)	2,127	91
4	Tesco (United Kingdom)	4,837	89
5	Schwarz group (Germany)	9,902	81

Source: Planet Retail
Country of headquarter is in parentheses

The growing concern by consumers around environmental issues, such as packaging waste and water utilization, are also being taken into account by the retail sector. Wal-Mart has pushed sustainability down to the producer level in the supply chain by implementing green logistics and moving toward its sustainability goals set in 2005 (Heying and Sanzero 2009). The main goals were to:

- Make use of 100% renewable energy.
- Create zero waste.
- Sell products that sustain Wal-Mart's resources and the environment.

To reach these three main goals, Wal-Mart is making use of a network of government agencies, nonprofit organizations, employees, and suppliers to help "green" its supply chains. It introduced green logistics by implementing a system that independently monitored overseas suppliers to make sure they met Wal-Mart's social and environmental standards (Wal-Mart 2010). This decision is having an impact on the company's growth and distribution methods (Plambeck 2007). The use of advanced technologies for distribution has affected the overall efficiency of their suppliers.

In the retail sector, the adoption of new technologies has improved coordination between partners of the food chain. For instance, Itradenetwork Company is one the global providers of on-demand supply chain management programs for the food industry (Itradenetwork 2010). In 2010, Metro signed a multiyear agreement with this company. With this innovative program, Metro Group will collaborate in real-time with hundreds of its supply chain partners. As of today, Metro Group purchases products from more than 45 countries around the world. This technology is part of the long-term strategy to capitalize on operational efficiencies by tracking food and its delivery to its customers according to their expectations.

Another initiative introduced by the retail sector to assure quality includes the development and use of private standards by leading firms. Private standards are often based on institutional and legal frameworks in which the company operates. The main driving forces behind these industries efforts are consumer demands and expectations for food safety and to some extent, environmental and animal welfare. The development of private standards is based on an increase in competition among retailers for consumer spending, a reduction in monitoring and inspection costs, and an increase in product sourcing from around the world (OECD 2006). Henson and Humphrey

(2009) identified four key drivers that are improving the control in agri-food supply chains, including:

- A reform of the regulatory system to respond to real or perceived risk in food production, transport, or processing
- Consumers' interest in food safety and quality, reinforced by company competitive strategies (i.e., product original, environmental aspect)
- The globalization and diversity of the food supply, which is imposing a new risk, and challenges that must be assessed through coordination and control of the value chain
- An increasing effort by the private sector to differentiate products by ensuring a higher level of food safety

These factors have generated a competitive environment in which retailers must supply food products to consumers with food safety standards. Globalization of the agri-food supply chain across multiple regulatory jurisdictions explains why retail food industries need to respond to this competitive environment. Over the years, several new global initiatives have been introduced regarding private standards. For example, European food retailers established GLOBAL GAP, soon followed by the establishment of the Global Food Safety Initiatives (GFSI), which set out the criteria for the development of private standards around food safety. Consumers in developed countries have increasingly indicated their willingness to pay for food product attributes that are largely related to production processes (organic, local, humane, and less environmentally damaging). In turn, there is considerable pressure on stakeholders across the agri-food system to restructure their production processes to meet such consumer preferences; hence, the expansion of the scope for "greening" the agri-food supply chain in recent years.

1.6 Conclusion

This chapter presented some of the core information related to agri-food supply chains and the global agri-food system as well as some examples taken from the sector. Technologies used in the agri-food industry were described followed by a snapshot of agri-food supply chains in Canada and around the world. This was followed by a description of the essential players and components of the global supply chain. Overall, the various players in the agri-food system face a diversity of drivers and concerns, primarily driven by consumers. Consequently, consumer trends related to food safety, quality, and sustainability are influencing players that are operating and interacting in food supply chains. Given the rise in concern about environmental sustainability of the food supply and agri-food system, there is an increasing trend toward attempts by large global retailers and processors to "green" the supply chain.

The notion of "greening" the supply chain refers to the ability of players downstream in the supply chain to require a certain level of environmental responsibility in the core business practices of their suppliers and vendors. If suppliers do not abide by the standards required by the buyer, the buyer company may be buying and using products that do not meet intended standards (Van der Grijp and den Hond 1999). Thus, there are impacts that go all the way back to the farm and primary agricultural production. As a result, it is important that all players in the chain be aware of the evolving demands for "greening" of their products. The discussions that follow in the next several chapters explore some recent green technologies and concepts under consideration or currently being applied in both primary agriculture and the food processing sector.

References

Agricultural and Agri-Food Canada. 2010. Structure and performance of the agriculture sector: Backgrounder. AAFC online at www.agr.gc.ca/. Accessed September 2010.

BASF. 2010. Cause of starch of potato commingling identified (press release). http://www.basf.com/group/pressrelease/P-10-421. Accessed September 24, 2010.

Cabinet Office of the United Kingdom. 2008. Food: An analysis of the issues. Cabinet Office of the UK, London www.cabinetoffice.gov.uk/media/cabinetoffice/stategy/assets/food/food_matters1.pdf. Accessed 14 Oct 2011.

Canada. Agriculture and Agri-Food Canada. 2011. *Overview of the Canadian agriculture and agri-food systems*. Ottawa: Agriculture and Agri-Food Canada.

Canada. Canadian Food Inspection Agency. 2008. Listeriosis outbreak. www.inspection.gc.ca/. Accessed 13 Oct 2011.

Canada. Statistics Canada. 2006. *Census of agriculture*. Ottawa: Statistics Canada.

Canadian Restaurant and Foodservices Association. 2010. *Foodservice facts 2010*. Toronto: Canadian Restaurant and Foodservices Association.

Christopher, M. 1992. *Logistics and supply chain management: Strategies for reducing costs and improving services*. London: Financial Times/Pitman.

Food and Agriculture Organization of the United Nations (FAO). 2010. FAOSTAT database. http://faostat.fao.org/. Accessed 20 Oct 2011.

Food and Agriculture Organization of the United Nations (FAO). Global Perspective Studies Unit. 2006. *World agriculture: Towards 2030/2050: Interim report*. Rome: Food and Agriculture Organization of the United Nations, Global perspective Studies Unit.

Globerman, S., and D. Shapiro. 2002. Infrastructure politique nationale et investissement étranger direct (document de travail no.37). Industrie Canada, Ottawa.

Henson, S., and J. Humphrey. 2009. *The impacts of private food safety standards on the food chain and on public standard-setting processes*. Rome: FAO/WHO.

Heying, A., and W. Sanzero. 2009. A case study of Wal-Mart's green supply chain management. www.apicsterragrande.org/Wal-Mart%20Sustainability.pdf. Accessed October 2010.

International Monetary Funds. 2010. World economic outlook database. Washington, DC. www.imf.org/external/pubs/ft/weo/2009/02/weodata/index.aspx. Accessed October 2010.

Itradenetwork. 2010. Corporate news, October 20, 2010. www.itradenetwork.com/html/news_articles/corp_news_article52.htm. Accessed October 2010.

McCullough, E.B., P.L. Pingali, and K.G. Stamoulis. 2008. *The transformation of agri-food systems: Globalization, supply chains and smallholder farmers*. Rome/London: FAO/EarthScan.

Mena, C., and G. Stevens. 2010. Delivering performance in food supply chains: An introduction. In *Delivering performance in food supply chains*, 1st ed, ed. C. Mena and G. Stevens. London: Woodhead Publishing.

Mentzer, J.T., W. de Witt, J.S. Keebler, S. Min, N.W. Nix, C.D. Smith, and Z.G. Zacharia. 2001. Defining supply chain management. *Journal of Business Logistics* 22(2): 1–25.

Nestlé. 2003. *The Nestlé people development review*. Zurich: Nestlé S.A.
Organization for Economic and Co-operation Development (OECD). 2006. *Working party on agricultural policies and markets: Final report on private standards and the shaping of the agro-food system*. Paris: OECD.
Pingali, P. 2006. Westernization of Asian diets and the transformation of food systems: Implications for research and policy. *Food Policy* 32(3): 281–291.
Plambeck, E. 2007. The Greening of Wal-Mart's Supply Chain, Supply Chain Management Review. www.scmr.com/article/CA6457969.html. Accessed October 2010.
Stevens, G.C. 1989. Integrating the supply chain. *International Journal of Physical Distribution and Materials Management* 19(8): 3–8.
U.S. Department of Agriculture (USDA). 2001. *Economic issues in agricultural biotechnology: Forces driving changes in the agricultural input industries: Plant breeding and biotechnology (AIB-762)*. Washington, DC: USDA, Economic Research Services.
Van der Grijp, N.M., and F. den Hond. 1999. *Green supply chain initiatives in the European food and retailing industry*. Amsterdam: Institute for Environmental Studies.
von Braun, J. 2007. *The world's food situation: New driving forces and required actions*. Washington, DC: International Food Policy Research Institute (IFPRI).
Wal-Mart. 2010. Global sustainability report: 2010 progress update. http://walmartstores.com/Sustainability/7951.aspx. Accessed October 2010.
World Bank, East Asia and Pacific Region. 2005. *Spread of avian flu could affect next year's economic outlook*. Washington, DC: World Bank.
World Organisation for Animal Health (WOAH). 2010. Food Safety-key themes. Paris. http://www.oie.int/index.php?id=83&L=0.

Chapter 2
Key drivers of the food chain

Corinne Gendron and René Audet

2.1 Agriculture as a market

2.1.1 The long history of food market disembeddedness

Because we rely on large companies and supermarkets to feed ourselves, we often forget that, unlike many other goods, food was not provided through markets during most of human history. The commoditization of food, which has led to its current status as merchandise, has indeed been the result of a long and complex process from the age-old practice of self-production, through the subsequent trade in small markets. By bringing out the specificity of food in the current economic framework, an examination of these origins helps us understand what can be coined as "the exceptionality of agriculture," that is, the fact that food is not like any other consumer good.

Approximately 10,000 years ago, the beginning of agriculture contributed to a deep transformation in the organization of social life, which evolved from the hunter-gatherer and nomadic forms of society to sedentariness. In 5550 BCE, the Sumerians started large-scale and intensive cultivation with irrigation techniques, allowing excess production that could be stored or sold, and reducing wild food to a marginal part of the usual diet of the population. Moschini (2008) explains this evolution as follows:

> The challenge of population pressure faced by an environment saturated with hunter-gatherers, catalysed by the changing climatic conditions at the end of the last Ice Age, led some people to attempt the cultivation of wild cereals in an effort to supplement their traditional sources of food (Cohen 1977). Thus was agriculture born about 12,000 years ago in the

C. Gendron (✉) • R. Audet
Department of Strategy and Social and Environmental Responsibility,
Management Sciences School, Quebec University at Montreal, Montreal, QC, Canada
e-mail: Gendron.Corinne@uqam.ca; audet.rene@uqam.ca

J.I. Boye and Y. Arcand (eds.), *Green Technologies in Food Production and Processing*, Food Engineering Series, DOI 10.1007/978-1-4614-1587-9_2,

Fertile Crescent, and independently in at least five other distant locations over the next few millennia (Bellwood 2005). A dwindling supply of large mammalian prey increased the attractiveness of early farming activities, and the ensuing Neolithic revolution saw the widespread transition of prehistoric humans from nomadic hunting-gathering bands to agriculture-based communities relying on a few successful domesticated species of plants and animals. Farming encouraged more permanent settlements, allowing the development of increasingly complex social structures and the ushering in of earlier civilizations.

Later on, numerous innovations led to increased productivity, freeing the population from other activities but also leading to specific configurations of social organization.[1] One could refer to commoditization and industrialization in relation to this period, but this would most certainly be an inappropriate anachronism given that society at that time was not what we call an "economic society," that is, a society primarily organized on the basis of modern economic rationality (Vergès 1989). The trade associated with the surplus was part of a more general social system in which other rationalities were used as organizing principles, whether religious or political.

This was obvious in the feudal period, when the serf had the right to cultivate his parcel of land, but was also required to give up a substantial part of the harvest to his lord according to the vassal relationship. It is also interesting to note that the serf was denied the right to hunt; that is, access to "wild food" was specifically reserved for the lord. At this time, commercial relations were enacted through two dimensions, which Braudel (1967) judicially distinguishes as small markets and great capitalism. Small markets were local spaces in which, in search of variety, people could trade their surplus with that of others. However, this was a marginal means of accessing food: A self-production system prevailed. For the elite, luxurious exchanges were a way to demonstrate supremacy more than to provide basic commodities, and in the Age of Discovery, agriculture also participated in the colonization process and was a determining factor in the organization of slavery.

This is why, although new farming techniques and space reorganization allowed for the production of a surplus that could be traded, it would be misleading to refer to the commoditization of food or the industrialization of agriculture with regard to this period, because agriculture and food markets were still organized according to ecological realities and embedded in social and political relations that cannot be reduced to economic transactions.

This is what Polanyi theorized as the concept of embeddedness, pointing out that food markets have not always been free markets, and that food staples have not always been considered to be commodities (2001). In Polanyi's view, although reciprocity served to sustain humane ways of sharing food and organizing society over

[1]The understanding of this dynamic may vary among authors, as illustrated by this comment by Moschini: "The path from these early days to the modern world eventually saw the gradual augmentation of technology in food production, which supported a population expansion that is continuing to this day. Some see that as a vicious circle and also lament that '… the transition from hunting and gathering to farming eventually resulted in more work, lower adult stature, worse nutritional condition and heavier disease burden' (Diamond 2002, p. 700). But civilization has also led to knowledge accumulation, culminating in the industrial revolution and the opening up of increasingly new opportunities" (Moschini 2008, pp. 331–332).

the millennia, the withholding and redistribution of food by a centralized authority (such as in the feudal system) also played a prominent role in many societies. These modes of economic exchange were always grounded—embedded—in social relations and nature's cycles. The land and the work force that helped to fertilize it and cultivate its fruits, were not market values, they were social institutions. On the eve of the Industrial Revolution, the concept of the self-regulated market shattered these institutions and put modern society on the path toward the commoditization of agriculture.

2.1.2 *The industrialization of agriculture*

The industrialization of agriculture can refer to different realities, starting with the mechanization of farming and the use of new technologies and modes of organization, through what has been called the "Agricultural Revolution" (as compared with the concomitant Industrial Revolution in England).[2] The turning point in this process was the integration of agricultural production into a new economic order organized on the basis of market rules and rationality instead of other logics, whether social or biological. This transformation of the status of agriculture as a "market" or "capitalist" activity—not merely economic but dedicated to trade with the aim of accumulation—was the result of many structural and ideological transformations that some authors have interpreted as the inexorable March of capitalism to encompass all human activities (Wallerstein 1974).

First among the transforming factors were new techniques and technologies—including the use of mechanized tractors and fertilizers—that led to an increase in productivity and, to some extent, freed production from unpredictable natural conditions. Second, the organization of land, symbolized by the enclosures in England (which are generally seen as one of the major starting points of industrial capitalism), also allowed for an increase in productivity but led to another dramatic change: the exodus of former self-sufficient peasants to cities, where they formed the new work force relying on markets to provide them with food. Last, it must not be forgotten that these transformations were anchored in the development of an ideology of industrial progress, closely associated with market dynamics as an optimal social regulation.

[2]Kerridge contradicts this understanding (1960). Referring to the thesis developed by Toynbee, he explains: "Toynbee rightly concludes that agricultural progress was relatively slow between 1700 and 1760. But he can hardly have been expected to have devoted much of his short working life to agricultural history, which to him was no more than a sideshow. He gives the matter short shrift and takes the obvious and easy course of turning away from his door all those facts that do not fit in with his preconceived notions that the period 1760–1843 witnessed 'an agrarian revolution [that] played as large part in the great industrial change at the end of the eighteenth century as does the revolution in manufacturing industries.'"

Thus, in Polanyi's view, the market appears to be an economic form of regulating social relations and human interaction with nature, a form that characterizes capitalism. Indeed, capitalism requires such a system based on the concept of a self-regulated market, at the expense of other economic models such as reciprocity and redistribution. Polanyi argues that at the peak of the industrial revolution, the concept of the self-regulated market shattered these other models. Trade and the search for profit became the fundamental rationale of economic activities in the new political economy of the eighteenth century, giving rise to "the market" as a social institution, different in this respect from local markets. Agriculture thus became part of the great capitalism referred to by Braudel. As a result, agriculture was subject to external pressures to become more productive and less expensive and, eventually, overrule natural cycles and processes. This evolution toward the disembeddedness of agriculture was later marked by the Green Revolution, and by the more recent process of globalization.

2.1.3 *From the green revolution to globalization*

The Green Revolution originated from the work of scientists who developed new varieties of crops, supported by the use of fertilizers and pesticides, both leading to increased agricultural productivity in response to the growing needs created by exponential population growth.[3] The Green Revolution, as a process, appeared at the end of the 1960s, and deepened until very recently. As Huffman (2004) explains:

> Overall, the productivity data suggest that the Green Revolution is best understood not as a one-time jump in yields, occurring in the late 1960s, but rather as a long-term increase in the trend growth rate of productivity. This occurred because successive generations of modern varieties were developed, each contributing gains over previous generations.
> Evenson and Gollin (2004) show that without the Green Revolution, crop yields in developing countries would have been 20% to 24% lower.

Applying Western fertilization and pesticide techniques to agriculture in the South, the Green Revolution accompanied the conversion of the war industry into the petrochemical industry, as best illustrated by Monsanto's evolution (Champion 2004). But agricultural production was also transformed by the growing practice of trade

[3]The demographic rationale of the Green Revolution is well illustrated by Huffman: "as the demand for food and fiber grew during the past 300 years, because of the Earth's expanding human population and rising per capita incomes, society met this demand first by increasing the land area under cultivation and later by improving crops so that their yields were higher. Before 1900, land was abundant almost everywhere, and in the United States, new lands were brought into production as the frontier moved across the country between 1700 and 1900. … By 1900, the frontier was closed in the United States, and this increased the urgency of finding new methods for increasing crop yields" (2004, p. 3).

that changed the conditions under which agricultural commodities were produced. Coleman et al. (2004) explain that since the end of World War II:

> [A]s agriculture has continued the movement from subsistence, peasant polyculture toward commercial farming, production has become more specialized. Processing companies buying from farmers have demanded more evenness in quality and standardization in size and the markets for these more specialized and standardized products are extending gradually further away from the locality in which crops or animals are grown. The greater emphasis on monocropping has come to rest on various expert systems to achieve economies of scale and higher productivity, including sophisticated farm equipment, chemicals to control pests and weeds, and to stimulate growth, and biological science to improve yields from plants and to raise output from animals. In some areas of livestock production like poultry and hogs, farmers are entering into quite specific contracts with processing companies that govern the nature and the quality of products to be delivered.

These developments have led to a deep transformation of the relationship between the farmer and the consumer, which has been progressively mediated by economic agents, whereas agricultural production has become part of a complex economic system encompassing science and technology, mechanical instruments, and worldwide distribution networks in a world marked by geopolitical tensions or conflicts and a new consumerist way of life. In this system, the self-producing farmer has progressively been replaced by a market producer whose conditions of production are dictated by influential economic actors in concordance with state-led development strategies.

> The production of agricultural commodities has gradually become enmeshed in linkages that are more globally extensive today than they were a half-century ago, whether through suppliers of inputs to agricultural production or through the processing, distribution and sale of agri-food products. Trade both in agricultural commodities and in processed food products has risen over the same period. This intensified economic competition on a more global scale has accentuated already powerful tendencies toward greater economic, technological and socio-cultural integration in the triad of the North America, Europe, and East and South-East Asia… (Goodman 2003).
>
> These more intense regional and economic relationships have developed through the greater presence of multinational firms in the sector, and important changes in technology used in the transportation and distribution of commodities and processed food products. Consequently, global economic relations are a more important factor in the lives of farmers and food consumers than they were a half-century ago (Coleman et al. 2004).

For these authors, globalization can be seen in direct continuity with industrialization and the Green Revolution in the process of agricultural disembeddedness.

> Globalising processes have accelerated changes in the countryside in most countries, changes already in train due to mechanization and chemical farming. More food is being produced on less land by fewer farmers than a half-century ago. Old patterns of relationships between farmers, the suppliers of their inputs and the buyers of their crops are shifting due to corporate restructuring. Political arrangements in the wealthier countries are being criticized and slowly changed in response to challenges from within and from those suffering the effects of subsidies in the Global South. Old orders are changing but there is scant evidence to date whether any new order will take their place (Coleman et al. 2004).

However, although globalization seems indeed to have intensified the process of commoditization initiated centuries ago, recent developments might be interpreted as a suspension, and even a turning point in this process. We would argue that following its accentuation through different phases of varying length and speed depending on the country, and its final intensification toward the end of the 1980s, the process of commoditization of agriculture ended with the last attempts made under the globalization process to handle food as an ordinary good. The negotiations aiming to liberalize international agricultural markets were defeated by protectionist strategies of the World Trade Organization (WTO), confirming a suspension of the rationale behind the commoditization of food.

2.1.4 Global trade and agricultural policy toward decommoditization?

In spite of the commoditization process that has been underway for centuries, among global trade issues, the agricultural sector has always been recognized as being somewhat particular (Aggarwal 2005; Trebilcock and Howse 2005). Along with the services and investment sectors, it was one of the last major economic sectors planned to be liberalized after the Uruguay Round of the WTO. After World War II, the United States refused to apply General Agreement on Tariffs and Trade (GATT) principles to the agricultural sector for political and strategic reasons, and was allowed to do so. Under the postwar reconstruction, Europe also relied heavily on subsidies and market protection in structuring its internal agri-food economy. Becoming competitors on a global scale, the EU (Eur.Union) and the United States were then drawn into a "war of subsidies" during the 1970s and 1980s—a process that brought about increasing price distortion in the agricultural market.

Although every national situation is different, this era saw agricultural policies being constructed around three axes that would eventually become the three "pillars" of the WTO Agreement on Agriculture. One is the traditional imposition of entry tariffs and taxes, which make imported products less competitive than domestic equivalents. For some imported products considered to be sensitive, a country can impose "tariff peaks"—meaning tariffs that are many times higher than regular tariffs—as a barrier to entry. Another policy area that has the effect of distorting prices is domestic support. All kinds of tax reduction for producers, public insurance programs for farmers, subsidies for production or for the purchase of machinery, etc., can be considered as domestic support measures. Finally, the most contested policy area under the GATT and WTO is that concerning export subsidies, which constitute public support to sectors oriented toward exportation. Although these can be justified as a means by which to counter the effects of entry barriers (of the type described in the preceding) in foreign markets, their impacts are being widely felt in developing countries, where the total amount of subsidies is a fraction of that of the United States or the UE. Overproduction in developed countries can also be subsidized to facilitate exportation to other countries, thus creating a form of "dumping" (Drope and Hansen 2006).

Thus, for different reasons—mainly political—agriculture has never been governed by a free market, and this has impeded the process of commoditization.

This has made multilateral policy negotiations in agriculture quite paradoxical. Although free trade originally meant opening up markets in the Global South to products from industrialized countries, free trade in agriculture refers to the reverse; that is, the opening of markets in the Global North to foodstuffs produced in developing countries. For the first time in the recent history of trade negotiations, the Uruguay Round tackled the liberalization of agriculture. The result, however, was rather modest. According to most experts on international trade law, the Uruguay Round Agreement on Agriculture, resulting from a bilateral bargain between the United States and the European Commission, consisted more in the legalization of protectionism in the agricultural sector than in its liberalization (Aggarwal 2005; Clapp 2006). Such a result was not going to be satisfactory to developing countries, and the next negotiating round would have to address the problem of agricultural protectionism more deeply.

2.1.5 North–south inequities in agricultural trade and the case for food security

In the agricultural sector, the huge subsidies and market protection policies enforced by the European Union and the United States (along with the majority of developed states) have had the most adverse impacts on developing countries, whose advantage based on low-cost agriculture is nullified by closed markets in the North. Instead of bringing real liberalization to the sector, as observed, the Uruguay Round Agreement on Agriculture legalized European and American protectionism. This is why the Doha Round was launched in 2001 with much anticipation on the part of developing countries who were hoping for "the redressal of the inequitable nature of existing provisions of the Agreement on Agriculture" (G77 2001). After more than 2 years of stalled negotiations at the WTO, it is now clear that the North–South inequities written into international trade law still constitute a great challenge to the notion of "equity" in global agricultural trade. Simultaneously, they impede a potentially formidable driver of the food chain: food security.

Food security is an interesting concept because it is based on the notion that food is more than a commodity—it is a vector of security or insecurity. Indeed, because a proper diet and the nourishment of communities, whether rural or urban, ensures greater stability and productivity, food is an essential driver of development. This is why the "G33 on special products and the special safeguard mechanism," a coalition of developing countries that emerged at the Cancun WTO Ministerial Conference in 2003, states that the "deepening global food crisis has clearly demonstrated that indeed the principles of food security, livelihood security and rural development are relevant and must indeed be a part of the equation in discussing the optimal solutions to [the Doha] Development Round" (G33 2008). If the Doha Round is ever concluded, multilateral trade agreements may well institutionalize mechanisms to ensure food security in developing countries, such as the possibility of designating "special products"—meaning products that are considered important for food security and rural development—that would benefit from better tariff protection, and a "special safeguard mechanism" that would allow countries to react with tariffs and quotas in the case of a sudden price drop or domestic market "flood" of competitive imported products (Lal Das 2005). Similar mechanisms could be available for developed countries under the denomination of "sensitive products" (Trebilcock and Howse 2005).

It is interesting to note that the content of the last negotiations occurring under the WTO, rather than being a pure translation of the commoditization rationale, indeed recognized the "exceptionality" of agriculture. Clearly, the food security issue has brought about the idea that agriculture may have functions other than simply being a commodity. Agriculture is increasingly presented as being "multifunctional," as shown by the new trends in food consumption and production. Thus, although we recognize that the globalization process is transforming agriculture through an intensification of the economic integration of its actors and activities, our understanding differs from that of Coleman et al. (2004) who see in globalization a deepening of the commoditization and disembedding process.

> In short, despite its rootedness in place and its dependence on the natural rhythms of the seasons, the life cycles of animals, and the climate, agriculture is changing rapidly in character as it becomes drawn into globalising processes in the economy, the world of politics, and culture. As an economic activity, it is losing its exceptionality and becoming one sector among others contributing to economic growth. Politically, the long-standing protective mantle of the nation-state is yielding to new forces, new rules and new constraints defined at regional and global levels. With nation-states yielding some authority, such cultural notions as farming being the backbone of the nation and these nations must be self-sufficient in food are being reconsidered and recast (Coleman et al. 2004).

It is mainly because of the exceptionality of agriculture that the trade liberalization process undertaken by the Uruguay Round failed, and that the globalization process took a new direction in accordance with this special status: a limitation of free trade primacy with regard to geo-strategic imperatives fuelled by the particular status of agriculture. This is not to say that we are entering a process of decommoditisation of food, but that the March towards commoditization has reached an end. This can be explained by a series of factors anchored in the nature of agriculture itself, wherein there have always been intrinsic limitations to the de-territorialization rationale of globalization, as clearly pointed out by Coleman et al.:

> [T]he immobility of land and the local circumstances of land use would seem to make farming a highly unlikely sector to be integrated into international and global processes. …
>
> With its implications for the declining constraints of physical location and its emphasis on de-territorialisation (Appadurai 1996), globalisation would seem to refer to processes not particularly relevant for understanding agriculture. After all, farming remains an activity intimately tied to particular rural communities, their soils, their microclimates, and their physical environments. Although buying and selling currencies might occur anywhere at any time, growing wheat or milking cows takes place in very particular localities, following regular patterns dictated by the seasons or basic cycles of life. Tied to the land and drawing upon centuries, if not millennia, of tradition and cumulated experience, farming, one of the ancient arts of human civilization, would seem to be somewhat beyond the reach of current globalising processes (Coleman et al. 2004).

As Coleman et al. (2004) explain, despite these characteristics, agriculture has not been spared the effects of globalization. However, we would go one step further. Inspired by a comment made by these authors suggesting that continuing globalization should not be seen as the only possible outcome of the current interplay of economics, politics and technology, we argue that the exceptionality of agriculture has actually changed the course of globalization, in part because of the struggle that its initial liberalization rationale brought about in this sector.

> The changes that we observe bring in their train a variety of avenues of contestation and struggle. The introduction of biotechnology, the growing of genetically modified organisms, and the incorporation of these into processed foods and animal feeds have given rise to field burnings, consumer boycotts, and political debates across the globe. Environmentalists have challenged "industrial farming" and its use of chemicals because of the effects on water courses and human health. Nation-states misuse global and regional trade rules to give "their" farmers an edge over competitors. Farmers continue to fight to stay on the land in most countries of the world. In short, the production and preparation of food promise to remain a prominent site of resistance to globalisation for the foreseeable future (Coleman et al. 2004).

Initially driven by universal free trade agreements, the economic integration that occurred during the first era of globalization is now entering a new age in which it obeys new logics and parameters. These parameters will both sanction the exceptionality and multifunctionality of agriculture and be supported by them. In this sense, agriculture plays an important role in the general restructuration of markets illustrated by the rise of responsible consumption and social responsibility strategies adopted by firms. In this new age, as the North–South tension is made more complex by commercial alliances rooted in the strategic interest of nations (as is the case with genetically modified organisms, GMOs), new issues such as environmental protection, greenhouse gas emissions, water supply, and foreseeable shortages are deeply changing the governance challenges and the rationale behind state-led negotiations and the strategies adopted by firms in the field of agriculture.

2.2 New trends: rediscovery of embeddedness through multifunctionality

Although agriculture has been a domain in which the human spirit has overcome nature's unpredictability and led to civilization and progress, today it seems to illustrate a total loss of control over nature's dynamic because it participates in disease, health problems, and social fracture. Agriculture is no longer presented as the instrument of progress and well-being, but rather as an area involving the convergence of economic, political, and social problems.

The publication of Rachel Carson's *Silent Spring* was a revelation for Western citizens concerning the dangers of chemicals for ecosystems (Hails 2002), and this revelation was followed by discoveries about the risk they present for human health as well. Considering the awakening provoked by what can be called the side effects of the Green Revolution, it is not surprising that the "Gene Revolution" has aroused suspicion. Even more recently, the environmental issue has brought up the problem of transportation, showing that the energy needed for food trade often exceeds the energy accumulated in food itself. In addition, issues of water supply and new data about fishery shortages (FAO 2006) have also led to concern.

Another issue that has been raised by the growing control of agricultural production by corporations relates to the use of land. Concurrent with the advent of urbanization, rural land is no longer seen as a space for social and human life, having been transformed into large dehumanized production areas in some countries, and being subject to severe conflict over usage in others. With regard to consumption, production processes have also been associated with health problems, especially obesity, heart disease, and even cancer. Moreover, the consumer has progressively been disconnected from the ecological and natural dimensions of food, which is increasingly packaged and transformed. Finally, food is increasingly seen as a major focal point of tension between the North and the South, as shown by the WTO negotiations, with tensions also arising between Europe and United States over the issue of GMOs.

This concentration of issues and the conflicts that have arisen over the last decade have been theorized in terms of the coexistence of four different paradigms to which countries refer in the governance of their agricultural sector: the dependent, competitive, multinational, and global production paradigms (Josling 2002). The dependent and competitive paradigms could be said to illustrate the first two economic periods of the last century. Coleman et al. (2004) define them as follows:

> A dependent paradigm organized around the core belief that agriculture fulfils basic food needs and provides national security, as well as social and political stability and rural employment and welfare, but requires government help to enable it to generate adequate incomes.
>
> A competitive paradigm that emphasizes agriculture as a sector that can hold its own against other sectors of the economy and that can thrive in a market economy and an international trade system (at least, where markets are permitted to operate free of distorting, dependent paradigm style policies).

In our view, we are currently experiencing competition between the two last paradigms,[4] a situation that is most likely to result in a combination of the two rather than the success of one over the other. These paradigms are also defined by Coleman et al. (2004) based on the work of Josling (2002):

> A multifunctional paradigm organized around the belief that agriculture is an integral part of the countryside and provides non-market goods that would be under-produced without some degree of government support.
>
> A globalised production paradigm that situates agriculture in potentially global food supply chains, where farmers are seen as supplying land and animal management services to an integrated vertical process from input supply and technology provision through marketing of the product.

As Coleman et al. (2004) add in a note, this last globalized paradigm does not exclude local food systems, which are part of the model: "Josling (2002) called this paradigm 'globalized' agriculture to emphasize the international nature of many of the supply chains. We follow this terminology while recognising that local and regional (short) supply chains are also part of the model. In some cases, the 'driver' of the chain is the ultimate consumer: in most cases it is the retailer that defines the

---

[4]The analysis offered by Coleman et al. presents current conflicts as a struggle between these four paradigms (2004, p. 119).

requirements to be met by the supply chain." This remark leads us to the reflection that we are indeed probably experiencing the exploration of a new articulation between the dependent and competitive paradigms, an articulation which will manifest itself through a globalized production paradigm encompassing a multifunctional paradigm, and that becomes possible only by assessing the particular nature of agriculture and confirming its exceptionality. In our view, it is this articulation that is illustrated by the numerous innovations prevailing in the agricultural field, some of which are presented in the following.

2.2.1 The turn to quality and alternative models of consumption

As shown by the diversification of restaurant food, the growing number of ethnic shops, the success of television shows about cooking, and the content of advertising, food consumption today refers to more than simply satisfying basic needs (if indeed it ever meant so little). Buying food, in many sectors of society, is a symbolic act that goes as far as an expression of identity or a political statement. Slow food, local food, labeling, community-sustained agriculture, etc.: All of these new trends mark a "turn to quality" in food consumption. The word "quality" here should not be restricted to the meaning of "better quality" or "different quality," but refers to the characteristics of food resulting notably from distinct processes. Indeed, the turn to quality relates to a search for better-quality products, but quality also includes a range of symbolic motivations behind the purchase of a food product: its origin (Is it from my region, from this area?), its mode of distribution (Will I have contact with the producer? Can I find it in specialized, health, or green food stores?), its mode of production (Is it antibiotic free, GMO free? Is it organic? Does it respect religious traditions—halal, kosher…), and values (Is it fair? Is this tuna dolphin safe?).

Thus, there is a great diversity of consumption practices existing under the label "turn to quality." Moreover, contradictions are numerous. For example, some authors have associated the turn to quality with progressive values such as fairness in consumer–producer relationships or ecological protection. Yet research has shown that buying local is not always a matter of fairness, and neither is it inescapably an ecological good. In a study of food consumption in five regions of England in the wake of the sanitary crisis of the 1990s, Michael Winter (2003) showed that what is sometimes interpreted as a search for quality can also appear as a fear of risk. He concludes that consumption is a complex act and that researchers should avoid false dichotomies such as capitalism on the one side and "buying local" on the other. A similar argument is made by Clare Hinrichs (2003) concerning local food distribution systems in Iowa, where localism seems to express a certain conservative reaction to globalization, a cultural defense against foreign food. Hinrichs thus argues for a broader vision of the turn to quality that could simultaneously integrate defensive conservative reactions and progressive sustainability values by putting them at the poles of a continuum where different degrees also exist.

This also implies that what is called the "embeddedness" approach to alternative agriculture should be reconsidered. According to David Goodman, the embeddedness concept favors only one aspect of social life—connectivity—and fails to see the social relations related to market dynamics (Goodman 2003). This approach thus tends to build a caricature that portrays society as being in opposition to the market. As drivers of the food chain, however, the turn to quality in consumption practices and the initiatives of alternative agriculture seem, rather, to involve a complex mix of the market, the social relationship and values, and also new modes of regulation.

2.2.2 *Environmental and sustainable agriculture certification*

One concern encompassed by the turn to quality trend is protection of the environment. Because, like other such issues, environmental protection cannot be assessed by examining the product itself, the consumer must rely on specialized distribution channels or certification programs. The Rainforest Alliance, to provide a first example, certifies shade-grown products in the tropics so as to help poor farmers sustain the forest as their home, as a source of income, and as a refuge for migrating birds. The Core Northeast Values initiative, with the collaboration of universities in the region, has developed a program to help farmers transition to ecological agriculture, which also involves certification and labeling. Producers from numerous regions are getting together to begin certifying local "terroir" products to protect their added value against imitation by larger competitors. Indeed, sustainable agriculture (and food) initiatives have widely adopted the certification and labeling strategy to promote their products. In doing so, they have contributed to the creation of a new regulatory framework in this sector—one that can be termed as "hybrid" (Audet 2010).

Certification and standardization existed well before most of these examples. Kosher and halal labels originating from faith-oriented organizations may have been among the first of the kind. However, governments have also played a part in the emergence of this phenomenon: The Dolphin Safe label was an initiative of the U.S. government to promote the use of a certain kind of (dolphin safe) net in tuna fishing, and the French government has long had its own system of certification and labeling for "origin appellations" in the winery, cheese and dairy, and meat products sectors, among others. Moreover, because a few sustainable agriculture labels have received a lot of attention from consumers and firms, governments have begun to introduce regulations to ensure consumer protection and foster the development of this sector. Organic agriculture may be the best example of this trend, as most governments now have regulations concerning organic certification and even organic standards (Willer et al. 2008).

In a bottom-up fashion, a hybrid regulation framework has emerged from the interlocking of all these initiatives, in the domain of agriculture as well as in many other sectors. This framework involves different types of organizations attending to certification, inspection, accreditation, promotion, standard-setting, etc., whether they originate from civil society or from government or intergovernmental agencies. Such a mix of private sector, civil society, and state actors brings a set of very complex issues to the growth of the food certification business. For example, labeling confers a new characteristic on a product, one that an unlabeled equivalent product does not have. According to free market promoters or Global South producers trying to export in northern markets, such a characteristic can become an unfair advantage, hence introducing "distortion" in the market. This is why WTO rules on technical barriers to trade state that certification schemes should aim at international harmonization, so that the advantage will be the same for certified products—such as organic products—all around the world, and certification does not become a technical barrier to trade (Audet 2010). As a result, hybrid regulatory frameworks increasingly interact with conventional national and international regulation frameworks, forcing them to adapt to the new reality of sustainable food production and distribution.

2.2.3 *Fair trade and organic agriculture*

Among types of sustainable agriculture certification, organic agriculture and fair trade are probably the best documented and most widely known. Although organic agriculture organizes agricultural production around techniques aimed at maintaining soil productivity and controlling pests without the use of chemicals, genetically modified organisms or other synthetic materials (antibiotics, growth hormones, etc.), fair trade, on the other hand, is a system of trade between marginalized producers in the South and consumers in the North, based on values related to equity. Both organic agriculture and fair trade use certification and labels to control the transparency of their respective food and staple chains and to make their products identifiable on market shelves. Historically, both initiatives conceived themselves as "movements," but fair trade and organic agriculture have evolved a great deal since their emergence in the 1960s. They now also share the challenge of conventionalization, that is, the process by which organic agriculture is being mainstreamed and its standards weakened (De Wit and Verhoog 2007), and by which fair trade is being extended to larger production models (such as plantations) and multinational agri-food corporations.

The conventionalization phenomenon is also somewhat challenging when it comes to analyzing sustainable agriculture, especially for those observers who tend to understand these initiatives as being in opposition to capitalism and free markets. It is in part in the literature on this subject that Polanyi's concepts of embeddedness and disembeddedness have grown roots. For example, Laura Raynolds (2000) and Elizabeth Barham (2002) have shown that the organic and fair trade movements were partly built on a critique of conventional agriculture, as both movements reject the specialization, intensification and "chemicalization" of agriculture. These authors demonstrate that productivism is a figure against which these movements define themselves, and that the "international organic and fair trade movements seek to create alternative trade circuits for items produced under more environmentally and socially sustainable conditions that simultaneously parallel and challenge the conventional global agro-food system. … Both movements work to re-embed production in natural and social processes and create an alternative agro-food system" (Raynolds 2000, p. 306).

Re-embedding means that the production and trade activities in organic and fair trade food chains are supposed to create or strengthen social relations of solidarity and redefine the role of farmers in natural cycles. Fair trade and organic agriculture hence hypothetically contribute to the creation of new social and economic institutions embedded in nature and social relations. This interpretation, however, can be challenged. In fact, one of the main characteristics of these social movements is their tendency to work inside the market. Of course, they have led to the development of alternative distribution routes and networks—although conventionalization is making these networks increasingly less alternative—but they have also had to mobilize expertise on product quality and sanitary standards, shipping and handling, marketing know-how, and above all, the business of standard-setting, evaluation, and certification. This has not only been true for fair trade and organic agriculture, but also for most initiatives related to sustainable agriculture and food.

2.2.4 *Social responsibility*

Partly in response to the protest movement and fair trade discourse, but more generally anchored in a global trend toward corporate social responsibility, agri-business corporations have begun to rethink the branding of their products, enhancing the social and environmental quality of the latter and sometimes initiating a more profound transformation of their strategic positioning. The case of Danone, among big players, and Liberty as an example of a smaller business, clearly illustrate how the discourse, as well as practices, development strategies, and public positioning have begun to be modeled around social issues, especially in the field of agriculture. Corporate social responsibility is often defined as voluntary initiatives going beyond the law and mechanisms for dialogue with stakeholders; however, Gendron (2009) has shown that this movement indeed reveals a deep transformation in business legitimacy rationale, forcing corporations to define their mission with respect to the common good and general interest of society. Corporations thus play a new role in the shaping of social issues, which shed light on their positioning and discourse like never before. One must add to this the fact that corporations are seen as being responsible for numerous contemporary problems, especially those of an ecological nature. Contrary to what is sometimes said, corporate social responsibility cannot be reduced to greenwashing or superficial change. When these practices are encountered, they are generally only first and wrong steps in a long path toward the real integration of new parameters in production, governance, and relations with communities and consumers or economic partners in the food chain. As shown by the "awakening" of some supermarkets that were denounced by environmental groups, corporate social responsibility is no longer an option, and even in areas that are not (as yet) regulated, firms are obliged to show their contribution to solving society's challenges.[5]

These corporate strategies are inspired by and in some ways embody an expression of the exceptionality of agriculture, and they build on the social and ecological embeddedness of the products themselves. They are thus participating in the construction of the new rationale that will predominate in agricultural governance in the coming years.

2.3 Conclusion

Considering the intensity, diversity, and persistence of struggles surrounding agriculture, it seems to be at the heart of the construction of a new rationale that will bring about or sustain a governance system that will move away from the ideal of liberalization. Although this hypothesis seems daring today, it is not absurd to think that the evolution of ecological issues and their growing importance at the international level will compel a governing system that challenges the paradigm of trade liberalization as the only or most powerful vector of public good.

In this sense, as it seems to bring together today's social and environmental problems, agriculture might also be the most fertile ground for governance innovation as well as a window into the most important cultural, political, and economic transformations of our societies. Following the arguments we have presented in this chapter, the key drivers of the food chain are not to be found in the increased commoditization of food, as was the case during the last centuries, but rather in the social, ecological, and even mythical reinvestment in food.

This reinvestment does not have to occur outside the market, because indeed it is transforming the market, which is increasingly being structured around normative parameters that cannot be reduced to the procedural imperatives of a free market. Agriculture is contributing to this transformation and to the rule of what might be called the second globalization; that is, an international integration organized on the basis of a rationale other than market liberalization. Multifunctionality will play an important role in the legitimization of this new order in the field of agriculture, and this is all the more probable because it will serve the interests of the dominant actors, as proved by the failure of the last round of WTO negotiations.

---

[5]By way of illustration, we present the following case: After being ranked last in 2009 by Greenpeace with regard to the issue of fisheries, the Metro supermarket adopted a new policy to avoid the distribution of endangered species, to which the environmental group responded favorably, granting them a fifth position in 2010.

Acknowledgments The authors would like to thank Johanna-Maud Egoroff and François Décary-Gilardeau for their help with the reference resources that inspired this chapter, Bernard Girard for his careful reading of the preliminary version of the text, and Nguyen and Murray associées for their work in editing the final version. René Audet would like to thank the Social Sciences and Humanities Research Council of Canada, postdoctoral scholarship program, for the financial support granted during the writing of this chapter.

References

Aggarwal, R. 2005. Dynamics of agriculture negotiations in the world trade organization. *Journal of World Trade* 39(4): 741–761.

Appadurai, A. 1996. Modernity at large cultural dimensions of globalization, Minneapolis: University of Minnesota Press.

Audet, R. 2010. L'institutionnalization de l'agriculture biologique vue par le sud: régulation inégalitaire vs tropicalization. In: *Développement durable et responsabilité sociale. De la mobilization à l'institutionnalization.* Montréal: Presses internationales Polytechnique.

Barham, E. 2002. Towards a theory of values-based labeling. *Agriculture and Human Values* 19(4): 349–360.

Bellwood, P. 2005. First Farmers—The Origin of Agricultural Societies. Oxford: Blackwell Publishing.

Braudel, F. 1967. *Civilization matérielle, économie et capitalisme (XVe-XVIIIe siècle)*, vol. 1. Paris: Armand Colin.

Champion, E. 2004. Analyse du discours de Monsanto 1970–2002: les dimensions sociales et environnementales dans le renouvellement de la légitimité institutionnelle de l'entreprise transnationale. Mémoire de maîtrise, Université du Québec à Montréal, Montréal.

Clapp, J. 2006. WTO agriculture negotiations: Implications for the global south. *Third World Quarterly* 27(4): 563–577.

Coleman, W., W. Grant, and T. Josling. 2004. *Agricultural in the new global economy.* Cheltenham: Edward Elgar Publishing.

Cohen, M. N. 1977. The Food Crisis in Prehistory—Overpopulation and the Origin of Agriculture. New Haven, CT: Yale University Press.

De Wit, J., and H. Verhoog. 2007. Organic values and the conventionalization of organic agriculture. *NJAS—Wageningen Journal of Life Sciences* 54(4): 449–462.

Drope, J.M., and W.L. Hansen. 2006. Anti-dumping's happy birthday? *The World Economy* 29(4): 459–472.

Evenson, R.E., D. Gollin. 2003. Assessing the impact of the green revolution, 1960 to 2000. Science 300: 758–762.

Food and Agriculture Organization of the United Nations (FAO). 2006. The state of the World Fisheries and Aquaculture 2006, Rome: FAO.

G33 on Special Products and the Special Safeguard Mechanism (G33). 2008. G33 Statement Informal TNC, Geneva, 26 June 2008.

Gendron, C. 2009. L'entreprise comme vecteur du progrès social: la fin ou le début d'une époque? *Cahiers de la chaire de responsabilité sociale et de développement durable—collection recherche (1-2009)* Université du Québec à Montréal, Montréal.

Goodman, D. 2003. The quality 'turn' and alternative food practices: Reflections and agenda. *Journal of Rural Studies* 19(1): 1–7.

Group of 77 and China (G77). 2001. Declaration by the Group of 77 and China on the Fourth WTO Ministerial Conference at Doha, Qatar, 22 Oct 2001.

Hails, R.S. 2002. Assessing the risks associated with new agricultural practices. *Nature* 418(6898): 685–688.

Hincrichs, C.C. 2003. The practice and politics of food system localization. *Journal of Rural Studies* 19(1): 33–46.
Huffman, W.E. 2004. Production, identity preservation, and labeling in a marketplace with genetically modified and non-genetically modified foods. *Plant Physiology* 134(1): 3–10.
Josling, T. 2002. Globalization and transnational corporations. In *Agricultural globalization, trade and the environment*, ed. J. Moss, G. Rausser, A. Schmitt, T. Taylor, and D. Zibermann. Dordrecht: Kluwer.
Lal Das, B. 2005. *The current negotiations in the WTO. Options, opportunities and risks for developing countries*. London/New York: Zed books/third world network.
Moschini, G. 2008. Biotechnology and the development of food markets: Retrospect and prospects. *European Review of Agricultural Economics* 35(3): 331–355.
Polanyi, K. 2001. *The great transformation: The political and economic origins of our time*. Boston: Beacon.
Raynolds, L.T. 2000. Re-embedding global agriculture: The international organic and fair trade movements. *Agricultural Human Values* 17(3): 297–309.
Trebilcock, M.J., and R. Howse. 2005. *The regulation of international trade*, 3rd ed. London/New York: Routledge.
Vergès, P. 1989. Représentations sociales de l'économie: une forme de connaissance. In *Les représentations sociales*, ed. D. Jodelet. Paris: Presses universitaires de.
Wallerstein, I. 1974. *The modern world-system I: Capitalist agriculture and the origins of the European world-economy in the sixteenth century*. New York: Academic.
Willer, H., M. Yussefi-Menzler, and N. Sorensen. 2008. *The world of organic agriculture. Statistics and emerging trends*. Bonn: IFOAM.
Winter, M. 2003. Embeddedness, the new food economy and defensive localism. *Journal of Rural Studies* 19(1): 23–32.

Part II
Life cycle/environmental impact assessment

Chapter 3
Life cycle assessment and the agri-food chain

John E. Hermansen and Thu Lan T. Nguyen

3.1 Introduction

It is well acknowledged that human activities may have adverse environmental effects and that measures are needed to mitigate such effects. Our consumption of different products is linked to environmental effects, the magnitude of which depends on the pattern of consumption. Life cycle assessment (LCA), a product-oriented environmental assessment tool, provides a systematic way to quantify the environmental effects of individual products or services from cradle to grave, thus allowing estimation of the environmental effects related to the consumption and use of different products. At the same time, this tool can be used efficiently to identify improvement options in relation to environmental effects associated with the production of a certain product. Life cycle assessment is thus supported by the United Nations Environmental Programme (UNEP) as an environmental management tool in addressing environmental challenges.

Life cycle thinking is also at the core of the integrated product policy (IPP) within the European Union (EU) (EC 2003), the aim being to reduce the environmental impact of our consumption. The IPP is thus a response to the World Summit on Sustainable Development in 2002 (WSSD 2002). It is a vital assumption of the IPP that the environmental performance of a product or service can be a factor in giving companies or their products a competitive edge; thus, it is a separate aim of the IPP to create the right framework for market conditions that favor environmental improvements in the product chain.

J.E. Hermansen (✉) • T.L.T. Nguyen
Department of Agroecology, Aarhus University,
Research Centre Foulum, Tjele, Denmark
e-mail: John.Hermansen@agrsci.dk; ThuLan.ThiNguyen@agrsci.dk

J.I. Boye and Y. Arcand (eds.), *Green Technologies in Food Production and Processing*,
Food Engineering Series, DOI 10.1007/978-1-4614-1587-9_3,

Following this rationale the EU Commission initiated a project to identify those products or services with the greatest environmental impact throughout their life cycle from cradle to grave (Tukker et al. 2006) as a background for policy measures to stimulate a reduction in environmental impact. The overall conclusion was that the three areas—food (and drink), private transport, and housing—each was responsible for 20–30% of the total environmental impact of the EU-citizens' overall consumption.

This highlights the importance of understanding the environmental impacts related to the consumption and use of different foods and the options to reduce these impacts. This is even more relevant because of the very big difference in the environmental impact among different foods (Mogensen et al. 2009), in which for example the contribution to global warming ranges from 0.1 to 20 kg CO_2 equivalents per kg of food ready at the retailer.

3.2 The concept of life cycle assessment

3.2.1 The general idea

On the one hand, the basic concept of LCA is to account for the use of resources (e.g., fuels, minerals, water, land) and the emissions of various air and water pollutants (e.g., carbon dioxide, nitrate, etc.) that occur during all stages in the production and use of a product; and on the other hand, to classify and group the effects of the resource use and emissions into relevant categories of importance for the evaluation of their environmental impact.

Figure 3.1 illustrates major processes involved in the production of pork. In the example, pig production takes place in Denmark but relies on feed production in South America and elsewhere, which requires input of fertilizer and energy, and pork consumption occurs outside Denmark. This type of food production and related trade is far from exceptional nowadays.

At each stage resources are used and emissions are generated that need to be accounted for through the life cycle inventory (LCI). The chain of most foods is very complex, consuming resources and producing a wide range of emissions. The effects of resource use and emissions are classified and characterized with respect to their environmental impact into various impact categories—often named midpoints—such as nonrenewable energy use, global warming, acidification, eutrophication, etc. This reflects how the use of resources and the emissions from production of food may affect the environment. In the food literature, the LCA often ends with these impact categories, but the LCA can also be taken further to estimate damage to the environment.

For example, Fig. 3.2 shows the overall scheme of a comprehensive impact assessment method, the Stepwise2006 (Weidema et al. 2007), linking various LCI results via 11 midpoint categories to damage categories concerning three safeguard subjects: humans, ecosystems, and resources. The classification of the LCI results,

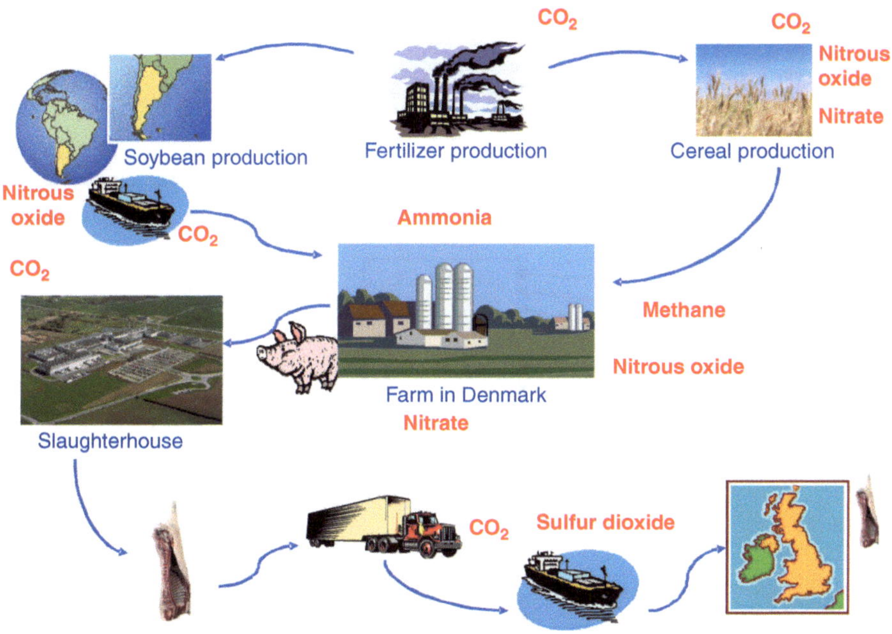

Fig. 3.1 Example of resource use and emissions to take into account in an LCA of pork (schematic)

for example, input from the technosphere, nature, and emissions (see Fig. 3.2) into midpoint categories is strongly supported by well-acknowledged standards (e.g., IPCC for global warming), whereas the further classification into damage categories is less well accepted. For decision support and communication, damage categories are very suitable, but uncertainty increases substantially, so an apparent dilemma exists for practitioners performing LCA.

Different classifications and methodologies exist in the units of the midpoints. For example, to display the result for acidification, the EDIP97 (Wenzel et al. 1997) uses the unit sulfur dioxide equivalent (kg SO_2e), whereas the EDIP2003 (Hauschild and Potting 2005; Potting and Hauschild 2005) uses area of unprotected ecosystem (m^2 UES) expressing the area of ecosystem that is moved from an exposure below to an exposure above the critical load in relation to maintaining the biodiversity of the area (Steen 1999; Goedkoop and Spriensma 2001; Jolliet et al. 2003; Bickel and Friedrich 2005).

As mentioned, many published LCA studies regarding food stop with the midpoint evaluation. Common midpoint categories used when analyzing the farming stage of food production are shown in Table 3.1, which also shows the characterization factors used when LCI results are translated into contributions to selected midpoint impact categories. For example, the major contributors to the global warming impact category are CO_2, N_2O, and CH_4, in which N_2O and CH_4 have a much higher impact per kg substance emitted than CO_2. Typically the contribution from N_2O and

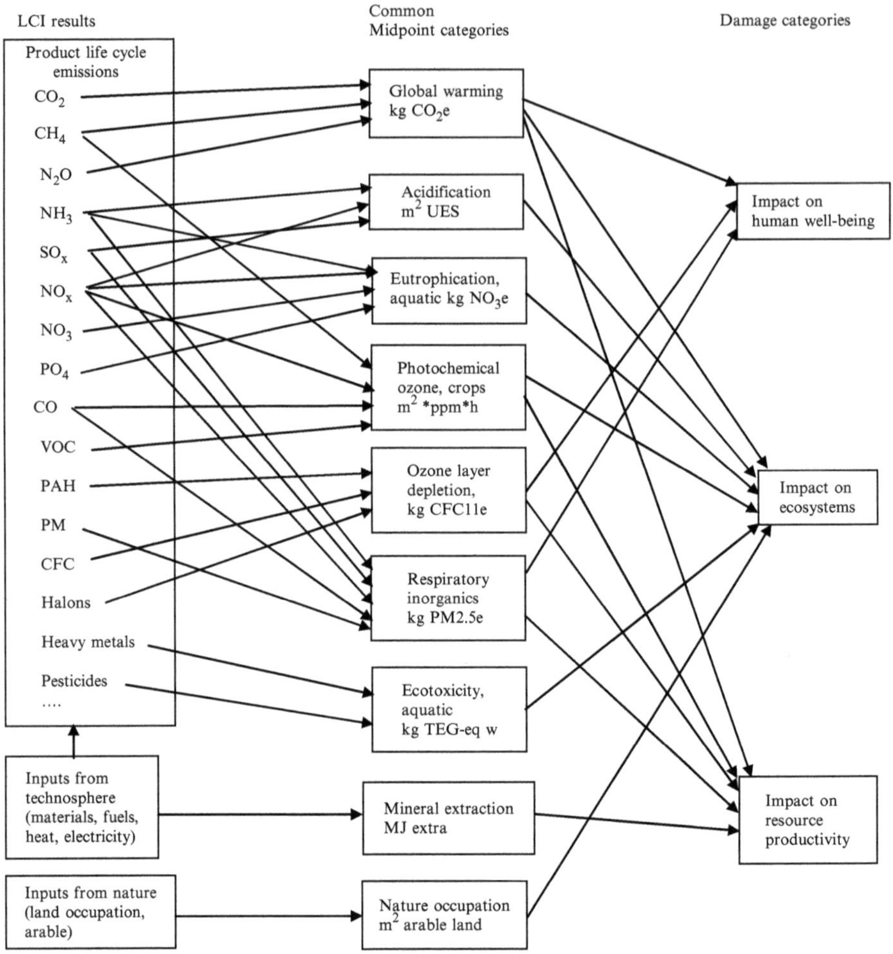

Fig. 3.2 General structure of an LCA model linking midpoint and damage categories

CH_4 for animal food products is more important for the total impact related to global warming than CO_2. Thus Dalgaard et al. (2007) estimated, for the pork chain illustrated in Fig. 3.1, that the emissions of N_2O were responsible for 44%, CH_4 for 32%, and CO_2 for 20% of the global warming impact of the pork produced.

When looking at all processes in the food chain, the number of relevant impact categories increases compared with the ones given in Table 3.1. Thus, Weidema et al. (2008) estimated the impacts that were important when considering meat and dairy consumption within the EU. In descending order, the following impact categories contributed to more than 10% of the impact of all consumptions (and services) within the EU: Ecotoxicity (aquatic), Eutrophication, Nature occupation, Acidification, Respiratory inorganics, Global warming, Respiratory organics, and Photochemical ozone.

Table 3.1 Selected impact categories with related units, contributing elements and characterization factors

Impact category (midpoint)	Unit	Contributing elements	Characterization factors
Acidification	kg SO_2e	SO_2	1
		NH_3	1.88
		NO_x[a]	0.70
Global warming[b]	kg CO_2e	CO_2	1
		CH_4	21
		N_2O	310
Eutrophication (nutrient enrichment)	kg NO_3e	NO_x	1.35
		P_2O_5	14.09
		NH_3	3.64
		NO_3^-	1
		PO_4^{3-}	10.45
		NH_4^+	3.6
		COD[c]	0.22
Land use	m^2a	Land occupation	1

Source: Thomassen et al. (2008a)
[a]NO and NO_2
[b]Assuming a 100-year time horizon
[c]Chemical oxygen demand: The amount of oxygen consumed in the oxidation of all the organic compounds, both biodegradable and recalcitrant, to carbon dioxide and water

3.2.2 *Approaches and complexities*

The previous section outlined a straightforward LCA, taking into consideration data requirement challenges for estimating emissions. However, apart from these technical issues, a number of assumptions have to be made when performing an LCA. ISO standards have been produced (ISO 14041 1998; ISO 14044 2006) to support transparency in these assumptions. According to these standards, an LCA includes the following stages:

- Defining goal and scope (functional unit, "type" of LCA)
- Outlining the model used (inventory, processes included, emissions accounted)
- Impact assessment
- Interpretation

This section emphasizes some methodological choices that may affect the results of the LCA, and that therefore must be stated explicitly when reporting LCA results. Two critical choices are apparent: whether to use an attributional or a consequential approach, and the handling of co-products through allocation or system expansion.

An attributional approach describes the emissions and resource flows within a chosen system attributed to the delivery of the functional unit. In contrast, a consequential approach estimates how emission and resource flows within a system change in response to the change in output of the functional unit (Ekvall and

Weidema 2004). It can also be rephrased as follows: Although the attributional approach describes the resource use and emissions that have occurred to produce the product in question, the consequential approach accounts for the resource use and associated emissions that arise to replenish stocks of the product that have been used.

Furthermore, in food production (particularly in livestock production) more than one product is commonly created. There is a need in these cases to allocate the environmental burden between different products. In principle, the ISO 14041 recommends that allocation should be avoided, if possible, by expanding system boundaries or dividing the unit process into subprocesses (ISO 14044 2006). If untenable, allocation for the system can be done in such a way that it reflects the physical or economic relationships.

The choice between the two methods for distributing the environmental loads between product and co-products (allocation or system expansion) depends on which type of LCA is to be used: attributional or consequential. As emphasized by Thomassen et al. (2008b), who investigated the difference in having an attributional or a consequential approach performed in life cycle assessment of milk, there is a strong connection between the choice of the approach used and how to handle co-products. With an attributional approach it is optional to use allocation or system expansion to handle co-products. However, when using the consequential approach, the only (rational) way to handle co-products is through system expansion.

As evidenced in the current literature, many studies have relied on an attributional view using economic allocation, which means distributing the emissions according to the market value of the different products produced. However, in most cases, LCA for agricultural and food products is used as a support tool for decision making—in "what would happen if?" situations—and in this respect, we argue that consequential LCA using the system expansion approach is most appropriate. This is sometimes called "the marginal approach" because basically, what is estimated is the change in resource use and emissions following the demand (and thus production) of a functional unit such as one extra kg, kWh, or MJ of a product. The subject and guidelines for estimating the marginal changes are described in more detail in Weidema et al. (2009).

3.3 Business and policy developments within life cycle assessment

As indicated, LCA is a tool that provides comprehensive insight into the environmental impact of the use of different products. Although many companies originally thought LCA might be a marketing communication tool , a survey by Frankl and Rubik (2000) showed that the most common use of the application was for product improvement and strategic choice support. It was difficult to communicate the results in detail because of various assumptions and impact categories. Halberg (2004) offers many examples of LCA use in the food business.

Recently, there has been a boost in the mutual interest from business and policy in LCA approaches exacerbated by the concern about global warming. There now seems to be a general consensus—in agreement with integrated product policy within the EU—that it is important to have a life cycle approach when considering this issue, and LCA is an appropriate tool to use for environmental analysis of food production systems instead of more partial tools such as "food miles." This is important because within the agriculture and food sector, emissions other than CO_2 (related to energy for transport) are even more important than CO_2 considering their contribution to the total impact on global warming. For example, among major contributors are emissions of N_2O that for many foods are the single most important contributor to global warming, and CH_4 is a particularly significant contributor for ruminant products.

There are some very important initiatives that need to be highlighted in this respect. In response to demands from business and policy in the United Kingdom, the British Standards Institute (BSI) has launched the development of a Publicly Available Specification (PAS 2050) for a method for measuring the embodied greenhouse gas (GHG) emissions from goods and services at the request of DEFRA (the Department for Environment, Food, and Rural Affairs) and the Carbon Trust (BSI 2008).

PAS 2050 is a carbon footprint analysis tool that considers all life cycle stages along the supply/value chain of a product (goods and/or services), that is, from raw materials to end of life, and includes the six greenhouse gasses under the Kyoto protocol. The PAS method aiming at quantifying embodied GHG emissions of goods and services is supposed to enable organizations such as businesses to effectively measure the climate change–related impacts of their goods and services, so they can use this information to improve their climate change–related performance. The PAS is intended as the first step toward a future internationally agreed method for organizations to measure the GHG emissions embodied in their goods and services. Some differences exist regarding PAS 2050 and some specifications in the ISO standard, mainly because of efforts to allow transparent implementation, including means of verification; therefore, when LCA results are supposed to be used in communication and for benchmarking, it is important that basic assumptions are stated explicitly.

Within the dairy sector remarkable initiatives are taken through the International Dairy Federation (IDF), which has launched the "Global Dairy Agenda for Action" in order to address the challenges of global warming related to dairy production. The Agenda for Action is a commitment by the world's dairy farmers, associations, and companies to reduce greenhouse gas emissions per unit of dairy production as part of its contribution to help address global warming. Most importantly this includes the promotion of the development of a standard methodology framework for assessing the carbon footprint of milk and dairy products based on the life cycle methodology (IDF 2010). Basically this initiative underpins the interest from the business side to contribute to the development of transparent standards within the LCA to be used for practical and benchmarking purposes. Concurrently, an initiative has been taken within the FAO. The FAO report "Greenhouse Gas Emissions from the Dairy Sector" covers all major milk production systems, from nomadic herds to intensified dairy operations in a life cycle approach (FAO 2010). The assessment is part of an ongoing program to analyze and recommend options for climate change mitigation. The next step is to use a similar approach to quantify GHG emissions associated with other major livestock species, including buffalo, poultry, small ruminants, and pigs.

The examples given in the preceding concentrate on global warming, which is in part caused by current interest in this issue and the need for common standards. This may seem to be a narrow view, taking the wide scope of LCA impact categories into consideration. Nevertheless, it is very important that these initiatives are built on the LCA, because this allows, assessments to be made of any effects related to other impact categories (e.g., acidification, eutrophication, photochemical smog, etc.) at the same time and on the same basis, and thus makes it possible for such impacts to be taken into consideration to avoid unintended side effects.

For the long-term development of the LCA methodology, it is worth mentioning the UNEP/SETAC initiative. In 2002, the United Nations Environmental Program (UNEP) and the Society for Environmental Toxicology and Chemistry (SETAC) launched The Life Cycle Initiative. This initiative aims to establish recommended methodologies and guidelines for impact assessment and thus enables users around the world to put life cycle thinking into practice. This initiative is exemplified by the UNEP-SETAC toxicity model (Larsen et al. 2008), and can be considered as a very important driver in methodological development and mutual recognition of methodologies applied in the LCA.

3.4 The issue of land use and biodiversity

Land use concerning food is a major issue currently not fully resolved in the LCA. The increased demand for food worldwide and the emerging demands for biofuels highlight the importance of considering this issue. FAO reports that the global demand for food is expected to continue to increase as a result of both population growth and rising incomes. The increase in the period from 2000 to 2050 is by a factor of approximately 1.5 for cereals and 2 for meat and milk (FAO 2006). The demand for biofuels driven by worldwide efforts to reduce fossil oil dependency is also growing. These increased demands, in combination, put heavy pressure on the availability of land resource and it is clear that this pressure has the potential to drive the clearance of natural habitats.

The occupation of land for food, and in particular for feed to support livestock production, is closely interrelated with impacts on biodiversity and, potentially, on emissions of CO_2 when land-use changes affect carbon stocks. The issue of land use has recently received increasing attention, both from academic and policy-oriented audiences to address the food, feed, and fuel dilemma. In current LCA studies, this land use impact category is often quantified as how much surface area in combination with a certain time period is required to produce one unit of output (Lindeijer 2000). This information is useful as a way to assess land use efficiency and productivity by making a comparison among different farming systems. However, because land is a limited resource, any increase in demand would inevitably lead to land use change, and this may decrease carbon stocks in vegetation and soils.

Recent reviews and discussions (Milà i Canals et al. 2006, 2007) have addressed this issue, acknowledging that it is important to include land use impacts in a LCA. A number of limitations and/or challenges, however, were raised such as off-site effects (identification of marginal land use), allocation of initial transformation impacts, assumption of natural relaxation, knowledge about the site-specific effects on soil quality, and biodiversity. Thus, better knowledge of the driving forces behind land-use change is essential.

Likewise, the food sector also poses a threat to global biodiversity through its use of new land for expanded pasture and crop area. Unfortunately, so far there have been no commonly accepted methods that reasonably translate land use pressures into loss of biodiversity, although a great deal of effort has been undertaken to do so (Lindeijer 2000; Mattsson et al. 2000; Weidema and Lindeijer 2001; Michelsen 2008). Schmidt (2008) pointed out that most existing methods addressing these impacts are either too coarse regarding the differentiation between different land use types, or too narrow regarding spatial coverage. Generic characterization factors for local species diversity in Central Europe were developed by Koellner and Scholz (2008), which can be used for marginal land use decisions within this specific region, but are recommended to be used just as a reference methodology for other regions, given that species diversity and the impact of land use on it can differ greatly from region to region. Thus, there is an urgent need to work out a coherent method for assessing the impacts of land use on biodiversity.

Given the fact that considerable efforts are being done to solve these issues, it will probably be typical to estimate the effects of land use on other impact categories to be included in the LCA of food. The inclusion would influence the magnitude of environmental impacts such as global warming, but to what extent remains to be explored. The issue is addressed in the next section, in which a case study comparing pork production in an indoor versus an outdoor system is presented.

3.5 Accounting for land use and land use change in estimating the GHG emissions related to livestock products: a case study on pig production in denmark

There are two major aspects of land use in relation to livestock production that can be considered in relation to GHG emissions: First, the occupation of land for feed production, which may take place at the farm or elsewhere in the world from the place where feeds are imported. Second, the (increased) demand for feed for livestock production may impose land use change, where native land is converted into cropland. In the PAS 2050 (BSI 2008) presently it is recommended that GHG emissions related to land use change are included in the carbon footprint of feeds if they are imported from countries where land use change does occur. In relation to this accounting choice, some default values are recommended. It should be emphasized that this is a preliminary way of addressing the issue, as acknowledged by the PAS 2050. The aim of this work was to exemplify the consequences of including land use aspects in the GHG balance of livestock production systems. To do so we introduced the term "opportunity cost" of land use in addition to land use change. It considers the value of land for alternative uses such as forestry, raising the question of what is the best way of using land for the same amount of output at least GHG cost. If the land were to revert to forest, it would effectively sequester carbon; therefore, the loss of carbon sequestration potential needs to be allocated to forage, cereal, and oilseed crops used for livestock. We compared pork production in an indoor versus an outdoor system using a life cycle approach. The conventional indoor pig production system was defined in a report on the environmental impacts of Danish pork production (Dalgaard et al. 2007). The outdoor system considered was a theoretical organic pig production system in which the sow herd was kept on grassland with access to small huts, and the fattening pigs were kept in indoor facilities as described in Halberg et al. (2010).

3.5.1 Materials and methods

The system boundary chosen for pork chain until farm gate in the two systems considered is presented in Fig. 3.3. In relation to feed composition, the consequential LCA allows us to make sound simplifications. For indoor pigs, we assumed that all feed was bought from outside and consisted primarily of cereals and soy meal. For outdoor pigs, we assumed that the feedstuffs included both bought-in and home grown feed. Home grown feed consisted of cereals, legumes, and grass and the bought-in feed was made up by cereals and soy meal in proportions that balanced the feed composition according to the pigs' nutritional requirements. The manure produced by the pigs in the two systems was considered to be available for growing crops, thus saving the use of fertilizers. Another assumption was that the "direct energy" consumed by the two systems was derived from diesel fuel and electricity for farm operation and animal housing, respectively.

The functional unit used to compare the two pig production systems was 1 kg meat live weight delivered from the farm. In our study, we used system expansion to handle the issue of allocation arising from the following subprocesses related to pig production: (1) soy meal production, (2) cash crop production in the outdoor system, and (3) indoor-pig manure utilization. For a detailed discussion on this particular subject see Nguyen et al. (2010a).

Essential assumptions regarding inputs, outputs, on-farm emissions and land use per 1,000 kg meat produced in the two systems are summarized in Table 3.2. A modeling technique for the estimate of CH_4 and N_2O emissions from the indoor and outdoor system is described in Nguyen et al. (2010a, b), respectively.

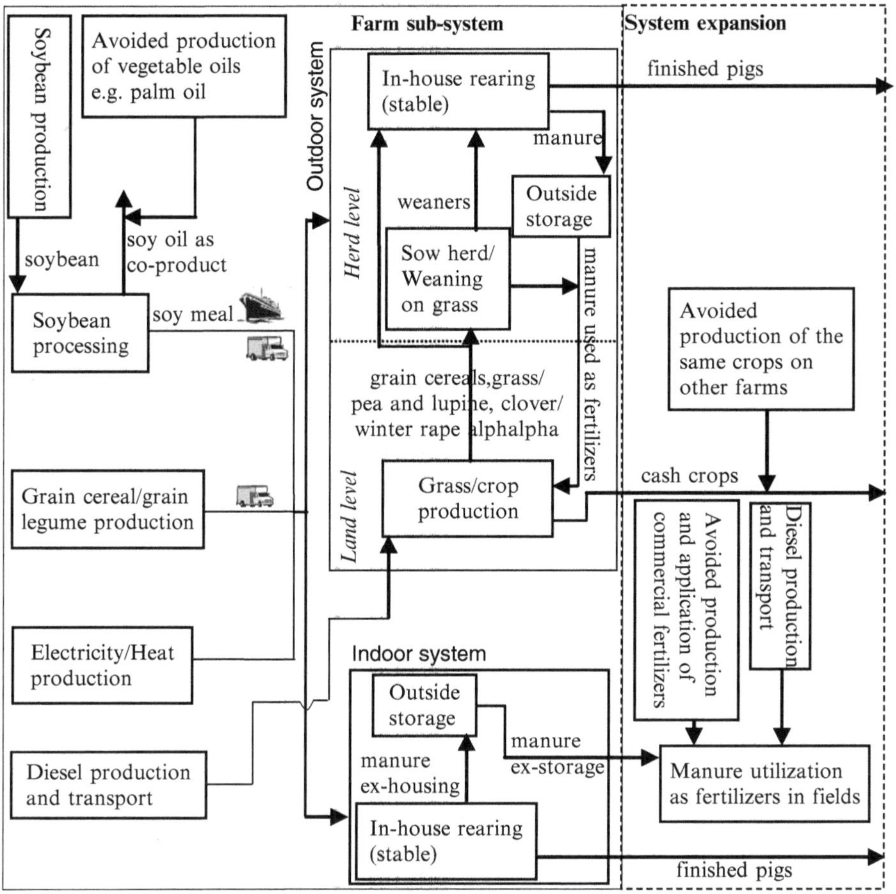

Fig. 3.3 Overview of the product chain of pig meat until farm gate in indoor and outdoor systems

3.5.2 GHG balance analysis: land-use–related GHG emissions

The most important protein feed that livestock production has to rely on is soy meal, which comes mainly from Latin America. Latin American rainforests are rapidly declining and one of the main causes is the expansion of soybean growing areas (WWF 2004). This creates a large source of carbon emissions. The clear connection between soy meal used in livestock production and land use change enables a straightforward translation from the area of land use for the feed into CO_2 emissions. The impact of land use related to home grown feed and bought-in cereals can be discussed in terms of the opportunity cost of rearing livestock or growing crops on land considering alternative uses of land for maximum carbon storage (e.g., forestry).

Table 3.2 Inventory for two different farm systems of pig production (reference flow = 1,000 kg meat live weight)

	Outdoor (organic)				Indoor (conventional)			
							Manure export	
	Herd level	Land level	Farm gate	Land use m^2a	Farm gate	Land use m^2a	Cost	Benefit (fertilizer displacement)
Inputs								
Bought-in feed, kg								
Grain cereals	1,807		1,807	5,836	2,620	5,339		
Soy meal	261		261	1,285	520	1,846		
Home grown feed, kg								
Grain cereals and legumes	1,104			3,367				
Grass (silage and grazed), DM	427			854				
Manure, N kg								
Collected indoors		32.3						
Deposited on grass		16.5						
Manure, P kg		10.1						
Direct (on-farm) energy use								
Electricity, kWh	186		186		229			
Heat, MJ	0		0		239			
Diesel (traction), MJ		1,167	1,167				168	11.3
Transport of bought-in feed tkm								
By truck	312		312		569			
By ship	3,134		3,134		6,187			
Outputs, kg								
Live pigs	1,000		1,000		1,000			
Grain cereals and legumes		1,104						
Grass (silage and grazed), DM		427						
Cash crops, DM		153	153	−590				
Manure (as nitrogen)	48.8				38.5			29
Manure (as phosphorous)	10.1				8.4			8.4
On-farm emissions kg								
CH_4			37.1		42.5			
N_2O			1.9		0.6		0.8	0.5

It remains a challenge to develop an acceptable method for assessing the effects of land use and land use change (LUC) on GHG emissions given the variation in carbon stocks, carbon sequestration potential, and the dynamic of land use change in different agro-ecological zones in the world. Searchinger et al. (2008) recently estimated GHG emissions from the production of corn ethanol in the United States using a worldwide agricultural model to account for land use change. Their calculation assumed the loss of all carbon in vegetation and 25% of soil carbon following land conversion from forest to cropland. Another study (Murty et al. 2002) found that, on average, the effect of land use change from forest to pasture on soil carbon was insignificant. The conversion also sacrifices the ongoing carbon sequestration that would take place each year if forest is not cleared. The information available from the two references enabled an estimate of the weighted average carbon emissions rate in kg CO_2/m^2/year for all forest systems, using a depreciation period of 30 years. We also assumed that the rate of carbon emissions from land conversion and the rate of carbon sequestration loss owing to using land for livestock and feed instead of reverting it to forest were of the same magnitude.

On this basis we estimated GHG implications based on the distinction of whether land use was for home grown feed (direct land use) or for bought-in feed (indirect land use). For the former case, we used the rate of 1.8 and 2.2 kg CO_2/m^2/year in Europe to estimate land opportunity cost concerning GHG savings potential, if grassland and cropland used to grow the pigs were to revert to forest, respectively. For the latter case, we considered the world weighted average emissions rate associated with the conversion of forest to cropland, 1.9 kg CO_2/m^2/year, would be acceptable for the estimate, but a check against the lowest and highest rate (1.2 and 2.9 kg in Africa and South/Southeast Asia, respectively) was also performed.

3.5.3 *Results and discussion*

The comparative GHG balance of the two pig systems is summarized in Table 3.3. When excluding land use impacts, pig meat from the outdoor system had a slightly lower GHG emission per kg produced than the indoor system: 3.2 versus 3.5 kg CO_2e/kg. For the outdoor system, a lower emissions rate generated by less input of bought-in feed, combined with emission savings potential from the sale of cash crop, made it a somewhat better choice than the indoor one in relation to GHG emissions.

If the potential land use change related to the use of soy meal was taken into account (as it might be according to the PAS 2050), the total estimated GHG emissions per kg pork produced increased considerably and the emissions related to the indoor system became 24% higher than those related to the outdoor system: 7.1 versus 5.7 kg CO_2e per kg pork. On the other hand, if we included land opportunity cost related to land occupation for growing bought-in cereals, the GHG emissions more than doubled and the outdoor system resulted in more or less the same GHG emission rate as the indoor system (16.9 versus 17.4 kg CO_2e per kg pork).

Table 3.3 Greenhouse gas emissions from the two pig systems with and without accounting for land-use related GHG emissions (g CO_2e/kg pig meat live weight)

Item	Outdoor (organic)				Indoor (conventional)		
	Without land use change	Land use change	Land opportunity cost: direct land use	Land opportunity cost: indirect land use	Without land use change	Land use change	Land opportunity cost
Bought-in feed incl. transport							
Soy meal	303	2,477			475	3,558	
Grain cereals and legumes	1,397			11,248	1,828		10,291
Mineral feed	9				6		
Home grown feed			9,125				
On-farm energy use	249				215		
On-farm emissions							
CH_4	927				1,063		
N_2O	568				408		
Manure export to cash crop farms					–452		
Cash crops production	−226		−1,318				
Sum	3,227	2,477	7,807	11,248	3,543	3,558	10,291
Sum, with land use change and land opportunity cost[a]	24,759 (lowest: 19,431, highest: 31,431)				17,391 (lowest: 12,015, highest: 24,124)		

[a]The lowest and highest estimate is derived from the lowest and highest value of carbon emissions rate used to account for GHG implications of land use for bought-in feed (1.2 and 2.9 kg CO_2/m^2/year), respectively

Finally, if the opportunity cost of the direct land use was taken into account, the outdoor system resulted in a higher GHG emission footprint than the indoor one (24.8 versus 17.4 kg CO_2e per kg).

In this case study, we used an outdoor organic pork system contrasted with an intensive pork system, because these systems differ in land use requirement among others. However, it is important to acknowledge that this is not a careful and comprehensive comparison between organic and conventional production but only an example. Nevertheless, the analysis highlights the importance of high productivity of the land occupied, and that products that have a high land use demand often have a very high impact on global warming. It is important, however, in this respect to distinguish between the use of arable land that may be used for competing purposes, and the use of semi-natural areas for such uses as grazing (wholly or in part) that does not markedly change the nature of these areas. In a broader context, our example can be regarded as reflecting the demand for land that could have been used better in another way.

Clearly, the LCA methodology enables the identification of differences in environmental impact of different products and the improvement product potentials, also with a particular reference to the increased pressure on natural resources such as land and its consequential effect on GHG emissions. Nevertheless, in the particular case of GHG emissions related to land use and its change, further research is needed to improve the methodology to obtain good estimates of the emissions occurring in different agro-ecological zones.

3.6 Conclusion

Life cycle assessment, "a methodological tool that applies life cycle thinking in a quantitative way on environmental analysis of activities related to processes or products" (European commission 2008) is now a well-established and well-acknowledged methodology—in policy making and business development—to quantify the environmental impact of our consumption. Furthermore, LCA can be used as a tool for investigating improvement options in the production of a particular product or service. It has well-developed methodologies for estimating a wide range of environmental impacts in relation to food production and is supported by different databases. Therefore, the LCA is very efficient as a tool for benchmarking and investigating improvement options in the food chain. However, it must be acknowledged that performing an LCA requires a number of assumptions, and different rationales for making these assumptions exist. These aspects may not be very important when the main aim of the LCA is to identify improvement options within a particular food chain. However, when using LCA for benchmarking among products or producers these assumptions and methodological choices are very important. Thus, it is imperative that in each case of benchmarking, the rationale for performing the LCA be made explicit. The practical consequence is that it should be clearly stated on what basis the LCA is performed when used for communication purposes.

References

Bickel, P., and R.Friedrich. 2005. *ExternE: Externalities of energy: Methodology 2005 update*. European Commission. Directorate-General for Research Sustainable Energy Systems, Brussels, Belgium (EUR 21951).

BSI. 2008. *PAS 2050:2008—specification for the assessment of the life cycle greenhouse gas emissions of goods and services*. UK: British Standards.

Dalgaard, R., N. Halberg, and J.E. Hermansen. 2007. Danish pork production. An environmental assessment. *DJF Animal Science* 82: 1–34.

EC (European Commission). 2003. Communication from the commission to the council the European parliament, integrated product policy. Building on environmental life-cycle thinking. COM(2003) 302 final. http://eur-lex.europa.eu/LexUriServ/site/en/com/2003/com2003_0302en01.pdf.

Ekvall, T., and B.P. Weidema. 2004. System boundaries and input data in consequential 18 life inventory analysis. *International Journal of Life Cycle Assessment* 9: 161–171.

European Commision. 2008. European Platform on LCA. LCA Info Hub. Introduction to LCA. http://lca.jrc.ec.europa.eu/lcainfohub/introduction.vm.

FAO. 2006. *Livestock's long shadow—environmental issues and options*. Rome, Italy: Food and Agriculture Organization of the United Nations.

FAO. 2010. *Greenhouse gas emissions from the dairy sector: A life cycle assessment*. Rome, Italy: Food and Agriculture Organization of the United Nations. http://www.fao.org/docrep/012/k7930e/k7930e00.pdf.

Frankl, P., and F. Rubik. 2000. *Life cycle assessment in industry and business: Adoption patterns, applications and implications*. Berlin: Springer.

Goedkoop, M., and R. Spriensma. 2001. *The eco-indicator 9. A damage oriented method for life cycle impact assessment: methodology annex*. Amersfoort, Netherlands: PRé consultants. www.pre.nl.

Halberg, N. 2004. Life cycle assessment in the agri-food sector. In: *Proceedings of the 4th international conference of life cycle assessment in the agri-food sector*, Bygholm, Denmark (Dias Report No. 61), 6–8 October 2003.

Halberg, N., J.E. Hermansen, I.S. Kristensen, J. Eriksen, N. Tvedegaard, and B.M. Petersen. 2010. Impact of organic pig production systems on CO_2 emission, C sequestration and nitrate pollution. *Agronomy for Sustainable Development* 30(4): 721–731.

Hauschild, M., and J. Potting.2005. Spatial differentiation in life cycle impact assessment—Steen gen, Denmark (Environmental News 80).

IDF (International Dairy Federation). 2010. The global dairy Agenda for action. http://www.dairy-sustainability-initiative.org/Public/Menu.php?ID=36. Accessed May 14, 2010.

ISO. 1998. *ISO 14041:1998 environmental management—life cycle assessment—goal and scope definition and inventory analysis*. Geneva, Switzerland: International Organisation for Standardisation.

ISO. 2006. *ISO 14044:2006 environmental management—life cycle assessment—requirements and guidelines*. Geneva, Switzerland: International Organisation for Standardisation.

Jolliet, O., M. Margni, R. Charles, S. Humbert, J. Payet, G. Rebitzer, and R. Rosenbaum. 2003. IMPACT 2002+: A new life cycle impact assessment methodology. *International Journal of Life Cycle Assessment* 8(6): 324–330.

Koellner, T., and R.W. Scholz. 2008. Assessment of land use impacts on the natural environment. Part 2: Generic characterization factors for local species diversity in Central Europe. *International Journal of Life Cycle Assessment* 13(1): 32–48.

Larsen, H.F., M. MacLeod, M. Margni, T.E. McKone, J. Payet, M. Schuhmacher, D. van de Meent, and M.Z. Hauschild. 2008. USEtox—The UNEP-SETAC toxicity model: Recommended characterisation factors for human toxicity and freshwater ecotoxicity in life cycle impact assessment. *International Journal of Life Cycle Assessment* 13(7): 532–546.

Lindeijer, E. 2000. Review of land use methodologies. *Journal of Cleaner Production* 8: 273–281.

Mattsson, B., C. Cederberg, and L. Blix. 2000. Agricultural land use in life cycle assessment (LCA): Case studies of three vegetable oil crops. *Journal of Cleaner Production* 8: 283–292.

Michelsen, O. 2008. Assessment of land use impact on biodiversity. *International Journal of Life Cycle Assessment* 13: 22–31.

Milà i Canals, L., R. Clift, L. Basson, Y. Hansen, and M. Brandão. 2006. Expert workshop on land use impacts in life cycle assessment (LCA). 12–13 June 2006 Guildford, Surrey, UK. *International Journal of Life Cycle Assessment* 11(5): 363–368.

Milà i Canals, L., C. Bauer, J. Depestele, A. Dubreuil, R. Freiermuth Knuchel, G. Gaillard, O. Michelsen, R. Müller-Wenk, and B. Rydgren. 2007. Key elements in a framework for land use impact assessment in LCA. *International Journal of Life Cycle Assessment* 12(1): 5–15.

Mogensen, L., J. Hermansen, N. Halberg, R. Dalgaard, J.C. Vis, and B.G. Smith. 2009. Life cycle assessment across the food supply chain. In *Sustainability in the food industry*, ed. C.J. Baldwin, 115–144. Ames, Iowa: Wiley/Blackwell. chapter 5.

Murty, D., M.U.F. Kirschbaum, R.E. McMurtrie, and H. McGilvray. 2002. Does forest con-version to agricultural land change soil organic carbon and nitrogen? A review of the literature. *Global Change Biology* 8: 105–123.

Nguyen, T.L.T., J. Hermansen, and L. Mogensen. 2010a. Fossil energy and GHG saving potentials of pig farming in the EU. *Energy Policy* 38: 2561–2571.

Nguyen, T.L.T., L. Mogensen, J. Hermansen, and N. Halberg. 2010b. Complexities in assessing the environmental impacts of livestock products. In *Environmental assessment and management in the food industry*, ed. U. Sonesson, J. Berlin, and F. Ziegler. Cambridge: Woodhead Publishing.

Potting, J., and M. Hauschild. 2005. Background for spatial differentiation in life cycle impact assessment—the EDIP2003 methodology. The Danish Environmental Protection Agency, Copenhagen, Denmark (Environmental project no. 996).

Schmidt, J.H. 2008. Development of LCIA characterisation factors for land use impacts on biodiversity. *Journal of Cleaner Production* 16: 1929–1942.

Searchinger, T., R. Heimlich, R.A. Houghton, F. Dong, A. Elobeid, J. Fabiosa, S. Tokgoz, D. Hayes, and T.H. Yu. 2008. Use of U.S. croplands for biofuels increases greenhouse gases through emissions from land-use change. *Science* 319(5867): 1238–1240.

Steen, B. 1999. *A systematic approach to environmental priority strategies in products development (EPS). Version 2000—Models and data of the default method.* Chalmers University of Technology, Environmental Systems Analysis, Gotheburg, Sweden (CPM report 1999:5).

Thomassen, M.A., K.J. van Calker, M.C.J. Smits, G.L. Iepema, and I.J.M. de Boer. 2008a. Life cycle assessment of milk production systems in the Netherlands. *Agricultural Systems* 96: 95–107.

Thomassen, M.A., R. Dalgaard, R. Heijungs, and I. de Boer. 2008b. Attributional and consequential LCA of milk production. *International Journal of Life Cycle Assessment* 13: 339–349.

Tukker, A., G. Huppes, J. Guinée, R. Heijungs, A. de Koning, L. van Oers, S. Suh, T. Geerken, M. Van Holderbeke, B. Jansen, and P. Nielsen. 2006. *Environmental impacts of products (EIPRO). Analysis of the life cycle environmental impacts related to the total final consumption of the EU-25.* Institute for Prospective Technological Studies, Sevilla, Spain (EUR 22284 EN).

Weidema, B.P., and E. Lindeijer. 2001. *Physical impacts of land use in product life cycle assessment.* Final report of the EURENVIRON-LCAGAPS sub-project on land use. Department of Manufacturing Engineering and Management, Technical University of Denmark, Lyngby, Denmark (IPL-033-01). Revised version dated 2001.12.31.

Weidema, B.P., M. Hauschild, and O. Jolliet. 2007. Preparing characterisation methods for endpoint impact assessment. http://www.lca-net.com/publications/. Accessed August 15, 2009.

Weidema, B.P., M. Wesnæs, J. Hermansen T. Kristensen N. Halberg, P. Eder, and L. Delgado. 2008. *Environmental improvement potentials of meat and dairy products.* Institute for Prospective Technological Studies, Sevilla, Spain (EUR 23491 EN).

Weidema, B.P., T. Ekvall, and R. Heijungs. 2009. Consequential LCA. Chapter for CALCAS deliverable D18 "Guidelines for applications of deepened and broadened LCA": http://www.lca-net.com/publications/.

Wenzel, H., M. Hauschild, and L. Alting. 1997. *Environmental assessments of products—vol 1: Methodology, tools and case studies in product development.* London: Chapman & Hall.

WSSD (World Summit on Sustainable Development). 2002. *Summit 2002, key outcome of the summit*. United Nations, Department of Public Information, Johannesburg, South Africa. http://www.un.org/jsummit/index.html.

WWF. 2004. ISTA Mielke, Oil World Annual 2004, Hamburg, May 2004 cited in Jan Maarten Dros, Managing the Soy Boom: Two scenarios of soy production expansion in South America. http://www.panda.org/downloads/forests/managingthesoyboomenglish_nbvt.pdf.

Chapter 4
Life cycle assessment of crop production

Frank Brentrup

4.1 Introduction

Agricultural and horticultural crop production is expected to supply society with sufficient raw materials for food, feed and bioenergy production. At the same time crop production must comply with the principles of sustainability (Commission of the European Communities 1999; UN-DSD 2000). According to the Brundtland report (WCED 1987), sustainability can be defined as "any development, which meets the needs of the present without compromising the ability of future generations to meet their needs." Sustainable development has to balance economic, social, and environmental aspects. For agricultural and horticultural crop production this means ensuring sufficient income for growers, producing high-quality products in sufficient quantities at affordable prices, while at the same time being environmentally benign.

This represents an increasing challenge because a growing world population will demand increasing food production that needs to be produced from a limited agricultural area (FAO 2003; 2006) (Fig. 4.1).

As a consequence agricultural productivity in terms of yield per hectare has to be improved to sustain sufficient global supplies of crop products. Intensive crop production using the complete range of available farm inputs is an important means to support this development because it allows the possibility of approaching the full yield potential of crops. However, at the same time intensive agriculture also raises environmental concerns and questions about its impact on the environment. Are agricultural productivity and sustainability compatible?

Plant nutrition is a vital component of crop production and plays an important role in both improving food security and reducing environmental impacts. Crops need nutrients to grow. They take up nutrients in the form of minerals irrespectively

F. Brentrup (✉)
Yara International, Research Centre Hanninghof, Dülmen, Germany
e-mail: frank.brentrup@yara.com

J.I. Boye and Y. Arcand (eds.), *Green Technologies in Food Production and Processing*, Food Engineering Series, DOI 10.1007/978-1-4614-1587-9_4,

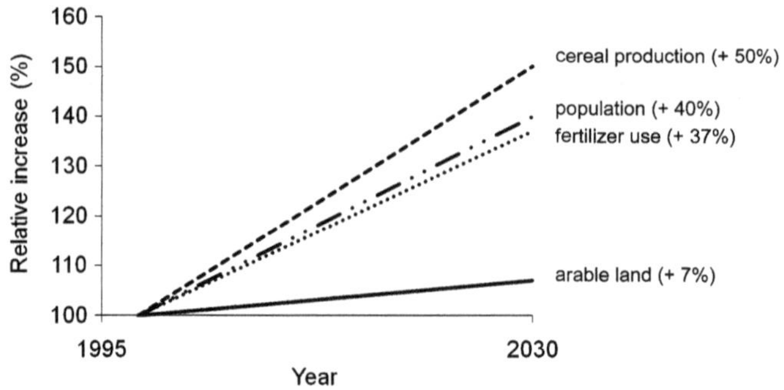

Fig. 4.1 Projected development of cereal production, world population, fertilizer use, and arable land (FAO 2003)

of their origin, which might be organic (e.g., manures, residues, soil organic matter) or mineral fertilizer. Erisman et al. (2008) estimate that today about 50% of the world's population can only be fed because of the use of mineral nitrogen fertilizers. Taking into account the additional future growth in global population accompanied by increasing demand for animal products, it will almost certainly be necessary to use more mineral plant nutrients in the future. Limited land resources further trigger the need for intensive crop production with optimum input of plant nutrients. Despite the fundamental benefits of mineral fertilizers in terms of food security, their production and use also have an environmental footprint. Energy use and losses of nitrogen compounds to the environment contribute to several environmental impacts such as eutrophication, acidification, and climate change.

To evaluate the environmental impact of crop production properly, it is necessary to have appropriate indicators in place. Particularly if there are choices between alternatives in crop production such as different farming concepts (organic, integrated, conventional), different production intensities (intensive, reduced), or options in the use of farming inputs (fertilizers, plant protection substances, application techniques), indicators are needed to evaluate the environmental impact of the alternatives on a scientific basis.

Life cycle assessment (LCA) provides a methodology to investigate environmental impacts in a holistic perspective. For crop production this means covering the wide range of different environmental impacts and including all the various activities involved, ranging from fertilizer production to tractor use in the field.

This chapter focuses on the environmental assessment of crop production using the LCA methodology. An economic or social assessment is not part of LCA.

4.2 Environmental assessment of crop production

The "environmental footprint" of crop production includes a wide range of different impacts such as nitrate leaching, ammonia volatilization, greenhouse gas emissions, or energy consumption, which itself may contribute to different environmental effects such as eutrophication, acidification, and global warming.

Many studies have shown that the nutrition of crops, especially with nitrogen, is of particular importance with regard to the environmental impact of crop production (IFA 2007). Therefore, the present chapter focuses on nitrogen-related impacts, although in an LCA of crop production other impacts such as those related to the use of pesticides, the consumption of energy, and the utilization of land also must be considered.

The following environmental effects often play an important role in environmental assessments of crop production.

4.2.1 Acidification

Acid deposition results mainly from anthropogenic emissions of sulfur dioxide (SO_2), NOx (NO, NO_2), and NH_3 and may damage acid-sensitive ecosystems at a regional to continental scale. In particular, forests and freshwater lakes are sensitive to acidification. Consequences range from defoliation and reduced tree vitality to declining fish stocks and decreasing biodiversity in lakes. Ammonia plays a significant role in the acidification of terrestrial and aquatic ecosystems. In 2005 ammonia was responsible for 35% of the acidification effect in EU15 (European Union member countries before enlargement: Austria, Belgium, Denmark, Finland, France, Germany, Greece, Ireland, Italy, Luxembourg, the Netherlands, Portugal, Spain, Sweden, and the United Kingdom) SO_2 and NOx for 28% and 37%, respectively (EEA 2008). Gaseous NH_3 emissions, such as those from the application of manure or mineral fertilizer, return to the surface through dry or wet deposition, partly combined with SO_2 as ammonium sulfate. Although most of the dry deposition of NH_3 takes place near the emission site, once combined with nitric or sulfuric acid in the atmosphere, the ammonia can be transported over distances of several thousand kilometers (EEA 1998).

4.2.2 Eutrophication

Eutrophication is the accumulation of nutrients in surface waters or land ecosystems that can lead to an undesired increase in biomass production and a shift in species composition. Eutrophication is particularly serious when it leads to algal blooms and subsequent oxygen-consuming decomposition processes, which may result in the death of fishes and other organisms. Ammonia emissions that deposit on water bodies and NO_3 leached from agricultural soils into surface water are two of the main anthropogenic sources of reactive nitrogen contributing to the eutrophication of aquatic ecosystems. For land ecosystems it is mainly the deposition of NOx and NH_3. Other major sources of nutrients to water systems are urban and industry wastes, particularly when wastewater is not, or is insufficiently, treated. These point sources emit also phosphorus (P), which is critical in the eutrophication of fresh water systems because they are generally P-limited. Eutrophication effects occur on the local level in small fresh water systems. However, problems also can appear on a regional or even continental scale if long-range transboundary emissions of NH_3 and NOx deposit on distant water bodies, or some leached N finally reaches rivers and coasts.

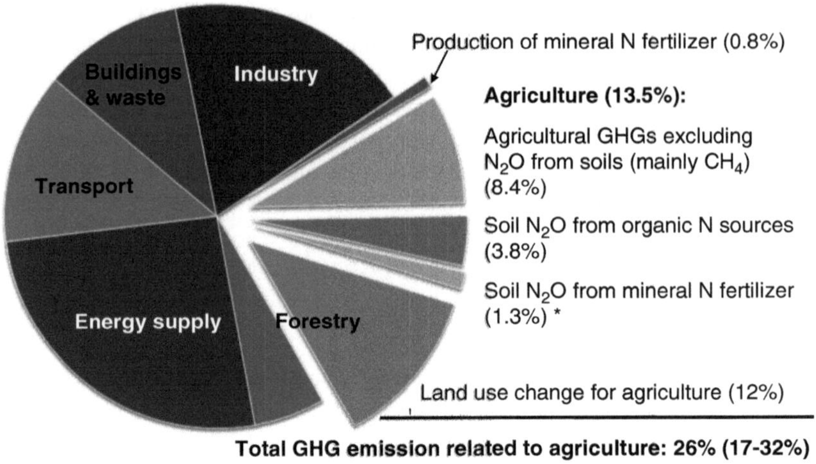

Fig. 4.2 Contribution of all activities related to agriculture to the global GHG emissions in 2004 (IPCC 2007; Bellarby et al. 2008) (*own calculations)

4.2.3 Global warming

Short-wave solar radiation heats the surface of the Earth, and the energy is radiated back through the atmosphere at longer wavelengths. Certain gases, for example CO_2 and N_2O, can absorb this longer wavelength radiation, trapping heat in the atmosphere. This greenhouse effect is essential for maintaining life on Earth. However, apparently because of human activity, concentrations of these greenhouse gases have increased, leading to an unnatural warming of the Earth's surface (global warming). This, in turn, will cause global and regional climatic changes with potentially severe consequences. The main anthropogenic contributors to the enhanced greenhouse effect are: CO_2, methane (CH_4), halogenated gases (e.g., CFCs), and N_2O. Agricultural GHG emissions according to IPCC (2007) have a share of 13.5% in the total global GHG emissions. There are also GHG emissions assigned to other sectors than agriculture that are directly linked to agricultural production, such as the production of mineral fertilizer and the loss of carbon because of land use change from natural area (mainly forest) into agricultural land. In total all GHG emissions related to agriculture amount on average to 26% of the total global GHG emissions with a range between 17% and 32% because of the uncertainty of the land use data. Of the N_2O emissions from agricultural soils, the largest share is related to organic N sources (Brentrup and Palliere 2008) (Fig. 4.2).

At the same time, agriculture can also mitigate global warming through carbon (C) sequestration and emission reduction, such as by adoption of improved farming practices. For instance, Burney et al. (2010) modeled the GHG emissions from global crop production, including emissions from fertilizer production, land use change, rice cultivation and arable soils, and estimated the net effect of improved agricultural practices between 1961 and 2005. The authors concluded that "while emissions from factors such as fertilizer production and application have increased, the net effect of higher yields has avoided emissions of up to 161 gigatons of carbon (GtC) (590 GtCO2e) since 1961." The analysis by Burney et al. (2010) shows "that investment in yield improvements compares favorably with other commonly proposed mitigation strategies. Further yield improvements should therefore be prominent among efforts to reduce future GHG emissions." These findings are supported by another recent study from West et al. (2010), who concluded that expanding croplands to increase agricultural crop production results in reduced carbon stocks and increased GHG emissions.

4.2.4 Particulate matter formation (dust) and photooxidant formation (summer smog)

There is increasing concern about the respiratory and related health problems resulting from particulate matter (PM) that contributes to atmospheric dust formation on a local to regional scale. The main sources for PM10 (particle size <10 μm) are stationary combustion plants, industry, and vehicles (EEA 1998). However, ammonia from agricultural sources may also contribute to PM formation as a secondary source after transformation in the atmosphere into ammonium sulfate or ammonium nitrate. Photooxidant formation of reactive chemical compounds such as ozone occurs by the action of sunlight on certain air pollutants. Photooxidants may be formed in the troposphere under the influence of UV light, through photochemical oxidation of volatile organic compounds (VOCs) and carbon monoxide (CO) in the presence of nitrogen oxides (NOx). Ozone is considered the most important of these reactive compounds, which may be injurious to human health and ecosystems and may also damage crops. International regulations to control ground-level ozone formation (e.g., from UNECE, EU) are in place to reduce these detrimental effects.

4.2.5 Land use

"Land use" describes the environmental impacts of utilizing and reshaping land for human purposes (Lindeijer et al. 1998; Köllner 2000). The environmental consequences of land use for arable farming or urban settlement are the decreasing availability of habitats and the decreasing diversity of wildlife species (EEA 1998). For assessing land use impacts it is necessary to measure the size of the area used for the specific purpose (e.g., to grow crops) and consider the ecological quality of the area under use. The ecological quality can be assessed by empirical investigations of species diversity (Köllner 2000) or by using a descriptive system of defined classes of land use types, which have been assigned to a certain ecological quality. An example for the latter approach is the Hemeroby concept (Kowarik 1999). Hemeroby is a measure for the "human influence on ecosystems" and it can therefore be used to characterize the environmental impact of different land use types (Brentrup et al. 2002a).

Hodgson et al. (2010) examined different land-use strategies for crop production with regard to their impact on biodiversity, which was in this example measured in terms of density and species richness of butterflies. The study compared a pure organic cropping system with a conventional system that was combined with a nature reserve. The conclusion was that "farming conventionally and sparing land as nature reserves is better for butterflies when the organic yield per hectare falls below 87% of conventional yield." On the other hand, the biodiversity of 1 ha of organic land is higher than that of 1 ha of conventional cropland. Therefore, it is crucial to consider both aspects at the same time, the area needed to produce a certain amount of crop products and the quality of the land-use type.

4.2.6 Depletion of abiotic resources

The issue related to the depletion of abiotic resources, such as fossil fuels or minerals, is their decreasing availability for future generations. In LCA, the consumption of abiotic resources is often aggregated to a summarizing indicator depending on the remaining reserves and the current consumption rates (Goedkoop and Spriensma 1999; Guinée 2001; Brentrup et al. 2002b). For LCA studies on agricultural crop production the consumption of fossil fuels and minerals such as phosphate, potash, and lime are of particular importance.

4.2.7 Toxicity

Besides dust and ozone formation, toxicity mainly includes the effects of plant protection substances and heavy metal emissions on humans (human toxicity) and ecosystems (ecotoxicity).

Plant protection substances are applied to control certain organisms (e.g., weeds, fungi, and insects) to improve the productivity of arable farming. However, via wind drift, evaporation, leaching, and surface runoff, part of the applied agrochemicals may impact upon terrestrial and aquatic ecosystems or even humans (Hauschild 2000). Different models have been developed to estimate the rate of unintended emissions of toxic substances, their fate in the environment and the final effects on ecosystems and humans (Guinée 1996; Jolliet and Crettaz 1997; Goedkoop and Spriensma 1999; Huijbregts 2001).

The agricultural use of mineral phosphate fertilizers and organic materials such as slurry, sewage sludge, or compost, may lead to emissions of heavy metals into soils. The contamination of these materials with heavy metals varies substantially depending on the origin of the raw material (P from rock, industrial, and household waste). For heavy metal emissions into soil, models developed by Goedkoop and Spriensma (1999) for human toxicity and Huijbregts (2001) for ecotoxicity are available for the estimation of their toxic potential. Both models take into account information on the environmental fate, the probable exposure of humans or ecosystems, and the potential toxic effects.

Depending on the production system under investigation, arable crop production may contribute to various but not necessarily all of the described environmental effects. Furthermore, the environmental impacts of crop production do not only occur during on-field activities such as tillage or fertilizer application, but also during the production and transportation of required farming inputs (e.g., fertilizers, plant protection substances, seeds, and machines).

The LCA methodology is especially designed for these kinds of environmental analyses of products and processes. The following section describes the general LCA concept and the specifics when applied to crop production.

4.3 Life cycle assessment methodology to investigate the environmental impact of crop production

LCA can be defined as an inventory and evaluation of all potential environmental impacts related to a product or process occurring throughout its life cycle, that is, from raw material extraction to disposal (cradle to grave). In practice, LCA studies do not always cover all life cycle stages of a product, but can be restricted to defined parts of it, for example in so-called "cradle to gate" or "gate to gate" studies. In LCA, impacts on natural ecosystems, human health, and natural resources are considered. However, today the LCA approach has also been adopted to calculate the so-called "carbon footprint" of products, which extracts the single impact category climate change out of the complete list of impact categories.

First attempts to analyze entire product systems were conducted in the 1960s in the United States with a focus on energy consumption (Curran 1996). The development of the recent LCA methodology started in the 1990s. Milestones of this development were the Society of Environmental Toxicology and Chemistry (SETAC) Code of Practice (Consoli et al. 1993) and the publication of ISO standards on LCA (ISO 14040–14049). According to the general concept and principles of the LCA methodology, LCA is divided into four phases: (1) goal and scope definition, (2) life cycle inventory, (3) life cycle impact assessment, and (4) interpretation.

4.3.1 Goal and scope definition

The first step in LCA is the *goal and scope definition*. Within this phase the system under investigation, its function, and its boundaries are described. In the following the production of 1 ton of winter wheat grain will serve as an example for an LCA study in crop production. Figure 4.3 shows the system needed to produce winter wheat as considered in this LCA study.

Fig. 4.3 The wheat production system

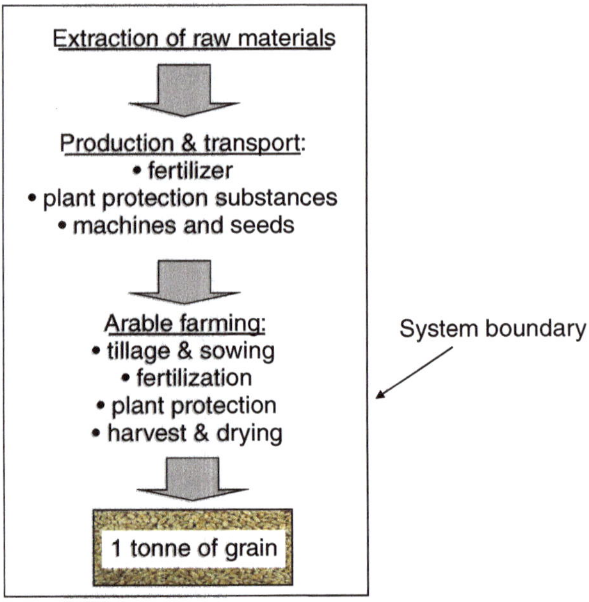

4.3.2 Life cycle inventory

The subsequent life cycle inventory (LCI) compiles resources (inputs) and emissions (outputs) connected to the examined system. To make the various inputs and outputs comparable, these data are related to a common functional unit, which will represent the main function of the system. In the wheat example the functional unit is 1 ton of winter wheat grain at the field border.

In contrast with most other sectors, horticultural and agricultural production most often takes place within an open natural setting, that is, without distinct borderlines of the environmental compartments of soil, water, and air. Many relevant agricultural emissions occur from the field and are dependent on environmental circumstances such as temperature, rainfall, or wind. On-field emissions in particular of N-containing compounds often contribute considerably to the final results of the LCA studies (Audsley 1997; Cederberg 1998; Küsters and Jenssen 1998; Andersson and Ohlsson 1999; Brentrup 2003). However, it is often difficult to derive exact rates of N released to the wider environment, because emission rates can vary greatly depending on soil type, climatic conditions, and agricultural management practices. Measurements of these emissions require considerable investment in terms of money and time, and in any case they show great variations (e.g., Isermann 1990, for NH_3) because they only reflect a snapshot of the specific conditions at the time of measurement. For LCA purposes, average potential emission rates adjusted to the conditions typical for the system under investigation are more appropriate. Therefore, methods have been proposed to easily calculate potential nitrogen emission rates taking into account important site-specific parameters (Brentrup et al. 2000; Bouwman et al. 2002; UNECE/EMEP 2007).

Table 4.1 List of environmental effects (impact categories) treated in LCA

General distinction	Impact category
Input-related categories	Depletion of abiotic resources (e.g., fossil fuels)
	Land use
Output-related categories	Climate change (global warming)
	Stratospheric ozone depletion
	Human toxicity, ecotoxicity
	Photooxidant formation (summer smog)
	Acidification
	Nitrification (eutrophication)

Another specific element of many agricultural LCIs is the need for land use data. Agricultural land use describes the continuous use of an area for agricultural production purposes or the conversion of another land use type into agricultural land. Land use is usually expressed in terms of area used for a defined time per functional unit. In addition to the quantity, the intensity of use should also be considered, that is, the remaining ecological quality of the land under use needs to be assessed (Brentrup et al. 2002a). Besides these direct land use effects, also indirect land use impacts may be relevant; for example, if the cultivation of bioenergy crops displaces food crops to other areas.

The data of the LCI per se do not allow comparisons to be made between different systems, for example, different ways to produce a tonne of wheat grain. Furthermore, the potential environmental impact of the various emissions and resource consumption is not considered in this phase. Therefore, the inventory data are further evaluated in the third step, the Life Cycle Impact Assessment (LCIA). Within the LCIA, the different inputs and outputs listed in the LCI are assigned to environmental effects, the so-called impact categories.

4.3.3 Life cycle impact assessment

The first step in LCIA is the characterization step. During characterization the inventory data are aggregated into indicators for each impact category. For instance, for the impact category climate change, CO_2, N_2O, and CH_4 emissions are aggregated into the impact category indicator CO_2-equivalents. Characterization is achieved through the use of characterization factors, which represent the potential of each emission to contribute to a specific environmental effect. The result of this first step is a list of indicator values for environmental effects, which gives the environmental profile of a product. Table 4.1 gives a list of the impact categories as proposed by SETAC (Udo de Haes et al. 1999a, b).

During the second step in LCIA, normalization, each of the indicator values is divided by a reference value, as for instance the respective indicator values per person in Europe (e.g., kg CO_2-equiv. per tonne of grain/CO_2-equiv. per capita in Europe). This normalization is performed to get information on the relevance of a product's impact in comparison with an independent reference value. Furthermore,

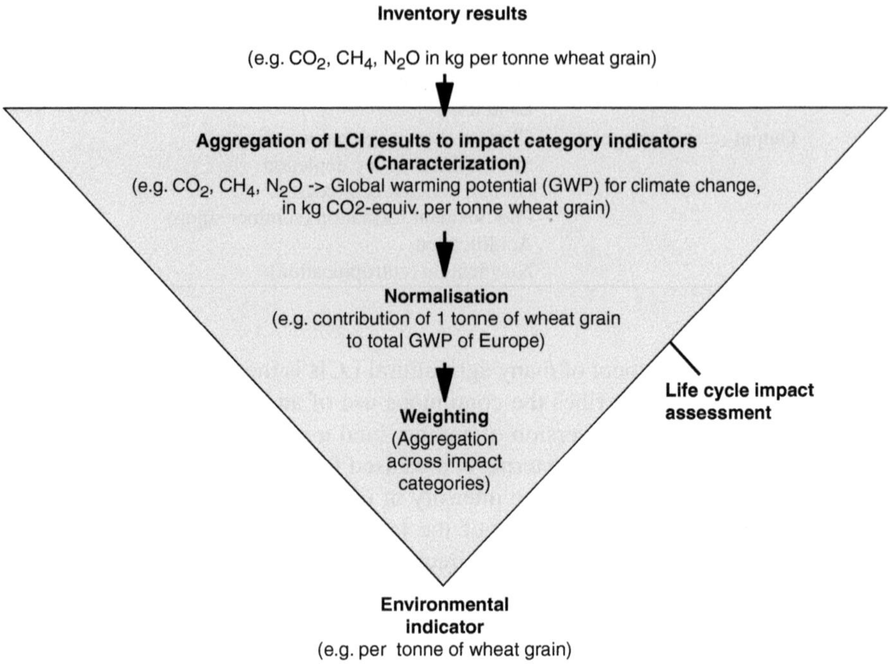

Fig. 4.4 The general life cycle impact assessment procedure

with normalization, the indicator values become dimensionless, which is a prerequisite for the final weighting step. The following weighting step aims at a final aggregation across all impact categories into one overall environmental indicator. Therefore, each normalized indicator value is multiplied by a weighting factor, which represents the potential of the respective impact category to harm natural ecosystems, human health, and resources in Europe. According to the ISO norms, normalization and weighting are optional steps in LCA, which are not mandatory. However, in LCA studies comparing alternative products or systems allow weighting of different environmental impacts, which is indispensable in the determination of the environmental preference of one or the other alternative. If weighting is not performed within LCA, users of LCA studies will tend to weigh the system's contribution to different environmental effects on their own. Instead of such a subjective individual way of weighting, a set of generic, study-independent weighting factors, as have for instance been proposed by Brentrup et al. (2004a), Goedkoop and Spriensma (1999), and others can help to interpret complex environmental data sets on a more transparent and documented basis. However, an evaluation and aggregation of different environmental effects will never be possible on a pure scientific basis and always includes more values and societal preferences than the other parts of LCA. Therefore, it is particularly important to present LCA results at different levels of detail, that is, as inventory data, as indicators for impact categories (environmental profile) and as normalization and fully aggregated weighting results. Figure 4.4 shows the general calculation procedure in LCA.

4.4 Example: results from the life cycle assessment study of wheat grain

The following section presents some selected results from the example study on winter wheat produced with different application rates of mineral nitrogen fertilizer. The study is based on the results of a long-term field experiment, the Broadbalk Experiment (Rothamsted, UK) (Brentrup et al. 2004b; Johnston 1994). Figure 4.5 shows the average yield response to fertilizer N. Up to 192 kg N/ha, a strong increase in grain yield from 2.1 to 9.3 t/ha can be observed, whereas at 240 and 288 kg N/ha no further yield increase occurred. The economic optimum N fertilizer rate (Nopt) is 210 kg N/ha. In agreement with IACR-Rothamsted, Nopt was calculated from a fitted quadratic response function and based on 0.57 Euro/kg N and 115 Euro/ton of wheat grain. The 192 kg N/ha treatment is closest to Nopt and therefore can be regarded as the optimum treatment in this trial for the years 1996–2000. In this chapter, the production intensity at Nopt is used as the reference treatment, with which the environmental impacts of the other treatments are compared.

4.4.1 Life cycle inventory results

This chapter does not include detailed LCI results of the wheat LCA study, because its aim is to present some selected results of the study so as to provide the reader with examples of what kind of outcomes can be expected from an LCA study on

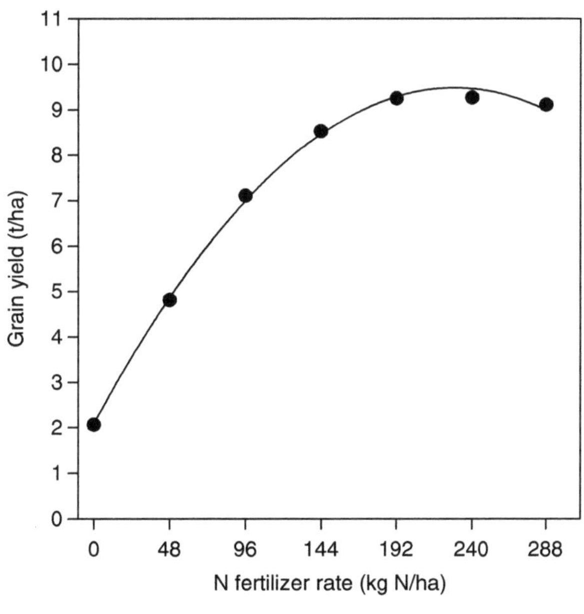

Fig. 4.5 Yield response of winter wheat to increasing N fertilizer rates (average for first wheat in rotation in 1996–2000)

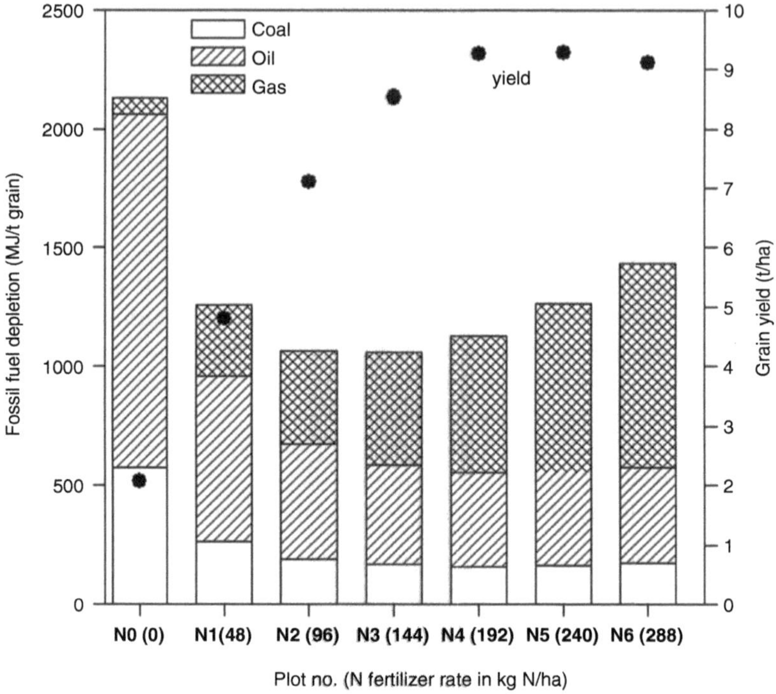

Fig. 4.6 Combined consumption of fossil fuels (MJ/t grain, bars) and yields (t/ha, dots) at increasing N fertilizer rates

crop production. Detailed LCI results and a comprehensive description of the underlying crop production system can be found in Brentrup et al. (2004b).

4.4.2 *Life cycle impact assessment results*

Four environmental effects from the wheat study are presented here in more detail: (1) consumption of fossil fuels, (2) land use, (3) climate change, and (4) eutrophication of water. The results for the other impact categories (consumption of minerals, human and ecotoxicity, acidification, eutrophication of land) are described and discussed in Brentrup et al. (2004a, b).

4.4.2.1 Consumption of fossil fuels

Figure 4.6 gives the combined consumption of fossil fuels in MJ per ton of grain for the 7 N fertilizer rates (bars). In addition, the yields are shown to illustrate the different productivity of the treatments, which strongly influences the fossil fuel consumption per ton of grain.

For N2 and N3 the energy consumption is lowest per tonne of grain (~1,060 MJ). That means it is possible to increase the yield from 7.11 (N2) to 8.53 t/ha (N3) by an additional application of 48 kg N/ha without increasing the energy use per ton of grain. The application of more than 144 kg N/ha or less than 96 kg N/ha results in an increasing energy use per ton of grain (N0, N1, N4, N5, N6). For N4 (192 kg N/ha) the consumption of natural gas and oil contribute 51% and 35%, respectively, in the total fossil fuel consumption per tonne of grain (coal = 14%). The more N fertilizer is applied, the higher is the share of natural gas in the consumption of fossil fuels. This is owing to the production of N fertilizer, which is based on natural gas, both as a feedstock (hydrogen source for ammonia synthesis) and as fuel.

4.4.2.2 Land use

The impact category "land use" deals with the degradation of natural land due to human utilization for agriculture, housing, roads, industry, etc. The calculation of a land use indicator not only includes the area used for a certain period of time but also the intensity of land use, for example, built-up area versus extensive pasture (Brentrup et al. 2002a).

However, for the compared wheat production systems the intensity of land use is uniform, because the wheat production in any treatment includes intensive soil preparation, plant protection measures, base fertilization with other nutrients than N, etc. Thus, for each N application rate the land area used for a certain period of time (in m^2*year/t grain) is multiplied by the characterization factor for intensive arable land use (0.8) (Brentrup et al. 2002a) to give the land use impact per tonne of grain for each treatment. Figure 4.7 reveals that the land use per tonne of grain is lower when the yield is higher. Therefore, the plots with the highest yields (N4, N5, and N6) show the lowest land use impacts (<900 m^2*year/t grain). In N0, the land is used most inefficiently (3,865 m^2*year/t grain).

4.4.2.3 Climate change

The Global Warming Potential (GWP) is used to express the contribution that gaseous emissions from arable production systems make to the environmental problem of climate change. Figure 4.8 reveals that the GWP per tonne of grain increases almost linearly with increasing inputs. The N_2O emissions, which are closely related to the N input, are responsible for this close relationship. Although the absolute emission rates of N_2O are much lower compared with those of CO_2, N_2O dominates the total GWP per ton of grain in all production intensities, except for N0. This is owing to the fact that 1 kg of N_2O has a GWP about 300 times higher than that of 1 kg of CO_2 (IPCC 2006). For N4, N_2O contributes 76% of the total GWP per ton of grain (CO_2 = 24%). Methane (CH_4) emissions are negligibly low for all N rates.

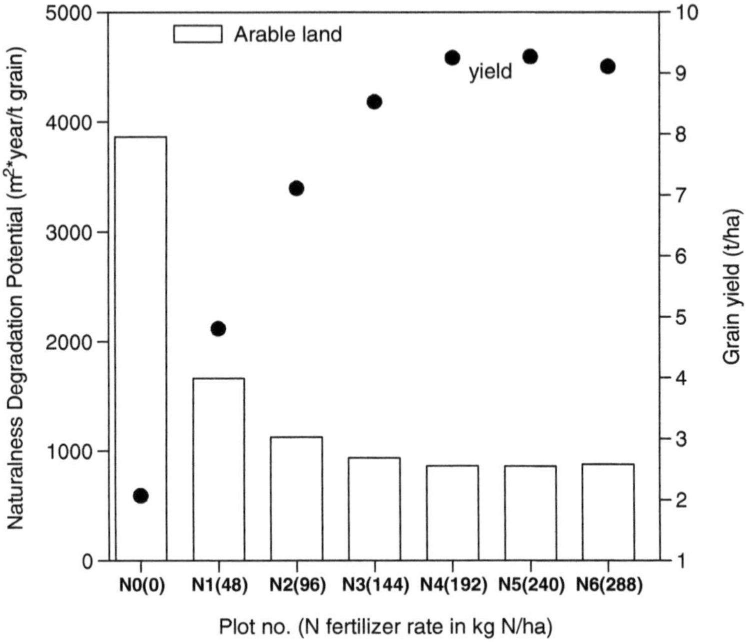

Fig. 4.7 Naturalness degradation potentials (m²*year/t grain, bars) and yields (t/ha, dots) at increasing N fertilizer rates

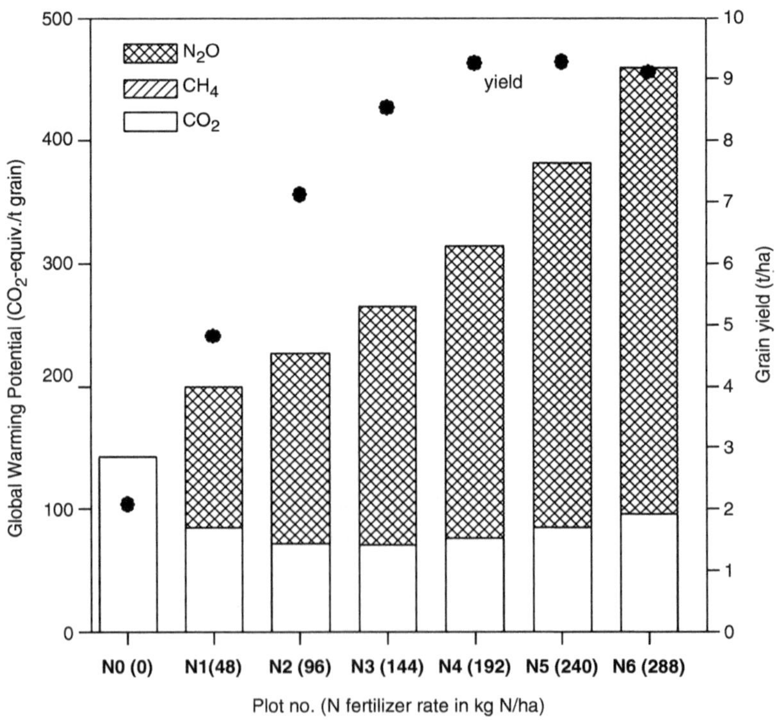

Fig. 4.8 Global warming potentials (kg CO_2-equivalents/t grain, bars) and yields (t/ha, dots) at increasing N fertilizer rates

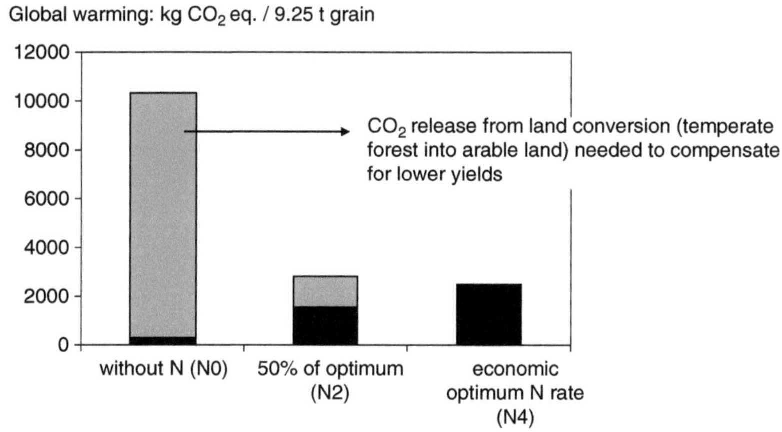

Fig. 4.9 Greenhouse gas emissions of wheat production in kg CO_2eq per 9.25 t grain (including production and transport of farming inputs, *and land use change*) at different N fertilizing intensities

However, crop production not only causes GHG emissions, but at the same time the crops fix about 1.6 tons of carbon dioxide per tonne of biomass through photosynthesis (Greef et al. 1993). At a yield of 18.5 t/ha (grain plus straw at economic optimum N rate) this amounts to 29.6 t CO_2 fixation per ha, that is, almost 12 times the CO_2 emissions per ha. If this biomass is used as biofuel (e.g., for direct incineration) and thereby avoids the use of fossil fuels, a substantial net saving of CO_2 emissions can be achieved. However, by far most of the grain is still consumed as food or feed and in this case the CO_2 fixation is only short to medium term and is usually not considered as a credit.

In view of the current and future demand for cereals it can be assumed that any reduction in the production intensity at one place needs to be compensated for by additional production at another place. In many cases this results in land use changes from non-agricultural (e.g., forest or wetlands) into agricultural land (IPCC 2007) (see Fig. 4.2). Figure 4.9 shows the additional CO_2 emission that occurs because of conversion of temperate forest into cropland and cultivation of cereals to compensate for the lower yields in the treatments N0 and N2. Bellarby et al. (2008) give data for CO_2 sequestration per area for different ecosystems. For a temperate forest the value is 560 t CO_2/ha, for arable land it is 300 t/ha, that is, the conversion from temperate forest into cropland would release 260 t CO_2/ha. In this study the total carbon loss was spread over a time period of 100 years; that is, the annual CO_2 loss owing to land use change is 2,600 kg CO_2/ha.

Without any nitrogen input the grain yield is so low (2.07 t/ha, N0) that the land required to compensate this yield loss would lead to CO_2 emissions that are more than four times higher than those of the economic optimum system. Taking further into account the scarce land reserves it becomes obvious that a "zero nitrogen" system is not a viable option. But also the N2 treatment shows a higher carbon footprint if the additional land use needed to compensate lower yields is considered. It can be concluded that intensive crop production aiming at most efficient utilization of

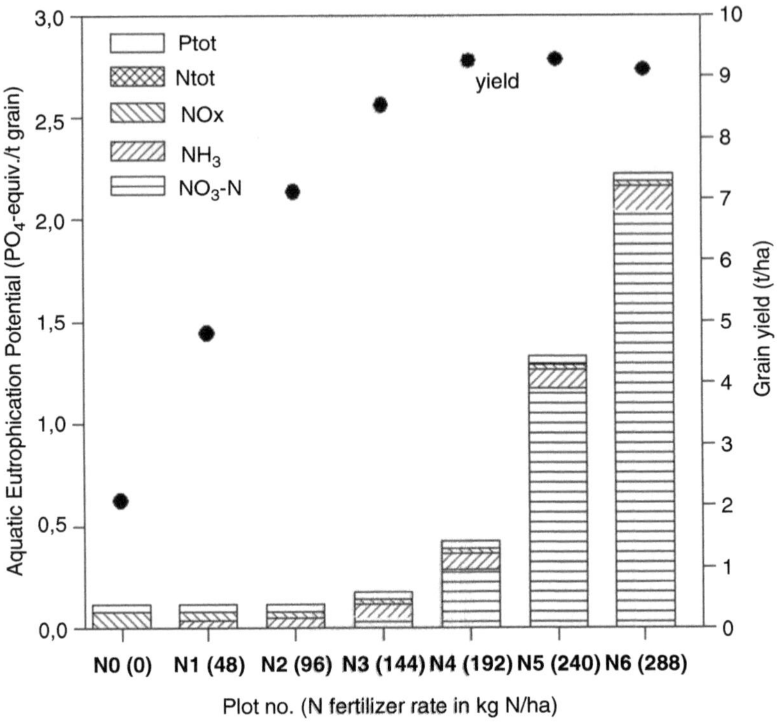

Fig. 4.10 Aquatic eutrophication potentials (kg PO_4-equivalents/t grain, bars) and yields (t/ha, dots) at increasing N fertilizer rates

resources, including agricultural land saves GHG emissions because natural land is potentially prevented from being converted into cropland.

A similar analysis was performed by Kindred et al. (2008). Their analysis shows

> that substantially different conclusions are reached about the effects of N fertilizer on GHG emissions from grain production, depending on whether (and how) indirect effects of LUC [land use change] are accounted for. Ignoring LUC suggests that N rates should best be cut to around 150 kg/ha less than the economic optimum in order to mitigate climate change. Whereas if LUC is accounted for, the best N rates for minimizing GHG emissions are much closer to the economic optimum, and sometime exceed it.

However, for further analysis of the wheat study presented in this paper, the CO_2 emissions resulting from potential land use changes will not be considered, because the assumption that yield differences need to be compensated was not part of the original LCA study (Brentrup et al. 2004b).

4.4.2.4 Eutrophication of water

Figure 4.10 shows the Aquatic Eutrophication Potential (AEP) for the different treatments. The figure clearly indicates that at N rates higher than N3 (144 kg N/ha), the eutrophication of water is dominated by NO_3 leaching (e.g., 92% of the total AEP for N6). In particular at production intensities above the optimum N rate (N4, 192 kg N/ha), nitrate losses via leaching result in high AEP values (2.22 kg PO_4-equiv./t grain for N6 versus 0.42 kg for N4). At N rates higher than 144 kg N/ha, airborne nutrient emissions, which deposit on surface waters (NOx, NH_3) and direct effluents of P (from P fertilizer production) contribute less to the AEP compared with nitrate (e.g., for N4: NOx and NH_3 = 24% of total AEP, Ptot = 9% and NO_3-N = 67%).

4.4.2.5 Normalization and weighting of the impact assessment results

Normalization and weighting aim at a further interpretation of the impact assessment results, which allow clearer conclusions on the environmental preference of one or the other options under comparison to be drawn. For instance, a high Global Warming Potential per ton of grain may represent only a small contribution to the total climate change effect in a given region, whereas a several times smaller eutrophication potential (per t grain) may represent a more important contribution to the overall eutrophication effect in the defined region. Thus, "the aim of the normalization of indicator results is to understand better the magnitude for each indicator result of the product system under study" (ISO 2000). During normalization the indicator results per functional unit (i.e., a ton of grain) are related to the respective indicator results for a defined reference area (e.g., Europe). Detailed normalization results can be found in Brentrup et al. (2004b).

However, from the normalization results a decision about the overall environmental preference of one or other production alternatives is still not possible. Although theoretically possible, the normalized values should not be summed up because information on the potential of the different effects to harm resource availability, natural ecosystems, and human health (weighting) is still not included. Therefore, the different impact categories were weighted based on the so-called distance-to-target principle (Brentrup et al. 2004a).

During the weighting step, each normalized indicator value is multiplied by a weighting factor, which represents the potential of the respective impact category to harm resources, natural ecosystems, and human health. The weighting factors used in this study for the different impact categories have been developed independently from any LCA case study and are based for each impact category separately on the ratio between a defined target indicator value and the current status of the impact (distance-to-target). The resulting index is called "EcoX" (Brentrup et al. 2004a).

In the present case study, the aggregated environmental impacts calculated for the N treatments of 48, 96, 144, or 192 kg N/ha were within a range of EcoX values of 0.16–0.22 per ton of grain (Fig. 4.11). The treatments receiving 0, 240, or 288 kg N/ha show 100–232% higher EcoX values compared with the lowest figure (N2, 96 kg N/ha). This result indicates that in high-yielding crop production systems (e.g., N4, 192 kg N/ha) economic and environmental considerations are not necessarily in conflict, whereas a significant undersupply or oversupply with nitrogen fertilizers (e.g., N0, N5, N6) leads to decreasing eco-efficiency in crop production.

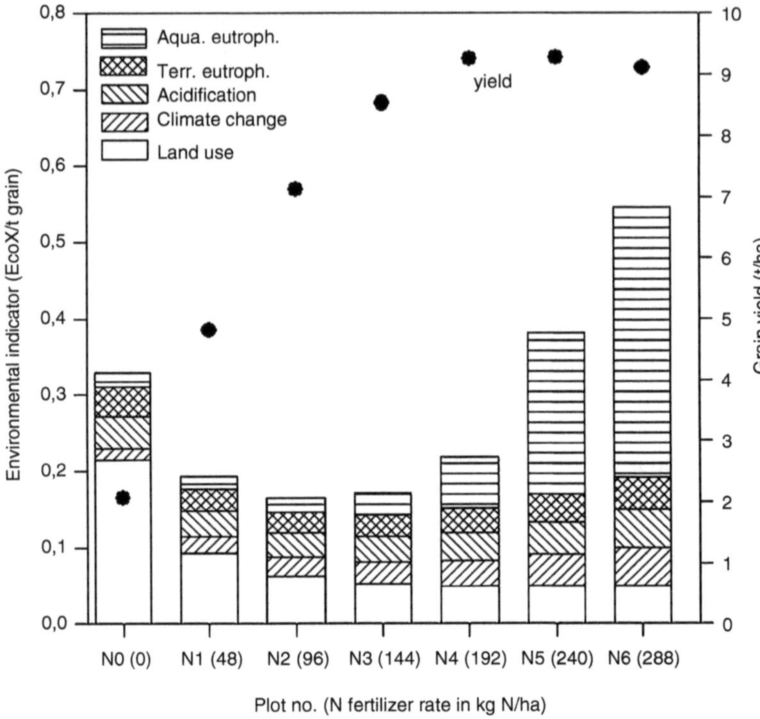

Fig. 4.11 Aggregated environmental indicator values (EcoX) per ton of grain (stacked bars) and yields (t/ha, dots) at increasing N fertilizer rates

Wheat production involves two environmental hot spots—land use and aquatic eutrophication. Thus, the greatest potential to minimize the environmental impact per tonne of grain is to achieve high yields per unit of land (i.e., high land use efficiency) and at the same time low NO_3 leaching rates, which are most responsible for aquatic eutrophication. At N3 (144 kg N/ha), the aggregated EcoX value for these two competing aspects is lowest. Consequently, other impacts than land use and aquatic eutrophication get a higher relative importance in this treatment compared with the other N rates.

The differences in the acidification potential are mainly determined by NH_3 emissions. Because ammonium nitrate is used as N fertilizer in the system under consideration, the NH_3 emissions are low compared with the use of other mineral and organic fertilizers (ECETOC 1994). Other LCA studies have shown that, for example, the use of urea or organic fertilizers (e.g., slurry) as N sources results in much higher acidification potentials (Küsters and Jenssen 1998). Therefore, in the wheat production system under investigation the release of acidifying emissions will be difficult to reduce. The same holds true for the contribution to terrestrial eutrophication because again the NH_3 emissions are a decisive factor for this indicator value.

For greenhouse gases the picture is different. Part of the global warming potential could be avoided by choosing an ammonium or urea based N fertilizer. In this case study significant amounts of N_2O are emitted during the production of nitric acid, which is part of ammonium nitrate production. However, as already mentioned a switch to urea or another non-nitrate fertilizer would lead to higher contributions to acidification and eutrophication, leading also to additional indirect N_2O emissions related to nitric acid production have been significantly reduced by 75–90% (Brentrup and Palliere 2008).

The impact categories related to abiotic resource consumption were excluded from this aggregation, because the problems related to the depletion of abiotic resources are substantially different from those related to the other impact categories. The impact categories other than the depletion of abiotic resources have direct effects on natural ecosystems (land use, acidification, eutrophication, ecotoxicity), or human health (human toxicity), or even both (climate change). In contrast, the depletion of abiotic resources, that is, the decreasing availability of raw materials for future generations, has no direct impact on human health or the shape of natural ecosystems. Nevertheless, the availability or nonavailability of resources such as fossil fuels or minerals imposes important challenges for the future.

4.5 Conclusions and outlook

Crop production, as all other human activities, has its environmental footprint, that is, horticultural and agricultural production of crops always has some unwanted side effects on the environment, such as nutrient losses, the release of greenhouse gases, or reduced biodiversity. The aim of sustainable development should be to balance the benefits and risks of human activities. For crop production this means to balance the tasks of supplying the society with sufficient food, providing the grower with sufficient income, and minimizing the burden to the wider environment.

There are different options available to the grower to produce the same crop. For instance, he or she can choose among different farm inputs (e.g., fertilizers, plant protection agents), tillage systems (e.g., direct seeding or conventional ploughing), or farming systems (organic versus integrated versus conventional). Finally, as much as possible, it is important for farmers, scientists, decision makers, and society to evaluate the sustainability of the various options on a scientific basis. For the environmental dimension of sustainability, LCA provides a methodology to investigate and compare different options in crop production.

LCA is a methodology that is designed for environmental analyses of products and processes. Life cycle assessment can be defined as an inventory and evaluation of all potential environmental impacts related to a product or process occurring throughout its life cycle, from raw material extraction to disposal (cradle to grave). In practice, LCA studies do not always cover all life cycle stages of a product, but can be restricted to defined parts of it, for example, in so-called “cradle to gate” or “gate to gate” studies.

Today, the structure and outline of the LCA methodology is clearly described in ISO norms (ISO 14040–14049). However, within this structure the way to calculate and interpret the available environmental indicators is not prescribed and different possibilities have been published. Therefore, it is important for any LCA study to openly describe all assumptions made and methods employed to derive aggregated indicators that describe the environmental profile of a crop production system.

The example presented in this chapter examined the environmental impact of winter wheat production with different levels of nitrogen fertilizer application. The study used an LCA approach, which is tailored to arable crop production systems and is described in detail in Brentrup et al. (2004a). From this LCA case study, it can be concluded that a good environmental performance in wheat production can be achieved by:

- Improving N use efficiency to reduce all sorts of N losses
- Maintaining high yields to use land most efficiently
- Applying nitrogen according to crop demand to minimize NO_3 leaching
- Using nitrogen fertilizers with low NH_3 volatilization rates (e.g., AN) to keep the acidification and terrestrial eutrophication potentials low
- Reducing N_2O emissions during fertilizer production and fertilizer use to reduce the global warming potential

References

Andersson, K., and T. Ohlsson. 1999. Life cycle assessment of bread produced on different scales. *The International Journal of Life Cycle Assessment* 4: 25–40.

Audsley, E. 1997. *Harmonisation of environmental life cycle assessment for agriculture: Final report.* Concerted Action AIR3-CT94-2028, Silsoe Research Institute, Silsoe, UK.

Bellarby, J., B. Foereid, A. Hastings, and P. Smith. 2008. *Cool farming: Climate impacts of agriculture and mitigation potential.* Amsterdam: Greenpeace International.

Bouwman, A.F., L.J.M. Boumans, and N.H. Batjes. 2002. Modeling global annual N_2O and NO emissions from fertilized fields. *Global Biogeochemical Cycles* 16(4): 1080.

Brentrup, F. 2003. *Life cycle assessment to evaluate the environmental impact of arable crop production.* Cuvillier Verlag Göttingen: Hannover University.

Brentrup, F., and C. Palliere. 2008. GHG emissions and energy efficiency in European nitrogen fertilizer production and use. In: *Proceedings of the international fertiliser society conference*, Cambridge, Dec 2008.

Brentrup, F., J. Küsters, J. Lammel, and H. Kuhlmann. 2000. Methods to estimate on-field nitrogen emissions from crop production as input to LCA studies in the agricultural sector. *The International Journal of Life Cycle Assessment* 5: 349–357.

Brentrup, F., J. Küsters, J. Lammel, and H. Kuhlmann. 2002a. Life cycle impact assessment of land use based on the Hemeroby concept. *The International Journal of Life Cycle Assessment* 7: 339–348.

Brentrup, F., J. Küsters, J. Lammel, and H. Kuhlmann. 2002b. Life cycle impact assessment of abiotic resource consumption—Conceptual considerations. *The International Journal of Life Cycle Assessment* 7: 301–307.

Brentrup, F., J. Küsters, J. Lammel, and H. Kuhlmann. 2004a. Investigation of the environmental impact of agricultural crop production using the Life Cycle Assessment (LCA) methodology. Part I: Development of an LCA method tailored to agricultural crop production. *European Journal of Agronomy* 20: 247–264.

Brentrup, F., J. Küsters, J. Lammel, P. Barraclough, and H. Kuhlmann. 2004b. Investigation of the environmental impact of agricultural crop production using the Life Cycle Assessment (LCA) methodology. Part II: Application of the LCA methodology to investigate the environmental impact of different N fertilizer rates in cereal production. *European Journal of Agronomy* 20: 265–279.

Burney, J.A., S.J. Davis, and D.B. Lobell. 2010. Greenhouse gas mitigation by agricultural intensification. *Proceedings of the National Academy of Sciences* 107(26): 12052–12057.

Cederberg, C. 1998. *Life cycle assessment of milk production—A comparison of conventional and organic farming*. SIK Report no. 643, The Swedish Institute for Food and Biotechnology (SIK), Gothenburg, Sweden.

Commission of the European Communities (CEE). 1999. *Directions towards sustainable agriculture COM (1999) 22 final*. Brussels: Commission of the European Communities.

Consoli, F., D. Allen, I. Boustead, J. Fava, W. Franklin, A.A. Jensen, N. de Oude, R. Parrish, R. Perriman, D. Postlethwaite, B. Quay, J. Séguin, and B. Vignon (eds.). 1993. *Guidelines for life cycle assessment: A code of practice*. Brussels: Society of Environmental Toxicology and Chemistry (SETAC).

Curran, M.A. 1996. The history of LCA. In *Environmental life cycle assessment*, ed. M.A. Curran. New York: McGraw-Hill.

Erisman, J.W., M.A. Sutton, J. Galloway, Z. Klimont, and W. Winiwarter. 2008. How a century of ammonia synthesis has changed the world. *Nature Geoscience* 1: 636–639.

European Chemical Industry Ecology and Toxicology Centre (ECETOC). 1994. Ammonia emissions to air in Western Europe. Technical Report no. 62, ECETOC, Brussels, Belgium.

European Environment Agency (EEA). 1998. *Europe's environment: The second assessment*. Copenhagen: EEA.

European Environment Agency (EEA). 2008. http://themes.eea.europa.eu/IMS/ISpecs/ISpecification 20041001122413/IAssessment1188820975411/view_content. Accessed 31 May 2010.

Food and Agriculture Organization of the United Nations (FAO). 2003. *World agriculture: Towards 2015/2030, an FAO perspective*. London: Earthscan Publications Ltd.

Food and Agriculture Organization of the United Nations (FAO). 2006. World agriculture: Towards 2030/2050, an FAO perspective. Interim Report, FAO, Rome.

Goedkoop, M., and R. Spriensma. 1999. *The Eco-Indicator 99: A damage oriented method of life cycle impact assessment methodology report*. Amersfoort: Pré Consultants B.V.

Greef, J.M., F. Hansen, G. Pasda, and W. Diepenbrock. 1993. Die Strahlungs-, Energie- und Kohlendioxidbindunglandwirtschaftlicher Kulturpflanzen—Ergebnisse und Modellrechnungen. *Berichte überLandwirtstchaft* 71: 554–566.

Guinée, J. 1996. Data for the normalization step within life cycle assessment of products. CML Paper no. 14 (revised version), Centre of Environmental Science (CML), Leiden, The Netherlands.

Guinée, J.B. 2001. *Life cycle assessment: An operational guide to the ISO standards*. The Netherlands: Centre of Environmental Science (CML), Leiden University.

Hauschild, M. 2000. Estimating pesticide emissions for LCA of agricultural products. In *Agricultural data for life cycle assessments*, vol. 2, ed. B.P. Weidema and M.J.G. Meeusen. The Hague: Agricultural Economics Research Institute (LEI). Report 2.00.01.

Hodgson, J.A., W.E. Kunin, C.D. Thomas, T.G. Benton, and D. Gabriel. 2010. Comparing organic farming and land sparing: Optimizing yield and butterfly populations at a landscape scale. *Ecology Letters* 13(11): 1358–1367.

Huijbregts, M.A.J. 2001. *Uncertainty and variability in environmental life cycle assessment*. The Netherlands: University of Amsterdam.

Intergovernmental Panel on Climate Change (IPCC). 2006. *2006 IPCC guidelines for national greenhouse gas inventories*, vol. 4. Agriculture, forestry and other land use. Hayama: IGES.

Intergovernmental Panel on Climate Change (IPCC). 2007. Technical summary. In *Climate change 2007: The physical science basis. Contribution of working group I to the fourth assessment report of the intergovernmental panel on climate change*, ed. S.D. Solomon, M. Qin, Z. Manning, M. Chen, K.B. Marquis, M.T. Averyt, and H.L. Miller. Cambridge: Cambridge University Press.

International Fertilizer Association (IFA). 2007. *Sustainable management of the nitrogen cycle in agriculture and mitigation of reactive nitrogen side effects*. Paris: IFA.
International Organization for Standardization (ISO). 2000. *Environmental management—Life cycle assessment—Life cycle impact assessment, International Standard ISO 14042:2000*. Geneva: ISO.
Isermann K (1990) Ammonia emissions from agriculture as part of its nitrogen balance and suggested solutions for their mitigation. Federal Agricultural Research Centre. Braunschweig, Germany. (in German)
Johnston, A.E. 1994. The Rothamsted classical experiments. In *Long-term experiments in agricultural and ecological sciences*, ed. R.A. Leigh and A.E. Johnston. Wallingford: CAB International.
Jolliet, O., and P. Crettaz. 1997. *Critical Surface-Time 95: A life cycle impact assessment methodology including fate and exposure*. Laussane: École Polytechnique Fédérale de Lausanne.
Kindred, D., P. Berry, O. Burch, and R. Sylvester-Bradley. 2008. Effects of nitrogen fertiliser use on greenhouse gas emissions and land use change. *Aspects of Applied Biology* 88: 53–56.
Köllner, T. 2000. Species-pool effect potentials (SPEP) as a yardstick to evaluate land-use impacts on biodiversity. *Journal of Cleaner Production* 8: 293–311.
Kowarik I (1999) Naturalness, nearness to nature and hemeroby as evaluation criteria. In *Handbook for nature conservation*, ed. W. Konold, R. Böcker, and U. Hampicke. Ecomed, Landsberg, Germany. (in German).
Küsters, J., and T. Jenssen. 1998. Selecting the right fertilizer from an environmental life cycle perspective. In: *International Fertilizer Industry Association (IFA) 1998 technical conference*, Merrakech, pp. 290–296.
Lindeijer, E.W., M. van Kampen, P.J. Fraanje, H.F. van Dobben, G.J. Nabuurs, E.P.A.G. Schouwenberg, A.H. Prins, N. Dankers, and M.F. Leopold. 1998. *Biodiversity and life support indicators land use impacts in LCA*. Delft: Public Works and Watermanagement, Ministry of Transport.
Udo de Haes, H.A., O. Jolliet, G. Finnveden, M. Hauschild, W. Krewitt, and R. Müller-Wenk. 1999a. Best available practice regarding impact categories and category indicators in life cycle impact assessment, Part I. *The International Journal of Life Cycle Assessment* 4: 66–74.
Udo de Haes, H.A., O. Jolliet, G. Finnveden, M. Hauschild, W. Krewitt, and R. Müller-Wenk. 1999b. Best available practice regarding impact categories and category indicators in life cycle impact assessment, Part II. *The International Journal of Life Cycle Assessment* 4: 167–174.
UNECE/EMEP. 2007. EMEP/CORINAIR Emission inventory guidebook—2007. Technical Report No 16/2007, EEA (European Environment Agency), Copenhagen, Denmark.
United Nations Division for Sustainable Development (UN-DSD). 2000. Agenda 21. Chapter 14: Promoting sustainable agriculture and rural development. United Nations (UN), New York, USA.
West, P., H. Gibbs, C. Monfreda, J. Wagner, C.C. Barford, S.R. Carpenter, and J.A. Foley. 2010. Trading carbon for food: Global comparison of carbon stocks vs. crop yields on agricultural land. *Proceedings of the National Academy of Sciences* 107(46): 19645–19648.
World Commission on Environment and Development (WCED). 1987. *Our Common Future (the Brundtland Report)*. Oxford: Oxford University Press.

Chapter 5
LCA of animal production

Xavier P.C. Vergé, Devon E. Worth, Raymond L. Desjardins, Brian G. McConkey, and James A. Dyer

5.1 Introduction

All life affects the environment, and ecosystems are constantly evolving in response to population pressures. Humans are no exception, and in our quest to house, feed, and entertain ourselves, we have developed an unequalled ability to modify our surroundings. With the development and exploitation of fossil fuel energy, in the span of about 100 years we have gone from having a relatively benign influence on the Earth system to having a significant global impact on the atmosphere, oceans, soils, and biodiversity. In a few decades humankind has become as powerful as geological forces. The use of fossil fuel energy has increasingly enabled us to override the natural cycles of nutrients and energy. In a sense, we have created our own "ecosystem" and have replaced ecosystem rules by economic processes. All environmental problems linked to human development are rooted in the fact that these two systems adhere to different economic versus ecological principles. Solutions will be found in a closer integration of anthropogenic processes with biogeochemical cycles. Because the economy is reliant on the Earth's ecosystem, the economic policy that will succeed is the one that respects the principles of ecology (Brown 2001).

X.P.C. Vergé (✉) • D.E. Worth • R.L. Desjardins
Agriculture & Agri-Food Canada, Ottawa, Canada
e-mail: vergex@agr.gc.ca

B.G. McConkey
Agriculture & Agri-Food Canada, Swift Current, Canada

J.A. Dyer
Agro-environmental Consultant, Cambridge, ON, Canada

J.I. Boye and Y. Arcand (eds.), *Green Technologies in Food Production and Processing*, Food Engineering Series, DOI 10.1007/978-1-4614-1587-9_5,

Understanding the origin of environmental degradation is a preliminary requirement to be able to find solutions that either minimize or eliminate adverse environmental impacts. Therefore, the first part of this chapter presents a general overview of recent changes in global agriculture, highlighting the main drivers of change and explaining the reasons why the agricultural footprint has become so large. Indeed, land and water use, biodiversity, and climate change are related to, and are affected by, agricultural activity. The second part of the chapter discusses the impact of global agriculture on the land, water, biodiversity, and climate.

This review aims to enhance our understanding of how agriculture, and more specifically animal production, affects the environment. However, a simple listing of impacts is of limited use to identify better management practices. Therefore, impacts are also discussed from a production system perspective. This is the objective of the third part of this chapter, which presents and discusses results from LCA studies carried out on five major animal production systems: dairy, beef, swine, poultry, and fish. The last part of the chapter discusses possible mitigation measures to minimize the environmental impact. As shown, the best management practices can only be found in a systems approach. Therefore, agricultural sustainability must be achieved through better integration of production systems into the ecosystem that supports them.

Finally, to put this chapter into a global perspective, we must note that feeding the human population relies on distribution systems and household incomes as much as it does on production. Current global food production is higher than food needs,[1] but the proportion of underfed people remains very high. In 1990 about 120% of the world's basic food needs (in terms of energy) were covered, which corresponds to an increase of more than 10% between 1970 and 1990 (Collomb 1999, 2002). Although the proportion of underfed people decreased during the same period, there were still about 845 million underfed people in 1990 (FAO 2009b). At the same time, about 80% of malnourished children in the developing world were in countries producing food surpluses. As Gardner and Halweil (2000) expressed the situation,

> For a variety of reasons, the century with the greatest potential to eliminate malnutrition instead saw it boosted to record levels

This chapter only presents and discusses food production in specific detail from animal sources and the use of LCA for estimating its environmental impacts. However, the importance of human issues has to be recognized. High food productivity without equitable food distribution and purchasing power is inefficient and leads to both social and environmental issues.

[1] Food demand has to be differentiated from food need, which depends on biological activity. Food need is therefore more directly linked to the human population in its diversity of age and gender but also level of activity. Food need could be considered as the level that must be reached, the lowest threshold in food production.

5.2 Food production

Food demand is a complex and dynamic process. It depends not only on population size, but also on human behavior such as habits, traditions, tastes, and other subjective factors (Collomb 1999, 2002). In addition, food availability can affect diet and food demand. Therefore, changes in food demand are the result of very diverse factors and numerous interrelationships. Although tastes, traditions, and cultures can affect food demand, the absolute quantity of food needed to feed a population is primarily driven by population. Therefore, in the paragraphs that follow we limit our discussion regarding food demand to population, although food demand is much more complex.

5.2.1 Human population

5.2.1.1 Change in global population

In 1961, the global human population was about 3.1 billion (Fig. 5.1). Over the past 50 years, the population has more than doubled and is now estimated at 6.8 billion people, an absolute gain of 3.7 billion people. It took only 12 years for the last billion to be added (Nellemann et al. 2009). This most recent rate of increase corresponds to about 230,000 persons per day. Over the next 40 years, the annual rate of

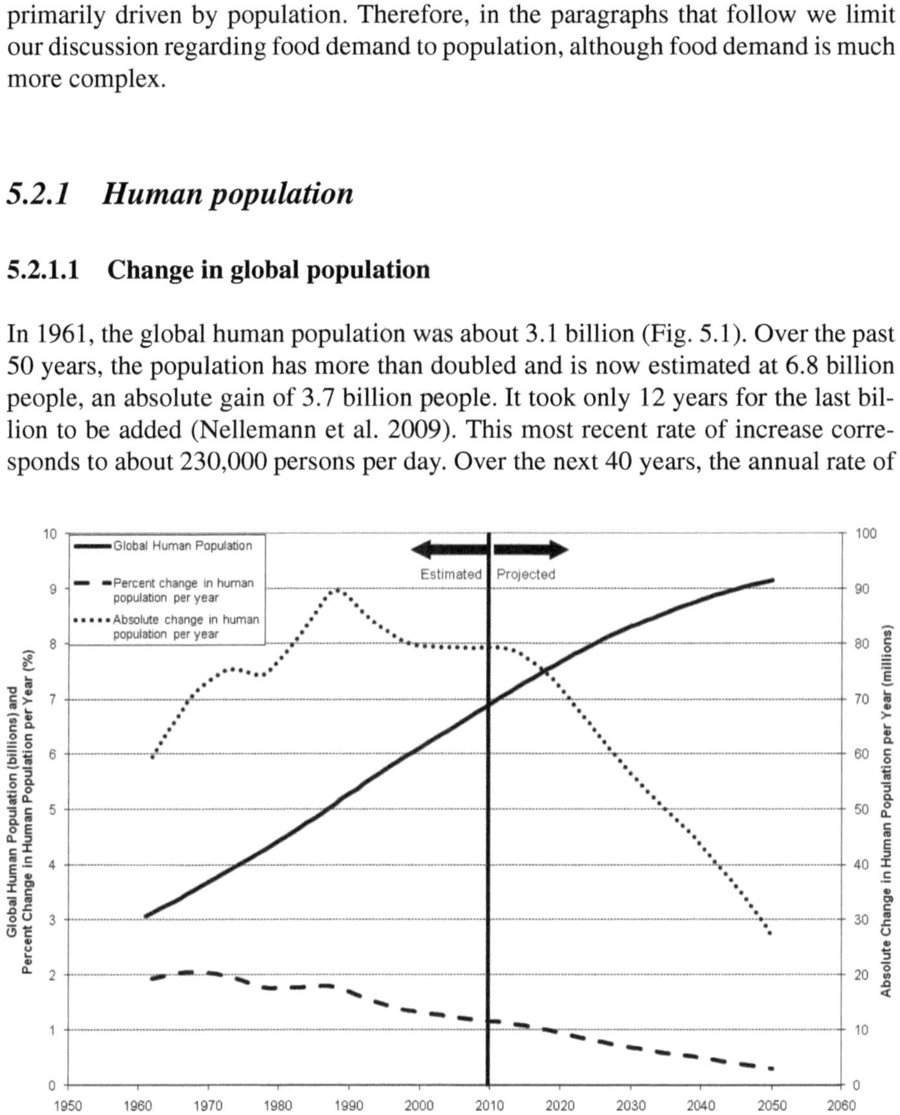

Fig. 5.1 Global estimated and projected population statistics, 1961–2050 (FAOSTAT 2010)

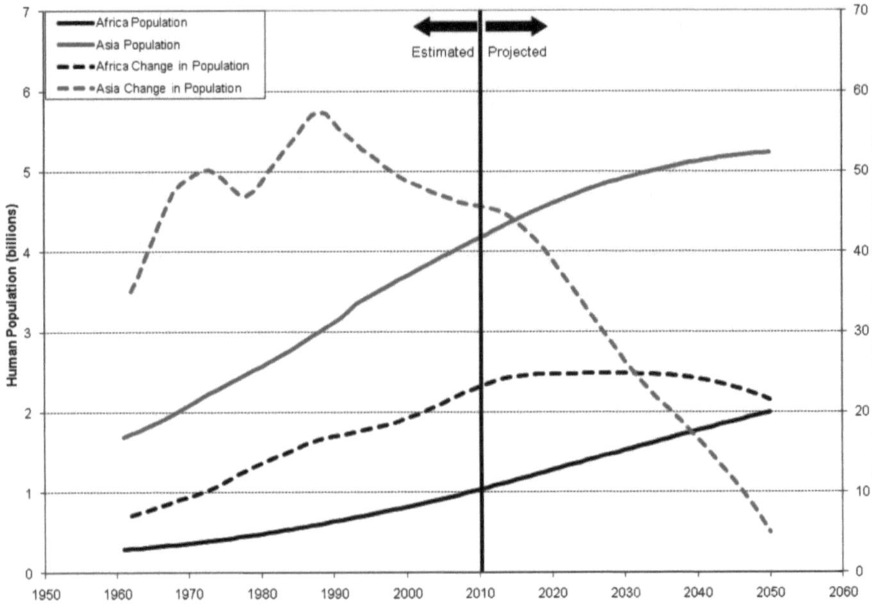

Fig. 5.2 Estimated and projected Asian and African population statistics, 1961–2050 (FAOSTAT 2010)

population increase is projected to decline from 1.2% to 0.3%. Despite a declining growth rate, the increase in population over this time is projected to be 2.3 billion people, as the global population climbs to 9.1 billion people by the year 2050. The majority of this increase is projected to occur in Africa (+1.0 billion) and Asia (+1.1 billion), with smaller increases in the Americas (+0.25 billion) and Oceania (+0.02 billion) offset by a small decrease in Europe (−0.04 billion).

The forces driving global population change differ from region to region. For instance, between 2010 and 2050 the absolute increase in population projected in Asia and Africa is about the same, or one billion people each. However, the growth rate for each continent is very different: increasing and then stabilizing in Africa, compared with constantly decreasing in Asia (Fig. 5.2). These differences are important to note because the future changes in agricultural production that will be necessary to feed this growing population will be achieved differently in each region. In this example, challenges will be greater in Africa not only because of the changes in technology, but also in the habits, traditions, and changes in social structure that will certainly occur.

5.2.1.2 Rural versus urban population

Urban population growth has an important social impact on food production. People living in large and densely populated cities tend to consume more protein from animal sources, higher fat content meals, and more sugar and cereals (Collomb 1999,

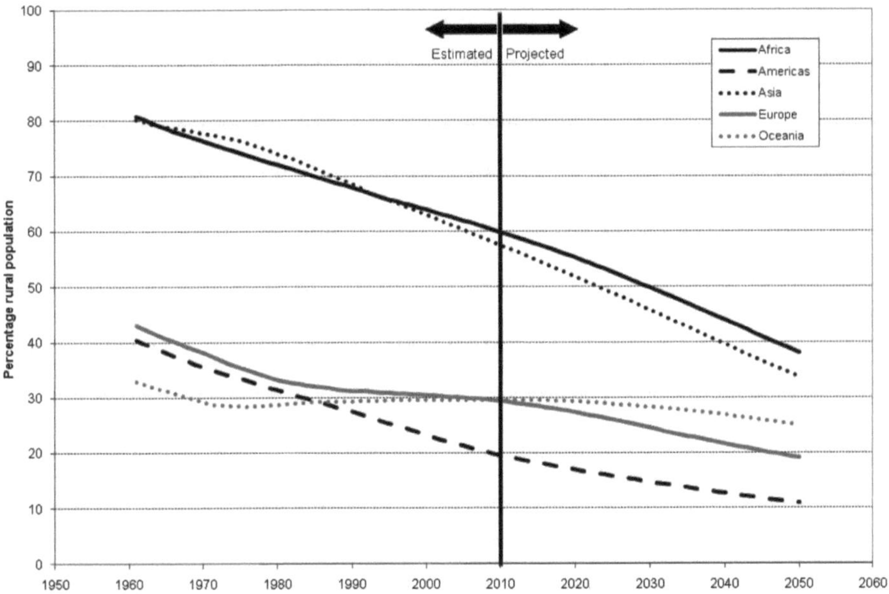

Fig. 5.3 Rural population as a percentage of total population for regions of the world (FAOSTAT 2010)

2002). Processed food is also much more common in the urban diet, which in turn leads to greater energy and material use but also losses through packaging wastes.

Currently, about 50% of the world population lives in a rural setting, but the urban/rural ratio is sharply divided by global regions. For instance, although 60% of the African and Asian population is rural, the rural population represents only 20% in the Americas (Fig. 5.3). Throughout the world, the rural population is projected to decline for the next 40 years, with the urban population representing approximately 90% of the population in the Americas, 62% in Africa, and 66% in Asia by the year 2050. The largest decreases in the proportion of rural populations are expected in Africa and Asia, where a significant population increase is predicted.

5.2.2 Food production from animal sources

Over the past 50 years the global growth rate in animal product consumption (meat, milk, and eggs) per capita was greater than any other agricultural products (FAO 2009a). Most of this growth occurred in China, where food energy consumption from livestock products was about 4.6 times higher in 2005 than it was 25 years earlier. Over the same 25-year period, energy derived from livestock products increased by more than 20% in Latin America and the Caribbean, about 7% in developed countries, and remained stable in sub-Saharan Africa.

Table 5.1 Consumption of livestock products (kg/person.year) by region, country group, and country in 1980 and 2005

	Meat		Milk		Eggs	
	1980	2005	1980	2005	1980	2005
Developed countries	73.6	82.1	197.6	207.7	14.3	13.0
Developing countries	14.1	30.9	33.9	50.5	2.5	8.0
China	13.7	59.5	2.3	23.2	2.5	20.2
Latin America and the Caribbean	41.1	61.9	101.1	109.7	6.2	8.6
India	3.7	5.1	38.5	65.2	0.7	1.8
Near East and North Africa	17.9	27.3	86.1	81.6	3.7	6.3
Sub-Saharan Africa	14.4	13.3	33.6	30.1	1.6	1.6
World	30.0	41.2	75.7	82.1	5.5	9.0

(After FAO 2009a)

Table 5.2 Meat consumption by region in 2000 and 2050 (projected)

	2000	2000–2050
	(kg/person-year)	% of change
Central and West Asia and North Africa	20	+65%
East and South Asia and the Pacific	28	+82%
Latin America and the Caribbean	58	+33%
North America and Europe	83	+7%
Sub-Saharan Africa	11	+100%

(After FAO 2009a)

According to FAO (2009a), consumption of animal products per capita could be broadly placed into three groups: the highest food consumption of animal products in the industrialized countries (mainly Europe and North America) with a daily intake of about 750 kcal/person in 2005, followed by the former centrally planned economies, as well as Latin America, the Caribbean, and East and Southeast Asia (mainly China), with a daily intake of almost 500 kcal/person, and finally by South Asia (mainly India) and Africa, with an average daily intake of 200 kcal/person (FAO 2009a). Note that these values only refer to the share of energy in diets coming from animal products.

The growth rate increase in animal product consumption was different depending on the type of products and among regions. Table 5.1 presents the change in the consumption of animal products per person (consumption per year and growth rate increase) over 25 years (1980–2005). To illustrate the variability, different country or country groups are presented in the developing world.

Egg consumption in China exceeded that in the developed countries, but meat and especially milk consumption were lower. Traditions are important considerations and those in European or North American diets are not the same as those in China, but the use of animal products will most likely keep increasing (Table 5.2). The same general observation can be made for India, where feed energy per capita derived from animal products was, on average, 2.5 times lower than in China and 3 times lower compared with developed countries.

In 2008, to support this consumption, the world cattle herd was estimated at about 1.34 billion head, the swine population was about 0.94 billion head, poultry were about 18.2 billion head, and sheep 1.1 billion head. This represents an increase of about 91%, 131%, 369%, and 8%, respectively, between 1961 and 2008 (FAOSTAT 2010). Milk production increased by 46% during the same period, and the dairy cow population went from 177 to 247 million (+39%). This is the result of increasing specialization in this industry, which led to higher milk yield per cow. Again, the change in milk yield per animal was region specific. In 1961, Africa and Asia had nearly identical milk yields; however, between 1961 and 2008, milk production in Asia increased by nearly 160%, whereas it remained constant in Africa.

Proteins derived from sea products (fish, crustaceans, and molluscs) account for about 15% of the global animal protein intake, but with very high regional variations. In the food-deficit countries this share goes up to nearly 20% (WHO/FAO 2003). Between 1950 and 2000, global fish production increased from 20 million to 130 million tons. These estimates include both sea catch and fish farming. Wild caught production peaked in 2000 at 96 million tons and declined slightly to 90 million tons in 2003. Conversely, aquaculture production increased from 20 to 42 million tons between 1993 and 2003 (Ryoichi et al. 2005).

5.2.3 *Changes in the animal production system*

In past decades the livestock production system has evolved, pushed on one side by the increasing demand for animal protein, and on the other side by high economic growth rates and technical innovations. Over time, this industry has become more and more specialized, and small farms with multiple production systems have progressively disappeared. More heavily influenced by market rules than by ecological processes, the face of agriculture in many regions has changed and has moved from classical family farms to large specialized factories. To maintain market competitiveness, specialization of animal production systems has occurred: adoption of high-energy diets (largely grain), genetic improvements (which led to international and highly oriented gene flows) (FAO 2007), and the introduction of new management techniques with high-density animal populations in confinement housing and the use of growth hormones and antibiotics. This production model is mainly driven by economic processes and focuses on short-term cost margins rather than ecological processes and based on long-term sustainability.

With the development of fast and efficient transportation and communication systems, agricultural activities were reorganized within each animal industry. Activities were broken down and decentralized, each of them often concentrated in specific geographical areas. There was a global shift in the geographic loci of the agricultural inputs and livestock operations. In concert with the specialization in animal production (shift from farm production to commodity production) the animal farming process has moved toward the industrial assembly line concept: imported animals fed with imported crops and additives, fossil energy to run the operation, and finally removal of animals from the land. This situation is the result of a necessary evolution owing to the rapid market change and strong economic pressure.

Globalization will most likely aggravate these trends, putting more and more pressure on all producing regions. Globalization has two major implications. First, in intensive systems the same basic production rules are applied everywhere without regard for the local situation. However, based on their natural resources (better soil, climate, etc.), social conditions (cheap labor cost, less environmental concerns, etc.), and government policies, some regions are more efficient producers than others. Second, globalization also characterizes the production processes: highly industrialized systems do not rely on local (on-farm) feed production. Because feed is the largest expense associated with animal production, intensive animal production systems are highly vulnerable to food market prices. These impacts put this production system in a very unstable and risky position. Although these changes in agricultural management have resulted in a strong increase in food production, important environmental issues have become more and more of concern. The following sections present and discuss the main issues related to animal production.

5.3 Impacts related to the animal production sector

5.3.1 Environmental issues

The changes in animal production systems that have occurred in only a few decades are now confronting us with new and important challenges. These changes began in an era in which economic growth was assumed to have no limits. In the 1960s, the industrial mode of food production increased at an annual rate of about 7%, mainly in developed countries, which was more than the population growth rate. At that time very little consideration was given to possible environmental consequences, notably from agriculture, which often are cumulative and were in their early stages. Many had never been seen or adequately appreciated in the past. As a result, from the beginning of the industrial era to the end of the 1970s, the environment was seen mainly as a supplier of raw material. However, in 1973, the first energy crisis occurred, forcing countries to think about their dependence on fossil fuel energy (Roberts 2004) and, more generally, that natural resources are limited.

An increasing awareness of the depletion of natural resources emerged around the middle of the 1980s. The risk of human dependence on finite resources was first pointed out by the newly founded Club of Rome, which published a report based on environmental modeling to simulate the interactions between ecological and human systems and, therefore, to identify the possible consequences of economic growth on Earth's resources (Meadow et al. 1972; Simmons 2000; Turner 2008). In the beginning of the 1970s, growing awareness that human activity was affecting the equilibrium of natural systems motivated governments to create Ministries of

Environment or reorient former government departments toward environmental issues (The United States and United Kingdom in 1970, Australia and Canada in 1971, and France in 1974). The risk that production, consumption, and pollution could push Earth's inherent resilience to the threshold and severely limit economic growth became better understood (Colombo 2001). Brown (2001) used a financial endowment as an analogy to describe ecological resources:

> The interest income from an endowment will continue in perpetuity as long as the endowment is maintained. If the endowment is drawn down, income declines. If the endowment is eventually depleted, the interest income disappears. And so it is with natural systems.

5.3.2 Land use

Although agriculture contributed to the economic growth during the middle of the last century, it also played an important role in waste production and natural resource depletion. Globally, all agricultural lands (including cropland, managed grassland and permanent ground cover with agro-forestry and bioenergy crops) occupy about 40–50% of the Earth's land surface (Smith et al. 2007). In 2007, animal production was estimated to use about one fourth of all ice-free land for pasture and about one third of all cultivated land for forage production (FAO 2009a).

5.3.2.1 Extensive production systems

In pasture-based systems, animal feed is mainly or entirely produced close to the farm. Therefore, an increase in animal production will primarily depend on pasture management, with the goal of increasing crop production and feed digestibility, so as to maximize weight gain and animal density. In these systems, the choice of plant species, as well as the time and duration of grazing, are extremely important management decisions.

Although grass-based production systems might be extensive in nature, their management can actually be intensive. Therefore, the environmental risk associated with extensive pasture-based management systems is overgrazing, in which livestock are stocked at a density that exceeds the carrying capacity of the rangelands or pasture. In such situations, the loss of vegetation can leave soil bare and vulnerable to erosion, soil carbon loss, and unable to recover (Dyer 1989; Zervas 1998; Conant and Paustian 2002). Water resources can also be negatively affected if overgrazing leads to chemical or biological contamination of leaching or runoff water, or through alterations to the water cycle.

Another important point to consider in pasture-based systems is deforestation. In this case new pasture lands are obtained at the expense of the surrounding forest and notably the tropical rainforest (Fearnside 1980, 1990).

5.3.2.2 Intensive production systems

Intensive animal production systems affect both the surrounding area and remote croplands. In these systems (often called "landless systems") animals are separated from the land that supports them either physically (when animals are confined to barns) or functionally (when animals are raised on a relatively large land area, but the carrying capacity of the land to produce food and dispose of waste is not considered). Either way, the local ecosystem is usually strongly disturbed. Because of its low commercial value and high cost of transportation, manure that is produced by confined feeding operations is generally applied in excess to nearby fields. This often results in compromised water quality owing to leaching and runoff of nutrients, as well as gaseous loss of nitrogen as ammonia, dinitrogen, and nitrous oxide (Menzi et al. 2010).

Livestock diets create secondary environmental impacts on the land used to grow feed crops. In intensive production systems, animal rations rely heavily on cereals or biofuel and vegetable oil co-products, and about 90% or more of animal feed is purchased off-farm (FAO 2009a). In developing countries the use of feed concentrates went from 240 million tons to 600 million tons between 1980 and 2005. Globally, the use of cereals for animal feeding represented about half of the world cereal production (Nellemann et al. 2009). Demand for livestock feed is a very important driver orienting crop production systems.

5.3.3 Biodiversity

Because animal production industries rely heavily on their surrounding environment, these systems have one of the strongest impacts on biodiversity. By its very nature, modern agriculture reduces biodiversity, as just one crop variety or animal breed is preferred over a complex of crops or breeds to mechanize operations and maximize efficiency (Altieri 1999). This is especially true for crop production, in which thousands of hectares can be covered by a single species. From an ecosystem perspective, such a monoculture is a completely unbalanced situation that relies heavily on fossil energy. Expansion of animal production often leads directly to deforestation and the loss of biodiversity. The heavy dependence on mechanized, high-input crop production for feed grains in animal diets leads to indirect impacts on biodiversity.

> Development tends to simplify ecosystems and to reduce their diversity of species. And species, once extinct, are not renewable. The loss of plant and animal species can greatly limit the options of future generations; so sustainable development requires the conservation of plant and animal species (Brundtland 1987).

However, it has to be noted that good management practices such as well-managed grazing systems may have beneficial environmental effects and can enhance biodiversity.

The second cause of the reduced biodiversity is that some animal production relies on the harvest of natural resources. For instance, overfishing is one of the main issues that the fishing sector is facing. In nearly 100 years, U.S. Chesapeake Bay oyster production has declined from about 45,500 t a year to 1,500 t because of overfishing. Similarly, the catches in the Gulf of Thailand dropped by more than 80% since 1963 and the Newfoundland cod fishery collapsed abruptly in 1992 (Brown 2001) and more than 35,000 fishers and plant workers became unemployed (Gien 2000). Mismanagement, lack of enforcement of conservation regulations and subsidies to fishing fleets (over US$27 billion annually) have left close to 30% of global fish stocks classed as collapsed (yielding less than 10% of their former potential). Some researchers estimate that virtually all commercial fisheries will have collapsed by 2050 unless urgent action is taken to bring far more intelligent management to global fisheries (UNEP 2010a).

Therefore, aquaculture is a growing source of protein in global diets. However, depending on how this industry is managed, it can have harmful effects on local ecosystems and on biodiversity as well. For example, shrimp farming developed at the expense of the mangroves is a threat to those ecosystems (Hein 2000). Along with aquatic habitat destruction, biodiversity impacts also include marine species becoming invasive or the accidental introduction of disease to local fish or invertebrate populations. Both the potential benefits and the environmental risks associated with aquaculture need careful consideration if this industry is to be sustainable.

Finally, animal production also reduces agricultural biodiversity through the selective breeding process. Since its origins, agriculture has always involved the systematic selection of plants and animals for desirable traits. Selection for the best performing crops and animals in local conditions has historically increased agrobiodiversity. Throughout history, the adaptation process to new requirements and situations has generated centers of agriculture biodiversity for many crops and livestock that are outside of their centers of origin (Fowler and Mooney 1990). So, why is animal production currently lowering agricultural biodiversity? The main reason is the industrialization of agriculture in response to global market competition. The Holstein-Friesian breed of dairy cattle is a typical example. This dairy breed is currently raised in about 128 countries and represented about three fourths of the global dairy semen export in 1991 (FAO 2007). Currently it represents about 93% of the whole dairy herd in Canada (AAFC-AAC 2008).

With market selection, the human agricultural-ecosystem has very few "ecological niches" in which rare animal breeds and heritage crop varieties can survive and continue to evolve. Globalization aggravates this situation by imposing the best breed along with its production system. In recognition of this recent decline FAO (2007) states that:

> Development and expansion of intensive livestock production systems have led to a reduction in diversity through the large-scale replacement of local breeds with a small number of globally successful breeds.

For example, although Holstein cattle now dominate the dairy industry, other dairy breeds, such as the Jersey, are in decline. With new reproductive biotechnologies, reproduction environments are controlled and have led to an unprecedented and highly oriented international gene flow. Of the 7,616 livestock breeds, 1,081 are currently recorded as "transboundary" (FAO 2007). Conversely, the 1996 Leipzig Global Plan of Action recognized that market forces could actually help conserve agro-biodiversity when fostered at a local or regional scale (Dyer 1998).

5.3.4 Climate change

Agricultural greenhouse gas (GHG) emissions originate through natural processes and crop production management practices. The three main gases attributed to agriculture are methane (CH_4), nitrous oxide (N_2O), and carbon dioxide (CO_2).

Methane is produced when organic matter is decomposed in an oxygen-free (anaerobic) environment. For livestock production, enteric fermentation and manure management are the two major sources of methane. Enteric fermentation is the digestive process that allows ruminants to digest cellulose. This is achieved through bacterial activity in the rumen and methane is produced as a byproduct of this digestive process. Manure produces methane when it is stored in liquid storage systems without being aerated.

Nitrous oxide is also emitted from stored manure. As opposed to methane, this emission process is partly aerobic, and needs oxygen. These agricultural emissions occur in solid storage systems and dry lots that are aerated. Nitrous oxide emissions also occur after field application of either manure or chemical nitrogen fertilizer, and in association with biological nitrogen fixation in pulse and legume crops.

Agricultural croplands can either emit or sequester carbon dioxide. The biological carbon cycle begins with carbon fixation in plants through photosynthesis, continues through the food chain, and ends with release of carbon dioxide back to the atmosphere by respiration or decomposition of plant and animal organic matter. The cycling of carbon from the atmosphere to the biosphere and back to the atmosphere is part of the annual carbon cycle and adds no new carbon to the atmosphere. Therefore, carbon dioxide from animal respiration does not need to be taken into account when studying the impact of animal production on climate change. Plant roots and harvesting residues do, on the other hand, accumulate atmospheric carbon in the soil. This long-term change in soil organic matter, or carbon sequestration, should be attributed to animal production systems because of the importance of animal diets in crop selection (Vergé et al. 2007). This linkage will be discussed further in the system boundaries section.

Carbon dioxide emissions also come from the use of fossil fuel energy in agricultural production systems. In Canada, the main fossil energy uses are for farm field work (tillage seeding and harvesting), the manufacture and supply of farming equipment and chemical fertilizer, electricity and heating fuel used in farm buildings, and local transport by farm-owned vehicles (Dyer and Desjardins 2009).

In 2005, global agriculture was estimated to be accountable for about 11% of all anthropogenic GHG emissions, equivalent to about 5.5 $GtCO_2e$ (between 5.1 and 6.1 $GtCO_2e$). The agricultural share of total anthropogenic CH_4 and N_2O emissions were about 47% and 58%, respectively (Smith et al. 2007). The two sources exclusively associated with animals, enteric fermentation, and manure management, were estimated to be 32% and 7%, respectively, corresponding to a total of about 2.2 $GtCO_2e$ in 2005. The third working group of the IPCC fourth assessment report, however, mentions that there is a wide range of uncertainty in the estimates of both the agricultural contribution and the anthropogenic total (Smith et al. 2007).

5.3.5 *Water use*

Compared with crop production, water use in the animal production system is generally very high because both the direct water consumption by animals and the water used to produce the animal feed have to be taken into account (Steinfeld et al. 2006). Animal water needs are higher in extensive systems than in intensive systems because the animals are more active. There is also indirect water use associated with animal production, notably for cooling equipment and cleaning barns and facilities, especially in intensive production systems that require imported feed and confined conditions.

Overall, livestock systems account for about 8% of water consumption worldwide, mainly for irrigating forage crops, with less than 1% as direct water consumption by farm animals (FAO 2009a). The world average water footprint per kg of meat was estimated to be 15,500 L for beef and 3,900 L for poultry with 4,800 and 6,100 L for pork and sheep, respectively (Hoekstra and Chapagain 2007).

Water quality can also be affected by the type of manure management practices. Nutrients in manure excreted by animals are estimated to be equivalent to the total amount of chemical fertilizers produced each year (Menzi et al. 2010). A large portion of manure nutrients reach the surface and underground waters and coastal marine ecosystems through leaching and runoff, leading to environmental problems such as eutrophication and soil salinization and potential human risks from pathogens, such as *E. coli* contamination (LeJeune 2003).

5.4 Environmental analysis and life cycle assessment

The need to better assess the impacts associated with the increasing pressure of production systems led to the development of indicators and methods of calculation adapted to the new environmental issues. The LCA method was first used to study impacts related to packaging and energy use, and then gradually developed a broader approach, taking soil, water, biodiversity, and human health into account (Jolliet et al. 2005).

There are now two primary types of life cycle analysis, Attributional Life Cycle Analysis (ALCA) and Consequential Life Cycle Analysis (CLCA) (Thomassen et al. 2008). ALCA describes the environmental impact that occurs because of the production of a certain amount (mass, volume, etc.) of the functional unit, whereas CLCA describes how environmental impacts change as the result of a modification to the amount of output from a system. Thus, ALCA describes a production process, whereas CLCA is based on building scenarios and sensitivity analysis. In the following section, we first present impact assessment studies for five animal production sectors including dairy cattle, beef cattle, swine, poultry, and fish using the ALCA approach. Some of the different concepts and proposed tools to be used in life cycle analysis are then discussed.

5.4.1 Life cycle assessment application to livestock commodities

5.4.1.1 Milk production

Life cycle assessment of the national and international dairy industry is probably the most advanced among all of the animal production industries. The International Dairy Federation (2009) recently conducted a review of the international scientific and gray literature to determine the environmental impact of milk production. In this study they report that throughout the entire life cycle of 1 kg of milk, an average of 1.2 kg CO_2e are emitted, with approximately 85%, or 1 kg CO_2e, being produced on-farm, from crop and milk production. The remaining 15% is attributed to transportation, processing, consumption, and disposal after the farm gate.

In a review of LCA studies, the carbon footprint for milk was found to range from 0.84 to 1.3 kg CO_2e per kg of milk (de Vries and De Boer 2010), which is lower than all other animal-based agricultural products. However, this small carbon footprint can partly be attributed to the high water content of milk. For this reason, it is common to report the carbon footprint (or other environmental impacts of milk) in terms of fat-corrected milk production or fat and protein-corrected milk production (Schau and Fet 2008). When only the protein content of milk is taken into account, the carbon footprint for milk production ranges from 24 to 38 kg CO_2e per kg of milk protein, which overlaps with the range of carbon footprints for protein production from poultry, pork, and egg production (de Vries and De Boer 2010). In Canada, the protein-based GHG emission intensity of milk was found to be lower than that for beef, but slightly higher than that for pork and poultry products (Dyer et al. 2010a).

5.4.1.2 Beef production

Quantification and generalization of the impact of beef production on the environment is complex because:

1. Often there are allocation issues associated with beef production as cull cows and surplus calves from the dairy industry are often transferred to the beef industry (Cederberg and Stadig 2003; Casey and Holden 2006)
2. There are a wide range of management techniques used to raise beef cattle from birth to slaughter, all of which entail separate feeding systems, land use requirements, and energy inputs (Vergé et al. 2008)

Because of the diversity in management strategies for beef production, the range of environmental impact estimated for beef production was greater than pork or poultry production (de Vries and De Boer 2010). Generally, the greatest impact was estimated for sucker beef, in which all the impacts from the mother are attributed to the meat product, whereas the lowest impact is for imported dairy calves, in which a significant portion of the impact from the mother is attributed to her primary product—milk.

Beef tends to have greater GHG emission intensities than pork and poultry production because of the relatively low fecundity of beef cattle and because the ruminant digestive system produces large amounts of CH_4 through enteric fermentation. For intensive production systems in the Midwestern United States, the smallest environmental impact (across all impact categories, including energy use, GHG emissions, eutrophication potential, and ecological footprint) from beef production was found to be from animals weaned directly to feedlots, with increasing impact for animals on background operations (nonbreeding cattle grazing farms) before being finished in feedlots, and for animals that are finished completely on pasture and hay (Pelletier et al. 2010). However, this analysis assumes a soil carbon neutral system, and when potential soil carbon sequestration of grazing lands is incorporated, the global warming impact of the grazing system is decreased to 15% less than the feedlot-weaned beef system.

Clearly, a standardization of the soil carbon treatment is needed to ensure consistency among studies.

It is somewhat surprising that a grass finishing system would have a greater energy use impact as compared with a feedlot system; however, it is understandable for this study, in which cattle are grazed on pastures that are periodically seeded and fertilized to maintain higher yields. Further, pastured cattle are fed a low-energy density hay diet over winter, which is harvested from fertilized fields (Pelletier et al. 2010). Combined, these factors make the grass-based system slightly more energy intensive as compared with the feedlot system. However, in grass systems in which the animals can graze for a greater portion of the year, or are not grazed on managed grasslands, this result might not hold true and the energy use for the grass system should be lower.

5.4.1.3 Pork production

Compared with dairy and beef production, pork production does not appear to have received as much attention in the literature. On a per kg of meat basis, pork production tends to have a smaller impact on land use, energy use, and climate change than beef production (de Vries and De Boer 2010). This is owing to the relatively high number of piglets per sow, the short time to reach maturity, and the young age at slaughter. However, for acidifying gases and eutrophication potentials there was no consistent trend in impact when comparing beef and pork production.

In France, organic, conventional, and "Red Label" pork production were compared using an LCA approach (Basset-Mens and van der Werf 2005). It was shown that per kg of pork (live weight), for climate change, nonrenewable energy use, and land use, the impact was least for animals that were raised under conventional conditions, which is broadly consistent with findings for the beef industry in the United States (Pelletier et al. 2010). However, in terms of acidifying gases and eutrophication potential, conventional pork production had a greater impact (Basset-Mens and van der Werf 2005). This study also highlights the importance of choosing an appropriate functional unit. When the environmental impact of pork production in Europe was expressed on a per ha of land used for pork production, organic agriculture had a lower impact as compared with conventional agriculture in terms of eutrophication, acidification, climate change, and nonrenewable energy use, largely because of less intensive production practices on the organic farm as compared with the production practices on the conventional farm.

Despite the fact that pork production tends to have a smaller climate change impact than beef production (Dyer et al. 2010a), pork production is still much less efficient than pulse crop production for protein. In a comparison of meals prepared using different pork feeding strategies (comparing soybeans with peas fed as protein) and a vegetarian meal in which the protein was derived from peas, the global warming impact of the vegetarian meal was roughly 30–50% less than the pork options, depending on the scenario (Davis et al. 2009). Similarly, the vegetarian meal had a smaller impact in terms of eutrophication, acidification, and land-use compared with pork production, but had a similar impact in terms of energy use (when the vegetarian meal was a pea burger meal).

5.4.1.4 Poultry production

Poultry production was found to have the smallest impact on land use, energy use, and global warming potential (de Vries and De Boer 2010). These results are consistent with a study in Canada that showed that poultry production is the least intensive means of producing animal protein based on GHG emissions (Dyer et al. 2010a). A cradle to farm gate analysis of the US broiler industry showed that the vast majority of environmental impacts associated with broiler production were embodied in the feed production phase (Pelletier 2008). This phase represented 80% of energy use, 82% of global warming potential, 98% of ozone depleting emissions, 96% of acidifying emissions, and 97% of eutrophying emissions. On average, emissions associated with on-farm inputs only contributed 9% of the environmental impact.

It should be noted in this analysis that the poultry manure that was land-applied effectively reduced the environmental burden of poultry production, because a credit was given for avoided fertilizer production. Although there would be some environmental impact associated with the management of manure (e.g., the energy used during transport and spreading, and local pollution from leakage during storage), these manure management issues were overshadowed by the environmental impact that was avoided by less synthetic fertilizer being manufactured.

5.4.1.5 Aquaculture

On a per capita basis, fish are not consumed in the same quantity in Western countries as other meats. For example, after correcting for food losses, on an annual basis and per capita, Canadians consume approximately 6.5 kg of fish, 13.5 kg of poultry, and 23.3 kg of red meat (beef and pork) (Statistics Canada 2009). However, the quantity of fish consumed by Canadians has been increasing over time, and with declining wild fish stocks it seems likely that farmed fish will increase as well.

The energy intensity of farmed salmon and wild caught cod in Norway is reported to be more efficient than farmed chicken (Ellingsen and Aanondsen 2006) and the energy return on investment (EROI) in salmon aquaculture in west coast Canadian salmon farming ranged from a low of 7.8% to a high of 17.8%, depending on the feed formulation (Pelletier and Tyedmers 2007). Therefore, the most efficient salmon farming compared well with poultry production, which was estimated to have an EROI of 17%. Similarly, farmed salmon compared well on the basis of global warming potential, with other types of terrestrial-based meat. The greenhouse gas emission intensity for farmed salmon ranged from 1.2 to 2.7 t CO_2e per ton of live weight salmon, which was approximately equal to that for poultry (1.8 t CO_2e), but better than for pork (3.2 t CO_2e), feedlot beef (14.5 t CO_2e), and sheep (18.5 t CO_2e) (Pelletier and Tyedmers 2007).

5.4.2 The use of life cycle assessment in animal producing systems

5.4.2.1 Life cycle assessment and intensity indicators

To estimate the environmental burden of a given product, LCA uses environmental impact factors calculated on a per unit basis, which is equivalent to the definition of intensity indicators. It is important to understand the limitations of this type of indicator for effective interpretation of the results they provide. Generally, intensity indicators compare the rate of emissions or wastes to the rate of production. Low values are associated with more environmentally friendly operations. For example, decreasing the GHG emissions from 2 kg CO_2e to 1 kg CO_2e to produce 1 kg of milk represents a very big improvement. This means that, keeping the same production levels, GHG emissions are reduced by 50%.

However, the reality is often a bit more complex than the picture presented in a calculation. For instance, achieving such a reduction in emissions per unit of product in milk production involves new feeding practices, changes in animal and herd management, or genetic modifications (Boadi et al. 2004) that have an impact on the milk composition. Therefore, although production may be twice as high, the milk does not contain the same nutritional value as the milk from the less productive cow.

Intensity indicators are based on relative calculations (i.e., the result of a ratio). Therefore, management practices that will keep production rates higher than the rate of increase of the associated waste products will be considered as beneficial and the production more environmentally benign. However, is it appropriate to consider a practice that leads to higher absolute waste production "environmentally friendly"? This is correct if the level of production is kept constant, meaning that better efficiency would lead to a reduction in absolute emissions. However, this analysis must also consider how higher efficiencies are achieved. Better efficiencies are generally obtained by a shift from extensive to intensive production systems and this change will then lead to the environmental effects previously mentioned (cf. Sect. 5.3). This again means that a holistic approach has to be used because positive results obtained for a single intensity indicator does not necessarily imply a reduced overall environmental impact.

Further, moving from extensive to intensive systems generally correspond with higher animal density and larger operations, which then means that environmental impacts are geographically concentrated on a smaller space. Because toxicity is not a question of quantity but of concentration, it is therefore possible to decrease the global environmental impact of a practice through intensification but at the expense of significantly increased, but localized, environmental impacts. The question is then whether global environmental issues can be solved by moving from a distributed production system, which has relatively minor local environmental impacts, to an intensified and centralized production system that has major local environmental impacts. The answer differs depending on the type of pollution. For GHG emissions, a "hot spot" could be preferable because emissions will eventually be diluted in the atmosphere. Therefore, the principles of area and concentration are not applicable and intensity indicators are well adapted. Conversely, they are important considerations for resident pollutions, such as overfertilization, nutrient losses, and eutrophication, for soil erosion and loss in soil fertility or decreasing biodiversity; thus, the use of intensity indicators is questionable.

Intensity indicators do not provide relevant information on the state of the ecosystems because the direct comparison between the production and the associated waste "bypass" the ecosystem supporting them. For instance, with new techniques and material, fish catches can increase and deplete stocks while minimizing the intensity indicator. Without more detailed information regarding ecosystem health, a new technique that is more efficient, but environmentally damaging, could be promoted.

Therefore, intensity indicators are very interesting for evaluating production efficiencies, but results have to be considered very carefully when they are used for evaluating environmental impacts and identifying new mitigation practices. They are not well adapted for assessing the sustainability of animal production.

Effective LCA is based on a range of different environmental indicators. Achieving sustainable agriculture requires that LCA focus on ecosystems with a shift away from a product point of view. The most important ecosystem parameter is the magnitude of the environmental effect (or the toxic concentration), which typically depends on the amount of waste per unit area. Therefore, LCA must consider the sites where industries are operating and the areas that provide the resources for that industry. Such a "functional area" unit must be used for each emission source. For example, GHG emitted directly from animals would be divided by only the area where animals are raised, whereas GHG emissions associated with manure should be based on the land on which it was applied. In agriculture, the state of the resource is itself often in need of assessment. In the IMPACT 2002+ methodology (Jolliet et al. 2003), no indicators were put forward to deal with stock depletion. For example, the aquatic ecotoxicity effect on loss in biodiversity has been considered as indicator but not the fish stock.

5.4.2.2 Objective of the system

Although the primary objective of agriculture is to feed people, the sector has always provided nonfood products such as wool or leather, or supplied draft animals for mechanical labor and transportation. Nowadays, the nonfood production includes bioenergy and, in a few cases, agrochemicals. Hence, both food and nonfood products must be considered when evaluating the environmental impact of agricultural activities. Agriculture also "produces" products that have little or no commercial value such as aesthetically pleasing landscapes or the preservation of rural traditions, which then strengthen social relationships within communities. Effects of these nontangible products are difficult to take into account but they should be considered and, if they are not quantified, they could at least be included in results analyses and discussions.[2]

Originally, LCA was developed for industry, in which the goal was the efficiency of production, rather than the assessment of environmental impacts. In assessing efficiency, the weak links and the most expensive steps are identified and modified to minimize inputs and maximize outputs. Although LCA has proved to be flexible enough to be applied as an environmental assessment tool, it needs to be complemented with an ecosystem perspective to be able to comprehensively evaluate the sustainability of production systems.

5.4.2.3 Functional unit

In LCA, products are compared based on their common function. Therefore, the unit of production output quantifies this function and differs from one study to another. The concept of functional unit used in LCA allows for a very wide range of comparisons because only the purpose of the product is actually considered and not its shape or matter.

When focusing on a specific production system (beef, dairy, swine, poultry, or fish), the functional unit can be a unit of product (kg of meat, milk, or dozen eggs). To assess the environmental impact of animal production as a whole industry requires a ubiquitous unit. Protein, as seen in the preceding, has been used because it is a common component of food and is essential to a healthy human diet. However, as previously discussed, agricultural systems do not just generate food products, but furnish a whole suite of associated co-products as well (leather, wool, labor, organic fertilizer, etc.). Therefore, it seems preferable to avoid using protein as the functional unit when a system as a whole is analyzed.

---

[2] These issues might be more easily managed if the French concept of terroir was used because spatial boundaries are based on human interactions with ecosystems. Terroir as defined in UNESCO (2005), is a "finite geographical region where a unique set of cultural traits, knowledge and practices have developed over time, and reflect the interaction between the natural and human environment" (freely translated from French by the authors).

Market price is a possible option because most agricultural commodities can be converted into a monetary value. However, the drawback of this option is that it is based on the dependence of market values and many external influences with little dependence on the products themselves. It is a floating reference, freely moving among commodities and over time. Therefore, with no fixed value, no true comparison can be made.

Two possible units that meet the requirements of a functional unit that is ubiquitous, directly linked to production, and easily accessible (measured and recorded regularly and frequently) are matter and energy. Energy is more broadly applicable than matter because all components (material, workforce, light, heat, etc.) can be converted into energy, which is not necessarily true for matter. As a functional unit, energy could be partitioned into renewable and fossil energy for both agricultural inputs and outputs and detailed for identifying hot spots in production processes.

5.4.2.4 System boundaries

A common problem of LCA is the establishment of the system boundary, or the sphere of influence. This is the extent up and downstream in the production process for which environmental impacts are attributed to the product being studied. There is no universal methodology to establish the boundaries of a system, except to ensure that all processes needed to fulfil the objective of the study have been included (Jolliet et al. 2005). Therefore, determining the system boundaries depends on the development of general rules that are broadly applicable rather than fixed limits applicable to all production systems.

In our case, the general objective is to estimate the environmental impact of agricultural activities and the capacity to sustain a specific animal production system. Therefore, LCA should consider both the production and the ecosystem perspectives when determining the system boundaries. From the production perspective the system has to be studied as a whole, including all production (or outputs) that correspond to a full system expansion. However, this also means that geographical boundaries have to be taken into account, because the ecosystem is spatially defined and impacts are related to this spatial unit. Therefore, to be able to evaluate the sustainability of animal production, boundaries have to be adapted to the farm, the factory, or the production sites being studied, including all impacted areas.

The Input–output method can be used as a tool to establish a system's boundary (Miller and Blair 1985; Jolliet et al. 2005). This method lists monetary flows between production processes. In our case, the focus is on agricultural production and therefore, to be as precise and exhaustive as possible, the list should be based on farm expenses detailing goods bought (electricity, fuel, fertilizers, feed, but also building materials, facilities, tractors or equipment, etc.) and farm activities. The list has then to be analyzed to determine the environmental impact associated with the identified goods and activities to avoid double counting.

Farm activities cause environmental impacts (positive or negative) if they physically modify the agro-ecosystem. For instance, tillage practices emit GHG through the use of fossil fuel and also impact soil biodiversity; application of manure or chemical fertilizer results in GHG emissions (CO_2 and N_2O) and can also result in soil compaction, on the other hand it can increase soil organic matter and enhance its structure and fertility. Deforestation also causes GHG emissions, soil quality concerns and biodiversity issues; conversely, a well-managed agro-forestry system will have positive effects on soil, atmosphere, and biodiversity and could also help to prevent flooding and soil erosion, loss of soil carbon and nutrients. All of these farm activities need to be considered when conducting a life cycle inventory (LCI). Other farming activities, such as milking dairy cattle, only involve the purchase of "goods" (which in this case is electricity).

Regarding goods, the environmental impact can be linked to external and remote lands (or ecosystems), such as in the case of the crops used in animal rations. As previously mentioned, it was estimated that, globally, livestock production uses about one fourth of all ice-free land for pasture and about one third of all cultivated land for forage production (FAO 2009a). Thus, when studying the environmental impact of animal production, impacts related to the land and crop production for the specific amount of crops included in the animal rations have to be included in the animal production system (Vergé et al. 2007, 2008, 2009a,b).

This approach avoids unfairly penalizing farmers that grow their own feed, as compared with farmers who do not own land and buy all their feed. This approach incorporates all impacts related to land involved in the production of livestock feed and can be used to develop efficient mitigation strategies. For example, by linking crop production to crop types and land management (Dyer et al. 2010b), it is possible to identify the diets and feeding strategies that have a smaller overall environmental impact.

5.4.2.5 Allocation

The same manufacturing or natural process can occur at the origin of multiple products, such as milk and meat from the dairy sector or meat and eggs from poultry layers. Impacts associated with these production processes need to be either distributed or allocated, to each of the individual products. Allocation is an important step in LCA studies and may have a significant impact on results and conclusions. For example, 70% of beef production in Sweden is derived from the surplus calves and cull cows of the dairy industry (Cederberg and Stadig 2003) and the allocation of environmental impacts from the cow affects both the dairy and beef industries.

There are four primary methods of allocating emissions:

1. No allocation: 100% of environmental impact is attributed to the primary product
2. Economic allocation: environmental impacts are attributed proportionally to the income derived from the products

3. Biological allocation: environmental impacts are attributed proportionally to the energetic requirements to produce each product
4. System expansion allocation: the system is expanded to estimate the emissions associated with an alternative method of producing the same product (i.e., in the case of surplus calves and cull cows, a beef system producing an equivalent amount of meat would be described)

According to the ISO 14044 standards (International Standard Organization 2006) established for LCA, insofar as is possible, allocation should be avoided by subdividing any one process into its subprocesses. When this cannot be done, system expansion is the preferred method of allocation (International Standard Organization 2006), and has been advocated as the most appropriate method for allocating milk and meat on a dairy farm (Cederberg and Stadig 2003).

Within ALCA, system expansion is the most desirable approach to associate environmental impacts with specific products, but other types of allocation procedures (financial, energetic, etc.) are most common (Thomassen et al. 2008), mainly because it is often difficult (or impossible) to expand the system to include a product that exactly mimics the co-product for which the system expansion analysis is attempting to correct. For instance, fish fillets are the primary product of aquaculture and farmed fish systems, but fish trimmings are also produced and some of the environmental impact must be attributed to them (Pelletier and Tyedmers 2007). However, there is no product that can mimic fish trimmings, so allocation is commonly based on economics or energy. In CLCA, system expansion is the only approach to allocation because it models the impact of a change in production, and therefore impacts other production systems through product displacement or avoidance.

Allocation can also be used to identify and evaluate a specific type of mitigation practice, such as recycling marginal products or wastes. For example, should manure from animal production be treated as a co-product, or as a waste product? If it is a co-product, then its environmental impact must be assigned to it specifically, but if it is a waste product from animal production, then its environmental burden is attributed entirely to the animal product. If manure is considered to be fertilizer (replacing synthetic fertilizers) that is applied to cropland, its impacts are attributed to itself rather than to the animal product. In general, agriculture produces component products that can potentially either harm the environment or be valuable. For LCA studies, this potential should be evaluated first.

5.5 Reducing the environmental impact of animal production systems on the environment

Animal production systems are varied and their production processes are complex. Globalization and trade liberalization have increased the length and complexity of supply chains for this industry. Longer supply chains have expanded the potential number and geographical extent of environmental impacts associated with animal products. This section will not list all possible mitigation practices that could potentially be applied. Instead, it will analyze the mitigation principle, present and discuss some examples, and propose possible ways for identifying and evaluating mitigation practices.

5.5.1 Mitigations practices and management objectives

Agriculture is a user of natural resources and an activity that can modify ecosystems. Its principal objective is to feed people, using management practices that achieve that goal sustainably. Management practices are the link between natural resources and food products—too much focus on only one of these elements leads to environmental disturbances.

The sector, including animal agriculture, has played an important and beneficial role, not only in social development, but also in ecosystem development. The grazing animal plays a unique ecological role. It maintains and enhances the structural heterogeneity of plant communities, which is vital to maintaining floral and faunal diversity (Rook and Tallowin 2003).

Agriculture reshapes the ecosystem with the goal of producing a few specialized products. It creates environmental problems when management pushes natural cycles and harvest beyond sustainable yields. Solving these issues requires mitigation practices and a change in management methods. Mitigation practices must be based on well-defined sustainable management objectives—knowledge and technologies are only tools that need to be controlled and directed to achieve the desired objectives. For example, as was previously described, manure can be either a harmful or a valuable product depending on its use. To some extent, it is management that gives environmental value and, in the case of manure, makes the difference between being a co-product or a waste.

Although the umbrella of environmental concerns may have a lot of commonality across regions, or even countries, the natural resources, the biophysical environment, geographical location, history, knowledge, and traditions differ locally. As a result, the most appropriate mitigation practices are not be the same either. In general, a practice will not harm the environment if it is well harmonized with the local ecosystem.

5.5.2 Beneficial management practices

The most important components of ecosystems are living beings because without them environmental issues, harmful impacts, pollution, or toxicity lose their meaning. Soil carbon is important for soil biology and plant growth but not for the mineral soil. Eutrophication does not hamper water, nor do greenhouse gases alter atmospheric composition by more than a fraction of a percent. It is the magnitude of pressures on living beings and their habitats that should be the real focus when assessing the benefits of management practices.

Also, when dealing with living beings and more specifically domesticated animals, welfare issues have to be considered. Because we are at the origin of agroecosystems, it is incumbent on humankind to consider and address the welfare issues throughout the animal life: birth and weaning stage, animal density and activity, transportation, and slaughter.

The objective of mitigation practices is fundamentally to have a better respect for the natural environment, but having a respect for the environment without respecting one of its living components is counterproductive and simply not consistent. The temptation to make animals responsible for environmental impacts such as global warming serves no useful purpose. Thus, since they appeared about 15 million years ago, ruminants have produced methane. So, fundamentally, animals are not the true cause of the current ecosystem disturbances, but rather it is the way they are managed. Adapting them to agricultural purposes is normal and needed, and can even be environmentally beneficial (Seré et al. 1995) but it has to be done within certain limits. Animals have to be used for what they are and can do. The basic attributes of each type of animal have to be first recognized and then promoted through their management.

For mitigating GHG emissions, diet manipulation is probably the best option for ruminants. However, there are two major concerns related to the adoption of this practice. The first concern is its applicability in developing countries in which GHG emissions are expected to increase rapidly in the coming decades. In many developing countries, improved feeds would not be affordable and would often undercut local cultural ecologies (Milich 1999). The second concern is connected with the animal itself: treating ruminants as monogastrics (e.g., in a diet of only 20% roughage) denies their most fundamental attribute. Besides, although this high-grain diet reduces enteric emissions, cattle cannot compete anyway with pork and poultry because they are not monogastrics. The strength of the ruminant is clearly not feed digestibility. The feed conversion rates (feed consumption to body weight gain ratio) for cattle are, in the best scenarios, two to three times higher than for pork and three to more than four times higher than for poultry. However, these two monogastric animals cannot be fed entirely with grass and forages. Cattle are grazing animals and are therefore the best commodity for solving the increasing land use competition issues (UNEP 2010b; Keyzer et al. 2005)—a consideration that is rarely incorporated into LCA (Garnett 2009), even though it could have a profound impact on the findings. For example, one of the driving causes of Amazonian deforestation is soybean production primarily for pork and poultry feed, not grazing (Nepstad et al. 2006). Through grazing, the ruminant digestive system can exploit forages produced on land that is unsuitable for the production of crops for human consumption. As a result, a large part of the ruminant diet can be derived from land that does not compete directly with crops that can be consumed directly by humans, as compared with pork or poultry. Furthermore, grazing can enhance biodiversity through an increase in mixed grass pastures and a decrease in annual monoculture cropping.

Beneficial management practices have to be approached from a global perspective because it is the whole system that one wants to balance. Therefore, mitigation measures for specific environmental effects must be oriented toward overall sustainability. Mitigation practices developed without taking a holistic approach may not be sustainable overall. For example, reducing methane emissions is important and beneficial from a climate change perspective, but if a methane-reduction practice leads to either high nitrous oxide or ammonia emissions through manure management, it is not an effective mitigation practice.

The same approach is true at a broader level of environmental management. Lowering GHG emissions at the expense of water quality or improving water quality at the expense of biodiversity are not viable solutions. Therefore, even mitigation practices that apply to a specific ecosystem component have to be conceived with consideration of the impact on the ecosystem as a whole or, in other words, applied in a holistic approach. This requires an understanding of ecosystem relationships and how those relationships may evolve over time. For example, copper is added to livestock feed as a stimulant for pigs, poultry, and ruminants (Kofoed 1980; Miller et al. 1986; Hegarty 1999; Eckel et al. 2008). In terms of production, this practice is beneficial but, from an ecosystem point of view, it is very questionable because in the long run manure applied to soil can lead to unsuitably high levels of soil copper concentrations.

So far, this section only considered the possibility of reducing emissions by modifying the practice at the origin of these emissions. However, there is also another type of mitigation practice through the capture and removal of pollutants already released. This kind of approach can be a very efficient way to remove CO_2 from the atmosphere, thus mitigating GHG emissions, through the implementation of agricultural practices that enhance soil carbon sequestration. These practices are very useful and have to be promoted. However, as important and useful as they are, this type of mitigation practice will not solve the environmental issues and should therefore be used in parallel with practices acting specifically on emission processes. In the case of GHG emissions, over time, the potential of carbon sequestration will decrease because the storage capacity of soils is limited. Therefore, first, it is a mitigation practice but not a true sustainable practice in the sense that it will not continue indefinitely (the rate of sequestration will decline and the storage capacity is finite) and, second, because carbon sequestration is limited in time, although benefits are very important this practices will not be able to solve the climate change issue as long as GHG emissions keep increasing.

5.5.3 Diversification in production systems

The fundamental origin of environmental issues related to intensive production systems is the specialization of all production processes towards a single purpose. All efforts are oriented toward a very specific product and the ecosystems around the industrial production site are poorly used. An artificial ecosystem responding to market forces is then "re-built" around the one mode of production. The productivity of modern agricultural systems is therefore highly dependent on external inputs (Altieri 1999).

Specialization is understandable in the context of market competition, but it only works in a stable environment because the more specialized the system, the more sensitive it becomes to very small environmental changes. Currently the agricultural environment is not stable: decreasing soil fertility, increasing risk of water pollution, climate change, and animal diseases, compounded by the financial and energy crises. Therefore, intensive animal production is under strong pressure and is unbalanced: with the fierce market competition, small family farms have difficulties to survive; for the same reasons and similar environmental issues, large farming operations are also in trouble; finally some of these industries already collapsed such as the Newfoundland cod fishery as previously mentioned. Animal production systems that are too specialized are, thus, both harmful to the environment and counterproductive to the viability of these industries.

Agricultural management practices based on more diversified production systems and multiple ecosystem components can alleviate some of the problems associated with more specialized systems. As much as possible, farmers should include the three following ecosystem components: animals, crops systems, and also forestry. Farmers are ecosystem users and, therefore, even if one of these components is not directly related to the main production, all of them should be considered and included in the farm activities because each of them occupies a specific and nonreplaceable position within the agro-ecosystem. In regard to agroforestry, the Brundtland report (1987) stated the following:

> Farmers can use agroforestry systems to produce food and fuel. In such systems, one or more tree crops are combined with one or more food crops or animal farming on the same land, though sometimes at different times. Well-chosen crops reinforce each other and yield more food and fuel than when grown separately. The technology is particularly suitable for small farmers and for poor-quality lands. Agroforestry has been practised by traditional farmers everywhere. The challenge today is to revive the old methods, improve them, adapt them to the new conditions and develop new ones.

Even if the animal-crop-forest association would be the most efficient (in terms of risk management and environmental impacts) at the farm scale, it could also be very interesting to use this approach at higher production levels. This means building close links between the three industries representing these different activities.

Many different links and relationships among tree crops, food crops, and farm animals can be found throughout the world. These interactions combine with specific farm environmental parameters (climate, natural resources, workforce, etc.) to cause the wide range of agro-ecosystems that can be found throughout the world. Consequently, there are possibly as many mitigation practices as there are different environmental situations. Hence, mitigation practices have to be adapted to every specific agro-ecosystem.

5.6 Conclusions

Over the past century, the animal production sector has changed dramatically. Small traditional family farms producing a variety of animal products have gradually been replaced by very large specialized factory farms. With globalization, agricultural activities have been reorganized, broken down, and spread over large regions and across national boundaries. To be as competitive as possible, animal production is now often concentrated in specific geographical areas, where the potential for profit is greatest.

Thus, industrialization has strongly influenced the agricultural system but, in turn, agriculture also reinforced it. Through mechanization and the harnessing of fossil fuels, farmers are now able to produce more with less people. As a result, the urban population has increased at the expense of the rural population. With the exponential growth in global population and the changing habits associated with an urban population, food demand has increased accordingly. What we eat and the way we eat, especially in developed countries, has evolved and the whole agri-food sector has changed and become more and more industrialized. However, the economic supply and demand rules are not adapted to manage the natural resources and the ecosystem limitations are becoming increasing apparent. The exponential increase in human population has acted to amplify these limitations and as a result the current production systems often place excessive pressure on the agro-ecosystems that support them. For instance:

1. In terms of "extraction," the use of natural resources is exceeding the ability of ecosystems to replenish the resource being consumed (deforestation, mining minerals for fertilizer production, etc.)
2. In terms of "outputs," emissions from modern production systems are potentially either toxic (heavy metals for example) or emitted in quantities that exceed the ecosystems' ability to process (phosphorus leached into water bodies, GHG emissions to the atmosphere, etc.)
3. Finally, with the new technologies and knowledge, new materials have been developed that outpace the evolution of nature: new materials are frequently not biodegradable, or are only biodegradable over the very long term and because they are not reused or recycled they become wastes

This situation leads to various environmental impacts: biodiversity losses, atmospheric pollution, water shortages and degradation of surface and subsurface water quality, decreasing soil fertility, etc.

In parallel, because ecosystems at the origin of the animal production system have been deeply disturbed by current management practices, the animal production system itself has become more and more fragile. Recent collapses of multiple ocean fishery stocks have demonstrated this ecosystem fragility and land-based animal production systems may face the same crisis.

To avoid this, ecosystem rules have to be followed and should become the framework of animal production systems. Therefore, agricultural management practices have to integrate the sustainability criteria and the recovery capacity of ecosystems. This means that the most appropriate practices will have to be specifically adapted to the natural and social situations, so that the best agricultural management practices will be as varied as the environmental conditions. Finally, a holistic approach has to be adopted because it is the whole system that one wants to eventually balance; however, the first ecosystem components to consider are living beings because they are at the heart of environmental issues.

To be able to apply this concept of animal management efficiently and as widely as possible, methods that can assess the degree of sustainability of various animal production practices must be developed. The LCA approach is an interesting method because it is adaptable to a variety of situations and uses a global approach for analyzing the production processes. However, this assessment tool must evolve to provide clear and precise information on the magnitude of impacts on ecosystems as the result of specific animal production systems. The objective of each LCA study has to be clearly defined and oriented toward the sustainability of ecosystems. The systems boundaries must be defined so as to encompass the whole production system and the type of indicators should be rethought and complemented with indicators providing relevant information on the state of the ecosystem. For that purpose, units other than those related to production have to be used. For example, spatial units are interesting in the case of resilient pollutions. Indeed, the magnitude of the environmental effect depends, in that case, on toxic concentrations and so on waste per unit of area. The concept of "functional area" proposed in this chapter is one of the new units that could be used to orient the LCA method to the assessment of the animal production sustainability.

With the development of ecosystem indicators, LCA can be a valuable tool to assess and develop management practices that minimize the environmental impacts of the animal production systems, to provide clear and relevant information to policy makers and help consumers to choose products that were made using sustainable management practices.

Acknowledgments A special thanks to Dr. Naveen Patni for his support during the writing of this chapter and for his advice concerning the environmental issues related to animal waste management. The authors would also like to thank Dr. Marie Boehm for her valuable comments during the writing of this chapter.

References

Agriculture and Agri-Food Canada/Agriculture et Agroalimentaire Canada (AAFC/AAC). 2008. Statistics of the Canadian dairy industry—L'industrie laitière canadienne en chiffre, AAFC-Dairy section (Catalogue no: A71-18/2008F-PDF).

Altieri, M.A. 1999. The ecological role of biodiversity in agroecosystems. *Agriculture, Ecosystems and Environment* 74: 19–31.

Basset-Mens, C., and H.M.G. van der Werf. 2005. Scenario-based environmental assessment of farming systems: The case of pig production in France. *Agriculture, Ecosystems and Environment* 105: 127–144.

Boadi, D., C. Benchaar, J. Chiquette, and D. Massé. 2004. Mitigation strategies to reduce enteric methane emissions from dairy cows: Update review. *Canadian Journal of Animal Science* 84: 319–335.

Brown, L.R. 2001. *Eco-economy: Building an economy for the earth.* New York: Earth Policy Institute, W.W. Norton.

Brundtland, G.H. 1987. *Our common future. Report of the world commission on environment and development.* Oxford, UK: Oxford University Press.

Canada. Statistics Canada. 2009. Food Statistics 2008. Statistics Canada, Ottawa (Catalogue no. 21-020-X).

Casey, J.W., and N.M. Holden. 2006. Quantification of GHG emissions from sucker-beef production in Ireland. *Agricultural Systems* 90: 79–98.

Cederberg, C., and M. Stadig. 2003. System expansion and allocation in life cycle assessment of milk and beef production. *The International Journal of Life Cycle Assessment* 8: 350–356.

Collomb, P. 1999. Une voie étroite pour la sécurité alimentaire d'ici à 2050. Economica: Paris. http://www.fao.org/DOCREP/003/X3002F/X3002F00.htm.

Collomb, P. 2002. *A narrow road to food security from now until 2050*. FAO Rome; 190.

Colombo, U. 2001. The Club of Rome and sustainable development. *Futures* 33(1): 7–11.

Conant, R.T., and K. Paustian. 2002. Potential soil carbon sequestration in overgrazed grassland ecosystems. *Global Biogeochemical Cycles* 16(4): 1143–1152.

Davis, J., U. Sonesson, D.U. Baumgartner, and T. Nemecek. 2009. Environmental impact of four meals with different protein sources: Case studies in Spain and Sweden. *Food Research International* 43(7): 1874–1884.

de Vries, M., and I.J.M. De Boer. 2010. Comparing environmental impacts for livestock products: a review of life cycle assessments. *Livestock Science* 128: 1–11.

Dyer, J.A. 1989. A Canadian approach to drought monitoring for famine relief in Africa. *Water International* 14(4): 198–205.

Dyer, J.A. 1998. *Niche market development and business planning: A handbook for growers of rare breeds and heritage crops*. Toronto: Seeds of Diversity Canada (SoDC) and Rare Breeds Canada (RBC).

Dyer, J.A., and R.L. Desjardins. 2009. A review and evaluation of fossil energy and carbon dioxide emissions in Canadian agriculture. *Journal of Sustainable Agriculture* 33(2): 210–228.

Dyer, J.A., X.P.C. Vergé, R.L. Desjardins, and D.E. Worth. 2010a. The protein-based GHG emission intensity for livestock products in Canada. *Journal of Sustainable Agriculture* 34(6): 618–629.

Dyer, J.A., X.P.C. Vergé, R.L. Desjardins, D.E. Worth, and B.G. McConkey. 2010b. The impact of increased biodiesel production on the greenhouse gas emissions from field crops in Canada. *Energy for Sustainable Development* 14(2): 73–82.

Eckel, H., U. Roth, H. Döhler, and U. Schultheiss. 2008. Assessment and reduction of heavy metal input into agro-ecosystems. In *Trace elements in animal production systems*, ed. P. Schlegel, S. Durosoy, and A.W. Jongbloed, 33–44. Wageningen: Wageningen Academic.

Ellingsen, H., and S.A. Aanondsen. 2006. Environmental impacts of wild caught cod and farmed salmon: A comparison with chicken. *International Journal of Life Cycle Assessment* 1: 60–65.

Fearnside, P.M. 1980. The effect of cattle pasture on soil fertility in the Brazilian Amazon: Consequence for beef production fertility. *Tropical Ecology* 21(1): 125–137.

Fearnside, P.M. 1990. Deforestation in Brazilian Amazonia. In *The earth in transition: Patterns and processes of biotic impoverishment*, ed. G.M. Woodwell. New York: Cambridge University Press.

Food and Agriculture Organization of the United Nations (FAO). 2007. *The state of the world's animal genetic resources for food and agriculture*. Rome, Italy: FAO-Commission on Genetic Resources for Food and Agriculture.

Food and Agriculture Organization of the United Nations (FAO). 2009a. *The state of food in agriculture: Livestock in the balance*. Rome, Italy: Electronic Publishing Policy and Support Branch-Communication Division, FAO.

Food and Agriculture Organization of the United Nations (FAO). 2009b. *The state of food insecurity in the world 2009: Economic crises, impacts and lessons learned*. Rome, Italy: Electronic Publishing Policy and Support Branch-Communication Division, FAO.

Food and Agriculture Organization of the United Nations (FAO). 2010. FAOSTAT statistical database. Rome, Italy. Accessed March 2010.

Fowler, C., and P. Mooney. 1990. *Shattering: Food, politics, and the loss of genetic diversity*. Tucson: University of Arizona Press.

Gardner, G., and B. Halweil. 2000. *Overfed and underfed: The global epidemic of malnutrition*. Washington, DC: Worldwatch Institute. Worldwatch paper; 150.

Garnett, T. 2009. Livestock-related greenhouse gas emissions: Impacts and options for policy makers. *Environmental Science and Policy* 12: 491–503.
Gien, L. 2000. Land and sea connection: The east coast fishery closure, unemployment and health. *Canadian Journal of Public Health* 91(2): 121–124.
Hegarty, R.S. 1999. Reducing rumen methane emissions through elimination of rumen protozoa. *Australian Journal of Agricultural Research* 50: 1321–1328.
Hein, L. 2000. Impact of shrimp farming on mangroves along India's East Coast. *Unasylva* 51(203): 48–55.
Hertwich, E., E. van der Voet, S. Suh, A. Tukker, M. Huijbregts, P. Kazmierczyk, M. Lenzen, J. McNeely, and Y. Moriguchi. 2010b. United Nations Environment Program (UNEP). Assessing the environmental impacts of consumption and production: Priority products and materials: A report of the working group on the environmental impacts of products and materials to the International Panel for Sustainable Resource Management. Organisation des Nations Unies pour l'éducation, la science et la culture (UNESCO). 2005. Rencontres Internationales Planète Terroirs—Actes. Paris, 10 Novembre 2005 (http://unesdoc.unesco.org/images/0015/001543/154388f.pdf).
Hoekstra, A.Y., and A.K. Chapagain. 2007. Water footprints of nations: Water use by people as a function of their consumption pattern. *Water Resources Management* 21: 35–48.
International Dairy Federation (IDF). 2009. Environmental/Ecological Impact of the Dairy Sector: Literature review on dairy products for an inventory of key issues. List of environmental initiatives and influences on the dairy sector. International Dairy Federation, Brussels, Belgium (Catalogue no. 436/2009).
International Standard Organization. 2006. ISO 14044. Environmental management—Life cycle assessment. Requirements and guidelines. ISO, Geneva, Switzerland.
Jolliet, O., M. Margni, R. Charles, S. Humbert, J. Payet, G. Rebitzer, and R. Rosenbaum. 2003. IMPACT 2002+M: A new life cycle assessment methodology. *International Journal of Life Cycle Assessment* 8(6): 324–330.
Jolliet, O., M. Saadé, and P. Crettaz. 2005. *Analyse du cycle de vie: Comprendre et réaliser un écobilan*. Lausanne, Suisse: Presse Polytechniques et universitaires romandes.
Keyzer, M.A., M.D. Merbis, I.F.P.W. Pavel, and C.F.A. van Wesenbeeck. 2005. Diet shifts towards meat and the effects on cereal use: Can we feed the animals in 2030? *Ecological Economics* 55: 187–202.
Kofoed, A.D. 1980. Copper and its utilisation in Danish agriculture. *Fertilizer Research* 1: 63–71.
LeJeune, J.T. 2003. *E. coli* O157 in dairy cattle: Questions and answers. Ohio State University Extension Fact Sheet. VME-16-03. Food Animal Health Research Program, Wooster, Ohio http://ohioline.osu.edu/vme-fact/0016.html. Accessed 15 June 2010.
Meadow, D.H., D.L. Meadow, J. Randers, and W.W. Behrens. 1972. *The limits to the growth: A report for the Club of Rome's project on the predicament of mankind.* New York: Universe Books.
Menzi, H., O. Oenema, C. Burton, O. Shipin, P. Gerber, T. Robinson, and G. Franceschini. 2010. Impacts of intensive livestock production and manure management on ecosystems. Chapter 9 in *Livestock in a changing landscape,* vol. 1: *Drivers, consequences and responses*, ed. H. Steinfeld, A.H. Mooney, F. Schneider, and E.L. Neuville, 139–163. Washington, DC: Island Press.
Milich, L. 1999. The role of methane in global warming: Where might mitigation strategies be focused? *Global Environmental Change* 9: 179–201.
Miller, R.E., and P.D. Blair. 1985. *Input–output analysis: Foundations and extensions*. Englewood Cliffs: Prentice-Hall.
Miller, W.P., D.C. Martens, L.W. Zelazny, and E.T. Kornegay. 1986. Forms of solid phase copper-enriched swine manure. *Journal of Environmental Quality* 15: 69–72.
Nellemann, C., M. MacDevette, T. Manders, B. Eickhout, B. Svihus, A.G. Prins, and B.P. Kaltenborn. 2009. The environmental food crisis. *The environment's role in averting future food crises: A UNEP rapid response assessment.* United Nations Environment Programme (UNEP)/GRID-Arendal, Birkeland Trykkeri AS, Norway.

Nepstad, D.C., C.M. Stickler, and O.T. Almeida. 2006. Globalization of the Amazon soy and beef industries: Opportunities for conservation. *Conservation Biology* 20: 1595–1603.

Pelletier, N. 2008. Environmental performance in the US broiler poultry sector: Life cycle energy use and greenhouse gas, ozone depleting, acidifying and eutrophying emissions. *Agricultural Systems* 98: 67–73.

Pelletier, N., and P. Tyedmers. 2007. Feeding farmed salmon: Is organic better? *Aquaculture* 272: 399–416.

Pelletier, N., R. Pirog, and R. Rasmussen. 2010. Comparative life cycle environmental impacts of three beef production strategies in the Upper Midwestern United States. *Agricultural Systems* 103(6): 380–389.

Roberts, P. 2004. *The end of oil: On the edge of a perilous new world*. New York: Mariner Books/Houghton Mifflin.

Rook, A.J., and J.R.B. Tallowin. 2003. Grazing and pasture management for biodiversity benefit. *Animal Research* 52: 181–189.

Ryoichi, Y., K. Masayasu, and M. Hiroshi. 2005. Science on sustainability 2006: A view from Japan. Report from the research on the scientific basis for sustainability (RSBS) project.

Schau, E.M., and A.M. Fet. 2008. LCA studies of food products as background for environmental product declarations. *International Journal of Life Cycle Assessment* 13: 255–264.

Seré, C., H. Steinfield, and J. Groenewold. 1995. *World livestock production systems: Current status, issues and trends*. Rome, Italy: Food and Agriculture Organization of the United Nations (FAO). FAO production and health paper; 127.

Simmons, M.R. 2000. Revisiting "The limits to growth": Could the Club of Rome have been correct, after all? Energy white paper. http://www.energybulletin.net/node/1512. Accessed March 2010.

Smith, P., D. Martino, Z. Cai, D. Gwary, H. Janzen, P. Kumar, B. McCarl, S. Ogle, F. O'Mara, C. Rice, B. Scholes, and O. Sirotenko. 2007. Agriculture. In *Climate change 2007: Mitigation. Contribution of working group III to the fourth assessment report of the intergovernmental panel on climate change*, ed. B. Metz, O.R. Davidson, P.R. Bosch, R. Dave, and L.A. Meyer. Cambridge, UK: Cambridge University Press.

Steinfeld, H., P. Gerber, T. Wassenaar, V. Castel, M. Rosales, and C. de Haan. 2006. *Livestock's long shadow: Environmental issues and options*. Rome, Italy: FAO.

Thomassen, M.A., R. Dalgaard, R. Heijungs, and I.J.M. De Boer. 2008. Attributional and consequential LCA for milk production. *International Journal of Life Cycle Assessment* 13: 339–349.

Turner, G. 2008. A comparison of the limits to the growth with thirty years of reality. Socio-economics and the environment in discussion—CSIRO. Working paper series 2009–09: June 2008.

United Nations Environment Program (UNEP). 2010a. Turning the tide on falling fish stocks—UNEP-Led green economy charts sustainable investment path. Press Release online: http://www.unep.org/resourceefficiency/News/PressRelease/tabid/428/language/fr-FR/Default.aspx?DocumentID=624&ArticleID=6566&Lang=en (last visit March, 2010).

Vergé, X.P.C., J.A. Dyer, R.L. Desjardins, and D. Worth. 2007. Greenhouse gas emissions from the Canadian dairy industry in 2001. *Agricultural Systems* 94: 683–693.

Vergé, X.P.C., J.A. Dyer, R.L. Desjardins, and D. Worth. 2008. Greenhouse gas emissions from the Canadian beef industry. *Agricultural Systems* 98(2): 126–134.

Vergé, X.P.C., J.A. Dyer, R.L. Desjardins, and D. Worth. 2009a. Greenhouse gas emissions from the Canadian pork industry. *Livestock Science* 121: 92–101.

Vergé, X.P.C., J.A. Dyer, R.L. Desjardins, and D. Worth. 2009b. Long term trends in greenhouse gas emissions from the Canadian poultry industry. *Journal of Applied Poultry Research* 18: 210–222.

World Health Organization (WHO)/Food and Agriculture Organization of the United Nations (FAO). 2003. Diet, nutrition and the prevention of chronic diseases. WHO/FAO Expert Consultation, Geneva, Switzerland (WHO Technical Report Series; 916).

Zervas, G. 1998. Quantifying and optimizing grazing regimes in Greek mountain systems. *Journal of Applied Ecology* 35: 983–986.

Chapter 6
Life cycle assessment of processed food

Yves Arcand, Dominique Maxime, and Reza Zareifard

6.1 Introduction

It is recurrently said that one of the first life cycle assessments (LCA) of a product was performed by a food company in the United States in the late 1960s. Indeed, the Coca-Cola Company applied the life cycle approach in 1969 to determine the better environmental choice between glass and plastic bottles, considering bottle manufacturing, transportation and use, as well as end of life (Hunt and Franklin 1996). It is less known that the following life-cycle based study, in 1972, as reported by these authors, was also dedicated to food packaging comparison (polystyrene foam vs. molded pulp meat trays). The early studies emphasized raw material demands, energy inputs, and waste generation flows. Attempts on more sophisticated analysis through the consideration of environmental impact on human health and ecosystem would come later in the evolution of LCA methodology, especially from 1990 when the Society of Environmental Toxicology and Chemistry (SETAC) started supporting worldwide research and development in the area of life cycle analysis.

Y. Arcand (✉) • R. Zareifard
Food Research and Development Centre, Agriculture and Agri-Food Canada, Saint Hyacinthe, QC, Canada
e-mail: Yves.Arcand@agr.gc.ca; Reza.Zareifard@agr.gc.ca

D. Maxime
Département de génie chimique, Interuniversity Research Centre for the Life Cycle of Products, Processes and Services, International Chair in Life Cycle Assessment, École Polytechnique de Montréal, Montréal (Québec), Canada
e-mail: Dominique.Maxime@CIRAIG.org

J.I. Boye and Y. Arcand (eds.), *Green Technologies in Food Production and Processing*, Food Engineering Series, DOI 10.1007/978-1-4614-1587-9_6,

Although numerous LCA studies on food packaging have been performed since the early days of LCA (Hunt and Franklin 1996; Gabathuler 1997), processed food or beverage products in themselves have far more rarely been assessed. The main reason is because environmental concerns at that time focused mainly on energy consumption within highly energy-intensive industries relying on fossils fuels or nonrenewable resources such as the oil and petroleum products industries, the chemical industry, or the metals manufacturing industry. Conversely, the agriculture and food industry sectors, whose core raw materials rely on renewable resources, were not seen as potentially harmful for the environment. However, since the mid-1980s, investigations of environmental impacts of food products became gradually more widespread, and the first published LCAs on food production and processing appeared in the early 1990s (Weidema 1993). Several factors have played contributory roles, mostly driven by the general awareness related to sustainable development, and triggered by key events such as the Brundtland report in 1987, the first report from the Intergovernmental Panel on Climate Change in 1990, and the 1992 Earth Summit Rio conference. First and foremost, the well-known low energy yield of food probably sparked off interest in an LCA approach to evaluate food products throughout their life cycle. Later, countrywide studies, generally commissioned by governments or policy makers, confirmed that the food products (especially meat and dairy products) consumed by citizens were some of the major contributors of the overall impact of nations in the developed world, similar to private transportation and housing (e.g., Weidema et al. 2005; Tukker et al. 2006). Such studies included all embedded impacts, from the agricultural production of raw materials to cooking operations at home and final waste management, via industrial processing, retailing, and all the transportation steps involved. Early studies identified some important issues, such as the unexpected consequences over time of farming intensification, and spread concerns over fertilizers and pesticide use, or emissions from animal breeding (e.g., greenhouse gases (GHG) from cattle or water contamination by nitrates from pig farming). A more recent factor of LCA development in the area of food is that requests have arisen from consumers. More and more "sustainable consumers" now request more transparent information about the environmental impacts generated by food products in order to make smart purchases, and the issue of environmental labeling is debated.

Thus, it is clear that environmental data are to be collected at each stage of the life cycle of a product, in order to assess step by step the environmental impacts not only of food items as sold to the consumer but also of all raw materials, ingredients, and intermediate products used from farm to fork. Historically, data was mostly collected on agricultural production systems, which is how LCA of farm products were first published. LCAs of cattle, pork, milk, crops such as corn and wheat, and oilseeds such as soya, as well as palm oil are the most widely studied using LCAs. However, very few of them extend the scope of the assessment up to the processing stage, or even further, because the first complete studies have rapidly revealed that most of the impacts arise from the agricultural step and not from subsequent ones (Hospido et al. 2005; Weidema et al. 2008; Sonesson 2009; Thoma et al. 2010). Hence, it was reasonable to focus on the production step, in order to refine data collection and to provide more accurate LCAs of agricultural products. However, for the reasons mentioned above, it is now appropriate to seek to produce and publish a more complete LCA including the food processing step. An additional benefit to be expected is a better and more detailed understanding of environmental impacts directly linked to food processing operations, as well as background impacts related to procurement of nonfood materials and ingredients (e.g., chemicals, cleaning agents, energy supplies).

After the farmer, the food industry is the second actor within the food chain, and it is the main provider of supplies to traders, i.e., the wholesaler or retailer. Basically, the food industry is where agricultural raw materials are processed to become packaged food products. However, one can hardly distinguish those sectors or industries engaged in primary processing of raw materials and those engaged in the production of finished—more sophisticated—food products dedicated to consumers. The former often supply the latter with basic food materials or ingredients that are further formulated and processed. Even though such a simplistic perspective is not obvious and could not be generalized, it helps understand the material flows within the food products' chain with respect to processing plants. It is also specifically relevant to life cycle assessment of food products, which involves inventorying inputs and outputs and relates them to specific processes. Typical examples, among others, of primary processing of raw materials are sugar and syrups manufacturing, grain milling manufacturing, and animal slaughtering. Outputs of such activities feed secondary processing plants belonging to the beverage sector, cereal products manufacturing, or meat processing plants, respectively. A transportation step may be involved between primary and secondary processing steps, depending on the sensitivity of products to perishability. Meat packing plants for instance are usually located next to slaughterhouses. On the other hand, powdered milk from a fluid milk plant has a longer shelf life and can be transported efficiently over long distances to food processors that require milk fat or protein ingredients. Beyond processors, there is a whole distribution industry involving exporters, importers, wholesalers, distributors, and retailers who bring food to the ever-increasing numbers of urban consumers. This involves mostly transportation and adequate packaging to extend shelf life. And these operations also leave an environmental footprint.

As the distance between the farmer and the consumer increases, most individuals loose touch with the food chain reality. When asked who are the most unsustainable stakeholders, the most popular answers are farmers using pesticides, transporters, processors, and packers. Yet, most comprehensive LCA studies have shown farmers (overusing water, land, and fertilizers) and consumers (wasting edible food and overusing energy when cooking) to be the greatest offenders. Therefore, more robust and transparent LCA studies have to be performed and publicly disclosed in order for there to be greater coherence of understanding.

The purpose of this chapter is not to get into highly technical considerations related to LCA, because it is not directed primarily to the LCA practitioner. It is intended rather to arouse the interest of professionals in food processing, such as engineers and technologists, and of people working in the food industry who would like to opt for an environmentally holistic approach to support decision-making and also for accounting and monitoring purposes using a life cycle perspective. The first two sections of the chapter present some important methodological considerations that shed light on the powerful potential of the LCA tool, but also on some challenging issues such as the burden required, the gathering of life cycle inventory (LCI) data, and impact assessment methods. Fortunately, these limitations of early-hour LCAs are somewhat alleviated nowadays, owing to ongoing scientific developments and to drivers that generalize the use of LCA studies. The last section of the chapter provides a focus on environmental issues by product groups within the food and beverage industry and reviews related LCA results found in the literature. This helps to shed light on the relevant life cycle inventory data that need to be collected in order to perform LCA of different food categories.

6.2 Methodological considerations for LCA in the food industry

Conducting an LCA is governed by the ISO 14040 series standards (ISO 2006a, b). Other guidelines exist and can be followed as well, provided that they are built upon minimum requirements and consensual rules of the ISO standards. For instance, the technical guide from the handbook of the European International Reference Life Cycle Data System (ILCD 2010a) is a very good reference that deserves attention. According to these standards, conducting an LCA is an iterative process involving four types of activities: (1) defining the goal and scope of the study; (2) collecting LCI data on inputs and outputs of every process involved; (3) assessing the life cycle impacts (LCIA) that characterize all resource usages and pollutant emissions of the processes involved; and (4) evaluating and interpreting the findings, along with sensitivity and uncertainty analyses, to support decision making.

The starting point of the goal and scope phase of an LCA is to state clearly the intended application of the study, because it will further influence the modeling approach to be chosen, which processes are to be considered and which can be omitted, whether specific requirements under ISO are to be met, etc. Table 6.1 presents an exhaustive list of potential applications of an LCA in relation to decision support and accounting/monitoring. As shown, an LCA can be intended for uses as diverse as corporate reporting (e.g., the scope 3 reporting for the Carbon Disclosure Project), forecast analysis of policy scenario (e.g., what is the effect on national GHG inventory if a government legislates 10% bioethanol in automotive fuel for 2015?), or a "classical" weak point analysis of a specific product (e.g., which are the main life cycle step contributors to the environment profile of 1 L of fluid milk processed by company X and shipped to wholesalers in country Y, in year 2012?).

Boundaries of the system under study are also defined during the goal and scope phase of the LCA, and allow scoping of the processes to be considered within the study. Boundaries are set in order to allow provision of the appropriate results for the intended application of the LCA. In practice, other numerous key questions should be answered during the goal and scope phase before reaching a robust definition of

Table 6.1 Most frequently used applications of LCA

Identification of Key Environmental Performance Indicators (KEPI) of a product group for Ecodesign/simplified LCA	Development of the "Carbon footprint," "Primary energy consumption," or similar indicator for a specific product
Weak point analysis of a specific product	Greening the supply chain
Detailed Ecodesign/Design-for-recycling	Providing quantitative life cycle data as annex to an Environmental Technology Verification (ETV) for comparative use
Perform simplified KEPI-type LCA/Ecodesign study	Clean Development Mechanism (CDM) and Joint Implementation (JI)
Comparison of specific goods or services	Policy information: Basket-of-products (or -product groups) type of studies
Benchmarking of specific products against the product group's average	Monitoring environmental impacts of a nation, industry sector, product group, or product
Green Public or Private Procurement (GPP)	Corporate or site environmental reporting including calculation of indirect effects in Environmental Management Systems (EMS)
Development of life cycle based Type I Ecolabel criteria	Certified supply type studies or parts of the analyzed system with fixed guarantees along the supply-chain
Development of Product Category Rules (PCR) or a similar specific guide for a product group	Accounting studies that according to their goal definition do not include any interaction with other systems
Development of a life cycle based Type III environmental declaration (e.g., Environmental Product Declaration [EPD]) for a specific good or service	Development of specific, average or generic unit process or LCI results data sets for use in specified types of LCA applications

Source: ILCD (2010a).

the system and setting up the boundaries (ILCD 2010a), but it is not the intent of this chapter to review them. According to the purpose of the study, boundaries may be narrowed from the cradle-to-grave system, i.e., the full life cycle, to cradle-to-shelf (retail), cradle-to-gate, or even gate-to-gate system boundaries (Fig. 6.1).

This setup is relative to the foreground system, throughout the food supply chain, although the supply and waste management background systems are obviously impacted from boundaries choice. When assessing the environmental footprint of one of its product, a food processor is likely—but not necessarily—to be concerned with boundaries limited to its sphere of influence, including suppliers: cradle-to-gate system or cradle-to-wholesale system. Gate-to-gate LCAs might appear of limited relevance, because all processes of the farming step(s) are excluded from the scope of the study. This is true obviously when assessing a food product, but it might be relevant if the LCA is intended for technology comparison. Indeed, provided that both technologies to be compared ensure the same ratio of agricultural raw materials input to food product output, two gate-to-gate systems can be modeled, analyzed, and compared without the need to include any upstream agricultural process. Neither LCI data of agricultural processes nor LCI data of upstream processes of agricultural step(s) (background process) are required, thus reducing the burden of LCI data collection. Typical examples of gate-to-gate systems are for the comparison of

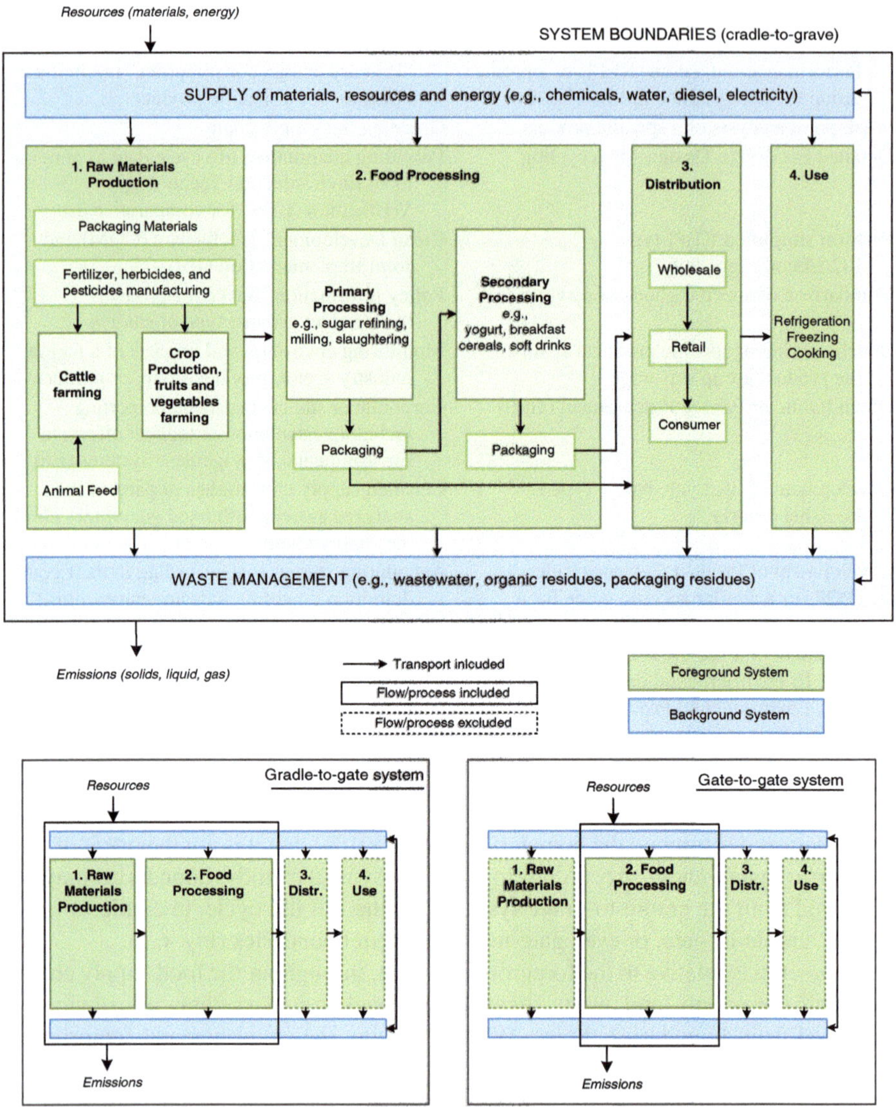

Fig. 6.1 System boundaries for food LCA study. Cradle-to-grave (*top*), cradle-to-gate (*bottom left*), and gate-to-gate (*bottom right*) system boundaries

flue gas cleaning technologies, energy sources (e.g., electric oven versus natural gas oven for baking), and filtration systems (e.g., plate and frame with filter-aid filtration of beer versus membrane filtration beer). However, it is worth noting that any new coproduct, any valorized byproduct, or any waste generated by one of the technologies has to be accounted for within the system modeled in order to account for the impacts it will generate (e.g., the impacts from waste management of the spent filter-aid) or those it will avoid (e.g., the avoided impacts of fertilizer use in the event of land spreading of the spent filter-aid). Last, an LCA related to food packaging should not be limited to a gate-to-gate analysis since any modification of packaging is very likely to affect downstream steps. For instance, transportation characteristics may be changed because different amounts of packaged products can be hauled in the same single truck, or product loss rate during distribution might be reduced if the new packaging is stronger. Furthermore, if product loss is involved, the LCA study should also include raw material production steps, including farming steps, in order to report results according to the functional unit.

One of the basic principles of the life cycle approach is that any product (or service, or process) fulfills a function. Hence, the system delineated within the boundaries setup is also offering functionality, and may even be multifunctional in the case of coproducts or valorized byproducts generation. A dairy farm is typically a multifunctional system, because milk and meat are the main outputs. As well, soymeal used for cattle feed is a coproduct of soybean oil processing plants, and cream is a coproduct of a fluid milk plant. Thus, when performing an LCA of fluid milk, several multifunctionalities have to be resolved step-by-step in order to allocate to the fluid milk only a part of all the upstream impacts, leaving a share of the dairy plant's impacts to the cream, a share of the dairy farm's impact to the meat, and a share of soybean oil processing plants' impacts to the soybean oil. Several allocation rules exist that are explained in ISO standards for LCA (ISO 2006a, b). Last, because the LCA methodology is quantitative, a baseline needs to be established that will be used to calculate the results of the LCIA phase accordingly. Thus, a functional unit (FU), which is chosen according to the main function of the product under consideration, is also defined in the goal and scope phase.

LCAs are performed in iterative loops of the four phases (goal and scope definition, LCI data collection and scaling to the functional unit, LCIA, and evaluation/interpretation) as a steering instrument. Practically, most adjustments are made toward the improvement of inventory data, until the required accuracy of the system's model and processes and the required completeness and precision of the inventory results have been attained, auguring a more accurate assessment of impact. Figure 6.2 depicts this iterative process.

During the LCI phase, every input flow (intermediate products and natural resources) and output flow (products, wastes, and pollutants emitted into soil, air, and water compartments) is identified, quantified, and scaled to the functional unit for every process that is identified to relevantly contribute to the system being studied. Any multifunctionality issue is resolved at this time. The LCI is eventually completed when the data compiled consist only in elementary substance flows from/to the ecosphere of natural resources and pollutants. Practically, during the first iteration of the LCA, a minimal set of data specific to the foreground systems is collected (primary or specific data from sites, farms, plants). Ideally, as many specific data as possible should be obtained in order to build up a model as representative as possible of geographical, temporal, and technological contexts of the system. This holds true even for background systems that are judged highly relevant, such as the composition of the electricity grid-mix. Then, the inventory is generally completed

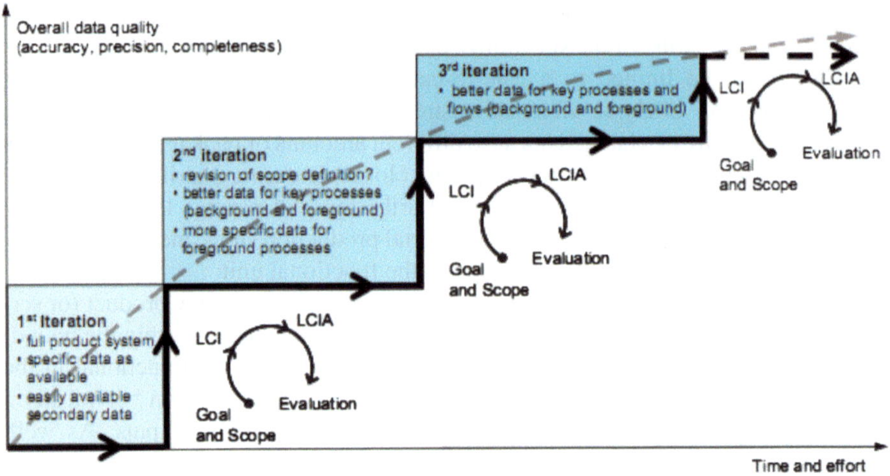

Fig. 6.2 Iterative nature of LCA (schematic) (From ILCD 2010a)

for missing foreground systems using generic data from existing databases. *Ecoinvent* (2011) is one of the most used generic LCI database worldwide. *Ecoinvent* is a European-oriented database that compiles exhaustive process datasets (>4,000), that have been peer-reviewed by experts in many fields (agriculture, forestry, pulp and paper, energy, transport, metallurgy, plastics, etc.). The goal of the first LCA iteration is usually to perform a screening assessment, requiring a limited time-effort burden. Actually, this screening LCA serves to spot the largest contributors to the overall impacts and to point out where the LCI should be specifically refined for the largest improvements in terms of accuracy, precision and completeness (see Fig 6.2). It is also advised during the screening LCA to test at least two different impact assessment methods.

An LCIA method is used to translate the emissions into air, water, and soil, as well as the resources consumed in terms of their contributions to different impacts on human health, natural environment, and natural resources uses. Impacts considered in an LCIA include climate change, ozone depletion, eutrophication, acidification, human toxicity (cancer and non–cancer related), respiratory inorganics, ionizing radiation, ecotoxicity, photochemical ozone formation, land use, and resource depletion. The thousands of emissions and resources obtained from the LCI phase are assigned to each of these impact categories. They are then converted into indicators using impact assessment models. Emissions and resources consumed, as well as different product options, can then be cross-compared in terms of the indicators. Here again, it is not the purpose of this chapter to go through impact assessment modeling and methods in details. For a state of the art review on this topic, the reader is advised to consult the specific volume of the ILCD handbook (2010b). It should be noted that all methods are not radically different, and that they share many common consensus. To date, IMPACT 2002+ (Jolliet et al. 2003) and

ReCiPe (ILCD 2010b) are probably the two most popular methods because they are regularly updated, they characterize the largest number of substances and provide the best coverage of impact categories, and their underlying models have been peer reviewed through scientific publications.

6.3 Limitations and perspective of life cycle assessment

6.3.1 Life cycle assessment limitations and improvements

It is very important to note that any LCA results are actually potential impacts and not real ones. This is an intrinsic characteristic of the LCA methodology, owing to the limited completeness of the inventory (in practice cut-off rules are applied to avoid dealing with an infinite number of flows to quantify), the use of generic data, and the limits on how the cause-to-effect chain can be truly modeled through impact assessment methods. The LCA science is still young and still in development. Hopefully, current and future improvements related to some of these weak points will lead to ongoing refinements towards a better representation of "reality"; that is, the consequences of anthropogenic activities on the ecosphere. More specifically, one can list the following limits and improvements:

- Available inventory databases do not cover completely or accurately all sectors of human activities. This is especially the case for agriculture and food processing activities databases. The former would involve a regional level of detail, mainly because of soil and climatic specificities; the latter because the food industry, historically, has not been a first-priority industry in the context of LCI database development as compared with much more impacting industries (e.g., energy, metals, transportation, building materials).
- Modeling the end of life management of products is still problematic. Data on recycling processes are still sparse; models to characterize the emissions from landfill sites must be improved; data on food losses at home have a poor representativeness (FAO 2011).
- Some impact pathways are difficult to model, and although approaches are proposed in the scientific literature, consensus does not always exist. Consequently, there are either missing characterization factors for certain substances inventoried, or the underlying uncertainty is very large, especially for elementary flows that are to be linked up to the ecosystem quality damage indicator, such as land use (Brandão 2010; McKone et al. 2011) and water use (Pfister et al. 2009; Jefferies et al. 2010). The consideration of indirect impacts of land use on carbon accounting is also a significant issue for the carbon footprinting of biofuels, biomaterials (e.g., polylactic acid used for food packaging), and so-called "green chemicals" sourced from biomass (Müller-Wenk and Brandão 2010).

However, LCI quality is improving rapidly thanks to national initiatives worldwide towards the development of regional databases (e.g., national *ecoinvent* databases based on the framework of future *ecoinvent* 3.0; French and U.S. agriculture LCI databases; the U.S. LCI database (2011)). Such databases offer a significant shift towards more geographical and technological representativeness, thus reducing the "genericity" of data and improving the accuracy of LCA results. Recently, water use inventory data have been made available thanks to the new Quantis Water Database, integrated with the *ecoinvent* database's framework (Quantis Water Database 2011) or through the data published by the Water Footprint Network (2011). Needless to say, the constantly increasing number of publications offers more water inventory data to LCA practitioners, including food LCA case studies (e.g., Pfister et al. 2008 for vegetable and fruit production; Heller et al. 2010 for milk production in the U.S.A.; Peters et al. 2010 for red meat production in Australia; and publications from the Water Footprint Network).

Hopefully, the regionalization and spatialization of impact assessment methods match the regionalization of life cycle inventories (Pfister et al. 2009 about water impacts; Manneh et al. 2010 about impacts on human health; Saad et al. 2010 about land use impacts on ecosystem quality). Some of these developments are already operational in some LCIA methods. For instance, the latest version of IMPACT 2002+ (v 2.2) is now extended, both to account for the characterization of water inventory data, and to account for regional impacts, because water use and consequences on human health and ecosystem are regional issues. A full integration of regional issues at the inventory level and at the impact assessment level will be achieved in the upcoming *ecoinvent* 3.0 databases and IMPACT WORLD method.

6.3.2 Economical and political drivers for development and use of LCA by the food industry

LCA is widely recognized as a tool by governments and public authorities. For instance, LCA is already used by the European Commission and Denmark for several years as a decision support of the "Integrated Product Policy" to develop policy guidance and prioritization of sectors and/or products (including food) with the highest potential for environmental improvement (European Commission 2010). The European Union, Canada, and other countries have implemented the extended producer responsibility, an environmental policy approach in which a producer's responsibility for a product is extended to the postconsumer stage of a product's life cycle (e.g., through its participation in the cost of collection and recycling of packaging materials and newsprints). This policy strategy acts as a driver for motivating the manufacturers to integrate life cycle thinking in the design of their packaging and products.

Another driver for the use of LCA is the support by governments for the provision of national LCI database to allow achieving better and faster LCA, and at lower cost, thus promoting its use by the industry in support of its choices, actions and communication. Both the Quebec government and France have recently funded such initiatives, with a specific funding for an agriculture LCI database in France.

Environmental labels on consumer products such as carbon footprint will be mandatory in some countries soon, as decided by public authorities. The French initiative for an agriculture LCI database results actually from the *Loi Grenelle* in France, which has decided to implement soon the display of environmental information on many consumer products (including food). The purpose is twofold: to allow consumers to make responsible choices by taking into account this new criterion, and, because the consumer will compare products, to promote a fair competition among brands that will pull up the environmental performance of products. A few proactive retailers already set up such a carbon footprint labeling ahead of regulations, hence establishing a sort of best practice (e.g., Tesco in the UK, Casino in France on products under their private brand). Similarly, some agrifood companies have also launched a communication process based on the disclosure of a full LCA, through the type III-Environmental Product Declaration (EPD, based on ISO 14025 and 14040 series ISO standards). EPDs are used as a product marketing tool, and provide an efficient way to quickly position the brand and the product both nationally and in the export market.

These proactive initiatives are driven, on one hand, by the positive image processors and retailers can convey to purchasers and consumers. On the other hand, performing an LCA reveals potential reductions of inputs of material, energy, and water, thus cost savings, simultaneous with the reduction of environmental damage. As well, LCA offers a complete accounting on which the company can build its sustainability report while monitoring and updating its sustainable developments strategy. All kinds of key environmental performance indicators (KEPI) can be derived, such as those from the Global Reporting Initiative (GRI), to communicate its environmental performance. Many companies in the sector of food and beverage publish such indicators in their reports to investors, buyers, and consumers. Many mention the use of LCA and even publish LCA results (e.g., Liberté, Pineridge Foods, Unilever, Coca-Cola, Nestlé, Fonterra, Danone).

6.3.3 Technical drivers for the development and use of LCA by the food industry

The above-mentioned improvements toward the availability of more comprehensive LCI data for LCA practitioners and towards a better representativeness of LCA results are technical drivers expected to increase the popularity of LCA. The public or commercial availability of generic LCI database containing datasets more representative of agricultural specificities of a given country will drastically reduce the burden for data collection, thus reducing the overall cost of an LCA of a food product. If relevant, each crop farming dataset and animal rearing dataset should be available for organic and conventional production, for intensive and extensive production, for different ecozones, for different animal feeding practices, etc.; in other words, for any environmental practice and ecological condition that is known to give rise to distinct environmental profiles for commercially processed food. Consequently, processors and retailers would find a benefit to argue for—and to prove—the environmental attributes of their lines of products if they have "greened" their supply chain.

A reported barrier to the use and the popularization of food LCA is also the inherent complexity to model food product system. To overcome this problem, product-group guidelines exist wherein specific rules, requirements, and guidance are provided. These guidelines are in line with ISO 14040 series LCA standards. Product Category Rules (PCRs) are examples of such guidelines, intended (but not mandatory) for type III EPD (see Table 6.1) of specific branded product. Schau and Fet (2008) reviewed existing LCA of food products with the objective of providing advice for food PCRs. PCRs are usually developed nationally.[1] Similar guidelines are actually under development in Quebec for LCA of dairy products, poultry meat, noncarbonated beverages, and packaging (as trans-sectoral guidelines). They will also cover social issues according to the United Nation Environment Program's guidelines for social life cycle assessment (UNEP 2009).

6.4 Addressing environmental issues specific to the food industry

Raw materials used in the food and drink industry are of agricultural origin from limited resources of land, water, and energy and are produced using manufactured inputs such as fertilizers, pesticides, and cleaning agents; it is therefore crucial to use them in a highly efficient manner. Environmental issues related to the food chain mainly include reduction of biodiversity (through land use change and monocultures), wastewater generation (including biological oxygen demand, nitrogen, phosphorus and sulfur availability, particles, and toxicity), generation of packaging wastes and of organic residues (including byproducts production) as well as air pollutants emissions (volatile organic compounds, particulate matter, greenhouses gases, odors, and refrigerant leaks). Nonenvironmental issues include animal welfare, workers' conditions, and the generation of acceptable social and economic conditions. The relative importance of these factors depends strongly on the food sector and the stakeholder within the food chain. For example, land use, reduction of biodiversity, water consumption, pesticides, and fertilizer leaching issues are strongly related to farmers. Issues related to processing are numerous and are strongly related to specific food sectors. For example, animal welfare is relevant for steps preceding and including animal killing within meat production, dairy products, and fisheries sectors only.

---

[1] See for instance, the *International EPD system* in Sweden, which acts as a central repository for most of European PCRs and EPDs (http://www.environdec.com).

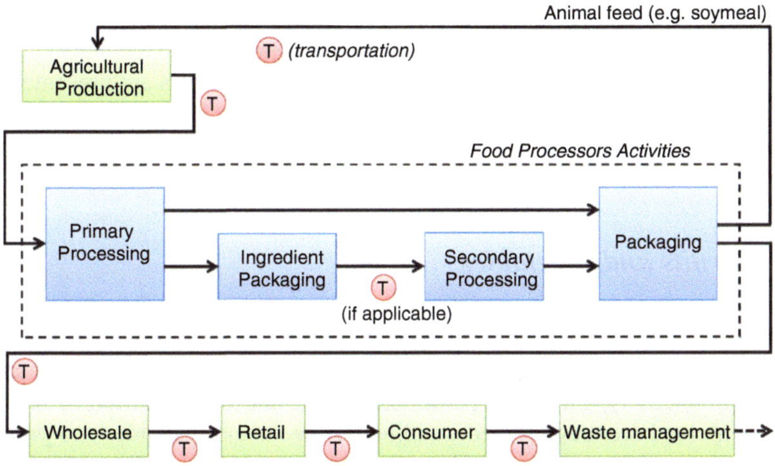

Fig. 6.3 Food processors activities

A generic flow diagram is presented in Fig. 6.3 that shows the major activities involved in food production, from the farm to the primary and secondary processing plants, followed by distribution up to the final destination.

As a general rule of thumb, freshly harvested plant products and recently killed animal produce are perishable materials that must be at least minimally processed to maintain their freshness. Initial cleaning is generally required to lower external microbial contamination, improve appearance, and ease further processing. Separation of the most perishable and nonedible parts is also done quickly (evisceration) to lower internal microbial contamination. Edible portions are then kept under conditions minimizing further microbial contamination (refrigerated, wrapped, marinated, acidified, dried, under controlled atmosphere, etc.). All these steps use energy, water, cleaning agents, and preservatives (sugar, salt, acids, etc.) and generate solid organic residues, packaging wastes, wastewaters, and air emissions (odors, particles, GHG, etc.). These may have some impact on the eutrophication, acidification, and/or hypoxia of water streams; the rapid filling of landfills, land use change (involving a modification of biodiversity), the production of ozone depleting or global warming substances (GHG, VOC, SOx, NOx), and even particles that can affect human health (particles, VOC, odors, etc.).

Further processing is both highly sector- and technology-specific. On one hand, heating is mostly done using fossil fuel combustion, which contributes to global warming and to acidification while depleting a nonrenewable resource. On the other hand, cooling is mostly done using electricity and refrigerants. Whereas refrigerants contribute to global warming and to ozone depletion, the local electricity grid has different impacts whether it is produced using hydroelectricity, coal combustion, or nuclear material. This adds a regional aspect to any LCA: the grid is mainly hydroelectric in Quebec, mainly fossil fueled in Alberta, and evenly mixed in Ontario (three different regions within Canada). As another example, drying can be done using heat, lyophilization, or desiccants (salt, sugar, chemicals) which, obviously, would generate different impacts even within the same plant.

In the following sections, major food manufacturing sectors are presented in an attempt to address and highlight the environmental issues involved in the processing and life cycle of different food products.

6.4.1 Fruits and vegetables

Fruits and vegetables (F&V) are an important part of everyone's diet. Before reaching the consumer, produce is grown on farmland or picked in natural habitats. To improve productivity, on-farm operations involve the use of different resources starting with land use, buying seeds, adding fertilizers and pesticides and providing sufficient water, light, and appropriate temperature throughout the entire growing season. If Mother Nature cooperates, the farmer will end the season harvesting the land to bring the produce to the processor or directly to the consumer. F&V can be consumed fresh, preserved, or as an ingredient of a more complex meal (e.g., soup, ready-to-serve meal). They are sensitive items that deteriorate more or less rapidly after harvesting. For this reason, even material sold as "fresh" is at least minimally processed and this often starts in the field.

Depending on the product and the farmer's income, harvesting can be done manually or using very sophisticated (and expensive) equipment. Some products are packaged in the field (e.g., cabbages) while others are piled in a truck (potatoes). It is common practice in some areas to cool certain fruits or vegetables in the field. Liquid nitrogen-cooled trucks can even provide transportation of fresh produce to the processing plant or directly to the market (Dauthy 1995). Harvested F&V might be stored shortly in simple stores without artificial cooling, or under refrigerated conditions, and for some instances in silos (e.g., potatoes, onions). Refrigerated storage is always preferable and in all cases a processing center needs a cold room for this purpose.

Upon arrival of produce at the processing center, the usual operations of cleaning, cutting and sorting are performed. Washing is used not only to remove field soil and surface microorganisms but also to remove fungicides, insecticides, and other pesticides, because their concentration is generally regulated. Washing water contains detergents or other sanitizers that can efficiently remove these residues. Sorting mainly involves two separate operations: (1) removal of nonstandard products and possible foreign bodies remaining after washing; and (2) quality grading based on variety, dimensional, organoleptic, and maturity stage criteria. Trimming of unnecessary material is also performed prior to packaging to minimize transportation costs. Because of its high perishability, it is common practice to process produce immediately as it is received from the field.

For fresh produce, the next step is packaging that is optimized to retain (or attain) their optimal qualities at the grocery store. Packaging protects sensitive material against unfavorable conditions (e.g., bruises, microorganisms, rodents; extremes temperature, humidity, gaseous substances, light). For example, hermetic packaging does not always improve shelf life because of accumulation of CO_2 and ethylene or because high humidity in the package may encourage microbial growth.

To slow down deterioration of produce per se, processors can use different techniques such as freezing, drying, acidifying (e.g., ketchup), fermenting (e.g., sauerkraut), or sweetening (e.g., jam). Canning can also be used to further increase the shelf life of F&V. These techniques involve material preparation (involving raw material losses) and energy or chemicals as preservative agents. As most F&V are also used as food ingredients, numerous processing techniques are often involved in food preparation, generating all kinds of impacts.

Some F&V require skin removal, which can be performed in various ways: mechanically, thermally, or chemically. Mechanical skin removal is performed with various types of equipment (e.g., blade peeling, scratching) depending on the result expected and the characteristics of the product. Skins can also be softened from the underlying tissues by submerging vegetables in a hot alkali solution. The vegetables with loosened skins are then conveyed under high velocity jets of water that wash away the skins and residual chemicals. In order to avoid enzymatic browning, this chemical peeling is followed by a short boiling in water or an immersion in diluted citric acid solutions. Other F&V with thick skins such as beets, potatoes, carrots, and sweet potatoes may be peeled using steam under pressure as they pass through cylindrical vessels. This softens the skin and the underlying tissue. When the pressure is suddenly released, steam under the skin expands and causes the skin to puff and crack. The skins are then washed away with jets of water at high pressure.

As described above, even minimally processed F&V have important environmental impacts. Trimming, peeling, and cutting require mechanical and thermal energy and sometimes large amounts of water to proceed and may generate tremendous amount of solid organic material. These steps also produce wastewater contaminated with strong alkalis, pesticides, detergents, and sanitizers (contributing to ecotoxicity) and with particulates and soluble organic material (contributing to eutrophication).

Most LCA studies performed on F&V conclude that food processors have a relatively low impact on all aspects along the food chain, the most important steps being cultivation (for most impacts) and consumption (impacts related to energy use) through shopping, transportation, cool storage, and the cooking of food (Pfister et al. 2008; Williams et al. 2008; Ingwersen 2010; Müller-Lindenlauf et al. 2010). Consumers and retailers are also responsible for most of the food waste (FAO 2011).

6.4.2 *Dairy*

Cows provide most of the raw milk consumed today. Animals are kept in herd grazing within extensive fields or kept in stalls eating formulated feeds within intensive dairy farms to improve their productivity. Because milk is a highly perishable product, it has to be refrigerated if not sold or processed immediately after milking. In industrialized countries, raw milk is transported to processing plants in refrigerated trucks to be processed into numerous products. Fluid milk, cream, cheese, butter, ice cream, dried milk powders, and isolated milk and whey compounds are the main dairy products. Heating, cooling, concentration, drying, and cleaning are five main operations in the dairy sector. These five main operations account for about 50% and 96% of dairy plant energy consumption in production of fluid milk and dry milk products, respectively (Ramirez et al. 2006).

Raw milk is first centrifuged to produce a fat rich and a fat poor phase. The fat rich phase is then pasteurized to be transformed into cream, butter, or used for ice cream. Part of the butter and the cream can be further concentrated into anhydrous milk fat using dryers. The low-fat phase is also pasteurized before being processed into numerous products such as fluid, fermented, condensed, or powdered milks, cheeses, casein, or whey protein concentrates and powders.

The main impacts of processing dairy products are related to energy consumption, because the industry relies heavily on cycling between high and low temperatures for pasteurization and storage (International Dairy Federation 2005). Raw material wastes are kept to a minimum through raw milk testing at the farm. Nevertheless, many processes generate highly charged organic wastewaters (whey, buttermilk, etc.). These are important contributors to eutrophication, but most modern dairy processors retrieve these valuable compounds for further valorization. As well, most dairy farms have implemented cleaning-in-place operations allowing the reuse of cleaning acids and alkali. Yet, dairy plant effluents remain one of the most challenging ones within the food and beverage industry, along with those from slaughterhouses and meat packing plants.

LCA of dairy products, especially fluid milk, are numerous and have been recently reviewed by the International Dairy Federation (2007, 2009, 2010a, b). The report confirms that most of the impacts come from on-farm operations. For fluid milk, 80% of the GHG produced comes from the farm while the remaining 20% is evenly distributed between milk hauling to plant, processing, packaging, distribution, retail, and consumer. About 50% of the on-farm GHG is owing to enteric fermentation and 33% to nitrous oxides (manure and fertilizers), the rest being generated from fossil fuels used by the farmer. In terms of the amount of energy used, 40% is attributed to the farm, 30% to processing, and 20% to packaging. Because it takes approximately 10 kg of milk to make 1 kg of cheese, >90% of the GHG impact of cheese can be attributed to on-farm activities. Most of the other impacts (water use, acidification, and eutrophication) originate from the feed production. Studies involving land occupation, ecotoxicology, and biodiversity loss, although being in their infancy, all show the important impacts of on-farm operations.

Heller et al. (2008, 2010) found that 70% of all GHG are produced on-farm (50% of the energy consumed) in the United States. Further break down shows that processing accounts for 35% of the energy used, producing 15% of GHG and being responsible for 15% of the acidification, 2% of the eutrophication, and 0.2% of the water used. Similar conclusions are reported in France, Norway, Brazil, and Italy (Eide 2002; Corson and Van der Werf 2008; Fantin et al. 2010; Maria de Léis et al. 2010).

Different dairy products generate different impacts because of the differences in their overall processing (production, processing, distribution). Because of the importance of the on-farm impacts, the cradle to grave complete LCA results of any processed product will strongly depend on the quantity of raw milk necessary to produce it. For example, it takes more cows to produce 1,000 kg of butter than 1,000 kg of fluid milk. Consequently, 1,000 kg of butter will be allocated more feed, more water, and more enteric fermentation from animals. This is why the International Dairy Federation (2009) report 1.5 times more nonrenewable energy to make 1 kg of yogurt than to make 1 kg of fluid milk (2 times for cream, 5 times for cheese, and butter and 6.5 times for powdered milk). Frozen products such as ice cream have a different impact profile. Garcia-Suarez et al. (2008) showed that the highest contributor to climate change impact for ice cream is neither the farm step (33% of overall impact) nor the processing step (2%) but the amount of energy used for keeping the final product frozen from the processor to the consumer (46%), that value depending strongly on product turnover and the percentage of shelves that are empty.

Some studies have reported reduced impacts per kg of products in larger processors (Eide 2002). This is probably because of more incentives to recycle wastes or turn them into byproducts and to more efficient cleaning procedures (it takes less detergent/sanitizers to clean a 10 kL/d process for a 1 day production than it is to clean a 5 kL/d process on 2 consecutive days of production). Eide (2002) assessed the cradle-to-grave impacts of milk, including the end of life management of packaging waste, considering three Norwegian dairies of different size. Results showed that the agricultural activities were the main hot spot, although the dairy processing, packaging, consumer phase, and waste management were also important parameters. The consumer was the main contributor to photooxidant formation and eutrophication impact categories; larger dairies were more efficient than the small and middle-sized industry; and the transport did not have any major influence.

6.4.3 Meat

As with fruits and vegetables, meat is one of the most ancient items on the human menu. Prior to our sedentary lifestyle, eating of meat was more diversified (more species) and less frequent. Regular provision of meat was one of the main reasons for human settlers to get herds. Nowadays, meat (as food) has evolved into a highly specialized industry in which fewer species are bred (mainly beef, pork, poultry, and fish). For land animals, the meat food chain starts at the farm where the animal is born and raised using feed, water, and more or less land. High productivity units also include the use of antibiotics, growth hormones, and high energy feeds to maximize the growth curve.

Meat can be consumed fresh, preserved, or used as an ingredient for more complex meals (e.g., ready-to-serve meals). It is a sensitive item that deteriorates very rapidly after the animal is killed. For this reason, even material sold as "fresh" is at least minimally processed immediately after killing.

Minimal meat processing involves killing the animal, bleeding, hide removal, evisceration, and carcass splitting. These operations are done very quickly to avoid excessive microbial/insect contamination. Excess fat is often trimmed at this point, immediately before treatments designed to remove surface contamination of fecal matter, dirt, hair, and blood. Other matter is washed off by methods that achieve varying degrees of decontamination. Methods include cold water or chemical sprays, warm water sprays (82°C), steam-vacuum sanitization, steam pasteurization tunnels, and any combinations of these (Young et al. 2001). For pigs the skin is seldom removed. Instead, the carcass is steam- or hot water-scalded then flamed to remove hairs. In many countries a specified deep leg temperature, typically 7°C, must be attained in a specified time, typically 12 h. At that time all other tissues of the carcass will be cooler than the specified site, with some approaching the temperature of the surrounding cold air. In some slaughterhouses, cold water is intermittently sprayed over carcass sides from nozzles above the suspension rails. While this has some cooling effect, the main advantage is maintenance of carcass weight. Boning, the separation of edible parts, can be done hot or cold. Poultry killing follows the same general procedure but the specific equipment is different. For example, washing is done by dipping the animal in hot and cold water containing some disinfecting agent (usually chlorine at 20 ppm) where permitted. Hot water dipping is used to ease feathers removal. To minimize contamination, regulation forces the use of multiple washing tanks with running water (>1 kg of freshwater per kg of animal processed).

Three types of wastes can be considered: solid organic material, contaminated wastewaters, and odors (European Commission 2005). Solid organic residues are made of manure, dirt, hair, hide, paunch content, offal, grease, and other nonedible parts. Wastewaters are filled with solid debris, blood, and microorganisms. Fortunately, the rendering industry uses a large part of these “wastes” in by-products such as leather, cosmetics, and the pet food industry to name a few (Diercxsens 2009).

Each year an estimated 45 million tons of plant protein are fed to U.S. livestock to produce approximately 7.5 million tons of animal protein for human consumption (USDA 2001). To produce this animal protein, about 28 million tons of plant protein from grain and 17 million tons of plant protein from forage are fed to the animals. The average energy input for all animal protein production systems is about ten times more per kg of animal protein produced than it is for kg of grain. Producing 1 kg of beef requires about 43 times more water than producing 1 kg of grain (Pimentel and Pimentel 1996). Livestock directly use only 1.3% of the total water used in agriculture. However, when the water required for forage and grain production is included, this dramatically increases the water requirement for livestock production. Producing 1 kg of fresh beef requires about 13 kg of grain and 30 kg of forage (USDA 2001). This much grain and forage requires a total of 43,000 L of water. On rangeland where an animal consumes about 200 kg of forage to produce 1 kg of beef, about 200,000 L of water are needed to produce the 1 kg of beef (Thomas 1987).

It is therefore not surprising that the first LCA on food of animal origin brought to attention the importance of embedded impacts. As methodologies were refined, extending boundaries and adding more impacts, the general conclusions pointed toward this sector being one of the most unsustainable of the food chain with, obviously, milk. Striking results come from studies presenting overall cradle-to-grave emissions from plant and animal foodstuff, as well as beverages (Tukker et al. 2006; Weidema et al. 2008). For the Food and Agriculture Organization of the United

Nations, Steinfeld et al. (2006) made an extensive review of environmental issues related to the livestock sector, emphasizing those related to land use (land transformation and land occupation), climate change, biodiversity, and water. Nguyen et al. (2010) also noted that the major issue currently not fully resolved in LCA regarding livestock impacts is land use.

Ohlsson (2010) reported results from Davis and Sonesson (2008) comparing popular produce and dairy products with the same quantity of different meat and found GHG levels below 100 g CO_2eq/kg of product for fruits and vegetables (e.g., onions, carrots, apples, lettuce), below 1,000 g CO_2eq for wheat and oat, between 1 and 2 kg for milk and chicken, between 2 and 4 kg for greenhouse grown cucumbers and tomatoes, near 5 kg for pork, and a whooping 19 kg CO_2eq/kg of beef. Although comparing the impact of the same amount of different foodstuff is debatable because the nutritional intake (the main function of food) is different, the level of discrepancy remains amazing. Another source comparing local and imported beef in Germany obtained values between 23 and 27 kg CO_2eq/kg beef (Müller-Lindenlauf et al. 2010).

Examining the distribution of the impacts along the food chain for pork in Sweden, Sonesson (2009) showed that about 90% of GHG emissions come from farm operations while 10% come from all other operations (slaughterhouse, transport, distribution, packaging, retail, consumer, and post-consumer). For beef, a rather similar figure is obtained by Tofield and Metcalfe (2009) who concluded that the amount of fertilizer used to grow feed and the enteric fermentation are responsible for 92% of all GHG emissions resulting from the production, processing, and retail of 1 kg of beef. One cow in the UK produces 2.3 t of GHG per year, which is equivalent, on average, to two European passenger cars!

Food of local origin is said to be ecofriendly because of the small transport distance required. Food from global provenance, on the other hand, is equated with high energy use. Schlich et al. (2008) showed that German beef has an important embedded energy (and GHG production). Argentinean beef, which is extensively bred on a large scale and transported by cargo ship and truck, has lower energy turnover and CO_2 release (Krause 2008). Comparing importation of poultry or cattle raised and slaughtered in South America and consumed in Europe, Prudêncio da Silva Jr. et al. (2008) showed that >60% of GHG produced comes from feed production, whereas less than 2% comes from poultry production and processing (slaughtering and freezing). Prudêncio da Silva Jr. et al. (2010) refined their work on GHG production and found that 95% of the CO_2 and 72% of the NO_X produced is embedded within animal feed. The study also compared two modes of poultry production in Brazil to conventional and organic (Label Rouge) production in France. In term of GHG produced, conventional French poultry generated 40% more GHG than its Brazilian counterpart, whereas the organic production generated 75% more GHG.

The introduction of land use impact has generally worsened the impacts of meat consumption. As an example, Cederberg et al. (2008) and Müller-Lindenlauf et al. (2010) concluded that, by including land use change, local beef production would have a significantly higher GHG production (40 kg CO_2eq/kg meat compared with 23 without land use change). But, adding this impact to imported South-American beef would bring the total GHG production to 55 kg CO_2eq/kg meat).

The case of Finnish broiler chicken has been investigated by Katajajuuri et al. (2008). All essential production phases from parent stock and production of farming inputs to product distribution and broiling in retail stores were included in the assessment. For each impact category, most of the environmental impact along the chain originated from housing of broiler chickens and cultivation of feed ingredients except for primary energy demand where broiling accounts for >50% of all energy used. Broiler housing and feed production had the most impact on water eutrophication and acidification caused by applied nutrient run-off and leaching, and through ammonia emissions from broiler chicken manure.

In general, meat coming from land animal has a significantly higher footprint than any other food simply because animals are inefficient energy and carbon transformers. Nilsson and Sonesson (2010) have compared vegetarian with different meat diets. Changing the average Swedish diet from beef to the same quantity of other popular meat sources (50% pork and 50% chicken) lowered the carbon footprint by 60% while becoming ovo-lacto vegetarian would lower it by 90%. Knowing that 76% of all human occupied territory in the United States is for agriculture and that >80% of this huge surface involves agricultural activities related directly or indirectly to the production of animal and/or animal feed (Costello et al. 2009), the reduction of beef consumption in the United States would help free large amounts of land to feed the projected increase of population (two billion more humans by 2050).

6.4.4 Seafood

Fish and shellfish are another source of food protein used by humans since prehistoric times. Fishing techniques were similar to hunting land animals using spear, baits, or traps, but recently fishermen have began to use farming techniques at sea or in basins (aquaculture). The top five consumed species worldwide are salmon, shrimp, tilapia, catfish, and crab (major consumption being in China and India according to FAO).

As with meat, seafood (including fish) can be consumed fresh, preserved, or used as ingredients for a more complex meal (e.g., soup). It is a sensitive item that deteriorates very rapidly after the animal is killed. For this reason, even material sold as "fresh" is at least minimally processed immediately after killing.

For wildlife fishes, the animal is first caught at sea, then transferred into the vessel where it may be chilled immediately or sorted prior to chilling. They are then either processed on the vessel or transported to a processing plant on land. Minimal processing involves head removal, bleeding, gutting, and sometimes filleting. This minimal processing has very short shelf life (a few days only refrigerated). To improve on this, the "cook and chill" process is often implemented where appropriate. After the eviscerated animal is washed and drained, it is immersed into a 10% sodium chloride solution for 1 h, drained, and immersed into a 5% tripolyphosphate solution and drained again. This is followed by heating for quick pasteurization (10 min at 70–90°C depending on regulation) then cooled rapidly just above freezing point (0–2°C). This ensures a longer shelf life. The procedure for non-shellfish products is similar to this technique.

Environmental issues in fish processing industries primarily include the following: water consumption and wastewater generation, solid waste generation and by-products production, emission to air, and energy consumption. Major sources of water consumption include: fish storage and transport; cleaning, freezing, and thawing; preparation of brines; equipment sprays; offal transport; cooling water; steam generation; and equipment and floor cleaning. For fish and fish meal processing energy is required for cooling, cooking, sterilizing, drying, evaporation, can cleaning, fork-lifting. Processes that involve heating, such as canning and fishmeal production need more energy than other processes (UNEP 1999). Emissions to air are exhausted into a vent or stack and emitted through a single point source to the atmosphere. The major air pollution sources in a typical seafood industry are from combustion sources like boilers and generators for electric power. A boiler is used for steam supply during precooking and the sterilization process. The examples of fuels used in the boilers are fuel oil, coal, and liquefied petroleum gases (UNEP 1999).

Odor is often the most significant form of air pollution in fish processing. Major sources include storage sites for processing waste, cooking by-products during fish meal production, fish drying processes, and odor emitted during filling and emptying of bulk tanks and silos. Fish quality may deteriorate under the anaerobic conditions found in onboard storage on fish processing facilities. This deterioration causes the formation of odorous compounds such as ammonia, mercaptans, and hydrogen sulfide (National Pollutant Inventory 1999).

LCAs of marine food yield similar conclusions as for animal foodstuffs. Aquacultures generally have higher impacts than wild catches. For example, in a multiculture of prawn, tilapia, crab; and milkfish in the Philippines, 97% of the land use impact, 95% of eutrophication, and 90% of acidification burden were owing to the production of feeds. Impacts of processing and transport to Western Europe was below 10% on average (Baruthio et al. 2008). Emanuelsson et al. (2008) have compared industrial and traditional shrimp fisheries in Africa (as imported to Europe) measuring ten different impacts. Industrial technologies have significantly more impacts in most categories including energy used, GHG produced, and natural habitat disturbance. Most impacts for both traditional and industrial come from the fishing step, while processing, packaging, storage, and transport to Europe are low contributors. Traditional fisheries have also significantly fewer by-catches, implying a lower impact on natural habitat.

Vold and Svanes (2008) have performed an LCA of cod in Norway (from cradle to consumer) measuring energy use and GHG emission. They found that more than 85% of the climate change impact is caused by fishing boat operations (diesel and refrigerant leaks), leaving less than 10% for the processing step, and less than 5% for the transportation and conditioning steps. In terms of energy consumption impact, the diesel used at sea is responsible for 50% of the overall impact, while processing and transport account for 30% and 20%, respectively, whereas packaging leads to marginal impact.

An extensive Norwegian study assessed the carbon footprint of 22 Norwegian seafood products transported differently to 8 national and international cities (Hognes 2010). Conclusions are: main contributors are fishing or aquaculture stage and transport to destination; wild catches generally have a lower carbon footprint than aquaculture; processing and packaging contribution is very low in absolute values (<1–2% but may account for up to 10% when all other stages have low GHG values). When compared with chicken, pork, and beef, the worst of fish displays a similar carbon footprint to chicken, which is half of pork and 15 times less than beef.

6.4.5 Grain and oilseed

Edible fats and oils processing involves a series of processes in which both physical and chemical changes are made to the raw material. Processing of oils is initiated by cleaning and milling followed by the extraction process to remove the fat or oil from the seed, bean, nut and fruit. Vegetable oil processing includes neutralization or refining, bleaching, and deodorization right after the extraction procedures.

Extraction of oil from materials of plant origin is usually done by pressing with the use of a continuous screw press or by extraction with volatile solvents. Before 1940, mechanical pressing was the primary method used. Mechanical pressing had limits because the oil recovery is poorer than with solvent extraction and the high temperatures generated damaged both the oil and the meal. The solvent method allows a more complete oil extraction at lower temperatures. Solvent extraction plants can be either batch or continuous. The continuous extraction plants can be percolation, immersion, or direct extraction plants. Generally, the oilseeds may be divided by oil content: above and below 20% oil content. In most cases, oilseeds with low oil content are subjected to both continuous and batch solvent extraction. High oil content seeds are normally extracted in two stages; first pressing and then solvent extraction (O'Brien 2006).

Mechanical pressing and solvent extraction with an organic solvent such as hexane are commonly used in oil extraction. A mechanical press can be used to squeeze oil out of oilseeds with high oil content. The efficiency of a hydraulic pressing system is very high because almost all the input energy can be recovered and reused during the next stroke. For oilseeds with low oil content, the solid-solvent extraction method is usually used. The removal of a large amount of solvent in the crude exact by distillation consumes a large amount of energy. Ultrafiltration or reverse osmosis with membranes can replace conventional distillation to minimize thermal damage to products and reduce energy consumption for solvent recovery in an oil mill (Wang 2009).

The grain and oilseed industry comprises establishments primarily engaged in milling grains and oilseeds to produce starch, oils, and to make breakfast cereal products. In the milling sector, a large portion of fuels is used for steam generation and a large portion of purchased electricity is used for motors and air compressors. Grain and oilseed milling, which is the largest energy consumer in the U.S. food industry, consumed 22.3% of total energy input of the whole food industry. The cost of purchased electricity in the grain and oilseed milling sector was 37% of the total energy expenses (United States Census Bureau 2006).

Oilseed processing requires a high energy demand. Wang (2009) reviewed the main products and processes implemented in the grain and oilseed milling sector and discussed the energy uses and conservation in the main unit operations, as well as energy utilization in processing onsite wastes. He concluded that the grain oilseed milling is a very energy intensive industry within the food industry. Wet milling undergoes a steeping operation followed by a series of separation operations to convert the starch, germ, fiber, and protein in the corn kernels into an array of marketable products. It is recommended that reduction of steeping time by pretreating of grain with enzymes and preheating to the steeping temperature and application of mechanical and membrane separation to dewater dilute slurries can save energy use in a wet milling. Energy conservation technologies for drying of grains and oilseeds should be an integrated part of the energy management project in the grain and oilseed milling sector.

Electricity and fuels such as natural gas and fuel oil are two major types of energy sources used in oilseed processing facilities. LCA analysis of soybean oil production by Li et al. (2006) showed that the energy consumption for preprocessing of soybeans before extraction is 2.38 MJ/kg of crude soybean oil produced. Among the total energy consumption, 1.19 MJ/kg is used for drying raw soybeans before mechanical cleaning, 0.60 MJ/kg is used for conditioning the soybeans to raise the moisture content and the temperature of the soybeans prior to flaking, and 0.51 MJ/kg is used in the expander to compress the soybean flakes into pellets for a more complete oil extraction. The total energy consumption for conventional extraction and separation with hexane is 3.19 MJ/kg of crude soybean oil. Most of the energy (1.38 MJ/kg) is used for hexane distillation from the extracted soybean oil. With potential energy recovery, the net energy consumption for extraction is 2.04 MJ/kg. The hexane loss is about 50 kg of hexane per 1,000 kg of crude soybean oil produced (Li et al. 2006).

In another LCA case study of soybean, Pollack and Greig (2010) reported that 99% of US soy meal coproduced from soybean oil production is used as animal feed, whereas only 1% is consumed for food and other industrial uses. From the total soybean oil production 83% is used for human consumption, 11% as biodiesel, and 6% for industrial uses. According to the US Soy association, there is no problem in cultivating soy for food, feed, and biodiesel. They also concluded that soybean oil and meal have favorable GHG and GWP because of the CO_2 sequestration (1.3 kg CO_2 eq/kg meal and 2.5 kg CO_2 eq/kg oil).

Kägi et al. (2010) presented the result of their comparison study for the LCA of several rice products: conventional and organic rice from Italy, conventional rice from the United States, and upland rice from Switzerland to see if there were significant differences in carbon footprint. Results indicated that organic rice has the highest carbon footprint per kg rice (3.75 kg CO_2eq/kg dry rice) followed by the US rice production and Italy, whereas the upland rice showed by far the best performance considering the GWP (1.60 kg CO_2 eq/kg dry rice). They have concluded that there were no significant differences between organic and conventional rice from Italy and also between conventional rice from the United States and Italy although transport distances vary a lot. The confidence intervals was reported to be 15–31% because of the high uncertainty of data, especially data from field emissions.

6.4.6 Sugar and confectionary

The sugar and confectionery product manufacturing sector is engaged in one or more of the following manufacturing activities: (1) processing sugarcane and sugar beet for new products such as sugar or chocolate, and (2) further processing sugar and chocolate for other products. Sugar is mainly produced from sugarcane and beet. About one-fourth of the world's sugar production comes from sugar beet and the production capacity was 40 million tons in 1999 (Erdal et al. 2007). Water consumption in sugar processing manufactures is relatively high. Approximately 10% of the sugar cane mass can be processed to commercial sugar and uses approximately 20 cubic meters of water per metric ton (m^3/t) of cane processed (MIGA 2006).

The basic sugar processes consist of slicing the agriculture product, diffusion, juice purification, evaporation, crystallization, and recovery of sugar. Sugar beet is first cleaned and washed to remove soil, stones, and organic matter from the beet. Cleaned and sliced beet is delivered to the extraction unit in which juice is extracted. Extraction of sugar from sliced beet requires heating to 75–80°C with steam (e.g., 110°C or higher). The resulting pulp is dewatered by mechanical pressing, followed by drying to produce dried pulp. The dried pulp is palletized for storage and transportation. The concentration of raw sugar juice is about 15%. The extracted raw juice is purified using lime-milk in order to remove impurities. The carbonation slurry is concentrated through filters and part of it is recycled while the rest is further concentrated to carbonation lime. The juice is then delivered to multiple-effect evaporators. The vapor generated in the evaporators is used for extraction, juice heating, crystallization, and other process heat. The energy needed to heat the first effect evaporator is supplied from a boiler. Concentrated juice at a concentration of 70% or higher from the last evaporator is delivered to multistage crystallization. Evaporating crystallization requires heating with steam (e.g., 120°C or higher). The crystallized sugar is recovered from the syrup by centrifugation. The syrup from the last crystallization stage is called molasses (Urbaniec et al. 2000; Grabowski et al. 2001; Krajnc et al. 2007).

Mashoko et al. (2010) performed a cradle-to-gate LCA of sugar produced in South Africa and evaluated the environmental impacts and energy consumption of the different life cycle phases of sugar production, which includes sugar cane farming, fertilizer and herbicide manufacturing, cane burning, sugar cane transportation, and sugar manufacture. They reported that an average of 8.46 t of cane is used to produce 1 t of sugar. The total fossil energy required for farming purposes is 372 MJ/t of sugar produced. The total transportation fossil energy required to produce a ton of raw sugar was calculated as 1893 MJ. During sugar manufacturing, coal is only used to start up boilers and to supplement bagasse supply in boilers during the off-season. The total energy use from coal was 1397 MJ/t of raw sugar produced, assuming a net calorific value of South African coal of 19.739 MJ/kg (Thomas et al. 2000).

The inventory and impact assessment results of the study conducted by Mashoko et al. (2010) showed that nonrenewable energy consumption is 5,350 MJ/t of raw sugar produced and 40% of this is owing to fertilizer and herbicide manufacturing. A reduction in fertilizer use for sugarcane farming could bring considerable savings in terms of fossil energy consumption and a reduction in GHG emissions.

Witayapairot and Yossapol (2009) studied the LCA of sugar production in Northeastern Thailand and reported that for the production of 1 t of sugarcane (equivalent to 60 kg sugar) almost 138 m^3 water is used at the sugarcane field and 9.7 kg CO_2eq is emitted. However for transportation >82 kg of CO_2eq is produced per ton of sugarcane, and about 15 kWh electricity is used at the processing plant for the production of 60 kg sugar. From the total energy used in sugar production, >30% is consumed for cane cultivation, 10% for transportation, and >50% used at the processing plant. Production cost in Thailand was found to be higher than in other producing countries although data is very limited to conduct a robust LCA of sugar production.

6.4.7 Bakery and tortilla production

The main ingredients for bakery products include flour, water, sugar, fat, egg, yeast, and salt. The ingredients should be formulated in accordance with the required rheological properties of dough and batter. In addition to formulation, processes including dough making, proofing, baking, cooling, and packaging are among the important steps in delivering a high-quality end product. Different process operating methods and conditions are required to produce different product varieties such as cakes, biscuits, cookies, tortillas, etc.

Generally, the production of bakery products consists of several common steps, including weighing, mixing, proofing, baking, cooling, and packaging. Each step plays an important role in achieving high and consistent product quality. In industrial bakeries, flour is stored in silos and transferred to production lines. In production lines, flour and other ingredients are mixed and kneaded to make dough or batter. Then the dough/batter is divided and molded. Some of the bakery products require a fermentation step to develop unique flavor and/or texture prior to the baking step at 180–300°C. The products are then cooled down before packing and delivery. Baking trays and molds are normally cleaned in hot water (Kannan and Boie 2003).

Baking is a complex unit operation involving simultaneous heat and mass transfer mechanisms. Heat is supplied to the dough through radiation, convection, and conduction. Starch uses water to form a gel and the physicochemical properties of dough are changed. The volume expansion of bakery products during baking is through two different mechanisms. Taking bread as an example, at the early stage of a baking process, the dough volume is expanded mainly due to yeast fermentation. When temperature is increased to around 55°C, yeasts are killed. Once the temperature reaches 60°C, any volume expansion is mainly caused by vapor pressure (Fan et al. 1999).

The estimated specific energy consumption at the plant could be in the range of 4.57–6.80 MJ/kg of processed flour depending on the type of fuels used in baking ovens and the type of backed products, size of the bakery, and the number of shifts. Baking ovens consume the largest portion of energy. About 73% of the total energy is consumed in baking, and 6% each for fermentation and cleaning (Kannan and Boie 2003).

A German study showed that bread from an industrial bakery is more energy efficient than local bakeries, which are better than homemade bread but only for low consumer distance transportation. Driving 0.6 km to get your bread is the break even point to home made bread (Müller-Lindenlauf et al. 2010).

In another LCA study of bread production in Germany, Reinhardt et al. (2003) reported that the energy consumption was about 4.6 MJ/kg of bread, of which almost 30% was consumed at the farm for crop production, 20% for flour production in milling process plants, and the remaining during baking process. Transportation by consumers from retail stores to their homes contributes significantly to energy use, climate change, ozone depletion, acidification, and eutrophication.

In another cradle-to-grave LCA of bread, Andersson and Ohlsson (1999) reported that processing has a minor effect on acidification and eutrophication, the main contributors being agriculture and transportation. Surprisingly, processing and transportation are responsible for >90% of photooxidant formation (the main contributor being processing). This is caused by the release of ethanol during the rising of the dough. Approximately 9 g of ethanol is formed per kg of white bread, of which 2–4 g is released during baking. Actually the pleasant smell of freshly baked breads is caused by alcohols, esters, acids, aldehydes, and ketones, which are strong photooxidants.

6.4.8 Other food

This food industry group comprises establishments engaged in manufacturing products with very different characteristics. Unit operations involved are rather complicated and diverse, but basic concepts of heat and mass transfer are often similar.

Not much information is available for the LCA of every individual food in this category, although there is mass production of these food materials. Overall, use of resources and emissions are considerable.

Mila i Canals et al. (2010) from Unilever industrial group reported three to five million tons production of CO_2eq/year for the production of Knorr products, which was estimated from global Knorr portfolio GHG emissions and represents approximately 1–4% of the Unilever carbon footprint. According to their analysis most environmental impacts occurred upstream, mainly 34% from the agricultural field, 10% in food processing and 4% in other manufacturing industries, 10% for packaging and waste, 6% for distribution, and 36% at the homes of consumer. They have suggested solutions in order to lower the climate change impact, targeting sustainable agriculture by producing low carbon crops, applying advanced drying and concentration technologies, and improving consumer habits and method of cooking at home.

6.5 Beverage

Fruit and vegetable juice manufacturing follows the same initial steps for fruit and vegetable preparation (prewashing, washing, and sorting) followed by crushing and preheating (using warm water at 55–60°C) to facilitate the extraction and dissolve pectin substances prior to extraction where a maximum of 80% of the initial fruit and vegetable is extracted using equipment adapted to each produce. Juice is then de-aerated, homogenized, and flash pasteurized prior to aseptic filling and rapid cooling. De-aeration process is performed under high vacuum, where the juice reaches its boiling temperature at about 35–40°C. Juice stability is assured by a flash pasteurization, which assures the destruction of natural microflora while keeping the initial organoleptic properties.

Flash pasteurization or high temperature short time (HTST) sterilization is performed at processing temperature of 130–150°C for about 8–12 s followed by cooling at 90°C, which is also the filling temperature in cans or bottles. For low pH beverages (less than 4.6), the full cans or bottles do not need further pasteurization because the bacteria that have potentially contaminated the products in the field or in the processing plant are easily destroyed at 90°C during filling because of natural juice acidity.

For bottles, it may be possible to avoid further sterilization if the following conditions can be respected: washing and sterilizing of receptacles, cap sterilization (with formic acid), filling and capping under aseptic conditions, in a space with UV lamps. For concentrated juices, membrane filtration and water removal under mild heat and intense vacuum is performed prior to filling.

Coltro et al. (2008) investigated the LCA of frozen concentrated orange juice produced in Brazil and showed that in the preparation of concentrated frozen orange juice, the heating and cooling processing operations alone were responsible for up to 30% of all energy used (versus 60% on farm activities and less than 5% during transport to Europe). Total energy consumption for the production of 1 t of frozen concentrated orange juice was reported as about 21,000 MJ, which includes the energy used for cultivation (70%), processing operation (25%), and transportation (5%).

In an LCA case study analyzing GHG emissions for beer production (cradle to consumer) in England, Tofield and Metcalfe (2009) reported that the GHG footprint was dominated by glass bottle manufacturing (50%) followed by processing (35%) and transportation, and not by the emission on the farm. Another study conducted by Pattara et al. (2010) indicated that the major contribution in terms of GHG (more than 70%) comes from packaging, mainly glass bottles, followed by the product distribution and then the agricultural operations.

6.5.1 Animal food

Pet foods are special foods that are formulated to meet nutritional needs of animals. The vast majority of domesticated animals being cats, dogs, birds, and fishes, pet foods generally consist of meats, meat by-products, cereal grains, vitamins, and minerals. While bird food is mostly grains, cat and dog foods are mainly available in three basic forms: namely, dry pet foods, semimoist pet foods, and moist pet foods. The conventional manufacturing processes of pet foods are similar to those of human foods.

The first step for the production of pet foods is raw material preparation. This starts with obtaining waste material from industries preparing food for human consumption. These are mostly animal and vegetable parts that are of good to excellent nutritional quality but that are not consumed (at least in sufficient proportion) by humans (such as offal, bones, cob).

The process includes grinding and mixing the fresh raw material into proper proportions. For moist pet food, the mixture is then cooked and canned. For dry pet food, the ground raw material is heated in a conditioner to melt the fat and then fed to an extruder. Extrusion combines several unit operations including mixing, cooking, kneading, shaping, and forming. Passage of the product through the extruder may induce enough heat to cook the product. Appropriate temperature within an extruder ranges from 112°C to 132°C, whereas the temperature of the product leaving the die is in the range of 120–127°C. The product is immediately cooled and cut into small-sized pieces.

Most pet foods are dried in forced-air convection conveyor dryers to prevent quality deterioration and microbial growth. Bite size wet material is spread onto a conveyor belt, and hot air is forced through the conveyor. Pet food dryers typically operate at an air temperature in the range of 100–180°C (Poirier 2003). Possible heat sources of dryers are gas burners, steam coils, and electric heaters, but the most common one is burners using a gaseous fuel, such as natural gas. After drying, pet foods are cooled down to 5–10°C of ambient temperature before being sent for storage or packaging. The coating step can be done by adding various liquids or powders, including liquid animal digests, fats, and tallow; and dried coatings, such as dried yeast, onto the outer surface of the products (Fritz-Jung et al. 2001; Ernst et al. 2002). Coatings can improve shape, appearance, and palatability of dry pet foods.

From a sectorial point of view, the pet food industry is one of the oldest recycling industries. Its use of byproducts from the human food industry prevents the need for final disposal, together with the resultant cost savings. This disposal reduces the price of meat for human consumption and the demand on the human food chain. Surprisingly, LCA of this industry sector has not been publicized. Through proper allocation, this industry allows for a strong organic material credit, which would reduce most impacts of human consumption such as energy use, land use, water use, eutrophication, acidification, greenhouse gases production, and so on. However, this industry still uses mechanical and heating energy, packaging, and transportation. In general, it can be said that the pet food industry is skilled in converting byproducts to added value products, which ensure proper nutrition for the pet population while reducing waste disposal of the global food industry.

6.6 Conclusion

One should be always careful with results of previous LCA studies, which should not be extrapolated into another context, or compared with more recent studies. LCA is quite a young science, which is constantly improving, especially in the area of data inventory (e.g., new regionalized datasets, data gaps make up), methodology (e.g., better modeling of end-of-life and waste management operations, consideration of market consequences), and impact assessment methods (e.g., improvement of impact pathways modeling, method regionalization, new impacts characterization such as those from water use). Consequently, an important momentum for LCA improvement is ongoing and it is expected that future LCAs will be (1) more robust, representative, and reliable; (2) carrying less uncertainty; (3) faster and cheaper to conduct, and (5) more comparable thanks to specific guidelines for different groups of products.

LCA is a powerful tool to help decision makers within the food industry in choosing one project versus another based on environmental criteria. Emphasis should be put on more comprehensive data on food processing operations in order to confirm through cradle-to-grave LCA studies that processors are unlikely to be blamed for their contribution to the overall life cycle impacts of food products. This would shed new light on the impact of processing within the entire food chain. Nevertheless, processors have the responsibility to green their supply chain, ensuring that (1) they are choosing raw materials, ingredients, and packaging materials embedding as little impacts as possible, and (2) the end-of-life impacts of packaging marketed with foodstuff are as low as possible. LCA is thus a valuable tool for food processors themselves and for the information it provides to all stakeholders of the food value chain of food, including consumers.

The communication of environmental attributes of products becomes an important selling point, with the benefit of an upward process of influence, because the environmental responsibilities are actually dispatched across the food supply chain, from retailers to food processors, and up to producers, for the final benefit of every citizen.

References

Andersson, K., and T. Ohlsson. 1999. Life cycle assessment of bread produced on different scales. *International Journal of Life Cycle Assessment* 4(1): 25–40.

Baruthio, A., J. Aubin, R. Mungkung, J. Lazard, and H.M. Van der Werf. 2008. Environmental assessment of Filipino fish/prawn polyculture using life cycle assessment. *Proceedings of the 6th international conference on LCA in the agri-food sector*. Zurich, November 12–14, 2008.

Brandão, M. 2010. Food, feed, fuel, forest or carbon sink: can LCA provide guidance on how to best use land for climate-change mitigation? In *EcoBalance 2010*, Tokyo.

Cederberg, C., K. Neovius, D. Eivind-Meier, A. Flysjö, and U. Sonesson. 2008. Life cycle greenhouse gas emissions from Brazilian beef. In *Proceedings of the 6th international conference on LCA in the agri-food sector*. Zurich, November 12–14, 2008.

Coltro, L., A.L. Mourad, S.P.M. Germer, T.A. Mendonça, and R.M. Kletecke. 2008. Energy use in the life cycle of frozen concentrated orange juice produced in Brazil. In *Proceedings of the 6th international conference on LCA in the agri-food sector*. Zurich, November 12–14, 2008.
Corson, M.S., and H.M.G. Van der Werf. 2008. Effect of structural and management characteristics on variability of dairy farm environmental impacts. In *Proceedings of the 6th international conference on LCA in the agri-food sector*. Zurich, November 12–14, 2008.
Costello, C., W. Michael Griffin, H. Scott Matthews, and C.L. Weber. 2009. Relating land use to consumed goods using input-output life cycle assessment method. *Life Cycle Assessment IX*. Boston, MA.
Dauthy, M.E. 1995. Fruit and vegetable processing. FAO Agricultural Services, Bulletin no. 119, FAO Corporate Document Repository.
Davis, J., and U. Sonesson. 2008. Life cycle assessment of integrated food chains: a Swedish case study of two chicken meals. *International Journal of Life Cycle Assessment*. doi:10.1007/s11367-008-003.
Diercxsens, P. 2009. Business strategy on waste prevention in the food industry. In *European conference on waste reduction*. Brussels, Belgium, October 14, 2009.
Ecoinvent. http://www.ecoinvent.org. Visited June 23, 2011.
Eide, M.H. 2002. Life cycle assessment of industrial milk production. *International Journal of Life Cycle Assessment* 7(2): 115–126.
Emanuelsson, A., A. Flysjö, M. Thrane, V. Ndiaye, J.L. Eichelsheim, and F. Ziegler. 2008. Life cycle assessment of southern pink shrimp products from Senegal: An environmental comparison between artisanal fisheries in the Casamance region and a trawl fishery off Dakar including biological considerations. In *Proceedings of the 6th international conference on LCA in the agri-food sector*. Zurich, November 12–14, 2008.
Erdal, G., K. Esengün, H. Erdal, and O. Gündüz. 2007. Energy use and economical analysis of sugar beet production in Tokat province of Turkey. *Energy* 32: 35–41.
Ernst, T.J., R.S. Lepp, and J.R. Jackson. 2002. Inhibition of *Tyrophagus putrescentiae* in pet food products. U.S. Patent 2002/0172740 A1.
European Commission. 2005. Integrated pollution prevention and control. Reference document on best available techniques in the slaughterhouses and animal by-products industries. Directorate General—Joint Research Centre, Institute for Prospective Technological Studies. Seville, Spain.
European Commission. 2010. Integrated Product Policy (IPP) web page at http://ec.europa.eu/environment/ipp/home.htm, and header "Identifying products with the greatest environmental impact and potential for environmental improvement" at http://ec.europa.eu/environment/ipp/identifying.htm. Visited June 23, 2011.
Fan, T., J.R. Mitchell, and J.M.V. Blanshard. 1999. A model for the oven rise of dough during baking. *Journal of Food Engineering* 41(2): 69–77.
Fantin, V., R. Pergreffi, P. Buttol, and P. Masoni. 2010. Life cycle assessment of Italian high quality milk production. *LCAFood 2010, VII international conference on life cycle assessment in the agri-food sector*. Bari, Italy, September 22–24, 2010.
Food and Agriculture Organization of the United Nations (FAO). 2011. *Global food losses and food waste: extent, causes and prevention*. Rome. www.fao.org.
Fritz-Jung, C., B. Singh, S. Bhatnagar, G.J. Woodward, K.L. Kettinger, D.R. Speck, J.A. Stoll, S.E. Woerz, and P. Zhang. 2001. Process for dry stable intermediate pet food composition. U.S. Patent 6,270,820 B1.
Gabathuler, H. 1997. The CML story: how environmental sciences entered the debate on LCA. *International Journal of Life Cycle Assessment* 2(4): 187–194.
Garcia-Suarez, T., S. Sim, A. Mauser, and P. Marshall. 2008. Greenhouse gas assessment of Ben & Jerry's ice-cream: communicating their 'Climate Hoofprint.' In *Proceedings of the 6th international conference on LCA in the agri-food sector*. Zurich, November 12–14, 2008.
Grabowski, M., J. Kleme, K. Urbaniec, G. Vaccari, and X.X. Zhu. 2001. Minimum energy consumption in sugar production by cooling crystallisation of concentrated raw juice. *Applied Thermal Engineering* 21: 1319–1329.

Heller, M., S. Cashman, K. Dick, D. Przybylo, W. Walter, and G. Keoleian. 2008. Life-cycle energy and greenhouse gas analysis of a large-scale vertically integrated organic dairy in the U.S. In *Proceedings of the 6th international conference on LCA in the agri-food sector*. Zurich, November 12–14, 2008.

Heller, M., J.S Gough, A.L. Kolodzy, B.A. Marshall, D. Wilson, and G.A. Keoleian. 2010. Life-cycle water use, nutrient cycling and solid waste generation of a large-scale organic dairy. In *LCAFood 2010*. Bari, Italy, September 22–24, 2010.

Hognes, E.S. 2010. Carbon footprint of Norwegian seafood. In *LCAFood 2010, VII international conference on life cycle assessment in the agri-food sector*. Bari, Italy, September 22–24, 2010.

Hospido, A., M.T. Moreira, and G. Feijoo. 2005. Environmental analysis of beer production. *IJARGE* 4(2): 152–162.

Hunt, R., and W. Franklin. 1996. LCA—How it came about. *International Journal of Life Cycle Assessment* 1(1): 4–7.

Ingwersen, W. 2010. Product category range of environmental performance for EPDs: the example of Costa Rican pineapple. In *LCAFood 2010, VII international conference on life cycle assessment in the agri-food sector*. Bari, Italy, September 22–24, 2010.

International Dairy Federation (IDF). 2005. Energy use in dairy processing. Bulletin no. 401, Brussels.

International Dairy Federation (IDF). 2007. Reduction of greenhouse gas emissions at farm and manufacturing levels. Bulletin no. 422/2007, Brussels.

International Dairy Federation (IDF). 2009. Environmental/Ecological impact of the dairy sector: literature review on dairy products for an inventory of key issues. List of environmental initiatives and influences on the dairy sector. Bulletin no. 436, Brussels.

International Dairy Federation (IDF). 2010a. Environmental issues at dairy farm level. Bulletin no. 443/2010, Brussels.

International Dairy Federation (IDF). 2010b. A common carbon footprint approach for dairy. The IDF guide to standard life cycle assessment methodology for the dairy sector. Bulletin no. 445/2010, Brussels.

International Reference Life Cycle Data System (ILCD). 2010a. *ILCD handbook: general guide for life cycle assessment—Detailed guidance*, 1st edition. Ispra (VA), Italy: European Commission, Joint Research Centre (DG JRC), Institute for Environment and Sustainability.

International Reference Life Cycle Data System (ILCD). 2010b. *ILCD handbook: analysing of existing environmental impact assessment methodologies for use in life cycle assessment*, 1st edition. Ispra (VA), Italy: European Commission, Joint Research Centre (DG JRC), Institute for Environment and Sustainability.

ISO. 2006a. ISO 14040: Environmental management—Life cycle assessment—Principles and framework, International standard, Ref ISO 14040:2006(E).

ISO. 2006b. ISO 14044: Environmental management—Life cycle assessment—Requirements and guidelines, International standard.

Jefferies, D., I. Muñoz, V. King, L. Mila I Canals, A. Hoekstra, M. Aldaya, and E. Ercin. 2010. A comparison of approaches to assess impact of water use in consumer products. In *LCAFood 2010, VII international conference on life cycle assessment in the agri-food sector*. Bari, Italy, September 22–24, 2010.

Jolliet, O., M. Margni, R. Charles, S. Humbert, J. Payet, G. Rebitzer, and R. Rosenbaum. 2003. IMPACT 2002+: A new life cycle impact assessment methodology. *International Journal of Life Cycle Assessment* 8(6): 324–330.

Kägi, T., D. Wettstein, and F. Dinkel. 2010. LCA of rice: confidence intervals for avoiding wrong conclusions. *LCAFood 2010, VII international conference on LCA in the agri-food sector*. Bari, Italia, September 22–24, 2010.

Kannan, R., and W. Boie. 2003. Energy management practices in SME—Case study of a bakery in Germany. *Energy Conversion and Management* 44(6): 945–959.

Katajajuuri, J.M., J. Grönroos, and K. Usva1. 2008. Environmental impacts and related options for improving the chicken meat supply chain. In *Proceedings of the 6th international conference on LCA in the agri-food sector*. Zurich, Switzerland, November 12–14, 2008.

Krajnc, D., M. Mele, and P. Glavic. 2007. Improving the economic and environmental performances of the beet sugar industry in Slovenia: increasing fuel efficiency and using by-products for ethanol. *Journal of Cleaner Production* 15: 1240–1252.

Krause F (2008) Energetic evaluation of different food supply chains: Example beef. Culliver: Göttingen. ISBN 978-3-86727-838-6.

Li, Y., E. Griffing, M. Higgins, and M. Overcash. 2006. Life cycle assessment of soybean oil production. *Journal of Food Process Engineering* 29: 429–445.

Manneh, R., M. Margni, and L. Deschênes. 2010. Spatial variability of intake fractions for Canadian emission scenarios: a comparison of three resolution scales. *Environmental Science and Technology* 44(11): 4217–4224.

Maria de Léis, C., V. Prudencio da Silva Jr., S.R. Soares, F.T. Olszensvski, and S. Meireles. 2010. Environmental assessment of two dairy systems in Brazil South. In *LCAFood 2010, VII international conference on life cycle assessment in the agri-food sector.* Bari, Italy, September 22–24, 2010.

Mashoko, L., C. Mbohwa, and V.M. Thomas. 2010. LCA of the South African sugar industry. *Journal of Environmental Planning and Management* 53(6): 793–807.

McKone, T.E., W.W. Nazaroff, P. Berck, M. Auffhammer, T. Lipman, M.S. Torn, E. Masanet, A. Lobscheid, N. Santero, U. Mishra, A. Barrett, M. Bomberg, K. Fingerman, C. Scown, B. Strogen, and A. Horvath. 2011. Grand challenges for life-cycle assessment of biofuels. *Environmental Science and Technology* 45(5): 1751–1756.

Milà i Canals, L., S. Sim, G. Neuer, K. Herstein, C. Kerr, G. Rigarlsford, T.G. Suarez, and H. King, H. 2010. Estimating the greenhouse gas footprint of Knorr products. In *LCAFood 2010, VII international conference on life cycle assessment in the agri-food sector.* Bari, Italy, September 22–24, 2010.

Müller-Lindenlauf, M., G. Reinhardt, S. Gärtner, J. Münch, and S. Häfele. 2010. Food import versus regional production. *LCAFood 2010, VII international conference on life cycle assessment in the agri-food sector.* Bari, Italy, September 22–24, 2010.

Müller-Wenk, R., and M. Brandão. 2010. Climatic impact of land use in LCA—Carbon transfers between vegetation/soil and air. *International Journal of Life Cycle Assessment* 15: 172–182.

Multilateral Investment Guarantee Agency (MIGA). 2006. World Bank Group. Environmental guidelines for sugar manufacturing. www.miga.org/documents/SugarManufacturing.pdf.

National Pollutant Inventory. 1999. Emission estimation technique manual for seafood processing industry. Environment Australia, 1999.

Nguyen, T.L.T., J.E. Hermansen, and L. Mogensen. 2010. What if land use aspects are included in GHG balance of livestock production systems? In *LCAFood 2010, VII international conference on life cycle assessment in the agri-food sector.* Bari, Italia, September 22–24, 2010.

Nilsson, K., and U. Sonesson. 2010. Changing diets—what is the influence on greenhouse gas (GHG) emissions of different consumption patterns? In *LCAFood 2010, VII international conference on life cycle assessment in the agri-food sector.* Bari, Italy, September 22–24, 2010.

O'Brien, R. 2006. Edible fats and oils processing and applications. In *Handbook of food science, technology, and engineering*, vol. 4, ed. Y.H. Hui. Boca Raton: CRC.

Ohlsson, T. 2010. Sustainability, food and the future. In *LCAFood 2010, VII international conference on life cycle assessment in the agri-food sector.* Bari, Italy, September 22–24, 2010

Pattara, C., A. Raggi, and A. Cichelli. 2010. Life cycle assessment and carbon footprint in the wine supply-chain. *LCAFood 2010, VII international conference on life cycle assessment in the agri-food sector.* Bari, Italy, September 22–24, 2010.

Peters, G., S. Wiedemann, H. Rowley, and R. Tucker. 2010. Accounting for water use in Australian red meat production. *International Journal of Life Cycle Assessment* 15(3): 311–320.

Pfister, S., F. Stoessel, R. Juraske, A. Koehler, and S. Hellweg. 2008. Regionalised LCIA of vegetable and fruit production: quantifying the environmental impacts of freshwater use. In *Proceedings of the 6th international conference on LCA in the agri-food sector.* Zurich, Switzerland, November 12–14, 2008.

Pfister, S., A. Koehler, and S. Hellweg. 2009. Assessing the environmental impacts of freshwater consumption in LCA. *Environmental Science and Technology* 43(11): 4098–4104.

Pimentel, D., and M. Pimentel. 1996. *Food, energy and society.* Niwot: Colorado University Press.

Poirier, D. 2003. Drying and cooling properties. In *Petfood technology*, ed. J.L. Kvamme and T.D. Phillips. Mount Morris: Watt Publishing.
Pollack, J.W. and A.L. Greig. 2010. Evaluating life cycle benefits of soybean production. In *LCAFood 2010, VII international conference on life cycle assessment in the agri-food sector.* Bari, Italy, September 22–24, 2010.
Prudêncio da Silva Jr., V., S.R. Soares, and R.A.F. de Alvarenga. 2008. Cradle to gate study of two differing Brazilian poultry production Systems. In *Proceedings of the 6th international conference on LCA in the agri-food sector*. Zurich, Switzerland, November 12–14, 2008.
Prudêncio da Silva Jr., V., H. van der Werf, and S.R. Soares. 2010. LCA of French and Brazilian broiler poultry production systems. In *LCAFood 2010*. Bari, Italy, September 22–24, 2010.
Quantis Water Database. http://www.quantis-intl.com/waterdatabase.php. Visited June 23, 2011.
Ramirez, C.A., M. Patel, and K. Blok. 2006. From fluid milk to milk power: energy use and energy efficiency in the European dairy industry. *Energy* 31: 1984–2004.
Reinhardt, G.A., J. Braschkat, A. Patyk, and M. Quirin. 2003. Life cycle analysis of bread production: a comparison of eight different options. In *4th international conference life cycle assessment in the agri-food sector*. Horsens, Denmark, October 6–8, 2003.
Saad, R., M. Margni, B. Wittstock, T. Koellner, and L. Deschênes. 2010. Land use within LCA: developing spatially explicit characterization factors for impacts on ecosystem services. In *SETAC Europe conference*. Seville, Spain, May 26, 2010.
Schau, E., and A. Fet. 2008. LCA studies of food products as background for environmental product declarations. *International Journal of Life Cycle Assessment* 13(3): 255–264.
Schlich, E., B. Hardtert, and F. Krause. 2008. Beef of local and global provenance: a comparison in terms of energy, CO2, scale, and farm management. In *Proceedings of the 6th international conference on LCA in the agri-food sector*. Zurich, Switzerland, November 12–14, 2008.
Sonesson, U. 2009. LCA for co-product: exploitation. Total Food 2009. Norwich, UK, April 22–24, 2009.
Steinfeld, H., P. Gerber, T. Wassenaar, V. Castel, M. Rosales, and C.D. Haan. 2006. *Livestock's long shadow: environmental issues and options*. Rome: Food and Agriculture Organization of the United Nations.
Thoma, G., J. Popp, D. Nutter, R. Ulrich, M. Matlock, D. Kim, Z. Niederman, C. East, N. Kemper, D. Shonnard, and F. Adom. 2010. Global warming potential of fluid milk consumed in the US: a life cycle assessment. In *LCAFood 2010, VII international conference on life cycle assessment in the agri-food sector*. Bari, Italy, September 22–24, 2010.
Thomas, G.W. 1987. Water: critical and evasive resource on semiarid lands. In *Water and water policy in world food supplies*, ed. W.R. Jordan. College Station: Texas A&M University Press.
Thomas, C., T. Tenant, and J. Rolls. 2000. The GHG indicator: UNEP guidelines for calculating greenhouse gas emissions for business and non-commercial organizations. UNEP Ozone and Action Unit, Paris. http://www.unepfi.org/fileadmin/documents/ghg_indicator_2000.pdf.
Tofield, B., and P. Metcalfe. 2009. Greenhouse gas emissions from agriculture, food and drink. Total Food 2009. Norwich, UK, April 22–24, 2009.
Tukker, A., G. Huppes, J. Guinée, R. Heijungs, A. de Koning, L. van Oers, S. Suh, T. Geerken, M. Van Holderbeke, B. Jansen, and P. Nielsen. 2006. Environmental impact of products (EIPRO). Analysis of the life cycle environmental impacts related to the final consumption of the EU-25. European Commission, Joint Research Centre (DG JRC), IPTS/ESTO, Technical Report EUR 22284 EN.
United Nations Environment Programme (UNEP). 1999. Industrial sector guide. Cleaner production assessment in fish processing industry. Danish environmental protection agency in cooperation with COWI Consulting Engineering and Planners AS.
United Nations Environment Programme (UNEP). 2009. Guidelines for social life cycle assessment of products.
United States Census Bureau. 2006. Annual survey of manufactures. Available at http://factfinder.census.gov.
United States Department of Agriculture (USDA). 2001. Agricultural statistics. US Department of Agriculture, Washington, DC.
United States Life Cycle Inventory Database (U.S. LCI). http://www.nrel.gov/lci/. Visited June 23, 2011.

Urbaniec, K., P. Zalewski, and X.X. Zhu. 2000. A decomposition approach for retrofit design of energy systems in the sugar industry. *Applied Thermal Engineering* 20: 1431–1442.

Vold, M., and E. Svanes. 2008. LCA as environmental improvement tool for products from line caught cod. In *Proceedings of the 6th international conference on life cycle assessment in the agri-food sector.* Zurich, Switzerland, November 12–14, 2008.

Wang L (2009) Energy conservation in grains and oilseeds milling facilities. In *Energy efficiency and management in food processing facilities*, ed. L. Wang. Boca Raton: CRC.

Water Footprint Network. 2011. http://www.waterfootprint.org/?page=files/Publications. Visited June 23, 2011.

Weidema, B.P. 1993. Life cycle assessments of food products. In *Proceedings of the 1st European invitational expert seminar on life cycle assessments of food products.* Lyngby, Denmark, November 22–23, 1993.

Weidema, B.P., A.M. Nielsen, K. Christianse, G. Norris, P. Notten, S. Suh, and J. Madsen. 2005. Prioritisation within the integrated product policy. Report no. 980 for the Danish Ministry of the Environment. http://www2.mst.dk/Udgiv/Publications/2005/87-7614-517-4/pdf/87-7614-518-2.pdf.

Weidema, B.P., M. Wesnæs, J. Hermansen, T. Kristensen, and N. Halberg. 2008. In *Environmental improvement potentials of meat and dairy products*, eds. Eder, P. and L. Delgado. Report EUR 23491 EN, European Commission, Joint Research Centre (DG JRC), Institute for Prospective Technological Studies. Sevilla, Spain.

Williams, A., E. Pell, J. Webb, E. Moorhouse, and E Audsley. 2008. Strawberry and tomato production for the UK compared between the UK and Spain. In *Proceedings of the 6th international conference on life cycle assessment in the agri-food sector*. Zurich, Switzerland, November 12–14, 2008.

Witayapairot, W., and C. Yossapol. 2009. Life cycle assessment of sugar production in Northeastern Thailand. In *International conference on green and sustainable innovation*. Chiang Rai, Thailand, December 2–4, 2009.

Young, O.A., S.J. Lovatt, N.J. Simmons, and C.K. Devinne. 2001. Carcass processing: quality controls. In *Meat science and application*, ed. O.A. Young, R.W. Rogers, Y.H. Hui, and W. Nip. Boca Raton: CRC.

Part III
Green technologies in food production

Chapter 7
Managing nutrient cycles in crop and livestock systems with green technologies

Jonathan Vayssières and Mariana Cristina Rufino

7.1 Introduction

Nutrients such as nitrogen (N) and phosphorus (P) are essential for the growth and development of organisms. Ecosystems regulate the flows and concentrations of nutrients through a number of complex processes including a diversity of species. Nutrients cycles have been substantially altered by human activities—mainly agriculture—over the past two centuries (Vitousek et al. 1997, Bouwman et al. 2009), with large positive and negative consequences for a range of ecosystem services and for human well-being (MEA 2005a).

The capacity of terrestrial ecosystems to absorb and retain the nutrients supplied either as fertilizers or from deposition has been undermined by the large extreme simplification of many ecosystems into large-scale, low-diversity agricultural systems (Lavelle et al. 2004). With the reduction of the buffering capacity of ecosystems such as riparian forests, wetlands, and estuaries, excess nutrients leak into groundwater, rivers, and lakes and are transported to coastal ecosystems (MEA 2005a).

In preindustrial times, the annual flux of N from the atmosphere to the land and aquatic ecosystems was roughly 110–210 MT N year^{-1} (i.e., 10^6 tons of N per year). Human activity contributed an additional 170 MT N year^{-1}, roughly doubling the

J. Vayssières (✉)
Livestock Systems Research Unit, Centre de Coopération Internationale en Recherche Agronomique pour le Développement (CIRAD), La Réunion, France
e-mail: jonathan.vayssieres@cirad.fr

M.C. Rufino
Plant Production Systems, Department of Plant Sciences, Wageningen University, Wageningen, The Netherlands

Sustainable Livestock Futures, International Livestock Research Institute, Nairobi, Kenya
e-mail: m.rufino@cgiar.org

J.I. Boye and Y. Arcand (eds.), *Green Technologies in Food Production and Processing*, Food Engineering Series, DOI 10.1007/978-1-4614-1587-9_7,

rate of creation of reactive N on the land surfaces of Earth (Galloway et al. 1995; Smil 1999). Phosphorus is also accumulating in ecosystems at a rate of 10.5–15.5 MT P year^{-1}, compared to a preindustrial rate of 1–6 MT P year^{-1}, mainly as a result of the use of phosphorus fertilizers in agriculture. Most of this accumulation occurs in soils (MEA, 2005a).

Nutrient concentration on agricultural land has allowed a large increase in food production, especially in industrial countries (Stewart et al. 2005), but at the cost of increased emissions of greenhouse gases and pollution in freshwater and coastal ecosystems. In contrast to the nutrient oversupply in the developed world, there remain vast areas of the Earth, notably in Africa and Latin America, where harvesting without external nutrient inputs has led to land degradation and depletion of soil fertility, with serious consequences for human food and nutrition (Smaling et al. 1997; Sanchez 2002).

The world's human population is expected to have increased about 40% by 2050 (United Nations Organization 2006). The demand for food will probably double as a consequence (Chaumet et al. 2009). Largest changes in food demand are expected in developing countries—especially in sub-Saharan Africa and Latin America—where the highest rates of population growth are projected. The tropics and neotropics will be the scene of major expansions in cultivated land and the increase in yields are expected to be largest (Chaumet et al. 2009). Several prospective studies conclude that agriculture intensification is the only option to respond to the future food demand and to limit conversion of land from natural vegetation to agriculture (e.g., Herrero et al. 2010, Wirsenius et al. 2010). Recent scenario studies that include projections of N fertilizer use indicate an increase of between 10% and 80% by 2020 depending on future food needs and the response of future farmers to soil degradation (Wood et al. 2004). Three out of four scenarios considered by the Millennium Ecosystem Assessment project that the global flux of N to coastal ecosystems will increase by 10–20% by 2030 (MEA 2005b). The concentration of N in rivers would not change much in most industrial countries, whereas a 20–30% increase is expected for developing countries. This would be the consequence of increasing N inputs to surface water mainly owing to increasing food production and associated inputs of N fertilizer, animal manure, atmospheric N deposition, and biological N fixation in agricultural systems (MEA 2005b). Consequently, producing food for the world population, but with limited impact on world ecosystems, will be a major global challenge with no single solution (Griffon 2006). Managing nutrients cycles in agriculture (crop and livestock systems) can certainly contribute to the goal. Conserving and using nutrients efficiently may help to increase crop yields and limit environmental impact. The underling questions of this study were: (1) What could be gained from managing nutrients cycles? and (2) What would be the productive and environmental goals and how can they be achieved in practice?

In this study we concentrated on N because it is one of the most limiting factors for crop production (Clark et al. 1999; Cassman et al. 2002), and a determinant for animal production, via the proteins and amino acids supply in the feed ration (Jarrige 1989). Nitrogen is present in multiple forms in agro-ecosystems, each of different life span, and it is subjected to numerous anthropogenic losses potentially harmful for the environment (Bouwman et al. 2009). Excessive flows of N contribute to eutrophication of freshwater and coastal marine ecosystems and acidification of freshwater and terrestrial ecosystems. Nitrogen losses contribute to global warming and, to some degree, also play a role in creation of ground-level ozone, destruction of ozone layer in the stratosphere (MEA 2005a).

In this chapter, we describe the sources of main N losses at the different steps of N cycling in agro-ecosystems. We suggest technical solutions to reduce the size of N losses, and illustrate the interest of analyses supported by integrated tools and indicators for two contrasted situations: a low-input system from the developing world and a high-input system from an industrial country. Finally, we discuss the research needs for better assessment of the likely gains to be attained at a global level, in land productivity and reduction of environmental impact by improving nutrient cycling management.

7.2 N cycling and main steps for losses

Losses of N take four main forms: ammonia (NH_3), nitrate (NO_3-), nitrous oxide (N_2O), nitrogen oxides (NO_x) and dinitrogen (N_2). Only the three first losses are discussed in this chapter because they are the most documented harmful losses. Figure 7.1 represents the main steps of N losses. Figures proposed by the International Panel on Climate Change (IPCC 1997, 2006) and FAO/IFA (2001) for regional and global inventories are given in bold and ranges of emissions rates encountered in the literature about specific studies are given in between brackets.

The three main steps for N loss are: (1) N gaseous emissions during manure handling, including manure collection in barns, storage and treatment; (2) N gaseous emissions after application and direct deposition during grazing; and (3) N losses through leaching and runoff after application and direct deposition onto soils. All emission rates given in this chapter are percentages of total N excreted (for manure handling) or of total N applied (for application or direct deposition).

7.2.1 N gaseous emissions during manure handling

Because of the low N use efficiency, about 90% of N ingested by livestock is excreted, mainly in the form of feces and urine. Apparent efficiency varies considerably between different species: from 5% for beef cattle to 34% for poultry (Van der Hoek 1998). Van der Hoek (1998) estimated a global figure of about 10% for N use efficiency by livestock, including pastoral systems. Most N in excreta mineralizes rapidly, especially for livestock diets rich in protein (Frank et al. 2002, Marini and Van Amburgh 2003). In urine, typically more than 70% of the N is present as urea and uric acid is the dominant N compound in poultry excretions (Whitehead 1995). Rapid degradation of urea and uric acid to ammonium leads to large NH_3 losses

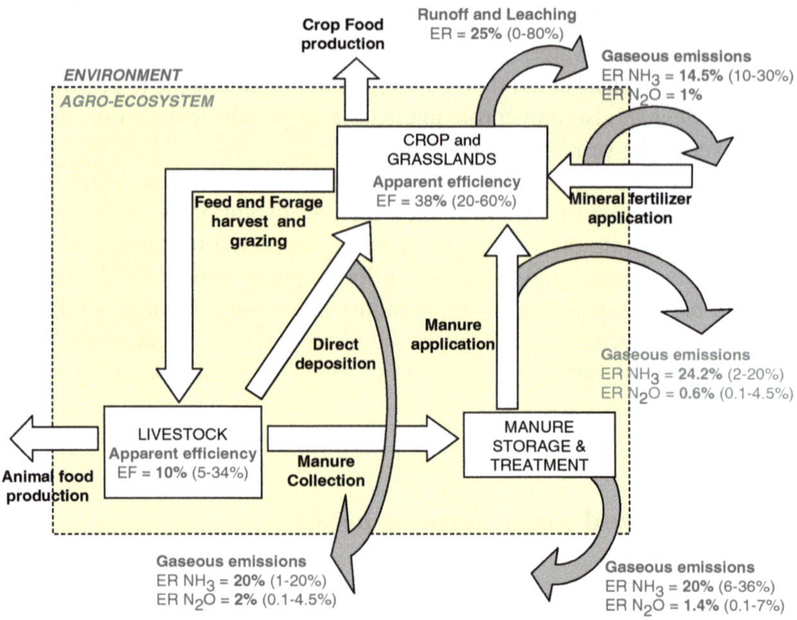

Fig. 7.1 Main steps of N losses (*arrows in grey*) occurring along the N cycle within agro-ecosystems (*arrows in white*) *EF* apparent nitrogen use efficiency, *ER* emission rate. Figures in bold are the ones proposed by IPCC (1997; 2006) and FAO/IFA (2001) for regional and global inventories and figures in between brackets are ranges encountered in the literature on specific studies. This range only considers studies conducted under temperate conditions and is mainly based on the review of Gac et al. (2006) that considered more than 160 references

through volatilization during manure handling. The emissions are subjected to many factors, particularly the concentration of ammonia, ambient temperature, moisture content, and the pH of the manure (Hartung and Phillips 1994). Bouwman et al. (1997) estimated losses of ammonia from excreta of different species: sheep and goats 6–10%; dairy cattle 26–29%; camels, buffalo, and horses 17–19%, and pigs and poultry 36%. Based on figures of Smil (1999) and Galloway et al. (2003), FAO (Steinfeld et al. 2006) estimates a global ammonia emission rate of 20% of N contained in manure.

Only a small fraction of the total N excreted is converted to N_2O during manure handling (Steinfeld et al. 2006). While the actual magnitude of the mineralization rate and resulting NH_3 emissions depends mainly on the manure composition, management, duration of the storage and temperature are the main drivers of N_2O emissions (Baggs and Philippot 2010). From manure heaps in Mali losses were smaller than 0.5% of initial N content (Predotova et al. 2010b, c). Under temperate conditions, N_2O emissions are negligible for liquid manure systems without natural crust cover (Harper et al. 2000; Lague et al. 2004). Amon et al. (2001) estimated N_2O emissions ranging from 0.27% to 1% for solid manure systems. Higher emission rates of up to 20% were observed for actively mixed cattle and swine deep bedding (Moller et al. 2000; Nicks et al. 2003). De Klein et al. (2010) estimated emission rates for N_2O of 1% for excreted sheep manure, and 2% for cattle, poultry and pigs. Default IPCC (2006) emission rates range from 0.01% for slurries to 7% for pig waste of deep-litter stables. Based on existing manure management in different systems and world regions, combined with default IPCC emissions rates, FAO proposes a global N_2O emission rate of 1.4% of N contained in manure (Steinfeld et al. 2006).

7.2.2 Gaseous N emissions after application and direct deposition

Under temperate conditions NH_3 volatilization rates after application depend on the manure type and quality ranging from 2% to 20% for solid manure and from 5% to 50% for slurry, all species considered (Gac et al. 2006). IPCC (1997) proposed a default N loss fraction for NH_3 volatilization of 20%, without differentiating between applied manure and directly deposited excreta. FAO/IFA (2001) estimated the N loss via ammonia volatilization from animal manure, after application to be 24.2% worldwide. Smil (1999) estimates this loss to be at least 15–20%.

Excreta deposited directly on the land are exposed to relatively large N loss rates, largely as ammonia, especially in tropical and dry situations (Vallis et al. 1985). Under temperate conditions, emission rates range from 1% to 8% of total N for grazing dairy cows and from 7% to 20% for outdoor reared monogastrics (Bussink and Oenema 1998; Gac et al. 2006). Wide variations in the quality of grazed forages and environmental conditions make N emissions from excreta on pastures difficult to estimate. FAO proposed a global volatilization rate of 40% for direct deposition (Steinfeld et al. 2006). The standard IPCC (1997) rate of 20% is more consistent with the literature.

Losses after manure application and excreta deposition on land are regarded as the largest livestock source of N_2O emissions by Steinfeld et al. (2006). The N_2O emissions depend on a variety of factors, particularly soil water filled spore space, organic carbon availability, pH, soil temperature, plant uptake and rainfall characteristics (Mosier et al. 2004). Under temperate conditions, observed emission rates ranges from 0.1% to 4.5% (Gac et al. 2006). They are of comparable magnitude for applied manure and for excreta directly deposited. IPCC (2006), makes a distinction between emission calculations for tropical and temperate soils and animal species. The current IPCC emission rate for manure application is 1% and in the range of 0.3–60% for direct deposition, recognizing a large uncertainty on the default rates. The FAO/IFA (2001) estimated the global N_2O loss rates at 0.6% for applied manure and 2% for deposited excreta.

The magnitude of N emissions after mineral fertilizer application depends on the fertilizer type. Bouwman et al. (1997) estimated that NH_3 emission losses from urea to be 15% and 25% in tropical and temperate regions, respectively. The NH_3 loss from ammonium bicarbonate (used as fertilizer) is higher: it may be 20% and 30% in tropical and temperate regions respectively. According to FAO/IFA (2001), which considered mineral fertilizer types used worldwide, the global NH_3 volatilization rate for mineral fertilizers is 14.5% and the N_2O emissions amount globally to 1% of the N applied. IPCC (2006) proposes a similar rate with reference to the work of Bouwman and Booumans (2000a, b).

7.2.3 N leaching and runoff after application and direct deposition

Leaching is another mechanism whereby N is lost from soils. Because of the relatively low N use efficiency by plants (38% Steinfeld et al. 2006, from 20% to 60%, Smil 1999) a large share of the applied or deposited N is not incorporated in the harvested plant tissues nor stored in the soil. It consequently enters in the "N cascade" (Galloway et al. 2003).

N is very mobile in soil solution, mainly as nitrate, and can be easily leached below the rooting zone to the groundwater. Organic forms of N can also be discharged into water courses through runoff and slides after heavy rain events. The importance of this process to N losses and pollution potential depends on soil and weather characteristics, the intensity, frequency and period of grazing, and the rate at which manure is applied (Steinfeld et al. 2006).

Brouwer and Powell (1998) estimated annual N leaching losses from millet fields receiving 9–10 $t\,ha^{-1}$ of manure to be on average 91 kg $N\,ha^{-1}$. However, in urban gardens of Niamey, Niger, Predotova et al. (2010a) estimated N leaching losses of 11.4 kg $N\,ha^{-1}$, for plots receiving an average of 10 $t\,ha^{-1}$ of manure. Kamukondiwa and Bergstrom (1994) measured N leaching losses of 17–39 kg $N\,ha^{-1}$ from lysimeters installed on a sandy soil in Zimbabwe cultivated with maize and wheat. Nyamangara et al. (2003) measured N leaching losses of 12–19 kg $N\,ha^{-1}$ for control plots cultivated with maize. These losses were increased with fertilization and ranged between 14–48 kg $N\,ha^{-1}$, depending on rainfall during the growing season (840 vs. 1397 mm) for a fertilizer rate of 60 kg $N\,ha^{-1}$. Losses ranged between 16 and 56 kg $N\,ha^{-1}$ for a fertilizer rate of 120 kg $N\,ha^{-1}$. Manure applications of 12.5 or 37.5 $t\,ha^{-1}$ had a small effect on N leaching. Losses ranged from 3% to 11% of applied N in second and third seasons, but 24–40% in the first season with large rainfall events. Relatively large amounts of N are lost through leaching in low-input African agriculture, where fertilizer use is far lower than the targeted 50 kg $N\,ha^{-1}$ (Vitousek et al. 2009).

N losses in runoff are usually below 5% of the applied fertilizer rate (Gangbazo et al. 1995). Overall N export (leaching and runoff) from agricultural ecosystems to water, as a percentage of N fertilizer input, ranges from 10% to 40% from loam clay soils to 25% to 80% for sandy soils (Carpenter et al. 1998). It is estimated that globally 25% of the N applied is lost and contaminates water resources (Galloway et al. 2004). Concerning N directly deposited to grazed areas, the N losses depend on the grazing intensity. Stocking rates, percent utilization of plants, and stubble heights are classically used to describe grazing intensity. With moderate intensity grazing practices, N losses are almost insignificant (Mosley et al. 1997). However, intensive grazing practices increase N losses in runoff and N leaching (Hooda et al. 2000;

Scrimgeour and Kendall 2002). Global N losses after deposition were considered negligible in the FAO's global assessment because of the large dominance of extensively grazed areas.

7.3 Technical solutions or green technologies to reduce N losses

Here we list a number of green technologies that can be used to reduce N losses, addressing a number of key processes in N cycle: (1) reducing N excretion in manure by modifying livestock diets, (2) improving manure handling, (3) limiting on-field N gaseous emissions, (4) limiting onfield leaching and runoff, and (5) improving N use efficiency by plants. Each stage can be addressed by specific technical options.

7.3.1 Reducing N excretion in manure

An important mitigation pathway lies in raising the low animal N assimilation efficiency (10% against some 40% for crops). It can be achieved through breeding, animal health including stress reduction, and especially through more balanced feeding, i.e., by optimizing protein intake to match the livestock requirements (Børsting et al. 2003) or modifying diets to increase the partitioning of excreta N to fecal N (Powell et al. 2010). Improved feeding practices also include grouping animals by gender and phase of production, and improving the feed conversion ration by tailoring feed to physiological requirements (Jarrige 1989).

Reducing the inorganic N content of manure leads to lower N_2O and NH_3 emissions from stables, during storage and after application to soil (Oenema et al. 2007). For cattle a proper balance in feed between degradable and non degradable proteins improves nutrient absorption and has been shown to reduce N excretion by 15–30% without affecting production levels (Tomlinson et al. 1996). For pigs a lower amount of crude protein supplemented with synthetic amino acids lowers N excretion up to 30%, depending on the initial composition of the diet (Jongbloed et al. 1997). Removing fiber and germ from corn is reported to reduce the level of N contained in urine and the feces by 39% (Sutton et al. 2001).

However, even when good practices are used, large quantities of mineral N still remain in the manure and technical options available at other stages have also to be considered to minimize N losses.

7.3.2 Improved manure handling

N losses from livestock housing and manure storage can be limited by immediate collection of manure, storage in covered containers, and adequate sizing of the storage facilities (Bussink and Oenema 1998). In low input systems, direct application or short storage time using a polyethylene cover reduced nutrient losses and preserved the quality of the manure as organic fertilizers (Rufino et al. 2007; Tittonell et al. 2010a). Appropriate storage capacity is of prime importance to prevent losses during the rainy season in places with frequent abundant rainfall event. The use of an enclosed tank can also limit the dilution process and then the overflow risk. Cover with a permanent lid reduces considerably (by 80–90%) gaseous loss (mainly NH_3) during storage (Sommer et al. 1993). Maintaining a natural crust on the slurry surface in an open tank is almost as effective (Smith et al. 2007).

The choice of N_2O emission mitigation options during storage are complex because important tradeoffs exist between methane and nitrous oxide emissions. For example, a management change from straw- to slurry-based systems may result in lower N_2O emission but increased CH_4 emission (Clemens and Ahlgrimm 2001).

Various treatment technologies (e.g., composting of solid manure, aeration of slurry) exist for concentrating the nutrients derived from separated phases to deliver products of higher fertilizer value and easier to transport, and to reduce local manure surpluses and then to reduce the potential for pollution of water resources through leaching and runoff (Burton et al. 2007). These technologies are mainly used for odor control (e.g., Westerman and Zhang 1997), and to reduce emissions of harmful gases (e.g., Vanotti et al. 2006). But most of these processes also lead to high N losses under the NH_3 gaseous form (Paillat et al. 2005). Addition of organic materials to reduce the size of the mineral pool of N has been suggested by Petersen et al. (1998) as an effective technology to reduce ammonia volatilization and gaseous emissions. Reducing the size of the substrate and avoiding aerobic conditions limit losses of ammonia and might stimulate slightly losses of N_2O, although this apparently depends more on available substrate as sources of energy (Petersen et al. 1998; Baggs and Philippot 2010).

7.3.3 Limiting on-field N gaseous emissions

In general, soil N emissions are strongly influenced by temperature and soil moisture, but also by agricultural practices (Bouwman et al. 2002a). N_2O emissions from slurry applications to grassland were reduced when slurry was stored for 6 months or passed through an anaerobic digester (e.g., for biogas production) prior to spreading (Amon et al. 2002) because there is less available C in the slurry applied to land. But these two technical solutions increase NH_3 volatilization. Rapid incorporation and shallow injection techniques for manure reduce N loss to the atmosphere by at least 50%, whereas deep injection into the soil essentially eliminates this loss (Rotz 2004). Another technical option for reducing N emissions during the application and deposition phase is the use of nitrification inhibitors (NIs). Monteny et al. (2006) cite examples of substantially reduced emissions. NIs can be used on pastures to act upon urinary N (Di and Cameron 2003). The rate of NH_3 emissions from mineral fertilizers can be reduced from on average 14.5% to 4% if injected anhydrous ammonia is used (as is widely done in the United States, Bouwman et al. 1997). Emissions of N_2O increase exponentially with N fertilizer rates, and the size of the losses depends on soil type, drainage conditions and fertilizer type (Bouwman et al. 2002b). Fertilization trials where optimal crop response and economic rates are sought must include estimations of gaseous emissions to capture environmental pollution.

7.3.4 *Limiting N leaching and runoff*

In general, the key to reducing on-field N losses is the fine-tuning of organic and mineral fertilizers application to the land with regard to environmental conditions, including timing, amounts and form of application in response to crop physiology and climate (Oenema et al. 2009). Synchronizing N application with the most nutrient-demanding growing period seems to be simple, but it has proved to be complicated in practice. With organic fertilizers, nutrient availability to plants varies with animal diets and manure quality, manure management practices, and climate (dilution by rains). Time-lag between N application and N mineralization can be significant, especially for manures in both high-input temperate and low-input tropical environments (Schröder et al. 2005, Nyamangara et al. 1999). Organic N stored in soils can be mineralized at times with low N uptake of crops and then be prone to leaching, e.g., in early spring in temperate areas and at the beginning of the rainy season in tropical areas (Chikowo et al. 2004). Spatial soil heterogeneity can be large in developing countries, especially in mixed crop-livestock systems (Ramisch 2005; Tittonell et al. 2005a,b). The challenge of "precision fertilization" (Chen et al. 2007) resides precisely in considering these spatio-temporal heterogeneities by combining different forms of N originating from fertilizers of different natures and origins (mineral or organic fertilizers; manure from ruminants or monogastrics; liquid, solid, or composted manure) to minimize N losses.

According to (Hill 1991), the presence and quality of the soil cover is also an important land management determinant of the intensity of leaching. For grazing systems, control of grazing season, intensity, frequency, and distribution of livestock can improve vegetation cover and reduce runoff and leaching. In temperate regions, excessive losses from manure can be limited by avoiding late fall and winter grazing (Stout et al. 1997). For cropping systems, a permanent ground cover, as in zero tillage cultural systems, may reduce considerably N leaching (Di and Cameron 2002).

7.3.5 *Improving N use efficiency by plants*

Improving N use efficiency by plants, similarly to animals, refers to increasing crop yield per unit of N inputs. It would result in a reduction of fertilizer applications and then of corresponding losses. It is possible via classical technologies such as irrigation, conservation agriculture (minimum tillage), genetic progress (including actively discussed transgenic plants), better planting methods, improved drainage, etc. N fixation by legumes also has an important role to play in the reduction of fertilizer application by incorporating legumes in grasslands and crop fields (intercropping) (Giller 2001). Optimizing practices and environmental parameters for plant cultivation would result in an increased efficiency that could reach 70%. The remaining 30% are inherent (unavoidable) loss from cultivated ecosystems (Raun and Johnson 1999).

7.4 Improving farm-scale N cycling and efficiency with integrated tools and indicators

The previous sections underline the importance and the variety of technical options available for mitigating N losses. Options are presented separately but they are in reality strongly related. For instance, reduced N emissions during manure storage can be compensated by higher N emissions after manure application and then efforts done at one stage may be in vain. Consequently, a holistic view of the N cycling in agro-ecosystems is needed. Simulation models can help in elaborating such a holistic view (Rotz 2004). This is illustrated with two modelling experiments completed in two contrasted situations. The first, in southern Africa, is about low-input systems in which soil fertility is decreasing and the priority is to improve farm productivity to respond to household food requirements. The second, in an ultraperipheral region of the European Union, is about high-input systems where the main problem is the reduction of N surpluses and resulting risks of fresh water pollution.

7.4.1 Exploring combinations of organic and inorganic sources of N to improve productivity and N use efficiency of zimbabwean crop-livestock farms

7.4.1.1 Context

Soils are inherently poor sands in vast areas of southern Africa (Sanchez 2002). Traditionally, farmers manage soil fertility to produce food combining local organic resources, litter, livestock manure, termitaria, and fallow periods. However, human population increase has led to strong competition for land and for organic resources needed to manage poor fertility. Most farmers in sub-Saharan Africa cultivate small pieces of land (usually smaller than 2 ha), use little mineral fertilizers, and own few or no livestock, which often leads to nutrient depletion (Vitousek et al. 2009). Fertilizer are needed to increase food production in Africa (Breman et al. 2001), although their effectiveness is often limited by soil degradation causing lack of response (Vanlauwe et al. 2010). Soil heterogeneity, reinforced through management, results in relatively small areas in the croplands used by farmers to secure crop yields (Zingore et al. 2007, Tittonell et al. 2008). To explore options to increase productivity of land, studies must consider the availability of inputs at farm scale, and the availability of organic inputs, which often in Africa goes beyond farm scale because of the local land tenure systems. The modelling exercise presented here uses experimental evidence from a communal farming system in Zimbabwe to analyze scope to increase food production, food self-sufficiency, and N use efficiency, and likely impacts to the environment.

7.4.1.2 Methodology

NUANCES-FARMSIM is a farm-scale model that links simulation results from four submodels. The coupling of the submodels allows representation of short- and long-term feedbacks that characterize interactions between crops and livestock, and their consequences on the natural resources of farming systems. Crop and soil modules are combined at field level in the model FIELD (Field-scale resource Interactions, use Efficiencies and Long-term soil fertility Development, Tittonell et al. 2010b). Combinations of crop and soils can be simulated for different field types. LIVSIM (LIVestock SIMulator, Rufino et al. 2009a) simulates animal production based on genetic potential, feed availability, and its quality. The dynamics of nutrients through manure collection, and storage are simulated by HEAPSIM (Rufino et al. 2007), estimating mass and nutrient cycling efficiencies of manure management. Weather and nutrient inflows constitute inputs to FARMSIM that are accounted for during the simulations, and modified for scenario analysis. Experimental data and calibrated process-based models were used to generate functional relationships that are built into the submodels of FARMSIM. Submodels incorporate processes and interactions in a descriptive fashion, and operate with different time steps: monthly for LIVSIM, HEAPSIM, and seasonal for FIELD. Details of this study can be found in Rufino et al. (2010).

7.4.1.3 Scenarios

For this study we selected a farming community of North-East Zimbabwe, consisting of 66 households, who classified themselves into four resource groups (RG): the livestock owners (RG 1 and 2) and the farmers without livestock (RG 3 and 4); groups 1 and 2, and 3 and 4 differed in land and labor availability. Farmers distinguish three types of fields in their farms: homefields, midfields, and outfields, which differ in fertility and receive different amounts of inputs. Homefields are the fields around the house, mid and outfields are more infertile fields located a few hundred meters to a few kilometers from the house. We considered two types of scenarios which were evaluated in time: (1) a baseline scenario in which we evaluated actual farmers' practices, and (2) targeted fertilization scenario, where inorganic fertilizer use is increased and organic resources are used more evenly across field types, increasing manure applications in outfields. Fertilizer rates of 60 kg N ha^{-1} and

30 kg P ha^{-1} used in the simulations were the most efficient rates in the experimental work of Zingore et al. (2007). Homefields of RG1 and RG2 received half of the fertilizer to simulate maintenance fertilization. In relatively poor homefields of RG3 and RG4, crop residues were incorporated into the soil and fertilizers were added to all fields. Crop residues from mid- and outfields of all RG were assumed to be grazed by free roaming livestock. Scenarios are evaluated through grain and livestock production, agronomic N efficiency, and partial N budgets.

7.4.1.4 Results

Increasing fertilizer use combined with organic resources spread at low rates across poor fields increased grain production in the outfields of wealthier farmers (RG 1 and 2) and in both home and outfields of poorer farmers (RG 3 and 4) (Figs. 7.2a, b and 7.3a, b). In the homefields of wealthier farmers, producing already close to attainable yield, the targeted fertilization increased agronomic efficiency (Figs. 7.2c and 7.3c) and reduced the partial N balance (Figs. 7.2e and 7.3e) compared with the baseline. N agronomic efficiencies were increased for all fields, and there was a positive trend in time for the targeted fertilization. Largest increases in agronomic efficiencies were observed in the responsive fields, i.e., the outfields of wealthy farmers and the homefields of poor farmers. N balances changed little or decreased slightly for all fields and resource groups.

The responses to fertilizers can be explained by the changes in soil carbon, resulting from the management of organic resources. Although soil C stocks decreased in most soils in the baseline because of continuous cultivation with little inputs, C stocks of homefields of wealthier farmers increased because of the concentration of fertilizers, manure, and crop residues (Fig. 7.4). In the targeted fertilization scenario, C stocks of all fields of wealthier farmers improved, as well as the homefields of poor farmers' fields. The long-term rate of degradation of outfields of poor farmers is slowed down in the targeted fertilization scenario.

Small improvements in soil C stocks resulting from investment in managing C and nutrient flows improved land production at the farm level (Table 7.1) and the food self-sufficiency ratio, creating grain surpluses for the poor farmers that can be marketed. However, the effect of fertilizer to improve land productivity is strongly associated with the utilization of common grazing lands to feed livestock and produce manure. Only a small proportion of excreted manure N (i.e., 15–17%) is recycled on farm and used for crop production. The rest is partly lost from grazing lands, and lost on-farm from stalls and during manure handling. Partial N budgets at the farm level are positive for wealthier farmers (RG1 and RG2) and very small for poorer farmers (RG3 and RG4), in the baseline. Increasing the use of fertilizers increases production but may increase the fraction of N lost and N removal from farm with the produce. This calls for evaluating strategies for improving food production also from the environmental and economic perspectives, so that farmers use the expensive fertilizers efficiently and do not pollute their environment. Other studies have showed that recycling in crop-livestock systems represents a small fraction

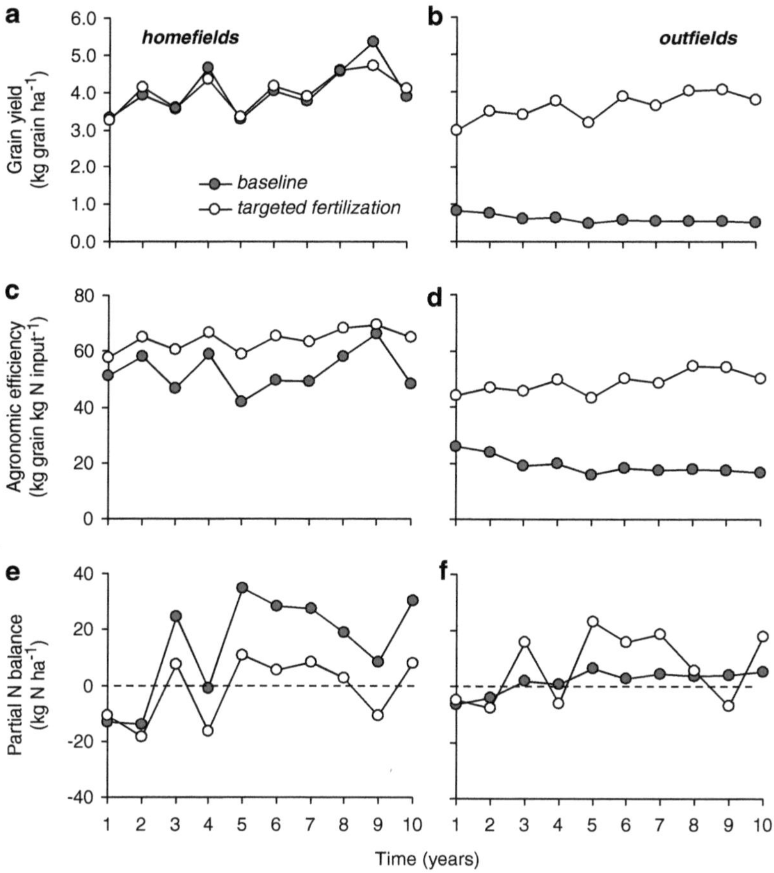

Fig. 7.2 Results from 10 year simulations using NUANCES-FARMSIM for two scenarios, a baseline and targeted fertilization for a relatively wealthier household: grain yield for (**a**) homefields and (**b**) outfields; agronomic efficiency for (**c**) homefields and (**d**) outfields; and partial N balances for (**e**) homefields and (**f**) outfields

(less than 10%) of total N flows (Rufino et al. 2009b). Improving cycling efficiency in a smallholder system is difficult to achieve, mainly because of labor shortages and because farmers often prioritize short term returns to their investments.

7.4.1.5 Conclusions

Smallholder farming agriculture in southern Africa must increase the size of the nutrient flows to improve food self-sufficiency and food security. This may bring environmental problems in the future, but so far agriculture does not create large surpluses of N, comparable with agriculture in the developed world (Vitousek et al. 2009). Yet, there is room to explore cost-effective options to manage soil fertility and contribute to improving food security.

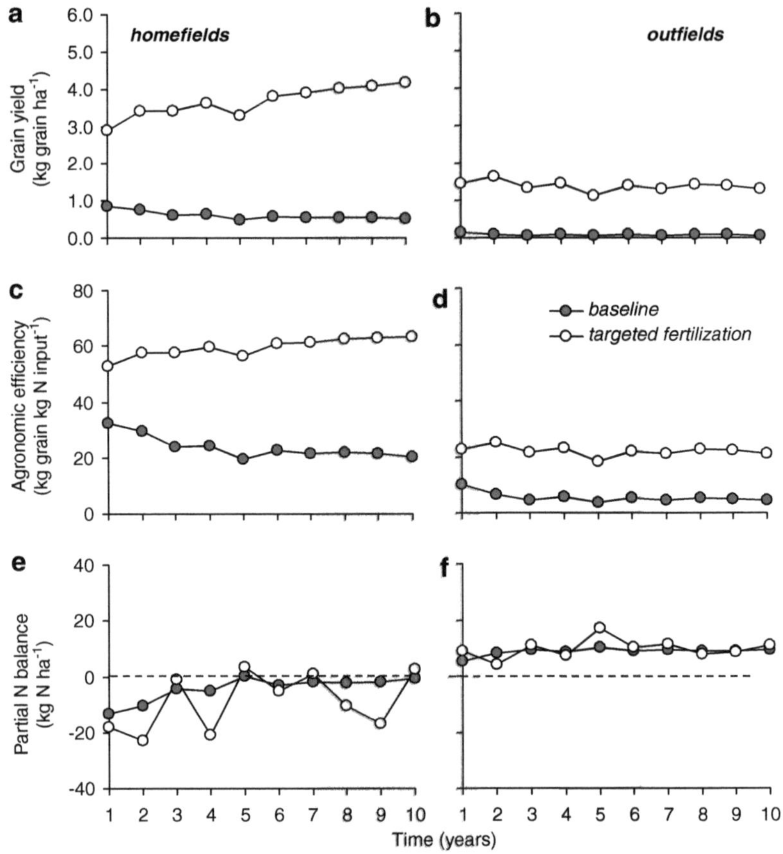

Fig. 7.3 Results from 10 year simulations using NUANCES-FARMSIM for two scenarios, a baseline and targeted fertilization for a relatively poorer household: grain yield for (**a**) homefields and (**b**) outfields; agronomic efficiency for (**c**) homefields and (**d**) outfields; and partial N balances for (**e**) homefields and (**f**) outfields

7.4.2 Exploring increased crop-livestock integration for reducing N surplus in dairy farms of la réunion tropical island

7.4.2.1 Context

Due to its isolation from continents and the limited availability of arable land, development of the agricultural sector of La Réunion Island is promoted with financial and technical support from the European Union. Since the early 2000s, subsidies on main inputs, such as concentrate feeds and mineral fertilizers, facilitated the quick intensification of dairy but did not stimulate recycling of on-farm byproducts (forages and organic fertilizers). In 2004, farms had an average productivity of 5,750 kg of milk cow^{-1} year^{-1}.

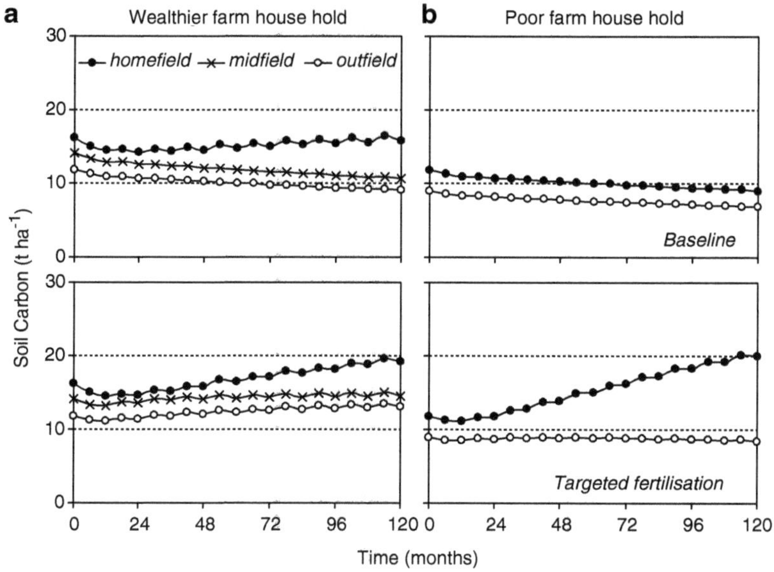

Fig. 7.4 Changes in soil carbon stocks for homefields, midfields, and outfields of relatively wealthier and relatively poorer farmers, under two scenarios: baseline, and targeted fertilization scenarios

Four thousand cows spread over 135 farms produced 22 million liters of milk. The local milk production covers about 30% of the consumption of the island (850 thousand inhabitants), the rest being covered by imports of powdered milk and cheese. This form of intensification based on high on-farm stocking rates (on average 4.6 Livestock Unit ha^{-1}) and on high input consumption (on average 11.6 kg of concentrate feeds fed cow^{-1} day^{-1}) generates a large farm-gate nutrient surplus (on average 477 kg N ha^{-1} $year^{-1}$) that raises environmental concerns (Vayssières et al. 2006). The farm model GAMEDE was developed to design, together with farmers, alternative management strategies that could reduce environmental risks associated with N surplus due to farming. Crop-livestock integration (CLI) at the farm level is considered as a promising strategy.

7.4.2.2 Methodology

GAMEDE is a dynamic simulation model representing the N cycle of a farm agro-ecosystem. The model integrates decision making and biophysical processes simulating systems interactions at a daily time step. The main system components are the farmer or decision maker, livestock, soils, and forage crops. Main biophysical modules are based on existing mechanistic models: MCP (Leteinturier et al. 2004), MOSICAS (Martiné 2003), INRATION (Faverdin et al. 2007), CNCPS (Fox et al. 2004), SEPATOU (Cros et al. 2003), and GRAZEIN (Delagarde et al. 2004). The integration of the modules is described in detail by Vayssières et al. (2009).

Table 7.1 Farm-scale indicators of productivity, calculated with results of simulations of NUANCES-FARMSIM for two scenarios (baseline and targeted fertilization), and for relatively wealthier and poorer households

Scenario	Land available (ha)	Cropped land (ha)	Livestock (heads $farm^{-1}$)	Household size (# of people)	Fertilizer N available (kg $farm^{-1} year^{-1}$)	Manure available (t $farm^{-1}$ $year^{-1}$)	Grain produced (kg $farm^{-1}$ $year^{-1}$)	Milk produced (kg $farm^{-1}$ $year^{-1}$)	Food self-sufficiency ratio[a] (dmnl)	Farm N balance[b] (kg $farm^{-1}$)	N balance (kg ha^{-1})
Baseline											
Wealthier farmers											
RG1	3.5	3.3	10	7	100	3.9	5.1	1526	3.3	126.6	36
RG2	2.2	2.0	5	5	45	2.0	2.9	830	2.6	55.7	25
Poor farmers											
RG3	1.9	1.7	0	6	35	0.2	0.6	0	0.4	7.7	4
RG4	0.9	0.8	0	4	13	0.1	0.2	0	0.2	−2.0	−2
Targeted fertilization											
Wealthier farmers											
RG1	3.5	3.3	12	7	174	4.5	12.6	1853	7.9	170.4	49
RG2	2.2	2.0	6	5	102	2.0	4.4	958	3.9	108.3	49
Poor farmers											
RG3	1.9	1.7	0	6	102	0.4	3.9	0	2.8	45.1	24
RG4	0.9	0.8	0	4	48	0.2	1.8	0	1.9	17.8	20

dmnl dimensionless

[a]Food self-sufficiency ratio = energy in produce/household energy needs, considering an average energy requirement of 9 MJ per capita per day, and an average energy content of 14 MJ per kg of grain, and 2.8 MJ per kg milk

[b]Farm N balance is partial balance including as inputs: grass N intake, crop residues N intake, fertilizer N, and as outputs: human excreta N, manure N lost (manure N excreted—manure N applied to crops), grain N sold

The organic and mineral fertilizer management module of GAMEDE stems from our own original work. It simulates daily N gaseous losses from fertilizers and the changes in the N content of organic fertilizers during their management (i.e., manure collection in barns, storage, composting, and application). Four possible manure management systems (MMS) are distinguished: (1) direct deposition during grazing, (2) liquid manure, (3) raw solid manure, and (4) composted solid manure. In practice, these MMS may coexist in a farm because farmers manage differently the excreta of livestock groups. Moreover, the same group of animals may contribute daily to the flow of manure of different MMS. The total excreta production of each animal group is divided between MMS according to the time livestock spent grazing or feeding in barns, and according to bedding practices. In GAMEDE, each MMS is conceptually a succession of handling stages represented as a chain of flows and stocks of variable length: the shorter chain corresponds to direct deposition during grazing, and the longer to producing composted solid manure. Whichever the MMS, the first flow of the chain is raw effluent production and the last one application on crops (or direct restitution to soils). Along the handling chain, the N and dry matter contents of the manure are updated daily according to the feeding and bedding practices (varying from one day to the next). Each fertilizer handling stage corresponds to an N emission flow calculated on the basis of emission rates (see Appendix 7.1). As the emission rates in the literature are mainly estimated with data from temperate climates, climatic correction indices were introduced to take into account the climate variability and tropical conditions of La Réunion. For details on the organic and mineral fertilizer management module, see supplementary material of Vayssières et al. (2009).

The indicator calculated with GAMEDE to assess CLI in farm agro-ecosystems is one indicator of the network analysis (Leontief 1951) originally used to quantify relationships within ecosystems (Fath and Patten 1999) and applied to analyze CLI in East African farming systems by Rufino et al. (2009b). Because a more tightly integrated system shows more internal recycling and less dependency on the external environment (Edwards et al. 1993), the degree of crop-livestock integration (DCLI) (dimensionless, dmnl) is calculated as follows:

$$DCLI = 1 - \frac{Nin}{TST}$$

where Nin (in kg N year^{-1}) is the total amount of N that is imported from the external environment into the system (as defined for the equation) and TST the total system throughflow (in kg N year^{-1}), i.e., the sum of all N throughflows in the system.

7.4.2.3 Scenarios

GAMEDE was used to simulate six farms representative of the diversity of management practices and the variety of soil and climate conditions encountered on the Island. For this exercise we selected farm 3 which may be considered the most illustrative case. For more simulations please refer to Vayssières et al. (2010). Four scenarios are distinguished.

The "*baseline scenario*" (scenario 0) represents current management practices of the farm. Ensiling of grass is the only method of forage harvest used on the farm. Livestock excreta are mainly managed as slurry, of which 70% is excreted indoors. Grasslands are fertilized mainly with large amounts of mineral fertilizers: 265 kg of mineral N and only 54 kg of manure N are annually spread per hectare of cut grassland. On-farm produced organic fertilizers are largely underused. The risk of slurry pit overflows is consequently high, particularly during the rainy season. The amount of concentrate feeds fed to cows is small (i.e., less than 10 kg FM of concentrate cow^{-1} day^{-1}) compared with other dairy systems on La Réunion. Feed self-sufficiency is consequently high (about 45%).

In the "*increased recycling of manure*" *scenario* (scenario 1), cubicles are built for lactating cows as advised by experts from the dairy industry. With cubicles, cows remain in the barns all day (and exercise areas are converted into grasslands for grass silage), rendering collection of manure more efficient. Because farm 3 is located in the rainiest part of the island, the slurry pit is covered to limit dilution and overflows. Large amounts of concentrated slurry are thus available for fertilizing grassland. The spreading of slurry on all the farm's grasslands (not currently practiced) is necessary to increase the total amount of slurry applied to limit overflows. The rates of application of slurry are increased to homogenize fertilization level among different farm plots. More manure and more concentrated manure on grasslands offer farmer 3 the opportunity to stop the use of mineral fertilizers without reducing forage production.

In the "*increased recycling of forages*" *scenario* (scenario 2), the surplus of on-farm produced grass silage resulting from increased recycling of manure (scenario 1) is used to feed animals and to reduce the amount of imported sugar cane straw for feed. Because of the silage surplus, substitution of sugar cane straw by grass silage is possible for heifers and dry cows. Moreover, farmer 3 currently does not adapt the amount of concentrate feeds fed to lactating cows according to their lactation stage. This practice was included in scenario 2.

In the "*improved herd reproductive performance*" *scenario* (scenario 3), a reduction of calving interval to 380 days is simulated, which was judged realistic by local experts. Keeping cows in barns facilitates detection of heat, given that farmers spend additional time observing their cows. Compared with current herd performance, this would correspond to a reduction of the calving interval by 18% for farm 3.

7.4.2.4 Simulation results

Figure 7.5 represents the evolution of N gaseous emissions and N losses due to slurry pit overflows over three years for scenarios 0 and 1. Table 7.2 shows the effect of scenarios 0 to 3 on farm self-sufficiency and its degree of CLI, on crop and animal productivity, and N flows. Values of Table 7.2 are means for the 2004–2006 period.

Increased recycling of organic fertilizers produced on-farm (scenario 1), largely improves fertilizer self-sufficiency (+29.7%) and abates N surplus (–22.6%). The increased use of slurry leads to a larger application of nutrients and organic matter to soils and increases the farm forage productivity by 3.9% (see Table 7.2). Better

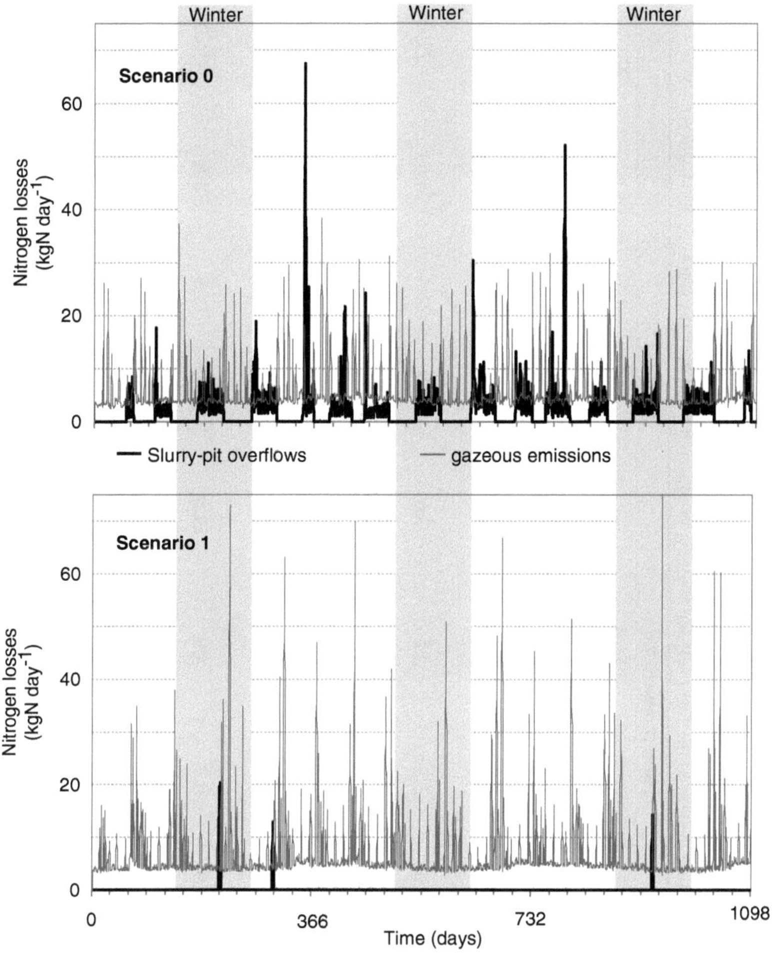

Fig. 7.5 Intra- and interannual variation of main N losses at the farm scale (farm 3, 2004–2006) simulated for two scenarios: the actual farming practices (scenario 0) and the increased recycling of manure on-farm produced (scenario 1)

management of slurry decreases N losses through overflows by 97%, but increases N gaseous emissions by 24.4%, mainly during application (see Fig. 7.5). Increased recycling of forages produced on-farm (scenario 2) increases farm feed self-sufficiency (+9.5%) and consequently contributes to N surplus abatements (−10%), because higher production of better quality forage allows reduction in imports of conserved forages. Adaptation of the amount of concentrate feeds fed to lactating cows according to their lactation stage increases the efficiency in the use of concentrate (+13.3%, see Table 7.2). Improvement of herd reproductive performance (scenario 3) reduces the time in which cows are unproductive, and therefore milk

Table 7.2 Contribution of different management in the farming system to the farm self-sufficiency, productivity and N flows (farm 3, 2004–2006)

	Farm self-sufficiency			Farm productivity			N flows			
Scenario	Degree of crop-livestock integration (dmnl)	Fertilizer self-sufficiency (dmnl)	Feed self-sufficiency (dmnl)	Forage productivity (10^3 UF ha^{-1} year^{-1})	Milk productivity (10^3 kgFM milk cow^{-1} year^{-1})	Feed conversion efficiency (dmnl)	Slurry pit overflows (kg N year^{-1})	N gaseous emissions (kg N year^{-1})	N efficiency (dmnl)	N surplus (kg N ha^{-1} year^{-1})
0. Baseline: values for actual practices	0.59	0.49	0.39	10.2	5.7	1.10	750	2050	0.19	547
1. Increased recycling of manure	13.1	29.7	0.0	3.9	2.5	2.6	–96.9	24.4	24.6	–22.6
2. Increased recycling of forages produced on-farm	3.1	–2.2	9.6	–1.6	–2.0	13.3	–1.8	–4.8	12.0	–10.0
3. 18% Improvement of herd reproductive performances	0.2	–1.8	2.3	–1.5	8.4	3.6	–1.3	–4.1	2.0	–3.4
Combination of the four scenarios	16.4	25.7	11.9	0.7	8.9	19.5	–100.0	15.5	38.6	–36.0

dmnl dimensionless, *FM* fresh matter, *UF* is the fodder unit defined by the UF/PDI feeding unit system characterizing the energy value of a considered feed

Except for the first line (scenario 0), which is expressed in absolute value, all values are percentages of variation of farm results compared to current farm performances (= scenario 0)

productivity is improved by 8.4%. Increased recycling of organic fertilizers and forages produced on-farm, combined with the improvement of herd reproductive performance, together increase the farm self-sufficiency and its degree of CLI (+16.4%). This leads to a decrease of the risk of N leaching. Slurry pit overflows and then runoff are minimized. The farm N use efficiency is finally improved by 38.6%, resulting in a N surplus below 350 kg N ha^{-1} $year^{-1}$.

7.4.2.5 Conclusions

As illustrated here for farm 3, simulation showed that changes in practices aimed at increasing CLI and farm self-sufficiency contribute to reduce the risk of nutrient leaching of La Réunion dairy farms even if increased recycling of manure enhances N gaseous losses at the same time. This modelling experiment underlines the role of indirect mitigation options, livestock feeding and reproduction in this case. Moreover additional indicators regarded in the complete analysis (Vayssières et al. 2010), also demonstrate the positive effect of increased N recycling on the farm economic efficiency.

7.5 Discussion: toward a global view of N losses and progress in world agro-ecosystems

As a consequence of the important role of food production in the alteration of the global N cycle, a number of studies have produced synthetic views of the terrestrial N cycle centered on agriculture (Bouwman et al. 2009). Smil (1999) accounted global N flows in relation to crop production for food and for feed. Van der Hoek (1998) developed a similar accounting but centered on animal production, including pastoral systems Global livestock excreta were estimed by Sheldrick et al. (2003). These studies were reused and detailed by FAO (FAO/IFA 2001; Steinfeld et al. 2006) to evaluate the global livestock sector impact on the environment. More recently, Liu et al. (2010) developed a high-resolution assessment of global N flows in cropland.

We reused the figures of these studies to build a view of the global agricultural sector according to the conceptual flows-stock model widely used to represent agro-ecosystems in both industrial (Petersen et al. 2006) and developing countries (Thorne 1998; Andrieu and Nogueira 2010; Alvarez et al. 2010). Figure 7.6 represents the simplified global N cycle for the whole agricultural sector for the mid 1990s (forestry not included). It quantifies the main N losses (uncontrolled flows) and its main N exchanges with other human activities (mainly food provision and mineral fertilizer consumption).

According to these figures, the global N losses from agro-ecosystems worldwide total 80 MT N $year^{-1}$ (i.e., 10^6 tons of N per year) which corresponds to 40 MT N-NH_3 and 3 MT N-N_2O via gaseous emissions, 17 MT N-NO_3- via leaching, and 20 MT organic N via runoff. It means that globally 15.4% of applied N in the form of mineral fertilizers and 30.4% of excreted N by domestic animals worldwide is

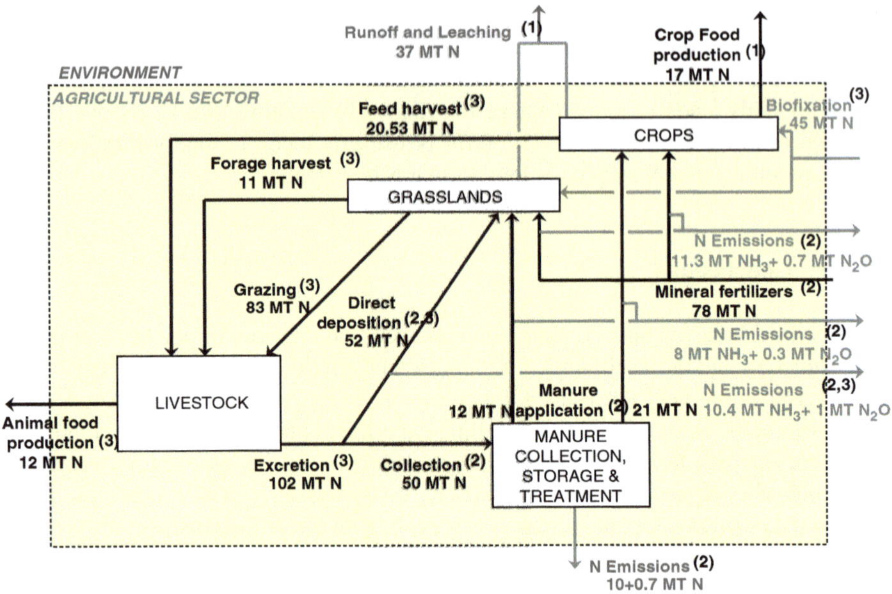

Fig. 7.6 Simplified global N cycle for the world agricultural sector for mid 1990s (flows are stated in MT N year^{-1}, i.e., 106 tons of N per year). [(1)] Smil, 1999; [(2)] Steinfeld et al. 2006; [(3)] Van der Hoek, 1998

lost in the form of harmful NH_3 and N_2O gases and not available for plants used to feed animals and humans. N losses are strongly linked to the presence of livestock activities in the cycle. Eighty five percent (85%) of total agricultural N losses, 72% of NH_3 and 74% of N_2O losses, are related to manure management. Beusen et al. (2008) estimated global NH_3 losses from agriculture to be: from housing and storage in mixed and pastoral systems 12–28%, from grazing 6% to 17%, from manure spreading in cropland and grassland 11–30, and from N fertilizer use in cropland 10–18%, and in grasslands 4–8%. Figure 7.6 shows that 44% of N available on earth (N in mineral fertilizers and N excreted by domestic animals) is lost before assimilation by plants. There is clearly much room for progress in the conservation of N in world agro-ecosystems.

In Sects. 7.3 and 7.4 we described several technical options for mitigating N losses to the environment at both process and system levels. We found no studies attempting to evaluate what would be the potential progress generated by the combination of all these technologies at a global level. Such a global evaluation is complicated by the diversity of agricultural systems and the variety of soils, climates, and socio-economical contexts encountered worldwide.

A good option to elaborate such a global view would be to use typologies proposed by FAO about world farming systems to identify representative regions and systems to be studied in priority. The analysis of these representative study cases would be extrapolated to evaluate what could be the global gains in food production and environmental aspects associated with improved N management in agro-ecosystems.

Different levels of detail exist in the proposed typologies, for instance 11 livestock systems are worldwide distinguished by Seré et al. (1995) and a typology of 72 farming systems is proposed by Dixon et al. (2001) just for the developing world. The appropriate level of detail will be defined by available data on nutrient losses and their potential reduction at both process (animal-manure-soil-plant) and system (farm) levels. While emission rates and factors regulating N losses are relatively well documented for temperate countries, references for tropical situations are clearly lacking (see Sect. 7.2). Moreover, the majority of technologies described in the literature are designed for temperate and intensive systems (see Sect. 7.2.), and they are not adapted for low-input systems encountered in tropical countries. Most farmers simply do not have access to these technologies at the moment.

Substantial progress can also be achieved by changing farming practices as discussed in Sect. 7.3. Integrated indicators proposed in Sect. 7.3—farm gate nutrient balances (Simon and Le Corre 1992; Nevens et al. 2006) and network analysis indicators (Rufino et al. 2009b)—are particularly helpful to analyze actual systems. They can be integrated in nutrient flow-stock models to simulate alternative farming systems and to explore different combinations of techniques and improved practices as illustrated in Sect. 7.3. In both tropical and temperate contexts, a better use of organic fertilizers will improve global N use efficiency and then increase the productivity of farming systems and decrease the risks to the environment from food production.

Here again, N dynamics in agro-ecosystems under temperate climates are relatively well documented but they are insufficiently studied in tropical countries. Simplified experimental methods such as the mass-based approaches may be helpful to develop a database in a more expedient and less costly manner (Robin et al. 2005; Hassouna et al. 2010). Nutrient flow models are nowadays relatively common in industrialized countries (e.g., Wastney et al. 2002 in New-Zealand; Rotz et al. 2005 in the United States; Chardon et al. 2007, in France; Oenema et al. 2007 in the Netherlands; Vayssières and Lecomte 2007 in La Réunion) but they are scarce in developing countries (e.g., Schaber 1997 in the Philippines; van Wijk et al. 2009 in Eastern Africa). While uncertainties on the size of the flows are probably large in tropical countries, some simplified models can be developed (with lower granularity), as a first step, to synthesize already locally available data and identify lacking data in the perspective of assessing the global contribution of improved nutrient management in the production of green food for the world human population.

7.6 Conclusion

As illustrated in this chapter for the nitrogen element (N), the sources of nutrient losses within the nutrient cycle in agro-ecosystems are numerous and these losses are substantial. For instance, it is estimated that 15.4% of applied N in the form of mineral fertilizers and 30.4% of excreted N by domestic animals worldwide is globally lost in the form of harmful gases (NH_3 and N_2O).

Appendix 7.1 Emission rates (ER, %) for the fertilizers considered in the GAMEDE whole dairy farm model and distinguishing the different management stages

		In barns			During storage			During treatment			On field		
Fertilizer		NH_3	N_2O	N_2	NH_3	N_2O	N_2	NH_3	N_2O	N_2	NH_3	N_2O	N_2
Bovine urine			–			–		–			Grazing: 15[17]	2[8]	6[b]
Bovine feces											Grazing: 3[17]		
Bovine slurry		30.6[a, 3]	0.2[12]	0.6[b]	3.1[3]	0.25[4]	0.75[b]	Mixing: 1.8[11]	0.15	0.45[b]	Spreading on grassland: 14[7] Spreading on sugar cane: 24.7[7]		
Bovine solid manure	Soft	29.4[a, 3]	0.5[4]	1.5[b]	3.2[3]				–		Spreading: 0.14[9]	0.9[6]	2.7[b]
								Swathing: 10[14, 15]	2[16]	9[16]			
	Strawy	40.7[a, 3]				2[8]	6[b]	First turning: 6[14, 15]	1[16]	5[16]	Spreading: 10[10]		
								Second turning: 1[14, 15]	1[16]	1[16]			
								Swathing: 0.11[14, 15]	0.01[16]	0[16]			
	Very strawy	32.8[a, 3]	2[8]	6[b]	9.2[3]			First turning: 0.04[14, 15]	0.003[16]	0.05[16]	Spreading: 6[10]		
								Second turning: 0.01[14, 15]	0.003[16]	0.08[16]			
Compost of bovine solid manure						0[2]						Spreading: 0[1]	
Pig slurry			–								Spreading: 12[13]	0.9[6]	
Poultry droppings						–			–		Spreading: 31.5[5]		
Mineral fertilizers											Spreading: 2[5]	1.25[8]	3.75[b]

Reference conditions: temperature = 10°C and rainfall = 0 mm

[a]ER relative to ammoniac N (others are relative to total N)

[b]ER estimated from the Webb (2001) equation: $ER_{N2} = 3$ a ER_{N2O}

[1] Amon et al. 1998; [2] Chadwick et al. 2002; [3] Dollé et al. 2000; [4] Dollé and Robin 2006; [5] EMEP-CORINAIR 2001; [6] Gac et al. 2006; [7] Génermont et al. 2003; [8] IPCC 1997; [9] Karisson and Salomon 2002; [10] Le Gall and Cabaret 1998; [11] Levasseur et al. 1999; [12] Marquis 2002; [13] Morvan and Leterme 2001; [14] Paillat et al. 2005; [15] Parkinson et al. 2004; [16] Robin et al. 2001; [17] Whitehead 1995

Numerous technical options are available to limit these losses and they are mainly related to animal husbandry and improved manure management techniques. But when applied together, these options interact, and then efforts done at one stage may be negatively compensated at another stage (e.g., limited emissions during collection may result in higher emissions during storage and after application). Consequently a holistic view of the nutrient cycling in the agro-ecosystems is needed, as illustrated in this paper with two farm-scale dynamic simulation models. Simulation results show that managing nutrient cycles can have a large impact on improving land productivity and food production in low-input systems, and reduce environmental impact of high-input systems. As shown with these two contrasted examples, agricultural systems encountered worldwide are diverse and current models are not generic enough to encompass this diversity. Science can contribute to generate simple and applicable tools that could help practitioners to evaluate system nutrient use efficiencies.

Tropical areas will be the field of major transitions in agriculture in the next decades, and they are paradoxically the least documented areas on biophysical and managerial processes regulating nutrient losses. Research must contribute to the understanding of nutrient cycles and opportunities to enhance green food production. The contribution of the developing agricultural sector to global environmental change is based mostly on speculations, because biophysical characterization of nutrient cycles grounded on field observations and measurements is lacking. This information is needed to evaluate with less uncertainty the global potential contribution of nutrient cycling and nutrient conservation in the production of green food for the world's human population.

References

Alvarez, S., P. Salgado, J. Vayssières, P. Tittonell, F. Bocquier, and E. Tillard. 2010. Modelling crop-livestock integration systems at a farm scale in a Highland region of Madagascar: A conceptual model. In *Proceedings of SAPT international conference*, Guadeloupe, French West Indies, 15–18 Nov 2010.

Amon, B., T. Amon, J. Boxberger, and C. Alt. 2001. Emissions of NH_3, N_2O, and CH_4 from dairy cows housed in a farmyard manure tying stall (Housing, Manure Storage, Manure Spreading). *Nutrient Cycling in Agroecosystems* 60: 103–113.

Amon, B., T. Amon, J. Boxberger, and A. Pöllinger. 1998. Emissions of NH3, N2O and CH4 from a tying stall for milking cows, during storage and farmyard manure and after spreading. In *Proceedings of 8th international conference on "Management strategies for organic waste in agriculture"*, Rennes, France, 26–29 May 19981998.

Amon, B., G. Moitzi, M. Schimpl, V. Kryvoruchko, and C. Wagner-Alt. 2002. Methane, nitrous oxide and ammonia emissions from management of liquid manures, Final Report 2002. On behalf of "Federal Ministry of Agriculture, Forestry, Environmental and Water Management" and "Federal Ministry of Education, Science and Culture".

Andrieu, N., and D.M. Nogueira. 2010. Modeling biomass flows at the farm level: A discussion support tool for farmers. *Agronomy for Sustainable Development* 30: 505–513.

Baggs, E.M., and L. Philippot. 2010. Microbial terrestrial pathways to N_2O. In *Nitrous oxide and climate change*, ed. K.A. Smith. Abingdon: Earthscan.

Beusen, A.H.W., A.F. Bouwman, P.S.C. Heuberger, G. Van Drecht, and K.W. Van Der Hoek. 2008. Bottom-up uncertainty estimates of global ammonia emissions from global agricultural production systems. *Atmospheric Environment* 42: 6067–6077.

Børsting, C.F., T. Kristensen, L. Misciattelli, T. Hvelplund, and M.R. Weisbjerg. 2003. Reducing nitrogen surplus from dairy farms: Effects of feeding and management. *Livestock Production Science* 83: 165–178.

Bouwman, A.F., and L.J.M. Boumans. 2002a. Emmisions of N_2O and NO from fertilised fields: Summary of available measurement data. *Global Biogeochemical Cycles* 16(4): 6.

Bouwman, A.F., and L.J.M. Boumans. 2002b. Modeling global annual N_2O and NO emissions from fertilised soils. *Global Biogeochemical Cycles* 16(4): 28.

Bouwman, A.F., A.H.W. Beusen, and G. Billen. 2009. Human alteration of the global nitrogen and phosphorus soil balances for the period 1970–2050. *Global Biogeochem Cycles* 23(GB0A04).

Bouwman, A.F., D.S. Lee, W.A.H. Asman, F.J. Dentener, K.W. Van Der Hoek, and J.G.J. Olivier. 1997. A global high-resolution emission inventory for ammonia. *Global Biogeochemical Cycles* 11(4): 561–587.

Breman, H., J.J.R. Groot, and H. Van Keulen. 2001. Resource limitations in Sahelian agriculture. *Global Biogeochemical Cycles* 11: 59–68.

Brouwer, J., and J.M. Powell. 1998. Increasing nutrient use efficiency in West African agriculture: The impact of micro-topography on nutrient leaching from cattle and sheep manure. *Agriculture, Ecosystems and Environment* 71: 229–239.

Burton, C., V. Jaouen, and J. Martinez, 2007. Traitement des effluents d'élevage des petites et moyennes exploitations. Guide technique à l'usage des concepteurs, bureaux d'études et exploitants. QUAE, Versailles.

Bussink, D.W., and J. Oenema. 1998. Ammonia volatilization from dairy farming systems in temperate areas: A review. *Nutrient Cycling in Agroecosystems* 51: 19–33.

Carpenter, S.R., N.F. Caraco, D.L. Correll, R.W. Howarth, A.N. Sharpley, and V.H. Smith. 1998. Nonpoint pollution of surface waters with phosphorus and nitrogen. *Ecological Applications* 8: 559–568.

Cassman, K.G., A. Dobermann, and D.T. Walters. 2002. Agroecosystems, nitrogen-use efficiency, and nitrogen management. *Ambio* 31(2): 132–140.

Chadwick, D.R., R. Matthews, R.J. Nicholson, B.J. Chambers, and L.O. Boyles. 2002. Management practices to reduce ammonia emissions from pig and cattle manure stores. In *Ramiran (2002). Proceedings of 10th international conference on "Recycling of agricultural, municipal and industrial residues in agriculture*," Strbskè Pleso, Slovakia, 14–18 May 2002.

Chardon, X., C. Rigolot, C. Baratte, A. Le Gall, S. Espagnol, R. Martin-Clouaire, J.-P. Rellier, C. Raison, J.-C. Poupa, P. Faverdin. 2007. MELODIE: a whole-farm model to study the dynamics of nutrients in integrated dairy and pig farms. In MODSIM 2007, *Proceedings of the International Congress on Modelling and Simulation: "land, water and environmental management: integrated systems for sustainability*", Christchurch, New Zealand, 10–13 Dec 2007.

Chaumet, J.M., F. Delpeuch, B. Dorin, G. Ghersi, B. Hubert, T. Le Cotty, S. Paillard, M. Petit, J.L. Rastoin, T. Ronzon, and S. Treyer. 2009. Agrimonde: Agricultures et alimentations du monde en 2050: scénarios et défis pour un développement durable. CIRAD/INRA.

Chen, G.F., L.Y. Cao, G.W. Wang, B.C. Wan, D.Y. Liu, and S.S. Wang. 2007. Application of a spatial fuzzy clustering algorithm in precision fertilisation. *New Zealand Journal of Agricultural Research* 50(5): 1249–1254.

Chikowo, R., P. Mapfumo, P. Nyamugafata, and K.E. Giller. 2004. Mineral N dynamics, leaching and nitrous oxide losses under maize following two years improved fallows on a sandy loam soil in Zimbabwe. *Plant and Soil* 259: 315–330.

Cros, M.J., M. Duru, F. Garcia and R. Martin-Clouaire. 2003. A biophysical dairy farm model to evaluate rotational grazing management strategies. *Agronomy* 23: 105–122.

Clark, M.S., W.R. Horwath, C. Shennan, K.M. Scow, W.T. Lantni, and H. Ferris. 1999. Nitrogen, weeds and water as yield-limiting factors in conventional, low-input, and organic tomato systems. *Agriculture, Ecosystems and Environment* 73: 257–270.

Clemens, J., and H.J. Ahlgrimm. 2001. Greenhouse gases from animal husbandry: Mitigation options. *Nutrient Cycling in Agroecosystems* 60: 287–300.

De Klein, C.A., M.R. Eckard, and T.J. van der Weerden. 2010. N2O emissions from the nitrogen cycle in livestock agriculture: Estimation and mitigation. In *Nitrous oxide and climate change*, ed. K.A. Smith. Abingdon: Earthsca.

Delagarde, R., P. Faverdin, C. Baratte, J.L. Peyraud. 2004. Prévoir l'ingestion et la production des vaches laitières : GrazeIn, un modèle pour raisonner l'alimentation au pâturage. In *Proceedings of the 11th International Symposium Rencontres Recherches Ruminants*, Paris, 8–9 Dec. 2004.

Di, H.J., and K.C. Cameron. 2002. Nitrate leaching in temperate agroecosystems: Sources, factors and mitigating strategies. *Nutrient Cycling in Agroecosystems* 64: 237–256.

Di, H.J., and K. Cameron. 2003. Mitigation of nitrous oxide emissions in spray-irrigated grazed grassland by treating the soil with dicyandiamide, a nitrification inhibitor. *Soil Use and Management* 19: 284–290.

Dixon, J., A. Gulliver, and D. Gibbon. 2001. *Farming systems and poverty: Improving farmers' livelihoods in a changing world*. Rome: FAO/World Bank.

Dollé, J.B., J. Capdeville, J. Martinez, and P. Peu. 2000. *Emissions d'ammoniac en bâtiments et au cours du stockage des déjections en élevage bovin*. Paris: Institut de l'Elevage.

Dollé, J.B., and P. Robin. 2006. Emissions de gaz à effet de serre en bâtiment d'élevage bovin. In *AFPF 2006, seminar on "Prairies, élevage, consommation d'énergie et gaz à effets de serres"*, Paris, 27–28 March 2006.

Edwards, C.A., T.L. Grove, R.R. Harwood, and C.J. Pierce Colfer. 1993. The role of agroecology and integrated farming systems in agricultural sustainability. *Agriculture, Ecosystems and Environment* 46: 99–121.

EMEP-CORINAIR. 2001. *Emission inventory guidebook*. Copenhagen: European Environment Agency.

FAO/IFA. 2001. *Global estimates of gaseous emissions of NH3, NO and N2O from agricultural land*. Rome: FAO.

Fath, B.D., and B.C. Patten. 1999. Review of the foundations of network environ analysis. *Ecosystem* 2: 167–179.

Faverdin, P., R. Delagarde, L. Delaby, F. Meschy. 2007. Réactualisation des équations du livre rouge: alimentation des vaches laitières. Quae, Versailles, pp. 23–55.

Fox D.G., L.O. Tedeschi, T.P. Tylutki, J.B. Russell, M.E. Van Amburgh, L.E. Chase, A.N. Pell, T.R. Overton. 2004. The Cornell Net Carbohydrate and Protein System model for evaluating herd nutrition and nutrient excretion. Animal Feed Science and Technology 112, 29–78.

Frank, B., M. Persson, and G. Gustafsson. 2002. Feeding dairy cows for decreased ammonia emission. *Livestock Production Science* 76: 171–179.

Gac, A., F. Béline, and T. Bioteau. 2006. Flux de gaz à effet de serre (CH4, N2O) et d'ammoniac (NH3) liés à la gestion des déjections animales: Synthèse bibliographique et élaboration d'une base de données. [Greenhouse gases and ammonia emissions from livestock wastes management: Bibliographic review and database development. Technical report.] CEMAGREF, Paris.

Galloway, J.N., J.D. Aber, J.W. Erisman, S.P. Seitzinger, R.W. Howarth, E.B. Cowling, and B.J. Cosby. 2003. The N cascade. *BioScience* 53: 341–356.

Galloway, J.N., F.J. Dentener, D.G. Capone, E.W. Boyer, R.W. Howarth, S.P. Seitzinger, G.P. Asner, C.C. Cleveland, P.A. Green, E.A. Holland, D.M. Karl, A.F. Michaels, J.H. Porter, A.R. Townsend, and C.J. Vöosmarty. 2004. Nitrogen cycles: Past, present, and future. *Biogeochemistry* 70: 153–226.

Galloway, J.N., W.H. Schlesinger, H. Levy II, A. Michaels, and J.L. Schnoor. 1995. Nitrogen fixation: Anthropogenic enhancement—Environmental response. *Global Biogeochemical Cycles* 9: 235–252.

Gangbazo, G., A.R. Pesant, D. Cluis, D. Couillard, and G.M. Barnett. 1995. Winter and early spring losses of nitrogen following late fall application of hog manure. *Canadian Agricultural Engineering* 37: 73–79.

Génermont, S., T. Morvan, J.M. Paillat, D. Flura, and H. Saint Macary. 2003. Volatilisation d'ammoniac après épandage de lisier en conditions tropicales. Cas des prairies d'altitude et de la canne à sucre à l'île de la Réunion. In *Modélisation des flux de biomasse et des transferts de fertilité—cas de la gestion des effluents d'élevage a` l'île de la Réunion*, ed. F. Guerrin and J.M. Paillat. Montpellier: CIRAD.

Giller, K.E. 2001. *Nitrogen fixation in tropical cropping systems*. Wallingford: CAB International.
Griffon, M. 2006. *Nourrir la planète—Pour une révolution doublement verte*. Paris: Odile Jacob.
Harper, L.A., R.R. Sharpe, and T.B. Parkin. 2000. Gaseous emissions from anaerobic swine lagoons: Ammonia, nitrous oxide, and dinitrogen gas. *Journal of Environmental Quality* 29: 1356–1365.
Hartung, J., and V.R. Phillips. 1994. Control of gaseous emissions from livestock buildings and manure stores. *Journal of Agricultural Engineering Research* 57: 173–189.
Hassouna, M., P. Robin, A. Brachet, J.M. Paillat, J.B. Dollé, and P. Faverdin. 2010. Development and validation of a simplified method to quantify gaseous emissions from cattle buildings. In *Proceedings of the XVIIth world congress of the international commission of agricultural and biosystems engineering (CIGR)*, Québec City, Canada 13–17 June.
Herrero, M., P.K. Thornton, A.M. Notenbaert, S. Wood, S. Msangi, H.A. Freeman, D. Bossio, J. Dixon, M. Peters, J. van de Steeg, J. Lynam, P.P. Rao, S. Macmillan, B. Gerard, J. McDermott, C. Sere, and M. Rosegrant. 2010. Smart investments in sustainable food production: Revisiting mixed crop-livestock systems. *Science* 327: 822–825.
Hill, M.J. 1991. *Nitrates and nitrites in food and water*. Chichester: Ellis Horwood.
Hooda, P.S., A.C. Edwards, H.A. Anderson, and A. Miller. 2000. A review of water quality concerns in livestock farming areas. *Science of the Total Environment* 250: 143–167.
IPCC. 1997. Revised 1996 IPCC guidelines for national greenhouse gas inventories—Reference manual (volume 3), Geneva. http://www.ipcc-nggip.iges.or.jp/public/gl/invs6.html.
IPCC. 2006. 2006 IPCC Guidelines for national greenhouse gas inventories—Agriculture, Forestry and Other Land Use (volume 4), Geneva. http://www.ipcc-nggip.iges.or.jp/public/2006gl/vol4.html.
Jarrige, R. 1989. *Ruminant nutrition: Recommended allowances and feed tables*. Versailles: INRA.
Jongbloed, A.W., N.P. Lenis, and Z. Mroz. 1997. Impact of nutrition on reduction of environmental pollution by pigs: An overview of recent research. *Veterinary Quarterly* 19: 130–134.
Kamukondiwa, W., and L. Bergström. 1994. Nitrate leaching in field lysimeters at an agricultural site in Zimbabwe. *Soil Use and Management* 10: 118–124.
Karisson, S., and E. Salomon. 2002. Deep litter manure to spring cereals—Manure properties and ammonia emissions. In *Ramiran (2002) 10th international conference on "Recycling of agricultural, municipal and industrial residues in agriculture."* Strbskè Pleso, Slovakia, 14–18 May 2002.
Lague, C., T.A. Fonstad, A. Marquis, S.P. Lemay, S. Godbout, and R. Joncas. 2004. *Greenhouse gas emissions from swine operations in Québec and Saskatchewan: Benchmark assessments*. Climate Change Funding Initiative in Agriculture (CCFIA), Canadian Agricultural Research Council, Ottawa.
Lavelle, P., D. Bignell, M. Austen, P. Giller, G. Brown, et al. 2004. Vulnerability of ecosystem services at different scales: Role of biodiversity and implications for management. In *Sustaining biodiversity and functioning in soils and sediments*, ed. D.H. Wall. Washington, DC: Island Press.
Le Gall, A., and M.M. Cabaret. 1998. *Mise au point de systèmes laitiers productifs et respectueux de l'environnement—Compte rendu de l'expérimentation conduite à la station Crécom de 1995 à 1998*. Paris: Institut de l'élevage.
Leontief, W.W. 1951. *The structure of American economy, 1919–1939: An empirical application of equilibrium analysis*. New York: Oxford University Press.
Leteinturier, B., R.Oger, D. Buffet. 2004. Rapport technique sur le nouveau module de croissance prairiale. Centre de Recherches Agronomiques de Gembloux, Belgique, pp. 37.
Levasseur, P., C. Boyard, J.C. Vaudelet, and P. Rousseau. 1999. Evolution de la valeur fertilisante du lisier de porcs au cours de la vidange de la fosse de stockage, influence du brassage. In *Proceedings of the 31th Journées Recherche Porcine en France, Institut Technique du Porc*, INRA, 2–4 Feb 1999.
Liu, J., L. You, M. Amini, M. Obersteiner, M. Herrero, A.J.B. Zehnder, and H. Yang. 2010. A high-resolution assessment on global nitrogen flows in cropland. *PNAS* 107(17): 8035–8040.
Marini, J.C., and M.E. Van Amburgh. 2003. Nitrogen metabolism and recycling in Holstein heifers. *Journal of Animal Science* 81: 545–552.
Marquis, A. 2002. Emissions de gaz à effet de serre par les animaux aux bâtiments. In *Proceedings of the 65th Congress de l'Ordre des Agronomes du Québec*.
Millennium Ecosystem Assessment. 2005a. *Ecosystems and human well-being Vol. I: Current state and trends*. Washington, DC: Island Press.

Millennium Ecosystem Assessment. 2005b. *Ecosystems and human well-being Vol. II: scenarios*. Washington, DC: Island Press.

Moller, H.B., S.G. Sommer, and B.H. Anderson. 2000. Nitrogen mass balance in deep litter during the pig fattening cycle and during composting. *The Journal of Agricultural Science* 137: 235–250.

Martiné, J.F. 2003. Modélisation de la production potentielle de la canne à sucre en zone tropicale, sous conditions thermiques et hydriques contrastées. Application du modèle. Ph.D. Thesis, INA-PG, Paris, 116 p.

Monteny, G.J., A. Bannink, and D. Chadwick. 2006. Greenhouse gas abatement strategies for animal husbandry. *Agriculture, Ecosystems and Environment* 112: 163–170.

Morvan, T., and P. Leterme. 2001. Vers une prévision opérationnelle des flux d'azote résultant de l'épandage de lisier: Paramétrage d'un modèle dynamique de simulation des transformations de l'azote et des lisiers. *Ingénieries* 26: 17–26.

Mosier, A., R. Wassmann, L. Verchot, J. King, and C. Palm. 2004. Methane and nitrogen oxide fluxes in tropical agricultural soils: Sources, sinks and mechanisms. *Environment, Development and Sustainability* 6: 11–49.

Mosley, J.C., P.S. Cook, A.J. Griffis, and J. O'Laughlin. 1997. Guidelines for managing cattle grazing in riparian areas to protect water quality: Review of research and best management practices policy. Report No. 15 Policy Analyses Group, College of Forestry, Wildlife, and Range Sciences, University of Idaho, Moscow, ID.

Nevens, F., I. Verbruggen, D. Reheul, and G. Hofman. 2006. Farm gate nitrogen surpluses and nitrogen use efficiency of specialized dairy farms in Flanders: Evolution and future goals. *Agricultural Systems* 88: 142–155.

Nicks, B., M. Laitat, M. Vandenheede, A. Desiron, C. Verhaege, and B. Canart. 2003. Emissions of ammonia, nitrous oxide, methane, carbon dioxide, and water vapor in the raising of weaned pigs on straw-based and sawdust-based deep litters. *Journal of Applied Animal Research* 52: 299–308.

Nyamangara, J., L.F. Bergström, M.I. Piha, and K.E. Giller. 2003. Fertilizer use efficiency and nitrate leaching in a tropical sandy soil. *Journal of Environmental Quality* 32: 599–606.

Nyamangara, J., M.I. Piha, and H. Kirchmann. 1999. Interactions of aerobically decomposed manure and nitrogen fertiliser applied to soil. *Nutrient Cycling in Agroecosystems* 54: 183–188.

Oenema, O., D. Oudendag, and G.L. Velthof. 2007. Nutrient losses from manure management in the European Union. *Livestock Science* 112: 261–272.

Oenema, O., H.P. Witzke, Z. Klimont, J.P. Lesschen, and G.L. Velthof. 2009. Integrated assessment of promising measures to decrease nitrogen losses from agriculture in EU-27. *Agriculture, Ecosystems and Environment* 133: 280–288.

Paillat, J.M., P. Robin, M. Hassouna, and P. Leterme. 2005. Predicting ammonia and carbon dioxide emissions from carbon and nitrogen biodegradability during animal waste composting. *Atmospheric Environment* 39: 6833–6842.

Parkinson, R., P. Gibbs, S. Burchett, and T. Misselbrook. 2004. Effect of turning regime and seasonal weather conditions on nitrogen and phosphorus losses during aerobic composting of cattle manure. *Bioresource Technology* 91: 171–178.

Petersen, S.O., A.M. Lind, and S.G. Sommer. 1998. Nitrogen and organic matter losses during storage of cattle and pig manure. *The Journal of Agricultural Science* 130: 69–79.

Petersen, S.O., S.G. Sommer, M.P. Bernal, C. Burton, R. Böhm, J. Dach, J.Y. Dourmad, C. Juhasz, A. Leip, R. Mihelic, T. Misselbrook, J. Martinez, F. Nicholson, H.D. Poulsen, G. Provolo, P. Sorensen, and A. Weiske. 2006. Recycling of manure and organic wastes—A whole-farm perspective. *Proceedings (Vol. I) of the 12th RAMIRAN international conference "Technology for recycling of manure and organic residues in a whole-farm perspective."* Aarhus, Denmark.

Powell, J.M., G.A. Broderick, J.H. Grabber, and U.C. Hymes-Fecht. 2010. Technical note: Effects of forage protein-binding polyphenols on chemistry of dairy excreta. *Journal of Dairy Science* 92: 1765–1769.

Predotova M., W.-A. Bischoff, and A. Buerkert. 2010a. Mineral-nitrogen and phosphorus leaching from vegetable gardens in Niamey, Niger. *Journal of Plant Nutrition and Soil Science* 174(1): 47–55.
Predotova, M., J. Gebauer, R.V.C. Diogo, E. Schlecht, and A. Buerkert. 2010b. Emissions of ammonia, nitrous oxide and carbon dioxide from urban gardens in Niamey, Niger. *Field Crops Research* 115: 1–8.
Predotova, M.E., A. Schlecht, and A. Buerkert. 2010c. Nitrogen and carbon losses from dung storage in urban gardens of Niamey, Niger. *Nutrient Cycling in Agroecosystems* 87: 103–114.
Ramisch, J.J. 2005. Inequality, agro-pastoral exchanges, and soil fertility gradients in southern Mali. *Agriculture, Ecosystems and Environment* 105: 353–372.
Raun, W.R., and G.V. Johnson. 1999. Improving nitrogen use efficiency for cereal production. *Agronomy Journal* 91: 357–363.
Robin, P., M. Hassouna, and J.M. Paillat. 2005. Multi-element combined methods (MECM) increase the reliability of emission factor measurement. In *Proceedings of the International Workshop on Pork production "Porcherie verte."* Paris, 25–27 May 2005.
Robin, P., J.M. Paillat, and S. Hacala. 2001. *Compostage des fumiers de litière accumulée de bovins. Estimation et caractérisation des pertes d'azote par volatilisation*. INRA, Versailles: Institut de l'élevage.
Rotz, C.A. 2004. Management to reduce nitrogen losses in animal production. *Journal of Animal Science* 82: 119–137.
Rotz, C.A., F. Taube, M.P. Russelle, J. Oenema, M.A. Sanderson, and M. Wachendorf. 2005. Whole-farm perspectives of nutrient flows in grassland agriculture. *Crop Science* 45(6): 2139–2159.
Rufino, M.C., P. Tittonell, M.T. van Wijk, A. Castellanos-Navarrete, R.J. Delve, N. de Ridder, and K.E. Giller. 2007. Manure as a key resource within smallholder farming systems: Analysing farm-scale nutrient cycling efficiencies with the NUANCES framework. *Livestock Science* 112: 273–287.
Rufino, M.C., M. Herrero, M.T. van Wijk, L. Hemerik, N. de Ridder, and K.E. Giller. 2009a. Lifetime productivity of dairy cows in smallholder farming systems of the highlands of Central Kenya. *Animal* 3: 1044–1056.
Rufino, M.C., P. Tittonell, P. Reidsma, S. López-Ridaura, H. Hengsdijk, K.E. Giller, and A. Verhagen. 2009b. Network analysis of N flows and food self-sufficiency: A comparative study of crop-livestock systems of the highlands of East and southern Africa. *Nutrient Cycling in Agroecosystems* 85: 169–186.
Rufino, M.C., J. Dury, P. Tittonell, M.T. Van Wijk, M. Herrero, S. Zingore, P. Mapfumo, and K.E. Giller. 2010. Competing use of organic resources, village-level interactions between farm types and climate variability in a communal area of NE Zimbabwe. *Agricultural Systems* 104: 175–190.
Sanchez, P.A. 2002. Soil fertility and hunger in Africa. *Science* 295: 2019–2020.
Schaber, J. 1997. *FARMSIM: A dynamic model for the simulation of yields, nutrient cycling and resource flows on Philippine small-scale farming systems*. Germany: Institut für Umweltsystemforschung, University of Osnabrück.
Schröder, J.J., A.G. Jansen, and G.J. Hilhorst. 2005. Long-term nutrient supply from cattle slurry. *Soil Use and Management* 21: 196–204.
Scrimgeour, G.J., and S. Kendall. 2002. Consequences of livestock grazing on water quality and benthic algal biomass in a Canadian natural grassland plateau. *Environmental Management* 29: 824–844.
Seré, C., H. Steinfeld, and J. Groenewold. 1995. World livestock systems. Current status, issues and trends. Animal production and health. Paper No. 127. FAO, Rome.
Sheldrick, W., J. Keith Syers, and J. Lingard. 2003. Contribution of livestock excreta to nutrient balances. *Nutrient Cycling in Agroecosystems* 66: 119–131.
Simon, J.C., and L. Le Corre. 1992. Le bilan apparent de l'azote à l'échelle de l'exploitation agricole: Méthodologie, exemples de résultats. *Fourrages* 129: 79–94.
Smaling, E.M.A., S.M. Nandwa, and B.M. Janssen. 1997. Soil fertility in Africa is at stake. In *Replenishing soil fertility in Africa*, eds. Buresh, R.J., P.A. Sanchez, and F. Calhoun. Madison, WI: American Society of Agronomy–Soil Science Society of America.
Smil, V. 1999. Nitrogen in crop production: An account of global flows. *Global Biogeochemical Cycles* 13(2): 647–662.

Smith, K., T. Cumby, J. Lapworth, T. Misselbrook, and A. Williams. 2007. Natural crusting of slurry storage as an abatement measure for ammonia emissions on dairy farms. *Biosystems Engineering* 97: 464–471.

Sommer, S.G., B.T. Christensen, N.E. Nielsen, and J.K. Schjørring. 1993. Ammonia volatilization during storage of cattle and pig slurry: Effect of surface cover. *The Journal of Agricultural Science* 121: 63–71.

Steinfeld, H., P. Gerber, T. Wassenaar, V. Castel, M. Rosales, and C. de Haan. 2006. *Livestock's long shadow*. FAO, Rome: Environmental issues and options.

Stewart, W.M., D.W. Dibb, A.E. Johnston, and T.J. Smyth. 2005. The contribution of commercial fertilizer nutrients to food production. *Agronomy Journal* 97: 1–6.

Stout, W.L., S.A. Fales, L.D. Muller, R.R. Schnabel, W.E. Priddy, and G.F. Elwinger. 1997. Nitrate leaching from cattle urine and feces in northeast USA. *Soil Science and Society of America Journal* 61: 1787–1794.

Sutton, A., T. Applegate, S. Hankins, B. Hill, G. Allee, W. Greene, R. Kohn, D. Meyer, W. Powers, and T. van Kempen. 2001. Manipulation of animal diets to affect manure production, composition and odors: State of the science. In *Animal agriculture and the environment*, ed. J.M. Rice, D.F. Caldwell, and F.J. Humenik. St. Joseph, MI: ASABE.

Thorne, P.J. 1998. *Crop-livestock interactions—A review of opportunities for developing integrated models*. Nairobi, Kenya: ILRI.

Tittonell, P., B. Vanlauwe, P.A. Leffelaar, E.C. Rowe, and K.E. Giller. 2005a. Exploring diversity in soil fertility management of smallholder farms in western Kenya: I. Heterogeneity at region and farm scale. *Agriculture, Ecosystems and Environment* 110: 149–165.

Tittonell, P., B. Vanlauwe, P.A. Leffelaar, K.D. Shepherd, and K.E. Giller. 2005b. Exploring diversity in soil fertility management of smallholder farms in western Kenya: II. Within-farm variability in resource allocation, nutrient flows and soil fertility status. *Agriculture, Ecosystems and Environment* 110: 166–184.

Tittonell, P., M. Corbeels, M.T. van Wijk, B. Vanlauwe, and K.E. Giller. 2008. Combining organic and mineral fertilizers for integrated soil fertility management in smallholder farming systems of Kenya—Explorations using the crop-soil model FIELD. *Agronomy Journal* 100: 1511–1526.

Tittonell, P., M.C. Rufino, B.H. Janssen, and K.E. Giller. 2010a. Carbon and nutrient losses during manure storage under traditional and improved practices in smallholder crop-livestock systems-evidence from Kenya. *Plant and Soil* 328: 253–269.

Tittonell, P., M. Corbeels, M.T. van Wijk, and K.E. Giller. 2010b. FIELD—A summary simulation model of the soil-crop system to analyse long-term resource interactions and use efficiencies at farm scale. *European Journal of Agronomy* 32: 10–21.

Tomlinson, A.P., W.J. Powers, H.H. Van Horn, R.A. Nordstedt, and C.J. Wilcox. 1996. Dietary protein effects on nitrogen excretion and manure characteristics of lactating cows. *Transactions of ASAE* 39: 1441–1448.

United Nations Organization. 2006. *World population prospects: The 2006 revision*. New York: UNO.

Vallis, I., D.C.I. Peake, R.K. Jones, and R.L. McCown. 1985. Fate of urea nitrogen from cattle urine in a pasture-crop sequence in a seasonally dry tropical environment. *Australian Journal of Agricultural And Resource* 36: 809–817.

Van der Hoek, K.W. 1998. Nitrogen efficiency in global animal production. *Environmental Pollution* 102: 127–132.

van Wijk, M.T., P. Tittonell, M.C. Rufino, M. Herrero, C. Pacini, N.D. Ridder, and K.E. Giller. 2009. Identifying key entry-points for strategic management of smallholder farming systems in sub-Saharan Africa using the dynamic farm-scale simulation model NUANCES-FARMSIM. *Agricultural Systems* 102: 89–101.

Vanlauwe, B., J. Kihara, P. Chivenge, P. Pypers, R. Coe, and J. Six. 2010. Agronomic use efficiency of N fertilizer in maize-based systems in sub-Saharan Africa within the context of integrated soil fertility management. *Plant and Soil* 339(1–2): 35–50.

Vanotti, M.B., A.A. Szögi, and C.A. Vives. 2006. Greenhouse gas emission reductions and carbon credits from implementation of aerobic manure treatment systems in swine farms.

In *Proceedings of the workshop on agricultural air quality: State of the science*. Potomac, MD, 5–8 June 2006.

Vayssières, J., F. Guerrin, J.M. Paillat, and P. Lecomte. 2009. GAMEDE: A global activity model for evaluating the sustainability of dairy enterprises. Part I—Whole-farm dynamic model. *Agricultural Systems* 101: 128–138.

Vayssières, J., and P. Lecomte. 2007. Modéliser les pratiques décisionnelles et les flux d'azote à l'échelle globale de l'exploitation : cas de l'élevage bovin laitier en contexte tropical insulaire. In *Proceedings of the International Symposium Rencontres Recherches Ruminants*, Paris, 5–6 Dec 2007.

Vayssières, J., Ph. Lecomte, F. Guerrin, F. Bocquier, and C. Verdet. 2006. Explaining the diversity of environmental performances according to a typology of farming practices combinations: The case of the dairy cattle breeding in Reunion Island. In *Ramiran 2006, Proceedings of the 12th international conference on "Technology for recycling of manure and organic residues in a whole-farm perspective."* Aarhus, Denmark, 11–13 Sept 2006.

Vayssières, J., M. Vigne, V. Alary, and P. Lecomte. 2010. Integrated participatory modelling of actual farms to support policy making on sustainable intensification. *Agricultural Systems* 104(2): 146–161.

Vitousek, P.M., H.A. Mooney, J. Lubchenco, and J.M. Melillo. 1997. Human domination of earth's ecosystems. *Science* 277: 494–499.

Vitousek, P.M., R. Naylor, T. Crews, M.B. David, L.E. Drinkwater, E. Holland, P.J. Johnes, J. Katzenberger, L.A. Martinelli, P.A. Matson, G. Nziguheba, D. Ojima, C.A. Palm, G.P. Robertson, P.A. Sanchez, A.R. Townsend, and F.S. Zhang. 2009. Nutrient imbalances in agricultural development. *Science* 324: 1519–1520.

Wastney, M.E., C.C. Palliser, J.A. Lile, K.A. Macdonald, J.W. Penno, and K.P. Bright. 2002. A whole-farm model applied to a dairy system. In *Proceedings of the New Zealand Society of Animal Production*, vol 62. NZ Society of Animal Production, 120–123.

Webb, J. 2001. Estimating the potential for ammonia emissions from livestock excreta and manures. *Environmental Pollution* 111: 395–406.

Westerman, P.W., and R.H. Zhang. 1997. Aeration of livestock manure slurry and lagoon liquid for odor control: A review. *Applied Engineering in Agriculture* 13(2): 245–249.

Whitehead, D.C. 1995. *Grassland nitrogen*. Wallingford: CAB International.

Wirsenius, S., C. Azar, and G. Berndes. 2010. How much land is needed for global food production under scenarios of dietary changes and livestock productivity increases in 2030? *Agricultural Systems*. doi:doi:10.1016/j.agsy.2010.07.005.

Wood, S., J. Henao, and M. Rosegrant. 2004. The role of nitrogen in sustaining food production and estimating future nitrogen fertilizer needs to meet food demand. In *Agriculture and the nitrogen cycle: Assessing the impacts of fertilizer use on food production and the environment*, ed. R. Arvin, J. Mosier, K. Syers, and J.R. Freney. Washington, DC: Island Press.

Zingore, S., H.K. Murwira, R.J. Delve, and K.E. Giller. 2007. Influence of nutrient management strategies on variability of soil fertility, crop yields and nutrient balances on smallholder farms in Zimbabwe. *Agriculture, Ecosystems and Environment* 119: 112–126.

Chapter 8
Environmental performance of organic farming

Christian Schader, Matthias Stolze, and Andreas Gattinger

8.1 Organic farming as a green technology

The environmental impacts of human activities have been increasing with growing populations and industrialization (Meadows et al. 1972). This led to the development of green technologies, which stand out for their positive environmental impacts or the avoided negative environmental impacts. Green technologies feature concepts like sustainability, "cradle-to-cradle" design, input reduction, innovation, and viability.

Organic farming suits this notion of a green technology. Organic farming has emerged in the course of the twentieth century as an environmentally friendly alternative to conventional agriculture (Niggli 2007; Vogt 2007). In the course of rapid structural change, conventional agriculture became increasingly capital intensive, input dependent, and specialized. Bound to strict rules regarding nutrient cycling and input avoidance, organic farming did not follow this path, which led to a significant gap between the two farming systems over time. Lampkin (1990) stresses, however, that several misconceptions exist regarding organic farming: commonly, organic farming is conceived as farming in the pre-1939 style or a production method that does not use chemicals, substitutes mineral fertilizers with organic fertilizers, and bans pesticides. However, the role of agro-ecosystem management and other progressive management practices is often ignored in such conceptions.

C. Schader (✉) • M. Stolze • A. Gattinger
Research Institute of Organic Agriculture (FiBL), Frick, Switzerland
e-mail: christian.schader@fibl.org; matthias.stolze@fibl.org; andreas.gattinger@fibl.org

J.I. Boye and Y. Arcand (eds.), *Green Technologies in Food Production and Processing*, Food Engineering Series, DOI 10.1007/978-1-4614-1587-9_8,

The international umbrella organization of organic agriculture, the International Federation of Organic Agriculture Movements (IFOAM), defines organic agriculture as:

> […] a production system that sustains the health of soils, ecosystems and people. It relies on ecological processes, biodiversity and cycles adapted to local conditions, rather than the use of inputs with adverse effects. Organic agriculture combines tradition, innovation and science to benefit the shared environment and promote fair relationships and a good quality of life for all involved (IFOAM 2009).

This definition highlights the "largely self-sustaining" (Köpke et al. 1997) nature of organic farming as a farming system. Thus, organic farming seems to meet the requirements for being a green technology. According to the above definition, organic farming is sustainable because it does not jeopardize needs of future generations, following the concepts of "cradle-to-cradle" design, input reduction, and sustaining soil fertility. Its worldwide rapid growth of 1.5 million hectares (ha) between 2006 and 2009 (Willer and Kilcher 2009) demonstrates its economic viability and its power to overcome self-imposed system restrictions by innovation.

Certification of organic farms is an important means to both establish credibility for consumers and guarantee a higher willingness-to-pay for organic produce compared to conventional products (Krystallis and Chryssohoidis 2005). Thus, certification generates additional farm income in order to make organic agriculture economically viable. Detailed standards, principles, and aims are set out by IFOAM in the periodically revised "IFOAM Norms." These contain the "IFOAM Basic Standards" and the "IFOAM Accreditation Criteria" as an international guideline for national standards in organic agriculture. According to Huber et al. (2010) 73 countries have implemented organic legislation, whereas 16 countries are currently in the process of drafting a legislation. In the EU, for instance, Council Regulation (EC) No. 834/2007[1] was also based on the IFOAM Basic Standards and provides a binding framework for EU Member States (IFOAM 2009). More detailed rules for the implementation of organic farming in the member states are set out in Commission Regulation (EC) No. 889/2008.[2] In Switzerland, the federal (country-wide) standards[3] have been developed according to Council Regulation (EEC) No. 2092/91[4] and were updated in 2010 according to Council Regulation (EC) No 834/2007.

---

[1] Council Regulation (EC) No. 834/2007 of 28 June 2007 on organic production and labelling of organic products, repealing Regulation (EEC) No. 2092/91, O.J. L 189/21 2007. This regulation was amended by Council Regulation (EC) No. 967/2008 of 29 September 2008, O.J. L 264/1 (2008).

[2] Commission Regulation (EC) No. 889/2008 of 5 September 2008 laying down detailed rules for the implementation of Council Regulation (EC) No. 834/2007 on organic production and labelling of organic products O.J. L 250/1, which was amended by Commission Regulation (EC) No. 1254/2008 of 15 December 2008, amending Regulation (EC) 889/2008 laying down detailed rules for implementation of Council Regulation (EC) No. 834/2007, O.J. L 337/80.

[3] Ordinance on Organic Farming. Verordnung des EVD vom 22. September 1997 über die biologische Landwirtschaft (SR 910.181).

[4] Council Regulation (EEC) No. 2092/91 of 24 June 1991 on organic production of agricultural products and indications referring there to on agricultural products and foodstuffs.

This chapter aims at comparing the environmental impacts of organic agriculture with those of conventional agriculture based on state-of-the-art literature. Furthermore, it aims at discussing methodological implications for the comparison of environmental impacts of farming systems.

Subsequent to this introduction, Sect. 8.2 focuses on the comparison of the environmental impacts of organic farming on biodiversity, climate change, resource depletion, ground and surface water pollution, air quality, and soil fertility with those of conventional farming. Section 8.3 discusses methodological implications of current research on the environmental impacts of farming systems by describing the most important problems of environmental assessments and by suggesting solutions for dealing with these problems. Finally in Sect. 8.4, conclusions are drawn regarding the impacts of organic farming.

8.2 Environmental impacts of organic farming

It is important to assess and quantify how green organic farming actually is. This is done by comparing organic agriculture with other farming systems, usually referred to as "conventional" agriculture. Acknowledging the fact that conventional agricultural practices are already diverse, it is difficult to determine its exact impacts. Moreover, agriculture as it is practiced in reality diverts from laws, standards, and regulations. Thus, there are basically two possible approaches for such a comparison:

- **Normative comparison**, i.e., comparing (minimum) standards: farming systems are assessed according to the environmental standards that have to be fulfilled. For conventional farming this is usually the environmental law or—if existing—cross-compliance standards.
- **Positive comparison**, i.e., comparing real farms: In reality farms divert from standards, laws, and regulations which they have to fulfil. For example, not all farms comply with minimum standards, such as an even nutrient balance. On the other hand other farms voluntarily tend to meet higher standards than they are obliged to.

Either type of comparison as well as blends of both can be found in literature. However, for drawing conclusions in this book chapter (i.e., to assess the environmental performance of organic farming as a green technology in practice), we will opt for the positive comparison.

As shown below, there are many studies identifying the positive and negative environmental effects of organic products or organic agriculture. Environmental impacts are grouped according to the types of natural resources concerned. In order to analyze the environmental impacts of organic farming on biotic and abiotic resources, first, biodiversity and landscape impacts will be looked at. After that climate change mitigation, resource depletion, ground and surface water pollution, air quality, and soil fertility will we reviewed. Social, economic, and ethological impacts of organic agriculture, such as provision of labor in rural regions, health benefits, and an increased animal welfare are not considered in this chapter.

8.2.1 Biodiversity and landscape

Biodiversity can be described according to four different levels. First, diversity within species (genetic diversity) includes the diversity of farm animals and crops. Genetic diversity enables species to adapt to changing environments (e.g., caused by climate change). Second, it is expressed at species level (encompassing faunal and floral diversity), most simply by monitoring the species in selected groups, such as birds or plants in a certain area. Third, biodiversity can be expressed in terms of the regional diversity of habitats and ecosystems in which species live (Christie et al. 2006). Fourth, ecosystem functions describe services delivered by functioning natural systems to humans. One of the key messages of a major study on "The Economics of Ecosystems and Biodiversity" (TEEB) was the "inextricable link between poverty and the loss of ecosystems and biodiversity" (TEEB. 2010).

Owing to agricultural activities, a great variety of ecosystems have been created which, overall, have enhanced biological diversity. On the other hand, agriculture negatively affects biodiversity directly through cultivation practices. Furthermore, it affects biodiversity indirectly through nitrogen emissions into the air and CO_2 emissions into the atmosphere. On land under intensive agricultural cultivation, biodiversity decreases significantly because of the high nutrient influx, high cutting frequencies on meadows, high stocking rates, use of pesticides, and modern methods of processing cut grass (Knop et al. 2006). In Alpine lowlands many diverse agricultural ecosystems have disappeared, while in the mountain regions two parallel trends are apparent: the intensification of productive areas and the abandonment of unproductive but ecologically valuable areas (Aeschenbacher and Badertscher 2008).

Biodiversity effects are among the most frequently studied environmental impacts of organic agriculture. Recent metastudies (Bengtsson et al. 2005; Fuller et al. 2005; Hole et al. 2005) show clear differences between organic and conventional farming systems. In very rare cases, organic production was found to have negative impacts, although this was outweighed by studies showing positive impacts. The differences vary among taxonomic groups, but for each species group large differences were found (Table 8.1). On average, about 50% greater species diversity was achieved on organic farms (Niggli et al. 2008).

Organic farming practices are most beneficial for birds, predatory insects, spiders, soil organisms, and the arable weed flora, while pests and indifferent organisms do not show different levels of abundance in the farming systems. Furthermore, differences in arable land between the farming systems are more pronounced than on grassland (Niggli et al. 2008).

Apart from differences at species-group level, structural differences at farm level are prevalent between organic and nonorganic farms (Gibson et al. 2007; Schader et al. 2008b). In addition, Boutin et al. (2008) identified higher species richness in seminatural habitats on organic farms compared with conventional farms.

Genetic biodiversity is influenced both positively and negatively by organic farming. On the one hand many organic farmers cultivate rare plant and animal species on their farms (e.g., because they are better adapted to local conditions); on the

Table 8.1 Number of studies analyzing the impacts of organic farming on biodiversity with respect to various *taxa* on the basis of 76 comparative studies[a]

	Impacts of organic farming		
Taxa	Positive	No difference	Negative
Plants	16	2	0
Birds	11	2	0
Mammals	3	0	0
Arthropods			
Beetles	15	4	5
Spiders	9	4	0
Butterflies	2	1	0
Bees	2	0	0
Other arthropods[b]	8	3	1
Bacteria, fungi and nematodes	12	8	0
Earthworms	8	4	2
Total	**87**	**28**	**8**

[a]Updated using studies from 2004 to 2008
[b]Mites, bugs, millipedes, flies, and wasps
Source: Hole et al. (2005), updated by Niggli et al. (2008)

other hand the restriction on admission of varieties hampers genetic diversity. Because there is insufficient scientific evidence on the impacts of organic farming on genetic biodiversity, we assume both farming systems to perform equally well.

Concerning **landscape** and habitat diversity, organic farming may perform better because of more diverse crop rotations (Norton et al. 2009) and higher implementation rates of structural elements such as hedges and fruit trees (Schader et al. 2009). However, landscape effects are very farm and site specific. Therefore, no general trend can be determined (Steiner 2006).

In a Swiss study, Schader (2009) analyzed the average habitat quality by combining an economic sector-representative farm group model (Sanders et al. 2005) with a species diversity model by Jeanneret et al. (2006) for different farm types, regions, and all of Switzerland. Figure 8.1 shows the average habitat quality in different farming systems for five species groups (that are on the red list of endangered species) and 11 indicator species as well as for an overall habitat quality indicator. Average habitat quality is expressed as a percentage of a hypothetical maximum habitat quality, which is achieved in an optimal habitat quality under optimal farm management conditions (=100%). The study showed that the average habitat quality over all species is 25% on organic farms and 16% on conventional farms. Schader attributed the 55% improvement of habitat quality to the on-average higher share of grassland, the on-average lower grassland intensity (lower stocking density, lower fertilization, fewer cuts) and the different farm type and regional distribution.

In summary, studies attribute the higher biodiversity in organic systems to the following factors: (a) ban on herbicides and artificial pesticides, (b) ban on mineral fertilizers, (c) more diverse rotations, (d) lower organic fertilization, (e) careful tillage, (f) a higher share of seminatural habitats in total UAA (Utilized Agricultural Area) (Bengtsson et al. 2005; Fuller et al. 2005; Hole et al. 2005).

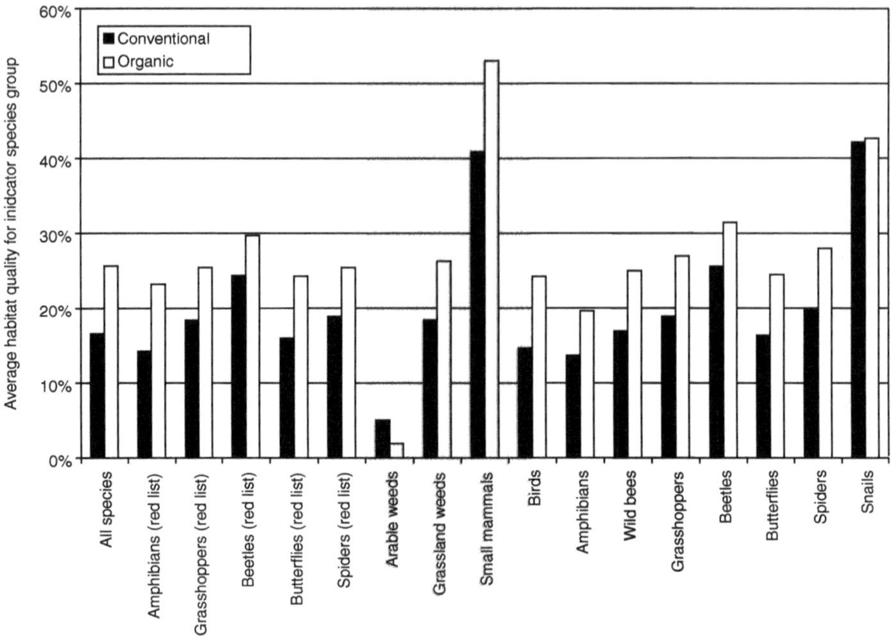

Fig. 8.1 Average habitat quality by species group (average over all regions and farm types, 2006/2007)

8.2.2 Resource depletion

Resource depletion is a problem of similar magnitude against the background of industrial civilizations' growing dependency on fossil fuels (Meadows et al. 1972). The debate about peak oil (i.e., the point in time when the maximum rate of global oil extraction has been reached) is currently increasing in intensity (Zittel and Schindler 2007). Agriculture, once a net energy producer has today become a net energy consumer for some commodities. Given the need for efficient resource use, energy use has become a standard environmental indicator (Pimentel et al. 2005; Frischknecht et al. 2007).

Nevertheless, many studies suggest that the whole food system (agricultural production, food processing, packaging, and distribution) makes up a large percentage of energy consumption (Ziesemer 2007). Studies comparing crop and animal products conclude that crop products have much higher energy efficiency per unit of digestible energy than produce from animal production (Pimentel and Pimentel 2005).

The impacts of organic agriculture on energy use can be analyzed on the basis of different functional units (Halberg 2008). While some studies use "area" as a unit (Haas et al. 2001), others take the weight of output from the farming system as a reference (Grönroos et al. 2006). Although the latter approach is in line with the standard procedure within life cycle assessments (Heijungs et al. 1992) and illustrates energy use per unit of food produced, it still has weaknesses when it comes to analyzing agricultural systems. Often, research on farming systems encompasses consideration of multiple outputs. Either these outputs need to be expressed in a single unit, or an allocation of the energy use has to be performed, or again, byproducts need to be deducted to enable a comparison across all products (Schader et al. 2008a). A product-related assessment additionally involves the determination of the functional unit. However, the scorings related to weight, volume, calories or protein might produce highly varied results.

Stolze et al. (2000) also concluded that organic farming systems perform better than conventional ones in terms of energy use. The energy use of growing permanent crops (olive, citrus, vineyards) and arable crops (grains, pulses, etc.) is, related to product output and area cultivated, lower in organic than in conventional farming systems. However, growing potatoes organically can require equal or more energy per product output than doing so conventionally. Also Lampkin's review (2007) identified that most product- and area-related energy use assessments of organic farming to date show a lower energy use per-ha. Due to the generally lower productivity of organic farming, per-ha comparisons reveal higher differences than product-based comparisons. Thus the choice of the appropriate functional unit is crucial when comparing organic and conventional agriculture (see Sect. 8.3.2.1 in the discussion section of this chapter). Haas et al. (2001) compared organic and conventional grassland farms in southern Germany. They found a 44–46% lower energy use per ha and per ton of milk. Thomassen et al. (2008b) also analyzed milk production and found that the energy efficiency of organic production was significantly higher compared to conventional production. They concluded that the use of concentrate feed in particular is a major driver and has potential for reducing energy use.

Grönroos et al. (2006) calculated that fossil energy use for organic rye bread and milk was lower by 13% for rye bread and 31% for milk—compared with conventional products. In a cradle-to-(farm-) gate perspective, the difference is even higher, with organic products consuming only 50% of the energy use of conventional products. Similar results were generated by Hoeppner et al. (2005) who compared the energy use throughout a rotation. Energy use and energy output were 50% and 30% lower, respectively, over organic rotations in a long-term field experiment, which results in higher energy efficiency (energy use per product) of organic farming compared with conventional farming.

Nemecek et al. (2005) demonstrated a lower energy use per ha and per product unit overall in organic systems for all major crops in Switzerland. This was done by analyzing data from long-term field experiments and generating subsequent calculations aimed at generalizing the results for Switzerland. An exception to this is potatoes, where a slightly higher energy use was calculated per ton of organic potatoes.

Figure 8.2 shows the differences in energy use per ha between organic and conventional farm types in Switzerland, based on a representative farm sample (Schader 2009). Pig and poultry farms have been excluded from the graph in order to maintain its readability for the other farm groups; because of the high stocking rates and

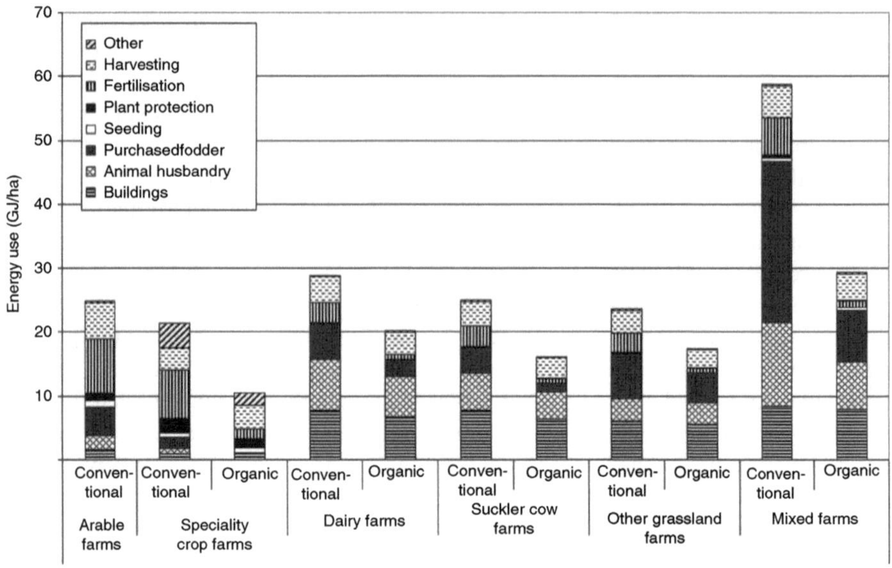

Fig. 8.2 Fossil energy use per ha on conventional and organic farms by farm type (2006/2007)

the high share of cereal-fed livestock, these farms have a calculative energy demand of 195 GJ/ha (not shown in Fig. 8.2). Beside pig and poultry farms, conventional mixed farms have the highest total energy use (60 GJ/ha). The average energy use (as a sum of all energy use components) of dairy, suckler cow, other grassland, arable and finally speciality crop farms ranges from 20 to 30 GJ/ha. Organic counterparts have an energy use which is about a third lower (10–20 GJ/ha), except for mixed farms where the average reduction in energy use amounts to 50%. Schader (2009) attributes the lower energy use of organic farms to lower quantities of purchased feedstuffs, particularly concentrates, lower stocking densities, the ban of mineral nitrogen fertilizers, and the absence of highly intensified specialized pig and poultry farms.

In summary, organic farming has a lower energy use per ha, and in most cases also per mass of product, than conventional farming. There are only a few exceptions on the crop-production side, notably potatoes, with organic systems displaying lower energy efficiency owing to low relative productivity levels. While milk production is more efficient on organic farms, poultry production has shown slightly lower energy efficiency. Thus, the quantitative advantage of organic farming depends crucially on the product, the geographic region, and the assumptions of the study.

Besides depletion of energy resources, in particular phosphorus (P) resources are exploited as phosphorus is applied on agriculture fields in large quantities. Because of the ban of easily soluble phosphate and potassium fertilizers and the widely closed nutrient cycles, organic farming leads to a lower depletion of nutrient resources. Positive impacts on P resource depletion are backed by life-cycle assessments (Nemecek et al. 2005).

There is little evidence on the impact of organic farming on water use efficiency (Stolze et al. 2000); however, because of lower yields organic farms might tend to use more water per unit of output. On the other hand, lower biomass production per area tends to make organic agriculture particularly attractive for areas with water scarcity. Furthermore, as animal production is most water intensive, the lower stocking rates of organic farms might lead to lower water use per ha.

8.2.3 Climate change

Climate change has been perceived for decades as a significant global environmental problem. Over the last years in particular environmental awareness has increased markedly within the general population, partly because of reporting on the international negotiations in relation to the Kyoto Protocol, and partly because of more visible impacts of climate change on ecosystems such as glaciers or Polar Regions. The higher incidence of natural disasters such as hurricanes, droughts, and floods has also contributed to the growing awareness of climate-change over the last few years (IPCC 2007). Recent studies estimate that the cost of current and projected levels of greenhouse gas emissions and the climate change caused by them exceeds their abatement costs (Stern 2007). Because agricultural production has an impact on all the three major greenhouse gases (CO_2, CH_4, and N_2O), it is perceived as a crucial and potentially cost-effective lever for mitigating climate change (Smith et al. 2007).

About 12–14% of global greenhouse gas emissions stem from the agricultural sector (Smith et al. 2007). However, while CH_4 and N_2O emissions are predominantly attributable to agriculture, the share of the agricultural sector in terms of CO_2 emissions caused mainly by burning fossil fuel is disproportionately small.

The literature review of Stolze et al. (2000) found that organic agriculture bears a lower CO_2 and NH_3 gas emission potential than conventional farming systems. However, because of lack of scientific literature, they concluded that there were no differences between these farming systems with respect to N_2O and CH_4.

As is the case for energy use, the impacts of organic agriculture on CO_2 emissions can be analyzed on the basis of different functional units (Halberg 2008). While some studies use "area" as a unit (Haas et al. 2001), most studies take the weight of output from the farming system as a reference. Since the performance of organic agriculture regarding CO_2 emissions is highly correlated to energy use, the same arguments apply as for the discussion of energy use in the above section. Unlike that of energy use though, net emissions of CO_2 (i.e., gross emissions subtracted by the sequestration rate) need to be taken into account. There are indications of better performance regarding carbon sequestration (Olesen et al. 2006; Niggli et al. 2009). Several long-term trials from the United States, Germany, and Switzerland (Mäder et al. 2002) show that organic farming systems are able to sequester on average 590 kg/ha per year more carbon from the atmosphere than the best performing conventional counterparts.

Table 8.2 Comparison of N_2O emissions per unit of area under conventional and organic management

References	Country	Method	Emissions in organic systems		
			Lower	Equal	Higher
Petersen et al. (2006)	Austria, Denmark, Finland, Italy, UK	Field measurement	x		
Chirinda et al. (2010)	Denmark	Field measurement		x	
Küstermann et al. (2008)	Germany	Modelling	x		
Flessa et al. (2002)	Germany	Field measurement	x[a]		
Sehy (2004)	Germany	Field measurement	x[a]		
Lynch (2008)	Canada	Field measurement	x		
Nemecek et al. (2005)	Switzerland	Modelling	x[b]		
Hansen et al. (2008)	Norway	Field measurement	x		

'X' indicates scientific evidence on higher, equal, or lower N_2O emissions under organic management

[a]No difference in output-related emissions

[b]Output related emissions are lower in organic systems

Source: Gattinger et al. (2010), Adapted

In summary, organic farming has lower CO_2 emissions per ha and, in most cases, also per ton of product than conventional farming. There are only a few exceptions on the crop-production side, notably potatoes, with organic systems displaying lower energy efficiency due to low relative productivity levels. Although milk production is more efficient on organic farms, poultry production has shown slightly lower energy efficiency. Thus, the quantitative advantage of organic farming depends crucially on the product, the geographic region, and the assumptions of the study. Furthermore, organic farming has a potential to mitigate climate change by reducing greenhouse gas emissions through their sequestration in soil.

About 75% of CH_4 emissions stem from enteric fermentation of ruminants. There are two different perspectives on the impact of organic farming on CH_4 emissions per unit of output. On the one hand, CH_4 emissions could be higher owing to less output (milk or meat) per livestock unit and time. On the other hand, many organic farms tend to keep dairy cows for more lactation periods than conventional farms, which lower the CH_4 emissions per unit of output during the growing up phase. When looking at the emissions per hectare (e.g., from an agri-environmental policy evaluator's perspective), organic farming has lower CH_4 emissions because of lower stocking densities.

So far, there are only a few studies available that compare N_2O-emissions from organic and conventional farming systems (Table 8.2). Chirinda et al. (2010) found no differences in N_2O-emissions between farming systems, while all other authors found lower N_2O-emissions per ha. Calculated per output quantity, N_2O-emissions in organic systems where equal to non-organic farming systems according to Flessa et al. (2002) and Sehy (2004) while Nemecek et al. (2005) modelled 18% lower N_2O-emissions in organic farming systems than in conventional ones.

Because in general, there is a linear relationship between N-Input und N_2O-release and in organic farming systems N-supply is up to 50% lower than in conventional farming systems, Gattinger et al. (2010) conclude that organic farming systems have a lower N_2O-emission potential than conventional farming systems.

In summary, data uncertainty concerning N_2O emissions from different fertilizers and from the soil does not allow general conclusions to be drawn on the impact of organic farming. N_2O is emitted from agricultural soils at specific periods of time, depending on nitrogen, carbon, and oxygen in the soil. Influencing factors on N_2O emissions are type and amount of nitrogen fertilization and water logging. In general, however, N_2O emissions could be lower owing to lower nitrogen fertilization rates and the applications of fertilizers with lower nitrogen concentration. On the other hand, N_2O emissions could be higher in organic systems per unit of output due to the higher land use.

8.2.4 Ground and surface water pollution

Eutrophication is defined as nutrient enrichment in sensitive ecosystems (UNECE 1999). Eutrophication entails various environmental impacts that cause both the loss of biodiversity and negative impacts on human health. Eutrophication leads to excessive growth of algae and excessive oxygen demand, with anaerobic conditions leading to foul smelling surface waters and fish death. These effects of eutrophication can be understood as societal costs, either in terms of abatement, purification, or restoration costs, or as damage costs if the negative impacts of eutrophication are not abated or fixed. Besides eutrophication, influxes of toxic substances particularly into surface water can pose a significant environmental threat because it can lead to severe harm to aquatic life.

The main environmental risks entailed by agricultural production in relation to ground and surface water pollution involve eutrophication with nitrogen and phosphorus and pesticide emissions. The leaching of mobile nitrates into ground and surface water and gaseous emissions such as ammonia (NH_3) from organic fertilizers are the major contributors to nitrogen eutrophication. Ammonia emissions affect ecosystems like forests, swamps and diverse meadows, which require low nitrogen levels. Furthermore, ammonia emissions into ecosystems cause acidification and the release of toxic substances including heavy metals.

Nitrate pollution in the lowlands has been the most severe environmental problem resulting from post-war policies (Gruber 1992). These policies provided incentives to run intensive, highly-yielding agricultural production involving heavy nutrient surpluses.

Phosphorus is relatively immobile in soils but can be emitted from agricultural systems to surface waters by erosion and run-off processes. While phosphorus rarely represents an environmental problem in rivers, it causes algae growth in lakes and seas, because normally phosphorus is the limiting nutrient for algae growth.

The decomposition of this additional plant material reduces the amount of oxygen. Finally, fauna die because of anaerobic conditions. Phosphorus emissions from agriculture give rise to high societal costs because of bad odors, costs of treatment, and the hindrance of recreational activities.

The reduction of nitrogen and phosphorus eutrophication demands efficient use of these nutrients (Herzog and Richner 2005). Evaluations have identified that the problem of nitrate leaching occurs predominantly in arable farming systems, although leaching can also occur from grassland receiving high fertilizer inputs. Therefore, Herzog and Richner (2005) suggest that farms should no longer be permitted to have a 10% nutrient surplus. Apart from systems that rely heavily on imported manures (e.g., horticultural systems), there is no nutrient surplus in organic systems, because nutrient import onto the farm is restricted for both feedstuffs and mineral fertilizer.

Several studies show that nitrogen leaching can be reduced by 40–64% through organic farming (e.g., Edwards et al. 1990; Younie and Watson 1992; Eltun 1995; Condron et al. 2000; Goulding 2000; Haas et al. 2001; Kirchmann and Bergström 2001; Mäder et al. 2002; Stopes et al. 2002; Auerswald et al. 2003; Pacini et al. 2003; Shepherd et al. 2003; Osterburg and Runge 2007).

In contrast, Nemecek et al. (2005) found higher eutrophication impacts for some organic crops compared to their conventional counterparts per kg output. In some places, these higher nutrient loads on arable land are attributed to the greater use of organic fertilizers in the organic system, because the life cycle assessments used by Nemecek et al. (2005) assume relatively high fertilization rates for organic farms.

Taking the data by Nemecek et al. (2005) and projecting them at sector level using statistical data and an economic model, Schader (2009) found on average 35% lower eutrophication rates. Figure 8.3 shows the average eutrophication with nitrogen for average conventional and organic farm types in Switzerland. As can be seen, nitrate eutrophication rates on different farm types vary drastically. Whereas farms specialized in animal production usually have nitrate leaching rates below 20 kg N-eq/ha, organic counterparts show a more than 50% reduction. Mixed farms have a higher nitrate leaching rate because of a higher share of arable land, whereas organic farms have a 20% lower eutrophication rate than their conventional counterparts. According to Schader (2009), there are only a very few organic specialized plant production farms, which makes it difficult to model nitrate leaching at sector level.

In the same study phosphorus eutrophication was also modelled, which showed a 10–20% decrease of phosphorus eutrophication on organic farm types, compared with their conventional counterparts (Fig. 8.4). Other studies on impacts of organic farming on eutrophication with phosphorus are scarce. However, acknowledging the fact that literature indicates significant efforts of organic farming to improve soil quality and to reduce erosion risk (see section on soil fertility below), phosphorus runoff can be assumed to be lower in organic systems (Shepherd et al. 2003; Schader 2009).

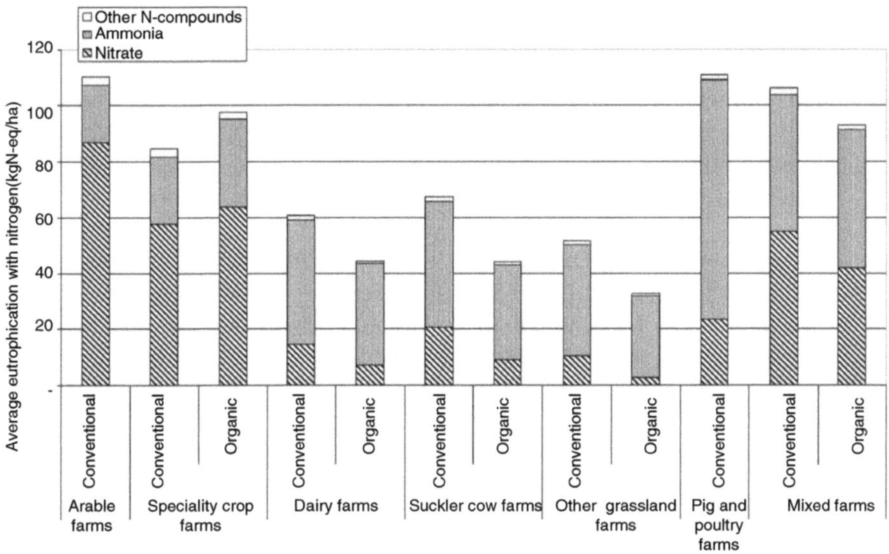

Fig. 8.3 Nitrogen eutrophication per ha on conventional and organic farms by farm type (2006/2007)

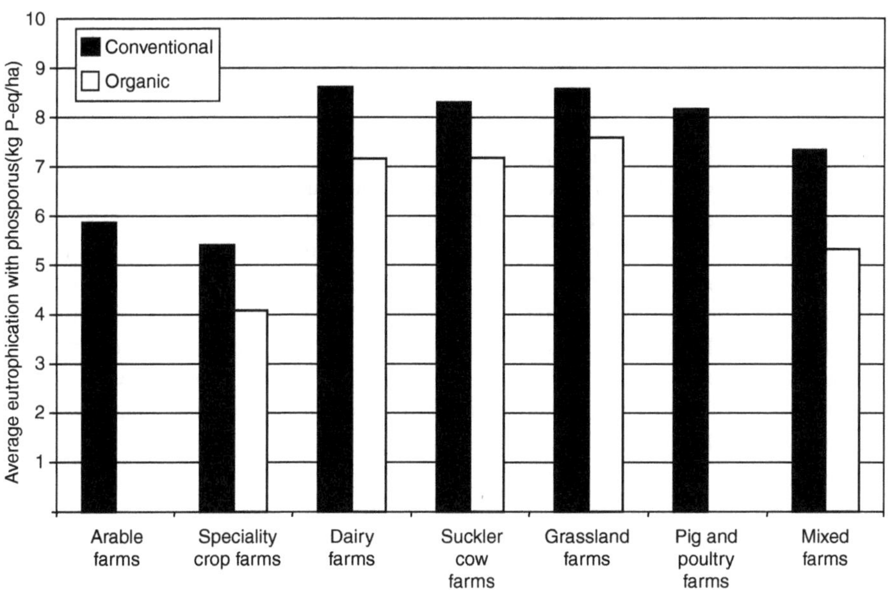

Fig. 8.4 Phosphorus eutrophication on conventional and organic farms by farm type (2006/2007)

Summing up, there are three facts underlining a lower eutrophication potential of organic farming:

- Organic farming systems have lower nutrient levels, which reduces the absolute quantity of nutrient loads that can be emitted from the system due to lower stocking rates and the ban of mineral nitrogen fertilizers.
- The quantity of directly available nitrogen is much lower in organically managed soils.
- Because nutrients cannot be imported easily into the systems, the opportunity cost (shadow price) of nitrogen losses is higher for organic farms than for conventional farms (Stolze et al. 2000). This implies a need for more efficient nutrient management in organic systems, although this does not eliminate losses. In addition, nitrate leaching can be high at the point of transition from the fertility building phase of the rotation to the cropping phase.

8.2.5 Air quality

The main environmental risks in terms of air pollution are entailed by agricultural production gaseous emissions such as ammonia (NH_3) from organic fertilizers and gaseous emissions from pesticides. Furthermore, ammonia emissions affect ecosystems like forests, swamps, and diverse meadows, which require low nitrogen levels.

There are only few comparisons between ammonia emissions from organic systems compared with conventional systems. In a study on representative life cycle assessment (LCA) approach by Schader (2009), ammonia emissions were found to be almost equal in both systems for each farm type. The fact, however, that specialized pig and poultry farms are hardly feasible according to organic standards, leads to a substantial reduction of stocking densities and thus to a reduction of ammonia emissions for the whole sector. Nevertheless the study assumed that fertilization rates were equal to the nutrient needs of the crops. In reality however, in organic systems nitrogen is applied in lower amounts, which also leads to reductions in ammonia emissions. Finally, opportunity cost of each kg of ammonia lost from the system is much higher on organic farms, because nitrogen cannot be fed into the system easily (Stolze et al. 2000). Therefore, organic farms should have a higher incentive to implement strategies for reducing the loss of nitrogen via ammonia emissions from the system.

Thus, per unit area, ammonia emissions are lower in organic systems because of lower stocking densities, whereas per livestock unit or milk/meat produced, emissions may be equal if productivity is in a similar range. Because most pesticides are banned in organic systems, there are substantially lower pesticide emissions from organic systems both per unit area and production unit.

8.2.6 Soil fertility

Soil is one of the major production factors of agriculture and therefore an essential natural resource for ensuring food security. Yet agricultural production at the same time poses a threat to this resource. According to Lal (2004) agricultural soils have lost a great share of their organic matter to wind and water erosion because of intensive agriculture, which is responsible for the loss of one third of fertile lands from 1955 to 1995 (Pimentel et al. 1995).

Organic agriculture encompasses a number of different activities within its system approach, which aim at increasing organic matter content in the soil. Most important is the ban of mineral fertilizers, which necessitates meeting the nutrient needs of the crops by organic fertilizers (Mäder et al. 2002). Furthermore, the fundamental importance of a crop rotation including short-term ley (i.e., nonpermanent meadows) supports the development of fertile soils (Pimentel et al. 2005).

These activities also favor biological activity in the soil. As has been reported above in the section on biodiversity, there is clear scientific evidence that soils under organic management have higher biological activity both in terms of species and general biomass (Mäder et al. 2002; Pfiffner and Luka 2007). Furthermore, organic farmers need to pay attention to soil fertility in the long term, because the ability of the soil to capture nutrients is crucial in organic systems (Köpke 2003).

Soil erosion has already been discussed in the context of phosphorus runoff, because erosion processes are the dominant driver for phosphorus losses. There are some studies reporting a beneficial effect of organic farming on soil erosion and soil structure (Siegrist et al. 1998; Stockdale et al. 2001; Shepherd et al. 2002). Organic farming performs much better in terms of soil biological activity than nonorganic farming. Soil erosion and organic matter content are also affected positively by organic practices.

8.3 Methodological implications for a comparison of farming systems

In this section we first present a methodological classification of studies reported according to nine characteristics. Second, we discuss implications for methodological choices when comparing organic and conventional production systems.

8.3.1 Classification of methodological characteristics

When comparing the environmental performance of organic farming with that of conventional farming, there are nine different methodological characteristics by which studies can be classified following the four main phases of life cycle assessment (Table 8.3). Five of these refer primarily to the phases "goal and scope definition" and

Table 8.3 Classification of methodological choices for comparing the environmental impacts of organic and conventional farming

LCA phase	Characteristic	Possible methodological choices
Goal and scope definition and life cycle inventory	Scope	Field, farm, region, sector-wide, global
	Functional unit	Area, mass, energy, protein, net value added
	Stage of processing	Raw product, processed and packed products and supply chains
	Life cycle perspective	Life-cycle perspective (cradle-to-grave, cradle-to-gate), economic perspective
	Normativeness	Based on statistics (positive view), based on normative models
Life cycle impact assessment and interpretation phase	Time perspective	Static view, dynamic view
	Scope of impacts	Single impact category, several impacts, only environmental impacts, all impacts relevant for sustainability of farming system
	View on sustainability	Weak sustainability, strong sustainability
	Units of impacts	Physical/biological units, relative effects, monetary units

"life cycle inventory," whereas the remaining four focus more on the phases "life cycle impact assessment" and "interpretation" of results.

First, the scope of the study can range from single fields, products, rotations, and farms to representative studies for whole sectors or even studies at the global level, although most studies are conducted at field or product level.

Second, the functional unit may either be related to area or production. Production can be measured by several units such as mass, energy content, protein content, or net value added.

Third, both raw products and processed products or entire supply chains can be assessed. Although standards for organic farming increasingly cover standards on processing (e.g., choice of additives), only raw agricultural products were assessed in the survey and in the bulk of studies found.

Fourth, studies can take on either a life cycle perspective or an economic perspective. Whereas the economic perspective usually covers only the resources used directly in the sector, a life cycle perspective covers also environmental impacts of inputs that occurred in earlier stages, e.g., fossil energy use for producing mineral nitrogen fertilizers (cradle-to-gate perspective). Even use and disposal phase can be covered, e.g., for specific inputs, buildings, or packaging materials (cradle-to-gate perspective). Commonly the impact categories climate change, resource depletion water and air pollution are assessed by agricultural life cycle assessments, usually based on the ISO Norms 14040 and 14044. Because of methodological difficulties, biodiversity and soil quality are only rarely assessed using LCAs.

Fifth, as already described in Sect. 8.2, the degree of normativeness of assumptions is different across studies. For example, a normative approach to assessing fertilization would be to assume that nutrient needs of plants are covered, whereas a positive approach is to calculate impacts based on the fertilizer amounts actually used. Most studies have a significant degree of normative assumptions in the models they use, due to a lack of empirical data.

Sixth, studies cover a static or dynamic perspective. All of the LCA studies followed a static approach, as the development of dynamic LCA models is still in its infancy.

Seventh, studies either analyze only a single environmental category, several, or they try to cover all relevant categories. Furthermore, social and economic impact categories are taken into consideration if aimng at a full coverage of sustainability.

Eighth, environmental impacts can either be aggregated to a single score or remain as several individual indicators. Whereas the first approach is in accordance with the notion of weak sustainability (i.e., full substitution of natural resources), the latter would be a strong sustainability approach.

Ninth, impacts can be expressed either in physical or monetary terms. Physical terms are used most frequently, whereas monetary terms are mostly used in economic comparisons. The latter approach is particularly useful if societal or farm-level costs and the environmental performance need to be compared with each other (e.g., using cost-benefit analysis).

8.3.2 *Discussion of methodological implications*

The large number of different methodological characteristics illustrates that finding appropriate methods for comparing farming systems is more difficult than for many other goods and services. Depending on the goal of the study and the geographic context, methodological choices can be very different (Hospido et al. 2010).

Life cycle assessments are a generally useful approach for assessing farming systems. Significant methodological advances in the assessment of environmental impacts of agriculture and food products were realized, owing to: (a) the introduction of life cycle thinking, which explicitly includes off-farm environmental impacts (e.g., owing to the production of fertilizers imported by the farm); and (b) the consideration of several environmental impact categories at once, instead of concentrating on a single impact category.

Yet, we see nine methodological problems with some of the current environmental impact assessments, particularly with LCAs, of farming systems. Building on the work by Reap et al. (2008a, b), we describe them and propose ways to address them in practical LCA work.

8.3.2.1 Consideration of the multifunctional character of agriculture

During the last decades the role of agriculture shifted from its mere production function to a multifunctional role. Multifunctionality acknowledges the fact that agriculture fulfils multiple roles in society. Apart from producing food and fibers, agriculture is carried out for landscape maintenance, conservation of natural resources, and cultural purposes (OECD 2001).

The functional unit needs to be chosen according to the function of agriculture which is addressed. Within a multifunctional setting several functions should be assessed. For addressing the function of generation of rural incomes, it makes sense to use "net value added" as a functional unit. Furthermore, in the case of agri-environmental policy evaluations (e.g., for direct payments), particularly agriculture's role to cultivate land is addressed. In such a case "area" is the most appropriate functional unit to consider.

If agriculture or farming systems as such are evaluated, it is necessary to consider all relevant functional units. Thus, the functional unit "area" can be seen as addressing the role of agriculture to maintain landscape, "digestible energy" addresses the function of providing food to the population, and finally "farm income" or "net value added" addresses the function of agriculture to provide rural livelihoods. However, multiple functional units are rarely considered, because this complicates both the research and the communication of the results. Furthermore, current ISO Standards do not offer sufficient guidance on multiple functional units.

8.3.2.2 Covering land use impacts

Land use impacts, for example, the change of biodiversity on agricultural land or soil fertility, can be related to an area more plausibly than to production (Schader et al. 2008c). This leads to the fact that many life cycle assessments neglect or even ignore these impact categories, although the categories are key indicators for comparing both farming systems. We argue that every full life cycle assessment should address biodiversity and soil fertility impacts if a comparison between organic and conventional agriculture is made and general conclusions on the systems' performance are drawn.

8.3.2.3 Heterogeneity of products

A mass-related functional unit is often used to visualize the environmental performance of a specific food product to the consumer. Food quality, however, is not considered appropriately in this functional unit as we speak of "heterogeneous products" when comparing conventional and organic products. The heterogeneity can be illustrated using the higher willingness-to-pay of consumers for organic produce (Krystallis and Chryssohoidis 2005), because the organic product may fulfil further functions and uses. As we know from many consumer studies, at least in Europe and North America, the choice of food products is often made irrespective of energy content. Often products are consumed even because of their low energy content.

Furthermore, many consumers are willing to pay more for organic products not only because of the environmental benefits but for expected health benefits. Although not a universally held opinion, there are indications that product quality of organic products is higher, particularly because of a lower risk of pesticide contamination and higher concentrations of nutritionally desirable fatty acids and antioxidants (Woese et al. 1997; Butler et al. 2008). We suggest that this higher product quality needs to be taken into account by choosing a monetary functional unit (e.g., consumer price).

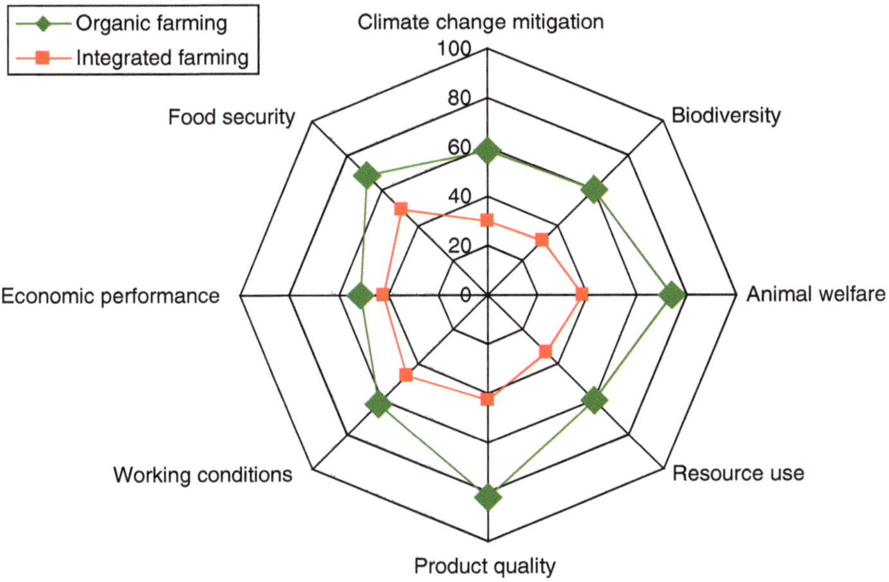

Fig. 8.5 Sustainability assessment of Swiss organic agriculture

8.3.2.4 Covering social and economic aspects of sustainable development

With the rise of the notion of sustainability (i.e., contribution to sustainable development) more comprehensive assessments of organic farming, including economic and social indicators, are demanded. The great complexity of a full coverage of sustainability, however, requires neglecting many details that would be taken into account in studies addressing specific environmental categories. Figure 8.5 shows results from a sustainability assessment using the nominal group technique as a participatory approach (Delbecq et al. 1975; Jeffreys 2002). The results show that neither organic nor conventional/integrated farming can be characterized as fully sustainable (=100%). Whereas conventional farming reaches sustainability scoring of 30–50%, organic agriculture scores better in all categories that were assessed. In particular regarding product quality, animal welfare, and biodiversity, organic agriculture was evaluated with nearly 80% sustainability scorings. Around 60% of sustainability score was given for the environmental indicators biodiversity, climate change mitigation, and resource use (Schader and Stolze 2010).

If a credible judgement on organic farming is to be made, all relevant impacts on sustainability need to be analyzed and made transparent. Whether approaches such as "social LCA" and "life cycle costing" are appropriate and feasible ways for this remains to be determined. It needs to be stressed, however, that not all impacts that cannot be quantitatively assessed are unimportant. We therefore suggest including qualitative assessments if quantitative data is weak. Furthermore, inter and transdisciplinary research needs to be reinforced in order to combine LCA methodologies with social, economic and psychological methods.

8.3.2.5 Regional variations in natural system capacity and environmental legislation

Both natural system capacity and environmental legislation varies regionally and by country. Therefore, it matters for some environmental impacts where the pollution occurs. Attributional LCAs tend to neglect this aspect by using "production mass" as a sole functional unit. This problem has been addressed in water footprinting methods, but is also relevant for eutrophication impacts, as the following example will illustrate.

A system inherent feature of organic agriculture is a lower intensity of production. Particularly stocking rates are restricted but also other restrictions apply, depending on the geographical context. These restrictions, on the one hand lead to lower environmental impacts per ha and lead to lower productivity per area, at least in developed countries (Badgley et al. 2006). If a policy aim is to reduce the eutrophication of groundwater in a particular region, policy-makers might try to reduce the amount of eutrophication in the region by lowering intensity of agriculture (e.g., by supporting the conversion of farmers to organic agriculture). The resulting production loss/gap might be compensated by moving production to regions with lower production cost (a) because of better site conditions or (b) because of less strict environmental legislation, causing indirect land use changes.

Thus, assessing impacts that are purely production-related can lead to wrong conclusions (i.e., a recommendation to intensify grain production if the gains in production mass outweigh the higher eutrophication). We therefore argue that production-related LCAs need to take into account indirect land use changes via consequential approaches, the capacity of natural systems, and define region-specific maximally allowed pollution rates, at least if applied at a larger scale or if results from a case study are generalized.

8.3.2.6 Consideration of the whole farm

A generic thought behind organic farming is the cradle-to-cradle design. Plant production activities provide animal feed and animal production activities provide organic fertilizer. One without the other is impossible and will therefore lead to flawed assessment results. Thus, there are several systems closely interlinked with each other on a mixed farm. Up to which point should emissions from organic fertilizers be allocated towards animal production and plant production? A consistent approach needs to be found, since results of different ways of allocation or system expansion can lead to very different results (Cederberg and Stadig 2003). If a whole farm approach is used, we suggest emissions from organic manure are attributed to animal production on the basis of the principle of economic allocation.

8.3.2.7 Normative assumptions

As was discussed above, the normativeness of assumptions creates a bias of results for the benefit for intensification of agriculture. For example, the higher the fertilizer amounts and the more toxic the pesticides are, the higher the risk of negative environmental impacts if substances are applied incorrectly. If risks of an improper use of the substances are not considered and good agricultural practice is assumed, a bias towards intensive agriculture is generated. We suggest taking into account the risks of toxic substances owing to improper use and over fertilization. Furthermore, empirical data need to be used as much as possible for rendering normative assumptions superfluous.

8.3.2.8 Resulting decisions of agents are not taken into account

There has been a long debate about attributional and consequential LCAs during the last years. With few exceptions, consequential LCAs are hardly ever applied in the agricultural field (Schmidt 2008; Thomassen et al. 2008a). However, the consequential perspective seems to be important in particular in agricultural LCAs. As indicated by the standards and restrictions of organic production, the management of organic farming systems differs systematically from that of nonorganic farming systems. Furthermore, agri-environmental schemes usually provide farmers low symbolic capital, which thus provides no incentive to change farmer's long-term behavior with respect to environmentally friendly farm management practices. Because organic systems are based more on the "naturalness" of the production, organic farmers might have a different system of cultural capital generation (Burton et al. 2008). Studies suggest that applying nature conservation measures is of higher symbolic capital for organic farmers (Stotten 2008), which might result in differences in uptake of agri-environmental schemes (Schader et al. 2008b). These differences in attitudes between organic and conventional farmers imply multiple and systematic effects on the environment (Morris et al. 2001; OECD 2004), which are disregarded in most of the LCA studies that were reviewed. Thus, we argue that if organic and conventional products are compared, a purely normative (standards-based) comparison leads to wrong conclusions. Positive aspects derived from empirical or modelled data and consequential LCAs should be used wherever feasible.

8.3.2.9 Bias of complexity and data availability

As discussed in relation to sustainable development, impact categories that are difficult to assess tend to be neglected or ignored. This results in the systematic trend that important environmental impacts are not considered. For instance, most assessments of the global warming potential do not include carbon sequestration in the soil, although carbon stores in agricultural soils have been identified as an important factor for climate change. We suggest including these factors as much as possible and applying in-depth sensitivity or uncertainty analysis.

Although the general use of the life cycle assessment method is seen as a significant improvement for assessing environmental impacts of agriculture, if the above implications are not taken into account, we risk biased results of current environmental impact assessments. We, therefore, recommend further research in order to consider the above mentioned problems, which need further methodological developments for a sound comparison of environmental impacts of organic and conventional products.

8.4 Conclusions

Figure 8.6 provides a qualitative overview of environmental impacts of organic farming relative to conventional farming on the basis of the review in Sect. 8.2 and taking into account the methodological discussion in Sect. 8.3. The dots represent the most frequently found result in literature. Because of regional differences, farm and management-specific impacts, and gaps in scientific measurement methodologies, there is a range of uncertainty (blue and grey color). However, several impacts can be determined relatively precisely, because their systematic influence dominates regional or farm-specific differences. The following section will summaries the conclusions regarding the different impact categories.

Impacts of organic farming on **biodiversity** range from much better to equal compared with nonorganic agriculture. **Genetic diversity** can be influenced both

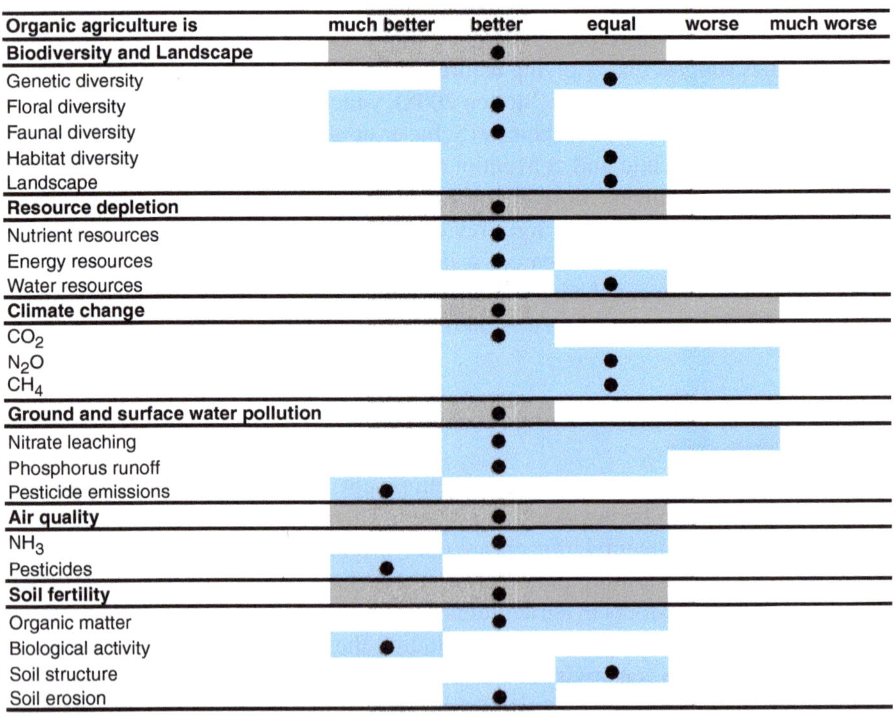

Fig. 8.6 Classification of environmental impacts and relative performance of organic farming compared to conventional farming (Source: Stolze et al. 2000, updated)

positively and negatively in organic systems. Because of lack of scientific evidence, we conclude that both systems perform equally well. According to most studies, organic agriculture clearly performs better for faunal **and floral species diversity** (Bengtsson et al. 2005). Concerning **landscape** and **habitat diversity**, organic farming may perform better because of more diverse crop rotations (Norton et al. 2009) and higher implementation rates of structural elements such as hedges and fruit trees (Schader et al. 2009). However, landscape effects are very farm and site-specific. Therefore, no general trend can be determined (Steiner 2006).

Regarding **resource depletion**, organic farming performs better regarding **nutrients** and **energy**, which confirms the evaluation done by Stolze et al. (2000). Compared with conventional farming, water consumption is not substantially affected by organic farming systems.

Stolze et al. (2000) found that organic agriculture bears a lower CO_2 and NH_3 gas emission potential than conventional farming systems. Furthermore, several long-term trials from the United States, Germany, and Switzerland (Mäder et al. 2002) show that organic farming systems are able to sequester on average 590 kg/ha per year more carbon from the atmosphere than the best performing conventional counterparts. On the basis of the suggested methodological implications, organic farming is likely to perform generally better in terms of CO_2 emissions. Regarding both CH_4 and N_2O emissions, there is not enough scientific evidence to make a final judgement. Furthermore, there are indications of better performance regarding CO_2 sequestration. Thus recent studies suggest a change in the appraisal that Stolze et al. made in 2000 from "equal" to "better."

Eutrophication of **ground and surface water** is very much dependent on what exactly is the subject of comparison. The impacts of nitrate leaching from organic farming can range from better to worse, compared to conventional agriculture. However, most of the studies analyzed found that organic farming performs better. Regarding pesticide emissions into ground and surface water, organic agriculture performs much better due to the ban on artificial pesticides.

Ammonia emissions into the **air** are lower in organic systems, mainly owing to the lower amounts of nitrogen in the system. However, depending on the assumptions, some studies show an equal performance of both systems. Because of the ban of artificial pesticides, air pollution is also lower.

Organic farming performs much better in terms of **soil** biological activity than nonorganic farming. Soil erosion and organic matter content are also affected positively by organic practices, although soil structure remains unaffected.

Both in organic and conventional farming there is potential for improving the environmental performance. Neither of the systems currently satisfies the principles of sustainability. However, organic agriculture on average performs better regarding most of the indicators than conventional systems. Furthermore, if social and economic indicators are taken into account, organic farming seems to render further benefits for society.

Methodologically, comparisons between organic farming need to be adequate to the aims of the study. As has been shown, there are several different levels of comparisons regarding scope, functional unit, stage of processing, life cycle perspective, normativeness, time perspective, scope of impacts, view on sustainability, and units of impacts. In our perception nine severe problems are commonly found in agricultural LCAs and studies using similar approaches. Besides heterogeneity of production systems, these problems are responsible for the highly fluctuating results found in the literature. We suggest reflecting methodological choices more carefully, to be aware of both the uncertainty of the results and the potentially misleading political consequences in the communication of LCA results on agricultural enterprises to consumers and policy makers. Future studies should furthermore integrate the analysis of environmental problems in the socio-economic context in order to come to a comprehensive view of sustainability of farming systems.

References

Aeschenbacher, S., and R. Badertscher. 2008. *Umweltziele Landwirtschaft – hergeleitet aus bestehenden rechtlichen Grundlagen*. Bern: Umweltwissen. Federal Office for the Environment (BAFU), Federal Office for Agriculture (BLW).

Auerswald, K., M. Keinz, and P. Fiener. 2003. Soil erosion potential of organic versus conventional farming evaluated by USLE modelling of cropping statistics for agricultural districts in Bavaria. *Soil Use and Management* 19: 305–311.

Badgley, C., J. Moghtader, E. Quinterio, E. Zakem, M.J. Chappell, K. Avilés-Vázquez, A. Samulon, and I. Perfecto. 2006. Organic agriculture and the global food supply. *Renewable Agriculture and Food Systems* 22(2): 86–108.

Bengtsson, J., J. Ahnström, and A.C. Weibull. 2005. The effects of organic agriculture on biodiversity and abundance: A meta-analysis. *Journal of Applied Ecology* 42: 261–269.

Boutin, C., A. Baril, and P.A. Martin. 2008. Plant diversity in crop fields and woody hedgerows of organic and conventional farms in contrasting landscapes. *Agriculture, Ecosystems and Environment* 123(1–3): 185–193.

Burton, R.J.F., C. Kuczera, and G. Schwarz, 2008. 'Exploring Farmers' Cultural Resistance to Voluntary Agri-environmental Schemes'. *Sociologia Ruralis* 48: 1, pp. 16–37.

Butler, G., J.H. Nielsen, T. Slots, C. Seal, M.D. Eyre, R. Sanderson, and C. Leifert. 2008. Fatty acid and fat-soluble antioxidant concentrations in milk from high- and low-input conventional and organic systems: Seasonal variation. *Journal of the Science of Food and Agriculture* 88(8): 1431–1441.

Cederberg, C., and M. Stadig. 2003. System expansion and allocation in life cycle assessment of milk and beef production. *The International Journal of Life Cycle Assessment* 6: 350–356.

Chirinda, N., M.S. Carter, K.R. Albert, P. Ambus, J.E. Olesen, J.R. Porter, and S.O. Petersen. 2010. 'Emissions of nitrous oxide from arable organic and conventional cropping systems on two soil types'. *Agriculture, Ecosystems and Environment* 136, pp. 199–208.

Christie, M., N. Hanley, J. Warren, K. Murphy, R. Wright, and T. Hyde. 2006. Valuing the diversity of biodiversity. *Ecological Economics* 58: 304–317.

Condron, L.M., K.C. Cameron, H.J. Di, T.J. Clough, E.A. Forbes, R.G. McLaren, and R.G. Silva. 2000. A comparison of soil and environmental quality under organic and conventional farming systems in New Zealand. *New Zealand Journal of Agricultural Research* 43: 443–466.

Delbecq, A.L., A.H. Van de Ven, and D.H. Gustafson. 1975. *Group techniques for program planning: A guide to nominal group and Delphi processes*, Longman higher education. Glenview: Scott, Foresman.

Edwards, C.A., R. Lal, P. Madden, R.H. Miller, and G. House. 1990. Research on integrated arable farming and organic mixed farming in the Netherlands. In *Sustainable agricultural systems*, ed.

C.A. Edwards, United States. Agency for International Development, et al. Ankeny: Soil and Water Conservation Society.
Eltun, R. 1995. Comparisons of nitrogen leaching in ecological and conventional cropping systems. *Biological Agriculture and Horticulture* 11: 103–114.
Flessa, H., R. Ruser, P. Dörsch, T. Kamp, M.A. Jimenez, J.C. Munch, and F. Beese, 2002. 'Integrated evaluation of greenhouse gas emissions from two farming systems in southern Germany. Special consideration of N2O emissions'. *Agriculture Ecosystems and Environment* 91, pp. 175–199.
Frischknecht, R., H.-J. Althaus, G. Doka, R. Dones, T. Heck, S. Hellweg, R. Hischier, N. Jungbluth, T. Nemecek, G. Reblitzer, and M. Spielmann. 2007. *Overview and methodology*, Final report ecoinvent v2.0, vol. 1. Duebendorf: Swiss Centre for Life Cycle Inventories.
Fuller, R.J., L.R. Norton, R.E. Feber, P.J. Johnson, D.E. Chamberlain, A.C. Joys, F. Mathews, R.C. Stuart, M.C. Townsend, W.J. Manley, M.S. Wolfe, D.W. Macdonald, and L.G. Firbank. 2005. Benefits of organic farming to biodiversity vary among taxa. *Biology Letters* 1: 431–434.
Gattinger, A., A. Müller, M. Häni, B. Oehen, and U. Niggli, 2010. *Klimaleistungen des Biolandbaus - Fakten und Hintergründe* Frick.
Gibson, R.H., S. Pearce, R.J. Morris, W.O.C. Symondson, and J. Memmott. 2007. Plant diversity and land use under organic and conventional agriculture: A whole-farm approach. *Journal of Applied Ecology* 44(4): 792–803.
Goulding, K.W.T. 2000. Nitrate leaching from arable and horticultural land. *Soil Use and Management* 16: 145–151.
Grönroos, J., J. Seppälä, P. Voutilainen, P. Seuri, and K. Koikkalainen. 2006. Energy use in conventional and organic milk and rye bread production in Finland. *Agriculture, Ecosystems & Environment* 117(2–3): 109–118.
Gruber, W. 1992. *Die schweizerische Agrarpolitik vor einer ökologischen Neuorientierung?* Freiburg: Universitätsverlag Freiburg.
Haas, G., F. Wetterich, and U. Köpke. 2001. Comparing intensive, extensified and organic grassland farming in southern Germany by process life cycle assessment. *Agriculture, Ecosystems and Environment* 83: 43–53.
Halberg, N. 2008. Energy use and green house gas emission in organic agriculture. *Paper presented at the international conference organic agriculture and climate change*, Enita of Clermont, France 17–18 April 2008.
Hansen, S., M.A. Bleken and B.K. Sitaula. 2008. Effect of soil compaction and fertilization practise on N_2O emission and CH_4 oxidation (Presentation). International Symposium "Organic Farming and Climate Change" 17 and 18 April 2008 at Clermont-Ferrand ENITA/France (http://www.abiodoc.com/fileadmin/uploads/Colloque/Diaporama/TH05/18_Hansen.pdf).
Heijungs, R., J.B. Guinée, G. Huppes, R.M. Lankreijer, H.A.U. de Haes, and A.W. Sleeswijk. 1992. *Environmental life cycle assessment of products*. Leiden: Centre of Environmental Science.
Herzog, F., and W. Richner. 2005. *Evaluation der ökomassnahmen—bereich stickstoff und phosphor*, Schriftenreihe der FAL, vol. 57. Reckenholz: Agroscope FAL Reckenholz.
Hoeppner, J.W., M.H. Entz, B.G. McConkey, R.P. Zentner, and C.N. Nagy. 2005. Energy use and efficiency in two Canadian organic and conventional crop production systems. *Renewable Agriculture and Food Systems* 21(1): 60–67.
Hole, D.G., A.J. Perkins, J.D. Wilson, I.H. Alexander, P.V. Grice, and A.D. Evans. 2005. Does organic farming benefit biodiversity? *Biological Conservation* 122: 113–130.
Hospido, A., J. Davis, J. Berlin, and U. Sonesson. 2010. A review of methodological issues affecting LCA of novel food products. *International Journal of Life Cycle Assessment* 15: 44–52.
Huber, B., O. Schmid, and G. Napo-Bitanem. 2010. Standards and regulations. *The world of organic agriculture—Statistics and emerging trends*, IFOAM, Bonn; FiBL, Frick; ITC, Geneva. 72–79.
IFOAM. 2009. International federation of organic agriculture movements (ifoam). http://www.ifoam.org/index.html. Accessed 5 Dec 2009.
IPCC. 2007. *Climate change 2007—Synthesis report*. Cambridge: Intergovernmental Panel on Climate Change (IPCC).

Jeanneret, P., D. Baumgartner, R. Freiermuth, and G. Gaillard. 2006. *Méthode d'évaluation de l'impact des activités sur la biodiversité*. Zürich-Reckenholz: Agroscope Reckenholz Tänikon (ART).

Jeffreys, I. 2002. A multi-objective decision-support system (modss) with stakeholders and experts in the hodgson creek catchment. In *A whole-farm and regional agroforestry decision-support system. A report for the rirdc/land & water Australia/FWPRDC joint venture agroforestry program*, eds. Harrison S., and J. Herbohn. Rural Industries Research and Development Corporation (RIRDC).

Kirchmann, H., and L. Bergström. 2001. Do organic farming practices reduce nitrate leaching? *Communications in Soil Science and Plant Analysis* 32(7): 997–1028.

Knop, E., D. Kleijn, F. Herzog, and B. Schmid. 2006. Effectiveness of the swiss agri-environment scheme in promoting biodiversity. *Journal of Applied Ecology* 43: 120–127.

Köpke, U. 2003. Conservation agriculture with and without use of agrochemicals. *Paper presented at the II. World congress on conservation agriculture "Producing in Harmony with Nature,"* Foz do Iguassu, Brazil, August 11–15, 2003.

Köpke, U., E.R. Keller, and G. Kahnt. 1997. Ökologischer landbau. In *Handbuch des pflanzenbaus 1. Grundlagen der landwirtschaftlichen pflanzenproduktion*, vol. Kapitel 9, ed. K.-U. Heyland, H. Hanus, E.R. Keller, et al. Stuttgart: Verlag Eugen Ulmer.

Krystallis, A., and G. Chryssohoidis. 2005. Consumers' willingness to pay for organic food: Factors that affect it and variation per organic product type. *British Food Journal* 107(5): 320–343.

Küstermann, B., M. Kainz and K.-J. Hülsbergen. 2008. Modeling carbon cycles and estimation of greenhouse gas emissions from organic and conventional farming systems. Renewable Agriculture and Food Systems 23(1): 38–52.

Lal, R. 2004. Soil carbon sequestration impacts on global climate change and food security. *Science* 304(5677): 1623–1627.

Lampkin, N.H. 1990. *Organic farming*. Ipswich: Farming Press Books.

Lampkin, N.H. 2007. Organic farming's contribution to climate change and agricultural sustainability. *Paper presented at the Welsh organic producers conference*, 18th Oct 2007.

Lynch, D. 2008. Greenhouse gas emissions from cropping in Atlantic Canada (Presentation). International Symposium "Organic Farming and Climate Change", 17 and 18 April 2008 at Clermont-Ferrand ENITA/France (http://www.abiodoc.com/fileadmin/uploads/Colloque/Diaporama/TH05/21_Lynch.pdf).

Mäder, P., A. Fließbach, D. Dubios, L. Gunst, P. Fried, and U. Niggli. 2002. Soil fertility and biodiversity in organic farming. *Science* 296: 1694–1697.

Meadows, D.H., D.I. Meadows, J. Randers, and W.W. Beherns III. 1972. *The limits to growth—A report to the club of Rome*. New York: Universe Books.

Morris, C., A. Hopkins, and M. Winter. 2001. *Comparison of the social, economic and environmental effects of organic, ICM and conventional farming—Final report to the countryside agency*. Cheltenham, IGER: CCRU.

Nemecek, T., O. Huguenin-Elie, D. Dubois, and G. Gaillard. 2005. *Ökobilanzierung von anbausystemen im schweizerischen acker- und futterbau*, FAL Schriftenreihe, vol. 58. Reckenholz: Eidgenössische Forschungsanstalt für Agrarökologie und Landbau (FAL).

Niggli, U. 2007. The evolution of organic practice. In *Organic farming: An international history*, ed. W. Lockeretz. Oxfordshire, Cambridge: CAB International.

Niggli, U., A. Fließbach, M. Stolze, J. Sanders, C. Schader, G. Wyss, O. Balmer, L. Pfiffner, and E. Wyss. 2008. *Gesellschaftliche leistungen der biologischen landwirtschaft*. Frick: Forschungsinstitut für Biologischen Landbau (FiBL).

Niggli, U., A. Fliessbach, P. Hepperly, and N. Scialabba. 2009. *Low greenhouse gas agriculture: Mitigation and adaptation potential of sustainable farming systems*. Rome: Food and Agriculture Organization of the United Nations (FAO).

Norton, L., P. Johnson, A. Joys, R. Stuart, D. Chamberlain, R. Feber, L. Firbank, W. Manley, M. Wolfe, and B. Hart. 2009. Consequences of organic and non-organic farming practices for field, farm and landscape complexity. *Agriculture, Ecosystems and Environment* 129(1–3): 221–227.

OECD. 2001. *Multifunctionality—Towards an analytical framework*. Paris: Organisation for Economic Co-operation and Development (OECD).
OECD. 2004. *Workshop on evaluating agri-environmental policies*. Paris: Organisation for Economic Cooperation and Development (OECD).
Olesen, J.E., K. Schelde, A. Weiske, M.R. Weisbjerg, W.A.H. Asman, and J. Djurhuus. 2006. Modelling greenhouse gas emissions from european conventional and organic dairy farms. *Agriculture, Ecosystems & Environment* 112(2–3): 207–220.
Osterburg, B., and T. Runge. 2007. *Maßnahmen zur reduzierung von stickstoffeinträgen in gewässer—eine wasserschutzorientierte landwirtschaft zur umsetzung der wasserrahmenrichtline*, vol. 307. Braunschweig: Bundesforschungsanstalt für Landwirtschaft (FAL), Landbauforschung Völkenrode—FAL Agricultural Research.
Pacini, C., A. Wossink, G. Giesen, C. Vazzana, and R. Huirne. 2003. Evaluation of sustainability of organic, integrated and conventional farming systems: A farm and field-scale analysis. *Agriculture, Ecosystems & Environment* 95(1): 273–288.
Petersen, S.O., K. Regina, A. Pöllinger, E. Rigler, L. Valli, S. Yamulki, M. Esala, and C. Fabbri, E. Syväsalo and F.P. Vinther. 2006. Nitrous oxide emissions from organic and conventional crop rotations in five European countries. Agriculture, Ecosystems and Environment 112: 200–206.
Pfiffner, L., and H. Luka. 2007. Earthworm populations in two low-input cereal farming systems. *Applied Soil Ecology* 37: 184–191.
Pimentel, D., and M.H. Pimentel. 2005. Sustainability of meat-based and plant-based diets and the environment. *The American Journal of Clinical Nutrition* 78: 660–663.
Pimentel, D., C. Harvey, P. Resosudarmo, K. Sinclair, D. Kurz, M. McNair, S. Christ, L. Shpritz, L. Fitton, R. Saffouri, and R. Blair. 1995. Environmental and economic costs of soil erosion and conservation benefits. *Science* 267: 1117–1123.
Pimentel, D., P. Hepperly, J. Hanson, D. Douds, and R. Seidel. 2005. Environmental, energetic, and economic comparisons of organic and conventional farming systems. *Bioscience* 55(7): 573–582.
Reap, J., F. Roman, S. Diuncan, and B. Bras. 2008a. A survey of unresolved problems in life cycle assessment—Part 1: Goal and scope and inventory analysis. *The International Journal of Life Cycle Assessment* 13: 290–300.
Reap, J., F. Roman, S. Diuncan, and B. Bras. 2008b. A survey of unresolved problems in life cycle assessment—Part 2: Impact assessment and interpretation. *The International Journal of Life Cycle Assessment* 13: 374–388.
Sanders, J., N.H. Lampkin, P. Midmore, and M. Stolze. 2005. A policy impact model for organic farming in Switzerland. In *ISOFAR: Proceedings of the conference "Researching Sustainable Systems", Adelaide 2005*, ed. U. Köpke, U. Niggli, D. Neuhoff, P. Cornish, W. Lockeretz, and H. Willer. Bonn: International Society of Organic Agriculture Research (ISOFAR).
Schader, C. 2009. *Cost-effectiveness of organic farming for achieving environmental policy targets in Switzerland*. Frick: Aberystwyth University, Wales. Research Institute of Organic Farming (FiBL).
Schader, C., and M. Stolze. 2010. *Nachhaltigkeitsbewertung der knospe-landwirtschaft—ergebnisse eines expertenworkshops*. Frick: Research Institute of Organic Agriculture (FiBL).
Schader, C., T. Nemecek, G. Gaillard, J. Sanders, and M. Stolze. 2008a. Using LCA data for agri-environmental policy analysis at sector level. *Paper presented at the 6th international conference on life cycle assessment in the agri-food sector 2008—Towards a sustainable management of the food chain*, Zurich, Switzerland, 12–14 Nov 2008.
Schader, C., L. Pfiffner, C. Schlatter, and M. Stolze. 2008b. Umsetzung von ökomassnahmen auf bio- und öln-betrieben. *Agrarforschung* 15(10): 506–511.
Schader, C., J. Sanders, T. Nemecek, N.H. Lampkin, and M. Stolze. 2008c. A modelling approach for evaluating agri-environmental policies at sector level. *Yearbook of Socioeconomics in Agriculture* 1: 93–132.
Schader, C., L. Pfiffner, C. Schlatter, and M. Stolze. 2009. Umsetzung von Agrarumweltmassnahmen auf bio- und konventionellen Betrieben der Schweiz. In: Mayer, J., Alföldi, T., Leiber, F., et al. eds. Werte—Wege—Wirkungen: Biolandbau im Spannungsfeld zwischen Ernährungssicherung, Markt und Klimawandel. Beiträge zur 10. Wissenschaftstagung Ökologischer Landbau, 11–13

Feb 2009, Zürich, 2009. Forschungsanstalt Agroscope Reckenholz-Tänikon (ART), Forschungsinstitut für biologischen Landbau (FiBL), ETH Zürich, Stiftung Ökologie & Landbau, 446–449.

Schmidt, J. 2008. System delimitation in agricultural consequential LCA. *The International Journal of Life Cycle Assessment* 13: 350–364.

Sehy, U. 2004. *N2O-Freisetzungen aus Ackerböden*: PhD Thesis. Technical University Munich.

Shepherd, M.A., R. Harrison, and J. Webb. 2002. Managing soil organic matter implications for soil structure on organic farms. *Soil Use and Management* 18: 284–292.

Shepherd, M., B. Pearce, B. Cormack, L. Philipps, S. Cuttle, A. Bhogal, P. Costigan, and R. Unwin. 2003. *An assessment of the environmental impacts of organic farming*. London: DEFRA, ADAS, Elm Farm, IGER.

Siegrist, S., D. Schaub, L. Pfiffner, and P. Mäder. 1998. Does organic agriculture reduce soil erodibility? The results of a long-term field study on loess in Switzerland. *Agriculture, Ecosystems & Environment* 69: 253–265.

Smith, P., D. Martino, Z. Cai, H. Gwary, H. Janzen, P. Kumar, B. McCarl, S. Ogle, F. O'Mara, C. Rice, B. Scholes, and O. Sirotenko. 2007. *Agriculture*, Climate Change 2007: Mitigation. Cambridge: Contribution of Working Group III to the Fourth Assessment Report of the Intergovernmental Panel on Climate Change. Cambridge University Press.

Steiner R.S. 2006. Landnutzungen prägen die Landschaft—konventionelle, bio-organische und bio-dynamische Anbaumethoden im Vergleich in ihrer Wirkung auf die Agrarlandschaft im Kanton Zürich.

Stern, N. 2007. *The economics of climate change—The Stern review*. London: Cambridge University Press.

Stockdale, E.A., N.H. Lampkin, M. Hovi, R. Keatinge, E.K.M. Lennartsson, D.W. Macdonald, S. Padel, F.H. Tattersall, M.S. Wolfe, and C.A. Watson. 2001. Agronomic and environmental implications for organic farming systems. *Advances in Agronomy* 70: 261–327.

Stolze, M., A. Piorr, A.M. Häring, and S. Dabbert. 2000. *Environmental impacts of organic farming in Europe*, Organic farming in Europe: Economics and policy, vol. 6. Stuttgart-Hohenheim: Universität Stuttgart-Hohenheim.

Stopes, C., E.I. Lord, L. Philipps, and L. Woodward. 2002. Nitrate leaching from organic farms and conventional farms following best practice. *Soil Use and Management* 18: 301–308.

Stotten, F. 2008. *Akzeptanzanalyse von naturschutzmaßnahmen im rahmen der schweizerischen agrarpolitik*. Aachen: RWTH Aachen.

TEEB (2010). *The Economics of Ecosystem and Biodiversity: Ecological and economic foundations*. In Kumar, P. (ed.), London and Washington: Earthscan.

Thomassen, M.A., R. Dalgaard, R. Heijungs, and I. de Boer. 2008a. Attributional and consequential lca of milk production. *The International Journal of Life Cycle Assessment,* 13(339–349).

Thomassen, M.A., K.J. van Calker, M.C.J. Smits, G.L. Iepema, and I.J.M. de Boer. 2008b. Life cycle assessment of conventional and organic milk production in the Netherlands. *Agricultural Systems* 96: 85–107.

UNECE. 1999. *Gothenburg protocol to abate acidification, eutrophication and ground-level ozone*. Gothenburg: United Nations Economic Commission for Europe (UNECE).

Vogt, G. 2007. The origins of organic farming. In *Organic farming: An international history*, ed. W. Lockeretz. Oxfordshire, Cambridge: CAB International.

Willer, H. and L. Kilcher. 2009. *The World of Organic Agriculture, Statistics and Emerging Trends 2009*: IFOAM, Bonn; FiBL, Frick; ITC, Geneva.

Woese, K., D. Lange, C. Boess, and K.W. Bögl. 1997. A comparison of organically and conventionally grown foods—Results of a review of the relevant literature. *Journal of the Science of Food and Agriculture* 74: 281–293.

Younie, D. and C. A. Watson. 1992. Soil Nitrate–N Levels in organically and intensively managed grassland systems. Aspects of Applied Biology, 30: 235–238.

Ziesemer, J. 2007. *Energy use in organic food systems*. Rome: Natural Resources Management and Environmental Department, Food and Agriculture Organisation of the United Nations (FAO).

Zittel, W., and J. Schindler. 2007. *Crude oil: The supply outlook*. Berlin: Energy Watch Group.

Chapter 9
Food transportation issues and reducing carbon footprint

Wayne Wakeland, Susan Cholette, and Kumar Venkat

9.1 Introduction

Transportation is the largest end-use contributor toward global warming in the United States and many other developed countries. The U.S. Department of Energy (DOE 2009) calculates that CO_2 emissions from transportation surpassed two billion metric tons in 2007. Yet a survey by Golicic et al.(2010) finds that fewer than 10% of Fortune 500 companies have addressed the environmental impacts of transportation, and even fewer are actively implementing improvements, despite the fact that such initiatives would also tend to reduce fuel usage and costs in the long run.

Transportation has a significant impact within the food and beverage sector because food is often shipped long distances and not infrequently via air. Heller and Keoleian (2000) estimate that diesel fuel use accounts for 25% of the total energy consumed within the U.S. food system. Pirog et al. (2001) report that nearly half of all fruit sold in the United States is imported, and that produce grown in North America travels an average of 2,000 km from source to point of sale.

Although the impact of transportation is important, full life cycle analyses indicate that for most foods transportation does not have the largest environmental impact. Some analysts, such as Weber and Matthews (2008), estimate that given the

W. Wakeland (✉)
Portland State University, Portland, OR, USA
e-mail: wakeland@pdx.edu

S. Cholette
San Francisco State University, San Francisco, CA, USA
e-mail: cholette@sfsu.edu

K. Venkat
CleanMetrics Corporation, Portland, OR, USA
e-mail: kvenkat@cleanmetrics.com

J.I. Boye and Y. Arcand (eds.), *Green Technologies in Food Production and Processing*, Food Engineering Series, DOI 10.1007/978-1-4614-1587-9_9,

typical household food basket, aggregate transportation accounts for just 11% of total carbon emissions associated with food production. We show in Sect. 9.5 that freight transport accounts for just 6% of overall emissions in the U.S. food sector, but its life cycle impact is greater in the case of plant-based foods that have relatively low production emissions. Therefore, it is still worthwhile to consider improving the food distribution system. There are often many options for delivering food to consumers, and these supply chain configurations can result in vastly differing energy and emissions profiles. In this chapter, we provide the background and tools for analyzing the energy intensity and resultant emissions of a food distribution system, evaluating tradeoffs and identifying opportunities for significant improvement. Note that we use the terms "carbon emissions" and "greenhouse gas emissions" interchangeably in this chapter, implying in both cases that all significant greenhouse gases emitted in a process are counted and reported as a single "carbon dioxide equivalent" figure.

9.2 Supply chain basics

Before we can further investigate transportation impacts, we must first introduce the concept of the supply chain: the sequenced network of facilities and activities that support the production and delivery of a good or service. Given the obvious importance of the supply chain, this field is rife with terminology and buzzwords, many of which are synonymous. For instance, supply chains are sometimes referred to as "demand chains" or "value chains." A supply chain starts with basic suppliers and extends all the way to consumers via stages. These stages may include such facilities as suppliers, factories, warehouses and other storage facilities, distribution centers, and retail outlets.

Figure 9.1 shows a sample supply chain, where the arrows denote the flow of a product toward the consumer. This figure depicts both inbound logistics (the delivery of raw materials and packaging to the manufacturer) as well as outbound logistics (the transportation and storage of the finished good to the end consumer). This chapter focuses on outbound logistics, colloquially known as "gate-to-kitchen" and "farm-to-fork" in the food and beverage industry. The emissions associated with outbound logistics vary by origin and type of food. Weber and Matthews (2008) estimate that food transportation may account for 50% of total carbon emissions for many fruits and vegetables, but less than 10% for red meat products. Although

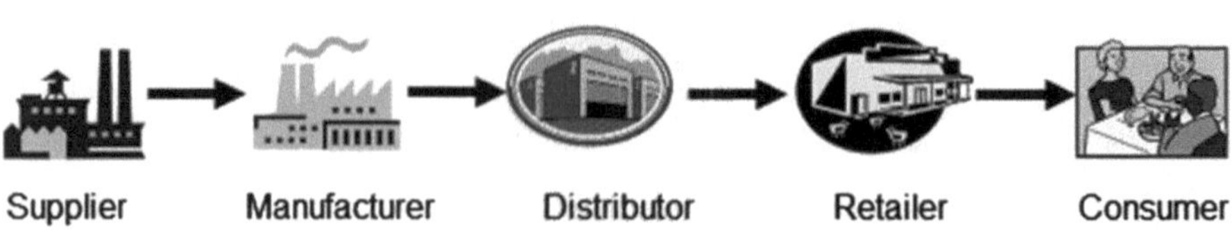

Fig. 9.1 A simple supply chain

inbound logistics can require substantial energy use, it is considered part of the production process and is discussed in earlier chapters.

Although the interrelationships between supply chain stages may be quite complex, all supply chains have one aspect in common—they end with a consumer. Supply chains for different products may be interlinked; one supply chain's end consumer may represent an intermediate node for another supply chain. Examples include a firm that buys components and assembles them into consumer items, and a soft drink producer that buys cylinders of compressed CO_2 to carbonate its products.

Much supply chain complexity results from the fact that few supply chains are completely controlled by one firm or vertically integrated. For example, producers and retailers are not typically owned by the same firm. Companies may outsource supply chain activities, especially transport and storage activities, which are handled more effectively by third party logistic (3PL) providers. Outside firms that form a part of a company's supply chain are channel partners. These partnerships require collaboration across organizations. We define supply chain management (SCM) as the coordination of business functions within an organization and between the organization and its channel partners. SCM strives to provide goods and services that fulfill customer demand responsively, efficiently, and sustainably.

SCM includes such functions as demand forecasting, purchasing (also known as sourcing), customer relationship management (CRM), and logistics. Logistics concerns the movement and storage of goods, services, and information. It is an umbrella term for such important functions as transportation, inventory management, packaging, and returns/reverse logistics. Some terminology will be helpful to understand who is doing what. The shipper initiates the movement of the product forward into the supply chain, the carrier is the party that does the actual moving of the product, and the consignee receives the product.

9.2.1 Transport modes

Within the developed world there are four basic transport modes for shipping large quantities of packaged products: water, rail, truck, and air. Trucking dominates, comprising more than 75% of the total U.S. freight transit bill. Trucking variables include truck type, ownership model (such as 3PL or company-owned fleet), and loading option (less-than-truckload or full-truckload). The dominant transport mode has shifted over time. The first transport revolution occurred when inland water transport replaced animal caravans. In the mid-1800s railroads displaced inland water as the dominant form of cargo transport, and in the mid-1900s trucking displaced railroads. Air cargo is a more recent and growing transport mode popular for short life cycle products such as flowers and luxury foods. The U.S. DOE (2009) estimates that air transport accounts for 9% of U.S. transportation fuel usage. Interestingly, water transport has started to make a comeback. In the United Kingdom (UK), for example, Tesco is relying on inland waterway barges for transporting

Table 9.1 Energy and emissions per ton-km

	MegaJoules per ton-km	kg CO_2e per ton-km
International water-container	0.2	0.14
Inland water	0.3	0.21
Rail[a]	0.3	0.18[a]
Truck[b]	2.7	1.8
Air[c]	10	6.8

Note that utilization and backhaul rates will affect all figures
[a]May depend on whether diesel or electric power is used
[b]Depends on size and type of truck, power source
[c]Includes effects from radiative forcing
Source: Based on data from Weber and Matthews (2008)

more of their beverage products. Short sea shipping, using ocean-going vessels for delivering cargo domestically, is popular in Europe and also holds promise for replacing many truck deliveries in the United States.

To compare transport modes with regard to energy usage and resultant emissions, we define a ton-km as the movement of 1 metric ton of cargo over 1 km. Table 9.1 shows that these modes have very different energy and emissions profiles. Caveats abound regarding the accuracy of these figures, but clearly air freighting is much more energy and emission intensive compared to other modes, especially water and rail. Of course, water and rail transport modes are contingent upon the availability of navigable water and established railroad tracks. An additional consideration is the potential need for supply chain responsiveness: air freight may be the only viable option for long-distance transport when customer orders require immediate fulfillment.

9.2.2 *Intermodal transport*

Before we choose one mode over another, we should consider intermodal transport. Defined as using more than one transportation mode to move a shipment between two points, an intermodal route might involve shipping cargo by water, then by rail, then by truck. Intermodal transport became practical with the advent of containerization, where products stay in the same container throughout their entire journey. Containerization was made possible through global standardization of container size and features, which dramatically reduced intermodal transfer times and significantly increased cost efficiency. From a sustainability viewpoint, the advantage of intermodal transport is that we can utilize more efficient modes for major transport corridors, and then shift to trucks for transport to remote destinations. Shippers can also use a 3PL provider to oversee the entire shipping process. One disadvantage of intermodal transport is its inherent complexity of coordination and the information technology support required to address that complexity. Another issue is the movement and repositioning of empty containers.

9.2.3 *Utilization and backhaul*

Many carbon analyzers base calculations on only transport mode and shipping distance. In our analysis, we will take into account additional factors, including vehicle utilization (how full the vehicle is) and backhaul (whether or not the vehicle carries freight on its return journey). Although fully laden vehicles use more fuel than nearly empty ones, most of the energy expended during a trip is used to move the vehicle and not its cargo. Underutilized vehicles waste energy, as do vehicles that return empty. Also, weight and volume limits must be respected, and all but the lightest and bulkiest cargo loads tend to "weigh out" rather than "cube out."

It can be difficult to determine utilization fractions and backhaul percentages, as these are likely to vary with each trip. Such information is even more challenging to obtain when transportation functions have been outsourced. However, some assumptions can be made. For example, vehicles chartered by 3PL providers are likely to have higher utilization fractions because they often carry cargo from multiple companies. Third party logistic providers are also likely to have higher backhaul rates, because they have more opportunities for obtaining return freight owing to their broader customer base.

9.2.4 *Warehousing*

Logistics involves not only the movement of goods, but their storage. Unless a product is custom ordered by an onsite client, it is likely that the product will enter storage at some point in its journey to the consumer. Such storage can occur at any supply chain stage: at the producer, distributor warehouse, and/or retailer stockroom. Intermediate supply chain stages range from pure storage centers to dedicated cross-dock facilities, in which cargo from upstream supplier trucks/railcars is transferred directly to outbound trucks/railcars destined for downstream stages. In addition to storage, warehouses can provide additional services: pick and pack (repackaging palletized products to smaller quantities destined for either retailers or end consumers), customs clearance, or even house product-finishing functions such as customizing goods to the local marketplace.

9.2.5 *Packaging*

Packaging decisions are inherently linked to the supply chain. Goods are frequently shipped in bulk and broken into consumer-sized quantities at a warehouse or other facility, and individual commodities are sometimes bundled into larger end-items, such as multipacks, and palletized. Packaging materials (pallets, boxes, totes, slipsheets, etc.) for both finished goods and intermediate support functions may be designed to be recyclable, compostable, or reusable. Non-landfilled packaging is highly desirable, but creates other challenges, such as the impact of reusable packaging in the reverse supply chain.

Packaging can often be reengineered to reduce package weight or bulk, which can translate into savings in raw materials, landfill impacts, and transport/storage energy use; but extra costs may be incurred elsewhere. For example, Safeway is working with Kimberly-Clark to pilot palletless deliveries of paper products. Although this may allow trucks to be packed with more products, labor costs are likely to increase in receiving, and pallets would still not have been totally eliminated from the supply chain because pallets would continue to be used at local warehouses.

9.3 What makes food supply chains special?

We have shown that supply chains can be long and complex. Food supply chains are some of the most difficult to manage as they must often address time constraints to avoid spoilage, as well as concerns about contamination, high weight-to-value ratios, fragility, unique packaging requirements, and the potential impact of food being wasted rather than consumed. We show here how these considerations affect outbound logistics.

One challenge relates to food production being inherently dependent on nature. Not only is the cultivation of many foods restricted geographically, but also temporally. Fruits, vegetables, and grains typically have fixed growing cycles with short and specific annual harvest periods. However, North American and European demand for many of these items is year round. There are three options for supplying fresh produce that is out of season locally: sourcing from distant growing areas, using long-term storage, or cultivating in a protected environment such as a greenhouse. Importing produce often results in lower overall emissions than harvesting and storing local produce for several months, as Hospido et al. (2009) and Milà i Canals et al. (2007) show for lettuce and apples, respectively. Indeed, energy needed for long-term cold storage can dominate a product's overall emissions profile. Carlsson-Kanyama (1998) shows, for example, that storage accounts for 60% of the carbon emissions associated with carrots. Higher emissions can result not only from the energy needed for climate control, but also from the inherent yield losses that occur during storage. Protected cultivation is even more energy intensive. Carlsson-Kanyama et al. (2003) show that tomatoes produced locally in Swedish greenhouses require ten times the energy as field-grown tomatoes imported from Southern Europe. Thus, long-distance supply chains, even though they are energy intensive, may yield the lowest overall footprint for providing out-of-season product to consumers.

A second challenge is related to situations where similar food commodities are produced locally as well as imported from distant locations, the emissions intensity of the production methods must be considered in any comparison of overall supply chain emissions. For example, Saunders and Barber (2007) find that milk solids produced locally in the United Kingdom generate 34% more emissions than the same product imported from New Zealand, even with transport included. This result reflects the more energy-intensive dairy production system in the United Kingdom.

A third challenge is that highly perishable foods require special handling to avoid yield loss and potential health issues. These foods often require cooling, refrigeration, or freezing during transport and/or storage. It may also be necessary to control other conditions, such as humidity, exposure to air, or contact with other items. These requirements increase energy usage and emissions.

A fourth challenge is that the location of facilities within a food supply chain can also affect emissions. For example, Sim et al. (2007) find that overall carbon emissions can be significantly reduced by locating processing and storage facilities in countries where more electricity is generated from renewable fuels or cleaner energy.

Fifth, when time is of the essence, as in the transport of highly perishable produce such as berries, air freight may be the only viable transport option. Air freighting may also be necessary in regions such as Africa where no other viable alternative exists for transporting produce to market. As previously shown, air freighting is highly energy intensive. Scholz et al. (2009) report that fresh salmon air freighted from overseas has about twice the environmental impact as frozen salmon transported by container ships over the same distance. The difference owing to transport modes is far more significant in this case than production choices such as wild versus farmed or organic versus conventional.

A sixth challenge is that safe food storage not only requires climate controls, but also a high degree of sanitation. In most developed countries, warehouses must be built and maintained to stringent guidelines to be certified as "food grade." In the United States, wood pallets may not be reused and may soon be phased out as unsanitary.

The process of packaging food is yet another challenge. Twede et al. (2000) emphasize that packaging beverage products is a high-speed automated process involving expensive equipment. Such capital investment and the need for a controlled environment favors centralizing packaging at the point of production, even if it might be more energy efficient to ship product in bulk. Food and beverage products typically require extensive packaging, which adds both weight and volume to the product. Additional energy and materials are required to create the packaging and transport it to the production site. Point (2008) performs a life cycle assessment of the Nova Scotia wine industry and finds that the largest contribution to emissions is owing to the production and transport of wine bottles.

9.4 Measuring transportation-related carbon emissions

This section presents the basics of performing a carbon audit and concludes with some examples from practice. Although other gases such as methane and nitrous oxide may contribute to global warming, aggregate greenhouse gas measures are typically reported in CO_2 equivalents (CO_2e), which is kgs of CO_2 emitted per kg of product. Carbon dioxide dominates, comprising 95% of total greenhouse gas emissions by volume (World Resources Institute 2004). Some points of confusion in carbon measures exist. For instance, U.S. documents report tons emitted, where most of the world measures in the SI (metric) units and reports in tons, as will we. Emissions are colloquially called "carbon emissions," which can lead to confusion as some older studies only weigh the carbon component of the gas, which is 30% of the total mass of CO_2. It is also now standard practice to report all significant greenhouse gas emissions as a single carbon emissions figure. The scope of the analysis depends on the purpose of the study. Scope 1 includes only direct emissions, whereas Scope 2 also includes indirect emissions from any consumption of purchased electricity, heat, or steam. Scope 3 is the broadest, including all other indirect emissions, such as the extraction and production of purchased materials and fuels, all outsourced activities, and waste disposal. Scope 3 can also include the substantial impact of radiative forcing from the contrails in tallying airplane emissions.

Scope must be carefully considered because incomplete framing (inappropriate scope) may lead to incorrect conclusions. For example, food miles are defined as the distance between the production source and the retail store, or "farm-to-fork." This metric has received substantial attention in the popular press and has been adopted by the business community. For example, the UK supermarket chain Tesco now provides food mile information. However, there is often no consideration of the energy used to transport supplies to the farm or the energy used for processing or storage. For example, Saunders et al. (2006) estimate that grass-fed lamb from New Zealand produces lower emissions overall than locally raised lamb fattened in a feedlot. Transport modes such as ocean and rail may actually be more efficient on a per-weight basis, even over long distances. Considering another dimension of sustainability, some African and South American farmers derive their livelihood from the service export markets. Tradeoffs between different facets of sustainability (environment, economics, equity; or planet, profits, people) can mean that carbon-based metrics may be misleading, especially in the context of incomplete framing.

Now that we understand supply chain basics and special logistical issues faced in food distribution, it would seem we are ready to collect data and enter this data into an analysis tool to derive the definitive answer to the question: How much carbon does our supply chain emit? Before we start broadcasting results with conviction and certainty, however, let us consider the following scenario. A person travels from San Francisco to New York and desires to purchase carbon offsets for the round-trip flight. The Internet has many free online carbon calculators, often with donation links for offsetting one's carbon footprint. Table 9.2 shows results from several calculators that target the typical U.S. consumer and report results in tons (the unit used in the United States). Even for a well-defined trip, the emissions reported and the recommended amount of carbon offsets to purchase vary widely. The amounts vary for both logical reasons (such as whether radiative forcing is included or not) and for obscure reasons (such as JetBlue's claim to being almost twice as carbon efficient as United Airlines). Note also that there are many other factors that may or may not have been considered by these carbon calculators, such as plane age and

Table 9.2 Divergent results of online carbon calculators—round trip from San Francisco to New York

	Tons CO_2e	Recommended offset	Implied $ per ton
Carbonfund.org	0.93	$9.34	$10.04
Adding radiative forcing	2.52	$25.22	$10.01
Terrapass.com			
Via JetBlue	1.462	$11.90	$8.14
Via Virgin	1.584		
Via United Airlines	2.215	$11.73	$5.30
Sustainabletravelinternational.org	1.86	$47.31	$25.44
Nativeenergy.com	2.055	$42.00	$20.44
Bonneville Education Foundation			
www.b-e-f.org	4.192	$56.00	$13.36

model, weather, utilization, and backhaul. Table 9.2 also shows that even when the amount of emissions being offset is about the same, the suggested donation is different. Different carbon calculators make different assumptions about the price per unit of offset.

So what lessons can be learned? First, it may not be realistic to expect highly accurate estimates of carbon emissions. It may be more important to strive for consistency and to avoid using different tools or techniques when comparing across scenarios. Johnson (2008) recommends that carbon footprints be defined sensibly and transparently because definitive standards have not yet emerged. It is also important to consider who will use the analyses and for what purpose. When providing consumers with recommendations for assessing personal transportation footprints, it may be appropriate to acknowledge the disparities between calculators and to provide a range of recommended offsets. There is a risk of backlash when savvy Internet users survey websites and discover how divergent the results can be. As an example of consumer skepticism, Rosenthal (2009) reports that in Sweden, where carbon labeling is starting to appear, the typical consumer reaction to carbon labels is bemusement.

In addition to providing consumers with information, carbon audits can provide useful insights for companies evaluating their operations. However, we believe that the figures from carbon audits should be viewed as guidelines rather than as precise and absolute truths. Given these caveats, we are now ready to consider some results from actual carbon audits performed for companies.

Case study: wine delivery to consumers

To illustrate the impact that supply chain design and implementation can have on carbon emissions, our first case considers the delivery of wine to the consumer. Energy usage associated with postproduction logistics is high for wine because the standard consumer packaging (a 750 mL glass bottle) is fragile, heavy, and bulky. Wine comprises 50% of the weight and less than 40% of the volume of a case of 12 bottles. Wine is also sensitive to temperature and must be stored in a controlled

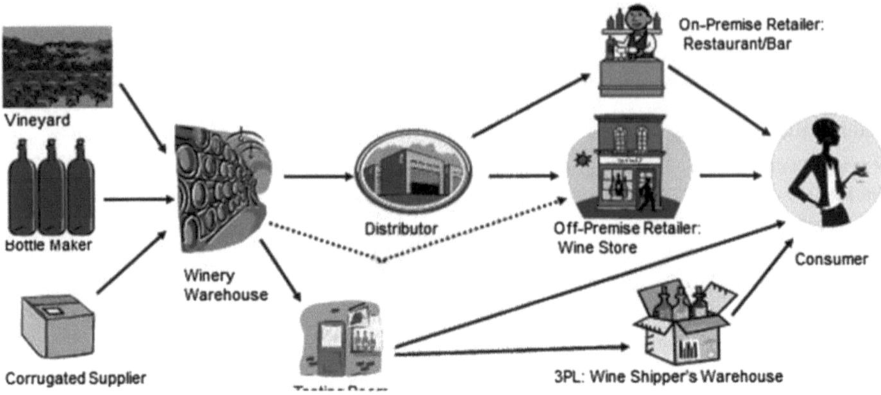

Fig. 9.2 An outbound logistics network for a winery

climate for all but the shortest periods. The outbound logistics network, depicted in Fig. 9.2, depicts paths with differing numbers of intermediary stages and several retail channels. The majority of wine sold in the United States is delivered within the framework of the 3-tier system, with product flowing from the winery to a distributor/wholesaler, then to a retailer before reaching the end consumer. However, in most states, wine can also be sold directly to the end consumer, either through tasting room sales, or through wine clubs and mailing lists where delivery is typically supported through a 3PL provider such as IBG or ShipCompliant, Inc. A few states such as California and Oregon allow wineries to self-distribute directly to both on-premise and off-premise retailers.

Cholette and Venkat (2009) use an online carbon calculator to model each of these options and provide a stage-by-stage view of resulting emissions for shipping a half case of wine to both local and cross-country consumers. The tool utilized, CargoScope (Venkat 2008), considers transportation distances, mode, temperature control, utilization, and backhaul rates for each link within the supply chain. Figure 9.3 provides a comparison of some of the different options investigated both for small and long-distance consumers. Not surprisingly, cross-country transport by rail (scenario D4) is more efficient than trucking (D1), which in turn is better than air freight (D2). Direct-to-consumer small package local delivery (L4) is the most efficient, in part because the overall transportation distance is minimized and the vehicles servicing the delivery area are highly utilized because of its compactness. Making a dedicated trip to the winery in a typical gasoline-powered car (L5a) results in 80 times more emissions than the least carbon-intensive method (L4). Most local commercial delivery configurations result in lower emissions than their long-distance counterparts, but there is a notable exception: long-distance delivery via rail (D4) is effectively equivalent to the standard, local 3-tier distribution scenario (L1).

Figure 9.3 also shows that the most energy-intensive transit link is often the last one–driving to the store. This is not surprising, because other studies, such as those by Browne et al. (2005) and Van Hauwermeiren et al. (2007), find that this link can

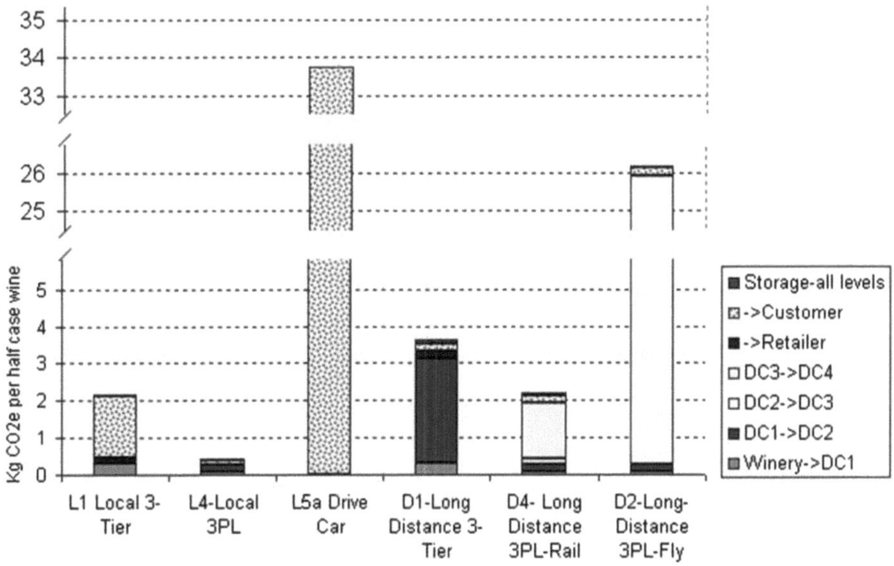

Fig. 9.3 Carbon emissions for local and long distance delivery scenario (Based on Cholette and Venkat 2009)

be the most carbon intensive even in European countries where consumers are traditionally more energy conscious than their U.S. counterparts. Because it is the least measurable and the most difficult to control, the retail-to-consumer link is typically outside the system boundary of most analyses. However, it may be worth considering the retail-to-consumer link when options include home delivery or when it may be possible to influence consumer shopping behavior. For example, if wine producers or retailers could provide incentives for consumers either to make larger purchases less often or to visit the store by bicycle, foot, or public transport, then the energy intensity of this last segment could be reduced. However, for supply chains involving multiple stages, such incentives may be impractical or difficult to implement and monitor.

Cholette and Venkat (2009) also find that no single supply chain configuration is ideal for all wineries. Large wineries that sell in volume to retailers, whereby a typical delivery would fill a reasonably efficient midsized truck, could consider self-distribution. For small wineries, where a typical delivery would fill a less-efficient light truck, 3-tier distribution would be more efficient than self-distribution. For retail store chains, the key to reducing carbon emissions can be to design their supply chain to maintain high vehicle utilization rates. For example, if a store chain moves sufficient volumes to be able to keep their fleet of delivery trucks fully utilized transporting goods to and from their distributors' warehouses, the third tier of the 3-tier distribution channel (a central warehouse for the retailer) may not be necessary, thereby saving considerable cost and reducing carbon emissions.

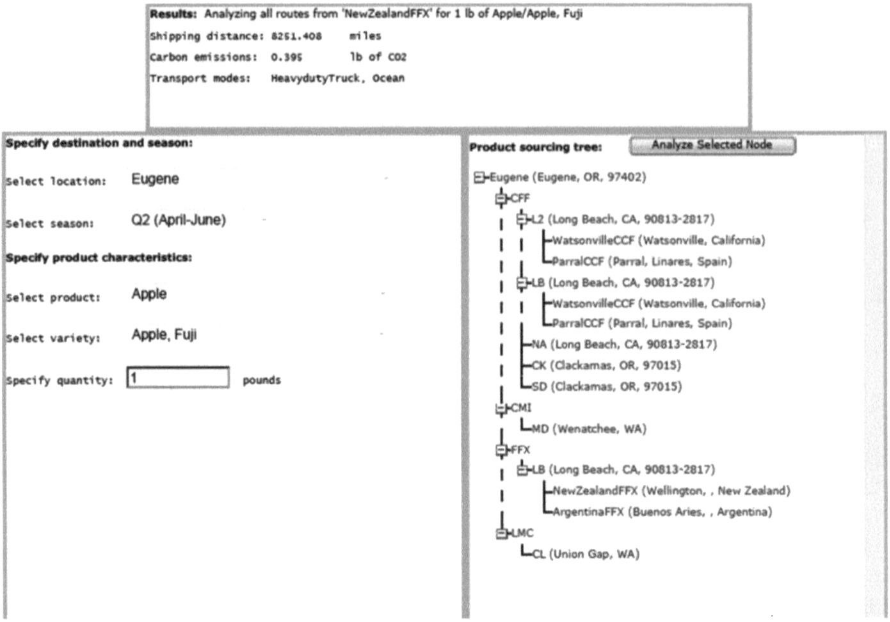

Fig. 9.4 Analysis of a supply chain for apples

Case study: an online supply chain emissions calculator

Organically Grown Co. (OGC), an Oregon-based regional distributor of organically produced fruits and vegetables, undertook a major initiative in 2008 to track the transport of every product that reached one of their three distribution centers located in northwestern United States. The company wanted to be responsive to customer requests for information about the environmental impacts of their products and to be a part of the sustainability discussion in the region.

About one third of the company's products are sourced from regional growers; the remainder are transported long distances from other states such as California and Hawaii, as well as from South America and other distant locations such as New Zealand. Transport modes include air, ocean, and land, as well as multimodal transport. Many products require temperature control during transit and storage. Products that have very short shelf lives often must be transported by air, whereas others can be sent refrigerated by ocean or land.

A web-based greenhouse gas emissions analyzer developed by CleanMetrics Corp. (2010) made use of OGC's database of thousands of suppliers and products, to compare the transport impacts of products sourced from different suppliers and locations. Figure 9.4 shows a typical result from the analysis of an international supply chain for apples, a typical crop that has a fixed harvesting cycle, which necessitates diverse sourcing in order to meet year-round demand. The destination

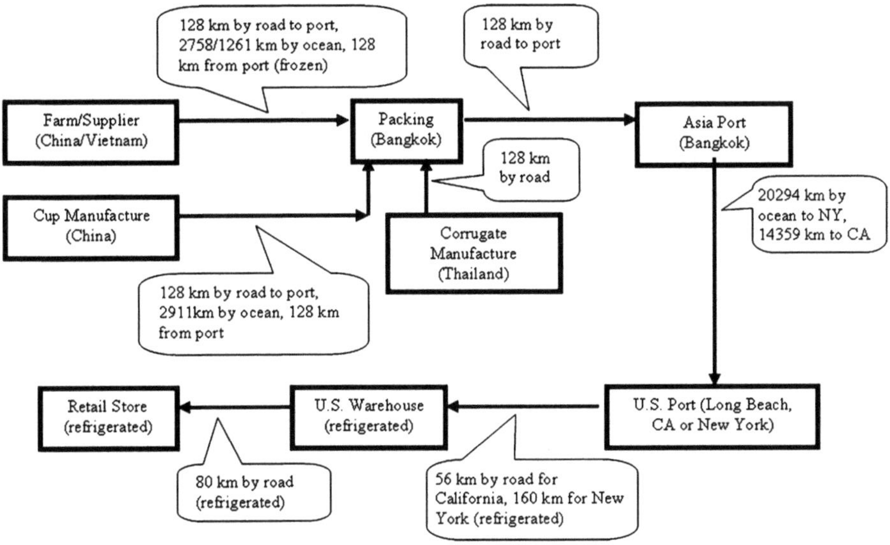

Fig. 9.5 A packaged fruit product supply chain

is a warehouse located in Eugene, Oregon, and the product is sourced from a variety of producers, ranging from growers that are local (Oregon) and regional (California and Washington) to international (Argentina and New Zealand). Results from these analyses provide OGC with visibility into product supply chains and information needed to incorporate transportation-related carbon footprint as one component of their food purchasing decisions.

Case study: an international supply chain for packaged fruit products

Sundia Corporation, based in Oakland, California, produces and distributes fresh fruits cut and packaged in plastic cups. Once packaged, the products do not require refrigeration until they are placed on supermarket shelves. This process allows the company to procure tropical fruits in Asian countries such as Vietnam and Thailand where they are grown, process the fruits close to the source location, and then ship the packaged products by ocean to stores in New York and California at a relatively low carbon cost. Figure 9.5 depicts the entire supply chain. A carbon footprint analysis of the supply chain shows that only about 30% of total emissions could be attributed to transport. The longest transport leg, the ocean segment from an Asian port to the United States, spanning 14,000–20,000 km, was responsible for just half of the transport-related emissions. The remaining 70% of total emissions came from growing, processing, and packaging the product. This is one of many recent industry examples where transport distances have not been good indicators of overall life cycle greenhouse gas emissions.

Case study: home delivery of groceries versus consumers driving to stores

Many national and regional grocery chains now offer convenient home delivery services. An example of this is the service provided by New Seasons Market in the Portland, Oregon metropolitan area. The company uses a fleet of delivery vans with the capacity to make up to ten deliveries on a single route. The vans are fueled with a blend consisting of 20% biodiesel, and they have separate cargo areas for unrefrigerated, refrigerated, and frozen goods. Deliveries are made directly from each of the company's retail stores to customers within a certain distance from that store. Customers typically place their grocery orders online and receive an estimated delivery time. Store employees shop for each customer and then load the filled shopping bags into a designated van. Delivery routes are calculated and mapped in advance using mapping and routing software such as MapPoint.

A carbon footprint study conducted by CleanMetrics (2010) for New Seasons Market compared the emissions produced by the delivery vans with the emissions that would have been generated had the customers driven from their homes to the nearest store. Using delivery data that included street addresses and delivery routes, the study calculated carbon emissions for both scenarios using actual driving distances and found that the delivery vans were more efficient by a factor of almost two. This result suggests that there may be significant potential for emissions reductions in the last leg of most supply chains where the product is finally delivered to consumers.

9.5 Putting transport emissions in context

Although transportation generally does not have the largest environmental impact in food supply chains, it can play a significant role depending on the specific supply chain and the modes of transport used. In this section, we put transportation-related greenhouse gas emissions in context by considering the major life cycle phases of food products in the U.S. food sector.

Based on food consumption and food waste data published by the United States Department of Agriculture (USDA 2008), we analyzed the typical life cycles of major food categories consumed in the United States, including meats, dairy, eggs, seafood, grains, nuts, vegetables, and fruits. This data set includes quantities of raw and processed food products delivered to retail locations, as well as percentages of food wasted at the retail and consumer levels. The analysis was conducted using the CarbonScope analytical tool (Venkat 2009). Typical cooking and waste disposal processes were assumed for the various food categories, with a standard freight transport distance of 1,500 miles by semi-trailer truck from the farm or processing facility to a typical retail location. Packaging, transport from retail stores to consumers' homes, and home refrigeration were not included in this analysis.

The results showed that freight transportation accounts for just 6% of the total carbon emissions (in millions of metric tons of CO_2e) for the food categories analyzed, indicated by the short bar in Fig. 9.6. Considering both animal-based and

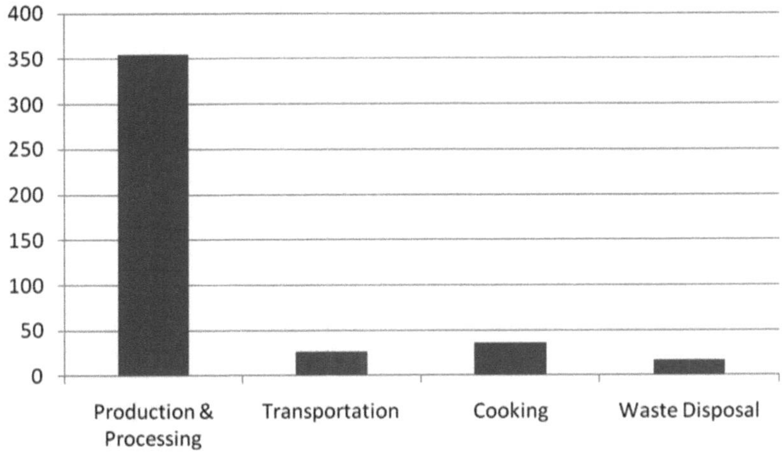

Fig. 9.6 Life cycle carbon emissions (millions of metric tones of CO_2e) for major food categories in the United States

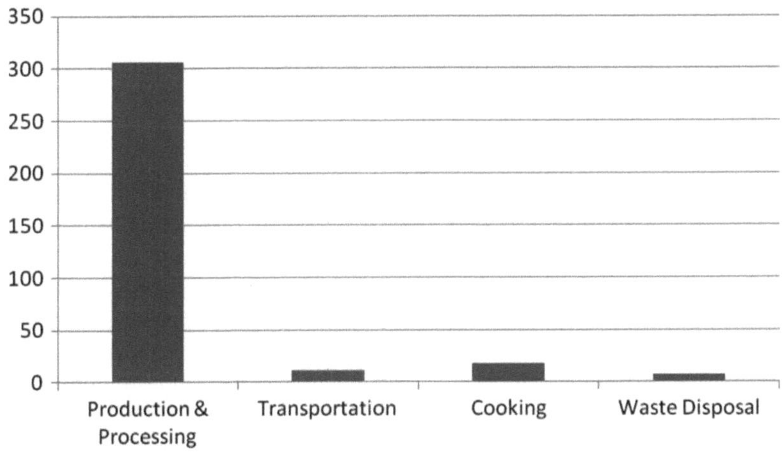

Fig. 9.7 Life cycle carbon emissions (millions of metric tons of CO_2e) for animal-based foods

plant-based foods, production and processing dominate with 81.6% of the emissions, followed by cooking, with 8.3%. Only waste disposal produces fewer emissions than transport, primarily because of the fact that more than 44% of the methane emissions from landfills are typically flared or converted to useful energy.

If we examine only the food products derived from animals (Fig. 9.7)—including meat, seafood, dairy and eggs—transportation contributes even less: just slightly more than 3% of total carbon emissions. For animal products, the high production-related carbon emissions dwarf emissions from all other factors.

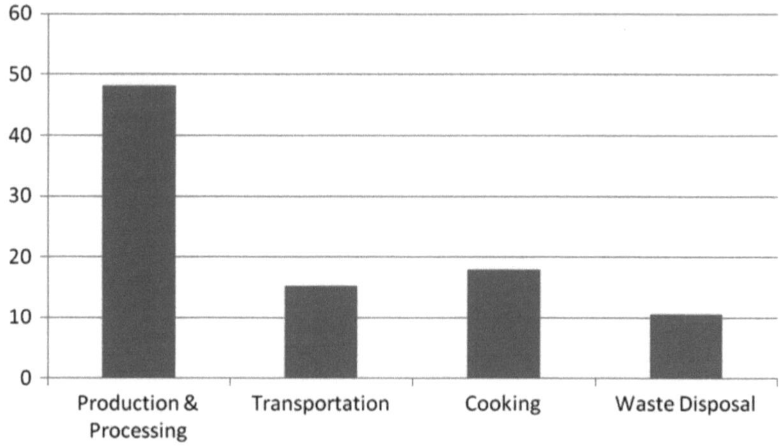

Fig. 9.8 Life cycle carbon emissions (millions of metric tones of CO_2e) for plant-based foods

On the other hand, if we consider only the plant-based products (Fig. 9.8)—including grains, nuts, fruits and vegetables—transportation contributes more than 16% of the life cycle emissions, because of the relatively low emissions from production. The preceding results indicate that the degree to which carbon emissions can be reduced by optimizing the distribution network depends on whether the food items are plant-based or animal-based.

If we replace our assumption of long-distance road transport with other distances and transport modes, the results will change significantly. For example, if air transport is used to deliver fresh imported foods from distant production locations, transportation will be a major contributor to the total life cycle emissions, regardless of production emissions. Ocean transport, on the other hand, generally produces low transport emissions per unit of freight. For foods that are imported via ocean, the road transport to and from the sending and receiving ports often generates emissions comparable to the [much longer] ocean segment, and therefore the total transport carbon emissions for ocean-related segments are likely to end up being comparable to the emissions associated with domestic truck transport for long distances.

9.6 Interactions and trade-offs

A classic challenge for supply chain managers is to strike the right balance between transportation and storage costs. Because costs and carbon emissions are correlated, Venkat and Wakeland (2006) examined the transportation and storage-related emission characteristics of a food supply chain that was adapted from Simons and Mason (2002). Figure 9.9 shows the carbon emissions per unit of final product. Venkat and Wakeland (2006) also determined the sensitivity of the total transportation and storage

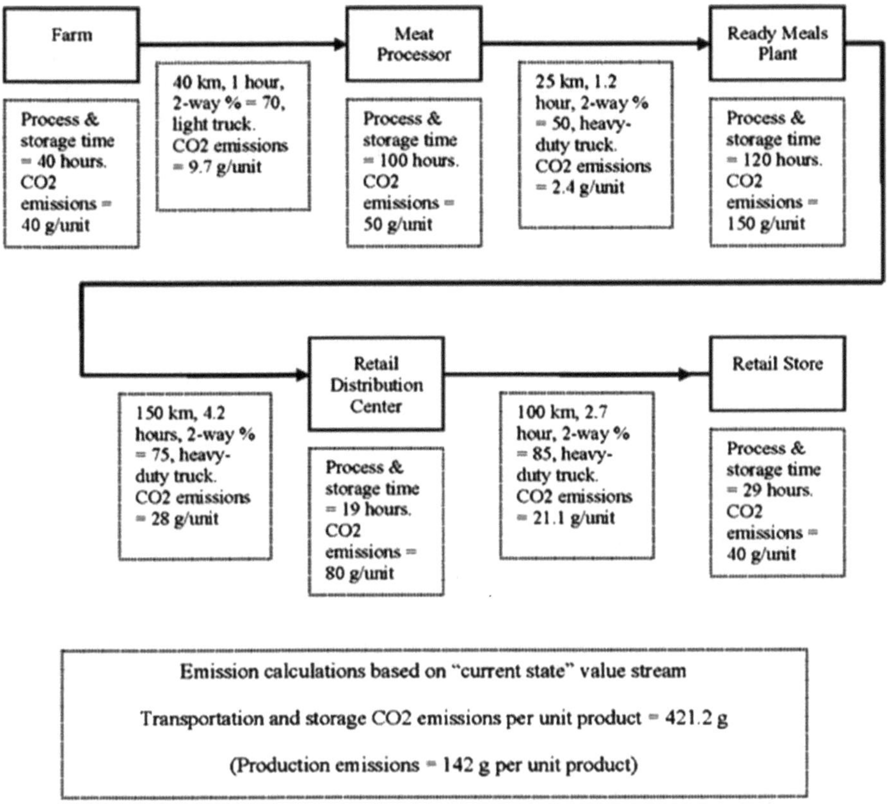

Fig. 9.9 Food supply chain-related carbon emissions (From Venkat and Wakeland 2006, adapted from Simons and Mason 2002)

emissions to the maximum distance between points in the supply chain. The total emissions would change by about 20% per 1,000 km (excluding any cold storage that might be required during transportation). Next, the impact of using a lean supply chain philosophy that emphasized small and frequent deliveries was analyzed. Results showed that for distances greater than 200 km, the increase in transportation-related carbon emissions would be greater than the reduction in storage-related carbon emissions.

Packaging design, production methods, and distribution are inherently interdependent. For example, the wine distribution scenarios explored in Sect. 9.4 consider wines bottled at the winery in standard 750 mL bottles. A case of 750 mL bottles is, by weight and volume, only about 50% product. As some wine producers shift to packaging in polyethylene terephthalate (PET) bottles, Tetra Pak, or Bag-in-Box formats, they realize energy and emissions savings by avoiding the transport of excess packaging. Recent carbon footprint analyses done for a packaging manufacturer show that from a carbon emissions standpoint, plastic is often a better material for making bottles than glass. For example, considering their full life cycle, a 360 mL PET bottle generates 41% fewer greenhouse gas emissions than a comparable glass bottle (Constar 2010). Full lifecycle includes the carbon emissions associated with transporting filled bottles to retail locations. Other winemakers have started to experiment with light-weighting (decreasing the amount of glass used in a bottle), although this can necessitate more careful handling in transport and storage. Furthermore, some wine companies have started to ship wine in bulk from Australia and bottle it in the United Kingdom, closer to the consumer market. Bulk shipping requires a firm to geographically distribute its operations and to be able to address preservation and contamination concerns. Of course, all alternative packaging formats are dependent on consumer acceptance.

9.7 Taking action

We consider what companies can do to cost-effectively reduce logistics-related carbon emissions without compromising quality and service levels. Sometimes a single project can provide a huge improvement opportunity. Jackson Family Wines recently consolidated its ten warehouses into a single energy-efficient distribution center (DC) with a rail spur connecting it to the Union Pacific Railroad. According to Bradley (2010), the new facility allowed the wine producer to stop shuttling inventory between storage locations and enabled increased usage of rail to transport its five million cases produced annually. Locating the DC near a key supplier supported the company's backhaul initiative as well: after transporting wine from the production facility to the DC, vehicles were able to pick up bottles from their nearby supplier on the return journey.

In other situations, a series of separate initiatives may be more appropriate than one large project. According to their website, by 2012 the UK-based retailer Tesco intends to reduce by half the emissions associated with delivering a case of goods. They plan to reach this goal through a variety of logistics improvements for moving goods, including switching to larger vehicles, partnering for backhaul opportunities, relying on rail for transport between DCs, and using barges to a greater extent.

A unifying factor in these initiatives is that they start with an initial study to benchmark the current system performance and to discover potential opportunities. Sometimes the results can be quite surprising, as seen in the following study. Recent research sponsored by the Oregon Transportation Research and Education Consortium (Pullman et al. 2009) compared a variety of packaging, food waste, and transportation scenarios for three food items: fresh or frozen chicken, raw potatoes, and processed diced tomatoes. These items were selected based on surveys and interviews of institutional food purchasers. Packaging alternatives consisted of waxed cardboard box versus plastic bag for chicken, cardboard versus reusable plastic container (RPC) for potatoes, and can versus plastic bag (in box) for tomatoes. Food waste scenarios considered uncooked and cooked waste that was either

Table 9.3 Example life cycle analysis of food waste, packaging, and distance

		Transportation			
		Local (within 100 miles)		Major distributor	
		Packaging		Packaging	
Processed tomatoes #10 can or equivalent		A Can (recycled)	B Bag-in-box	A Can (recycled)	B Bag-in-box
Food waste	None	1.56	1.39	1.61	1.44
(disposed before	50% Compost, before cooking	2.22	1.89	2.38	2.00
or after cooking)	50% Landfill, after cooking	4.09	3.76	4.20	3.86

From Pullman et al. (2009)

partially composted or 100% landfilled. Transportation alternatives included local versus long distance and fresh versus frozen.

For each scenario, full life cycle carbon assessments were done to determine the embodied carbon (the total greenhouse gas emissions generated by the product life cycle within a system boundary of interest and reported in kg of CO_2e). The CarbonScope analytical tool (Venkat 2009) was employed to do the analysis. The primary standard used for the product life cycle greenhouse gas emissions calculations was PAS 2050 (BSI Group 2008). PAS 2050 relies on the ISO 14040 series of standards (International Standards Organization 2009) and the 2006 IPCC Guidelines for National Greenhouse Gas Inventories (Intergovernmental Panel on Climate Change 2009).

Pullman et al. (2009) conclude that food waste has a large adverse impact on the environment (especially if landfilled, as is often the case with cooked vegetables and meats); an unexpected finding that can be easily acted upon. Even raw vegetables that are composted rather than consumed increase carbon impact (see Table 9.3). Institutions may want to consider policies to encourage the use of packaging that reduces food waste (both before and after cooking). Similarly, "buy local" efforts for the three products evaluated earlier would make sense and would contribute to carbon reduction. Transportation-related carbon impacts were the most significant for frozen meat items.

Overall, "food miles" do not matter as much as other considerations when determining the carbon impact of food production, consumption, and disposal (except perhaps for fresh food that is air freighted). Minimizing food waste and composting the unavoidable food waste could have a much larger benefit than switching from a distant supplier to a local supplier. Also, when analyzed carefully, one must conclude that plastic packaging generally has a smaller environmental footprint than steel, paper, or glass, because of its low volume (thinness) and light weight. Considerations may, however, be required on other health impacts of plastic packaging (e.g., physiological impacts on animals who consume plastics in landfills, etc.).

As we have seen from previous examples, items transported by highly utilized larger trucks result in lower emissions per unit than when transported in smaller vehicles. Such operational realities have led some analysts to posit a natural ecology of scale must exist. Schlich and Fleissner (2005) provide examples of how the international sourcing of juices and lamb is less energy intensive on a per-unit basis than the local (German) equivalent. They calculate that energy and cost savings arise from large-scale production and distribution, as well as from the comparative natural advantages some countries have in growing or raising food. For example, Brazil's climate is naturally more conducive to growing fruit than the climate in most of Europe, and Brazilian juice production typically occurs on a much large scale than European producers are able to support. Consequently, Schlich and Fleissner (2005) found that the lower emissions owing to producing juice in Brazil rather than Europe more than offset the transportation-related emissions associated with transporting the juice from Brazil to Europe. Such analyses tend to be controversial, in part because the answer is not what many people expect. Also, these results do not mean that local production can never compete with long distance sourcing regarding energy use and carbon emissions. The key is for local operations to find ways to be efficient and to achieve the economies that tend to be associated with large scale operations. Local production also provides other benefits that are not as easily incorporated into analyses, such as local employment, rural conservation, and consumer trust in origin and local brands.

It is not necessarily the case that only giant conglomerates can be highly efficient. It is possible for smaller firms to mimic the efficiency of larger entities. For example, farmers in European wine regions have often formed cooperatives for the production and marketing of their products. Indeed, Schlich (2010) documents the Hessische Bergstraße cooperative within the Rhine Valley, which is comprised of more than 500 family-owned vineyards. Despite the small size of these individual winemakers, the Hessische Bergstraße cooperative coordinates production and distribution in a way that allows these farmers to capture efficiencies of scale normally associated with a much larger producer.

When aggregation occurs even further downstream in the supply chain, producer collectivism is not required. The following example explores how the logistics needed to support California farmers' markets could be improved, while preserving the small scale and independent nature of both the farms and the markets. Such improvements will not only yield financial benefits, but will also improve the energy intensity associated with bringing these products to consumers.

The proliferation of farmers' markets in the last few years has been a boon to consumers looking to purchase fresh, locally grown fruits and vegetables. Worthen (2010) documents that California now has more than 500 markets, which are attended by nearly 3,000 farmers. However, this growth in the number of markets has negatively impacted many farmers, who report having to make multiple trips to visit more markets in order to sell the same amount of produce. Jog (2010) profiles Schletewitz Farms, one such fruit producer located near Fresno, California, that sells year round to nine greater Bay Area markets, each 200–350 km away. Such distances are typical, as many of the certified market suppliers are located in the Central Valley, whereas most markets are found in population centers closer to the coast, as seen in Fig. 9.10, which depicts a representative subset of farmers' markets and the farms that service them. Likewise, a map of vendors found at the Jack

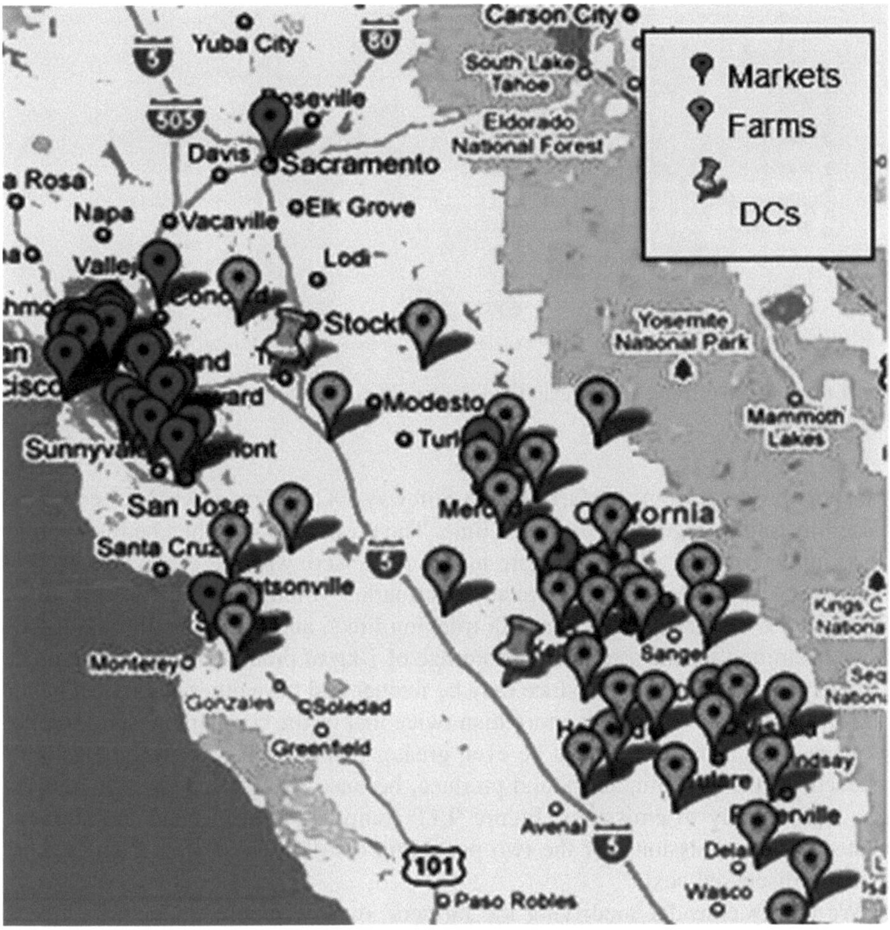

Fig. 9.10 Subset of farmers' markets, farms and candidate DCs (From Jog 2010). The Northern DC belongs to Safeway. The other DC represents a plausible consolidation point for Central Valley Farmers

London Square Farmers Market in Oakland shows that farmers come from as far north as Yuba City and as far south as San Diego, almost 800 km away (Jog 2010).

On a per unit basis, supply chains relying on extensive use of small capacity vehicles are likely to be less energy efficient than their larger-scale counterparts. To illustrate this, we consider Sanger Farm, a regular vendor at the San Francisco Ferry Market. We assume that they nearly fill (90%) a large pickup truck with produce and that they are able to completely sell the produce at the market (a best-case scenario). Using CargoScope, we calculate that the round-trip journey to transport the produce from Sanger Farm to San Francisco results in .45 kg CO_2e emitted per kg of produce.

We next consider a neighboring farm that may sell high volumes of produce through a supermarket such as Safeway. In this supply chain scenario, a midsized commercial truck (which has about ten times the capacity of a pickup truck) would

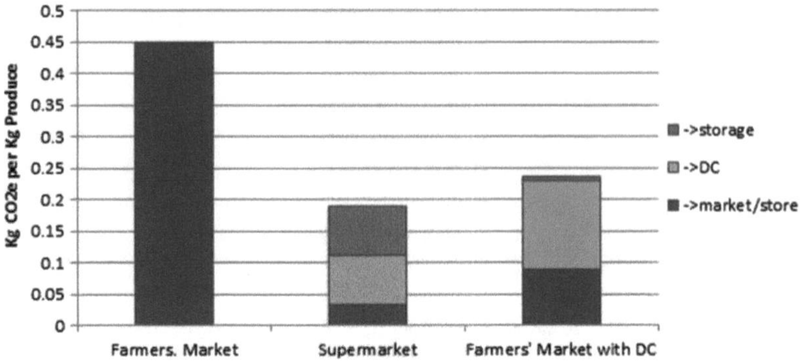

Fig. 9.11 Emissions from farm to retail for different sales channels (calculations performed using CargoScope)

transport produce from the farm to the Safeway DC in Tracy, where it could be placed in cold storage for a period of time. The produce would then be transported by a similar truck to a Safeway store in San Francisco, where it could take up to a week to be sold. Considering typical supermarket volumes and dwell times, we assume 90% utilization rates for both trucking links, and determine that .19 kg of CO_2e is emitted for the transport and storage of 1 kg of produce from farm to retail gate. The farmers' market produce may be fresher and have other ecological benefits, but its carbon footprint is more than twice that of the comparable supermarket produce. This difference would be even greater if the dwell time at the DC were reduced for the supermarket-bound produce, because refrigerated storage contributes significantly to emissions. Figure 9.11 compares the transport and storage-related carbon emissions for the two preceding distribution scenarios and a third scenario discussed next.

We finally consider modifying the farmers' market supply chain to include a consolidation DC that is near an interstate highway, but still relatively close to many Central Valley farms (such as in the vicinity of the small town of Cantua Creek). Rather than individually driving their produce across the state to farmers' markets in the Bay Area, participating farmers would deliver produce to the DC, where it would be stored for a period of a few days and then consolidated into midsized commercial trucks, each destined for a particular farmers' market. Market stalls could be staffed by employees of the DC or individuals hired at the local markets Assuming the same high utilization rates would hold, Fig. 9.11 shows that this newly designed supply chain would be much less carbon emissions intensive, approaching the relatively low levels of emissions provided by the supermarket supply chain.

Jog (2010) explores the potential for a farmers' market consolidation DC at a regional level. Although it would be impractical to gather historical sales data for every farmer at every market, Jog creates a simplified approximation by modeling the underlying network, which includes 135 Bay Area markets and more than 1,000 farms that sell produce at these markets. Jog's analysis treats the farms within a given zip code as a single larger farm that is equivalent in volume. The result is slightly more than 200 equivalent farms. Using this approximation tends to slightly underestimate the total travel distances. Considering one type of undifferentiated produce, assuming uniform production rates for each of the equivalent farms, and classifying each market into one of three sizes, Jog models this network as a transportation problem. The optimization problem was to find the minimum total travel distance that could meet the market demand.

Without a consolidation DC, nearly 400 trucks would be needed to supply the markets over the course of a week, with an average round trip of slightly more than 200 km and a total travel distance of 84,000 km. It should be noted that this optimal solution greatly understates actual transit, as farmers typically travel more than 300 km per trip and visit several markets per week. In this minimized solution, farmers would visit only two or three markets per week, and markets would have at most a few vendors. Of course, such a lack of diversity would be counter to the mission of a farmers' market.

Next, Jog adds the Cantua Creek consolidation DC to the model and finds that the optimum in this case occurs when nearly 60% of the produce is routed through the DC, which reduces the overall distance traveled by 20%. Furthermore, market diversity would be supported, because trucks arriving to the market from the consolidation DC would carry produce from multiple farmers, a benefit for small-scale farmers who produce specialized products. Such farmers would be able to reach more consumers than is currently possible, because it would be economically feasible to ship fractional truckloads of their goods to different markets by consolidating them with goods from other small-scale farmers.

Of course, the reality of funding and implementing such a consolidation DC, supporting transport from the DC to the markets, and arranging for staffing would be much more complex than building models and calculating potential benefits. Perhaps the DC could be created by the farmers as a cooperative, with collaborative staffing and support, as well as partnerships with organizations such as the California Federation of Certified Farmers' Markets. Alternatively, the DC could be created as a joint venture funded by a major reseller. Another option might be to allow produce to be sold at the DC or the market on a consignment basis. At present, California farmers' markets allow only direct producers. The concepts described above would tend to reduce consumer contact with growers, which has been one of the primary benefits of farmers' markets. However, the current trend toward more and more farmers' markets, without corresponding improvements to the underlying support structures, will not be sustainable in the long run for most small family farms and could ultimately deter participation of just the sort of vendors these markets were created to showcase.

9.8 Conclusion

The transportation-related carbon footprint varies from a few percent to more than half of the total carbon footprint associated with food production, distribution, and storage. Supply chains are complex and varied, and food supply chains are especially challenging because of seasonality, freshness, spoilage, and sanitary considerations. Measuring transportation-related carbon footprint involves careful choice of the scope of the analysis, and there is much uncertainty in the results. Caution is warranted regarding the absolute numbers from carbon assessments, so it may be best to focus primarily on relative comparisons.

The winemaker case study showed that a local 3PL approach had the lowest carbon footprint, and that the highest carbon footprint resulted from consumers driving to the winery. This does not mean, however, that local 3PL would be the best solution for all wineries. The case study involving organic fruit and vegetable supply showed how food carbon data can be provided to consumers in order to support their food purchasing choices, and the case study regarding packaged fruit indicated that transportation distances are not always a good indicator of total carbon footprint. Another case study indicated that home delivery can cut the transportation-related carbon footprint almost in half compared to consumers driving to the store individually.

Supply chain planners must carefully consider the trade-off between transportation-related energy cost & carbon footprint and storage-related energy cost & carbon footprint. Also, the frequent small deliveries called for by lean manufacturing practices, although optimizing efficiency within a facility, can increase overall carbon footprint. Packaging is another important consideration, and the use of plastics rather than glass tends to lower carbon footprint. Benefits for the environment and health are further accrued by plastic recycling.

To reduce carbon footprint, suppliers are consolidating their operations, increasing their use of rail and water transit, and increasing transport efficiency by filling trucks and considering backhaul opportunities. Food waste is another potentially significant contributor to carbon emissions, which could potentially be reduced via alternative packaging options. Our research also indicates that food-miles, a metric that many consider to be of primary importance, do not actually correlate very well with overall carbon footprint. Finally, although farmers' markets have many desirable attributes, they unfortunately tend to have a much higher carbon footprint than conventional food distribution. This differential could be lessened considerably by using consolidation DCs close to the farms.

Whether comprising a large or small share of a product's total emissions, transportation is an unavoidable step in the supply chain for nearly every food product. However, the economic and environmental impacts of food transportation can be moderated. We reiterate that although food transportation decisions can sometimes be considered separately from other issues, this is not always appropriate. Food transportation and storage involves trade-offs that necessitate taking an overall system perspective.

References

Bradley, P. 2010. An inside look at Jackson Family Wines' new eco-friendly warehouse. http://www.dcvelocity.com/articles/20100222inside_an_eco_friendly_dc/. Accessed 15 May 2010.

Browne, M., C. Rizet, S. Anderson, J. Allen, and B. Keita. 2005. Life cycle assessment in the supply chain: A review and case study. *Transport Reviews* 25(6): 761–782.

BSI Group. 2008. PAS 2050:2008—Specification for the assessment of the lifecycle greenhouse gas emissions of goods and services. http://www.bsigroup.com/en/Standards-and-Publications/Industry-Sectors/Energy/PAS-2050/. Accessed 18 Mar 2009.

Carlsson-Kanyama, A. 1998. Climate change and dietary choices—How can emissions of greenhouse gases from food consumption be reduced? *Food Policy* 23(3–4): 277–293.

Carlsson-Kanyama, A., M.P. Ekström, and H. Shanahan. 2003. Food and life cycle energy inputs: Consequences of diet and ways to increase efficiency. *Ecological Economics* 44(2–3): 293–307.

Cholette, S., and K. Venkat. 2009. The energy and carbon intensity of wine distribution: A study of logistical options for delivering wine to consumers. *Journal of Cleaner Production* 17(16): 1401–1413.

CleanMetrics. 2010. http://www.cleanmetrics.com. Accessed 18 Sep 2010.

Constar. 2010. Sustainability analysis for PET food jars. http://www.constar.net/pdf/Convert-It_Analysis_Example.pdf. Accessed 24 June 2010.

DOE. 2009. U.S. CO_2 emissions from energy sources 2008 flash estimate. http://www.eia.doe.gov/. Accessed 20 Jan 2010.

Golicic, S., C. Boerstler, and L. Ellram. 2010. Greening transportation in the supply chain. *MIT Sloan Management Review* 51(2): 47–55.

Heller, M., and G. Keoleian. 2000. *Life cycle-based sustainability indicators for assessment of the U.S. food system [CSS00-04]*. Ann Arbor: Center for Sustainable Systems, School of Natural Resources and Environment.

Hospido, A., L. Milà i Canals, S. McLauren, et al. 2009. The role of seasonality in lettuce consumption: A case study of environmental and social aspects. *International Journal of Life Cycle Assessment* 14: 381–391.

Intergovernmental Panel on Climate Change. 2009. 2006 IPCC guidelines for national greenhouse gas inventories. http://www.ipcc-nggip.iges.or.jp/public/2006gl/index.html. Accessed 18 Mar 2009.

International Standards Organization. 2009. ISO 14040: 2006 Environmental management–life cycle assessment–Principles and framework (2). http://www.iso.org/iso/iso_catalogue/catalogue_ tc/catalogue_detail.htm?csnumber=37456. Accessed 18 Mar 2009.

Jog, S. 2010 Optimizing the logistics of farmer's markets: A carbon footprint study. Master's thesis, San Francisco State University, California.

Johnson, E. 2008. Disagreement over carbon footprints: A comparison of electric and LPG forklifts. *Energy Policy* 36(4): 1569–1573.

Milà i Canals, L., S.J. Cowell, S. Sim, and L. Basson. 2007. Comparing domestic versus imported apples: A focus on energy use. *Environmental Science and Pollution Research* 14(5): 338–344.

Pirog, R., T. Van Pelt, K. Ensayan, and E. Cook. 2001. *Food, fuel and freeways: An Iowa perspective on how far food travels, fuel usage, and greenhouse gas emissions*. Ames: Leopold Center for Sustainable Agriculture.

Point, E. 2008. Life cycle environmental impacts of wine production and consumption in Nova Scotia, Canada. Master's thesis, Dalhousie University, Halifax.

Pullman, M., R. Fenske, and W. Wakeland. 2009. *Food delivery footprint: Addressing transportation, packaging, and waste in the food supply chain*. OTREC Research Report, RR-08-154, Oregon Transportation Research and Education Consortium, March 2009.

Rosenthal, E. 2009. To cut global warming, Swedes study their plates. *The New York Times*, 23 Oct 2009.

Saunders, C., and A. Barber. 2007. *Comparative energy and greenhouse gas emissions of New Zealand's and the United Kingdom's dairy industry*. Christchurch: Lincoln University. AERU Research Report No. 297.

Saunders, C., A. Barber, and G. Taylor. 2006. *Food miles—Comparative energy/emissions performance of New Zealand's agriculture industry*. Christchurch: Lincoln University. AERU Research Report No. 285.

Schlich, E. 2010. From vineyard to point of sale: Allocation of energy use and CO_2 emission to entire supply chains of wine. In *4th annual meeting of American association of wine economists (AAWE)*. Davis, CA, June 2010.

Schlich, E., and U. Fleissner. 2005. The ecology of scale: Assessment of regional energy turnover and comparison with global food. *International Journal of Life Cycle Assessment* 10(3): 219–223.

Scholz, A., U. Sonesson, and P. Tyedmers. 2009. Catch of the freezer. *The New York Times*, 8 Dec 2009.

Sim, S., M. Barry, R. Clift, and S.J. Cowell. 2007. The relative importance of transport in determining an appropriate sustainability strategy for food sourcing. *International Journal of Life Cycle Assessment* 12(6): 422–431.

Simons, D., and R. Mason. 2002. Environmental and transport supply chain evaluation with sustainable value stream mapping. In *Proceedings of the 7th logistics research network conference*, Birmingham, UK.

Twede, D., R. Clarke, and J. Tait. 2000. Packaging postponement: A global packaging strategy. *Packaging Technology and Science* 13(1): 105–115.

United States. Department of Agriculture. 2008. Loss-adjusted food availability: Spreadsheets. http://www.ers.usda.gov/Data/foodconsumption/FoodGuideSpreadsheets.htm. Accessed 10 Sep 2010.

Van Hauwermeiren, A., H. Coene, G. Engelen, and E. Mathijs. 2007. Energy lifecycle inputs in food systems: A comparison of local versus mainstream cases. *Journal of Environmental Policy Plan* 9(1): 31–51.

Venkat, K. 2008. CargoScope web software tool. http://www.cleanmetrics.com/html/cargoscope.htm. Accessed 18 Mar 2009.

Venkat, K. 2009. CarbonScope web software tool. http://www.cleanmetrics.com/html/carbonscope.htm. Accessed 18 Mar 2009.

Venkat, K., and W. Wakeland. 2006. Is lean necessarily green? In *50th anniversary conference of the international society for the systems sciences*, Sonoma, CA, July 2006.

Weber, C.L., and H.S. Matthews. 2008. Food-miles and the relative climate impacts of food choices in the United States. *Environmental Science and Technology* 42(10): 3508–3513.

World Resources Institute. 2004. Greenhouse gas protocol, a corporate accounting and reporting standard. http://www.ghgprotocol.org/files/ghgp/public/ghg-protocol-revised.pdf. Accessed 20 June 2009.

Worthen, B. 2010. Farmers' markets see risk from growth. *Wall Street Journal*, 11 Feb 2010. http://online.wsj.com/article/SB10001424052748704533204575048000839268376.html. Accessed 10 Mar 2010.

Part IV
Green technologies in food processing

Chapter 10
Supercritical and near-critical CO_2 processing

Eric J. Beckman

10.1 Introduction

The critical temperature, or Tc, is that temperature above which a substance cannot be liquefied, regardless of the pressure applied. At the critical point (Tc, Pc), the liquid and vapor phases adopt equal densities and merge, forming a single phase whose physical properties (a) are intermediate to those of the respective liquids and gases, and (b) vary smoothly as the fluid density is varied. Because operation above the critical temperature precludes the formation of a liquid–vapor meniscus, supercritical fluids can be separated from products of value via a simple depressurization.

Although there are several compounds whose critical temperatures are sufficiently mild (<100°C) to allow easy access to the supercritical regime, most have received little attention owing to various process or safety issues. Propane, for example, exhibits solvent behavior similar to that of hexane, but it is rarely considered as a viable process fluid owing to the combined issues of high process pressure and inherent flammability. Fluoroform is a more polar solvent (than propane) with relatively low toxicity, but its high cost and significant climate change potential (a common factor among the hydrofluoroalkanes) has inhibited its use outside of academia. Nitrous oxide exhibits similar characteristics, with the added disadvantage that its oxidizing potential has produced some unexpected explosive events during laboratory work. As such, when one thinks of supercritical fluid technology, one quickly zeroes in on the use of carbon dioxide as the solvent of choice.

E.J. Beckman (✉)
Chemical Engineering Department, University of Pittsburgh, Pittsburgh, PA, USA
e-mail: Beckman@pitt.edu

J.I. Boye and Y. Arcand (eds.), *Green Technologies in Food Production and Processing*, Food Engineering Series, DOI 10.1007/978-1-4614-1587-9_10,

CO_2 is the supercritical fluid of choice because its critical temperature is readily accessible (~304 K), because it is a relatively benign material itself (nonflammable, relatively low toxicity), and because it is naturally abundant and hence inexpensive. Unfortunately, CO_2 is a feeble solvent, making it difficult to use for processes where substantial solubility of high molecular weight and/or polar compounds is required. In extraction processes, two strategies have been employed to overcome CO_2's poor solvent power—the use of cosolvents (also known as entrainers) and the design of auxiliaries (surfactants, chelating agents). A brief discussion of entrainers appears below; a discussion of auxiliary design appears in the next section.

The use of cosolvents dates to the earliest work on the use of carbon dioxide in extractions, in which methanol was added (typically at levels less than 10 mol%) to improve CO_2's ability to solvate polar solutes. Some key questions arise during use of any cosolvent with CO_2: (1) Does the use of the cosolvent cancel the benign advantages accrued by use of CO_2? and (2) Does the cosolvent partition preferentially to the matrix or product during the process? With respect to the former, it must be remembered that construction and operation of a high pressure process is inherently more expensive than an analogous one-atmosphere setup, and hence the preservation of the "green" advantages of CO_2 (preclusion of cosolvent use) may be paramount in the decision to move forward with a CO_2-based process. Regarding the latter, Johnston and colleagues (Shim and Johnston 1989; Shim and Johnston 1991) showed long ago that the partition coefficient of compounds dissolved in CO_2 is a strong function of operating conditions—if an entrainer preferentially partitions to either the extractant or the matrix during processing, once again the "benign" advantages of a CO_2 process may be lost. Finally, if operation under supercritical conditions is required (versus dense liquid), it must be remembered that addition of a cosolvent will alter the critical temperature—in the first approximation the critical temperature of a mixture is a mole fraction average of the individual T_cs (Reid et al. 1987). Addition of even small amounts of high T_c cosolvents (like methanol) rapidly raises the T_c of the mixture.

The use of carbon dioxide as a solvent or raw material has been investigated somewhat continuously in academia and/or industry since 1950; interest in the use of CO_2 in these roles has intensified during the past 20 years as large-scale plants using CO_2 have been brought on line. Extraction of compounds from natural feedstocks for use in (or as) foodstuffs, nutraceuticals, and pharmaceuticals has been the most resilient application area in the field since the 1980s. Reviews on natural products extraction appear frequently; some of the more recent are listed in the reference section (Brunner 1994; McHugh and Krukonis 1994; Brunner 2005; Herrero et al. 2006; Reverchon and De Marco 2006). As noted above, specific interest in CO_2 where food is concerned is magnified by its perceived "green" properties—in addition, unlike water, the supercritical regime of CO_2 is readily accessible. As such, if one is interested in the extraction of a nonpolar, relatively low molecular weight material such as fatty acids (Sahena et al. 2009), there is substantial previous literature available on appropriate conditions and on transport issues related to design and scale-up (Valle and De La Fuente 2006). If, on the other hand, one is interested in a more complex solute, fundamental information allowing for prediction of phase behavior or the design of useful auxiliaries is still lacking.

Whereas the bulk of CO_2-based applications in the food industry to date have targeted extractions, there is the possibility for other CO_2-based technologies to cross over from polymer science (foaming, additive blending), catalysis (hydrogenation), and biotechnology (enzymatic reactions); these will be discussed later. Further, much of the most exciting work in CO_2 technology since 1990 has involved the design of auxiliaries (surfactants, chelating agents)—these might find use in food processing as well.

10.2 Physical properties of CO_2

The pVT (pressure-volume-temperature) properties of CO_2 have been known since the 1930s (Michels and Michels 1936; Michels et al. 1936); extensive data sets are available in the literature and on the web in the form of correlations of density, viscosity, dielectric constant, etc., as functions of temperature and pressure (Angus et al. 1976). CO_2's critical pressure (and hence its vapor pressure in the "near-critical," or liquid regime) is significantly higher than analogous values for alkane, fluoroalkane, or hydrofluoroalkane fluids. CO_2's anomalously high critical pressure is but one result of the effect that CO_2's strong quadrupole moment exerts on its physical properties. While the high critical pressure is problematic, the most unfortunate outcome of the effect of quadrupole moment on physical properties was the premise, first advanced during the late 1960s, that CO_2 might prove to be a solvent whose strength would rival or surpass that of alkanes and ketones (Giddings et al. 1969). Because early models employed to calculate CO_2's solvent power relied on a direct relationship between the Hildebrandt solubility parameter (δ) and the square root of the critical pressure [$(P_c)^{1/2}$], the solubility parameter of CO_2 was overpredicted by 20–100%, leading to early inflated claims as to the potential for using CO_2 to replace conventional organic solvents.

For a supercritical fluid the solvent power, in the first approximation, is proportional to solvent density, and hence SCFs have been noted as "tuneable solvents" in the past (McHugh and Krukonis 1994). As such, researchers have employed CO_2 (and other SCFs) to fractionate mixtures by either molecular weight (polymers/oligomers) or polarity, or both. It must be emphasized that when using CO_2 for such purposes, one must always be cognizant of CO_2's feeble solvent power—fractionation can only occur if the mixture (or its most valuable components) can be solubilized to the extent needed for economical operation.

10.3 Environmental and safety advantages to use of CO_2 in processes

Carbon dioxide is non-flammable, a significant safety advantage in using it as a solvent. It is also naturally abundant, with a TLV (threshold limit value) for airborne concentration at 298 K to which it is believed that nearly all workers may be repeatedly exposed day after day without adverse effects of 5,000 ppm (Lide 1996; Praxair 1999), rendering it less toxic than many other organic solvents (acetone, by comparison, has a TLV of 750 ppm, pentane is 600 ppm, chloroform is 10 ppm (Praxair 1999; Lide 1996)). Carbon dioxide is relatively inert towards reactive compounds, another process/environmental advantage (byproducts owing to side reactions with CO_2 are relatively rare), but CO_2's relative inertness should not be confused with complete inertness. For example, an attempt to conduct a hydrogenation in CO_2 over a platinum catalyst at 303 K will undoubtedly lead to the production of CO, which could poison the catalyst (Minder et al. 1996). The same reaction run over a palladium catalyst under the same conditions will by contrast produce lesser amounts of CO as a byproduct (Solymosi et al. 1985; Erdohelyi et al. 1986), and hence knowledge of CO_2's reactivity is vital to its use in green chemistry.

Carbon dioxide is clearly a "greenhouse gas," but it is also a naturally abundant material. Like water, if CO_2 can be withdrawn from the environment, employed in a process, then returned to the environment "clean," no environmental detriment accrues. However, while CO_2 could in theory be extracted from the atmosphere (or the stack gas of a combustion based power plant), most of the CO_2 employed in processes today is collected from the effluent of ammonia plants or derived from naturally occurring deposits (for example, tertiary oil recovery as practiced in the United States (Lake 1989; Chakma et al. 1991)). Because industrially available CO_2 is derived from manufactured sources, if CO_2 can be isolated within a process one could consider this a form of sequestration, although the sequestered volumes would not be high. Ultimately, one should consider the source of CO_2 used in a process in order to adequately judge the sustainability of the process.

CO_2's combination of high TLV and high vapor pressure means that residual CO_2 left behind in substrates is not a concern with respect to human exposure—the same can certainly not be said to be true for many man-made and naturally-occurring organic compounds. Because there is effectively no liability due to "residual" CO_2 in materials following processing, CO_2 is not considered a solvent requiring process reevaluation by the U.S. FDA. Only water also enjoys this special situation. Indeed, most of the commercial operations employing CO_2 as a solvent were initiated to take advantage of CO_2's particular advantages in products designed for intimate human contact (such as food), or CO_2's non-VOC (volatile organic compound) designation (such as the foaming of thermoplastics). The commercialization of fabric cleaning using CO_2 was based both from CO_2's advantages in human-contact applications *and* situations where emissions appear unavoidable.

10.4 Environmental and safety disadvantages inherent to use of CO_2 in a process

Because CO_2's vapor pressure at room temperature is over 60 bar, use of CO_2 in a process clearly requires high-pressure equipment, creating a potential safety hazard relative to the same process operated at one atmosphere operation. In addition, uncontrolled release of large quantities of carbon dioxide can asphyxiate bystanders owing to air displacement. These issues have not impeded the commercialization of CO_2-based processes, nor is it likely they will do so in the future. It should be remembered that the low density polyethylene polymerization process, first commercialized in the 1940s and still in operation today (Rodriguez 1996), runs continuously at 2,000–3,000 bar and 520 K with a highly flammable component, and hence safe operation of a 100–200 bar CO_2-based plant is readily achievable using current technology. Whether to use liquid or supercritical CO_2 is a choice that actually involves safety as well as chemistry considerations. While use of supercritical CO_2 almost always involves use of higher pressure (to achieve the same solubility of a given substrate as in the liquid case), other factors should also be considered. First, supercritical CO_2 will exhibit a higher compressibility than liquid CO_2, and hence the supercritical fluid will be better able to absorb excess heat evolved from an exothermic reaction whose rate suddenly exceeds typical operating conditions. On the other hand, use of saturated liquid CO_2 (in the presence of the vapor phase) would allow boiling to be used as a means to absorb excess heat. Use of supercritical CO_2 (versus liquid) could avoid complications owing to a phase separation occurring upon a departure from established temperature or pressure conditions within a given reactor. For example, if one is employing a mixture of oxygen, substrate, and liquid CO_2 in a particular process, a sudden drop in pressure owing to a perturbation in the process could lead to formation of a flammable gaseous phase—use of a supercritical mixture could avoid this problem as no vapor–liquid separation will be encountered. Indeed, it should also be remembered that the T_c of a mixture of CO_2 and other materials will differ from that of pure CO_2 [see, for example, Reid, et al. (1987) for useful correlations], and hence T-p conditions sufficient for supercritical operation with pure CO_2 may create a liquid in the case of the mixture.

10.5 Process design: are CO_2-based plants inherently uneconomical?

The number of processing plants operating worldwide that employ supercritical CO_2 is above 100 and growing steadily. Most of the current plants use CO_2 to process food in some way (extraction or fractionation), yet other types of plants have been or are being brought on stream (for example, fluoropolymer synthesis by DuPont, hydrogenation by Thomas Swan, coatings by Union Carbide, polyurethane processing by Crain Industries). Despite this steady growth, there is a general sense (or unease) within both the academic and industrial communities that there are elements connected to the design and construction of CO_2-based plants that effectively block greater use of the technology.

Several authors have reviewed aspects of process design and costing of "supercritical" plants (Cygnarowicz-Provost 1996; Gani et al. 1997; Smith and Huse 1998; Perrut 2000; Hauthal 2001); these reviews typically focus on a specific industry. For example, Perrut reports that for the case of extraction, the relative cost of a supercritical plant scales as $(V*Q)^{1/4}$, where V is the column volume and Q the flow rate. Hence the rule of thumb here is that one should operate at high concentration if possible, thus minimizing both V and Q.

Each of the authors who has reviewed process design using supercritical CO_2 emphasizes that one needs access to the relevant fundamental parameters in order to complete and optimize the design. Such parameters include both the relevant thermodynamic model for the mixture(s) in question with the appropriate binary interaction parameters, reaction data (rate constants, heats of reaction, Ahrrenius constants), and transport constants (densities, diffusivities, and viscosities). Note that these parameters are exactly the same as would be required to design a 1-atmosphere process, and hence there is nothing inherently "foreign" about a CO_2-based process that inhibits design and costing. Indeed, high pressure alone is not sufficient to explain the perceived difficulty of CO_2-based process scale-up, given that hydroformylation operates at 200–300 bar at large scale, whereas low density polyethylene is produced at over 2,000 bar. If one has access to the necessary basic information, one can employ software such as ASPEN to accomplish the process design, and ICARUS to handle the costing.

Hence, we must conclude that, if the inhibition in the scale-up of CO_2-based processes is real rather than perceived, then it must be caused by a lack of the fundamental parameters needed for process design, plus other factors that would inhibit the commercialization of any "new" technology. For example, it is relatively difficult at present to predict the effect of molecular structure on phase behavior in CO_2 of molecules that exhibit any substantial degree of complexity. Carbon dioxide exhibits both nonpolar tendencies (low dielectric constant) and "polar" properties (Lewis acidity, strong quadrupole moment), and hence predictions of phase behavior are not straightforward (as in the case of alkanes or alkenes). Research (Zhang et al. 2000; Blas and Galindo 2002; Byun et al. 2000; Colina et al. 2002) has shown that the Statistical Associating Fluid Theory (SAFT) can provide good descriptions of the phase behavior of complex mixtures including CO_2, yet the complexity of this model and/or lack of suitable parameters may currently limit its use industrially. Group contribution models have been applied to CO_2 solutions somewhat narrowly, generally targeting a single class of solutes (Keshtkar et al. 1997; Jaubert and Coniglio 1999; Diaz et al. 2000; Espinosa et al. 2000; Artal et al. 2001). What appears to be needed is a means to easily predict the properties of mixtures involving CO_2, such that confident predictions of process requirements and costs can be made using conventional process software such as ASPEN.

10.6 Operating a process economically with CO_2: heuristics

While use of CO_2 as a solvent is often considered to be "green," operation of any process at high pressure typically involves higher costs than the analogous process operated at one atmosphere. If such a process is considered "green," but cannot be created and operated economically, then the process will be of academic interest only

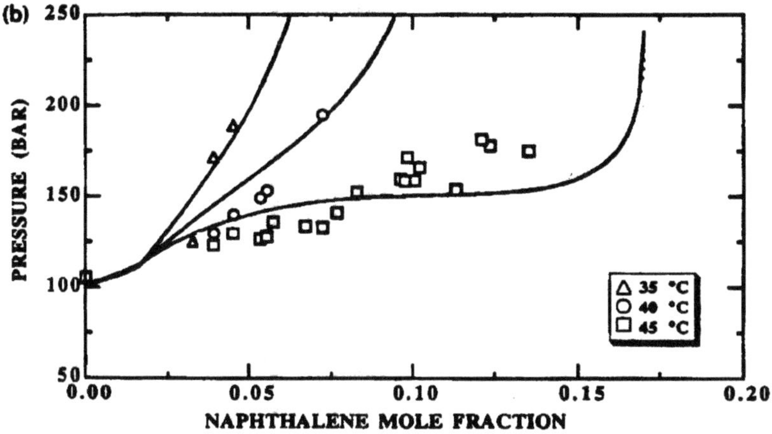

Fig. 10.1 Solid–fluid phase behavior (McHugh and Krukonis 1994): CO_2-naphthalene

and its potential green benefits unrealized. There are some simple “rules of thumb” that one can use to render the cost of a CO_2-based process as low as possible.

10.6.1 Operate at high concentration

One way in which to minimize the cost of a CO_2-based process is to minimize the size of the equipment. Given that CO_2 is typically proposed as a solvent (rather than a reactant), the most obvious means by which to minimize equipment size is to minimize the amount of solvent (CO_2) flowing through the process. Consequently, one should try to choose or design substrates such that they exhibit high solubility in CO_2. In addition, those processes where CO_2 is employed as the minor component (use of CO_2 as a plasticizer in foaming, for example) are likely to be favored economically.

Another aspect of this issue is reflected in the typical phase behavior of compounds in CO_2 (Figs. 10.1 and 10.2). Note that in the typical phase diagram of a crystalline solid in CO_2, an essentially pure solid phase exists in equilibrium with a solution. Given that the solid phase cannot be processed, one obviously makes use of the solution, where naturally CO_2 is the major component. For the case of liquid–liquid phase behavior, a CO_2-rich phase exists in equilibrium with a substrate-rich phase. However, because CO_2 has been shown to lower the viscosity of solutions substantially, one can actually pump and process the substrate rich phase. Further, one can operate at lower pressure in addition to at higher concentration. Consequently, it may be beneficial to employ systems where liquid–liquid phase behavior occurs rather than liquid–solid. Efficient operation of a process is both economically favorable and more environmentally friendly.

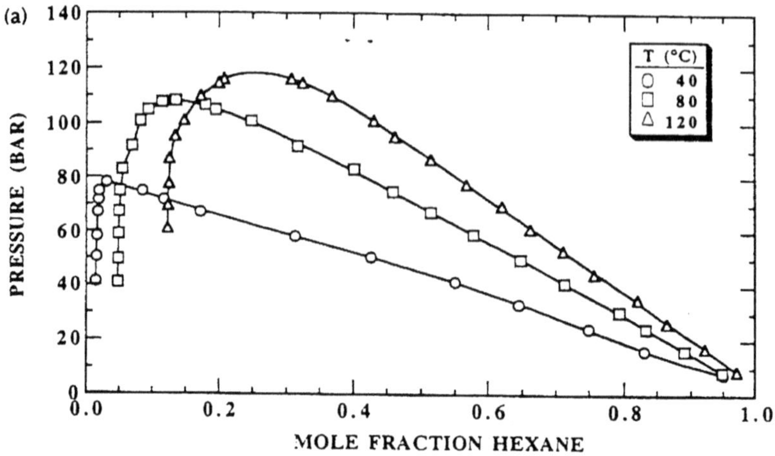

Fig. 10.2 Liquid–liquid phase behavior: CO_2-hexane (McHugh and Krukonis 1994)

10.6.2 Operate at as low a pressure as possible

Operation of a process at high pressure is more expensive than at 1 atm, owing to equipment design and construction, as well as the additional safety features that are necessary. Further, the capital cost of a high-pressure process is not linear with pressure, because the pressure ratings of certain vital equipment (flanges, for example) are available in discrete steps (60, 100 bar, for example). In addition, the number of companies with experience in high-pressure process design drops dramatically as the operating pressure rises above 200 bar.

Clearly, these caveats strongly recommend operating at the lowest pressure possible. One means by which to accomplish this is in the chemical design of reactants and/or substrates. It has been known for a number of years that certain functional groups are more "CO_2-philic" (thermodynamically more CO_2-friendly) than others. Use of CO_2-philic functional groups in the design of substrates or catalysts can greatly lower the needed operating pressure, although it should be remembered that their use could easily raise raw material costs.

Given that carbon dioxide is a relatively feeble solvent, a classic technique for lowering operating pressure (or raising operating concentration) is to employ cosolvents. Methanol and ethanol are most commonly used (Brunner 1994; McHugh and Krukonis 1994; Eckert et al. 2000; Lucien and Foster 2000), but a wide range of organic solvents have been employed in this fashion, usually at concentrations below 40%. Regarding whether the use of cosolvent/CO_2 mixtures is green, one must make a determination on a case-by-case basis. For example, in a conventional chemical process, one must decide whether it is more efficient to use a low pressure process with 100% organic solvent or a high pressure process using only 5–10% organic solvent (for example) with the balance CO_2. To date, the typical answer has been to opt for the low pressure, solvent-based process. However, if the solvent (owing to the nature of the process) is to be emitted to the atmosphere, there are examples where the choice has been to opt for the CO_2/cosolvent route. In the UniCarb coatings process (Chinsoo et al. 1990, 1991; Hoy et al. 1992) developed by Union Carbide during the 1980s and 1990s, CO_2 was employed to replace one component of a solvent mixture used in spray coating, creating a CO_2/cosolvent based process. The foaming of thermoplastics such as polystyrene (Suh 1991) is often conducted using a mixture of CO_2 and an alkane, a more efficient route than employing either 100% alkane or 100% CO_2. One can also employ relatively lower process pressures by operating in the two-phase regime (gas–liquid) rather than employing pressures high enough to maintain a single phase; more about this option will be described in a later section.

Another somewhat obvious route to lowering the operating pressure is by operating at subambient temperatures. Here, however, one must balance the advantage gained by reducing the operating pressure with other impacts, such as the energy cost for cooling and any reduction in reaction rate owing to reduced temperature. Whereas dropping the temperature is an obvious mechanism to reduce the operating pressure, there are others that have received far less attention. For example, the identification of a minimum boiling azeotrope where CO_2 is the majority component could provide a solvent that is both green and exhibits a vapor pressure far lower than that of pure CO_2. Azeotropes are desirable in that process steps requiring flashing of the material (or small leaks) won't change the composition of the solvent. Azeotropes can be maximum boiling (where the vapor pressure of the mixture is higher than either of the pure component vapor pressures) or minimum boiling (the opposite, and here desired situation) (Khoury 1995). Although addition of a second component might lessen the sustainability of the solvent, a solvent that is mostly CO_2 may be better than one than contains no CO_2, and the reduction of the pressure through use of a minimum boiling azeotrope might lower the operating pressure sufficiently to allow economical scale-up of the process. Some CO_2-based azeotropes have been identified (Rowlinson and Sutton 1955; Ohgaki and Katayama 1977; Decaire et al. 1994a, 1994b) as a result of research by chlorofluorocarbon (CFC)-producing companies in a search for alternative refrigerants. Consequently, most of the known CO_2 azeotropes are mixtures with fluorocarbons (it is also known that ethane forms an azeotrope with CO_2). Because azeotropes typically form between compounds whose boiling points are separated by 50 K or less, the number of potential azeotrope-forming cosolvents for CO_2 is likely limited, but this could provide an interesting route to solvents that are both green and versatile.

10.6.3 Recover products without high-pressure drops

It has been mentioned in the literature that use of CO_2 as a solvent is advantageous because reduction of the pressure to 1 atmosphere results in the complete precipitation of any dissolved material, rendering product recovery easy. This may be true, but use of such a route for product recovery raises costs, as one must then either

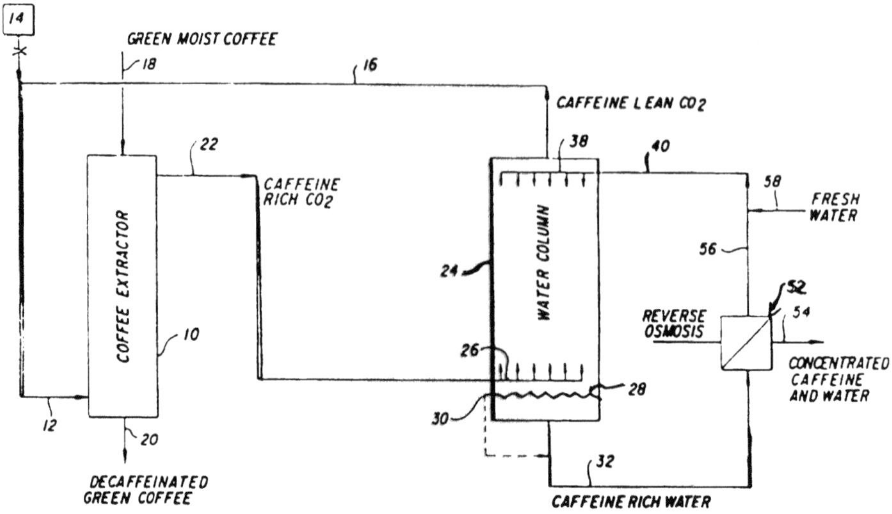

Fig. 10.3 Process schematic for coffee decaffeination using CO_2 (McHugh and Krukonis 1994)

recompress the CO_2 prior to re-use or compress make-up CO_2. As gas compression is energy-intensive and expensive, a greener route to product recovery is desirable.

One example of product recovery without a high-pressure drop is liquid-liquid extraction against water. A liquid-liquid extraction between an organic and aqueous phase inevitably cross-contaminates the phases, normally requiring remediation of one, and probably both phases. In the case of a water-CO_2 extraction, however, the inevitable cross-contamination is benign (carbonated water!). Indeed, the CO_2-based coffee decaffeination process employs a water-CO_2 extraction to recover the caffeine, allowing the CO_2 to move in loop at relatively constant pressure (Fig. 10.3). Further, the cross-contamination here is actually beneficial, because the low pH in the "CO_2-contaminated" water allows for a higher partition coefficient for caffeine, while the "water-contaminated" CO_2 is a better extractant for caffeine than pure CO_2.

10.6.4 Operate the process continuously if possible

The rationale for operating in a continuous mode is that the equipment can be smaller while maintaining high productivity. While this is usually straightforward for liquid substrates, it can be much more difficult for the processing of solids at high pressure. Indeed, there currently does not exist a viable means for introducing and removing solids continuously from a high pressure (100 bar +) process. Those commercial CO2-based processes that employ solids use either batch or semibatch mode. This does not mean that such processes are unimportant—on the contrary, several have been commercialized, including extraction of caffeine from coffee beans and tea leaves, certain acids from hops, and various components from spice plants (Zosel 1974; Prasad et al. 1981; Palmer and Ting 1995; Montero et al. 1996; Tiwari 2007; Jusforgues et al. 1998; Perrut 2000).

In the late 1980s, Chiang and colleagues at the University of Pittsburgh developed a process (LICADO) for the cleaning of coal that employed a biphasic mixture of CO_2 and water (Chiang and Klinzing 1986). Here, the coal was introduced to the process continuously as a slurry in water. If the use of a water slurry of solid substrate is tolerable, this is a useful means by which to introduce solids continuously into a high-pressure process. The Chiang work notwithstanding, there remains no truly efficient means by which to inject and remove granular solids from a high-pressure system (screw feeders have been tried with limited success). There are clearly a number of areas (such as food processing) where continuous injection and removal of solids would greatly enhance the economic viability of a CO_2-based process, yet lack of the mechanical means by which to accomplish this relegates the process to batch or semibatch operation. Note that the chemical basis for continuous polyurethane foam production using liquid CO_2 as the blowing agent was established in the early 1960s, whereas commercialization only occurred after development of the proper equipment in the early 1990s.

A clever example of the use of phase behavior trends to accomplish continuous processing, as well as to recover products without large pressure drops, is shown by Charpentier and colleagues (Charpentier et al. 1999, 2000) in the examination of the continuous polymerization of fluorinated monomers in carbon dioxide. Here, the monomers are soluble in CO_2 (as are many vinyl monomers) whereas the polymers are insoluble (also a relatively general trend). Thus monomer can be continuously recycled through the continuously stirred tank reactor while the polymer precipitates and is collected.

10.6.5 Recover and reuse homogeneous catalysts and CO_2-philes

The discovery of CO_2-philes in the early 1990s allowed for the exploration of a number of processes in CO_2 that had been heretofore untenable owing to CO_2's feeble solvent power. Highly CO_2-soluble surfactants and catalyst ligands became available, leading to a number of important discoveries regarding chemistry in carbon dioxide. However, the new CO_2-philes are significantly more expensive than their CO_2-phobic counterparts, and hence it is important to the economics of a CO_2-based process that any CO_2-philes used in the process be recycled as extensively as possible. Note that recycle of CO_2-philes not only makes good economic sense, but is also more sustainable than the case where the CO_2-philes are simply disposed.

Recovery and recycle of homogeneous catalysts is important whenever such catalysts are employed, because the metals employed in such catalysts are typically expensive. In the case of a CO_2-based process, the ligands are also likely to be expensive (they must be designed to exhibit high CO_2 solubility), and hence the need for effective catalyst recycle is even more important.

In summary, attention must always be paid to the economic viability of processes employing CO_2 as reactant and/or solvent—while CO_2-based processes are generally thought to be "green," their benefits will never be realized if the cost of such processes dwarfs conventional analogs.

10.7 What is CO_2 "like," solvent-wise and how can we use this information for design?

Although in some situations (such as fatty acid extraction), the nature of the solute is such (relatively low molar mass, non-polar) that neat CO_2 is an effective solvent, there are many cases (both in food and chemical production) where this is not the case. Unfortunately, CO_2 is a feeble solvent (Kirby and McHugh 1999; Kauffman 2001); although it can solubilize low-molecular weight, volatile compounds at pressures below 100 bar, polar and high-molecular weight materials are usually poorly soluble at tractable pressures. In traditional nonpolar organic solvents, such a situation does not impede processing, because one simply adds auxiliaries (surfactants, chelating agents, affinity ligands, etc.) to permit extraction of polar targets into nonpolar solvents. For the case of CO_2, traditional auxiliaries are not soluble (Consani and Smith 1990), a situation that nearly eliminated active consideration of CO_2 as a process solvent in the late 1980s.

In the late 1960s Giddings suggested a simple correlation between solubility parameter and critical pressure that indicated that CO_2's solvent power should be similar to that of pyridine (Giddings et al. 1968, 1969). However, the strong quadrupole moment of carbon dioxide affects CO_2's pVT properties (including the critical pressure) without influencing its solvent strength. Consequently, early calculations of the solubility parameter were invariably inflated. This was actually confirmed by the very study that proposed that CO_2's solubility parameter should approach that of pyridine; polymers that would dissolve in pyridine were not soluble in carbon dioxide. Subsequent calculations performed during the early 1980s (Allada 1984) using CO_2's equation of state strongly suggested that CO_2's solubility parameter should approach that of normal alkanes. However, experimental work by Heller's group on the phase behavior of polymers performed during that time (Orr et al. 1983) clearly demonstrated that CO_2's solvent power is inferior to that of n-alkanes—very few polymers tested by Heller showed any significant solubility in carbon dioxide at moderate (less than 200 bar) pressures. Experimental work by Johnston's group (McFann et al. 1994) suggested that solubility parameter was not the best means by which to characterize the solvent power of compressible fluids such as carbon dioxide. Johnston suggested instead that polarizability/volume is a better measure of solvent power; by this standard CO_2 is judged to be a feeble solvent, in line with experimental evidence.

An ability to a priori design highly CO_2-soluble ("CO_2-philic") compounds would advance the "greening" of demanding yet important applications in various industries. It is somewhat ironic that while molecular attributes required to allow dissolution in water (our "other" green solvent) are relatively obvious, a full understanding of the solvent character of CO_2 remains a challenge. CO_2's solvent power has in the past been likened to that of toluene based on FT-IR spectroscopy, to hexane and to pyridine based on thermodynamic solubility parameter calculations, and to acetone based on hydrogen bond accepting tendency. These descriptions have all been discarded over the years as experimental data have revealed them to be oversimplified. CO_2 exhibits "multiple solvent personalities," in that one of CO_2's three MOSCED (Modified Separation of Cohesive Energy Density) solvent parameters (Thomas and Eckert 1984) suggests a resemblance to perfluorohexane, another to acetone, and the third to acetonitrile. CO_2 clearly has great potential as a green solvent, but a thorough fundamental understanding of its behavior is needed to take full advantage of its properties. This is true for both the design of CO_2-soluble auxiliaries (surfactants, chelating agents) and to allow true prediction of phase behavior from the molecular structure of target solutes. Without such fundamental information, use of CO_2 in processing remains more art than science.

In 1992, DeSimone and colleagues (1992) published the first report of a truly "CO_2-philic" material, when they showed that a poly(perfluoroalkyl acrylate) with over 2,500 repeat units was miscible with CO_2 at pressures below 150 bar. By contrast, Heller and colleagues (Heller et al. 1985) had earlier noted that typical non-fluorinated polymers with fewer than 25 repeat units were insoluble in CO_2 at 200 bar. Subsequent work showed that attachment of fluorinated "ponytails" to chelating agents, surfactants, and catalyst ligands generally enhanced the solubility of such compounds in CO_2 (Yee et al. 1992). The problem of CO_2's weak solvent strength seemed to have been solved; fluorination = CO_2-philicity.

Unfortunately, the use of fluorinated ponytails to achieve CO_2 solubility is relatively expensive, and towards the end of the 1990s, it also became somewhat environmentally problematic. More interesting, however, were the reports in the literature showing that the heuristic, "fluorination = CO_2-philicity" was not quite correct. Some, but interestingly not all fluorinated alkane, acrylate, and ether compounds are miscible with CO_2 at much lower pressures than their nonfluorous counterparts. Attempts to explain the CO_2-philic character of fluorinated CO_2-philes have focused on determining whether there exist any specific interactions between CO_2 and these molecules. Yee et al. (1992) used Fourrier Transform Infra-Red spectroscopy (FTIR) to investigate mixtures of CO_2 and hexafluoroethane, finding no evidence of specific attractive interactions between the F atoms and CO_2. The authors consequently attributed the observed enhanced solubility of fluorocarbons to weak solute–solute interactions. However, when Dardin et al. (Dardin et al. 1998) compared ^1H and ^{19}F NMR chemical shifts of n-hexane and perfluoro-n-hexane in CO_2, they observed a chemical shift in the C6F14 spectra, which they ascribed to C6F14-CO_2 van der Waals interactions. By contrast, Yonker and Palmer (2001) showed (using ^1H and ^{19}F NMR) that neither fluoromethane (CH3F) nor trifluoromethane (CHF3) exhibit significant specific attractive interactions with CO_2.

Theoretical studies have also resulted in contradictory findings. Using restricted Hartree-Fock level ab initio calculations, Cece et al. (1996) suggested that there exist specific interactions between CO_2 and the fluorines of C2F6. Han and Jeong (1997), however, disagreed with these results, noting that Cece et al. did not take into account basis set superposition error (BSSE) corrections during their calculations.

Using similar ab initio calculations, but accounting for BSSE corrections, Diep et al. (1998) reported no evidence of CO_2–F interactions in perfluorinated compounds. Raveendran and Wallen (2003) computationally investigated the effect of stepwise fluorination on the CO_2-philicity of methane in an effort to address the existence of F–CO_2 interactions. In partially fluorinated systems, the fluorine atom acts as a Lewis base towards electron deficient carbon atom of CO_2, and the hydrogen atoms, having increased positive charge due to the neighboring fluorine, act as Lewis acids towards the electron rich oxygen atoms of CO_2.

Fried and Hu (2003) used MP2 calculations (6-31++G** basis set) in an effort to identify the nature of specific interactions between CO_2 and fluorinated substituent groups on polymers. They reported that quadrupole–dipole interactions are important contributors to the total energy of interaction. In experimental studies by McHugh's group (Rindfleisch et al. 1996; McHugh et al. 2002), the favorable miscibility of fluorocarbons has also been attributed to dipole–quadrupole interactions. The authors noted that fluorination imparts solubility to the polymer provided that polarity is also introduced to the polymer via such fluorination. Too high a level of fluorination produces an adverse effect on miscibility because of dominance of dipole–dipole interactions between the polymer chains.

Clearly, there is considerable controversy in the literature surrounding the origin of the miscibility of some fluorinated polymers in CO_2, yet there do seem to be some interesting lessons to be learned from this work regarding CO_2-phile design, namely:

- The presence of fluorine creates molecules with weak self-interaction, rendering miscibility with CO_2 possible at lower pressures.
- Electronegative fluorine may exhibit specific interactions with CO_2's electron poor carbon, lowering miscibility pressures.
- The presence of fluorine will affect the acidity of neighboring protons, allowing for the possibility of specific interactions between these protons and CO_2's oxygen atoms.

10.8 Nonfluorous CO_2-philes: the role of oxygen

Given the environmental and cost issues related to the use of fluorine, designs of CO_2-philic auxiliaries since 2000 has focused on intriguing literature concerning interactions between oxygen-containing functional groups and CO_2. For example, Kazarian et al. (1996) reported the existence of Lewis acid–Lewis base interactions (via FT-IR) between CO_2 and the oxygen of a carbonyl. In this and other studies, it was shown that the carbonyl oxygen interacts with the carbon atom of CO_2, wherein the geometry and strength of the interaction may vary depending on adjacent groups. The use of oxygen-containing functional groups appeared particularly advantageous in CO_2-phile design because it allows for creation of specific interactions with CO_2 while minimizing the strength of self-interactions of the solute.

We subsequently (Fink et al. 1999; Kilic et al. 2003) showed that addition of carbonyl-containing functional groups lowers the miscibility pressures of silicones in CO_2; the combination of weak self-interaction (silicones) and CO_2:carbonyl groups interaction is clearly favorable. The particular utility of acetates in lowering miscibility pressures in CO_2 was demonstrated by Raveendran and Wallen (2002b) and by our group (Potluri et al. 2002) using acetate-functional saccharides and polysaccharides. To our CO_2-phile molecular "wish list" of weak self-interaction and oxygen:CO_2 interactions we added high flexibility/low softening point, assuming that this characteristic would enhance the entropy of mixing of the compound with CO_2. We then designed a series of ether-carbonate copolymers (Sarbu et al. 2000) that exhibited lower miscibility pressures than the fluorinated polyethers we had employed as CO_2-philes for over a decade. Nevertheless, our set of guidelines remained simply guidelines; true a priori design was not possible. In particular, we noted that very small changes to structure led to dramatic and unpredictable changes in phase behavior, a frustrating situation. For example, it has been known (Rindfleisch et al. 1996) for almost a decade that polymethyl acrylate (PMA) and polyvinyl acetate (PVAc) exhibit miscibility pressures in CO_2 that differ by hundreds of bar—this result would not be predicted by any group contribution thermodynamic model currently in use without purely empirical adjustments. While polyvinyl acetate exhibits relatively accessible miscibility pressures, we have found that the addition of a single methylene unit (polyallyl acetate) creates a material that is for all intents and purposes, insoluble. Again, traditional thermodynamic models provide no guidance here.

We believe that at least part of the answer to these puzzles lies in CO_2's ability to act as both Lewis acid and Lewis base, coupled with subtle effects of neighboring substituents on the acidity of certain protons. For example, Raveendran and Wallen (2002b), in an analysis of interactions between acetate groups and CO_2, found that the acidity of the methyl acetate protons allows for binding of CO_2 through both its carbon and oxygen atoms. We have found experimentally (Baradie et al. 2004) that copolymers of vinyl acetate (VAc) and tetrafluoroethylene (TFE) exhibit lower miscibility pressures than either of the homopolymers. Not surprisingly, calculations made (using MP2/aug-cc-pVDZ level of theory) on the geometry and strength of the interactions between CO_2 and various dyads (TFE-VAc, VAc-VAc, etc.) in the copolymer showed that the presence of the difluoromethylene groups in the backbone render neighboring protons more acidic.

In summary, combinations of theory and experiment are rapidly advancing our ability to design cost-effective CO_2-philic materials, and hence to predict how various materials will behave in solution with CO_2. We would propose the following molecular characteristics for a "CO_2-phile."

10.8.1 *Flexible, high free volume materials*

Eastoe et al. (2001) have previously demonstrated that increasing free volume through functional group changes will enhance solubility of compounds in CO_2. Elevating free volume and flexibility can, for example, be accomplished through branching and use of ether linkages in the main chain.

10.8.2 Weak self-interactions

O'Neill et al. (1998) previously noted that most of the CO_2-philes known exhibit relatively weak self-interaction, as evidenced by low cohesive energy density. We have found, for example, that while tertiary amines interact more strongly with CO_2 than do carbonyls, the stronger self-interaction of the amine-containing compounds actually elevates their miscibility pressures in CO_2 versus oxygen-containing analogs.

10.8.3 Multidentate interactions between CO_2 and solute functional groups, where interactions involve both the carbon and oxygen in CO_2

Oxygen-containing functional groups are advantageous in that they interact with CO_2, adjust the acidity of neighboring protons, and add comparatively little to the strength of self-interaction of the solute.

Eastoe and coworkers have designed a number of low molecular weight surfactant systems that take advantage of some of these design parameters—while miscibility pressures remain high, they have shown that surfactants are soluble and allow for water solubilization, a key feature (Eastoe et al. 2006).

10.9 Chemistry in CO_2: hydrogenations

In the chemical process industry, oxidations, hydrogenations, and hydroformylations are carried out routinely; use of CO_2 as a solvent in such processes has been extensively investigated, owing to CO_2's ability to solvate large amounts of either O_2, H_2, or CO. Because the food industry is primarily interested in hydrogenations, in this section we focus on the use of H_2 as reactant. Further, while homogeneous catalysis is very important to the fine chemical and pharmaceutical industry, because it is somewhat less important to food processing, we opt to focus primarily on heterogeneous catalysis. Discussion of oxidation and hydroformylation in CO_2, as well as homogeneous catalysis, can be found elsewhere (Beckman 2003).

Hydrogenation is widely used in industry at scales ranging from grams per year to tons per hour. Hydrogenation is conducted at large scale in either the gas or liquid phase; further, whereas gas phase reactions are performed over a solid catalyst (heterogeneous catalysis), liquid phase reactions are conducted in either two (homogeneous catalyst, liquid and gas each present) or three (heterogeneous catalyst, liquid and gas each present) phase modes. Finally, heterogeneous catalysis is conducted in batch, continuous slurry, and fixed bed reactor configurations, although the latter is less common than the former two. Despite the broad range of potential reactor configurations and reactions, we will restrict this discussion to those hydrogenations currently carried out in the liquid phase—addition of a supercritical solvent to a gas-phase reaction will simply dilute the reactant concentrations, reducing the rate significantly.

10.10 Liquid-phase hydrogenations: advantages to use of supercritical solvents

A number of hydrogenations (synthesis of unsaturated fatty acids, reduction of fatty esters to alcohols) are conducted commercially in organic solvent, and replacement of these solvents with benign carbon dioxide will reduce both liability (reduced flammability, potential toxicity issues) and the potential for VOC emissions owing to fugitive losses. In addition, use of any supercritical fluid in a liquid-phase hydrogenation process can significantly alter the relative importance of fundamental processes governing the rate expression. In a three-phase hydrogenation, the rate can be governed purely by the kinetics of the reaction, but more likely will depend on the rate at which hydrogen diffuses from the gas phase to the active sites on the catalyst. The overall rate of transport is itself governed by three resistances in series: (1) the resistance to transport of H_2 across the gas–liquid interface, (2) the resistance to transport of H_2 through the liquid to the surface of the catalyst, and finally (3) resistance to transport of H_2 within the pores of the catalyst. Given that the overall rate is related to the sum of the resistances in series (Satterfield 1980), one term can easily dominate the expression for the overall rate. Use of a supercritical fluid solvent (as opposed to a traditional liquid) eliminates the gas–liquid interface, as low T_c gases such as H_2, O_2, and CO are completely miscible with fluids above their critical point. However, this does not necessarily mean that the reaction will be kinetically controlled, because one must deal with the remaining two resistances to transport (bulk liquid to solid surface, interpore diffusion). Because the diffusion constant is embedded in each of these resistances, the use of a supercritical fluid can also aid in their elimination, although simply switching from a conventional liquid to a supercritical fluid solvent for hydrogenation by no means guarantees that the reaction rate will depend solely on the underlying kinetics.

It should be noted that significant effort is expended in hydrogenation reactor design to ensure that H_2 is well dispersed in the liquid phase—effective sparging greatly increases the contact surface area between the phases and hence the rate at which H_2 diffuses into the liquid. If use of a supercritical fluid allows for a reactor redesign (for example, plug-flow versus continuous-stirred tank given that gas sparging is unnecessary), then it may be possible to enhance the selectivity of the reaction through reactor design improvement, reducing waste.

Indeed, selectivity is a major concern in any chemical process—hydrogenation is no exception. It is well known that solvents affect the yield and selectivity of various hydrogenation reactions, whereby "one very useful, although fallible, generality is that in a series of solvents, the extremes in selectivity will be found at the extremes of the dielectric constant ..." (Rylander 1985). The supercritical fluids most often employed as hydrogenation solvents, propane and CO_2, exhibit dielectric constants at the lower end of the scale (1.5–1.7), and we might expect to see an effect on selectivity if a polar solvent is replaced by CO_2. In addition, the physical properties of supercritical fluids are readily varied over a significant range through changes to pressure and temperature, and it may be possible to affect selectivity by altering these variables. Finally, addition of CO_2 or operation above the critical point of the reactant mixture could aid in coke removal from the catalyst, prolonging its life or maintaining favorable selectivity (Baptiste-Nguyen and Subramaniam 1992; Ginosar and Subramaniam 1994, 1995; Subramaniam and Ashraf 1995; Subramaniam and Arunajatesan 1999; Subramaniam 2001). Clearly, enhancing selectivity of a reaction will ultimately reduce the volume of byproducts generated, and potentially the volume of waste emanating from a particular process.

Harrod and colleagues (van den Hark et al. 1999; van den Hark and Harrod 2001) have successfully performed the hydrogenation of fats and oils using supercritical propane; propane was employed to allow for solubility of both the substrates (whose solubility in CO_2 is poor) and hydrogen, which is completely miscible with any supercritical fluid. The homogeneous propane/H_2/substrate mixture was fed into a packed bed containing a commercial Pd catalyst—extremely high reaction rates were indeed achieved (gas–liquid transport resistance being eliminated) and the concentration of trans fatty acids (an undesirable byproduct) was reduced. Hence, the green advantages to this reaction would include reduced waste content and smaller and more efficient reactors. However, the use of propane is problematic, and it is not clear whether the process advantages because of faster reaction rate balance the disadvantages deriving from use of a flammable solvent and the problems inherent to high-pressure process design/development. Further, the catalyst deactivated quickly, an important problem for both economic and sustainable reasons. Tacke et al. (1996) also investigated the hydrogenation of fats and oils (over a supported Pd catalyst), although they employed CO_2 as the supercritical solvent. Again, rates were shown to be significantly higher in the supercritical case (sixfold increase in space-time yields), and selectivity and catalyst lifetime were also improved. Each of these features contributes to enhancing the green potential of the process, while the need for high pressure operation detracts both from the cost and the sustainability (energy, unit operation complexity). Macher and Holmqvist (2001) also examined the hydrogenation of an oil in supercritical propane; similar results to those found by Harrod were obtained. King and coworkers (Andersson et al. 2000; King et al. 2001) examined the hydrogenation of vegetable oil and fatty acid esters over nickel catalysts using both CO_2 and propane as supercritical solvents, and under conditions where either one or two fluid phases existed in the reactor. This approach is interesting, because it ultimately could prove a useful engineering solution to the problem of solubilizing substrates in CO_2 at moderate operating pressures.

Indeed, Chouchi et al. (2001) examined the hydrogenation of pinene (over Pd/C) in supercritical CO_2. They found that the rate of the reaction was significantly faster in the two-phase regime (i.e., lower pressures) than when the pressure was raised to the point where only a single fluid phase existed. The reason for this seems clear; the Chouchi study was performed by charging a known amount of each of the ingredients to the reactor, then pressurizing with CO_2. The partitioning of compounds between phases (in the two-phase system) must have been such that the concentration of reactants in the lower phase was higher than under single-phase conditions. In other words, raising the pressure to create a single phase simply diluted the reactants, lowering the rate. Note that the concentration of CO_2 in the lower phase (in the two phase system) was likely to be substantial, because CO_2 should interact favorably with a volatile, low molecular weight compound such as pinene. Further, the concentration of hydrogen in the lower phase must also have been substantial to support the high rate observed, and hence we see that CO_2 can swell an organic substrate significantly and carry substantial amounts of hydrogen into a "swollen" liquid phase. CO_2 could therefore function as a "reversible diluent," much in the same way that it is employed as a "reversible plasticizer" in polymer science. In this case, addition of CO_2 at relatively low pressures would enhance solubility of H_2 in the substrate, raising rates while not impacting process costs precipitously.

Subramaniam et al. (2002) have authored a comprehensive review on process design issues inherent to catalytic processes performed in carbon dioxide; interested readers should consult this paper. A key future research issue that will impact heterogeneous hydrogenations in CO_2 is the lifetime of the catalysts, particularly the widely used palladium catalysts. The literature contains examples of successful hydrogenations over Pd in CO_2, and also examples where the rapid formation of CO led quickly to catalyst poisoning and deactivation. Subramaniam's group has recently presented a rationale for the seemingly contradictory results in the recent literature. They showed (using high pressure FT-IR) that CO forms very quickly (within minutes) on Pd in a mixture of CO_2 and H_2, and then over much longer times alters its mode of binding to reduce catalyst activity. Temperature is a key parameter in this process, where temperatures above 343 K seem to greatly accelerate the process. Longer residence times (as would be experienced in batch reactors or CSTRs) also enhance the rate of poisoning.

In summary, hydrogenation in supercritical fluids has been extensively investigated over the past 15 years, and it is clear that hydrogenation reactions can be successfully conducted in CO_2 and other fluids. The primary rationale for use of a supercritical solvent in hydrogenation reactions is the elimination of transport limitations to reaction through enhancement of the solubility of hydrogen at the reaction locus. Hydrogen is poorly soluble in conventional hydrocarbon liquids and water, and use of CO_2 as the solvent has been shown to enhance H_2 solubility and hence improve the efficiency of the reaction. Attaining kinetic control over the reaction can lead to reduced byproduct formation and lower energy input, although in the case of typically exothermic hydrogenations, energy removal is more important than energy addition.

A key point that arises if one examines the literature is that one doesn't need to create a single phase (of SCF, substrate, and hydrogen) to create a situation where transport limitations can be eliminated. For example, one can attain kinetic control over the reaction simply by ensuring that a significant amount of CO_2 is present in the liquid phase (maintaining a gas phase of CO_2/H_2). Here the CO_2 functions as a diluent (and viscosity reducer) that enhances the solubility of hydrogen in the lower phase.

The enhanced hydrogen solubility more than makes up for the dilution effect from the CO_2. Whereas elimination of the resistance owing to transport of H_2 into the liquid phase does not by definition create kinetic control over the reaction (resistances owing to diffusion to and within the catalyst also exist), the previous work has shown that the solubility of H_2 in the liquid is typically the limiting factor. The use of CO_2 as the "H_2 solubility enhancing diluent" could have broad ramifications on the practicality for conducting hydrogenations in supercritical fluids, in that it could make the use of benign (and non-flammable) CO_2 more viable. The poisoning of noble metal catalysts via the formation of CO from CO_2 and H_2 could seriously impact the economic viability of hydrogenation processes conducted in carbon dioxide.

10.11 Polymerization and polymer processing

Research studies into polymer/CO_2 mixtures where CO_2 is the minor component (i.e., CO_2 as diluents and/or foaming agent) are numerous; perhaps much more numerous than those on dilute solutions of polymers in CO_2. Commercial processes have been created around such systems owing to the favorable thermodynamics—it is quite possible that analogies could be created within the food processing industry (foamed food products are routinely generated, although steam is most often used as the porogen). As such, it is useful to examine advances in this area with an eye towards envisioning possible analogies.

In Fig. 10.4, we see a generic phase diagram of a polymer and a SCF (Wiesmet et al. 2000), showing the various phase separation envelopes and the behavior both above and below the solvent critical temperature. As can be seen in Fig. 10.4, the liquid–liquid phase envelope is asymmetric (owing to the large disparity in size between polymer and solvent) with the liquid-liquid critical point shifted towards the 100% solvent axis. This is important—it means that solubilization of low concentrations of polymer in solvent will require the highest pressures. Swelling of the polymer by the solvent (moving to the right along the x-axis in Fig. 10.4) requires significantly lower pressures. Thus, in certain polymer-SCF mixtures, one can observe very high degrees of swelling (>25% in polyacrylate-CO_2 mixtures, for example) at pressures of 100 bar and below (Wissinger and Paulaitis 1991a, b). The relatively low pressures required to elicit high degrees of swelling may be one reason why applications where CO_2 is the minor component have been successfully commercialized, while those employing dilute polymer solutions have not.

As noted above, CO_2 will swell many polymers extensively, even those normally considered "CO_2-phobic." As shown in the generic phase diagram (see Fig. 10.4), this is because of the asymmetry of the liquid–liquid phase envelope, itself arising from the disparity in size (and hence vapor pressure) of the solvent and solute. Swelling a polymer with CO_2 will drop its viscosity significantly (depending upon temperature, by orders of magnitude). This large drop in viscosity allows for a number of CO_2-enhanced processes. For example, Berens et al. (1992) demonstrated that the swelling of a polymer by carbon dioxide enhances the rate of infusion of

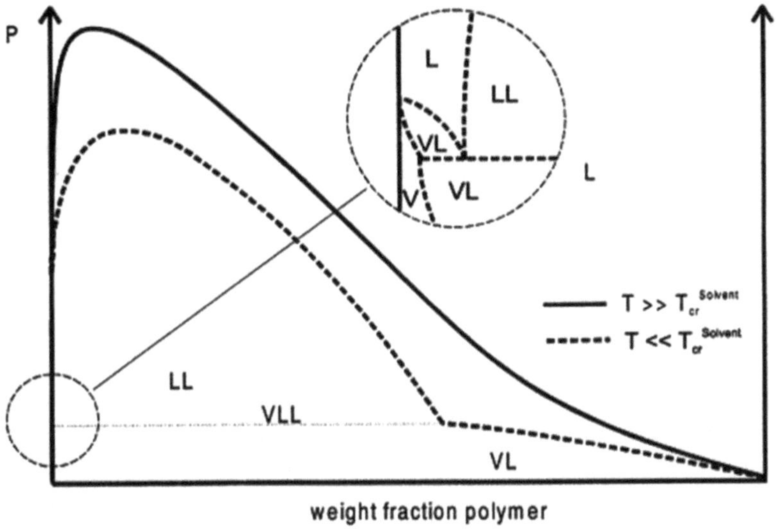

Fig. 10.4 Qualitative P-x diagram of a polymer-CO_2 binary mixture, both above and below the critical temperature of the solvent. Figure includes liquid-liquid (LL), vapor–liquid (VL), and three phase vapor–liquid–liquid (VLL) types of phase envelopes (Wiesmet et al. 2000)

model compounds. Kazarian et al. (1997) later exploited this effect in a novel way; they showed that one can greatly enhance the kinetics of mixing of a CO_2-incompatible dye with a polymer. In this work, the dye and polymer are thermodynamically compatible, but the rate of infusion of the polymer by the dye is glacially slow. CO_2 plasticizes the polymer (while not actually dissolving very much, if any, of the dye), lowering the viscosity and allowing fast blending. The dying of fabric and fibers using CO_2 has been extensively examined in Europe and the US (Bach et al. 1998, 1999, 2002; Argyle and Propp 1999; Eggers et al. 1999a, b, c; Schollmeyer et al. 1999; Montero et al. 2000; Poddevin et al. 2000; Smith et al. 2000; Hendrix et al. 2001); here again the dye and polymer are thermodynamically compatible while the dye is sparingly soluble in CO_2. Consequently, the dye partitions preferentially into the swollen polymer, in which the CO_2 diluent enhances the kinetics of the thermodynamically favorable process. It is interesting to note that Johnston's group (Shim and Johnston 1989, 1991) outlined the fundamentals for such a process several years ago using a silicone polymer, CO_2, and toluene as the model "infusant." One wonders if there aren't analogies to such "infusant" based processes in the food industry?

Major challenges remaining in this process are in many ways "mechanical." How does one design a treatment chamber that allows fast charging, fast sample changeover, and rapid dying? Is there sufficient thermodynamic and transport information available to model and hence scale-up the process? Note that this situation is analogous to that for continuous polyurethane production using carbon dioxide—the chemical challenges were overcome long before the mechanical issues were settled. A further challenge would include redesigning conventional dyes to allow for higher CO_2 solubility, which would provide for more even coating.

Applying the concept of carbon dioxide as "reversible plasticizer," Shine and Gelb (1998) showed that one could mix a thermally labile bioactive compound (here a vaccine) into polycaprolactone. Howdle et al. (2001) expanded this work into the tissue engineering field. Here, CO_2 was used to swell an aliphatic polyester, depressing its T_g to well below room temperature. A temperature and shear-sensitive enzyme was then mixed with the swollen polymer; upon depressurization the enzyme was found to be dispersed throughout the now foamed polymer, and to have retained its activity. Such a process allows the blending of temperature-sensitive compounds with polymers without the need for additional solvent-based processing.

The extrusion based foaming of polymers is conceptually simple, yet requires complex analysis to fully understand the system. In the case of polystyrene, a fluid is injected into the extruder, where the pressure and temperature are sufficient (ostensibly) to create a single-phase mixture of blowing agent and polymer. Mixing is enhanced through strategic screw design. Following mixing, the melt is cooled (in some cases in a second, tandem extruder) to build melt strength, because the addition of the fluid greatly lowers the melt viscosity. The die is cooler still. Upon exiting the die, the rapid pressure drop creates a supersaturated solution, wherein small pores containing CO_2 nucleate and grow (nucleating agents are often added to stimulate this process). The pores grow until the rapidly rising viscosity of the polymer (owing to cooling and loss of blowing agent) restricts further expansion. In conventional extruded foam, the cells are of order 100–1,000 μm in diameter. Microcellular foam (Pierick 1999), formed in much the same way albeit with higher concentration of CO_2 in the polymer melt, exhibits cells 50 μm and below in size.

The generation of foamed thermoplastics using CO_2 as the sole blowing agent is most definitely "green" processing, as the CO_2 replaces either organic or hydrofluorocarbon agents that would otherwise directly enter the atmosphere. A number of researchers have investigated the fundamentals of foam formation using high pressure CO_2, and several important conclusions have arisen (Goel and Beckman 1995; Arora et al. 1998; Stafford et al. 1999):

- The number of cells nucleated during a pressure quench in a CO_2-swollen polymer depends directly upon the degree of swelling of the polymer. Swelling, in turn, rises as pressure rises and as temperature falls. To create more cells one must adjust conditions to ensure higher degrees of swelling.
- The growth of cells is dependent upon the degree to which CO_2 diffuses into the nuclei, and also the degree to which CO_2 expands as pressure drops. At the same time, growth is inhibited by the retractive force of the polymer melt, which increases as the temperature drops and CO_2 diffuses from the melt. Hence, to make smaller cells, one must restrict growth soon after nucleation by vitrifying the system before the pressure drops to the point where CO_2 begins to expand significantly. If one desires to make a large number of very small cells, then in theory one should start with a high degree of swelling of the polymer by CO_2, and vitrify the material as soon as possible after nucleation of pores. Unfortunately, very high

degrees of swelling lower the melt strength (related to viscosity) significantly, and hence pores tend to coalesce during growth (Sparacio and Beckman 1998).

- Our understanding of the fundamental processes that control foam morphology derives in large part to fundamental studies performed in academia and industry during the late 1980s and early 1990s. For example, early studies of the effect of pressure on the swelling of polymers by CO_2 by Berens and Huvard (1989), Liau and McHugh (1985), and Wissinger and Paulaitis (1991a, b) paved the way for future work on polymer foaming. Wang et al. (1982) first explored the behavior of the glass transition of a polymer versus CO_2 pressure; this was followed by a seminal study by Condo and Johnston (1994). Fundamental studies of the viscosity of polymer-CO_2 melts, for example, were performed by Kwag et al. (1999), and also by Royer et al. (2000). These studies provided the data that made later studies of foam formation more tractable. As such, extension of CO_2-based foaming into the food industry will require analogous basic research into the interaction between CO_2 and the matrices of choice.

A large number of patents have been issued for both the foaming of polymers with CO_2 and the use of CO_2 to dye textiles. For the case of polymer foaming, the technology has achieved commercial status, both macrocellular foam formation (Dow, for example) and microcellular foam formation (Trexel has licensed technology developed at MIT by Nam Suh's group (Park et al. 2000)). The textile work has been advanced to the pilot stage in Germany and in the US.

10.12 Enzymatic chemistry in CO_2

At first glance, enzyme/CO_2 mixtures appear as ideal reaction systems for the performance of green chemistry. Enzymes are naturally derived catalysts that are highly selective, while CO_2 is a naturally abundant, benign solvent. However, research into enzymatic reactions in CO_2 has dropped precipitously since the mid-1990s, and no commercialization of such processes is currently anticipated. The reasons for this are straightforward and scientifically based, deriving from the substantial research performed in this area during the 1990s.

Enzymes are naturally derived catalysts, proteins whose primary, secondary, and tertiary structure has evolved to create a catalyst that is highly selective and very active under a set of narrowly defined conditions. Enzymes themselves are green catalysts, and their means of production (fermentation) is also typically a green process. In nature, enzymes perform their catalytic function in water, yet Klibanov (and others) showed that enzymes would function adequately (not as well as in water) in organic media provided that a small amount of water remains bound to the enzyme (Klibanov 1997). Further, while lipases (and other analogous enzymes) naturally perform hydrolysis reactions in an aqueous environment, these same enzymes were shown to perform esterification in an organic environment. Because enzymes don't dissolve in the organic solvents under consideration, enzymatic chemistry in organic solvents is governed by heterogeneous reaction kinetics. This, however, is not a drawback, because catalyst recovery is easier than for a homogeneous system. Given this background, enzymatic reactions in CO_2 seemed an ideal combination of green solvent with green catalyst.

During the early 1990s, a number of enzymes were evaluated in carbon dioxide, primarily in support of esterification reactions (Mesiano et al. 1999). For the most part, activities were very low, much lower than for the same reaction conducted in a conventional organic solvent. In addition, rates in CO_2 were substantially lower than rates in other compressible fluids (ethane, propane, fluoroform). In some key publications, Russell and colleagues outlined the reason for CO_2's low activity—apparently carbon dioxide reacts with primary amine residues (primarily from lysine) to form carbamic acid and/or ammonium carbamates (Kamat et al. 1992, 1995). This derivatization was observed experimentally, and is apparently responsible for the reduced activity of many enzymes in CO_2 (note that not all enzymes suffer from this reduced activity, consistent with the fact that enzymes exhibit a range of protein sequences, secondary and tertiary structures). Carbamate formation is reversible, as removal of the enzyme from CO_2, followed by examination of the rate in either water or another organic solvent reveals no change in inherent activity. Even bubbling of gaseous carbon dioxide through a suspension of enzyme in organic solvent can produce the reversible drop in activity. Consequently, interest in enzymatic chemistry using enzyme powder in CO_2 diminished greatly.

At this same time, advancements in the design of CO_2-philic surfactants allowed for the possibility of performing enzymatic chemistry in the aqueous core of micelles formed in carbon dioxide, a situation that would eliminate the problems owing to carbamate formation (polar solvents destabilize the carbamates). Indeed, work by Randolph and Johnston (Johnston et al. 1996), as well as Beckman's group (Ghenciu et al. 1998), showed that one could solubilize an enzyme in the core of a micelle, and then recover the protein via depressurization. However, CO_2 dissolves in water and forms carbonic acid, and not surprisingly the pH within the micelles was shown to be less than 3.0. While Johnston showed that one could buffer such a system to pH's from 5.0 to 6.0 (Holmes et al. 1999), the ionic strength required was far higher than would normally be recommended for use with an active enzyme. Thus, realization of the full "green" potential of enzyme-CO_2 systems was again blocked by technical realities.

Other issues to note regarding use of enzymes in CO_2 include the need by the enzyme for a certain amount of bound water and the equilibrium nature of many of the reactions. Although CO_2 is usually considered a nonpolar solvent, it will solubilize approximately 2,500 ppm water at moderate pressures (100 bar, room temperature). Because enzymes will not function in organic media if stripped of all of their water, care must be taken to prevent CO_2 from dehydrating the enzyme. In addition, many of the enzymatic reactions that one might wish to perform in CO_2 are governed by equilibrium, and hence one must examine means by which to remove the byproduct or product from the neighborhood of the enzyme.

A final obstacle to use of enzymes in supercritical fluids lies in the poor solubility of many of the polar substrates that one might wish to transform. For example, while many of the literature studies performed during the early 1990s examined esterifications, the starting material (carboxylic acid) was usually not particularly soluble in CO_2 (hardly surprising given what is known about CO_2).

The previous paragraphs make plain the technical hurdles that would need to be overcome to render enzymatic chemistry in CO_2 generally practical and useful. Either enzymes must be identified (or developed through a directed evolution-like process) that do not form carbamates with CO_2 (or where carbamate formation does not impede activity) or a way must be found to buffer a CO_2/water mixture without resorting to an ionic strength that will harm the enzyme. Conversely, identification of enzymes that thrive at low pH or high ionic strength would also be worthwhile in this regard.

10.13 Reactions at interfaces and/or multi-phase mixtures

Reactions at interfaces (or transport across interfaces to facilitate reaction) in CO_2-based systems have been proposed as a useful means by which to support green chemistry in carbon dioxide while easing separation problems post-reaction. Indeed, if one can effectively segregate catalyst, reactants, and products in various phases in the reactor, downstream separation is certainly easier. However, one is now also faced with thermodynamic (phase behavior) and transport limitations to reaction. A key proviso in attempting to use a biphasic system (with CO_2) to perform green chemistry is that the continuous component of each phase (CO_2 and the second component) should either be environmentally benign (and hence cross-contamination is irrelevant) or should be immiscible over essentially the entire concentration regime. Thus, only the components of interest (reactants, products) are moving across the phase boundary.

Reactions making use of the CO_2/water biphasic mixture have long been proposed as green alternatives to conventional reactions. Each of these solvents is inherently benign, they are immiscible over a broad range of concentrations, and the inevitable cross-contamination that occurs on phase contact does not require remediation. Eckert's group (Chandler et al. 1998) first examined the use of a conventional phase transfer catalyst in a CO_2/water mixture and found that despite the lack of "CO_2-philic" ligands, the tetraalkyl ammonium bromide was effective at catalyzing the reaction across the interface. While Eckert employed a phase transfer catalyst, Johnston and colleagues (and later Tumas) enlarged the interfacial surface area through creation of an emulsion (Jacobsen et al. 1999). The enhanced surface area in the emulsion greatly enhanced the rate of the model reactions performed by these two groups (Fig. 10.5). Hancu and Beckman (2001) also examined the use of added surfactant to enhance reactivity in a CO_2/water biphasic system. Here, CO_2 dissolves in aqueous hydrogen peroxide, forming percarbonate (through two distinct mechanisms). The percarbonate ion (basic conditions are employed) then reacts with an alkene at the interface, forming the epoxide. The addition of surfactant to this system substantially enhanced the reaction rate, as did the addition of a phase transfer catalyst. The usual caveat in CO_2/water biphasic mixtures is that the low pH can cause problems for some reactions (Bonilla et al. 2000).

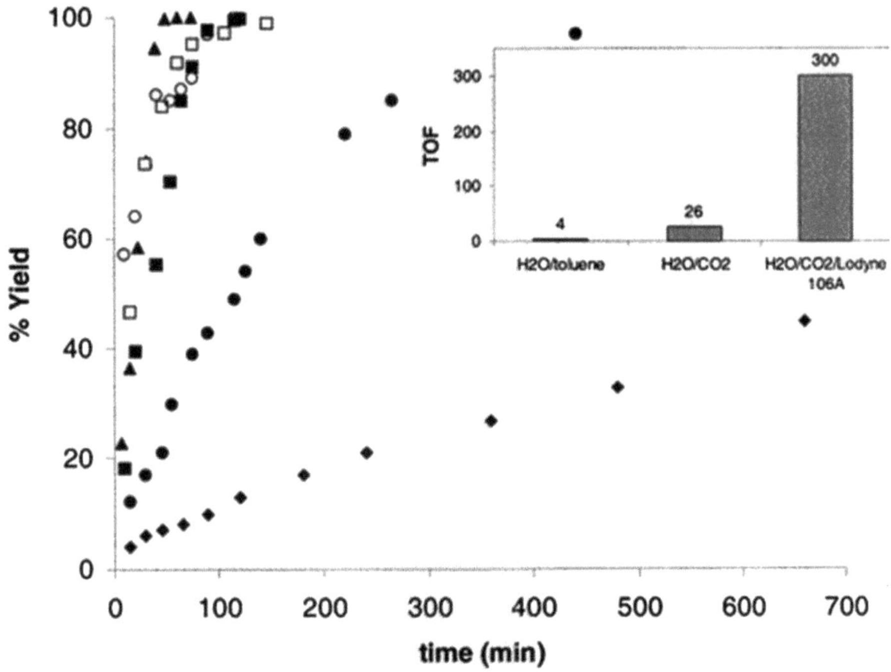

Fig. 10.5 Time profile of formation of ethyl benzene from hydrogenation of styrene performed in biphasic water/toluene (♦), biphasic water/CO_2 (•) and in emulsions using PFPE MW = 2,500 (▲), PFPE MW = 740 (■), Lodyne 106A (○), or PBO-PEO (□) as surfactants. Reaction conditions: 50/50 wt.% water/CO_2, 1.5% surfactant, 80 mM styrene, 1 mol% catalyst (to substrate), Rh/L = 1/6, 40 C, 4,000 psi. TOF values at 50% conversion are given as a comparison for biphasic H_2O/toluene, H_2O/CO_2, and H_2O/CO_2 emulsion systems (Jacobsen et al. 1999)

10.14 Summary

From its inception, research into the use of CO_2 as a process solvent has been motivated by the desire to create process schemes that are greener than their predecessors. From the first coffee decaffeination work in the late 1970s, the use of CO_2 has expanded into an ever widening series of scientific subdisciplines, from biotechnology to microelectronics processing to polymer synthesis. Applications in food processing were some of the first examples of large scale manufacturing processes employing CO_2, and CO_2's benign character will likely continue to prompt work in this area. Future advances, however, will be predicated on a better understanding of CO_2's molecular behavior in solution, so as to be able to truly design processes.

References

Allada, S.R. 1984. Solubility parameters of supercritical fluids. *Industrial & Engineering Chemistry Process Design and Development* 23: 344.

Andersson, M.B.O., J.W. King, and L.G. Blomberg. 2000. Synthesis of fatty alcohol mixtures from oleochemicals in supercritical fluids. *Green Chemistry* 2: 230.

Angus, S., B. Armstrong, and K.M. de Reuck. 1976. *Carbon dioxide*. Oxford: Pergamon Press.

Argyle, M.D., and W.A. Propp. 1999. Method and apparatus for the application of textile treatment compositions to textile materials. USPTO, US Patent No. 5,709,910.

Arora, K.A., A.J. Lesser, and T.J. McCarthy. 1998. Preparation and characterization of microcellular polystyrene foams processed in supercritical carbon dioxide. *Macromolecules* 31: 4614.

Artal, M., J. Munoz Embid, I. Velasco, C. Berro, and E. Rauzy. 2001. Representation for binary mixtures of *n*-alcohols+sub and supercritical CO_2 by a group-contribution method. *Fluid Phase Equilibria* 178: 119.

Bach, E., E. Cleve, E. Schollmeyer, M. Bork, and P. Korner. 1998. The dyeing of polyolefin fibers in supercritical carbon dioxide part 1: Thermomechanical properties of polyolefin fibers after treatment in CO2 under dyeing conditions. *Journal of the Textile Institute* 89: 647.

Bach, E., E. Cleve, E. Schollmeyer, T. Vardag, and P. Korner. 1999. The dyeing of polyolefin fibers in supercritical carbon dioxide part 2: The influence of dye structure on the dyeing of fabrics and on fastness properties. *Journal of the Textile Institute* 89: 657.

Bach, E., E. Cleve, and E. Schollmeyer. 2002. Past, present and future of supercritical fluid dyeing technology—An overview. *Review of Progress in Coloration and Related Topics* 32: 88.

Baptiste-Nguyen, S., and B. Subramaniam. 1992. Coking and activity of porous catalysts in supercritical reaction media. *AIChE (American Institute of Chemical Engineers) Journal* 38: 1027.

Baradie, B., M.S. Shoichet, Z. Shen, M. McHugh, L. Hong, Y. Wang, J.K. Johnson, E.J. Beckman, and R.M. Enick. 2004. Synthesis and solubility of linear poly(tetrafluoroethylene-*co*-vinyl acetate) in dense CO2: Experimental and molecular modeling results. *Macromolecules* 37(20): 7799–7807.

Beckman, E.J. 2003. Green chemical processing using CO2. *Industrial and Engineering Chemistry Research* 42(8): 1598–1602.

Berens, A.R., G.S. Huvard. 1989. Interaction of Polymers with Near-Critical Carbon Dioxide, in *Supercritical Fluid Science and Technology, American Chemical Society Symposium Series*, 406(14): pp 207–223 http://pubs.acs.org/isbn/9780841216785.

Berens, A.R., G.S. Huvard, R.W. Korsmeyer, and F.W. Kunig. 1992. Application of compressed carbon dioxide in the incorporation of additives into polymers. *Journal of Applied Polymer Science* 46: 231.

Blas, F.J., and A. Galindo. 2002. Study of the high pressure phase behaviour of CO_2+*n*-alkane mixtures using the SAFT-VR approach with transferable parameters. *Fluid Phase Equilibria* 501: 194–197.

Bonilla, R.J., B.R. James, and P.G. Jessop. 2000. Colloid-catalysed arene hydrogenation in aqueous/supercritical fluid biphasic media. *Chemical Communications* 11: 941–942.

Brunner, G. 1994. *Gas extraction*. Darmstadt, FRG: Steinkopff Verlag.

Brunner, G. 2005. Supercritical fluids: Technology and application to food processing. *Journal of Food Engineering* 67(1–2): 21–33.

Byun, H.S., K. Kim, and M.A. McHugh. 2000. Phase behavior and modeling of supercritical carbon dioxide–organic acid mixtures. *Industrial & Engineering Chemistry Research* 39: 4580.

Cece, A., S.H. Jureller, J.L. Kerscher, and K.F. Moschner. 1996. Molecular modeling approach for contrasting the interaction of ethane and hexafluoroethane with carbon dioxide. *The Journal of Physical Chemistry* 100: 7435.

Chakma, A., M.R. Islam, and F. Berruti (eds.). 1991. *Enhanced oil recovery*. New York: American Institute of Chemical Engineers.

Chandler, K., C.W. Culp, D.R. Lamb, C.L. Liotta, and C.A. Eckert. 1998. Phase-transfer catalysis in supercritical carbon dioxide: Kinetic and mechanistic investigations of cyanide displacement on benzyl chloride. *Industrial & Engineering Chemistry Research* 37: 3252.

Charpentier, P.A., K.A. Kennedy, J.M. DeSimone, and G.W. Roberts. 1999. Continuous polymerizations in supercritical carbon dioxide: Chain-growth precipitation polymerizations. *Macromolecules* 32: 5973.

Charpentier, P.A., J.M. DeSimone, and G.W. Roberts. 2000. Decomposition of polymerisation initiators in supercritical CO_2: A novel approach to reaction kinetics using a CSTR. *Chemical Engineering Science* 55: 5341–5349.

Chiang, S.H., and G.E. Klinzing. 1986. Process for removing mineral matter from coal. USPTO, U.S. Patent No. 4,613,429.

Chinsoo, L., K.L. Hoy, and M.D. Donohue. 1990. Supercritical fluids as diluents in liquid spray application of coatings. USPTO, U.S. Patent No. 4,923,720.

Chinsoo, L., K.L. Hoy, and M.D. Donohue. 1991. Supercritical fluids as diluents in liquid spray application of coatings. USPTO, U.S. Patent No. 5,027,742.

Chouchi, D., D. Gourgouillon, M. Courel, J. Vital, and M.N. da Ponte. 2001. The influence of phase behavior on reactions at supercritical conditions: The hydrogenation of α-pinene. *Industrial & Engineering Chemistry Research* 40: 2551.

Colina, C.M., C.K. Hall, and K.E. Gubbins. 2002. Phase behavior of PVAC–PTAN block copolymer in supercritical carbon dioxide using SAFT. *Fluid Phase Equilibria* 553: 194–197.

Condo, P.D., and K.P. Johnston. 1994. In situ measurement of the glass transition temperature of polymers with compressed fluid diluents. *Journal of Polymer Science Part B: Polymer Physics* 32(3): 523–533.

Consani, K.A., and R.D. Smith. 1990. Observations on the solubility of surfactants and related molecules in carbon dioxide at 50°C. *The Journal of Supercritical Fluids* 3(2): 51–65.

Cygnarowicz-Provost, M. 1996. Design and economic analysis of supercritical fluid extraction processes. Supercritical Fluid Technology in Oil and Lipid Chemistry, pp. 155–179.

Dardin, A., J.M. DeSimone, and E.T. Samulski. 1998. Fluorocarbons dissolved in supercritical carbon dioxide. NMR evidence for specific solute–solvent interactions. *The Journal of Physical Chemistry B* 102: 1775–1780.

Decaire, B.R., P.B. Logsdon, E.A.E. Lund, R.R. Singh, and I.R. Shankland. 1994a. Azeotrope-like compositions of trifluoromethane, carbon dioxide and sulfur hexafluoride. USPTO, US Patent 5,275,751.

Decaire, B.R., P.B. Logsdon, E.A.E. Lund, R.R. Singh, I.R. Shankland, D.P. Wilson, and R.H.P. Thomas. 1994b. Azeotrope-like compositions of trifluoromethane and carbon dioxide or hexafluoroethane and carbon dioxide. USPTO, US Patent 5340490.

DeSimone, J.M., Z. Guan, and C.S. Elsbernd. 1992. Synthesis of fluoropolymers in supercritical carbon dioxide. *Science* 257: 945.

Diaz, S., S. Espinosa, and E.A. Brignole. 2000. Modeling and simulation tools for supercritical fluid processes. *Computer-Aided Chemical Engineering* 8: 319.

Diep, P., K.D. Jordan, J.K. Johnson, and E.J. Beckman. 1998. CO_2–Fluorocarbon and CO_2–hydrocarbon interactions from first-principles calculations. *The Journal of Physical Chemistry A* 102: 2231.

Eastoe, J., A. Paul, S. Nave, D.C. Steytler, B.H. Robinson, E. Rumsey, M. Thorpe, and R.K. Heenan. 2001. Micellization of hydrocarbon surfactants in supercritical carbon dioxide. *Journal of the American Chemical Society* 123: 988.

Eastoe, J., S. Gold, S. Rogers, P. Wyatt, D.C. Steytler, A. Gurgel, R.K. Heenan, X. Fan, E.J. Beckman, and R.M. Enick. 2006. Designed CO2-philes stabilize water-in-carbon dioxide microemulsions. *Angewandte Chemie* 118(22): 3757–3759.

Eckert, C.A., D. Bush, J.S. Brown, and C.L. Liotta. 2000. Tuning solvents for sustainable technology. *Industrial & Engineering Chemistry Research* 39: 4615.

Eggers, R., J. von Schnitzler, and K. Truckenmuller. 1999a. Method for the dyeing of yarn from a supercritical fluid. USPTO, US Patent No. 5,938,794.

Eggers, R., J. von Schnitzler, and G. Worner. 1999c. Process for the dyeing of a textile substrate. USPTO, US Patent No. 5,972,045.
Eggers, R., J. von Schnitzler, R. Huber, and G. Worner. 1999b. Process for dyeing a textile substrate in at least one supercritical fluid. USPTO, US Patent No. 5,958,085.
Erdohelyi, A., M. Pasztor, and F. Solymosi. 1986. Catalytic hydrogenation of CO_2 over supported palladium. *Journal of Catalysis* 98: 106.
Espinosa, S., S. Diaz, and E.A. Brignole. 2000. Optimal design of supercritical fluid processes. *Computers & Chemical Engineering* 24: 1301.
Fink, R., D. Hancu, R. Valentine, and E.J. Beckman. 1999. Toward the development of "CO_2-philic" hydrocarbons. 1. Use of side-chain functionalization to lower the miscibility pressure of polydimethylsiloxanes in CO_2. *The Journal of Physical Chemistry B* 103: 6441–6444.
Fried, J.R., and N. Hu. 2003. The molecular basis of CO_2 interaction with polymers containing fluorinated groups: Computational chemistry of model compounds and molecular simulation of poly[bis(2,2,2-trifluoroethoxy)phosphazene]. *Polymer* 44: 4363.
Gani, R., G. Hytoft, and C. Jaksland. 1997. Design and analysis of supercritical extraction processes. *Applied Thermal Engineering* 17: 889.
Ghenciu, E.G., A.J. Russell, and E.J. Beckman. 1998. Solubilization of subtilisin in CO2 using fluoroether-functional amphiphiles. *Biotechnology and Bioengineering* 58: 572.
Giddings, J.C., M.N. Myers, L. McLaren, and R.A. Keller. 1968. High pressure gas chromatography of nonvolatile species. *Science* 162: 67.
Giddings, J.C., M.N. Myers, and J.W. King. 1969. Dense gas chromatography at pressures to 2000 atmospheres. *The Journal of Chromatographic Science* 7: 276.
Ginosar, D.M., and B. Subramaniam. 1994. Coking and activity of a reforming catalyst in near-critical and dense supercritical reaction mixtures. *Studies in Surface Science and Catalysis* 88: 327.
Ginosar, M., and B. Subramaniam. 1995. Olefinic oligomer and cosolvent effects on the coking and activity of a reforming catalyst in supercritical reaction mixtures. *Journal of Catalysis* 152: 31.
Goel, S., and E.J. Beckman. 1995. Nucleation and growth in microcellular materials: Supercritical CO2 as foaming agent. *AIChE (American Institute of Chemical Engineers) Journal* 41: 357.
Han, Y.K., and H.Y. Jeong. 1997. Comment on "Molecular modeling approach for contrasting the interaction of ethane and hexafluoroethane with carbon dioxide". *The Journal of Physical Chemistry A* 101: 5604.
Hancu, D., Beckman, E.J. 2001. Generation of hydrogen peroxide directly from H2 and O2 using CO2 as the solvent. *Green Chemistry* 3(2): 80–86.
Hauthal, W.H. 2001. Advances with supercritical fluids. *Chemosphere* 43: 123.
Heller, J.P., D.K. Dandge, R.J. Card, and L.G. Donaruma. 1985. Direct thickeners for mobility control of carbon dioxide floods. *Society of Petroleum Engineers Journal* 25: 679.
Hendrix, W.A., G.A. Montero, C.B. Smith, and D.L. Butcher. 2001. Method for introducing dyes and other chemicals into a textile treatment system. USPTO, US Patent No. 6,261,326.
Herrero, M., A. Cifuentes, and E. Ibañez. 2006. Sub- and supercritical fluid extraction of functional ingredients from different natural sources: Plants, food-by-products, algae and microalgae. *Food Chemistry* 98(1): 136–148.
Holmes, J.D., K.J. Ziegler, M. Audriani, C.T. Lee, P.A. Bhargava, D.C. Steytler, and K.P. Johnston. 1999. Buffering the aqueous phase pH in water-in-CO2 microemulsions. *The Journal of Physical Chemistry B* 103: 5703.
Howdle, S.M., M.S. Watson, M.J. Whitaker, M.C. Davies, K.M. Shakesheff, V.K. Popov, F.S. Mandel, and D.J. Wang. 2001. Supercritical fluid mixing: Preparation of thermally sensitive polymer composites containing bioactive material. *Chemical Communications* 1: 109–110.
Hoy, K.L., K.A. Nielsen, and L. Chinsoo. 1992. Supercritical fluids as diluents in liquid spray application of coatings and spraying from an orifice. USPTO, U.S. Patent No. 5,108,799.
Jacobsen, G.B., C.T. Lee, and K.P. Johnston. 1999. Enhanced catalyst reactivity and separations using water/carbon dioxide emulsions. *Journal of Organic Chemistry* 64: 1201–1207.

Jaubert, J.N., and L. Coniglio. 1999. The group contribution concept: A useful tool to correlate binary systems and to predict the phase behavior of multicomponent systems involving supercritical CO_2 and fatty acids. *Industrial & Engineering Chemistry Research* 38: 5011.
Johnston, K.P., K.L. Harrison, M.J. Clarke, S.M. Howdle, M.P. Heitz, F.V. Bright, C. Carlier, and T.W. Randolph. 1996. Water-in-carbon dioxide microemulsions: An environment for hydrophiles including proteins. *Science* 271: 624.
Jusforgues, P., M. Shaimi, and D. Barth. 1998. *Chromatographic Science Series* 75:403.
Kamat, S., J. Barrera, E.J. Beckman, and A.J. Russell. 1992. Biocatalytic synthesis of acrylates in organic solvents and supercritical fluids: I. Optimization of enzyme environment. *Biotechnology and Bioengineering* 40: 158.
Kamat, S., G. Critchley, E.J. Beckman, and A.J. Russell. 1995. Biocatalytic synthesis of acrylates in organic solvents and supercritical fluids: III. Does carbon dioxide covalently modify enzymes? *Biotechnology and Bioengineering* 46: 610–620.
Kauffman, J.F. 2001. Quadrupolar solvent effects on solvation and reactivity of solutes dissolved in supercritical CO_2. *The Journal of Physical Chemistry A* 105: 3433.
Kazarian, S.G., M.F. Vincent, F.V. Bright, C.L. Liotta, and C.A. Ecker. 1996. Specific intermolecular interaction of carbon dioxide with polymers. *Journal of the American Chemical Society* 118: 1729.
Kazarian, S.G., N.H. Brantley, B.L. West, M.F. Vincent, and C.A. Eckert. 1997. In situ spectroscopy of polymers subjected to supercritical CO2: Plasticization and dye impregnation. *Applied Spectroscopy* 51: 491–494.
Keshtkar, A., F. Jatali, and M. Moshfeghian. 1997. Evaluation of vapor—liquid equilibrium of CO_2 binary systems using UNIQUAC-based Huron—Vidal mixing rules. *Fluid Phase Equilibria* 140: 107.
Khoury, F.M. 1995. *Predicting the performance of multi-stage separation processes*. Boca Raton, FL: CRC Press.
Kilic, S., S. Michalik, Y. Wang, J.K. Johnson, R.M. Enick, and E.J. Beckman. 2003. Effect of grafted Lewis base groups on the phase behavior of model poly(dimethyl siloxanes) in CO_2. *Industrial & Engineering Chemistry Research* 42: 6415.
King, J.W., R.L. Holliday, G.R. List, and J.M. Snyder. 2001. Hydrogenation of vegetable oils using mixtures of supercritical carbon dioxide and hydrogen. *Journal of the American Oil Chemists' Society* 78: 107.
Kirby, C.F., and M.A. McHugh. 1999. Phase behavior of polymers in supercritical fluid solvents. *Chemical Reviews* 99: 565.
Klibanov, A.M. 1997. Why are enzymes less active in organic solvents than in water? *Trends in Biotechnology* 15: 97.
Kwag, C., C.W. Manke, and E. Gulari. 1999. Rheology of molten polystyrene with dissolved supercritical and near critical gases. *Journal of Polymer Science Part B: Polymer Physics* 37: 2771.
Lake, L.W. 1989. *Enhanced oil recovery*. Englewood Cliffs, NJ: Prentice Hall.
Liau, I.S., and M.A. McHugh. 1985. High pressure solid polymer-supercritical fluid phase behavior. In *Supercritical fluid technology*, ed. J.M.L. Penninger, M. Radosz, M.A. McHugh, and V.J. Krukonis. Amsterdam: Elsevier.
Lide, D.R. (ed.). 1996. *CRC handbook of chemistry and physics*, 76th ed. Boca Raton, FL: CRC Press.
Lucien, F.P., and N.R. Foster. 2000. Solubilities of solid mixtures in supercritical carbon dioxide: A review. *The Journal of Supercritical Fluids* 17: 111.
Macher, M.B., and A. Holmqvist. 2001. Hydrogenation of palm oil in near-critical and supercritical propane. *European Journal of Lipid Science and Technology* 103: 81.
McFann, G.J., S.M. Howdle, and K.P. Johnston. 1994. Solubilization in nonionic reverse micelles in carbon dioxide. *AIChE (American Institute of Chemical Engineers) Journal* 40: 543.
McHugh, M.A., and V.J. Krukonis. 1994. *Supercritical fluid extraction*. Boston: Butterworth.
McHugh, M.A., I.H. Park, J.J. Reisinger, Y. Ren, T.P. Lodge, and M.A. Hillmyer. 2002. Solubility of CF_2-modified polybutadiene and polyisoprene in supercritical carbon dioxide. *Macromolecules* 35: 4653.

Mesiano, A., E.J. Beckman, and A.J. Russell. 1999. Supercritical biocatalysis. *Chemical Reviews* 99: 623.
Michels, A., and C. Michels. 1936. Series evaluation of the isotherm data of CO2 between 0 and 150 C and up to 3000 atmospheres. *Proceedings of the Royal Society of London Series A* 153: 201.
Michels, A., C. Michels, and H. Wouters. 1936. Isotherms of CO2 between 70 and 300 atmospheres (amagat densities between 200 and 600). *Proceedings of the Royal Society of London Series A* 153: 214.
Minder, B., T. Mallat, and A. Baiker. 1996. Enantioselective hydrogenation in supercritical fluids. Limitations of the use of supercritical CO2. 3rd international symposium on high-pressure chemical engineering. *Process Technology Proceedings* 12: 139–144.
Montero, G.A., T.D. Giorgio, and K.B. Schnelle. 1996. Scale-up and economic analysis for the design of supercritical fluid extraction equipment for remediation of soil. *Environmental Progress* 15: 112.
Montero, G.A., C.B. Smith, W.A. Hendrix, and D.L. Butcher. 2000. Supercritical fluid technology in textile processing: An overview. *Industrial & Engineering Chemistry Research* 39: 4806.
O'Neill, M.L., Q. Cao, M. Fang, K.P. Johnston, S.P. Wilkinson, C.D. Smith, J.L. Kerschner, and S.H. Jureller. 1998. Solubility of homopolymers and copolymers in carbon dioxide. *Industrial & Engineering Chemistry Research* 37: 3067.
Ohgaki, K., and T. Katayama. 1977. Isothermal vapor–liquid equilibrium data for the ethane–carbon dioxide system at high pressure. *Fluid Phase Equilibria* 1(1): 27–32.
Orr, F.M., J.P. Heller, J.T. Taber, and R.J. Card. 1983. CO2 as solvent for oil recovery. *Chem Tech* 13(8): 482–487.
Palmer, M.V., and S.S.T. Ting. 1995. Applications for supercritical fluid technology in food processing. *Food Chemistry* 52: 345.
Park, C.B., N.P. Suh, and D.F. Baldwin. 2000. Method for providing continuous processing of microcellular and supermicrocellular foamed materials. USPTO, US Patent No. 6,051,174.
Perrut, M. 2000. Supercritical fluid applications: Industrial developments and economic issues. *Industrial & Engineering Chemistry Research* 39: 4531.
Pierick, D. 1999. The mucell molding technology: microcellular foam. *Paper presented at Molding '99 New Orleans*, March 1–3.
Poddevin, N., J. Fages, and R. Guidoin. 2000. Method for manufacturing and treating textiles. USPTO, US Patent No. 6,120,558.
Potluri, V.J., J. Xu, R.M. Enick, E.J. Beckman, and A.D. Hamilton. 2002. Peracetylated sugar derivatives show high solubility in liquid and supercritical carbon dioxide. *Organic Letters* 4: 2333.
Prasad, R., M. Gottesman, and R.A. Scarella. 1981. Decaffeination of aqueous extract. USPTO, U.S. Patent No. 4,246,291.
Praxair Material Data Safety Sheet, P-4574-H, May 1999. http://www.praxair.com/praxair.nsf/AllContent/D3CBF4EF4186073485256A860080D221/$File/p4574j.pdf.
Raveendran, P., and S.L. Wallen. 2002a. Sugar acetates as novel, renewable CO_2-phile. *Journal of the American Chemical Society* 124: 7274.
Raveendran, P., and S.L. Wallen. 2002b. Cooperative C–H···O hydrogen bonding in CO_2–Lewis base complexes: implications for solvation in supercritical CO_2. *Journal of the American Chemical Society* 124: 12590.
Raveendran, P., and S.L. Wallen. 2003. Exploring CO_2-philicity: Effects of stepwise fluorination. *The Journal of Physical Chemistry B* 107: 1473.
Reid, R.C., J.M. Prausnitz, and B.E. Poling. 1987. *The properties of liquids and gases*. New York: McGraw-Hill.
Reverchon, E., and I. De Marco. 2006. Supercritical fluid extraction and fractionation of natural matter. *The Journal of Supercritical Fluids* 38(2): 146–166.
Rindfleisch, F., T.P. DiNoia, and M.A. McHugh. 1996. Solubility of polymers and copolymers in supercritical CO_2. *The Journal of Physical Chemistry* 100: 15581.
Rodriguez, F. 1996. *Principles of polymer systems*. New York: McGraw-Hill.

Rowlinson, J.S., and J.R. Sutton. 1955. The statistical thermodynamics of solutions of non-spherical molecules. II. Liquid–vapour equilibrium and the formation of azeotropes. *Proceedings of the Royal Society of London Series A* 229: 396–404.

Royer, J.R., Y.J. Gay, J.M. DeSimone, and S.A. Khan. 2000. High-pressure rheology of polystyrene melts plasticized with CO2: Experimental measurement and predictive scaling relationships. *Journal of Polymer Science Part B: Polymer Physics* 38: 3168.

Rylander, P.N. 1985. *Hydrogenation methods*. Orlando, FL: Academic.

Sahena, F., I.S.M. Zaidul, S. Jinap, A.A. Karim, K.A. Abbas, N.A.N. Norulaini, and A.K.M. Omar. 2009. Application of supercritical CO_2 in lipid extraction—A review. *Journal of Food Engineering* 95(2): 240.

Sarbu, T., T. Styranec, and E.J. Beckman. 2000. Non-fluorous polymers with very high solubility in supercritical CO_2 down to low pressures. *Nature* 405: 165.

Satterfield, C.N. 1980. *Heterogeneous catalysis in practice*. New York: McGraw-Hill.

Schollmeyer, E., E. Bach, E. Cleve, M. Bork, M. Steinhauer, and J.P. Korner. 1999. Process and device for treating textile substrates with supercritical fluid. USPTO, US Patent No. 5,953,780.

Shim, J.J., and K.P. Johnston. 1989. Adjustable solute distribution between polymers and supercritical fluids. *AIChE (American Institute of Chemical Engineers) Journal* 35: 1097.

Shim, J.J., and K.P. Johnston. 1991. Molecular thermodynamics of solute-polymer-supercritical fluid systems. *AIChE (American Institute of Chemical Engineers) Journal* 37: 607.

Shine, A.D., and J. Gelb. 1998. Microencapsulation process using supercritical fluids. USPTO, US Patent 5,766,637.

Smith, C.W., and G. Huse. 1998. In *Supercrital fluid cleaning*, ed. J. McHardy and S.P. Sawan. Westwood, NJ: Noyes Publishers.

Smith, C.W., G.A. Montero, and W.A. Hendrix. 2000. Method of dyeing hydrophobic textile fibers with colorant materials in supercritical fluid carbon dioxide. USPTO, US Patent No. 6,048,369.

Solymosi, F., Erdöhelyi, A., Lancz, M. 1985. Surface interaction between H2 and CO2 over palladium on various supports. *Journal of Catalysis* 95(2): 567–577.

Sparacio, D., and E.J. Beckman. 1998. Generation of microcellular biodegradable polymers using supercritical carbon dioxide. In *Solvent-free polymerizations and processes: Minimization of conventional organic solvents*, ACS symposium series, vol. 713, ed. M. Hunt and T. Long. Washington, DC: American Chemical Society.

Stafford, C.M., T.P. Russell, and T.J. McCarthy. 1999. Expansion of polystyrene using supercritical carbon dioxide: Effects of molecular weight, polydispersity, and low molecular weight components. *Macromolecules* 32: 7610.

Subramaniam, B. 2001. Enhancing the stability of porous catalysts with supercritical reaction media. *Applied Catalysis A: General* 212: 199.

Subramaniam, B., and V. Arunajatesan. 1999. Coking of solid acid catalysts and strategies for enhancing their activity. *Studies in Surface Science and Catalysis* 126: 63.

Subramaniam, B., and J. Ashraf. 1995. "*In situ mitigation of porous catalysts with supercritical reaction media" in innovations in supercritical fluids*. ACS Symposium Ser. 608. American Chemical Society, Washington DC, USA. http://pubs.acs.org/isbn/9780841233249.

Subramaniam, B., C.J. Lyon, and V. Arunajatesan. 2002. Environmentally benign multiphase catalysis with dense phase carbon dioxide. *Applied Catalysis B: Environmental* 37: 279.

Suh, K.W. 1991. Polystyrene and structural foam. In *Polymeric foams*, ed. D. Klempner and K.C. Frisch. New York: Oxford University Press.

Tacke, T., S. Wieland, and P. Panster. 1996. Hardening of fats and oils in supercritical CO2. *3rd international conference on high pressure chemical engineering*, Zurich, Process Technology Proceedings 12.

Thomas, E.R., and C.A. Eckert. 1984. Prediction of limiting activity coefficients by a modified separation of cohesive energy density model and UNIFAC. *Industrial & Engineering Chemistry Process Design and Development* 23: 194.

Tiwari, K.K. 2007. Processes using supercritical fluids. *Chemical Industry Digest* 9: 71.

Valle, J., and J. De La Fuente. 2006. Supercritical CO_2 extraction of oilseeds: Review of kinetic and equilibrium models. *Critical Reviews Food Science Nutrition* 46(2): 131–160.
van den Hark, S., and M. Harrod. 2001. Hydrogenation of oleochemicals at supercritical single-phase conditions: Influence of hydrogen and substrate concentrations on the process. *Applied Catalysis A:General* 210: 207.
van den Hark, S., M. Harrod, and P. Moller. 1999. Hydrogenation of fatty acid methyl esters to fatty alcohols at supercritical conditions. *Journal of the American Oil Chemists' Society* 76: 1363.
Wang, W., E.J. Kramer, and W.H. Sachse. 1982. Effects of high-pressure CO2 on the glass transition temperature and mechanical properties of polystyrene. *Journal of Polymer Science: Polymer Physics Edition* 20: 1371–1384.
Wiesmet, V., E. Weidner, S. Behme, G. Sadowski, and W. Arlt. 2000. Measurement and modelling of high-pressure phase equilibria in the systems polyethyleneglycol (PEG)–propane, PEG–nitrogen and PEG–carbon dioxide. *The Journal of Supercritical Fluids* 17: 1.
Wissinger, R.G., and M.E. Paulaitis. 1991a. Glass transitions in polymer/CO_2 mixtures at elevated pressures. *Journal of Polymer Science Part B: Polymer Physics* 25: 2497.
Wissinger, R.G., and M.E. Paulaitis. 1991b. Glass transitions in polymer/CO_2 mixtures at elevated pressures. *Journal of Polymer Science Part B: Polymer Physics* 29: 631.
Yee, G.G., J.L. Fulton, and R.D. Smith. 1992. Fourier transform infrared spectroscopy of molecular interactions of heptafluoro-1-butanol or 1-butanol in supercritical carbon dioxide and supercritical ethane. *Journal of Physical Chemistry* 96: 6172.
Yonker, C.R., and B.J. Palmer. 2001. Investigation of CO2/fluorine interactions through the intermolecular effects on the ^{1}H and ^{19}F shielding of CH3F and CHF3 at various temperatures and pressures. *The Journal of Physical Chemistry A* 105: 308.
Zhang, Z.Y., J.C. Yang, and Y.G. Li. 2000. Prediction of phase equilibria for CO_2–C_2H_5OH–H_2O system using the SAFT equation of state. *Fluid Phase Equilibria* 169: 1.
Zosel, K. 1974. Process for recovering caffeine. USPTO, U.S. Patent No. 3,806,619.

Chapter 11
Green separation technologies in food processing: supercritical-CO_2 fluid and subcritical water extraction

John Shi, Sophia Jun Xue, Ying Ma, Yueming Jiang, Xingqian Ye, and Dianyu Yu

11.1 Introduction

Phytochemical compounds (so-called bioactive components) have generated significant interest as they can be incorporated into high value products including functional foods, nutraceuticals, pharmacologicals, and fragrances. They have been studied for different functional activities including their antioxidant, antimicrobial, antihypertensive, anticancer, and anti-inflammatory properties, and for the prevention of neurodegenerative diseases (e.g., Parkinson, Alzheimer, and Huntington), etc. A number of unit operations in food engineering are related with changing the composition of solutions, mixtures, and solids through methods that do not involve chemical reactions. In food engineering, these particular unit operations are normally referred to as separation processes and are directed toward separating a substance into its component parts, or concentrating a particular substance from another using various materials. Application of separation technologies is used either to recover

J. Shi (✉) • S.J. Xue
Guelph Food Research Center, Agriculture and Agri-Food Canada, Guelph, ON, Canada
e-mail: john.shi@agr.gc.ca; jun.xue@agr.gc.ca

Y. Ma
College of Food Science and Engineering, Harbin Institute of Technology, Harbin, China
e-mail: maying@hit.edu.cn

Y. Jiang
South China Botanical Garden, Chinese Academy of Sciences, Guangzhou, China
e-mail: ymjiang@scbg.ac.cn

X. Ye
School of Biosystems Engineering and Food Science, Zhejiang University, Hangzhou, China
e-mail: psu@zju.edu.cn

D. Yu
College of Food Science and Engineering, Northeast Agricultural University, Harbin, China
e-mail: dyyu2000@yahoo.com.cn

J.I. Boye and Y. Arcand (eds.), *Green Technologies in Food Production and Processing*, Food Engineering Series, DOI 10.1007/978-1-4614-1587-9_11,

high value components from agricultural commodities, being an important operation for the production of food products such as oil and proteins, or especially for the development of health-promoting food ingredients and high value-added food products such as antioxidants and flavors. Separation processes such as extraction, concentration, purification, and fractionation of bioactive components (phytochemicals) from agricultural materials are the main processes used to obtain high value end-products. These products may become wellness ingredients for use in functional foods and nutraceuticals. Thus the unit operations comprising these separation processes play a key function in the total procedures for value-added food processing. Many potential high-value products can be developed from natural resources by different separation technologies and processes. Carotenoids including lycopene, β-carotene, astaxanthins, and lutein make up a world market nearing $1 billion with a growth rate of about 3%. Therefore, efforts to utilize natural agricultural materials for the production of high value-added products, especially health-promoting foods and ingredients, are of great interest to the food and biotechnology industries.

Traditional extraction techniques employ large amounts of sometimes toxic organic solvents for removing targeted components from plant materials. In many countries, to address safety, health, and environmental issues, strict regulations regarding the use of organic solvents (i.e., hexane) are forcing the food industry to search for alternative processes. Furthermore, increasing consumer awareness of the presence and use of organic chemical solvents in foods nowadays and the desire to buy natural products is encouraging the food and natural products industry to develop and commercialize "green" technologies. Recently, there has been a lot of interest from the scientific community and increasing industrial demands to perform research and development on extraction and separation technologies with the goal of eliminating the use of organic chemicals, because these products are being increasingly used in producing functional ingredients and natural products (e.g., nutraceuticals or supplements). Supercritical-CO_2 fluid extraction technology and subcritical water extraction are possible alternative methods providing environmentally friendly "green" processing techniques for new or improved food processing applications. These "green" environmentally friendly separation technologies and processes provide detailed information on the equipment selections, system design, and methods for extraction and purification of multiple classes of phytochemicals from plant materials, while retaining intact their bioactivity, or even improving their bioactivity and functionality. These newer techniques have made impressive advances in obtaining adequate segregations of components of interest with maximum speed, minimum effort, and minimum cost at as large a capacity as possible in production-scale processes.

11.2 "Green" separation process design

Worldwide, there is pressure for industry to adopt new sustainable separation processes that do not require the use of environmentally damaging organic solvents. Separations are interphase mass transfer processes because they involve heat, mass, and phase transfers, as well as chemical reactions among components from the plant materials. The engineering processes for separation systems include modeling, simulations, optimization control studies, and thermodynamic analyses. The principles of mass conservation and component transfer amounts are used to analyze and design industrial processes. Proper understanding of molecular properties and the use of thermodynamic approaches constitute powerful tools for the design of successful separation processes. Moreover, the design of a separation process is strongly dependent on the phase equilibrium scenario, which is highly sensitive to changes in operating conditions. Thus, phase equilibrium engineering plays a key role in the synthesis and design of these processes. Phase equilibrium engineering is the systematic application of phase equilibrium knowledge to process development. The applied knowledge necessary for system design comprises data banks, experimental data, phenomenological phase behavior, thermodynamic analysis, and mathematical modeling procedures for phase equilibrium process calculations, mass and heat transfer analyses, the characteristics of the targeted components, and effects of processing conditions. The following issues of engineering properties are important to consider for process optimization and commercialization.

11.2.1 Technical requirements for a separation system

The design of a separation process depends on the separation to be performed and the properties of the material used, as well as the targeted bioactive components. An important consideration in determining the appropriateness of a separation technique and system is the actual purity requirements for the end products. In most organic solvent (toxic-chemical-free) separation technologies and systems, a combination of new available techniques is necessary for system optimization. Product design is related to reasonable separation and purification steps, and economic feasibility and raw material selection. "Green" separation processes are environmentally friendly processes that result in less air pollution and industrial waste (e.g., energy, greenhouse gases emission, reduction of waste water production). Separation operations are interphase mass transfer processes because they involve certain heat, mass, and phase transfers, as well as chemical reactions among food components. The engineering properties of the targeted food components via separation systems include separation modeling, simulations, optimization control studies, and thermodynamic analyses. The principles of mass conservation and component transfer amounts are used to analyze and design industrial processes. Molecular intuition and a thermodynamic approach constitute powerful tools for the design of a successful separation process. The following issues may be involved in the consideration of a possible system design:

(a) Knowledge of phase equilibrium, mass transfer rate, and solubility data are important for scale-up of the extraction process and equipment.
(b) Information on and experimental data for the effects of processing conditions on the physicochemical properties and degradation of the bioactivity of compounds are important for designing a suitable extraction system and procedure.

(c) Proper solvent selection is based on the solubility characteristics of the desired compounds (targeted compounds), which should largely dissolve in the extraction solvent, ideally to achieve as pure a substance as possible.
(d) A pump for transporting the solvent is required for recycling. An additional pump may be required for cosolvent incorporation during extraction.
(e) An extractor, which will be responsible for charging a solid material into a high pressure and potentially high temperature zone, and a separator, which may involve a change of pressure and temperature, and a sample collectors.
(f) The capacities of the heat exchanger and condenser should also be considered in designing the system. For example, the heat capacity of water is very large, so considerable effort is necessary to remove the excess stored energy.
(g) With the choice of a suitable pump and extraction capacity, only the dimensioning of the extraction vessel and optimizing of the processing parameters are required.
(h) Economics and safety should always be considered and indeed may be the final determiners in designing a separation system.

11.2.2 Food quality and separation systems

The major issues related to product quality after separation are the impacts of processing on bioactivity of extracts and the nutritional value of the end products, as well as the quality characteristics. To meet food safety regulations, no toxic-chemical solvent residues are permitted in the end food products (e.g., "green" food products). Nutrition and health regulations must be met. Some important requests include high stability of nutrients and bioactive components, processes operating at low temperatures to reduce thermal effects, processes that exclude light to reduce light induced (UV) irradiation effects, and processes that exclude oxygen to reduce oxygen effects. The final product must maintain uniformity and quality consistency, and a purity that can meet food grade or pharmaceutical grade requirements.

11.2.3 Scaling up technology for industrial production

Scaling up of a natural product separation process from the laboratory development stage to the commercialized production scale is, by no means, a simple affair. The scale-up of an innovative separation process for large scale manufacture is essential in order to operate the separation in a reproducible and consistent manner for commercial purposes and to avoid potential exposure to biological or chemical post-translational modifications, which could result in poor product quality. Avoidance of enormous variations from process to process, thus necessitates careful attention to details at all stages of product development. When the technology in a food process is designed for industrial-scale production, an important area for consideration is the balance of capital and operating costs as the scale of the separation operation increases. The process of scale-up of a separation technology also involves optimization with respect to increasing the efficiency of each stage, giving rise to increased demands on the accuracy of the online quality control. The process must be determined to be reliable for the industrial scale production of food or pharmaceutical grade ingredients that will be used in a wide variety of applications including food, nutraceuticals, pharmaceuticals, and cosmetic products.

11.2.4 New technology development

Extraction of health-promoting components from plant materials has usually been accomplished by conventional extraction processes such as solid-liquid extractions employing methanol, ethanol, acetone, or hexane, and also through steam distillation or evaporation processes to remove solvents from the extracts. Currently, the demand for natural bioactive compounds is increasing because of their use by the functional food and pharmaceutical industry. Thus, there has been increasing interest in the use of environmentally friendly "green" separation technologies able to provide high quality-high bioactivity extracts while precluding any toxicity associated with the solvents. Some of the motivations to employ "green" separation technologies as a viable separation technique are: (a) tightening government regulations on toxic-chemical solvent residues and pollution control, (b) consumers' concern over the use of toxic-chemical solvents during processing, and (c) Increased demand for higher quality products that traditional processing techniques cannot meet.

One of the most important considerations in developing new extraction processes is the safety aspect. In this sense, a variety of processes involving extractions with supercritical-CO_2 fluid extraction, membrane-based separation, molecular distillation, and pressurized low-polarity water extraction, etc., are generally recognized as "green" separation technology and are considered clean and safe processes to meet requirements (Ibáñez et al. 1999; Herrero et al. 2006; Chang et al. 2008). They have been developed and are regarded as emerging innovative separation technologies that meet food quality and safety requirements. These processes can be used to solve some of the problems associated with conventional organic solvent-oriented separation processes. Operation parameters and other factors related to the quality of the original plant, including its geographic origin, the harvesting date, its storage, and any pretreatment processes before extraction also influence the separation operations and the final composition of the extracts obtained.

11.3 Supercritical-CO_2 fluid extraction

11.3.1 Principles and properties of supercritical fluid extraction

A gas or a liquid is normally used as the extraction solvent for supercritical fluid extraction. When a gas or liquid is compressed under pressure and heated past its critical point, it enters a phase called the "supercritical phase" and the extraction medium is called a "supercritical fluid" in this state. The critical temperature (Tc) and pressure (Pc) at which this happens are unique to each pure substance. In the supercritical state, the supercritical fluid possesses many of the properties of both gas and liquid. For example, the liquid-like density of a supercritical fluid provides its high solvent power whereas the gas-like viscosity and diffusivity, with zero surface tension, enhance the transport properties of the supercritical fluid solvent during extraction. Therefore, this unique property enables the superficial fluid to penetrate into porous solid materials more effectively than can a liquid solvent, resulting in faster mass transfer, and therefore gives faster and greater extraction yields.

The extraction process can be easily adjusted by pressure and temperature. However, operating pressures and temperatures above the critical points would affect the properties of the supercritical fluid such as its density, viscosity, diffusivity, heat capacity, and thermal conductivity, and would enhance the ability of the supercritical fluid to penetrate and extract targeted molecules from source materials. One of the main characteristics of a supercritical fluid is the possibility of modifying the density of the fluid by changing its pressure and/or its temperature. Since density is directly related to solubility (Raventós et al. 2002; Shi et al. 2009a, b), by altering the extraction pressure, the solvent strength of the fluid can be modified. Besides, the solvent power of superficial fluid significantly depends upon its density. The solvent extracting ability increases with density increase at a given temperature, or increases with temperature rise at a given density.

Many solvents are candidates for supercritical fluid extraction. The most desirable super critical fluid (SCF) solvent for extraction of natural products for food and medicines is carbon dioxide (CO_2), and the extraction process is called supercritical-CO_2 fluid extraction. The advantageous characteristic traits of CO_2 are inertness, non-flammability, non-corrosiveness, low cost, easy availability, odorless, tasteless, and environmentally friendly, with relatively mild critical conditions of pressure (Pc) at 7.38 MPa and temperature (Tc) at 31°C (Fig. 11.1). Therefore, its near-ambient critical temperature makes it ideal for thermolabile natural products (Mendiola et al. 2007). Because it is also a gas at room temperature, once the extraction is completed, a substantial elimination of CO_2 is achieved without residues by simply decompressing the system, yielding a solvent-free extract. However, CO_2 is not a prefect solvent for high molecular weight and polar compounds. Small amounts (ranging up to 20 mol%) of polar or nonpolar cosolvents called modifiers can be incorporated to increase the solubility of such compounds during supercritical-CO_2 fluid extraction.

Carbon dioxide, therefore, has favorable properties including the ease of changing selectivity by the addition of relatively small amount of modifiers such as ethanol and other polar solvents (e.g., water). As a result, CO_2 has been considered the most desirable supercritical fluid for extracting natural products for food and medicinal uses (Shi et al. 2007a, b, c, d; Kasamma et al. 2008; Shi et al. 2009a, b; Yi et al. 2009). Other supercritical fluids such as ethane, propane, butane, pentane, ethylene, ammonia, sulphur dioxide, water, chlorodifluoromethane, etc. are also used in supercritical fluid extraction processes.

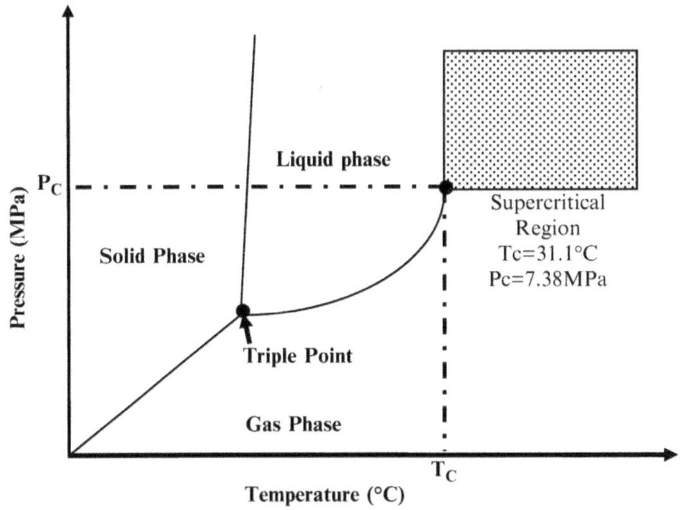

Fig. 11.1 Supercritical pressure–temperature diagram for carbon dioxide

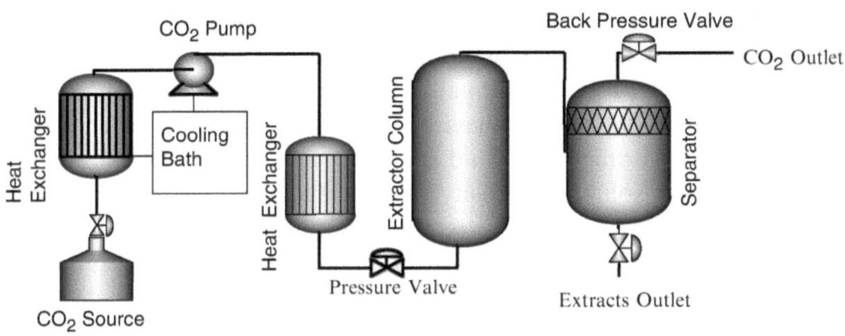

Fig. 11.2 Schematic diagram of a typical single stage supercritical fluid extraction system with CO_2

Basically, the supercritical-fluid extraction system consists of pumps for delivering solvent and cosolvent throughout the system and for raising the pressure of the recycled solvent, a high-pressure extractor, a pressure reduction valve, heat exchangers, compressors, and one or more separators in which the extract is collected and the solvent (e.g., CO_2) is depressurized and removed (Fig. 11.2). The extraction cells and the separators are commonly equipped with independent controls for temperature and pressure; thus the fractionation of the extracted compounds can be carried out by stepwise decompression. Therefore, different compounds can be collected in separators, depending on their differential solubility in the supercritical fluid. Additionally, it is possible to install a system to recycle the fluid employed. Raw materials are usually ground and charged into a temperature controlled extractor, forming a fixed bed, which is usually the case for batch and single stage mode (Shi and Zhou 2006; Kasamma et al. 2008).

The process of supercritical fluid extraction requires intimate contact between the packed beds formed by a ground solid substratum (fixed-bed of extractable material) with a supercritical-CO_2 fluid that is fed into the extractor through a high-pressure pump. During the supercritical fluid extraction process, the solid phase comprised of the solute and the insoluble residuum (matrix) is brought into contact with the fluid phase, which is the solution of the solute in the supercritical-CO_2 fluid (solvent). The extracted material is then conveyed to a separation unit via a pressure reduction valve. At reduced temperature and pressure, the extract precipitates in the separator while the CO_2, free of extract, is recycled to the extractor.

The physicochemical properties of supercritical-CO_2 fluid greatly influence the solvent strength and the solubility of targeted compounds in the fluid. In a supercritical-CO_2 fluid extraction process, the physicochemical properties of supercritical-CO_2 fluid, such as the density, diffusivity, viscosity, and dielectric constant, can be controlled by varying the operating conditions of pressure and temperature, or both in combination (Tena et al. 1997; Kasamma et al. 2008; Shi et al. 2007a, c; Shi et al. 2009b). Thus, the separation process can be affected by simply changing the operating pressure and temperature to alter the solvating power of the solvent. After modifying CO_2 with a cosolvent, the extraction process can significantly enhance the selectiveness and separation power, and in some cases can even extend its solvating powers to polar components (Shi et al. 2009a).

Many challenges to the design and development of commercially viable supercritical fluid extraction processes for natural products still remain. These include a better understanding of the phase behavior and solubility of multi-bioactive component mixtures in supercritical-CO_2 fluid and the generation of much needed fundamental data, including solubility, density, and interfacial tension, as well as changes in mass transfer phenomena under different operating conditions such as temperature, pressure, flow rate, and the effects of the composition and proportion of incorporated cosolvents (Lucien and Foster 2000; Marcus 2006; Shi et al. 2009a, b). Solvent power is related to the density of the supercritical-CO_2 fluid, and it can be varied at different operating conditions, mainly by varying temperature, pressure, and flow rate. Generally, density decreases with increasing temperature at a constant pressure, and pressure has more pronounced effects on the changes in density at a constant temperature. Shi et al. (2007a, d) observed that, with an increase in density, the solvating power of CO_2 increased lycopene extraction from tomato skins. Moreover, the region above the critical point provides the greatest density changes, and thus for even the slightest changes in pressure and temperature, the density of CO_2 will effectively change the operating variables (pressure and temperature) in this zone. In the region below the supercritical point, the density of CO_2 changes, but it does not vary substantially with the changes in pressure and temperature (Shi et al. 2009a, b).

Targeted compounds will have different solubilities in supercritical-CO_2 fluid and that will greatly influence the extraction efficiency and bioactivity of the extracts (Lucien and Foster 2000, Shi et al. 2009a; Yi et al. 2009). The impact of solubility enhancement on selectivity has been assessed, and some opportunities for improving the selectivity of extraction have been highlighted and extensively studied (Yi et al. 2009). Generally, elevation of the temperature leads to an increase in solubility of the targeted components in supercritical-CO_2. However, there is also an increased concern for the stability of bioactive extracts because the bioactivity of natural extracts might degrade when subjected to high processing temperatures.

Most of the investigations on solubility have been concerned with binary systems consisting of a single solute in contact with only supercritical-CO_2. In contrast, solubility data derived from multi-component systems have not been well established. This is important because intermolecular interactions between components can significantly alter the selectivity of the supercritical-CO_2. For example, in some studies, the addition of a cosolvent (modifier) to supercritical-CO_2 resulted in an enhancement of the solubility of the targeted component (Sauceau et al. 2004; Yi et al. 2009). Food grade solvents such as water, lipids, and ethanol can be used as cosolvents to increase solubility and enhance extract yield. Furthermore, not only the composition of the cosolvent, but the proportion of cosolvents involved in extraction is also considered a factor for system and process design and for optimization of the operating conditions, as excess cosolvents may cause either negative or negligible and un-economical effects (Shi et al. 2009b). Vasapollo et al. (2004) found that the presence of vegetable oil as cosolvent improved yields and contributed to the stability of lycopene. Shi et al. (2009a) investigated the effects of ethanol, water, and canola oil modifiers on the profile of lycopene extractions, and the results showed that extraction efficiency was improved by the addition of any of the three modifiers, and that yields increased with increasing amounts of modifier (from 5% to 10%). However, the rate of yield increase was lower when the proportion of ethanol concentration was increased from 10% to 15%.

11.3.2 Process systems and industrial applications

Changes in food processing practices and new opportunities for innovative food products have spurred interest in supercritical-CO_2 fluid extraction. There are many advantages of using supercritical fluids instead of conventional organic solvents. These include achieving higher purity extracts, absence of toxic solvent residues, single step processing, reduced operating costs, selective fractionation, faster separation, environmental friendliness, and physiological compatibility. In addition, the oxygen free operating system prevents oxidation, and the use of low temperatures minimizes thermal degradation of sensitive materials.

Because of consumer perception of the negative impacts of chemical solvent extractions and the increasing demands for natural products from natural sources having no toxic chemical contamination, there is worldwide pressure for industry to adopt new sustainable processes to produce natural products for food or nutritional supplement applications. Under such pressure, supercritical fluid-CO_2 technology has been developed and successfully used to extract essential oils, functional fatty acids, antioxidants, and other bioactive compounds, including the extraction and fractionation of carbohydrates (Glisic et al. 2007; Shi et al. 2007a, b, c; Montañés et al. 2008; Mitra et al. 2009; Montañés et al. 2009; Sanchez-Vicente et al. 2009; Shi et al. 2010a). It has particularly found relevance in those food and pharmaceutical applications that process and handle complex, thermosensitive, bioactive components, and is increasingly being applied in the areas of nutraceuticals, flavor, and production of other high-value items. Examples of these applications include the purification of solid materials; separation of tocopherols and antioxidants; removal of pesticide residues from herbs; production of medicines and food products; the detoxification of shellfish; the concentration of fermented broths, fruit juices, essential oils, spices, and coffee; and the separation of caffeine, etc. (Perrut 2000; González et al. 2002; Kasamma et al. 2008; Martinez et al. 2008; Miyawaki et al. 2008; Liu et al. 2009a, b; Shi et al. 2009a; Yi et al. 2009; Herrero et al. 2010).

Supercritical-CO_2 fluid extraction is governed by four key steps; extraction, expansion, separation, and solvent conditioning. The steps are accompanied by four generic primary components: extractor (high pressure vessel), pressure and temperature control system, separator, and pressure intensifier (pump). The major process parameters are temperature, pressure, and flow rate. Most commercial operations for supercritical-CO_2 extraction of solid materials operate on a batch system (see Fig. 11.2). Once the supercritical-CO_2 and the feed reach equilibrium in the extraction vessel, through the manipulation of pressure and temperature to achieve the ideal operating conditions, the extraction process proceeds. The mobile phase, consisting of the supercritical-CO_2 fluid and the solubilized components, is transferred to the separator where the fluid is reduced by decreasing the pressure of the system. The extract precipitates in the separator while the supercritical-CO_2 fluid is either released to the atmosphere or recycled back to the extractor.

Recently, industry has focused on "fractional separation" where the natural materials are extracted under relatively severe pressure and temperature conditions to remove all the desired components. The resulting fluid extract is then passed through a series of 2, 3, or 4 separator vessels in which the operating parameters (temperatures and pressures) in each vessel are set to selectively precipitate one specific component of interest (Fig. 11.3). This can create a range of unique fractions with new application potentials. For example, this fraction system is able to extract/fractionate a number of different carotenoids (e.g., β-carotene, lycopene), oleoresins, and colorants from tomatoes. High antioxidant activity phenolic compounds have also been extracted from rosemary leaves with supercritical-CO_2 (Chang et al. 2008, Huang et al. 2010; Shi et al. 2010b; Xiao et al. 2010). Rizvi and Bhaskar (1995) evaluated the feasibility of supercritical fluid processing of milk fat (i.e., fractionation, scale-up, and economic aspects), and reported that the scale-up of a supercritical extraction processes was a practical approach.

In the case where highly volatile components are being extracted, a multistage configuration may have to be employed as shown in Fig. 11.4 (Shi and Zhou 2006; Kasamma et al. 2008). The processes described above are semibatch continuous processes, where the supercritical-CO_2 flows in a continuous mode while the extractable solid feed is charged into the extraction vessel in batches. In a commercial-scale processing plant, multiple extraction vessels are sequentially used to enhance process performance and output. The semicontinuous extractor designs allow intermittent loading and unloading of solid material through lock-hopper vessels, reducing downtime and improving production efficiency. The need to improve the design into truly continuous

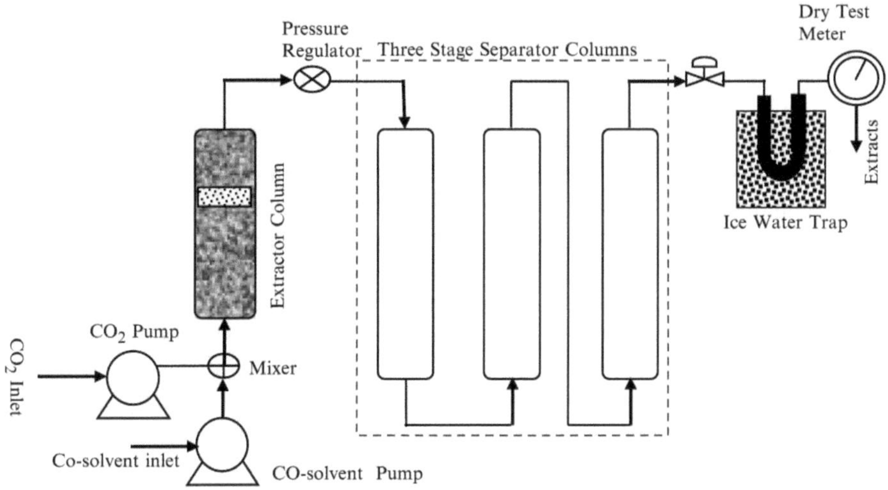

Fig. 11.3 Schematic diagram of supercritical fluid extraction system used to fractionate bioactive components from plant matrix using supercritical carbon dioxide

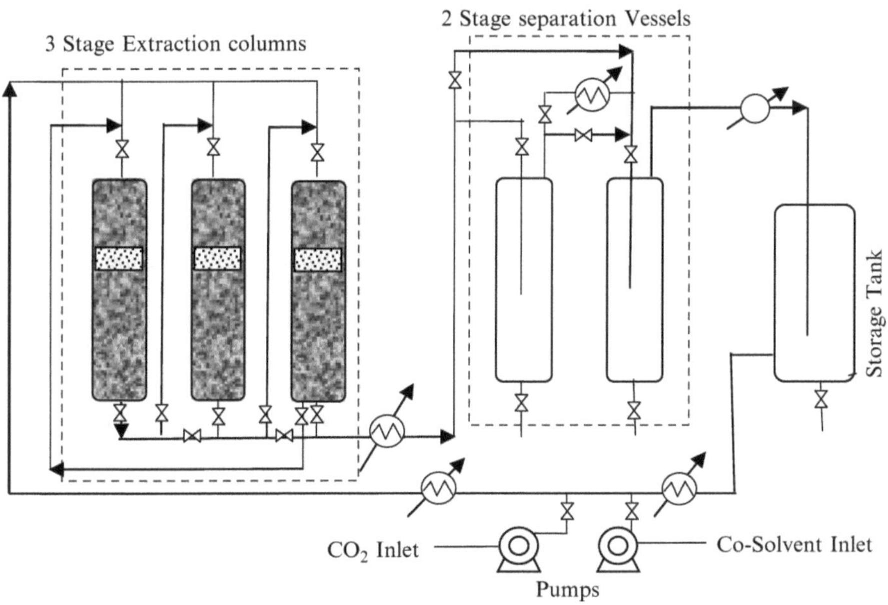

Fig. 11.4 Schematic diagram of commercial scale multi-stage supercritical fluid extraction system used to fractionate bioactive components. (The symbols "Ⓦ", "⧖" represent pressure valves and heat exchangers, respectively.)

modes is growing. Supercritical-CO_2 fluid extraction could be cost effective under large scale production conditions.

Large scale supercritical-CO_2 fluid extraction has become a practical process for the extraction of high-value products from natural materials. The solvating power of supercritical-CO_2 fluids is sensitive to temperature and pressure changes. Thus the extraction parameters may be optimized to provide the highest possible extraction yields with maximum bioactivity for health-promoting components targeted in bioactive extraction processes (Kasamma et al. 2008; Chen et al. 2009; Yi et al. 2009). Although a temperature rise in the extraction process generally can increase the solubility of components in supercritical-CO_2 fluids, the conditions under which thermally labile targeted compounds are negatively affected should be considered (Shi et al. 2007a, b, c). The intensity and the length of heat processing affect the health-promoting properties of bioactive components. Therefore, ideally the extraction time and temperature should be minimized. Minimizing such conditions also leads to a more economically viable process (Shi and Zhou 2006; Kasamma et al. 2008). Excessively high flow rates may reduce the contact time between the solute and the solvent and restrict the fluid flow in the sample if it becomes compacted. The optimal flow rate appears to vary with the targeted molecule. Relatively high flow rates may have a negative effect on extraction of some components. Under the same condition of flow rates and operating temperature, the pressure rise can significantly increase extraction yields. Most supercritical-CO_2 applications extract natural products or bioactive compounds at pressures ranging between 20 and 50 MPa, and at temperatures between 40°C and 80°C. Because of the poor solubility of some bioactive substances in supercritical-CO_2, food and/or pharmaceutical grade modifiers, ranging from 3% to 20%, are sometimes added to help in extracting the more polar compounds from plant materials.

One of the main processing aspects that should be considered in supercritical fluid extraction is extraction optimization. The use of optimum conditions for the different variables influencing extraction could significantly enhance the recovery or extraction yield of target compounds. With the aim to effectively optimize these variables (mainly extraction temperature, pressure, time, type, and percentage of modifiers, sample size, etc), different approaches have been applied. Appropriate experimental designs and statistical modeling should therefore be used to optimize these processes as compounds of interest will each have their unique characteristics and are likely to be extracted under different specific temperature and pressure combinations. Understanding the specific characteristics and physicochemical properties of the targeted bioactive compounds, and the effects of the parameters influencing extraction efficiency, bioactivity, cost, and efficiency, and determining and optimizing the parameters required to maximize yields and bioactivity of the targeted components, can help in establishing optimal conditions for large-scale processing (Shi and Zhou 2006; Kasamma et al. 2008).

Shi et al. (2009a, b) found optimized conditions for lycopene extraction to achieve higher yields with maximum antioxidative properties were 70°C and 30 MPa with a flow rate of 1.5 m/min using a one L separation cell. Optimum process parameters of 56°C and 26 MPa with an extraction time of 4 h was reported for the extraction of passion fruit seed oil with a good yield of 25.8% and yields of

89.4% and 72%, respectively, for high unsaturated fatty acids and linoleic acid (Liu et al. 2009a). Aromatic plants are mostly used as raw materials to extract natural antioxidants. Yépez et al. (2002) obtained high yields of odorless and flavorless extracts with high antioxidant activity from coriander (*Coriander stivum*) under moderate conditions of 45°C and 17.7 MPa. Ribeiro et al. (2001) found that the high antioxidant activity in extracts from lemon balm (*Melissa officinalis L.*), was achieved by optimizing the extraction at 10°MPa and 35°C for 4 h. Hadolin et al. (2001) found that extraction conditions of 60°C and 20 MPa produced the most concentrated vitamin E-rich oil from *Silybum marianum* with a relatively high extraction yield. Perretti et al. (2007) demonstrated the fractionation of fish oil containing a high fraction enriched in ω-3 fatty acids and with a suitable EPA/DHA ratio. These results highlight the possibility of modifying the fatty acid ethyl ester composition of extracts by optimizing the extraction conditions in terms of pressure, temperature, and supercritical carbon dioxide flow rate.

Extraction and fractionation of carbohydrates by supercritical fluid extraction have been recently reported. The extractions involve the use of CO_2 with a relatively low amount of polar modifier to effectively fractionate carbohydrate extract formed by lactose and lactulose. Montañés et al. (2008) used a full factorial experimental design to evaluate the effects of extraction pressure, temperature, proportion of modifier and flow rate. They later reported the use of 100 bar, 100°C, and 0.2 m/min with 4% modifier as optimum conditions of processing (Montañés et al. 2009). Supercritical fluid extraction has also been widely employed for the extraction of aromatic compounds. The application of mild pressures and temperatures (100 bar and 40°C) allowed the highest concentration of aroma compounds from sugar cane (Gracia et al. 2007).

One of the major advantages of superficial-CO_2 fluid extraction is to eliminate the refining process, which is otherwise required to remove undesirable compounds before consumption when conventional oil extraction techniques are used. If some valuable compounds are contaminated, they can also be lost during the refining process. By using supercritical fluid extraction, extracts can be obtained that are enriched with the particular compounds of interest, for example wheat germ oil (Eisenmenger and Dunford 2008) and rice brain oil (Chen et al. 2008).

Supercritical fluid extraction as a "green" processing technology has attracted a great deal of interest in recent times and is increasingly being used in food, natural products and pharmaceutical applications. It is of particular interest because of its environmental benefits. Furthermore, it provides flexibility as conditions can be optimized and various modifiers can be added to obtain selective fractionation of target compounds.

11.4 Subcritical water extraction

11.4.1 Principles and instrumentation

Subcritical water extraction, also known as pressurized low-polarity water extraction, or superheated water extraction, has recently become a popular green processing

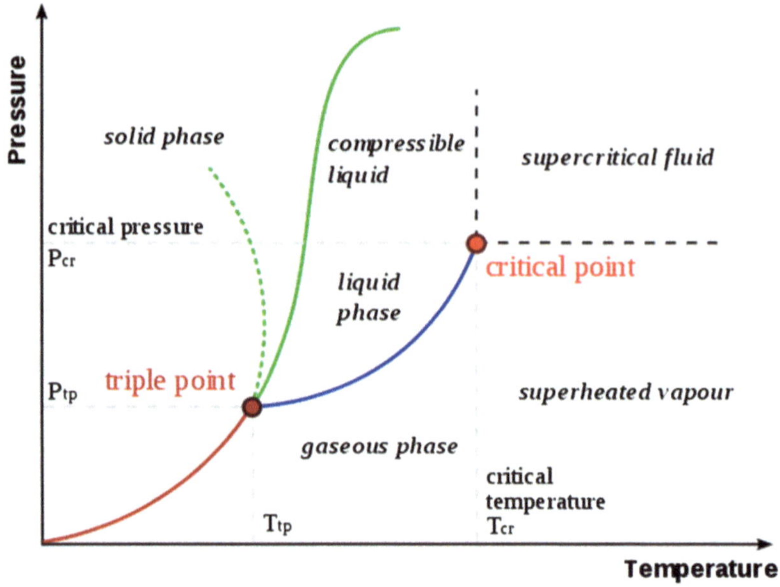

Fig. 11.5 Water phase temperature–pressure diagram

technology and is emerging as a promising extraction and fractionation technique to replace traditional extraction methods. The subcritical water extraction is not only used for concentration of bioactive compounds such as antioxidants, lignans, and anthocyanin from natural materials (Li-Hsun et al. 2004; Ju and Howard 2005; Cacace and Mazza 2006), but it has also been used in sample preparation to extract organic contaminants from foodstuffs for food safety analysis and solids/sediments for environmental monitoring purposes (Herrero et al. 2006).

The subcritical water extraction process is an environmentally friendly technique that can provide higher extraction yields from solid plant materials (Luque de Castro and Jiménez-Carmona 1998). Subcritical water extraction is based on the use of water as an extractant in a dynamic mode, at temperatures between 100°C and 374°C and under pressure high enough to maintain the liquid state leading to increased extraction efficiency through improvements in mass transfer and changes in the polarity of water. The critical temperature and pressure of water are shown as a phase diagram in Fig. 11.5 (Tc = 374°C, Pc = 221 bar).

Water is a polar solvent, but its polarity decreases if heated from 25°C to 200°C under pressure to maintain its liquid state, making it possible to use it to extract lower polar or non-polar compounds. This unique property of water is the fundamental principal of subcritical water extraction. Water changes dramatically when its temperature rises because of the breakdown in its hydrogen-bonded structure with temperature. The high degree of association in the liquid causes its relative permittivity (more commonly called its dielectric constant) to be very high under ambient conditions. But as the temperature rises, the hydrogen bonds break down, and the dielectric constant falls.

The most outstanding feature of water as an extraction solvent is the easy manipulation of its dielectric constant (ε, dimensionless). In fact, this parameter can be changed within a wide range just by changing the temperature under moderate pressure. Thus, at ambient temperature and pressure, water has a dielectric constant around 80, making it an extremely polar solvent. This parameter is drastically lowered by raising the temperature under moderate pressure. For example, subcritical water at 250°C and a pressure over 40 bar has $\varepsilon = 37$, which is similar to that of ethanol and allows for the leaching out of low-polarity compounds. At 205°C, water's dielectric constant has fallen so that it is equal to that of methanol (i.e., 33) at ambient temperature. Thus, between 100°C and 200°C superheated water behaves like a water-methanol mixture. Partly because of its fall in polarity with temperature, superheated water can dissolve organic compounds to some extent, especially if they are slightly polar or polarizable like aromatic compounds. Water can, therefore, be used as extraction solvent to extract polar, moderately polar, and non-polar compounds by adjusting the extraction temperature from the range of 50–275°C.

The extraction is, thus, based on the changing dielectric constant of water with temperature, which modifies the polarity of water making it capable of extracting different compounds, such as hydrophilic or lipophilic compounds. Also, the use of pressure of 5 MPa would be high enough to prevent the water from vaporizing at temperatures from 100°C to 250°C. Once pressure is high enough to keep water in a liquid state, additional pressure is not necessary as it has limited influence on the solvent characteristics of water. Increasing the water temperature from 25°C to 250°C causes similar changes in dielectric constant, surface tension, and viscosity (Kronholm et al. 2007; Brunner 2009). Thus a subcritical water extraction process can easily dissolve organic compounds such as phytochemicals that are normally insoluble in ambient water.

As shown in Fig. 11.6, the system of subcritical water extraction consists of a water reservoir, oven, extraction cell, restrictor (valve), and vial (extracts collector). The water reservoir is coupled to a high pressure pump that introduces the water as a solvent into the system, the oven where the extraction cell is placed and extraction takes place, and a restrictor to maintain the pressure. The extracts are collected in a vial placed at the end of the extraction system. A coolant device may be required for rapid cooling of the resultant extract. Because of the unique properties of low-polarity water solvent, the extraction has a disproportionately high boiling point for its mass, a high dielectric constant, and high polarity. As the temperature rises, there is a marked and systematic decrease in permittivity, an increase in the diffusion rate and a decrease in the viscosity and surface tension. In consequence, the more polar target materials with high solubility in water at ambient conditions are extracted most efficiently at lower temperatures, whereas moderately polar and non-polar targets require a less polar medium, induced by elevated temperature.

The subcritical water extraction process has the ability to selectively extract different classes of compounds, depending on the temperature used. The selectivity of subcritical water extraction allows for manipulation of the composition of the extracts by changing the operating parameters, with the more polar extracted at lower temperatures and the less polar compounds extracted at higher temperatures (Basile et al. 1998; Ammann et al. 1999; Clifford et al. 1999; Miki

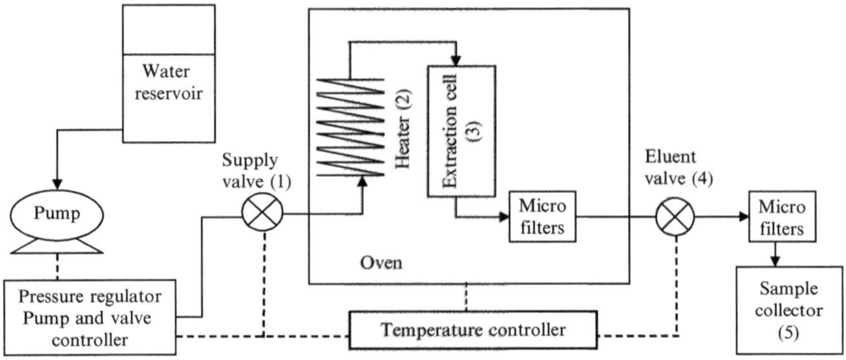

Fig. 11.6 Schematic diagram of subcritical water extraction system. The electrical connections are marked by dashed lines, while the path of subcritical water is shown by solid line with arrow. The high pressured water passes through a supply vale (1) into a heating coil (2) and into an extraction cell (3). The microfilters are placed before and after an eluent valve (4). The extract is collected in a sample collector (5)

et al. 1999; Kubatova et al. 2001, Soto Ayala and Luque de Castro 2001). Ibáñez et al. (2003) studied the effects of temperature, over a wide range of temperatures (i.e., 25–200°C), on recovery of antioxidant compounds from rosemary by using subcritical water extraction. The results showed a clear effect of water temperature on extraction yield, which increased at higher extraction temperatures. A decrease in the capability of water to dissolve the most polar compounds was observed at high extraction temperatures (200°C), while a high concentration of nonpolar compounds such as carnosic acid was obtained. In addition to the antioxidants from rosemary, a high yield of aroma compounds from rosemary was also achieved. The effects of temperature, pressure, flow rates, and extraction time significantly influenced the extraction selectivity, yield, and bioactivity of the extracts. Optimization of processing conditions for subcritical water extraction is important for large scale applications. Ozel et al. (2003) found that optimized conditions for improving the extraction yields of essential oil from *Thymbra spicata* were obtained at 150°C and 60 bar with a flow rate of 2 m/min for 30 min. Ozlem and Mazza (2008) also reported that flow rate, flow direction, and sample size greatly influenced the yield and properties of saponins and cyclopeptides from cow cockle seed.

The solubility of a material in an organic compound is often many orders of magnitude higher than its solubility in water at ambient temperature for two reasons. One is the polarity change, and the other is that the compound has low solubility at ambient temperature. A subcritical water extraction process will have a high positive enthalpy of solution, and thus a large increase in solubility with temperature. Because of the greater solubility of some organic compounds in superheated water, this medium could be considered as a replacement of conventional organic solvents for the extraction of some compounds. It is important to note, however, that some additional (generally undesirable) reactions of the compounds being processed may also occur (e.g., hydrolysis, oxidation, etc.) owing to the use of such high temperatures in processing.

11.4.2 Industrial applications

Using a subcritical water extraction process provides a number of advantages over traditional extraction techniques (i.e., hydrodistillation, organic solvents, solid–liquid extraction). Some advantages include, shorter extraction times, higher quality of the extracts (especially for essential oils), lower costs of the extracting agent, and an environmentally compatible technique. Because water is perhaps the most environmentally friendly solvent available at high purity and low cost, subcritical water extraction has been widely used to extract different compounds from several vegetable materials. It also has been exploited for the extraction of avoparcin in animal tissue (Curren and King 2001), fungicides in agricultural commodities (Pawlowski and Poole 1998), fragrances from cloves (Rovio et al. 1999), antioxidative components from sage (Ollanketo et al. 2002), anthocyanins and total phenolics from dried red grape skin (Ju and Howard 2005), saponins from cow cockle seed (Güçlü-Üstündag et al. 2007), and other bioactive components from plant materials (Ong and Len 2003). Some additional successful applications of this technique are for the extraction of essential oils from various plant materials (Khajenoori et al. 2009; Mortazavi et al. 2010), sweet components from *Siraitia grosvernorii*, lactones from kava roots, antioxidant compounds from microalgae *S. platensis* (Ibáñez et al. 1999, 2003; Herrero et al. 2004, 2006), and Ginkgo biloba and biophenols from olive leaves (Japón-Lujána and Luque de Castro 2006).

Several comparison studies have been made between subcritical water extraction and traditional extraction such as by Soxhlet extraction. Clifford et al. (1999) demonstrated that the amount of eugenol and eugenyl acetate recovered from clove (*Syzygium armaticum*) by using subcritical water at 150°C was similar to that obtained using Soxhlet extraction and hydrodistillation. These compounds have similar antioxidant properties as α-tocopherol. Jimenez-Carmona et al. (1999) compared subcritical water extraction with hydrodistillation methods for the extraction of essential oils from marjoram. The results showed that the quality of the oil obtained was better than that from steam distillation, because it contained more of the oxygenated compounds and fewer terpenes. The yield was also slightly higher than that from steam distillation, in spite of the fact that all the terpenes were not extracted. This may be because, at the higher temperature and under pressure, the plant material was more effectively penetrated. However, about twice the amount of water is required than for steam distillation, but energy costs are much less than for steam distillation. The energy required to heat a given mass of water from 30°C to 150°C under pressure is one fifth of that needed to boil water from 30°C at atmospheric pressure. Furthermore, it is possible to recycle most (three quarters) of the heat, whereas it is difficult to recycle heat in steam distillation. Thus, in spite of the fact that twice as much water is needed, only one tenth of the energy of a steam distillation is required. Subcritical water extraction has been suggested as a method to extract valuable health-promoting compounds from plant materials.

11.5 Summary

One of the most important trends in the food industry today is the demand for "natural" foods and ingredients that are free from toxic-chemical additives. The growing interest in natural foods has raised the demand for natural health-promoting products of nonsynthetic origin. High-value functional substances can be obtained from biological materials by various purification and separation methods from plant materials or byproducts. The challenges in the separation processes are to both meet food regulation guidelines and to conduct the separation effectively and economically. Public health, environmental, and safety issues are the major concerns in the use of organic solvents in food processing. Supercritical-CO_2 fluid extraction and subcritical water extraction are considered as "green" and environmental friendly separation technologies that have emerged as attractive alternatives to traditional methods for the concentration of bioactive compounds. The separation problems encountered in the production of soluble materials have a number of aspects that influence the nature of the extraction technique chosen. An important factor is the solubility and stability of the extracted product during extraction. Moreover, the extraction power and polarity of the CO_2 fluid can be adjusted by incorporating small amount of cosolvents during supercritical fluid extraction, and simultaneously, the selectivity also can be enhanced. Water is the most environmentally friendly and cheapest solvent, and it can be used to extract non-polar compounds (such as some antioxidants) by increasing its temperature over 100°C under applied high pressure. In the case of unstable products, a rapid separation process is necessary in order to avoid significant loss of products. One of the greatest advantages of supercritical-CO_2 fluid extraction and subcritical water extraction is their rapidity, with quicker extraction times than traditional methods. Supercritical-CO_2 fluid extraction is particularly favorable for the extraction of thermally labile bioactive substances and processing can be easily controlled by adjusting temperature and pressure.

The demand for safe, natural and high value-added products is redirecting the focus of the food and pharmaceutical industries into seeking the development of new and clean technologies for their products. Emerging "green" processing technologies such as supercritical-CO_2 fluid and subcritical water extraction processes have been widely used in different fields, such as for the extraction of essential oils, food ingredients, natural products, pharmaceutical and cosmetic products, by-product recovery as well as for food toxicology and eco-toxicology studies. Extracts from natural sources are key elements in the manufacturing of health-promoting functional foods and ingredients, thus, the development and use of "green" separation processes and technology will likely continue to be widely employed in the processing of bioactive components, especially for use as supplements for health-promoting foods.

References

Ammann, A., D.C. Hinz, R.S. Addleman, M. Wai Chien, and B.W. Wenclawiak. 1999. Superheated water extraction, steam distillation and SFE of peppermint oil. *Fresenius' Journal of Analytical Chemistry* 364: 650–653.

Basile, A., M.M. Jimenez Carmona, and A.A. Clifford. 1998. Extraction of rosemary by superheated water. *Journal of Agricultural and Food Chemistry* 46: 5204–5209.

Brunner, G. 2009. Near critical and supercritical water. Part I. Hydrolytic and hydrothermal processes. *The Journal of Supercritical Fluids* 47(3): 373–381.

Cacace, J.E., and G. Mazza. 2006. Pressurized low polarity water extraction of lignans from whole flaxseed. *Journal of Food Engineering* 77: 1087–1095.

Chang, C.H., C.C. Chyau, C.L. Hsieh, Y.Y. Wu, Y.B. Ker, H.Y. Tsen, and R.Y. Peng. 2008. Relevance of phenolic diterpene constituents to antioxidant activity of supercritical CO_2 extract from the leaves of rosemary. *Natural Product Research* 22: 76–90.

Chen, C.R., C.H. Wang, L.Y. Wang, Z.H. Hong, S.H. Chen, W.J. Ho, and C.M.J. Chang. 2008. Supercritical carbon dioxide extraction and deacidification of rice bran oil. *The Journal of Supercritical Fluids* 45: 322.

Chen, J., J. Shi, C. Yi, S. Xue, Y. Jiang, and X. Ye. 2009. Effects of supercritical CO_2 fluid parameters on chemical composition and yield of carotenoids extracted from pumpkin. *Food Science and Technology/Lebensmittel-Wissenschaft und-Technologie* 43: 39–44.

Clifford, A.A., A. Basile, and S.H.R. Al Saidi. 1999. A comparison of the extraction of clove buds with supercritical carbon dioxide and superheated water. *Fresenius' Journal of Analytical Chemistry* 364: 635–637.

Curren, M.S.S., and J.W. King. 2001. Ethanol-modified subcritical water extraction combined with solid-phase microextraction for determining atrazine in beef kidney. *Journal of Agricultural and Food Chemistry* 49: 2175–2180.

Eisenmenger, M., and N.T. Dunford. 2008. Bioactive components of commercial and supercritical carbon dioxide processed wheat germ oil. *Journal of the American Oil Chemists' Society* 85: 55–61.

Glisic, S.B., D.R. Misic, D. Stamenic, I.T. Zizovic, R.M. Asamin, and D.U. Skala. 2007. Supercritical carbon dioxide extraction of carrot fruit essential oil: Chemical composition and antimicrobial activity. *Food Chemistry* 105(1): 346–352.

González, J.C., O.I. Fontal, R. Mercedes, J. Vieytes, M. Vieites, and L.M. Botana. 2002. Basis for a new procedure to eliminate diarrheic shellfish toxins from a contaminated matrix. *Journal of Agricultural and Food Chemistry* 50(2): 400–405.

Gracia, I., J.F. Rodriguez, M.T. Garcia, A. Alvarez, and A. Garcia. 2007. Isolation of aroma compounds from sugar cane spirits by supercritical CO_2. *The Journal of Supercritical Fluids* 43: 37–42.

Güçlü-Üstündag, O., J. Balsevich, and G. Mazza. 2007. Pressurized low polarity water extraction of saponins from cow cockle seed. *Journal of Food Engineering* 80: 619–630.

Hadolin, M., M. Skerget, Z. Knez, and D. Bauman. 2001. High pressure extraction of vitamin E-rich oil from *Silybum marianum*. *Food Chemistry* 74: 355–364.

Herrero, M., E. Ibáñez, J. Señoráns, and A. Cifuentes. 2004. Pressurized liquid extracts from Spirulina platensis microalga: Determination of their antioxidant activity and preliminary analysis by micellar electro-kinetic chromatography. *Journal of Chromatography A* 1047(2): 195–203.

Herrero, M., A. Cifuentes, and E. Ibáñez. 2006. Sub- and supercritical fluid extraction of functional ingredients from different natural sources: Plants, food-by-products, algae and microalgae: A review. *Food Chemistry* 98: 136–148.

Herrero, M., J.A. Mendiola, A. Cifuentes, and E. Ibáñez. 2010. Supercritical fluid extraction: Recent advances and applications. *Journal of Chromatography A* 1217: 2495–2511.

Huang, W., Q. Deng, B. Xie, J. Shi, F. Huang, B. Tian, R. Liu, Q. Huang, and S. Xue. 2010. Purification and characterization of an antioxidant protein from ginkgo biloba seeds. *Food Research International* 43(1): 86–94.

Ibáñez, E., A. Oca, G. de Murga, S. Lopez-Sebastian, J. Tabera, and G. Reglero. 1999. Supercritical fluid extraction and fractionation of different pre-processed rosemary plants. *Journal of Agricultural and Food Chemistry* 47: 1400–1404.

Ibáñez, E., A.K. Tovaa, J. Señoráns, S. Cavero, G. Reglero, and S.B. Hawthorne. 2003. Subcritical water extraction of antioxidant compounds from rosemary plants. *Journal of Agricultural and Food Chemistry* 51: 375–382.

Japón-Lujána, R., and M.D. Luque de Castro. 2006. Superheated liquid extraction of oleuropein and related biophenols from olive leaves. *Journal of Chromatography A* 1136: 185–191.

Jimenez-Carmona, M.M., J.L. Ubera, and M.D. Luque de Castro. 1999. Comparison of continuous subcritical water extraction and hydrodistillation of marjoram essential oil. *Journal of Chromatography A* 855: 625–632.

Ju, Z.Y., and L.R. Howard. 2005. Effects of solvent and temperature on pressured liquid extraction of anthocyanins and total phenolics from dried red grape skin. *Journal of Agricultural and Food Chemistry* 51: 5207–5213.

Kasamma, L., J. Shi, and G. Mittal. 2008. Response surface analysis of lycopene yield from tomato using supercritical CO_2 fluid extraction. *Separation and Purification Technology* 60: 278–284.

Khajenoori, M., A. Haghighi Asl, and F. Hormozi. 2009. Proposed models for subcritical water extraction of essential oils. *Chinese Journal of Chemical Engineering* 17: 359–365.

Kronholm, J., K. Hartonen, and M.L. Riekkola. 2007. Analytical extractions with water at elevated temperatures and pressures. *Trends in Analytical Chemistry* 26(5): 396–412.

Kubatova, A., A.J.M. Lagadec, D.M. Miller, and S.B. Hawthorne. 2001. Selective extraction of oxygenates from savory and peppermint using subcritical water. *Flavour and Fragrance Journal* 16: 64–73.

Li-Hsun, C., C. Ya-Chuan, and C. Chieh-Ming. 2004. Extracting and purifying isoflavones from defatted soybean flakes using superheated water at elevated pressure. *Food Chemistry* 84: 279–284.

Liu, G., X. Xu, Q. Hao, and Y. Gao. 2009a. Supercritical CO_2 extraction optimization of pomegranate (*Punica granatum* L.) seed oil using response surface methodology. *Food Science and Technology/Lebensmittel-Wissenschaft und-Technologie* 42: 1491–1495.

Liu, S., F. Yang, C. Zhang, H. Ji, P. Hong, and C. Deng. 2009b. Optimization of process parameters for supercritical carbon dioxide extraction of Passiflora seed oil by response surface methodology. *The Journal of Supercritical Fluids* 48(1): 9–14.

Lucien, F.P., and N.R. Foster. 2000. Solubilities of solid mixtures in supercritical carbon dioxide. *The Journal of Supercritical Fluids* 17: 111–134.

Luque de Castro, M.D. and Jiménez-Carmona, M.M. 1998. Potential of water for continuous automated sample-leaching. *Trends in Analytical Chemistry* 17(7): 441–447.

Marcus, Y. 2006. Are solubility parameters relevant to supercritical fluids? *The Journal of Supercritical Fluids* 38: 7–12.

Martinez, M.L., M.A. Mattea, and D.M. Maestri. 2008. Pressing and supercritical carbon dioxide extraction of walnut oil. *Journal of Food Engineering* 88: 399–404.

Mendiola, J.A., M. Herrero, A. Cifuentes, and E. Ibáñez. 2007. Use of compressed fluids for sample preparation: Food applications. *Journal of Chromatography A* 1152(1–2): 234–246.

Miki, W., K. Nakahara, T. Fujii, K. Nagami, and K. Arai. 1999. *Process for producing essential oil via treatment with supercritical water and essential oil obtained by treatment with supercritical water*. Int. Patent Appl. WO99/53002A1.

Mitra, P., G.S. Ramaswamy, and K.S. Chang. 2009. Pumpkin (*Cucurbita maxima*) seed oil extraction using supercritical carbon dioxide and physicochemical properties of the oil. *Journal of Food Engineering* 95: 208–213.

Miyawaki, T., A. Kawashima, and K. Honda. 2008. Development of supercritical carbon dioxide extraction with a solid phase trap for dioxins in soils and sediments. *Chemosphere* 70(4): 648–655.

Montañés, F., N. Corzo, A. Olano, G. Regiero, E. Ibáñez, and T. Fornari. 2008. Selective fractionation of carbohydrate complex mixtures by supercritical extraction with CO_2 and different cosolvents. *The Journal of Supercritical Fluids* 45(2): 189–194.

Montañés, F., A. Olano, G. Regiero, E. Ibáñez, and T. Fornari. 2009. Supercritical technology as an alternative to fractionate prebiotic galactooligosaccharides. *Separation and Purification Technology* 66(2): 383–389.
Mortazavi, S.V., M.H. Eikani, H. Mirzaei, M. Jafari, and F. Golmohammad. 2010. Extraction of essential oils from *Bunium persicum* Boiss. using superheated water. *Food and Bioproducts Processing* 88: 222–226.
Ollanketo, M., A. Peltoketo, K. Hartonen, R. Hiltunen, and M.L. Riekkola. 2002. Extraction of sage (*Salvia officinalis* L.) by pressurized hot water and conventional methods: Antioxidant activity of the extracts. *European Food Research and Technology* 215: 158–163.
Ong, E.S., and S.M. Len. 2003. Pressurized hot water extraction of berberine, baicalein and glycyrrhizin in medicinal plants. *Analytica Chimica Acta* 482: 81–89.
Ozel, M.Z., F. Gogus, and A.C. Lewis. 2003. Subcritical water extraction of essential oil from *Thymbra spicata*. *Food Chemistry* 82: 381–386.
Ozlem, G. U., Mazza, G. 2008. Extraction of saponins and cyclopeptides from cow cockle seed with pressurized low polarity water. *LWT-Food Science and Technology* 41(9): 1600–1606.
Pawlowski, T.M., and C.F. Poole. 1998. Extraction of thiabendazole and carbendazim from foods using pressurized hot (subcritical) water for extraction: A feasibility study. *Journal of Agricultural and Food Chemistry* 46: 3124–3132.
Perretti, G., A. Motori, E. Bravi, F. Favati, L. Montanari, and P. Fantozzi. 2007. Supercritical carbon dioxide fractionation of fish oil fatty acid ethyl esters. *The Journal of Supercritical Fluids* 40: 349–353.
Perrut, M. 2000. Supercritical fluid applications: Industrial developments and economic issues. *Industrial and Engineering Chemistry Research* 39(12): 4531–4535.
Raventós, M., S. Duarte, and R. Alarcón. 2002. Application and possibilities of supercritical CO2 extraction in food processing industry: An overview. *Food Science and Technology International* 8: 269–284.
Ribeiro, M.A., M.G. Bernardo-Gil, and M.M. Esquivel. 2001. *Melissa officinalis L*: Study of antioxidant activity in supercritical residues. *The Journal of Supercritical Fluids* 21: 51–60.
Rizvi, S.S.H., and A.R. Bhaskar. 1995. Supercritical fluid processing of milk fat: Fractionation, scale-up, and economics. *Food Technology* 49(2): 90–97.
Rovio, S., K. Hartonen, Y. Holm, R. Hiltunen, and M.L. Riekkola. 1999. Extraction of clove using pressurized hot water. *Flavour and Fragrance Journal* 14(6): 399–404.
Sanchez-Vicente, Y., A. Cabañas, J.A.R. Renuncio, and C. Pando. 2009. Supercritical fluid extraction of peach (*Prunus persica*) seed oil using carbon dioxide and ethanol. *The Journal of Supercritical Fluids* 49: 167–173.
Sauceau, M., J.J. Letourneau, B. Freiss, D. Richon, and J. Fages. 2004. Solubility of eflucimibe in supercritical carbon dioxide with or without a cosolvent. *The Journal of Supercritical Fluids* 31(2): 133–140.
Shi, J., and X. Zhou. 2006. Solubility property of bioactive components on recovery yield in separation process by supercritical fluid. In *Functional food ingredients and nutraceuticals: Processing technologies*, ed. J. Shi. Boca Raton: CRC Press.
Shi, J., Y. Kakuda, X. Zhou, G. Mittal, and Q. Pan. 2007a. Correlation of mass transfer coefficient in the extraction of plant oil in a fixed bed for supercritical CO_2. *Journal of Food Engineering* 78: 33–40.
Shi, J., L.S. Kassama, and Y. Kakuda. 2007b. Supercritical fluid technology for extraction of bioactive components. In *Functional food ingredients and nutraceuticals: Processing technologies*, ed. J. Shi, 3–43. Boca Raton: CRC Press.
Shi, J., G. Mittal, E. Kim, and S. Xue. 2007c. Solubility of carotenoids in supercritical CO_2. *Food Reviews International* 23: 341–371.
Shi, J., X. Zhou, and L.S. Kassama. 2007d. Correlation of mass transfer coefficients in supercritical CO_2 separation process. *Drying Technology* 25(2): 335–339.
Shi, J., M. Khatri, J. Xue, G. Mittal, Y. Ma, and D. Li. 2009a. Solubility of lycopene in supercritical CO_2 fluid affected by temperature and pressure. *Separation and Purification Technology* 66: 322–328.

Shi, J., C. Yi, J. Xue, Y. Jiang, Y. Ma, and D. Li. 2009b. Effects of modifier on lycopene extract profile from tomato skin using supercritical-CO_2 fluid. *Journal of Food Engineering* 93: 431–436.

Shi, J., S. Xue, Y. Jiang, and X. Ye. 2010a. Supercritical-fluid extraction of lycopene from tomatoes. In *Separation, extraction and concentration processes in the food, beverage and nutraceutical industries*, ed. S. Rizvi, 619–639. Philadelphia: Woodhead Publishing Limited.

Shi, J., C. Yi, X. Ye, S. Xue, Y. Jiang, Y. Ma, and D. Liu. 2010b. Effects of supercritical CO_2 fluid parameters on chemical composition and yield of carotenoids extracted from pumpkin. *Food Science and Technology/Lebensmittel-Wissenschaft und-Technologie* 43(1): 39–44.

Soto Ayala, R., and M.D. Luque de Castro. 2001. Continuous subcritical water extraction as a useful tool for isolation of edible essential oil. *Food Chemistry* 75: 109–113.

Tena, M.T., M. Valcarcel, P. Hidalogo, and J.L. Ubera. 1997. Supercritical fluid extraction of natural antioxidants from rosemary: Comparison with liquid solvent sonication. *Analytical Chemistry* 69: 521–526.

Vasapollo, G., L. Longo, L. Rescio, and L. Ciurlia. 2004. Innovative supercritical CO_2 extraction of lycopene from tomato in the presence of vegetable oil as cosolvent. *The Journal of Supercritical Fluids* 29: 87–96.

Xiao, J., B. Tian, B. Xie, E. Yang, J. Shi, and Z. Sun. 2010. Supercritical fluid extraction and identification of isoquinoline alkaloids from leaves of *Nelumbo nucifera* Gaertn. *European Food Research and Technology* 231(3): 407–414.

Yépez, B., M. Espinosa, S. López, and G. Bolaños. 2002. Producing antioxidant fractions from herbaceous matrices by supercritical fluid extraction. *Fluid Phase Equilibria* 197: 879–884.

Yi, C., J. Shi, J. Xue, Y. Jiang, and D. Li. 2009. Effects of supercritical fluid extraction parameters on lycopene yield and antioxidant activity. *Food Chemistry* 113(4): 1088–1094.

Chapter 12
Electrodialysis in food processing

Martin Mondor, Denis Ippersiel, and François Lamarche

12.1 Introduction

Electrodialysis is an electromembrane process whereby the ionic composition of liquids may be modified by the use of ion permeable membranes under the influence of a direct electrical potential. The membranes used in conventional electrodialysis carry either a fixed positive charge and are called anion exchange membrane or a fixed negative charge and are then called cation exchange membrane. The fixed charge carried by the membranes allows them to facilitate passage of ions of the opposite charge (counter-ions) and repel ions of the same charge (co-ions) a principle known as Donnan exclusion.

A field of electrodialysis uses a special type of membrane known as bipolar. Bipolar membranes (BPM) consist of a laminate of cation and anion exchange layer. Upon the application of an electrical potential these membranes will assure the transport of electrical charges via protons and hydroxyl ions produced by water splitting. These membranes can be used to adjust stream pH value.

Advantages given by the use of this technology include: product purification with no dilution, rapid and controlled salt removal from a product stream, efficient removal of low charged molecular species with high yield, ions substitution from adjacent solution, and pH variation and adjustment with no addition of external solutions.

M. Mondor (✉) • D. Ippersiel • F. Lamarche
Agriculture and Agri-Food Canada, Food Research and Development Centre,
Saint-Hyacinthe, QC, Canada J2S 8E3
e-mail: martin.mondor@agr.gc.ca; denis.ippersiel@agr.gc.ca; francois.lamarche@agr.gc.ca

J.I. Boye and Y. Arcand (eds.), *Green Technologies in Food Production and Processing*,
Food Engineering Series, DOI 10.1007/978-1-4614-1587-9_12,

12.2 Conventional ED

12.2.1 Basics principles of conventional electrodialysis (ED)

In conventional electrodialysis the electrical charges on ions allow them to be driven through solutions of water swollen membranes when a direct current is applied at the electrodes, such that positively charged cations will migrate toward the negatively charged cathode and negatively charge anions will migrate toward the positively charged anode. Under the influence of an electrical potential, cations will permeate cation-exchange membranes (identified as C, Fig. 12.1) but will be retained by anion exchange membranes (identified as A, Fig. 12.1) and negatively charged anions will permeate anion exchange membranes but will be retained by the negatively charged cation-exchange membranes (see Fig. 12.1). With a stack construct in a dilution-concentration configuration wherein cations and anions exchange membranes are alternating, ion permeation through membranes will result in an ion concentration increase in compartments known as concentrate or brine and a decrease in ion concentration in the adjacent compartment known as the diluate.

Many components are necessary for the proper and efficient operation of an electrodialysis plant. However, the key element is the electrodialysis stack, which consists

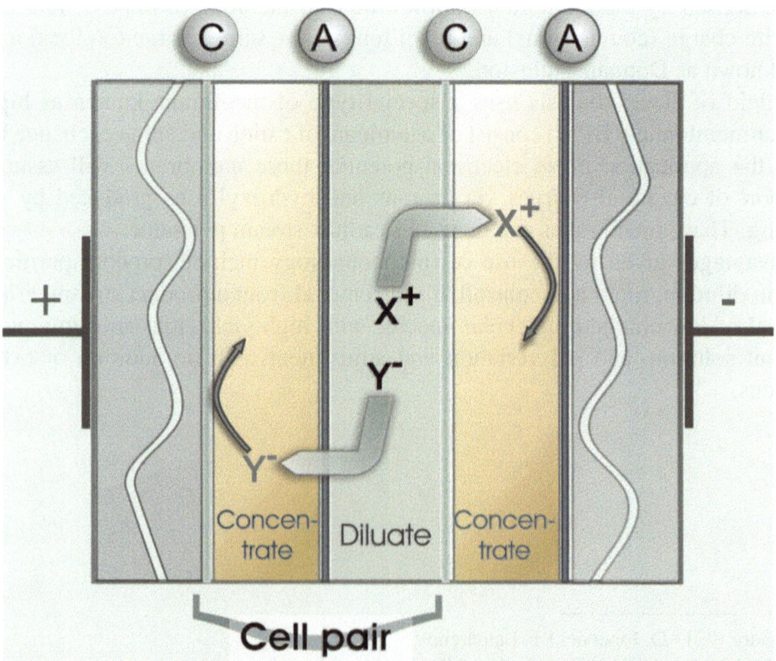

Fig. 12.1 Conventional electrodialysis in a dilution-concentration configuration. A: anionic membrane. C: cationic membrane

of an alternating repeating unit known as a cell-pair with electrodes rinsing solution at both ends (see Fig. 12.1). The material most commonly used for the cathode electrode is stainless steel. Whereas graphite, platinum coated titanium, or niobium, being more stable to oxidation, are used as anode material. In an electrodialysis stack a cell-pair is a sequence consisting of an anion exchange membrane, concentrate compartment, cation exchange membrane, and diluate compartment; it is typical to have up to 200 cell pairs stacked between the electrodes (Strathmann 2004). Membranes stacking results in efficient use in terms of energy consumption, because total ions transfer from a diluate to a concentrate solution is proportional to the current passing through a cell-pair multiplied by the number of cell-pairs. This also implies that electrodes cost is minimized, because a large membrane area may be associated with a relatively small electrode area and the energy consumed by electrode reactions are therefore negligible compared with the energy used for ion transfer.

12.2.2 Monopolar membranes

Membranes used in electrodialysis are thin sheets or films of anion or cation exchange resins reinforced with synthetic fabrics, which are necessary to give a mechanical strength. The fabrics used in most commercial ion-exchange membranes consist of hydrophobic polymers such as polystyrene, polyethylene, or polysulfone crosslinked with divinylbenzene. The degree of crosslink determines the polymer degree of swelling and improves membrane mechanical strength. The material they are made from determines to a large extent the mechanical, chemical, and thermal stability of the membranes. An increase in the degree of cross-linking results in an increase in membrane selectivity, stability, and electrical resistance. Concentration and type of fixed charges on the polymer determine membrane permselectivity and electrical resistance. Types of fixed charges most commonly used on cationic membranes include the sulfonic acid group $-SO_3^-$, which is completely dissociated over nearly the entire pH range and the carboxylic acid group $-COO^-$, which is virtually undissociated in the pH range <3. Fixed charges on the anion exchange membranes include: quaternary ammonium group $NR3^+$, which is completely dissociated over the entire pH range, and secondary ammonium group $-NH_2R^+$, which is only weakly dissociated. Concentration of fixed charges influences the mechanical stability and permselectivity of the membrane because of the impact on their degree of swelling (Strathmann 2004). Cationic and anionic exchange membrane can be discriminated using a few drop of 0.05% solution of either methyl orange or methylene blue that will stain the membrane deep blue or yellow respectively.

The properties of ion-exchange membrane determine to a large extent the technical feasibility and the economics success of an industrial process. Parameters such as electrical resistance, permselectivity, ion-exchange capacity, solvent transfer, resistance to chemical attack, and good mechanical and form stability are all important electrodialytic properties that must be considered.

Parameters determining the different properties of ion-exchange membranes are often contrary to each others. Fundamental to the economic use of electrodialysis is the electrical resistance of the ion-exchange membranes. Membrane electrical resistance determining the energy requirements of the electrodialysis process is a function of the ion-exchange capacity and ion mobility within the membrane. Fixed ionic charges on the membrane matrix determine the electrical resistance such that a high concentration of fixed charges results in low electrical resistance but, in general, causes a high degree of swelling and a poor mechanical stability. Membrane mechanical strength may be improved by increasing the degree of cross-linking but this also increases membrane electrical resistance (Strathmann 2004).

Permselectivity, which is the membrane ability to allow the passage of counter-ions while preventing the passage of co-ions, is a property more important than the membrane ion-exchange capacity, which represents the numbers of fixed charges by dry mass of membrane (Young 1974). In theory, most of the electric current in a membrane is carried by counter-ions while membrane concentration of co-ions is relatively low. High membrane permselectivity is important because it has a direct impact on process efficiency.

Solvent transfer by electroosmosis and osmosis will affect the maximum attainable concentration of the concentrate. Under the influence of an applied electrical current solvent is transferred through the membrane in the hydration ions shelf, a phenomenon known as electroosmosis. It was estimated that water transport through a membrane represents 0.05–0.2 dm^3 of water per mole of migrating ion passing through each membrane (Davis et al. 1997). In addition osmosis, which consists in water transfer from a dilute to a concentrate solution, may occur even in the absence of an electrical field.

12.2.3 Operating variables

As in other processes using membranes, solutions used in electrodialysis must be exempt of suspended particles; therefore a pretreatment is often required. Filtration of feed solution can be applied to remove particulate material that may otherwise foul the surface of the ion-exchange membranes (Fidaleo and Moresi 2006). Sometimes, other procedures such as flocculation or precipitation are required to remove ions present near their saturation limits. Presence of species such as calcium and magnesium ions, suspended solids, silica, and organic matter has to be addressed. As mentioned previously, electrodialysis is based on ion migration caused by the application of an electrical potential; however, uncharged species will move by diffusion and high molecular weight, and less mobile ions will not be transferred at a high rate. Streams pressure when operating the stack must all be equal and be equilibrated with the pressure of the diluting stream to prevent transfer through pressure gradient.

As a first determination the limiting current density is the first operating variable that has to be determined. Electric current in solutions is carried by both cations and anions according to their transference number, which is a function of ions mobility.

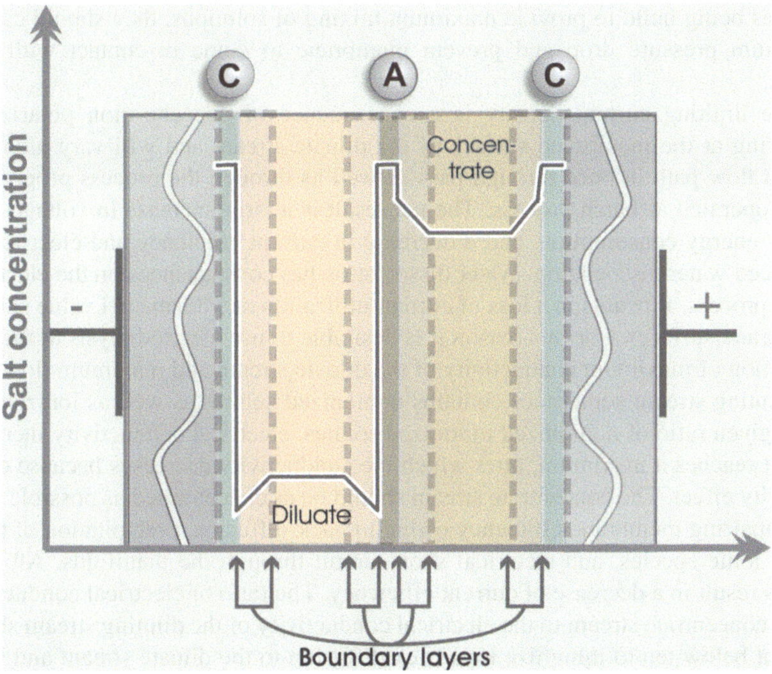

Fig. 12.2 Salt concentration profile during conventional electrodialysis. A: anionic membrane. C: cationic membrane

However, in ion-exchange membranes, current is mainly carried by the associated counter-ions, which have a transference number very close to one. At the surface of a membrane facing the diluate stream there is a lower concentration of ions in solution because of the lower ions transference numbers in the membrane (Fig. 12.2). On the membrane side facing the concentrate stream there is an associated accumulation of ions at the membrane surface because of the lower transference number of ions in solution. The net result is the establishment of a concentration gradient at both sides of the membrane surface, a situation known as concentration polarization.

Ions are moving from the bulk phase to the boundary layer at a rate related to the applied electric current. During electrodialysis a steady state is obtained when ions from the bulk phase replenish the ions removed from the boundary layer through the membrane (see Fig. 12.2). If ion concentrations at the membrane surface of the diluting stream are reduced to zero, there are no more ions available to carry the electric current through the membrane. The limiting current density is attained when this situation is reached. Concentration polarization may be partly reduced by generation of a turbulent flow at the membrane surface. The utilization of a spacer and a relatively high flow rate will result in a good mixing of the various streams and will reduce the phenomenon of concentration polarization. However, at the membrane surface boundary layer an almost static solution will always be present. In industrial electrodialysis stacks, spacers are typically between 0.3 and 2 mm thick.

Besides being build to provide maximum mixing of solutions, they should cause a minimum pressure drop and prevent membrane to come in contact with each others.

The limiting current density is a consequence of concentration polarization occurring at the membrane surface of the diluate stream and will vary along the stream flow path in once through pass as well as through the process progression when operated as batch process. The net result is a large increase in voltage drop, higher energy consumption, and a decrease in current efficiency and electric field enhanced water dissociation. Water dissociation has consequences on the electrodialysis process. It results in a loss of current utilization and drastic pH value shifts at membrane surfaces. For this reason it is desirable to use electrodialysis at the concentration of maximum conductivity of the diluate stream and maximum flow rate.

Diluting stream sometimes contains nonionized solutes as well as ionized one. For a given ratio of nonionized to ionized solutes, electrical conductivity increases until it reaches a maximum, after which the conductivity decreases because of the viscosity effect. The concentrate stream should be as concentrated as possible without impairing membrane efficiency owing to back-diffusion, precipitation of transferred ionic species, and electrical short circuit through the manifolds. All these effects result in a decrease of current efficiency. The ratio of electrical conductivity of the concentrate stream to the electrical conductivity of the diluting stream should be kept below ten to minimize ions back diffusion to the diluate stream and water transfer by osmosis. As the electrodialysis process progress the electric conductivity of the diluate stream reduces, because ions are transported to the concentrate stream and there is an associated drop in current density. This drop in current density is advantageous, because it reduces the possibilities of operating the stack at or above the limiting current density. For this reason it is usually preferable to operate the stack with a constant voltage rather than at constant current in batch operation. Moreover, because the voltage between both electrodes is constant when operating at constant voltage but the resistance of the solutions flowing through a stack is changing along the flow path from the entrance to the exit of the stack, resulting in changes in current density along the flow path, it is also desirable to operate the stack at constant voltage in once-through continuous operation.

Process temperature should be as high as possible without impairing membrane and product integrity, because it will result in a lower electrical resistance and viscosities of the fluids. The process is usually carried at 80% of the limiting current density to prevent any undesirable's consequences resulting if operating above the limiting current density.

The mode of operation of an electrodialysis stack in an industrial process is a function of the composition of the feed stream and degree of product stream ion depletion or enrichment required, which can be monitored online by conductivity measurement. Feed solutions in an electrodialysis process can be treated on a once-through continuous basis feed and bleed basis or batch, which is the simplest mode of operation. In a continuous mode of operation, the stream to be processed is passed through the stack only once; in commercial applications ions removal rarely exceeds 50% in a single pass. To obtain the desired degree of removal some type of staging is required. This implies that the treated solution has to go through a series of stacks until the required degree of depletion or concentration is obtained. In feed and bleed mode of operation, some of the solution that has already been processed is returned with the feed so that steady state conditions result. Recycling of solutions is less efficient than once-through flow in terms of equipment utilization and energy consumption, because the same solution has to be pumped and electrodialyzed repeatedly.

In batch operation, streams to be treated are placed in a reservoir, circulated through the stack, and recycled to the reservoir until the desired degree of ions removal or enrichment is attained. Batch operation has the advantage that the process cycle can be adjusted toward the desired end product and be independent of feed concentration, which may vary from batch to batch. In the stack, rate of ions transported from diluate stream to the concentrate stream is a function of the residence time of the different streams (flow rate), membrane area, and the applied current density. Stream flow velocities will determine the residence time of a solution in the stack. High flow velocities are required to assure efficient stream mixing so that concentration polarization at the membrane surface will be reduced, thus allowing it to work at a higher current density. The current density used will determine the required membrane surface area for a given capacity. However, working at high flow velocities implies a shorter residence time, therefore reducing the degree of ions depletion and concentration that can be obtained in a single pass in a stack.

12.3 Bipolar membranes ED

The invention of bipolar membrane has greatly extended the potential of electrodialysis. Upon the application of a direct electrical potential these membranes allow the electrodissociation of water molecules into proton and hydroxyl ions without the generation of O_2 and H_2 gases produced at the electrodes when utilizing electrolysis, which is the conventional method. The overvoltage required by the generation of these gases consumes about half the required electrical energy for this process. Bipolar membrane can therefore be advantageously used to adjust stream pH value without the external addition of acids or bases, which are often sources of impurities (Davis et al. 1997).

12.3.1 Basic principles of bipolar membrane (BPM)

Bipolar membrane consists of an anionic and cationic membrane laminated on top of each other. The junction between both membranes where water dissociation takes place is known as the transition region or transition layer. With a thickness in the order of 10 nm or less it results in a membrane with a minimum electrical resistance. Important properties of BPM are: low electric resistance, high selectivity for counter-ions, excellent stability at extreme pH values, high water dissociation, and high water permeability of the monopolar layers.

As with conventional electrodialysis the application of an electric field across the membrane results in the removal of ions contain in the transition layer. When all ions present in the transition layer are removed, further transport of electric charges can only be accomplished by protons and hydroxyls ions generated by water dissociation at the interface between the anion and cation exchange membrane. A theoretical potential of 0.83 V at 25°C is required to achieve water splitting. The actual potential drop across a BPM is in the range of 1.2–2 V for current density between 500 and 1500 A/m^2. The energy required is the sum of the energy for water dissociation and to transport ions from the feed solution, and protons and hydroxyl ions from the transition region of the bipolar membrane into the acid and base solutions (Strathmann 2004). Water will diffuse from the surrounding solutions into the BPM such that the transition region is continuously replenished with water molecules. The H^+ and OH^- ions produced further migrate from the transition layer through the cation exchange layer toward the cathode for the H^+ and through the anion exchange layer toward the anode for the OH^-. Water dissociation in BPM is accelerated 50 million times by the strong electric field in the transition region when compared with the rate of water dissociation in aqueous solution (Strathmann 2004).

12.3.2 Operating variables

Membrane configuration used in a stack depends on the process application. Because it is possible to combine different types of membranes into a stack, various membrane configurations may be used. The production of an acid and a base from the corresponding inorganic or organic salt is the main process application using bipolar membrane electrodialysis (BPMED).

In a three compartment cell, the basic unit known as a cell triplet (Fig. 12.3) consists of an anion exchange membrane, a cation-exchange membrane, a bipolar membrane, a salt compartment, an acid compartment, and a base compartment that can be repeated in the stack (Thate et al. 2000).With this configuration three distinct streams are obtained. As in conventional electrodialysis with the application of a direct electric current, ions contained in a salt stream migrate through the adjacent monopolar membrane and combine with either protons or hydroxyl ions generated by the BPM. The net result is the production of an acid and a base stream as well as a depleted salt stream. As in conventional electrodialysis changes in concentration of the various stream is a function of the applied current density, membrane area, and flow velocities. Maximum obtainable concentration of 2–3 N for strong acid and 4–5 N for base are achievable with this configuration. Permselectivity of the bipolar and monopolar membrane to H^+ and OH^- limit the maximum achievable concentration.

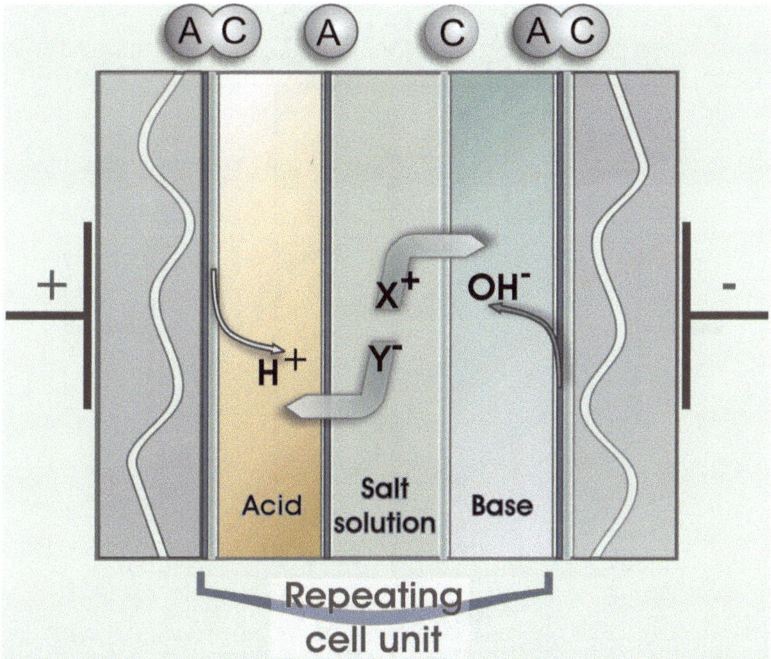

Fig. 12.3 Bipolar membranes electrodialysis with a three-compartment cell. AC: bipolar membrane. A: anionic membrane. C: cationic membrane

Other membrane configurations may be used where it is not practical to use the cell triplet configuration. Stack configuration using a two compartment cell with cationic membrane and a BPM is used preferentially when the salt of a weak acid having low conductivities has to be processed to recuperate a pure base and a mixture of acid and salt in the other compartment (Fig. 12.4a). A cell configuration using anionic and bipolar membrane is used to process weak base to produce a pure acid solution and a mixture of base and salt (Fig. 12.4b).

When compared with conventional electrodialysis, BPMED is a process run at a higher current density and the voltage drop across a cell unit is also higher (Strathmann 2004). This implies that a limited number of cell units can be used in a stack because of the amount of heat generated that has to be dissipated. In BPMED, limiting current density is reached when water is depleted from the transition layer because of lower water diffusion in the ion-exchange membrane. When water transport in the BPM is insufficient, membrane resistance increases drastically, and membranes will dry out and become inefficient if this condition occurs for a prolonged period of time. For economical and technical reasons, current density used is significantly lower than the water transport limited current density (Strathmann 2004).

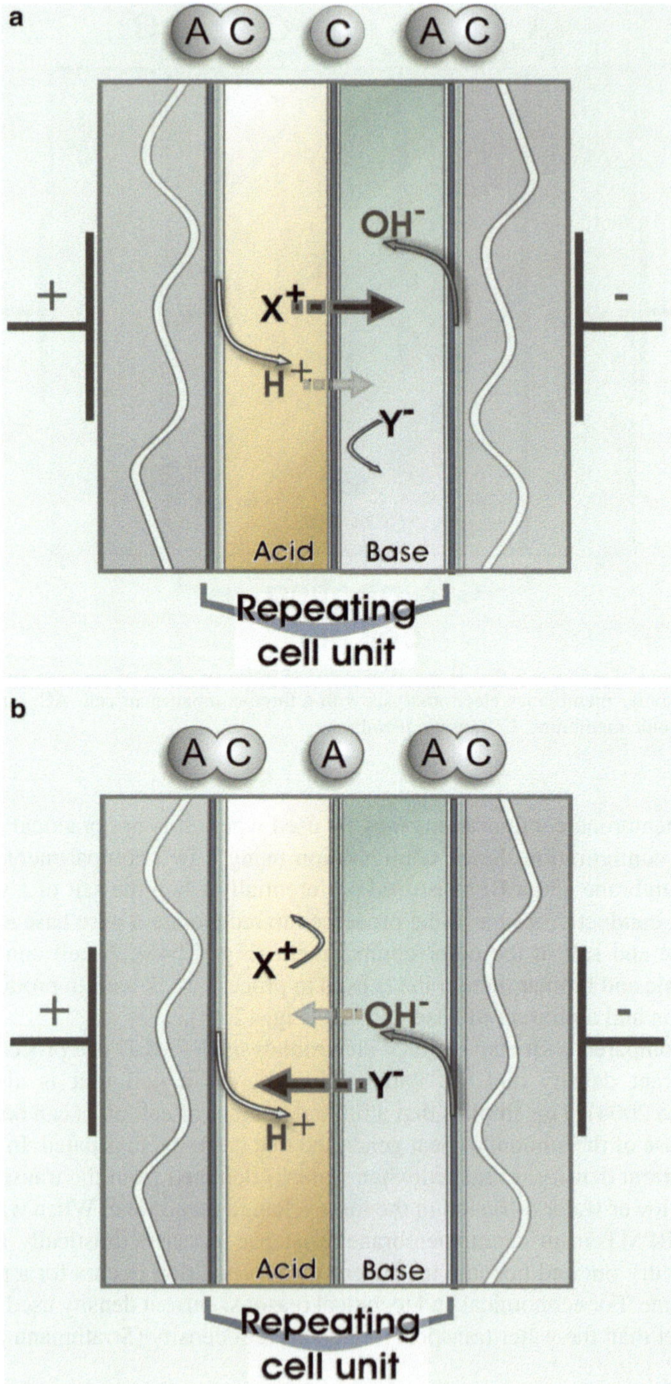

Fig. 12.4 Bipolar membranes electrodialysis with a two-compartment cell. (**a**) Bipolar-cationic membranes configuration. (**b**) Bipolar-anionic membranes configuration. AC: bipolar membrane. A: anionic membrane. C: cationic membrane

12.4 Green technologies applications of conventional and bipolar ED in food processing

The most important use of conventional electrodialysis is the desalination of brackish water for the production of potable water. However, other applications are gaining increasing importance with large-scale industrial installations including some applications in the food industry (e.g., whey and molasses demineralization, tartaric stabilization of wine, and deacidification of fruit juices) (Fidaleo and Moresi 2006). In the domain of food industries, electrodialysis with bipolar membranes was also recently used on an industrial scale for the production of organic acid (Pourcelly and Bazinet 2009). In addition to the aforementioned industrial scale applications of conventional electrodialysis and of electrodialysis with bipolar membranes in the food industry, there are also some applications currently under development that have potential for future industrial applications. In the following sections the industrial scale applications of both technologies in the food sector, as well as a few selected applications under development, will be briefly presented with the emphasis on the applications having the most potential to help environmental protection.

12.4.1 Demineralization of food solutions

The high mineral content of some food solutions and the need for their demineralization have made conventional electrodialysis increasingly important with large-scale industrial installations. One application is the demineralization of whey, the liquid by-product that is obtained from cheese making. For every kilogram of cheese produced about 10 kg of whey are left over, resulting in large volumes of this byproduct. There exist two types of whey, namely sweet whey (pH <6) and acid whey (pH <4). Whey contains 4.5–5.0% lactose, 0.04% nitrogen components, and approximately 0.5–0.8% ash. Its economical potential is influenced by its high water content (94–95%) (Fidaleo and Moresi 2006). Also, because whey has a biological oxygen demand varying between 30 and 45 kg/m^3, it cannot be directly discharged as wastewater in the sewage. In a dried form, about 10% of the whey solids consist of salts (mainly KCl), prohibiting its use in, for instance, infant foods or other applications requiring low salt content (Lopez Leiva 1988).

Conventional electrodialysis with a typical cationic–anionic two-compartment configuration (see Fig. 12.1) was shown to be an efficient technology to demineralize whey with up to 90% demineralization (Lopez Leiva 1988). In order to achieve the desired level of demineralization, the whey is circulated in the diluate compartments between the anionic and the cationic membranes. Under the influence of the electrical field, anions (Y^-) move in the direction of the anode permeating through the adjacent anionic membranes, whereas cations (X^+) move in the direction of the cathode permeating through the adjacent cationic membranes, resulting in whey demineralization. A demineralization in the order of 90% is considered to be the practical economic limit because of the decreasing electrical conductivity of the whey contains in the diluate stream with demineralization.

The optimal performance of whey demineralization by conventional electrodialysis will depend on many factors related to whey composition and to the desired level of demineralization. A majority of plants using electrodialysis use pre-concentrated whey (28–38%) as feed. By increasing the whey concentration, the increase in electrical conductivity enhances the efficiency without negative impact from the concomitant increase in viscosity (de Boer and Robbertsen 1983). Temperature also influences the rate of demineralization, and the optimal temperature is a compromise between high demineralization rate (30–40°C) and low bacteriological development (10°C) (Lopez Leiva 1988).

When compared with the traditional demineralization process using ion-exchange resins, conventional electrodialysis has several advantages from an environmental and energy point of view. Greiter et al. (2002) estimated that the volume of wastewater generated by the conventional electrodialysis process use to demineralize 45 m^3 of nanofiltrated whey is in the order of 1.25 m^3 as compared with 3.7 m^3 for ion-exchange resins that is to say approximately three times less. In addition, ash content and organic load of the effluent generated by the conventional electrodialysis process is in the order of 8.1 and 8.4 kg/m^3, respectively, which is significantly less than the 36.3 and 26.0 kg/m^3 reported for the process using ion-exchange resins (Greiter et al. 2002). As a result, the disposition of the electrodialysis effluent is less problematic from an environmental point of view than for the ion-exchange resin effluent. The electrodialysis also requires less energy than the ion-exchange resin process. The overall energy demand for demineralization of whey by conventional electrodialysis is in the order of 12.8 kWh/m^3, which includes the energy demand for pumping, electrodialysis desalting, and reduction of the organic load, whereas it is in the order of 35.4 kWh/m^3 for the ion-exchange resin and includes the energy demand for pumping, production of regenerants, and reduction of organic load (Greiter et al. 2002). For the aforementioned reasons, conventional electrodialysis is the method of choice for industrial demineralization of whey. However, as previously mentioned the maximum demineralization achievable by electrodialysis is in the order of 90%, whereas it is possible to remove as much as 99% of the mineral using traditional ion-exchange resins. Another drawback of electrodialysis is membrane fouling; this phenomenon can alter irreversibly the membrane integrity and can provide a drastic decrease in process effectiveness. Recently, some work has been dedicated to the control or minimization of membrane fouling during demineralization of whey (Ayala-Bribiesca et al. 2006; Casademont et al. 2009).

A second application of conventional electrodialysis is for the removal of melassigenic ions from beet or cane sugar syrups. Beet or cane molasses are the main solutions used in the sugar-manufacturing process. Despite an average sugar content of 50% (w/w), further sugar recovery is hampered by the presence of impurities ranging from suspended materials to inorganic salts and color substances (Fidaleo and Moresi 2006). Among these impurities, alkali metals are highly melassigenic by holding sugar in the molasses and preventing it from being recovered as crystalline sugar (Elmidaoui et al. 2002). In order to remove these impurities from beet or cane sugar syrups before crystallization, conventional electrodialysis with a typical cationic–anionic two-compartment configuration (see Fig. 12.1) was used (Tako and Brahim 1993; Thampy et al. 1999; Elmidaoui et al. 2002; Lutin et al. 2002).

As aforementioned for the demineralization of whey, the feed, which consists of preconcentrated juice with 30–40 °Brix is fed in the diluate compartments between anionic and cationic membranes. Various cationic and anionic membranes were used in past studies aiming for the removal of melassigenic ions from beet or cane sugar syrups, but Elmidaoui et al. (2002) and Lutin et al. (2002) reported that the use of Neosepta AXE 01 anionic membranes (in combination with CMX-Sb cationic membranes) enabled work with a feed temperature ranging between 50°C and 60°C, which was a significant improvement when compared with other type of anionic membranes that were limited to 40°C. Demineralization rates reported in the literature vary between 50% and 80%. Energy consumption reported by Lutin et al. (2002) for the electrodialysis process was in the order of 1.1 kWh/m^3 for ion transport and pumping, whereas 0.02 kg HCl/m^3 and 0.06 kg $NaOH/m^3$ were required to clean the membranes (Lutin et al. 2002; Fidaleo and Moresi 2006). Following the demineralization step, the juice can be further concentrated and crystallized to sugar crystals.

More recently, Elmidaoui et al. (2006) have conducted the electrodialysis treatment in batches at the industrial level (Sunabel Corp.) on solutions sampled at various points of the process, i.e., juice (16 °Brix), syrup (60–70 °Brix), and mother liquor B (78 °Brix). The electrodialysis configuration used is the one aforementioned and previously described by Elmidaoui et al. (2002). Electrodialysis was carried out on a volume of 30 L and under a constant current density of either 9, 14, or 15 mA/cm^2 for juice, syrup, and mother liquor B, respectively. A temperature of 55°C was used for all three solutions. Because of the high viscosity of syrup and mother liquor B, the demineralization was conducted on solutions at 45 °Brix obtained by dilution with juice. The demineralization rate, which was defined as the ratio between the decrease in conductivity after time "*t*" and initial conductivity, was fixed at 70%. For the three solutions, pH remains almost stable (8.74–8.44 for juice, 8.45–8.10 for 45 °Brix syrup, and 8.61–8.40 for 45 °Brix mother liquor B), coloration (ICUMSA) decreases (1,714–1,498 for juice, 6,303–5,925 for 45 °Brix syrup, and 11,031–10,349 for 45 °Brix mother liquor B), and final product purity increases (85.99–92.50% for juice, 83.42–92.55% for 45 °Brix syrup, and 76.59–85.88 for 45 °Brix mother liquor B), indicating an improvement in the quality of the treated solutions. Based on the initial content of the main melassigenic ions (Na^+, K^+ and Ca^{2+}), flow rate, and operating costs, without pumping, Elmidaoui et al. (2006) concluded that the lower investment was obtained by applying the demineralization step to the juice followed by mother liquor B and by the syrup. In addition, it is expected that membrane fouling will be less significant for the juice than for the syrup and mother liquor B; as well the frequency of membrane replacement would also be lower, mainly because of the lower Brix observed for the juice.

12.4.2 Wine stabilization

Tartaric acid (H_2T) and potassium ions (K^+) are natural constituents of grapes that can be found at a level varying between 1–3 and 0.8–1.5 kg/m^3, respectively (Fidaleo and Moresi 2006). Tartaric acid can dissociate into tartrate (HT^-) and bitartrate ions

(T^{-2}), which in turn may precipitate in the form of potassium hydrogen tartrate (KHT) and calcium tartrate (CaT), with the latter being practically insoluble in aqueous solutions. In the typical pH range of wines (3–4) KHT is predominant over CaT, because tartrate ions are found in higher proportion than the bitartrate ions. Tartrate crystals precipitation in bottle is undesirable especially, because these crystals may be confused with fragments of broken glasses thus inducing wine bottle refusal in consumer (Fidaleo and Moresi 2006).

To overcome this problem, excess of KHT is traditionally removed by cooling the wine to −4°C over several days to induce KHT precipitation prior to bottling (Maujean 1994). Another approach is directed to impair the crystallization process, especially when KHT content is closely exceeding the equilibrium one, by enriching wine with antiprecipitation additives, such as metatartaric acid, carboxycellulose, or yeast mannoproteins (Fidaleo and Moresi 2006). Despite its quite general application, the cold stabilization technique is not only expensive and time-consuming, but also its complexity does not allow a precise control of the final KHT concentration. Also, this operation can affect wine quality because of the simultaneous precipitation of polysaccharides and polyphenols together with KHT crystals (Vernhet et al. 1999; Gonçalves et al. 2003). Thus, the use of other KHT removal techniques like ion-exchange resins (Mourgues 1993) and electrodialysis (Escudier et al. 1993, 1997; Moutounet et al. 1994; Saint-Pierre et al. 1995) have been alternatively proposed for tartrate stabilization of wines. Treatment by ion-exchange resins consists of equilibrating the wine with a cation exchange resin that replaces the potassium ions by hydrogen or sodium ions (Mourgues 1993; Gonçalves et al. 2003). However, in addition to being not allowed in the European Union because it alters the ionic balance of wine, as well as its organoleptic properties, the treatment by ion-exchange resins also requires a step for resins regeneration, which results in the generation of large amount of effluents having a negative impact on the environment. On the other hand, the approach using electrodialysis is approved for commercial use in the European Union, it allow flexibility over the extent of removal of KHT, it does not alter the organoleptic properties of wine, and it is more friendly from an environmental point of view.

Tartaric stabilization of wines by conventional electrodialysis is actually performed industrially using a typical cationic–anionic configuration (see Fig. 12.1). In order to achieve the desired level of K^+, the wine is circulated in the diluate compartments. Under the influence of an electrical field, the anions (Y^-) including tartaric, lactic, and malic acids move in the direction of the anode permeating through the anionic membranes, whereas the cations (X^+) mainly K^+ but also Mg^{+2}, Ca^{+2}, and Na^+ move in the direction of the cathode permeating through the cationic membranes. The concentrate compartments are filled with a salt solution (i.e., NaCl or KHT). Because the permeation of the aforementioned anions is slower than for the aforementioned cations, a slight decrease of wine pH is usually observed (0.15–0.25 pH units). When the wine conductivity has reached the predetermined equilibrium value, it is automatically discharged and replaced with another batch of wine. The concentration of KHT in the concentrate compartments is regulated by dilution with deionized water in order to avoid its precipitation onto the membranes

(Moutounet et al. 1994). In terms of membrane fouling, it has been observed that anionic membranes are more prone to fouling than cationic membranes, especially when red wines are treated. However, once per day rinsing of the membrane stack can minimize fouling and extend its lifetime. The electric energy consumption including the pumping energy ranges from 0.5 to 1.0 kWh/m^3, which is ten times lower than the energy requirement of the cold stabilization procedure. An investment of about US$400,000 and operating costs of US$0.60/L of wine treated would enable treatment of about 40,000 m^3 of wine per year (Fidaleo and Moresi 2006). The ED-treated wines are generally stable, their basic characteristics (i.e., pH, acidity, sugar content, alcohol level) as well as organoleptic properties are practically unaltered and meet the regulation (Fidaleo and Moresi 2006).

12.4.3 Separation of organic acids produced by fermentation

A considerable number of organic acids are produced by fermentation but usually low concentrations (0.2–1 N) are obtained. Operation of the fermentor near neutral pH value by the addition of a base to neutralize the formed organic acids is required to increase yield. In the conventional approach, $Ca(OH)_2$ is usually used for that purpose. In addition to maintaining the pH near neutral, it enables at the same time precipitation of the organic acids as their calcium salts, which results in an improved productivity because high product concentration is responsible for low productivity, owing to product inhibition. The precipitated calcium salts are then recuperated and a strong acid such as H_2SO_4 is added to regenerate the organic acids resulting in the formation of waste—$CaSO_4$ sludge. In addition of not being economically favorable, the conventional process is not environmentally friendly. Because the H^+ and OH^- ions are the only ones necessary for this separation, the continuous input of other ions and output of the corresponding waste results in environmental pollution (Huang and Xu 2006).

Many processes have been suggested as potential alternatives to the traditional calcium salts precipitation approach for the recuperation/purification of organic acids from fermentation broths. One of these processes that is often mentioned in the literature is the so-called membrane recycle bioreactors (Enzminger and Asenjo 1986). In this approach, the fermentation broth is continuously recycled through a cross-flow microfiltration unit that allows continuous removal of the organic acid. The microfiltration permeate containing the organic acid is treated using a conventional electrodialysis configuration (see Fig. 12.1). The permeate solution is fed into the diluate compartments. The result is a dilute stream depleted of organic acid that is recycled back into the bioreactor, and a concentrated stream containing the organic acid to be supplementary refined (Fidaleo and Moresi 2006). In addition to helping environmental protection, this approach also has the advantages of improving the organic acids formation rate in the bioreactor, because product inhibition is greatly reduced.

Electrodialysis with bipolar membranes also can be successfully combined with the fermentation technology to improve the productivity of the process and minimize the negative impact on the environment. A specific application reported by Mani (1991) is the regeneration of acid and base during the production of organic acids through fermentation. For the combined process, NaOH can be used in replacement of the $Ca(OH)_2$ to neutralize the fermentation broth. In this case, the organic salts would mainly remain in their soluble form and the salts stream can be fed to a three compartment cell (see Fig. 12.3). In this case, the organic anions ($RCOO^-$) would permeate through the anionic membranes and form the corresponding organic acids (RCOOH) by combining with the H^+ ions generated by the bipolar membranes, whereas a base (NaOH) that can be used for the neutralization of the fermentation broth would be obtained from the combination of the OH^- ions and the Na^+ that permeate through the cationic membranes. This combined process enables operation of the fermentor at relatively low product concentrations, when compared with the process without electrodialysis with bipolar membranes, which results in an improved productivity. A second advantage is that the recovered organic acids are directly obtained in high concentration so that the subsequent purification via crystallization or other techniques is relatively inexpensive. Also, there is minimal amount of effluents generated by the combined process.

The reader interested in more detailed information about the use of conventional electrodialysis as well as electrodialysis with bipolar membranes in the fermentation industry is referred to the work of Fidaleo and Moresi (2006) who presented a comprehensive review on the subject. Detailed information about the application of conventional or bipolar membranes electrodialysis for the production of specific organic acids in the food industry can also be found in the literature: acetic acid (Nomura et al. 1988, 1994; Zhang et al. 1993; Chukwu and Cheryan 1999; Pourcelly and Bazinet 2009), citric acid (Novalic et al. 1995, 1996; Datta and Bergemann 1996; Sappino et al. 1996; Moresi and Sappino 1998, 2000; Ling et al. 2002), lactic acid (Hongo et al. 1986; Boyaval et al. 1987; Nomura et al. 1987, 1991; Ishizaki et al. 1990; Lee et al. 1998; Boergardts et al. 1998; Kim and Moon 2001; Madzingaidzo et al. 2002), malic acid (Sridhar 1987; Benafi-Bako et al. 2004), propionic acid (Weier et al. 1992; Boyaval et al. 1993; Zhang et al. 1993) and amino acids (such as lysine) (Kikuchi et al. 1995; Mani 2000; Grib et al. 2000; Lee et al. 2003).

12.4.4 Deacidification of fruit juices

Fruit juices are appreciated by consumers for their aroma and flavors, but the high acidity of some of them limits their use as an ingredient in the formulation of various preparations such as beverages, ice creams, marmalades or cocktails (Vera et al. 2007). As a result, some acidic fruit juices are to be deacidified to facilitate their use in food products. Organic acids such as citric present in high concentration in lemon, orange, pineapple, and passion fruit juices or malic in apple and grape juices are responsible for fruit juices acidity.

A conventional method used for deacidification of fruit juices is the use of anion-exchange resins: particular ions (such as OH^-) within the polymeric resin are exchanged with the citrate and/or malate ions present in the fruit juice, which is passed through the resin. Even if some anion-exchange resins are authorized for use in the food industry and have been shown to be efficient for deacidification of fruit juices, their use leads to changes in the organoleptic properties of the juice and results in the generation of a large volume of effluents during the regeneration step, which is a significant drawback from an environmental point of view (Vera et al. 2003a, b). In practice, for economical and environmental reasons, resins are regenerated to about two thirds of their total capacity; but even if this approach results in the use of lower amount of regenerant, it is conceivable that it would also reduce the deacidification efficiency of the resin by allowing a small proportion of the citrate and/or malate ions that are being taken up by the resin to slip through without exchange.

Removal of citric acid from fruit juices was also achieved through the precipitation of calcium citrate by the addition of $CaCO_3$ or $Ca(OH)_2$ to the juice. In this case, the chemical reactions taking place in the juice are the following (Vera et al. 2003a):

$$2H_3Cit + 3CaCO_3 \rightarrow Ca_3(Cit)_2 + 3H_2O + 3CO_2 \tag{12.1}$$

$$2H_3Cit + 3Ca(OH)_2 \rightarrow Ca_3(Cit)_2 + 6H_2O \tag{12.2}$$

Even if such procedures have not so far conferred sensory dislike to the juices, the use of $CaCO_3$ is not recommended because the release of CO_2 induces foaming and poor pH control, whereas that of $Ca(OH)_2$ may cause some precipitation in the final product. Furthermore, the consumer is often averse to any chemical addition in natural products, making this approach less attractive (Vera et al. 2003a; Fidaleo and Moresi 2006).

Use of conventional electrodialysis (see Fig. 12.1), was suggested to reduce the acidity of several fruit juices, such as lemon, orange, pineapple, grape (Adhikary et al. 1983), mandarin orange (Kang and Rhee 2002), and clarified passion fruit (Vera et al. 2003b) juices. In order to achieve the desired level of deacidification, the juice is circulated in the diluate compartments. Under the influence of the electrical field, the anions (Y^-) including the citrate and/or malate ions move in the direction of the anode permeating through the anionic membranes, whereas the cations (X^+) including H^+ move in the direction of the cathode permeating through the cationic membranes, which should result in the fruit juice deacidification. The concentrate compartments are filled with dilute NaOH solution. Results reported by Kang and Rhee (2002) for mandarin orange juice and by Vera et al. (2003b) for clarified passion fruit juice indicated that even if a decline in the titratable acidity was obtained, there was also a decrease in the juice pH. This would be because of the fact that conventional electrodialysis using the aforementioned configuration removes both H^+, and in these particular applications, the citrate ions, enhancing the dissociation equilibrium of the weak citric acid (Vera et al. 2003b).

As a result, a more suitable approach for deacidification of fruit juices could be the use of electrodialysis with bipolar membranes, because the H^+ and OH^- ions resulting from the water splitting can be used for the deacidification, and the production of organic acids (Voss 1986; Bailly 2002; Vera et al. 2003a, 2009), leading to economical and environmental benefits. A typical configuration with bipolar and anionic membranes is used for this purpose (see Fig. 12.4b). The juice is fed in the base compartments receiving the OH^- generated by bipolar membranes, which results in partial neutralization of H^+ ions present in the juice. At the same time, the citrate and/or malate ions are extracted from the juice by permeating through the anionic membranes and they combine, in the adjacent compartment, with the H^+ ions generated by the bipolar membranes to produce citric and/or malic acids. For the removal of citric acid from passion fruit juice, this configuration was tested by Vera et al. (2009) on pulpy unclarified juice, on juice clarified by tangential microfiltration, on twice-concentrated clarified juice, and on centrifugated juice. For all juices, the deacidification was carried out up to pH 4.5 with satisfactory results. In terms of electrodialysis performance it was lower with the unclarified and with the concentrated juices because of fouling of the anionic membranes, which resulted in higher membrane resistance. Also, whatever the juice pretreatment, all deacidified juices were of high quality in terms of physicochemical properties, only slight changes of color were observed (Vera et al. 2009). Even if electrodialysis with bipolar membranes is more expensive than the conventional techniques in terms of equipment cost, the organic acid that is simultaneously produced during the deacidification could reduce the overall cost of this technique. Therefore, electrodialysis with bipolar membranes could be an environmentally friendly alternative to the conventional techniques for the deacidification of fruit juices (Vera et al. 2003a, b).

12.4.5 Production of plants protein isolates with emphasis on soy

Plant proteins have made up a higher proportion of the human diet in recent years. This increasing demand for plant proteins is related to a growing interest in reducing the consumption of animal products for health, social and economic reasons. A large proportion of the plant proteins used in the food industry is in the form of protein isolates (at least 90% protein on a dry weight basis). Among the different sources of plant protein ingredients, soy is the most important accounting for some 68% of global plant protein consumption (USDA 2009). Because soy protein isolates are by far the plant protein isolates that are the most used as ingredients in the food industry, in the following section the emphasis will be put on comparison of the traditional isoelectric precipitation process with the processes using electrodialysis with bipolar membranes for production of soy protein isolates. However, in general, the processes described below as well as the main conclusions would also apply for the production of protein isolates from other plant sources.

Soy protein isolate is traditionally prepared by isoelectric precipitation process (Mounts et al. 1987; Lusas and Rhee 1995). It consists in solubilizing proteins from flakes or flours at a 1:10 to 20 solids/solvent ratio, in 30–50°C water adjusted to pH 8–11 with a base, such as NaOH. Follow a centrifugation step to remove the insoluble fibres. Soy proteins present in the supernatant are then precipitated out by acidification to their isoelectric point (pH 4.2–4.5) using a mineral acid, such as HCl, and recuperated by a second centrifugation step. The resulting precipitate containing the proteins is then suspended in water to remove the minerals and sugars, in order to increase protein content, and after reconcentration by applying a third centrifugation step, it is neutralized to pH 7 with a dilute base, such as NaOH. The resulting soy protein dispersion is fed to a spray-dryer to produce an isolate. The isolates produced by this approach have poor functional properties because of protein denaturation, as demonstrated by Petruccelli and Anon (1994) and Wagner et al. (2000), and a high phytic acid content (1–3% w/w), which alters the solubility of the isolates especially at low pH (Omosaiye and Cheryan 1979; Brooks and Morr 1985). This process also has a negative impact on the environment because it requires large amount of water (extraction, washing of precipitate and neutralization steps) and generates large volume of effluents (isoelectric precipitation and washing steps). The effluent generated following the isoelectric precipitation step is especially problematic and constitutes a serious water pollution threat (high biochemical oxygen demand), unless properly processed, because it contains whey-like proteins that remain soluble for the interval of pH 4.2–4.5 and salt (NaCl), which is formed because of the use of NaOH during the extraction step, and HCl during the isoelectric precipitation step. The whey-like proteins found in this effluent are difficult to recover because of low solids concentration varying from 1% to 3% (Meena and Usha 1990), which makes their isolation not economically viable.

A process using electrodialysis with bipolar membranes was proposed to produce soy protein isolates (Bazinet et al. 1996). The global approach is similar to the one aforementioned that is to say it can be divided in the same main steps namely proteins extraction, proteins isoelectric precipitation, washing of the precipitate, neutralization to pH 7 and spray-drying. However, in this case protein isoelectric precipitation is achieved using electrodialysis with bipolar membranes with a bipolar-cationic configuration (see Fig. 12.4a). More specifically the soy protein extract is fed into the acid compartments receiving the H^+ generated by the bipolar membranes, which come into contact with the proteins stream, bringing them to their isoelectric point and resulting in their precipitation (Bazinet et al. 1998a). At the same time, cations present in the proteins stream, namely the K^+ naturally present in soy and the Na^+ resulting from the addition of NaOH during the extraction step, permeate through the cationic membranes and form the corresponding bases with the OH^- generated by the bipolar membranes. These bases can be used for the extraction step or for the neutralization step prior to the spray-drying. Centrifugation can then be used, as in the conventional process for separation of the precipitated proteins, which are then washed, neutralized, and spray-dried as is the case for the conventional process. Process variables investigated by Bazinet et al. (1997a) for the

BP-C configuration were: current density of 25 mA/cm^2, soy protein concentration ranging between 15 and 60 g/L, and temperature ranging between 10°C and 35°C. As pointed out by Bazinet et al. (1997a, b) and by Pourcelly and Bazinet (2009), the soy protein concentration and the temperature are the two primary factors influencing the energy required to generate the H^+ and OH^- and for the ions transport through the cationic membranes. By expressing the energy required for the production of 1 kg of protein isolate (excluding the pumping energy), a drop in required energy is observed with an increase in soy protein concentration of the acidified stream (0.693 kWh/kg for a soy protein concentration of 15 g/L with 0.06 M KCl versus 0.294 kWh/kg for a soy protein concentration of 30 g/l with 0.06 m KCl versus 0.129 kWh/kg for a soy protein concentration of 60 g/l with 0.06 M KCl—temperature of 20°C—adapted from Bazinet et al. 1997a). In parallel, a drop in energy consumed occurs as the temperature of the soy extract is increased (0.728 kWh/kg for a temperature of 10°C versus 0.455 kWh/kg for a temperature of 20°C versus 0.371 kWh/kg for a temperature of 35°C) for a acidified stream of soy protein with a concentration of 30 g/l and 0.06 M KCl (Bazinet et al. 1997b).

From an environmental point of view, the approach using electrodialysis with bipolar membranes for isoelectric precipitation of proteins has significant advantages, when compared with the traditional approach. Because it does not use any added acids during the process to adjust the pH and because the base used during the extraction step can be regenerated, the volume of effluent generated by the protein precipitation step is less than for the traditional approach. In addition, because the soy protein extract is partially demineralized during the protein precipitation step, salt content of the effluent is less than for the traditional process. In addition, soy protein isolates produced by this process have a lower salt content than soy protein isolates produced by the traditional isoelectric process. However, both precipitation processes result in supernatants with similar soluble protein content (whey-like proteins: 1.5–1.7%) (Bazinet et al. 1998b).

The main limitation to the replacement of the traditional isoelectric precipitation step by the isoelectric precipitation using electrodialysis with bipolar membranes is because of fouling of the electrodialysis cell by precipitated proteins. This results in an increase of cell resistance (decrease of the system efficiency), a loss of protein (yield decrease), and it complicates soy protein extract passage in the cell. An online centrifugation step at the exit of the electrodialysis cell has been proposed by Bazinet et al. (1998a) to allow the recovery of precipitated proteins and as a result minimize the aforementioned limitations. Pourcelly and Bazinet (2009) suggested that there would also be a need to optimize the hydrodynamic design of the electrodialysis cell in order to minimize fouling.

More recently, Mondor et al. (2004a, b), Alibhai et al. (2006), Skorepova and Moresoli (2007) and Ali et al. (2010) have combined electrodialysis using bipolar membranes with ultrafiltration–diafiltration for production of soy protein isolates with low phytic acid content. When compared with the traditional approach, the protein extraction step remains the same, the isoelectric precipitation step and the washing step are replaced by a combination of electrodialysis with bipolar membranes (to adjust the pH of the soy extract to 6–7, i.e., the pH at which phytic acid is mainly found in its free form) and ultrafiltration–diafiltration (to concentrate and remove some minerals and sugars); the neutralization to pH 7 (if required) and spray-drying steps are also not modified. From an environmental point of view, the main advantage is that the volume of water required in the process is reduced because there is no need to resuspend the proteins in a solvent (dilute NaOH) before the spray drying, as is the case with the two previously discussed approaches, because there is no proteins precipitation involved in the current approach. Also the amount of proteins in the permeate is in the order of only 0.016% (adapted from Skorepova and Moresoli 2007—assuming the permeate density is equals to the one of water at 22°C), when a 100 kDa ultrafiltration membrane is used to process to VCR 4.5 a 1.8% total solid soy extract (which is comparable to the feed generated using an extraction ratio of 1:15 with an extraction yield of 30% based on the flakes weight) for which the pH was adjusted to 6 with electrodialysis with bipolar membranes (Skorepova and Moresoli 2007). This is significantly lower than the values aforementioned for the supernatant generated during the precipitation step for the conventional and for the electrodialysis with bipolar membranes, and not surprising, because it is well known that most of the whey-like proteins are retained by the ultrafiltration membrane (Kumar et al. 2003; Ali et al. 2010). As for the isoelectric precipitation using electrodialysis with bipolar membranes, there are no added acids during the pH adjustment step, which is an advantage compared with the traditional approach. Another advantage, when compared with the isoelectric precipitation approach using electrodialysis with bipolar membranes, is the minimization of cell fouling, because for the pH range of 6–7 most of the soy proteins remain soluble. In terms of product functionality, the combination of electrodialysis with bipolar membranes to adjust the pH of the extract to 6 with ultrafiltration (VCR 5)-diafiltration (VD 4) resulted in a soy protein isolate with improved solubility characteristics by as high as 25%, when compared with soy protein isolate produced by the traditional approach. Improvement in proteins solubility was most important between pH 2 and 4.5 (Ali et al. 2010). This improvement in solubility was attributed to a lower phytic acid/protein ratio as well as to a higher whey-like protein content observed for the isolates produced by the combined approach than for the one produced by the traditional process. The main limitation of the combined process is the declining permeate flux observed during ultrafiltration.

In addition to the production of soy protein isolates, it was also reported that electrodialysis with bipolar membranes would also allow the fractionation of the major soy protein fraction, i.e., 7S and 11S (Pourcelly and Bazinet 2009) (when performed at 10°C), as well as the recovery of protein and coagulant from soy tofu whey (Bazinet et al. 1999a, b). Again the approach would consist of the isoelectric precipitation of the proteins. The reader interested in more detailed information about both applications is referred to the work of Pourcelly and Bazinet (2009).

12.4.6 Dairy proteins

A large number of existing or potential applications of electrodialysis have been identified in the dairy industry. One application is the demineralization of whey, which was discussed in details in a previous section. Demineralization of whey is the largest area of use of conventional electrodialysis in the food industry.

A potential application is the production of acid caseinate by electrodialysis with bipolar membranes (Bazinet et al. 1999a, b, 2000a, b). The approach is basically the same than for the production of plants protein isolates, which was discussed in a previous section. It consists of circulating skim milk in the acid compartments receiving the H^+ generated by the bipolar membranes (see Fig. 12.4a), in order to precipitate the casein proteins by bringing them to their isoelectric point. Again the advantage from an environmental point of view is that the volume of effluent generated by the proteins precipitation step is less than for the traditional approach, which generates large volume of chemical effluents because of the addition of bases and acids during treatments.

Using electrodialysis with bipolar membranes to gradually decrease the pH of a whey solution reconstituted from a commercial whey protein isolate (WPI, BiPRO), Bazinet et al. (2004a) reported some interesting results in terms of whey proteins fractionation, when compared with chemical acidification. When a 10% whey protein solution was used, a β-lactoglobulin-enriched fraction (β-lg) was produced with a total protein purity of 98% and a content of 97.3% of β-lg and 2.7% of α-lactalbumin (α-la). Even if the chemical acidification approach resulted in a product with similar β-lg and α-la content than the approach with electrodialysis with bipolar membranes (97.7% and 2.1%, respectively), the total protein purity was lower for the chemical acidification process with 94.3% of total protein and also it did not allow precipitation of a large amount of whey protein, whereas 53.4% of β-lg was precipitated by the electrodialysis technique. At concentration higher than 10%, the conductivity of the whey protein solution became the limiting factor. In order to solve this problem, Bazinet et al. (2004b) electroacidified a 20% whey protein solution with conductivity control (i.e., when pH 5.0 or a conductivity of 350 μS/cm was reached, conductivity was maintained constant at 350 μS/cm by the addition of 1.0 M KCl). By this approach, a 27% protein precipitation was obtained at pH 4.6, which was higher than the value of 21% reported without conductivity control but still significantly lower than the value of 53.4% observed with the 10% whey protein solution.

As for the production of plant protein isolate, the main limitation to the production of acid caseinate and whey protein fractions using electrodialysis with bipolar membranes appears to be fouling of the electrodialysis cell by precipitated proteins. The reader interested in having more details about both aforementioned applications of electrodialysis with bipolar membranes is referred to the review of Pourcelly and Bazinet (2009), on the subject.

12.4.7 Regeneration of wastewater

Another application of electrodialysis with bipolar membranes that is of interest for environmental protection is the regeneration of acids and bases from the corresponding salts solutions. Because a considerable part of industrial wastewater is salt solutions that are formed either directly during processing or because of neutralization of acid/base effluents, regeneration of acids and bases from salts solutions can help environmental protection by decreasing the amount of salts rejected into rivers (Koter and Warszawski 2000).

A typical electrodialysis configuration used for the regeneration of acids and bases from salts solutions is the three-compartment cell illustrated in Fig. 12.3. In this configuration, when an electrical potential is applied across the electrodialysis stack, the cations (X^+) move in the direction of the cathode permeating through the cationic membranes and combine with the OH^- ions, generated by the bipolar membranes, to form the corresponding base (i.e., XOH). At the same time, the anions (Y^-) move in the direction of the anode permeating through the anionic membranes and combine with the H^+ ions, generated by the bipolar membranes, to form the corresponding acid (i.e., HY). In most cases, the feed stream is a mixture of salts, which means that a mixture of the corresponding acids and bases will be produced in their respective stream. However, the products composition will depend of the transport properties of ions across the monopolar membranes (Huang and Xu 2006). In the food industry, a significant part of these applications seem to be directed to the regeneration of acids and bases that have been used in a chemical process either for neutralization or regeneration of ion-exchange resins (Davis and Laterra 1987).

Typical operating conditions for a three-compartment cell were reported by Mani (1991): a current density of 50–150 mA/cm^2, which corresponds to a unit cell voltage of 1.6–2.5 V, temperature up to 50°C, and a minimum conductivity of the feed solution in the order of 35,000 μS/cm. However, the specific values for the aforementioned operating conditions will vary with the type of cell and membranes used, as well as with the nature of the different streams. The salts in the feed solution must also be in their soluble form and the presence of multivalent ions impurities such as Ca^{2+}, Mg^{2+}, Fe^{3+}, etc. must be kept below 2 ppm in order to avoid the formation of insoluble impurities inside the cell stack and to avoid membranes poisoning. Using the Aquatech cell and applying the aforementioned guidelines, Mani (1991) reported that the overall process has a current efficiency of at least 80% at a current density of 100 mA/cm^2. At this performance level the cell could generate a metric ton of sodium hydroxide from the corresponding salt solution using 1,400–2,000 kWh, which represents a fraction of the costs encountered commercially for the production of sodium hydroxide; that is to say between 2,800–3,500 kWh/ton (Nagasubramanian et al. 1977). In general, acid concentration produced by this technology will reach 1–2 N for strong acids and 3–6 N for weak acids, whereas base product concentration will reach a value between 3 and 6 N.

As pointed out by Mani (1991) and by Pourcelly and Bazinet (2009), for the regeneration of weak acids (i.e., organic acids), a two-compartment cell combining bipolar membranes and cationic membranes (see Fig. 12.4a) could be more suitable from and energy point of view than the aforementioned three-compartment cell. The reason is that the pure acid, which is formed by the combination of the anions (Y^-) with the H^+ ions generated by the bipolar membranes, being weakly dissociated has low conductivity, and as a result a high resistance is observed in the acidic compartment throughout all the treatment if a three-compartment cell is used. By working with a BP-C two-compartment cell, this problem can be minimized by circulating the salt solution in the acidic compartments between the bipolar and the cationic membranes. As previously explained, a relatively pure base solution (i.e., XOH) will be formed and retained in the basic compartments, whereas a mixture of salt (X^+Y^-) and weak acid (HY) will be formed and retained in the acidic compartments. When operated in a batch mode, as the electrodialysis progress, salt (X^+Y^-) concentration is decreasing while acid (HY) is being formed. Even if the resistance of mixed acid/salt solution found in the acidic compartment is increasing as a function of the electrodialysis progression, it is lower than for the corresponding acid solution generated in the acidic compartment for a three-compartment cell and thus the two-compartment cell process is more efficient from an energy point of view. The process using a two-compartment cell is particularly advantageous when the salts stream is a concentrated solution (i.e., 2–5 N). When the conversion of the salt stream reaches 95%, the electrodialysis is stopped and the residual X^+ ions can be removed using cation-exchange resins, if required. A similar approach with a two-compartment cell combining bipolar membranes and anionic membranes (see Fig. 12.4b) could be advantageously used to regenerate weak bases from the corresponding salts solution.

12.5 Application of ultrafiltration–electrodialysis integrated process for the separation of peptides

As presented in previous sections, conventional electrodialysis with ion-exchange membranes is a technique with a multitude of potential applications in the food industry. However, this versatile technique also shows some serious limitations that restrict its application (Van der Bruggen et al. 2004). An inherent limit of the technology is related to the size of the ionic species (i.e., that in order to migrate through the ion-exchange membranes, the ionic species must have a radius or a molecular weight that does not exceed a certain limit allowed by the membrane porosity). In the contrary case, the ion-exchange membranes do not allow their passage because of steric rejection (Aider et al. 2008a, b, c).

The idea to replace one or more ion-exchange membranes with a porous membrane acting as a molecular barrier proved to be profitable, because it was possible to separate charged molecules of molecular weight higher than allowed by the traditional ion-exchange membranes (Ruckenstein and Zeng 1998; Galier and Roux-de Balmann 2004). Depending on the molecular weight of the solutes, nanofiltration, ultrafiltration, or microfiltration can be used as porous membrane. By exploiting the electrokinetic properties of the biological molecules, and by choosing optimum conditions of the medium such as pH, ionic strength, duration of treatment, electric field value, and appropriate electrodialysis cell configuration, it was possible to carry out separation that appeared impossible with an electrodialysis with conventional ion-exchange membranes (Causserand et al. 1994).

Recently, the use of a conventional electrodialysis cell, in which some ion-exchange membranes were replaced by ultrafiltration membranes (UF-ED integrated process) was reported for the isolation of polyphenols from tobacco extract (Bazinet et al. 2005) and from green tea infusion (Labbé et al. 2005), for the separation of chitosan oligomers (Aider et al. 2008b, 2008c, 2009), as well as for the fractionation of bioactive peptides (Poulin et al. 2006, 2007, 2008; Firdaous et al. 2009). In the present section, only the application of this technique for the separation of peptides will be discussed; however, the basic principle is the same for the isolation of other charged species with a molecular weight that would not allowed them to permeate through conventional ion-exchange membranes.

Three situations are to be considered for the separation of peptides using electrodialysis with ultrafiltration membrane (1) the situation in which cationic peptides are to be separated from a large number of hydrolyzed protein fractions, (2) the situation in which anionic peptides are to be separated from hydrolyzed protein fractions, and (3) the situation in which both cationic and anionic peptides are to be separated from neutral molecules. For the separation of cationic peptides (situation 1), an electrodialysis cell consisting of an anionic membrane located at one extremity of the stack to separate the electrode rinsing solution from the hydrolyzed protein solution, one UF membrane to separate the hydrolyzed protein solution from the KCl receiving solution, and one cationic membrane located at the other extremity of the stack to separate the KCl solution from the electrodes rinsing solution can be used. The sequence consisting of a hydrolyzed protein compartment-UF membrane-KCl product compartment-cationic membrane can be repeated in order to increase the membrane surface area available for the separation (Poulin et al. 2007). Under the influence of the electrical field, the cationic peptides are transported through the UF membranes into the product compartments while the K^+ and Cl^- migrate out of the product compartments. In the second case, where the anionic peptides are to be separated from the hydrolyzed protein fractions, one just needs to change the position of the hydrolyzed protein solution with that of the KCl solution. Finally, if both cationic and anionic peptides are to be separated from neutral molecules (situation 3), two UF membranes are placed between the ion-exchange membranes instead of one as is the case with the first two aforementioned situations. The feed solution containing the peptides is fed into the compartment between the two UF membranes, and KCl solutions are fed in each of the compartments receiving the peptides.

Aiming for the separation of cationic peptides from a 0.25% (w/v) β-lg tryptic hydrolysate aqueous solution for which the pH was adjusted to 5, Poulin et al. (2008) studied the impact of the feed solution flow rate on the peptides fractionation. They reported that irrespectively of the flow rate value, only 7 of 23 peptides identified in the raw hydrolysate had migrated in the KCl solution. Although the same peptides were found regardless of the feed flow rate, for two of these peptides with limited electrophoretic mobility, their relative migration was influenced by this parameter with an increase in the feed flow rate resulting in a decrease in their migration. The relative migration was not influenced by the feed flow rate for the other five cationic peptides. The relative migration (in percent; relative to peptide in the feed solution) in the KCl solution of the seven cationic peptides of interest varied between 1.18%

and 19.98% after 180 min of operation with an electrical field of 2.75 V/cm. No significant fouling of the UF membrane was reported after 12 electroseparation runs. The same authors (Poulin et al. 2007) demonstrated that the impact of the electrical field on the relative migration in the KCl solution of the seven cationic peptides is more significant than the ones of the feed flow rate, because by operating under an electrical field of 11.0 V/cm they were able to achieve a relative migration varying between 2.6% and 39.8% depending on the cationic peptides of interest.

Using the aforementioned electrodialysis configuration with two UF membranes placed between the ion-exchange membranes (situation 3), Firdaous et al. (2009) investigated the fractionation of an alfalfa white protein concentrate hydrolysate by electrodialysis with ultrafiltration membranes. The hydrolysate solution was fed between the two UF membranes, whereas KCl solutions (KCl 1 and KCl 2) were circulated in the product compartments to recuperate anionic and cationic peptides, respectively. The effect of pH of the hydrolysate solution (three and nine) on the migration of peptides was studied in these conditions. Among a total of 70 peptides contained in the hydrolysate, only 8 peptides were found in the separated adjacent recovery solutions (KCl 1 and KCl 2). Among these eight migrating peptides, four significant anionic peptides were recovered in the KCl 1 solution and one antihypertensive peptide identified as VW was recovered in KCl 2. The maximum transmission rate recorded for a peptide was 18.23%.

When compared with traditional techniques, the integrated UF-ED approach would have better productivity and lower cost than the chromatographic/ion-exchange resin techniques, it would have better selectivity than the conventional pressure-driven processes (ultrafiltration, nanofiltration), and it would have a lower tendency to foul (Lapointe et al. 2005; Firdaous et al. 2009). However, to our knowledge, no direct comparison was made between the UF-ED approach and the traditional chromatographic/ion-exchange resin techniques in terms of selectivity. Also, one may assume that there will be transport of water through the UF semipermeable membranes separating the hydrolysate solution and the receiving solution(s), from the solution that is dilute in solute to the solution that is concentrated, a phenomenon known as osmosis. Consequently, we may expect that the concentration of peptides in the receiving solutions will be limited by the gradual transfer of water from the hydrolysate solution, which could represent a limitation for the application of the UF-ED integrated approach at the industrial scale unless this phenomenon could be controlled. In terms of the environmental impact, the integrated UF-ED approach can help environmental protection when compared with ion-exchange resin techniques that result in the generation of large volumes of effluents because of the need to neutralize or regenerate the ion-exchange resins.

12.6 Concluding remarks and perspectives

Since the first industrial application of conventional electrodialysis in the food industry in 1960 (demineralization of cheese whey), a number of potential applications of conventional electrodialysis in the food sector has been developed. Also the industry has seen the introduction of electrodialysis with bipolar membranes, which uses electricity to generate H^+ and OH^- from water dissociation at the bipolar membranes interface. In addition to a few potentially valuable applications, this process is now used on an industrial scale for organic acids production. However, applications of both conventional and bipolar membranes electrodialysis are still in their infancy because quite a limited number of the novel applications studied so far in the laboratory and pilot plant have seen application at the industrial scale. This is certainly because of the high cost of membranes and equipment, as well as to the short monopolar membrane lifetime, which can be no longer than 1 year for applications with high fouling potential (Fidaleo and Moresi 2006).

However, as illustrated in previous sections, some of the applications of both conventional and bipolar membranes electrodialysis have a great potential to reduce pollution problems and thus help environmental protection. With the strengthening of environmental policies this should bring further interest toward electrodialysis applications and help to minimize the membranes and equipment costs, especially with the membrane market that is now open to new Far East manufacturers. In addition, the lifetime of monopolar membranes is continuously increasing because of research in polymer science (Pourcelly and Bazinet 2009). Also, the integration of electrodialysis with pressure-driven processes and with biochemical units, such as fermentation, seems to be a practical strategy for sustainable growth because such integrations can achieve not only a synergetic effect for a specific separation but also an optimization on utilization of materials, energy, and space (Xu and Huang 2008). Consequently, we expect the aforementioned factors to foster electrodialysis applications in the food sector for the coming years.

References

Adhikary, S.K., W.P. Harkare, K.P. Govindan, and A.M. Nanjundaswamy. 1983. Deacidification of fruit juices by electrodialysis. *Indian Journal of Technology* 21: 120–123.

Aider, M., D. de Halleux, and L. Bazinet. 2008a. Potential of continuous electrophoresis without and with porous membranes (CEPM) in the bio-food industry: Review. *Journal of Membrane Science* 19: 351–362.

Aider, M., S. Brunet, and L. Bazinet. 2008b. Electroseparation of chitosan oligomers by electrodialysis with ultrafiltration membrane (EDUF) and impact on electrodialytic parameters. *Journal of Membrane Science* 309: 222–232.

Aider, M., S. Brunet, and L. Bazinet. 2008c. Effect of pH and cell configuration on the selective and specific electrodialytic separation of chitosan oligomers. *Separation and Purification Technology* 63: 612–619.

Aider, M., S. Brunet, and L. Bazinet. 2009. Effect of solution flow velocity and electric field strength on chitosan oligomer electromigration kinetics and their separation in an electrodialysis with ultrafiltration membrane (EDUF) system. *Separation and Purification Technology* 69: 63–70.

Ali, F., D. Ippersiel, F. Lamarche, and M. Mondor. 2010. Characterization of low-phytate soy protein isolates produced by membrane technologies. *Innovative Food Science and Emerging Technologies* 11: 162–168.

Alibhai, Z., M. Mondor, C. Moresoli, D. Ippersiel, and F. Lamarche. 2006. Production of soy protein concentrates/isolates: Traditional and membrane technologies. *Desalination* 191: 351–358.

Ayala-Bribiesca, E., M. Araya-Farias, G. Pourcelly, and L. Bazinet. 2006. Effect of concentrate solution pH and mineral composition of a whey protein diluate solution on membrane fouling during conventional electrodialysis. *Journal of Membrane Science* 280: 790–801.

Bailly, M. 2002. Production of organic acids by bipolar electrodialysis: Realizations and perspectives. *Desalination* 144: 157–162.

Bazinet, L., M. Boulet, D. Ippersiel, R. Labrecque, F. Lamarche, and R. Toupin. 1996. Systematic study on the preparation of a food grade soyabean protein. *Report for the Canadian Electricity Association n 9326 U 987*, Research and Development, Montreal.

Bazinet, L., F. Lamarche, R. Labrecque, and D. Ippersiel. 1997a. Effect of KCl and soy protein concentrations on the performance of bipolar membrane electro-acidification. *Journal of Agricultural and Food Chemistry* 45: 2419–2425.

Bazinet, L., F. Lamarche, R. Labrecque, and D. Ippersiel. 1997b. Effect of number of bipolar membranes and temperature on the performance of bipolar membrane electroacidification. *Journal of Agricultural and Food Chemistry* 45: 3788–3794.

Bazinet, L., F. Lamarche, and D. Ippersiel. 1998a. Bipolar-membrane electrodialysis: Applications of electrodialysis in the food industry. *Trends in Food Science and Technology* 9: 107–113.

Bazinet, L., F. Lamarche, and D. Ippersiel. 1998b. Comparison of chemical and bipolar membrane electrochemical acidification for precipitation of soybean proteins. *Journal of Agricultural and Food Chemistry* 46: 2013–2019.

Bazinet, L., F. Lamarche, D. Ippersiel, and J. Amiot. 1999a. Bipolar membrane electro-acidification to produce bovine milk casein isolate. *Journal of Agricultural and Food Chemistry* 47: 5291–5296.

Bazinet, L., D. Ippersiel, and F. Lamarche. 1999b. Recovery of magnesium and protein from soy tofu whey by electrodialytic configurations. *Journal of Chemical Technology and Biotechnology* 74: 663–668.

Bazinet, L., D. Ippersiel, C. Gendron, J. Beaudry, M. Mahdavi, J. Amiot, and F. Lamarche. 2000a. Cationic balance in skim milk during bipolar membrane electroacidification. *Journal of Membrane Science* 173: 201–209.

Bazinet, L., D. Ippersiel, R. Labrecque, and F. Lamarche. 2000b. Effect of temperature on the separation of soybean 11S and 7S protein fractions during bipolar membrane electroacidification. *Biotechnology Progress* 16: 292–295.

Bazinet, L., D. Ippersiel, and B. Mahdavi. 2004a. Fractionation of whey protein by bipolar membrane electroacidification. *Innovative Food Science and Emerging Technologies* 5: 17–25.

Bazinet, L., D. Ippersiel, and B. Mahdavi. 2004b. Effect of conductivity control on the separation of whey protein by bipolar membrane electroacidification. *Journal of Agricultural and Food Chemistry* 52: 1980–1984.

Bazinet, L., Y. DeGrandpré, and A. Porter. 2005. Electromigration of tobacco polyphenols. *Separation and Purification Technology* 41: 101–107.

Benafi-Bako, K., N. Nemestothy, and L. Gubicza. 2004. A study on applications of membrane techniques in bioconversion of fumaric acid to L-malic acid. *Desalination* 162: 301–306.

Boergardts, P., W. Krischke, and W. Troesch. 1998. Process for purifying dairy wastewater. US Patent 5,746, 920.

Boyaval, P., C. Corre, and S. Terre. 1987. Continuous lactic acid fermentation with concentrated product recovery by ultrafiltration and electrodialysis. *Biotechnology Letters* 9: 801–806.

Boyaval, P., J. Seta, and C. Gavach. 1993. Concentrated propionic acid production by electrodialysis. *Enzyme and Microbial Technology* 15: 683–686.

Brooks, J.R., and C.V. Morr. 1985. Effect of phytate removal treatments upon the molecular weight and subunit composition of major soy protein fractions. *Journal of Agricultural and Food Chemistry* 33: 1128–1132.

Casademont, C., P. Sistat, B. Ruiz, G. Pourcelly, and L. Bazinet. 2009. Electrodialysis of model salt solution containing whey proteins: Enhancement by pulsed electric field and modified cell configuration. *Journal of Membrane Science* 328: 238–245.

Causserand, C., J.-P. Lafaille, and P. Aimar. 1994. Transmission of bio-molecules through porous membranes triggered by an external electric field. *Journal of Controlled Release* 29: 113–123.

Chukwu, U.N., and M. Cheryan. 1999. Electrodialysis of acetate fermentation broths. *Applied Biochemistry and Biotechnology* 77–79: 485–499.

Datta, R., and E.P. Bergemamm. 1996. Process for producing of citric acid and monovalent citrate salts. US Patent 5,532,148.

Davis, T.A., and A. Laterra. 1987. Onsite generation of acid and base with bipolar membranes. In *Paper presented at the 48th Annual Meeting of the International Water Conference*, Pittsburgh, PA.

Davis, T.A., J.D. Genders, and D. Pletcher. 1997. *A first course in ion permeable membranes. The electrochemical consultancy*. UK: Alresford Press LTD.

de Boer, R., and T. Robbertsen. 1983. Electrodialysis and ion-exchange processes: The case of milk whey. In *Progress in food engineering*, ed. C. Cantarelli and C. Peri. Switzerland: Foster-Verlag AG.

Elmidaoui, A., F. Lutin, L. Chay, M. Taky, M. Tahaikt, and R. Alaoui Hafidi. 2002. Removal of melassigenic ions for beet sugar syrups by electrodialysis using a new anion-exchange membrane. *Desalination* 148: 143–148.

Elmidaoui, A., L. Chay, M. Tahaikt, M.A. Menkouchi Sahli, M. Taky, F. Tiyal, A. Khalidi, and R. Alaoui Afidi. 2006. Demineralisation for beet sugar solutions using an electrodialysis pilot plant to reduce melassigenic ions. *Desalination* 189: 209–214.

Enzminger, J.D., and J.A. Asenjo. 1986. Use of cell recycle in the aerobic fermentative production of citric acid by yeast. *Biotechnology* 8: 7–12.

Escudier, J.-L., B. Saint Pierre, J.L. Batlle, and M. Moutounet. 1993. Procédé et dispositif automatique de stabilisation tartrique de vins. FR Patent 2709308-A1.

Escudier, J.-L., M. Moutounet, B. Saint-Pierre, and J.-L. Batlle. 1997. Stabilisation tartrique des vins par membranes: résultats et développements technologiques. In *Proceedings of 11ème Colloque Viticole et Œnologique*, Montpellier.

Fidaleo, M., and M. Moresi. 2006. Electrodialysis applications in the food industry. In *Advances in food and nutrition research*, vol. 51, ed. Steve Taylor, 265–360. San Diego: Academic.

Firdaous, L., P. Dhulster, J. Amiot, A. Gaudreau, D. Lecouturier, R. Kapel, F. Lutin, L.-P. Vézina, and L. Bazinet. 2009. Concentration and selective separation of bioactive peptides from an alfalfa white protein hydrolysate by electrodialysis with ultrafiltration membranes. *Journal of Membrane Science* 329: 60–67.

Galier, S., and H. Roux-de Balmann. 2004. Study of biomolecules separation in an electrophoretic membrane contactor. *Journal of Membrane Science* 241: 79–87.

Gonçalves, F., C. Fernandes, P.C. dos Santos, and M.N. de Pinho. 2003. Wine tartaric stabilization by electrodialysis and its assessment by the saturation temperature. *Journal of Food Engineering* 59: 229–235.

Greiter, M., S. Novalin, M. Wendland, K.-D. Kulbe, and J. Fischer. 2002. Desalination of whey by electrodialysis and ion-exchange resins: Analysis of both processes with regard to sustainability by calculating their cumulative energy demand. *Journal of Membrane Science* 210: 91–102.

Grib, H., D. Belhocine, H. Lounici, A. Pauss, and N. Mameri. 2000. Desalting of phenylalanine solutions by electrodialysis with ion-exchange membranes. *Journal of Applied Electrochemistry* 30: 259–262.

Hongo, M., Y. Nomura, and M. Iwahara. 1986. Novel methods of lactic acid production by electrodialysis fermentation. *Applied and Environmental Microbiology* 52: 314–319.

Huang, C., and T. Xu. 2006. Electrodialysis with bipolar membranes for sustainable development. *Environmental Science and Technology* 40: 5233–5243.

Ishizaki, A., Y. Nomura, and M. Iwahara. 1990. Built-in electrodialysis batch culture, a new approach to release of end product inhibition. *Journal of Fermentation and Bioengineering* 70: 108–113.

Kang, Y.J., and K.C. Rhee. 2002. Deacidification of mandarin orange juice by electrodialysis combined with ultrafiltration. *Nutraceuticals & Food* 7: 411–416.

Kikuchi, K., T. Gotoh, H. Takashashi, S. Higashino, and J.S. Dranoff. 1995. Separation of amino acids by electrodialysis with ion-exchange membranes. *Journal of Chemical Engineering of Japan* 28: 103–109.

Kim, Y.H., and S.-H. Moon. 2001. Lactic acid recovery from fermentation broth using one-stage electrodialysis. *Journal of Chemical Technology and Biotechnology* 76: 169–178.

Koter, S., and A. Warszawski. 2000. Electromembrane processes in environment protection. *Polish Journal of Environmental Studies* 9: 45–56.

Kumar, N.S.K., M.K. Yea, and M. Cheryan. 2003. Soy protein concentrates by ultrafiltration. *Journal of Food Science* 68: 2278–2283.

Labbé, D., M. Araya-Farias, A. Tremblay, and L. Bazinet. 2005. Electromigration feasibility of green tea catechins. *Journal of Membrane Science* 254: 101–109.

Lapointe, J.-F., S.F. Gauthier, Y. Pouliot, and C. Bouchard. 2005. Fouling of a nanofiltration membrane by a -lactoglobulin tryptic hydrolysate: Impact on the membrane sieving and electrostatic properties. *Journal of Membrane Science* 253: 89–102.

Lee, E.G., S.-H. Moon, Y.-K. Chang, I.-K. Yoo, and H.N. Chang. 1998. Lactic acid recovery using two-stage electrodialysis and its modelling. *Journal of Membrane Science* 145: 53–66.

Lee, H.-J., S.-J. Oh, and S.-H. Moon. 2003. Recovery of ammonium sulphate from fermentation waste by electrodialysis. *Water Research* 37: 1091–1099.

Ling, L.-P., H.-F. Leow, and M.R. Sarmidi. 2002. Citric acid concentration by electrodialysis: Ion and water transport modelling. *Journal of Membrane Science* 199: 59–67.

Lopez Leiva, M.H. 1988. The use of electrodialysis in food processing. Part 2: Review of practical applications. *Lebensmittel- Wissenschaft und-Technologie* 21: 177–182.

Lusas, E.W., and K.C. Rhee. 1995. Soy protein processing and utilization. In *Practical handbook of soybean processing and utilization*, ed. D.R. Erickson. St. Louis, MO: AOCS Press, Champaign, IL; United Soybean Board.

Lutin, F., M. Bailly, and D. Bar. 2002. Process improvements with innovative technologies in the starch and sugar industries. *Desalination* 148: 121–124.

Madzingaidzo, L., H. Danner, and R. Braun. 2002. Process development and optimisation of lactic acid purification using electrodialysis. *Journal of Biotechnology* 96: 223–239.

Mani, K.N. 1991. Electrodialysis water splitting technology. *Journal of Membrane Science* 58: 139–146.

Mani, K.N. 2000. A process for simultaneous production of amino acid hydrochloride and caustic via electrodialytic water splitting. Eur Pat Appl EP 1,016,651.

Maujean, A. 1994. Traitement par le froid artificiel des vins en relation avec leur stabilisation vis-à-vis des troubles cristallins tartriques. In *Les acquisitions récents dans les traitements physiques du vin*, ed. B. Doneche. Paris: Lavoisier Tec & Doc.

Meena, G., and B. Usha. 1990. Soy protein isolate: Production technology, functional properties, nutritive value and applications—a review. *Indian Food Pack Nov-Dec* 90: 5–22.

Mondor, M., D. Ippersiel, F. Lamarche, and J.I. Boye. 2004a. Production of soy protein concentrates using a combination of electroacidification and ultrafiltration. *Journal of Agricultural and Food Chemistry* 52: 6991–6996.

Mondor, M., D. Ippersiel, F. Lamarche, and J.I. Boye. 2004b. Effect of electro-acidification treatment and ionic environment on soy protein extract particle size distribution and ultrafiltration permeate flux. *Journal of Membrane Science* 231: 169–179.

Moresi, M., and F. Sappino. 1998. Effect of some operating variables on citrate recovery from model solutions by electrodialysis. *Biotechnology and Bioengineering* 59: 344–350.

Moresi, M., and F. Sappino. 2000. Electrodialytic recovery of some fermentation products from model solutions: Techno-economic feasibility study. *Journal of Membrane Science* 164: 129–140.

Mounts, T.L., W.J. Wolf, and W.H. Martinez. 1987. Processing and utilization. In *Soybean: Improvement, production and uses*, ed. J.R. Wilcox. WI: American Society of Agronomy, Madison.

Mourgues, J. 1993. Utilisation des résines échangeuses d'ions. *Rev des Oenologues* 19: 51–54.

Moutounet, M., J.-L. Escudier, and B. Saint-Pierre. 1994. L'électrodialyse, adaptation à la stabilisation tartrique des vins. In *Les acquisitions récents dans les traitements physiques du vin*, ed. B. Doneche. Paris: Lavoisier Tec & Doc.

Nagasubramanian, K., F.P. Chandla, and K.J. Liu. 1977. Use of bipolar membranes for generation of acid and base: An engineering and economic analysis. *Journal of Membrane Science* 2: 109–124.
Nomura, Y., M. Iwahara, and M. Hongo. 1987. Lactic acid production by electrodialysis fermentation using immobilized growing cells. *Biotechnology and Bioengineering* 30: 788–793.
Nomura, Y., M. Iwahara, and M. Hongo. 1988. Acetic acid production by an electrodialysis fermentation method with a computerized control system. *Applied and Environmental Microbiology* 54: 137–142.
Nomura, Y., K. Yamamoto, and A. Ishizaki. 1991. Factors affecting lactic acid production rate in the built-in electrodialysis fermentation an approach to high speed batch culture. *Journal of Fermentation and Bioengineering* 71: 450–452.
Nomura, Y., M. Iwahara, and M. Hongo. 1994. Production of acetic acid by *clostridium thermoaceticum* in electrodialysis culture using a fermenter equipped with an electrodilyzer. *World Journal of Microbiology and Biotechnology* 10: 427–432.
Novalic, S., F. Jagschitz, J. Okwor, and K.D. Kulbe. 1995. Behaviour of citric acid during electrodialysis. *Journal of Membrane Science* 108: 201–205.
Novalic, S., J. Okwor, and K.D. Kulbe. 1996. The characteristics of citric acid separation using electrodialysis with bipolar membranes. *Desalination* 105: 277–282.
Omosaiye, O., and M. Cheryan. 1979. Low-phytate, full-fat soy protein product by ultrafiltration of aqueous extracts of whole soybeans. *Cereal Chemistry* 56: 58–62.
Petruccelli, S., and M.C. Anon. 1994. Relationship between the method of obtention and structural and functional properties of soy protein isolates. 1. Structural and hydration properties. *Journal of Agricultural and Food Chemistry* 42: 2161–2169.
Poulin, J.-F., J. Amiot, and L. Bazinet. 2006. Simultaneous separation of acid and basic bioactive peptides by electrodialysis with ultrafiltration membrane. *Journal of Biotechnology* 123: 314–328.
Poulin, J.-F., J. Amiot, and L. Bazinet. 2007. Improved peptide fractionation by electrodialysis with ultrafiltration membrane: Influence of ultrafiltration membrane stacking and electrical field strength. *Journal of Membrane Science* 299: 83–90.
Poulin, J.-F., J. Amiot, and L. Bazinet. 2008. Impact of feed solution flow rate on peptide fractionation by electrodialysis with ultrafiltration membrane. *Journal of Agricultural and Food Chemistry* 56: 2007–2011.
Pourcelly, G., and L. Bazinet. 2009. Developments of bipolar membrane techmology in food and bio-industries. In *Handbook of membrane separations: Chemical, pharmaceutical, food, and biotechnological applications*, ed. A.K. Pabby, S.S.H. Rizvi, and A.M. Sastre. Boca Raton, FL: CRC Press Taylor and Francis Group.
Ruckenstein, E., and X. Zeng. 1998. Albumin separation with cibacron blue carrying macroporous chitosan and chitin affinity membranes. *Journal of Membrane Science* 142: 13–26.
Saint-Pierre, B., J. Batlle, J.-L. Escudier, and J.-L. Batlle. 1995. Obtention de la stabilité tartrique des vins au moyen d'un procédé d'électrodialyse contrôlé. In *Proceedings of 5ème Symposium International d'oenologie*, Bordeaux.
Sappino, F., M. Mancini, and M. Moresi. 1996. Recovery of sodium citrate from aqueous solutions by electrodialysis. *Italian Journal of Food Science* 8: 239–250.
Skorepova, J., and C. Moresoli. 2007. Carbohydrate and mineral removal during the production of low-phytate soy protein isolate by combined electroacidification and high shear tangential flow ultrafiltration. *Journal of Agricultural and Food Chemistry* 55: 5645–5652.
Sridhar, S. 1987. Method for recovery of L-malic acid. *Ger Offen DE* 3: 542–861.
Strathmann, H. 2004. *Ion-exchange membrane separation processes*. Amsterdam, The Netherlands: Elsevier.
Tako, M., and M. Brahim. 1993. Demineralization of molasses by electrodialysis. *International Sugar Journal* 95: 243–247.
Thampy, S.K., P.K. Narayanan, G.S. Trivedi, D.K. Gohil, and V.K. Indushekhar. 1999. Demineralization of sugar cane juice by electrodialysis. *International Sugar Journal* 101: 365–366.

Thate, S., G. Eigenberger, and H.-J. Rapp. 2000. Introduction. In *Handbook on bipolar membrane technology*, ed. A.J.B. Kemperman. Enschende, The Netherlands: Twente University Press.

USDA. 2009. Soy stats 2009, a reference guide to important soybean facts & figures. http//:www.Soystats.com.

Van der Bruggen, B., A. Koninckx, and C. Vandecasteele. 2004. Separation of monovalent and divalent ions from aqueous solution by electrodialysis and nanofiltration. *Water Research* 38: 1347–1353.

Vera, E., J. Ruales, M. Dornier, J. Sandeaux, F. Persin, G. Pourcelly, F. Vaillant, and M. Reynes. 2003a. Comparison of different methods for deacidification of clarified passion fruit juice. *Journal of Food Engineering* 59: 361–367.

Vera, E., J. Ruales, M. Dornier, J. Sandeaux, R. Sandeaux, and G. Pourcelly. 2003b. Deacidification of the clarified passion fruit juice using different configurations of electrodialysis. *Journal of Chemical Technology and Biotechnology* 78: 918–925.

Vera, E., J. Sandeaux, F. Persin, G. Pourcelly, M. Dornier, G. Piombo, and J. Ruales. 2007. Deacidification of clarified tropical fruit juices by electrodialysis. Part II. Characteristics of the deacidified juices. *Journal of Food Engineering* 78: 1439–1445.

Vera, E., J. Sandeaux, F. Persin, G. Pourcelly, M. Dornier, and J. Ruales. 2009. Deacidification of passion fruit juice by electrodialysis with bipolar membrane after different pretreatments. *Journal of Food Engineering* 90: 67–73.

Vernhet, A., K. Dupre, L. Boulange-Petermann, V. Cheynier, P. Pellerin, and M. Moutounet. 1999. Composition of tartrate precipitates deposited on stainless steel tanks during the cold stabilization of wines. Part I. White wines. *American Journal of Enology and Viticulture* 50: 391–397.

Voss, H. 1986. Deacidification of citric acid solutions by electrodialysis. *Journal of Membrane Science* 27: 165–171.

Wagner, J.R., D.A. Sorgentini, and M.C. Anon. 2000. Relation between solubility and surface hydrophobicity as an indicator of modifications during preparation processes of commercial and laboratory-prepared soy protein isolates. *Journal of Agricultural and Food Chemistry* 48: 3159–3165.

Weier, A.J., B.A. Glatz, and C.E. Glatz. 1992. Recovery of propionic and acetic acids from fermentation broths by electrodialysis. *Biotechnology Progress* 8: 479–485.

Xu, T., and C. Huang. 2008. Electrodialysis-based separation technologies: a critical review. *AIChE iournal* 54(12): 3147–3159.

Young, P. 1974. An introduction to electrodialysis. *Journal of Society of Dairy Technology* 27: 141–150.

Zhang, S.T., H. Matsuoka, and K. Toda. 1993. Production and recovery of propionic acid and acetic acids in electrodialysis culture of *propionibacterium shermanii*. *Journal of Fermentation and Bioengineering* 75: 276–282.

Chapter 13
Enzyme-assisted food processing

Benjamin K. Simpson, Xin Rui, and Jiang XiuJie

13.1 Introduction

Enzymes are biological catalysts and they function to speed up rates of biological reactions in living organisms. Living organisms manufacture the same functional classes of enzymes to enable them to carry out their normal metabolism. Because foodstuffs are derived from living organisms, they also contain naturally present enzymes that catalyze various reactions in foods. In addition to the action of endogenous enzymes, certain food processing operations also have enzymes deliberately added to foodstuffs as processing aids to transform foods into finished products.

The action of enzymes in foods may result in products with extended shelf life and improved textures and flavors, and enable a variety of food products to be fabricated from the harvested produce. The use of enzymes in food processing has been increasing over the past several decades because their action is based on greener chemistry compared with most traditional methods. They are relatively more selective and specific, thus they enable food products to be produced in forms with fewer co-products or contaminants on a more consistent basis. They are relatively nontoxic and thus are preferred to traditional food processing approaches based on chemicals. They are

B.K. Simpson (✉)
Food Science and Agricultural Chemistry Department, McGill University
(Macdonald Campus), Quebec, Canada H9X 3V9
e-mail: benjamin.simpson@mcgill.ca

X. Rui
Bioresource Engineering Department, McGill University
(Macdonald Campus), Quebec, Canada H9X 3V9
e-mail: xin.rui@mail.mcgill.ca

J. XiuJie
Hei Long Jiang Ba Yi Agricultural University College of Food Science, Da Qing, PR China
e-mail: Xiujiejiang@gmail.com

J.I. Boye and Y. Arcand (eds.), *Green Technologies in Food Production and Processing*,
Food Engineering Series, DOI 10.1007/978-1-4614-1587-9_13,

active in low concentrations and under relatively milder reaction conditions, thus their use does not require expensive corrosion-proof stainless steel vessels. Their protein nature also permits their ready inactivation after they have been used to attain the desired transformation in the food material. Also, they can be immobilized onto stationary support materials to permit their recovery and re-use to cut down processing costs.

Developments in enzyme engineering is also easing the design and production of superior enzymes with improved activity and stability under conditions that hitherto have not been very conducive for enzyme use, such as improved stability and activity at extreme temperatures and pHs, as well as tolerance to organic solvents and inhibitors. These developments provide food manufacturers with the requisite tools to produce wholesome food products in safe and stable forms to help improve global food security.

Food enzymes are used in several applications such as baking, manufacture of alcoholic and nonalcoholic beverages, production of cheese and other dairy products; as well as in the making of eggs and egg products, seafood products, meats and poultry products, cereal products, and candies and confectionaries.

Nevertheless, the presence and action of enzymes in foods are not always desirable. For instance, endogenous enzymes in fresh fruits, vegetables, meats, and poultry, as well as seafood, may increase undesirable changes in food texture, flavor, appearance, and/or safety. The naturally occurring food enzymes may cause destruction of essential food components such as vitamins and essential fatty acids, and they may convert molecules from their useful and innocuous forms into their toxic counterparts.

For those enzymes intentionally added to foods as processing aids, it is necessary to stop their action after the desired products are formed, otherwise their continued activity in the food products could adversely affect food texture, color, flavor, and other food qualities. Thus, it is necessary for food technologists and food manufacturers to implement useful strategies to prevent or reduce the harmful post-harvest and/or post-processing effects of food enzymes.

13.2 Food enzyme sources

Food enzymes have been traditionally derived from plant, animal, and microbial sources. Examples of plant enzymes are the amylases and glucanases, papain, bromelain, ficin, and lipoxygenase (LOX); the animal enzymes types include pepsins, trypsins, chymotrypsins, catalase, pancreatic amylase, pancreatic lipase, and rennet; and examples of the microbial enzymes are amylases, pullulanases, cellulases, lactases, pectinases, invertases, glucose isomerase, catalase, microbial lipases, and proteases.

Most of the industrial food enzymes are derived from safe microorganisms because microorganisms are easier and faster to cultivate, and their use as enzyme source is not subject to seasonal fluctuations and various political and agricultural decisions that influence the general use of animals or plants. Although all classes of enzymes should occur in most microorganisms, their use as sources of food enzymes is restricted to only a few species that are generally recognized as safe (GRAS species) such as *Aspergillus* species, *Saccharomyces cerevisiae*, *Bacillus aureus,* or *Kluyveromyces* species, or rigorously appraised for their safety for food use.

Traditionally food enzymes do not often make good candidates for effective use in food processing because the activity and/or stability of most of them tend to be hampered by temperature and pH extremes that may be required for particular processes. The availability of traditional enzymes may also not be consistent or cost effective. These limitations with traditional food enzymes are being addressed by recent advances in enzyme engineering that are enabling the design and production of recombinant enzymes (Table 13.1) with superior catalytic properties to better suit various food processing applications (Olempska-Beer et al. 2006).

Table 13.1 Sources of industrial food enzymes—selected traditional and recombinant forms

Source	Enzyme	Some food applications
Plant sources		
Papaya	Papain	Meats, baked goods, brewing
Pineapple	Bromelain	Meats, baked goods
Animal sources		
Ruminants	Chymosin (rennet)	Dairy (cheese)
Mammals (guts)	Lipase	Dairy products, oleo products, baked goods
Microbial sources		
Bacillus spp.	Alcalase	Waste utilization, bioactive compounds
	α-Amylase	Baked goods, beverages, and high glucose syrups
	β-Amylase	Baked goods, beverages, and high glucose syrups
	Pectinase (alkaline)	Fruit juices, coffee, and tea
	α-D-galactosidase	Soybean pretreatment
Aspergillus niger	Protease	Soy sauce
	Pectinase (acidic)	Fruit juices
	Lipase	Dairy products, oleo products, baked goods
	Glucose oxidase	Baked goods, beverages, dried Food mixes
Aspergillus niger, Candida guilliermondii, Kluyveromyces marxianus	Fructosidase	Fructose production
Candida cylindracea, Penicillium spp., *Rhizopus* spp., *Mucor* spp.	Lipase	Dairy products, oleo products, baked goods
Aspergillus spp., *Penicillium* spp., *Aureobasidium* spp.	Fructosyl transferase	Fructo-oligosaccharides production

(continued)

Table 13.1 (continued)

Source	Enzyme	Some food applications
Streptoverticillium mobaraense	Transglutaminase (Tgase)	Meats, seafood, dairy products
Pleurotus ostreatus	Laccase	Beverages (wine, fruit juice, beer); jams, baked goods
Recombinant enzymes[a]		
Aspergillus oryzae	Aspartic proteinase	Meats (tenderizer)
	Chymosin	Dairy (cheese)
	Glucose oxidase	Baked goods, beverages, dried Food mixes
	Laccase	Beverages (wine, fruit juice, beer); jams, baked goods
	Lipase	Dairy products, oleo products, baked goods
	Pectin esterase	Beverages (wine, fruit juice, beer)
	Phytase	Baked goods, animal feeds,
Aspergillus oryzae, Pichia pastoris	Phospholipase A_1	refining of vegetable oils
Bacillus lichenformis	α-Amylase	Baked goods, beverages, and high glucose syrups
	Pullulanase	Baked goods, beverages, and high glucose syrups
Bacillus subtilis	α-Acetolactate Dehydrogenase	Dairy, flavor generation
Escherichia coli K-12	Chymosin	Dairy (cheese)
Fusarium venenatum	Xylanase	Cereals, baked goods, animal feed
Kluyveromyces maxianus var. lactis	Chymosin	Dairy (cheese)
Pseudomonas fluorenscens Biovar 1	α-Amylase	Baked goods, beverages, and high glucose syrups
Trichoderma reesei	Pectin lyase	Beverages (wine, fruit juice, beer)

[a]Olempska-Beer et al. (2006)

These recombinant enzymes are being designed and made to make them function better at extreme pH and temperature, or survive better in nonaqueous solvents for use in manufacturing new molecules for food and related uses.

Some of the advantages with recombinant enzymes include the capacity to produce them in relatively purer forms, in higher yields, and on a sustained and consistent basis at low cost. Their purity and overproduction also minimizes extensive quality assurance requirements. The facility to produce recombinant enzymes also allows new and useful enzymes from hazardous organisms to be produced in useful and safe recombinant forms in safer microorganisms.

13.3 Major food enzyme groups

The major food enzymes include types from the oxidoreductase, transferase, hydrolase, and isomerase groups of enzymes. Examples of the oxidoreductase enzymes of importance in food processing include the glucose oxidases (GOX), ascorbic acid oxidases, lipoxygenases (LOX), polyphenol oxidases (PPO), xanthine oxidase, and peroxidases. Well-known examples of the transferase enzymes encountered in foods and food processing include the transglutaminases (TGases), fructosyl transferases (FTases), cytodextrin glycosyl transferase (CGTase), and amylomaltase. Common food enzymes from the hydrolase family of enzymes are the proteases, carbohydrases, lipases, and nucleases. The best known example of an isomerase put to use in the food industry is glucose isomerase.

13.3.1 Oxidoreductases

The oxidoreductases (Ors) catalyze redox reactions in substrate molecules. They are widespread in plants, animals, and microorganisms. Some of the ORs of interest and significance in foods are glucose oxidase (GOX), polyphenol oxidases (PPO), peroxidase and catalase, lipoxygenase (LOX), xanthine oxidase (XO), ascorbic acid oxidase (AAO), and sulfhydryl oxidase (SO). Others include alcohol dehydrogenase, aldehyde dehydrogenase, lactate dehydrogenase, glutathione dehydrogenase, laccase, and lactoperoxidase.

Glucose oxidase (GOX) is produced commercially from various fungi such as *Aspergillus niger*, *Penicillium chrysogenum*, and *P. vitale*. The enzyme catalyzes the oxidation of β-D-glucose via D-glucono-δ-lactone to D-gluconic acid. Glucose oxidase is used in the food industry to improve bread dough texture, de-sugar egg white and prevent Maillard type browning reactions, and eliminate O_2 from the head spaces of beverages (e.g., wines) and dried packaged foods to lessen enzymatic browning and off flavor formation.

Polyphenol oxidases are ubiquitous in nature and are present in plants, animals, and microorganisms. The enzyme is produced on a commercial scale from mushrooms and fungi such as *Alternaria tenuis* and *Thermomyces lanuginosus,* and from bacteria, such as *B. licheniformis* and *B. natto*. Polyphenol oxidases are Cu^{2+}-containing enzymes and enhance enzymatic browning by hydroxylating and/or oxidizing phenolic compounds via intermediates such as hydroxyquinones and quinones that subsequently polymerize into dark-colored melanins. These browning reactions are responsible for the undesirable discoloration in freshly cut vegetables and fruits (e.g., potatoes, lettuce, apples, avocadoes, and bananas), and "blackspot" formation in raw shrimps, lobsters, and crabs.

In the food industry, PPO induced browning is desirable in products such as chocolate, cocoa, coffee, prunes, raisins, tea and tobacco. Polyphenol oxidase is also put to nonfood uses in pharmaceutical and biomedical industries as components of biosensors to distinguish between codeine and morphine, to treat Parkinson's disease, phenylketonuria (PKU) and leukemia, and for wastewater treatment to remove phenolic pollutants.

Lipoxygenases (LOX) are also found in plants, animals, and microorganisms, and are obtained from soy flour for commercial use. Lipoxygenases oxidizes essential polyunsaturated fatty acids, for example, linoleic, linolenic, arachidonic, eicosapentaenoic acid (EPA), and docosahexaenoic acid (DHA). Lipoxygenase also oxidizes carotenoids and degrades them into various breakdown products such as hydrocarbons, alcohols, carbonyl compound, and epoxides. These effects of LOX produce off flavors in foodstuffs, as well as free radical intermediates that can destroy biomolecules and essential food components such as proteins, astaxanthin, β-carotene, and vitamin A. Thus, the action of LOX in foods is generally undesirable, particularly the generation of harmful free radicals to cause the destruction of essential biomolecules such as carotenoids and other adverse effects in foods such as color loss in salmonids during frozen storage, and off-flavor development in fatty foods. However, the enzyme is put to beneficial use in bread making for the bleaching of flour and for improving the viscoelastic properties of dough. Lipoxygenase is also involved in flavor generation in plants via the formation of 2-hexenal, an aldehyde. The aldehyde thus formed may be oxidized further to the corresponding carboxylic acid by aldehyde dehydrogenase or reduced by alcohol dehydrogenase to the corresponding alcohol. The carboxylic acids and alcohols may interact and form esters to impart flavors to the plant material.

Peroxidases and catalases are heme iron containing enzymes that occur widely in plants, animals, and microorganisms. They use H_2O_2 as co-reactant for the oxidation of other peroxides to form H_2O and other oxidized products, for example, alcohols. Examples of peroxidases are glutathione peroxidase and cytochrome c peroxidase. Catalase is a specific peroxidase that catalyzes the breakdown of hydrogen peroxide. In the food industry, heat inactivation of peroxidases is used as an index of the efficacy of blanching. Catalase is used in the food industry to eliminate H_2O_2 from food products, for example, to remove residual H_2O_2 after its use for cold pasteurization of milk. It is also used together with GOX to de-sugar eggs and remove head space O_2 from beverages and packaged dried foods. Peroxidases are also used to get rid of phenolic contaminants from industrial waste waters. In foods, peroxidases participate in undesirable browning and off-flavor development in fruits and vegetables.

Xanthine oxidase (XO) is a molybdo-flavoprotein enzyme found largely in mammalian milk and liver. XO catalyzes the formation of uric acid from the oxidation of hypoxanthine and xanthine. The enzyme has been implicated in the spontaneous development of oxidized flavors in raw milks, a phenomenon that may be prevented by treating raw milks with trypsin (Lim and Shipe 1972).

Ascorbic acid oxidase (also known as L-ascorbic acid oxidase) catalyzes the oxidation of vitamin C to the less active form, dehydroascorbic acid.

$$2\text{L - ascorbic acid} + O_2 \rightleftharpoons 2\ \text{dehydroascorbic acid} + 2H_2O$$

In general, this effect of ascorbic acid oxidase is undesirable, except perhaps in the rare cases in which vitamin C is ingested in excess to cause toxicity. Sources of ascorbic acid oxidase include squash and the bacterium *Aerobacter aerogenes*.

Lactate dehydrogenase (LDH) occurs widely in plants, animals and microorganisms such as *Lactococcus, Enterococcus, Streptococcus,* and *Pediococcus*. The enzyme catalyzes the interchanges between lactic acid and pyruvic acid. Nicotinamide adenine dinucleotide (NAD^+) participate in the forward reaction (see equation that follows) as a hydrogen acceptor while NADH participates as a hydrogen donor in the reverse reaction from pyruvic acid to lactic acid.

$$\text{Lactic acid} + \text{NAD}^+ \rightleftharpoons \text{Pyruvic acid} + \text{NADH} + \text{H}^+$$

The lactic acid formed is an essential component of dairy products such as sour milk products, yogurt, kefir, and cottage cheese. Formation of lactic acid also lowers the pH of the product and promotes curdling of casein in fermented milks; it also contributes to the sour flavor of sourdough breads, and is used in brewing to lower the pH and add "body" to beer.

In general, the levels of LDH increase during the breakdown of body tissues, thus LDH levels are used as part of the diagnosis of health disorders such as hemolysis, meningitis, acute pancreatitis, and HIV (human immunodeficiency virus).

Sulfhydryl oxidase (SO) enzymes catalyze the formation of disulfide bonds via the oxidation of thiol groups into disulfides as follows:

$$2\text{RSH} + \text{O}_2 \rightarrow \text{RS} - \text{SR} + \text{H}_2\text{O}_2$$

Sulfhydryl oxidases have been found to occur in the milks of mammals (Burgess and Shaw 1983), egg albumen (Hoober et al. 1999; Thorpe et al. 2002), and also microorganisms (de la Motte and Wagner 1987). The SO enzymes from cow's milk, pancreas, and kidney are said to be Fe dependent (Sliwkowski et al. 1984), whereas others from the small intestine and the skin have been shown to have Cu^{2+} ions (Yamada 1989).

The enzyme has been used for flavor control in ultra-high temperature (UHT) processed milk (Kaufman and Fennema 1987).

13.3.2 *Transferases*

The transferases catalyze the transfer of groups from a donor substrate molecule to the acceptor molecule to form a new product, as summarized by the following equation:

$$\text{A} - \text{X} + \text{B} \rightarrow \text{A} + \text{B} - \text{X}$$

where AX is the donor and B is the acceptor. Examples of the transferases are transglutaminases (TGases), transglycosidases, amylomaltases, and transphosphorylases. Some of the transferases such as TGases, and various glucanotransferases (e.g., fructosyl transferase, cyclodextrin glycosyl transferase, and amylomaltase), are used extensively in the food industry.

The transferase fructosyl transferase (FTase) has been found to occur in plants, for example, onions (Robert and Darbyshire 1980) and asparagus (Shiomi et al. 1979), and in microorganisms such as *Streptococcus mutans* (Wenham et al. 1979), *Fusarium oxysporum* (Gupta and Bhatia 1980, 1982), *Bacillus subtilis* (Homann and Seibel 2009), and *Aspergillus* spp (Prapulla et al. 2000). FTase catalyzes the synthesis of fructose oligomers, or fructose oligosaccharides (FOS). Fructo-oligosaccharides account for about one third of the sweetness of sucrose, but they are nondigestible by humans (Oku et al. 1984; Yun et al. 1995), and serve as good sources of dietary fiber for the growth of beneficial intestinal microflora. Thus, FOS are applied as low calorie sweeteners and as food ingredients with health benefits (Lee et al. 1992). Fructose oligosaccharides are also used as sugar substitutes in foods such as reduced calorie jam, ice cream, and confectionery products. The enzyme from *B. subtilis* has been used to synthesize fructose disaccharides such as xylsucrose, galactosucrose, and 6-deoxysucrose for use as sweeteners or bulking agents in food products (Cheetham et al. 1989).

Cyclodextrin glycosyl transferases (CGTases) catalyze coupling, cyclization, disproportionation and hydrolysis reactions. The capacity of these enzymes to form cyclic compounds from linear polysaccharide chains is significant because the cyclic molecules formed can trap or encapsulate other molecules to form inclusion composites for applications in foods, cosmetics, and fine chemicals (Shahidi and Han 1993). Cyclodextrins (or CDs) with varying units of glucose joined together by α, $(1 \rightarrow 4)$ glycosidic bonds have been reported. They are the 6 glucose CDs (or α-CD), the 7 glucose CDs (or β-CD) and the 8 glucose CDs (or γ-CD). CGTases are present in microorganisms including *Bacillus*, *Klebsiella*, *Micrococcus*, and *Clostridium* (Bonilha et al. 2006; Menocci et al. 2008).

Another transferase enzyme is amylomaltase, also known as 4-α-glucanotransferase. It catalyzes the transfer of units between α-1,4-glucans or from one α-1,4-glucan to glucose. The enzyme has been found to occur in microorganisms, for example, *Clostridium butyricum* and *E. coli* (Goda et al. 1997; Tafazoli et al. 2010), and in plants, for example, potato (Sträter et al. 2002) and barley (Yoshio et al. 1986). The enzyme acts on both malto-oligosaccharides and amylose to form medium to highly polymerized water-soluble cycloamyloses (Takaha et al. 1996; Terada et al. 1997) with potential for use in the food, pharmaceutical, and fine chemical industries.

Transglutaminases (TGases) are transferases that catalyze acyl transfer reactions between lysine and the γ-carboxyamide groups of glutamines in proteins (Motoki and Seguro 1998). Such acyl transfers result in extensive strong and resistant covalent cross-link formation between biomolecules such as proteins. The products usually have modified properties with respect to mechanical properties and solubility in aqueous systems. Advantage is taken of this attribute of the enzyme in commercial food processing to improve the texture of gluten-free breads for celiac disease patients (Moore et al. 2004, 2006), preparing imitation surimi-type products and fish balls, and binding small pieces of meat into larger pieces of meat. The enzyme is also used as a binding agent to improve the texture, firmness, and elasticity of high protein foods (e.g., surimi or ham), and emulsified meat products (e.g., sausages and hot dogs). TGases are also used to improve the texture of low-grade meats, improve the consistency of dairy products, and make firmer noodles (Kim and Kim 2009).

13.3.3 *Proteases*

Proteases (also known as proteinases or proteolytic enzymes) are enzymes that catalyze the hydrolytic cleavage of peptide bonds in proteins and peptide molecules to form lower molecular weight peptides and amino acids. The greater bulk of the enzymes currently used in the food industry are proteases (Layman 1986; Godfrey and West 1996). In addition to their use in food processing, proteases are also used in the animal feed, detergent, leather and textiles, photographic, and related industries (Kalisz 1988). Proteases are also used as a digestive aid to improve digestion and utilization of proteins by certain individuals who cannot readily digest these molecules.

Two types of proteases are distinguished based on their mode of action as either endoproteases or exoproteases. Endopeptidases act randomly to cleave peptide bonds within protein molecules to form smaller peptides. Examples are the digestive proteases, pepsin, rennet, chymosin, trypsin, and chymotrypsin. Other proteases are papain, bromelain, ficin, elastase, and thermolysin. Exopeptidases remove amino acids successively from the terminals of the protein molecule to produce free amino acids and the residual polypeptide molecule. Thus, two types of exoproteases are distinguished, those that cleave amino acids at the carboxyl or C-terminal of the protein (or carboxypeptidases) and those that cleave amino acids from the amino or N-terminal (or aminopeptidases). Examples of carboxypeptidases are cathepsins A and C, carboxypeptidases A, A2, B, and C (known as metallo carboxypeptidases), and angiotensin converting enzyme (ACE). Examples of the aminopeptidase are aminopeptidase A or glutamyl aminopeptidase, aminopeptidase B or arginyl aminopeptidase, leucine aminopeptidase, and the methionine aminopeptidases 1 and 2 (also known as METAPS 1 and 2).

Proteases are also classified based on the nature of their active sites into four subgroups, namely the acid, serine, sulfhydryl, and metallo-proteases. Their distinctive features are summarized in the following subsections.

13.3.3.1 The acid proteases

The acid proteases are sometimes referred to as aspartic acid proteinases or aspartic acid proteases because of the nature of their active sites that have one or more side chain carboxyl groups from aspartic acid that the enzymes must have for functional activity. They tend to show optimal activity within the acid pH range. Examples the acid proteases are the pepsins, chymosins, rennets, gastricins (from animals), cyprosin and cardosin (from plants), and several others from microorganisms such as *Rhizomucor miehei*, *R. pusillus*, and *Aspergillus niger* (Mistry 2006). This group of enzymes is typified by the susceptibility to inhibition by pepstatin. They are used extensively in dairy processing to curdle milk in cheese making. They are able to do this by cleaving critical peptide bonds between phenylalanine and leucine residues in κ-casein of milk to form micelles that coagulate and precipitate out as the curd. The specific cleavage of the critical bonds by these proteases is crucial and ensures that continued proteolysis is averted in the finished products to prevent excessive texture softening, bitterness and/or "off-odors" in the product(s), defects that could proceed in the products when other proteases are used.

13.3.3.2 Serine proteases

The serine proteases are characterized by the presence of a serine residue in their active sites; they are also generally inhibited by the serpins, organophosphates such as di-isopropylphosphofluoridate (DIFP), aprotinin and phenyl methyl sulfonyl fluoride (PMSF). The animal types include trypsins, chymotrypsins, thrombin, and elastase; examples of plant serine proteases are cucumisin, macluralisin, and taraxilisin (Kaneda and Tominaga 1975; Rudenskaya et al. 1995); and the microbial forms include subtilisin and proteinase K (Couto et al. 1993). They are generally optimally active within the pH range of 7–11, and are used extensively in the food and animal feed industries to produce protein hydrolysates from substrates such as whey, casein, soy, keratinous materials, and scraps from meat and fish processing discards (Dalev 1994; Wilkinson and Kilcawley 2005). They are also put to use to enhance flavor development during ripening of dairy products. Serine proteases are also used for regioselective esterification of sugars, resolution of racemic mixtures of amino acids, and production of synthetic peptides for use as pharmaceutical drugs and vaccines (Guzmán et al. 2007). Other industrial uses of serine proteases are for the treatment and bating of leather to remove unwanted pigments and hair from leather (Adigüzel et al. 2009). They are also used to treat industrial effluents and household waste to lower the viscosity and biological oxygen demand (BOD), and increase the fluidity and flow of solid waste byproducts (such as hair, feather, skin, hooves, and bones), and for the treatment of "stickwater" from slaughterhouses and fish processing plants (Dalev and Simeonova 1992). They are also used in the degumming and the detergent industries (Godfrey and West 1996), and in medicine for the treatment of burns, wounds, and abscesses (Lund et al. 1999).

13.3.3.3 Sulfhydryl proteases

The sulfhydryl proteases (or proteinases) are also referred to as thiol or cysteine proteases (or proteinases) because of the presence of free sulfhydryl group(s) in their active sites. This group of enzymes is sensitive to inhibition by thiol reagents (e.g., alkylating agents and heavy metal ions). They occur in plants, animals, and microorganisms. Examples of plant forms include papain, bromelain, chymopapain, and ficin; examples of the animal types are cathepsins B and C, calpains and caspases; and the microbial types have been described in species such as *Streptococcus* (Whitaker 1994), thermophilic *Bacillus* (Do Nascimento and Leal Martins 2004), and *Pyococcus* sp. (Morikawa et al. 1994). Their catalytic activities are enhanced by cysteine and 2-mercaptoethanol (ME), as well as dithiothreitol (DTT) and its epimer dithioerythritol (DTE); and they are inhibited by cystatin, iodoacetate, *p*-chloromercuribenzoic acid, leupeptin and N-ethylmaleimide. They tend to be optimally active at slightly acidic to neutral pH (i.e., pH 6–7.5), and are relatively heat stable—a feature that accounts for their use in meat tenderizers. They do so by degrading myofibril protein and collagen fibers to make meats tender. In commercial applications, papain and bromelain are more frequently used because the higher hydrolytic capacity of ficin can cause excessive softening and mushiness in treated meats (Wells 1966).

Sulfhydryl proteases are also used in the brewing industry to remove haze and improve beverage clarity. Haziness may form in the product from the aggregation of larger peptide and polyphenol molecules into high molecular weight complexes, a development that is especially enhanced during chilling of beverages. Sulfhydryl proteases are also used in baking to improve dough elasticity and firmness by increasing their viscoelastic, gas-retaining, and thermosetting properties (Mathewson 1998).

13.3.3.4 Metalloproteases

Metalloproteases are so-called because they must have certain metal ions (e.g., Ni^{2+}, Mg^{2+}, Mn^{2+}, Ca^{2+}, Zn^{2+} and Co^{2+}) in their active sites for catalytic activity. Examples include urease that has nickel (Ni^{2+}) in its active site. Urease catalyzes the hydrolysis of urea as shown in the following equation:

$$(NH_2)_2CO + H_2O \rightarrow CO_2 + 2NH_3$$

The NH_3 produced from the hydrolysis acts to neutralize gastric acid in the stomach. The enzyme occurs in microorganisms (in for example, bacteria, yeast), as well as in plants and animal tissues (e.g., liver, blood, and muscle). Urease occurs in high quantities in the Jack beans, soybeans, and in the bacterium *Bacillus pasteurii* and in intestinal bacteria such as *Eubacterium aerofaciens, E. lentum,* and *Peptostreptococcus productus*. It is also found in high levels in yeasts (e.g., *Candida curvata, C. bogoriensis,* and *Rhodotorula* species), and in the blue-green algae *Anacystis nidulans* and *Anabaena doliolum*. The enzyme is also found in large quantities in and other plant seeds, as well as in certain some animal tissues.

An example of Mg^{2+}-containing metalloprotease is inorganic diphosphatase that cleaves diphosphates into inorganic phosphates, and phosphatidylinositol triphosphate 3-phosphatase that catalyzes the hydrolysis of phosphatidylinositol 3,4,5-trisphosphate into the bisphosphate and inorganic phosphate.

An example of calcium (Ca^{2+})-containing metalloprotease is lactonase that hydrolyzes lactones into their corresponding hydroacids. Lactonase also participates in the metabolism of galactose and ascorbic acid.

Glutamate carboxypeptidase is a Zn^{2+}-containing metalloprotease, and it catalyzes the breakdown of N-acetylaspartylglutamic acid into N-acetylaspartate and glutamic acid, and another Zn^{2+}-containing type is glutamyl aminopeptidase (or aminopeptidase A) that catalyzes the removal of glutamate and aspartate residues from the N-termini of polypeptides. This reaction by glutamyl aminopeptidase is important because of its link with the degradation of angiotensin II into angiotensin III that have an effect on high blood pressure. Another important Zn^{2+}-containing metalloprotease is the *insulin-degrading enzyme* (IDE) found in humans and causes degradation of low molecular weight polypeptides such as insulin, glucagon, and endorphins. Collagenases also contain Zn^{2+} in their active site and they catalyze the breakdown of peptide bonds in collagen.

Methylmalonyl coenzyme A mutase is a Co^{2+}-containing metalloprotease and it participates in the isomerization of methylmalonyl-CoA to succinyl-CoA during the extraction of energy from proteins and fats (Kobayashi and Shimizu 1999).

Metalloproteases derived from animals include carboxypeptidases A and B, collagenases and gelatinases; microbial types include thermolysin (used in the food industry to produce the low calorie sweetener, aspartame), lysostaphin, and acidolysin; and plant types are exemplified by the Zn^{2+}-containing endoproteases from grass pea. The metalloproteases are inhibited by agents such as phosphoramidon, dimercaprol, porphine, ethylenediaminetetraacetic acid (EDTA), vitamin B_{12}, and 1,10-phenanthroline.

Commercially available metalloenzymes include Neutrase (Novo Nordisk 1993) and Thermoase (Amano Enzymes), and some of the commercial applications include the synthesis of peptides for use as low calorie sweetener, or drugs for the food and pharmaceutical industries, respectively. Thermolysin and vimelysin are both used to synthesize the dipeptide, aspartame (Murakami et al. 1998). Thermolysin is also used in combination with other proteases to produce food protein hydrolysates and flavor-enhancing peptides, and for accelerating the ripening of dry sausages. They are also used to digest slaughterhouse waste, and in brewing to aid filtration of beer and the production of low-calorie beers.

13.3.4 Carbohydrases

Carbohydrases are a group of enzymes that catalyze the synthesis or hydrolytic breakdown of polysaccharides. Examples include the amylases, maltases, and lactases that degrade starch, maltose, and lactose, respectively. Others are invertases that break down sucrose into glucose and fructose, glucanases that hydrolyze glucans, cellulases that degrade celluloses into low molecular weight cellodextrins, cellobiose, and/or glucose, and pectinases that break down pectin into galacturonic acid. Carbohydrases are widespread nature and occur in plants, animals, and microorganisms. Amylases, pectinases, β-galactosidases, and chitinases are all carbohydrases present in living organisms (plants, animals, and microorganisms). Microorganisms from which commercial carbohydrases are produced include bacteria such as *Bacillus licheniformis,* various fungi including *Aspergillus niger*, *A. oryzae*, *Trichoderma reesei,* and yeasts such as *Kluyveromyces lactis* and *Saccharomyces cerevisiae*. Carbohydrases also occur in marine algae.

Based on their mode of action, carbohydrases may be classified as endo- and exo-carbohydrases. The endo-carbohydrases, for example, α-amylase, cleave the internal glycosidic bonds of polysaccharides, whereas the exo-forms such as β-amylase break down carbohydrates via successive removal of maltose units from the nonreducing end of their substrates. Alternatively, they may be classified into the starch hydrolyzing types versus the nonstarch hydrolyzing types. The starch hydrolyzing types, for example, amylases, glucanases, cellulases, and maltases catalyze the hydrolysis of amylose and amylopectin, and are used principally to facilitate the production of high glucose syrups, alcoholic beverages, and baked goods. The nonstarch hydrolyzing forms break down polysaccharides other than amylose and amylopectin. Examples are the cellulases, maltases, galactosidases and lactases, pectinases, invertases, chitinases, and fructosidases.

Examples of the commonly used carbohydrases include the amylases that are extensively applied in the starch, beverages, and the sugar industries. A recombinant heat stable α-amylase derived from *B. licheniformis* by Novozymes is used for the continuous liquefaction of starch at high temperatures (105–110°C). This same heat-stable recombinant enzyme is used to liquefy starch during brewing and mash distillation, and in sugar manufacture to hydrolyze the starch in sugar cane juice. The α-amylases are also put to non-food uses in the textiles industry for de-sizing of fabrics and in detergents to facilitate the removal of starch-based stains from fabrics. A type of amylase known as amyloglucosidase (AMG) catalyzes the hydrolysis of both α-1,4- and α1,6- glycosidic bonds and removes glucose units from liquefied starch sequentially. It is derived commercially from *A. niger,* and used to produce high glucose syrups from starches and dextrins.

Glucans are carbohydrates made up of glucose units linked together by glycosidic bonds. Examples of glucans include amylose starch (with α-1,4 bonds), cellulose (β-1,4 bonds), zymosan (β-1,3 bonds), glycogen (α-1,4- and α-1,6 bonds), amylopectin starch (α-1,4- and α-1,6 bonds), and pullulan (with both α-1,4- and α-1,6 bonds). They are ubiquitous in nature and perform structural, defense, and growth functions. They also serve as storage forms of chemical energy (Thomas et al. 2000). Glucanases (also carbohydrases) catalyze the hydrolysis of glucans into glucose oligomers, trisaccharides, such as maltotriose and nigerotriose, disaccharides like maltose, trehalose, and cellobiose, and glucose—a monosaccharide. Several microorganisms produce glucanases to digest glucans and provide them with nutrients and energy source (Abd-El-Al and Phaff 1968; Kikuchi et al. 2005). Plant glucanases occur in seeds, tobaccco, rice, acacia and others (Thomas et al. 2000; Elortza et al. 2003; Leubner-Metzger 2003; Hrassnigg and Crailsheim 2005; González-Teuber et al. 2010); whereas animal types have been recovered from insects, sea urchin embryos, nematodes, mollusks, crayfish, shrimp, earthworms

(Lee et al. 1992; Bachman and McClay 1996; Beschin et al. 1998; Sritunyalucksana et al. 2002; Kozhemyako et al. 2004; Kikuchi et al. 2005). Glucanases are produced commercially from bacteria to break down glucans in cereals during brewing to aid filtration and prevent clogging of filters.

Pullulanases, also known as debranching enzymes, are specific glucanases that break down pullulans. Two types of pullulanases are distinguished; the type I pullulanases that cleave the α-1,6 bonds, and the type II pullulanases that break down both the α-1,4 and α-1,6 linkages. The enzymes occur in bacteria such as *B. cereus* and *Klebsiella* sp. (Pugsley et al. 1986; Nair et al. 2006). Plant sources of these enzymes include rice, wheat and maize (Dinges et al. 2003; Yamasaki et al. 2008; Liu et al. 2009).

Because of their ability to digest glucans into low molecular weight glucose oligomers and glucose units, glucanases and pullulanases are used extensively in industry to produce animal feeds, beverages (fruit juices, tea, beer, and wine), baked goods, leather, and textiles. They are also used in detergents, oil extraction, and for waste treatment.

Various amylase mixes are used to hydrolyze carbohydrates. For example, Viscozyme is produced and marketed by Novozymes from the fungus *A. aculeatus* and used widely in food processing to degrade nonstarch polysaccharides in plant materials, thereby reducing the viscosity of such materials, improving clarity and flow properties, and increasing yields.

Another group of carbohydrases is the galactosidases. They catalyze the hydrolysis of galactosides such as lactose, lactosylceramides, ganglioside, and glycoproteins into monosaccharide units. Galactosidases are classified as α- and β-galactosidases based on their mode of action. The α-forms (also known as α-gal) catalyze the hydrolysis of terminal α-galactosyl residues from α-galactosides, including the disaccharide melibiose into galactose and glucose. Alpha-gal is used as a dietary supplement to assist the breakdown of complex carbohydrates into simpler forms to facilitate digestion and reduce intestinal gas build up and bloating. The enzyme may also be used to break down complex carbohydrates to increase the fermentable sugar and subsequently, the alcohol content in fermented beverages. The β-gals hydrolyze β-galactosides such as lactose, various glycoproteins, and ceramides into simpler forms.

Lactase refer to specific β-galactosidase enzyme that catalyzes the breakdown of milk lactose exclusively into its constituent monosaccharides, galactose, and glucose units. Lactases are produced in commercial quantities from the yeasts *Kluyveromyces lactis* and *K. fragilis,* and from fungi such as *A. niger* and *A. oryza* (Mustranta et al. 1981). Other lactase producing microorganisms are *Streptococcus cremoris, Lactobacillus bulgaricus,* and *Leuconostoc citrovorum* (Ramana Rao and Dutta 1978). Lactases are also found in the mucous lining of the small intestines of cows, rabbits, and humans quite early in intrauterine life. Lactase levels tend to be higher in younger animals and human infants and the levels generally decline with age (Heilskov 1951). The enzyme is commercially produced from *K. lactis* by Novozymes as Lactozym and widely used in the dairy industry to produce fermented milks, ice cream, milk drinks, and lactose-reduced milks for lactose-intolerants and household pets (e.g., cats) that cannot utilize lactose. Lactase treated milks tend to be sweeter than their untreated counterparts because the monosaccharides glucose and galactose produced in the treated milks from the hydrolysis of lactose are sweeter than lactose. Crystallization of lactose in condensed milk can also be prevented by treating the milk first with lactase. Lactase-treated milks also ferment faster than untreated milks, because lactose ferments at a slower pace than the glucose and galactose in the treated milks. Ice cream made with lactase-treated milk (or whey) is substantially depleted of lactose crystals, thus the undesirable "sandy" feel because of lactose crystallization is prevented.

The carbohydrases referred to as maltases catalyze the breakdown of the disaccharide maltose into its constituent glucose units. Maltase is found in vertebrate animals, including humans, and in plants and microorganisms. In humans, digestion of starch commences with ptyalin in the mouth followed by pancreatic amylases in the gut to form maltose. Intestinal maltase next breaks down the maltose into glucose for utilization by the body to produce energy or for storage in the liver as glycogen. Maltase (from barley malt) is used in brewing to increase the glucose content for conversion into alcohol. Maltase from *A. oryzae* is used in the beet sugar and molasses distilleries; and in the industrial production of glucose syrups (McWethy and Hartman 1979).

Cellulase enzymes catalyze the hydrolysis of the β-1,4- glycosidic bonds in cellulose. Cellulases are produced by bacteria (e.g., *Clostridium thermocellum, Acidothermus cellulolyticus)* (Tucker et al. 1988), fungi (*T. reesei* and *Fusarium roseum*) (Sidhu et al. 1986) and protozoa (e.g., *Ostracodinium obtusum bilobum*) (Coleman 1985). Cellulases are also produced in plants (e.g., cress and pepper) (Ferrarese et al. 1995), and in animals (e.g., termites, crustacea and insects) (Watanabe and Tokuda 2001). Native cellulose is highly crystalline and resistant to hydrolysis. Its complete digestion requires preliminary unraveling of the crystalline molecules into simple (or linear) cellulose chains by endocellulases. The linear cellulose chains then serve as substrates for exocellulases to form smaller oligosaccharides (such as cellobiose and cellotetrose), that are subsequently hydrolyzed further by cellobiases into β-D-glucose units.

Cellulase is used in coffee processing to enhance drying of the beans, and in laundry detergents, as well as the textile and pulp and paper industries. Cellulases are used as digestive supplements, and also to facilitate in the biotransformation of feedstock into biofuels. The enzyme is marketed commercially as *celluclast* by Novozymes and is used to digest cellulose into the fermentable sugars cellobiose and glucose, or to reduce the viscosity of soluble cellulosic materials.

Invertases and sucrases are hydrolytic enzymes that catalyze the hydrolysis of sucrose into fructose and glucose. The difference between the two is that invertases split the O–C (fructose) bond, whereas sucrases break the O–C (glucose) bond. The equimolar mixture of glucose and fructose formed from the hydrolysis is known as invert sugar or artificial honey. Invertase is produced from *Saccharomyces cerevisiae* or *S. carlsbergensis* for commercial use, although bees also make the enzymes to produce honey from nectar. Invertases and sucrases are also found in bacteria and in plants such as carrots, tomatoes, and the seeds, stems, roots, and tubers of several higher plants (Cooper and Greenshields 1961, 1964; Elliott et al. 1993; Bogs and Geider 2000; Quiroz-Castañeda et al. 2009). Novozymes produces the enzyme and markets it commercially under the brand name Maxinvert, for extensive use in the confectionery industry to produce chocolate-covered berries and nuts, soft-centered candies, marshmallows, and fudges. Yeast invertase is also commercially available from Enzeco for the inversion of sucrose into fructose and glucose.

Pectinases (also known as pectic enzymes) are carbohydrases that hydrolyze pectins into D-galacturonic acid residues. Pectins occur in fruits and can contribute to the viscosity and haziness in beverages. Pectic enzymes break down pectins into smaller molecules to enhance their solubility and remove haze in beverages. Members of the pectinase family of enzymes include pectinesterase (PME) and polygalacturonase, and they are widespread in avocado, carrots, and tomato, and in some bacteria (e.g., *Bacillus polymyxa* and *Erwina aroideae*), fungi (e.g., *A. niger, Penicillium digitatus,* and *Fusarium oxysporum*), and yeasts (e.g., *Saccharomyces fragilis*). Pectin methyl esterase catalyzes the hydrolysis of pectins via the removal of methoxyl groups to form pectic acids, i.e.,

$$\text{pectin} + H_2O \rightleftarrows \text{pectic acid} + \text{methanol}$$

Pectin lyase or pectolyase catalyzes the removal of methyl esters in pectins from the non-reducing termini. Food-grade pectolyase is produced from fungi and used to remove residual pectins in wines and ciders. The enzyme is also used in combination with cellulase in plant cell culture to break down plant cell walls to produce protoplasts.

Polygalacturonases (PGases) break down the glycosidic bonds that join galacturonic acid residues in pectins. Two types of PGases are distinguished based on their mode of action as either endo- or exo-PGases. The endo-forms hydrolyze the 1,4-α-D-galactosyluronic acid residues of pectins in a random fashion, while their exo-counterparts degrade pectins by terminal action to produce galacturonic acid residues.

Novozymes produces a variety of pectinase for several applications in the food industry. Examples of the Novozyme pectinases are Pectinex produced from *A. niger* for fruit for vegetable juice processing and the maceration of plant tissues, Peelzym produced from *A. aculeatus* for hands-free peeling of citrus fruits, and NovoShape derived from *A. aculeatus* and *A. oryzae* for maintaining the shape and structure of whole portions or bits of fruits in a product.

Fructosidases or inulases are carbohydrase enzymes that hydrolyze inulins or fructans into smaller units. Inulin is a linear plants polysaccharide consisting of about 35 fructose molecules linked together by β-1,2 glycosidic bonds, and terminating in a glucose unit. There is considerable interest in this group of enzymes for use by the food industry to produce fructose from fructans (Muñoz-Gutiérrez et al. 2009). The enzymes occur in plants and are produced by microorganisms such as *Aspergillus*, *Saccharomyces fragilis,* and *S. marxianus*.

13.3.5 Lipases

Lipases are a large group of hydrolytic enzymes that catalyze the hydrolysis of ester bonds in lipids/fats such as triglycerides, phospholipids, and cholesteryl esters. Lipases are classified into triacylglycerol lipase (or true lipases), phospholipases, sterol esterase, and retinyl-palmitate esterase—based on the nature of their substrate molecules.

There are several categories of lipase specificity. Lipases that are specific for fatty acids tend to prefer specific classes of fatty acids such as short-chain, long-chain, or polyunsaturated. Some lipases exhibit positional specificity and display regioselectivity in differentiating between the external, primary (*sn*-1 and *sn*-3 positions) and internal, secondary (*sn*-2 position) ester bonds. For example, several microbial lipases, gastric lipases, and phospholipases show a preference for the external ester bonds. Some display partial stereospecificity such as preference for either *sn*-1 or *sn*-3 position of triglycerides, whereas some others show enantioselectivity and can distinguish between enantiomers of chiral molecules (Hou 2002). For triglycerides, lipases hydrolyze ester bonds in triglycerides to form free fatty acids and glycerol.

13.3.5.1 Conventional sources of lipases

The pancreas and serous glands from ruminants, including young calf and lamb, and pig pancreas have served as significant sources of lipases for flavor development in the dairy industry for several decades now. Microbial lipases have since received a lot of attention in lipase research and have been exploited considerably in numerous applications because of the comparative ease of their production, genetic manipulation, and wider range in properties. Microorganisms now constitute the major sources of commercial lipases.

Lipases also occur in plants. However, there is a relative lack of information on plant lipases compared with their mammalian and microbial counterparts. Plant lipases that have been studied in some depth include triacylglyceride lipases, phospholipases, glycolipases, and sulfolipases. Plant lipases have a unique ability to cleave all three fatty acids from triglycerides, and distinct substrate specificities and capacity to synthesize structured lipids. In food processing and preservation, an understanding of the properties of plant lipases is crucial for food quality and freshness, particularly in oilseeds, cereals, oily fruits, and for the production of edible oils (Mukherjee and Hills 1994).

Phospholipases catalyze mostly water-insoluble substrates and are involved in digestive, regulatory, and signal transduction pathways, among others. Their classification as phospholipases A_1, A_2, B, C, and D is based on the nature of the phospholipid ester bond they hydrolyze. Several of these enzymes have many disulfide bonds and require Ca^{2+} ions for optimal catalytic activity. They are also relatively small in size and have molecular weights ranging from 13 to 30 kDa, although a few have been characterized with molecular weights ≥80 kDa (Lopez-Amaya and Marangoni 2000a). Like lipases, phospholipase activity causes deterioration of seafood quality (Lopez-Amaya and Marangoni 2000b).

13.3.5.2 Some applications of lipases

Dairy products

In dairy products, lipases play an important role in the development of flavor. For example, membrane lipases can cause rancidity in fresh milks. In processed milk and milk products, this may not be a major problem because the lipases do not survive pasteurization treatments. In some dairy products, lipolysis is desired to produce the flavors normally associated with those products, such as Roquefort cheese, butterfat, cheddar cheese, and blue cheese. In these products, there is supplementation with microbial lipases to enhance ripening. During ripening process, the added lipases hydrolyze milk fat into free fatty acids. Different types of lipases may contribute unique flavors. In this regard, short-chain (i.e., C4-C6) fatty acids releasing lipases produce so-called sharp flavors in dairy products, whereas lipases furnishing long-chain fatty acid (>C12) tend to induce more subtle flavors (Schmid and Verger 1998). For example, cheddar cheese has higher amounts of butyric acid (C4:0) compared with milk fat (Bills and Day 1964). For this purpose, lipases from *Penicillium roqueforti, A. niger, R. arrhizus,* and *Candida cylindracea* are found to be very useful (Ha and Lindsay 1993). Although free fatty acids produce distinctive flavors in cheeses and other dairy products, controlling of lipolysis after the desired modification in the product has been achieved is useful because excessive lipolysis could induce undesirable rancid flavors in the products.

Oleo products

Oleo products are food spreads (ranging from margarines to specialty fats, shortenings, and lard substitutes) prepared mostly from vegetable oils and used as butter substitutes. Oleo products manufacture makes use of the capacity of lipases to catalyze interesterification reactions to modify acylglycerols from fats and oils such as in the making of cocoa butter equivalents. Cocoa butter has 1-palmitoyl-2-oleoyl-3-stearoyl-glycerol (POS) or 1,3-distearoyl-2-oleoyl-glycerol (SOS) as its main components, and is widely used in making chocolates and candies. Palm oil mid fraction with 1,3-dipalmitoyl-2-oleoyl-glycerol (POP) as its prevalent triglyceride, has been used as substrates to imitate cocoa butter by mixing it with co-substrates such as tristearin (SSS). In the presence of lipases, interesterification reactions proceed to interchange fatty acids in *sn*-1- and *sn*-3- positions to form POS and/or SOS. Some commercially available fungal lipases, for example, lipozyme produced from *Mucor miehei* by Novozymes, and Enzymatix F3 produced from *Rhizopus* sp., by Gist-brocades are capable of effecting such product-simulating interesterifications.

Rhizopus arrhizus lipase in the immobilized form has been used to produce cocoa butter equivalent from oils, and the substrate conversion rate is increased considerably from the addition of defatted soya lecithin (Mojovic et al. 1993). Lipase catalyzed interesterifications are also used to fabricate shortenings depleted in trans fats. Shortenings are usually produced via hydrogenation of vegetable oils and the process produces trans-fats as co-products. Lipozyme TLIM from *Thermomyces lanuginosus* is capable of producing trans-fat free shortenings from sunflower oil or fully hydrogenated soybean oil (Li et al. 2010). Free trans-fat shortening has also been produced from palm and rice bran oils by using the same lipozyme TLIM.

13.3.6 Isomerases

The isomerases catalyze intra molecular rearrangement of substrates into their isomers. Although the literature is replete with information on several isomerases, the main one with food use is glucose isomerase (in its immobilized form) (Chaplin and Bucke 1990). The enzyme is produced from microorganisms such as *B. coagulans, Actinoplanes missouriensis*, *Streptomyces albus,* and *S. olivochromogenes.*

D-fructose is sweeter than D-glucose, and is important in the food industry as a sweetener. The free form of the enzyme is costly, thus it is immobilized to permit re-use and make it cost effective for the isomerization of D-glucose into fructose. The making of high fructose corn syrup (HFCS) involves the initial liquefaction of corn starch with α-amylase before the isomerization with glucose isomerase. The fructose yield in HFCS is temperature dependent, and yields of about 55% are obtained at 95°C compared with the 42% recovered at the conventional production temperature ranging from 60°C to 70°C. Thus, the high thermal stability of glucose isomerase is useful for this conversion.

13.4 Enzymes in food and feed manufacture

Enzymes have been used since ancient times for food transformation and product formulations. Enzymes used as food processing aids include various proteases, lipases, carbohydrases, transferases, isomerases, and oxidoreductases. Common examples are papain, bromelain, rennets, pepsins, trypsins, chymotrypsins, pancreatic lipases, lysozyme, various amylases, cellulases, glucose oxidase, polyphenol oxidases, transglutaminases, and glucose isomerases. These enzymes produce certain desirable changes in foods, including improvements in texture, consistency, solubility, appearance, flavor, and yields, among other advantages. Enzymes have been used as processing aids in the manufacture of baked goods, dairy products, meats and fish products, and various beverages (Table 13.2).

Table 13.2 Selected food products and the enzymes involved in their transformations

Baked goods	Amylases	To effect breakdown of starches to increase reducing sugars to improve fermentation, increase loaf volume, prevent staling, and increase product shelf life
	Cellulases and xylanases	To improve conditioning, mixing time, and machinability of doughs
	Oxidases	To improve texture/structure of dough, increase dough volume, and reduce the need for emulsifiers and chemical oxidants (e.g., bromates)
	Proteases	To break down protein molecules in the dough and improve dough handling, reduce dough mixing time, and control dough retraction, and also enhance flavor development; and to break down gluten and protect individuals that are gluten intolerant
	Lipases	To modify fat to reduce dough stickiness, improve stability, and increase volume of doughs; and also enhance flavor
Dairy products	Proteases	To act on milk proteins to modify texture and solubility properties of milk and other dairy products; accelerate cheese ripening, and improve flavor intensity
	Lipases	To transform the fat in milk and other dairy products to produce creamy texture, accelerate ripening, enhance flavor, and increase emulsifying properties
	Lactases	To hydrolyze lactose and prevent development of "sand" taste in frozen ice cream and yogurts, and protect individuals who are lactose intolerant
	Transglutaminases (TGases)	To improve texture and curtail syneresis in yogurts, and improve the mouthfeel properties of sugar-free low-calorie food products
	Catalases	To breakdown residual H_2O_2 used for cold pasteurization of milk in certain countries
Starch products	Amylases, glucoamylases, and pullulanases	To liquefy starch to produce low-viscosity dextrose syrups comprised of maltodextrins and other oligosaccharides; and to saccharify the maltodextrins/oligosaccharides into glucose syrups that are high in reducing sugars (glucose, maltose, and isomaltose)
	Glucose isomerases (immobilized)	To isomerize some of the glucose high glucose syrups (HGS) into fructose—to result in high fructose syrups (HFS) that are sweeter than glucose and HGS. HFS and HGS are both used as sweeteners in the food and beverage industries
	Carbohydrases and protease mixtures	For the breakdown and removal of starches and proteins from oil seeds during oils extraction, to facilitate the recovery and improved oil yields
	Proteases	To modify grain proteins and improve their utility as animal feed

(continued)

Table 13.2 (continued)

Alcoholic beverages	Amylases including glucoamylases (heat-stable forms)	To liquefy grain starches to increase fermentable sugar (glucose and maltose) content and wort yields; ease clarification and filtration of the beverage; enhance yeast growth and accelerate fermentation rates
	Cellulases and pectinases	To break down celluloses and pectic substances in plant materials to facilitate liquefaction, increase fermentable sugar contents, and reduce viscosity for easier handling by yeasts for faster fermentation
	Proteases	To break down proteins in plant materials to improve solubility and flavor, facilitate filtration, and prevent post-product haze formation for increased clarity and stability
	Xylanases and glucanases	To break down complex carbohydrates (xylans and glucans) to improve filtration and clarity, and reduce residual carbohydrate levels during the manufacture of light beers
	Mix of glucanases and proteases	To enhance beer clarity and storage stability, increase solubility and nutritional value, and reduce viscosity haziness in beverages
(a) Fruit juices (nonalcoholic beverages)	Pectinases	To break down and remove pectins to increase yields, facilitate clarification, and improve stability; for peeling operations in some citrus products
	Mix of cellulases and glucosidases	To facilitate maceration of high carbohydrate-containing fruits (e.g., mangoes, bananas, and papaya) and vegetables (e.g., carrots and ginger) for better recovery of juices
	Carbohydrase enzyme mix	To improve maceration to facilitate and enhance juice extraction, increase juice yields, and enhance color and flavor of juices
	Naringinases	To break down naringin and reduce bitterness in citrus juices
(b) Coffee and tea products	Glucose oxidases (GOX)	For the removal of glucose and head space O_2 from bottled/carton packaged soft drinks and dry mixes
	Pectinases	To assist in maceration of tea leaves to improve strength and color
	Tannases	To remove tannins
	Polyphenoloxidases	To improve enzymatic browning and enhance color of the products
Meats and seafood products	Proteases (heat stable forms; e.g., papain, ficin, and bromelain)	To modify texture and induce tenderness in meats and squid, chewability and digestibility, to reduce bitterness and improve flavor as well as nutritive value, produce hydrolysates from meat scraps, underutilized fish species and fish processing discards; enhances flavors in fermented herring (matjes)
	TGases	To improve texture in meats and seafood products, form restructured meats from trimmings and surimi-type products some soft fleshed fish species, form "umami" flavors for use as additives to meat products

(continued)

Table 13.2 (continued)

	Mix of proteases and carbohydrases	To de-skin fish, and aid the breakdown of roe sacs to improve roe yields
Egg products	Lipases	To eliminate egg (whole, liquid or yolk) lipids to preserve its foamability
	Phospholipases	To modify whole egg or egg yolk to increase its emulsifying and gelling properties
	GOX	To de-sugar powdered eggs and curtail enzymatic browning
	Catalases	To break down residual H_2O_2 in egg products or egg ingredients (into H_2O and O_2) after partial sterilization of the eggs with H_2O_2. used to sterilize the product
Dietetic foods	Pancreatins and glutaminases	To facilitate digestion and utilization of wheat proteins (gluten) by individuals who are intolerant to gluten
	Lactases	To break down milk lactose to enhance safe utilization by lactose intolerants
Protein products	Proteases	To hydrolyze proteins derived from yeasts (e.g., single cell proteins), meats, and vegetable such as corn, soy, pulses (e.g., peas and lentils), and wheat to produce highly soluble and low viscosity protein hydrolysates with intense/improved flavors and high nutritive value for use as food ingredients and flavor enhancers; proteases (e.g., chymotrypsins) are also used in food to reduce bitterness resulting from bitter peptides
	Carbohydrase mix	To break down complex carbohydrates in high protein–containing vegetables, and facilitate their removal for the extraction of vegetable proteins in "pure" form and in high yields
Candies and confectioneries	Amylases and invertases	To produce high maltose and high glucose syrups for the manufacture of hard candies, soft drinks, and caramels; increase sweetness and aid soft cream candy manufacture; and for the recovery of sugars from candy scraps
	Amylase, invertase, and protease mix	To improve sweetness, texture, and stability of candied fruits
	Lipases	To modify butterfat to increase buttery flavors and reduce sweetness in candies and caramels
Animal feed and pet care	Cellulases and pectinases	To facilitate silage production
	Alkaline proteases	To decolorize whole blood from abattoirs for use as animal feed
	Carbohydrase, phytase, and protease mix	To increase metabolizable energy and protein utilization; reduce viscosity and liberate bound phosphorous
	Proteases	To reduce pain and inflammation and increase wound healing

13.4.1 Baked goods

Baked goods (breads, biscuits, cakes, etc.) are produced from flours derived from starchy crops such as cereals, grains and tubers like wheat, corn, rice, barley, sorghum and millet, potatoes, cassava, and others. Amylases break down the starches in the flours into low molecular weight dextrins that are subsequently utilized by yeasts to produce dough for the baking process. The action of these enzymes in baked goods manufacture results in improvements in products' texture, volume, taste, color, handling, and anti-staling properties. For example, the actions of glucose oxidase and lipase enzymes in dough strengthen the texture and enhance elasticity of the doughs, and this facility of the enzymes obviates the need for adding chemicals such as bromates and ascorbate for that purpose. Other enzymes employed in baking include lipoxygenases for bleaching flour and strengthening doughs, various proteases for degrading and weakening gluten to improve plasticity of doughs for the purpose of making products such as biscuits; pentosanases or xylanases for improving machinability of the doughs to make them more flexible, easier-to-handle as well as increase loaf volume; and asparaginases to prevent the acrylamide formation by breaking down asparagine in the flours to cut down its availability for reaction with sugars to form acrylamide at elevated temperatures.

13.4.2 Dairy products

Dairy products manufacture involves use of several enzymes, including lipases, lactases, and proteases such as rennets. The proteases cause milk caseins to aggregate and precipitate as clots or curds during cheese making; both proteases and lipases enhance flavor development and formation during cheese ripening, and also in products such as coffee whiteners, snack foods, soups and food spreads; and lactases facilitate the production of lactose-free milks and dairy products for the benefit of lactose intolerant individuals. Catalases are used to degrade residual H_2O_2 that is used in some countries to preserve milk and whey (Burgess and Shaw 1983).

13.4.3 Meat and fish products

Enzymes (proteases) are used in the meat industry to improve meat tenderness. The tenderization process entails changes in muscle proteins structure arising from the breakdown of connective tissue proteins. The proteolytic enzymes responsible for the tenderization include endogenous types in the post mortem animal such as cathepsins, or intentionally added exogenous forms such as papain and bromelain. In some meats, the action of the endogenous enzymes alone is inadequate to achieve the desired tenderization within a useful time period. The exogenous enzymes that are used to assist in the tenderization process are invariably heat stable proteases from microbial and plant sources, to permit reasonable stability and activity at cooking temperatures. These meat tenderizing proteases also tend to have broader specificities. The enzymes for tenderization may be sprinkled over the entire surface of the carcass; alternatively, lumps of the meat may be immersed in the enzyme protease solution for a period of time. These approaches invariably result in uneven distribution of the enzymes in the meats and result in unequal tenderness in different parts of the meat. More effective techniques include distributing the enzyme throughout the meat by injections of the protease solutions into the live animal before slaughter, or via post mortem injections over several parts of the carcass.

In the fishing industry, enzymes are used for various purposes such as for de-skinning fish using a mixture of proteases and glycolytic enzymes (Haard and Simpson 1994; Gildberg et al. 2000). A variety of proteases are also used to accelerate fish fermentations into fish sauce and fish paste (Haard and Simpson 1994), and to develop and improve flavor of fermented products, such as fermented herring, also known as matjes (Simpson and Haard, 1984). Proteases have also been found to be useful in tenderizing squid meat (Raksakulthai and Haard 1992).

13.4.4 Beverages

Several enzymes are employed to assist the manufacture of alcoholic and nonalcoholic beverages. Thermostable amylases, glucanases, and pullanases are used to hydrolyze starches from cereals, grains, tubers, and fruits to increase fermentable sugar levels in the beverage. Heat-stable bacterial phytases are used to hydrolyze insoluble phytins into soluble phytic acid, which lowers the pH of the beverage to better suit microbial and enzymatic activity in the fermentation medium. Various proteases are also used to digest large molecular weight proteins (e.g., albumins and globulins) into smaller and soluble peptides to reduce haze formation in the products.

Fruit and vegetable juice processing also involves the use of enzymes such as pectinases for peeling the fruits, and for other effects including maceration, removal of pectins, viscosity reduction, and clarification of the juice. Amylases and amylo-glucosidases are used to hydrolyze starches present in high starch–containing fruits to prevent haze formation in these products. Cellulases and hemicellulases are used to digest cellulose and hemicellulosic materials present in fruit cell walls, whereas naringinase is used to reduce bitterness in certain juices such as grapefruit juice.

Enzymes such as polyphenoloxidases and peroxidases play a useful role in the manufacture of tea, coffee, and cocoa for their use as beverages. These enzymes promote enzymatic browning reactions during the fermentation of these products to produce the dark colors normally associated with these products. For coffee, the dark color is further enhanced by roasting of the beans. Amylases are used to liquefy starches in these products to permit free flow in the beverages derived from cocoa and chocolate, and tannases are used as aids in tea manufacture to break down tannins to enhance the solubilization of tea solids. Enzymes such as invertase, lipoxygenases, peroxidases, and proteases also play a useful role in flavor formation and development in these beverages, whereas glucose oxidase is also used to remove head-space O_2 from the bottled products.

13.4.5 Candies and confectioneries

Various hydrolytic enzymes are used to facilitate the manufacture of candies and confectioneries. For example, amylases and invertases are used to make chocolate-covered soft cream candies, and they are used with proteases to assist the recovery of sugars from candy scraps (Cowan 1983). An enzyme mix of pectinases, cellulases, invertase, and proteases are used to make candied fruits (Mochizuki et al. 1971), and amylases are also employed for the production high maltose and high glucose syrups as sweetener for use in hard candies, soft drinks, and caramels. Lipolytic enzymes are used to modify butterfat to increase buttery flavors in candies and caramels and reduce sweetness in these products (Burgess and Shaw 1983).

13.4.6 Animal feed and pet care

Harvesting and processing byproducts (or discards) from fruit and vegetables, livestock and fish are rich in useful biomolecules and nutrients such as carbohydrates, proteins, lipids, vitamins and minerals. Some of these components such as complex carbohydrates and structural proteins cannot be utilized or digested readily as food. In this regard, various enzymes are added to these materials to transform them into more digestible and useful forms, for better feed efficiency. Examples of the enzymes used to degrade the complex carbohydrates, proteins, phytates, etc., to aid animal feed manufacture include microbial amylases, cellulases, glucanases, xylanases, phytases, and proteases.

Cellulases and pectinases are also used to aid silage production, whereas alkaline proteases are used to decolorize bloods from livestock slaughtering for use as good protein sources in animal feeds.

Enzymes are also used in pet care to relieve the effects of dry or scaly hair coats, allergies, digestive and immune disorders, skin and weight problems, bloating, and other disorders that cause distress for household pets. Enzymes are also used to enhance wound healing, rectify pain, inflammations and soreness suffered by highly active or arthritic pets.

A summary of food products manufactured via is provided in Table 13.2.

13.5 Controlling enzymatic activity in foods

Enzymatic activity in foods is not always desirable. Examples of the undesirable effects of enzymes in food include (1) extended enzymatic activity in foods beyond the desired transformation that can adversely impact food quality; (2) the presence of endogenous enzymes such as polyphenol oxidases, lipoxygenases, lipases, phospholipases, peroxidases, and others in agricultural materials that can cause undesirable changes in the appearance, texture, and flavors in foods, (3) transformation of innocuous molecules naturally present in raw materials into toxic compounds (biogenic amines) in food products; or (4) destruction of essential food components. Thus, controls for enzymatic activity in food stuffs to prevent their undesirable effects are needed. Enzyme activity and behavior in foods are influenced by several factors such as temperature, pH, water activity (A_w), inhibitors, chelating agents, and reducing agents; and advantage is taken of these effects to develop controls for enzymatic activity in foods.

13.5.1 Temperature effects

Temperature can affect both the activity and stability of enzymes. An increase in temperature generally increases the average kinetic energy of enzyme molecules and their substrates, and increases reaction rates (up to the optimum temperature for the enzyme's activity and/or stability). Beyond the temperature optimum, further increases in temperature are not useful for increasing the rates of enzyme catalysis. This is because the higher temperatures cause disruption of the forces that maintain the integrity of the molecules needed for stability and normal catalytic functions. Thus, most enzymes undergo denaturation and inactivation at higher temperatures and lose catalytic activity. This heat inactivation phenomenon may be irreversible or reversible depending on the intensity and/or extent of the heat treatment, and advantage is taken of this high temperature effect on enzymes to control undesirable enzymatic reactions and protect foods from enzyme-induced food quality loss through thermal processes such as sterilization, blanching, ultra high temperature (UHT), high temperature short term (HTST), pasteurization, and related thermal treatments. Nevertheless, thermal processing of foods has the drawback of causing destruction of heat-labile essential foods components (e.g., some vitamins and essential oils), thus the need for alternative and effective nonthermal methods.

Low temperatures slow down the average kinetic energy of enzyme molecules. The low thermal energy makes enzyme molecules sluggish, they collide less frequently and less effectively with their substrates; thus, reaction rates are slower at low temperatures, and use is made of the low temperature effects on enzymes by using approaches such as refrigeration, icing, or chilling, refrigerated sea water (RSW) storage and freezing to slow down the undesirable effects of enzymes after harvest.

Low temperature treatments are generally used to control enzymatic activity in fresh vegetables, eggs, seafood, and meats; however, other processed foods such as food spreads (margarine and butter), vacuum-packaged meats and seafood products, pasteurized milk, cheeses, and yogurt also benefit from the useful effects from low temperature treatments. Low temperature treatments generally slow down enzymatic activity, but they do not completely inactivate the enzymes. Enzymatic activity still takes place, albeit at much reduced rates. Once the food material is removed from the low temperature environment, enzyme activity may restart and produce their undesirable effect in foods. Thus, refrigerated and chilled storage in particular may not be considered as effective methods to use for long-term storage of fresh foods.

There are certain enzymes that have remarkable stability and activity of extreme temperatures, and may survive the traditional temperature treatments and induce spoilage in foods. For such enzymes, other procedures are needed to inhibit their action.

13.5.2 *Effect of pH*

Enzyme activity in foods is also affected by pH. Enzymes tend to be destabilized and inactivated irreversibly at extreme pH values, and use is made of this effect by pH to control enzymatic activity in foods. For example, adding acids such as citrate, lactate, or ascorbate to shrimp, fresh fruits, and vegetables, creates acidic conditions in which polyphenol oxidase is inactive and unable to elicit their undesirable discolorations in the food products. In the same way, when a mixture of glucose, catalase, and glucose oxidase is applied to raw shrimp, the glucose is oxidized to gluconic acid to lower pH and inactivate polyphenol oxidase and prevent the undesirable discolorations they cause. Sometimes vinegar or lemon juice is sprinkled on fresh fruits and vegetables to stop the browning in these products. Also, lowering the pH of citrus juice achieves inactivation of pectin esterase (Owusu-Yaw et al. 1988); and lipoxygenase activity in soy flours is significantly reduced at pH $\leq$5.0 (Thakur and Nelson 1997).

13.5.3 *Effect of inhibitors*

Enzyme inhibitors act to slow down enzyme catalysis. They are able do this in various ways; for example, by binding with the enzyme, or by eliminating co-reactants from the enzyme catalyzed reaction milieu, such as removal of O_2 from polyphenol oxidase catalyzed reactions. They may also strip the enzyme of an essential cofactor or prosthetic group, and alter the pH to destabilize the enzyme. Thus, a range of compounds are available for inhibiting enzyme activity in foods. They include reducing agents such as sulfites and cysteine, metal ion chelators such as EDTA, carboxylic acids and phosphates, acidulants such as phosphoric acids, citric acid, sorbates, benzoates, and molecules such as polypeptides, organic acids, and resorcinols that bind directly to enzymes and prevent their activity. A number of food-grade protein inhibitors derived from egg whites, bovine/porcine plasma and potato flour are all used to control undesirable proteolysis in foods. Egg white contains the inhibitor ovomucoid that hinder serine protease activity; bovine and porcine plasmas have α_2-macroglobulin, a broad spectrum protease inhibitor; and potato flour has a serine protease inhibitor (potato serine protease inhibitor or PSPI). Extracts from these food materials containing these inhibitors have been used to prevent surimi texture softening because of proteolysis (Weerasinghe et al. 1996).

13.5.4 Effect of water activity (A_w)

Enzymes generally require some minimum amount of moisture to maintain their integrity and activity. Moisture availability, described as water activity (A_w) in foods may be modified by several methods including drying, freezing, addition of humectants (e.g., salt, sugar, honey, maple syrup, gelatin, glycerol, and other polyols), or by lyophilization (Tejada et al. 2008). The enzymes require moisture to provide a thin film of moisture (bound water) to maintain the appropriate integrity for functional activity. Moisture is also needed to solubilize substrates and also serve as the reaction medium to enhance the interaction of the enzyme with the substrate molecules and/or to participate in the reaction as co-reactant. A reduced A_w induces conformational changes, destabilization of enzyme integrity, and loss of catalytic activity. In general, most enzymes require $A_w \geq 0.85$ to have catalytic activity, the exceptions being lipases that may actually remain active and gain activity at A_w values as low as 0.3, and even as low as 0.1 (Loncin et al. 1968). The unusual behavior of lipases is attributed to an interfacial phenomenon that proceeds better in a reduced moisture environment. Foodstuffs prepared based on A_w reduction to control the enzymes (and microbial) activity in foods include the use of sugar in food spreads such as jams and jellies, the use of salt in pickled vegetables, the addition of glycerol in cookies and liqueurs, and incorporation of gelatin in confectioneries and candies.

13.5.5 Effect of irradiation

Irradiation or cold pasteurization of foods involves the use of ionizing radiations to treat foods to preserve them. Food irradiation uses gamma rays, x-rays, and accelerated electron beams to control enzyme activity. However, the dosage needed to achieve complete and irreversible inactivation of enzymes may be too high and could induce undesirable effects of its own (e.g., nutrient loss) in foods. The technique is used to a limited extent in meats, seafood, fruits, and in cereal and grains for long-term preservation. However, the procedure is not very appealing to consumers, and therefore not extensively used in food processing.

13.5.6 Effects of pressure

High pressure treatment or high pressure processing (HPP) is a nonthermal method used to process foods. At high pressures, enzyme molecules (and microorganisms) suffer inactivation due to conformational changes in their 3D structures (Cheftel 1992). Different enzymes display different sensitivities to pressure. Some may be very sensitive to pressure and become inactivated at relatively low pressures (few 100 MPa), whereas some others may withstand much higher pressures. The pressure effects on enzymes may or may not also be reversible, depending on the intensity and the duration of the pressure treatment. Those enzymes that can tolerate extreme pressures may be deactivated using suitable pressure treatments in combination with other barriers such as temperature, pH and/or inhibitors (Ashie and Simpson 1995; Ashie et al. 1996; Sareevoravitkul et al. 1996; Katsaros et al. 2010).

The relative advantages with high pressure processing include the minimal adverse effects it has on the attributes of food such as flavor, appearance, and nutritive value. High pressure treatments keep foods fresher and highly nutritious, improve texture, and maintain the appearance and taste of foods. The HPP approach also protects heat-labile essential food components better, and extends product shelf life with minimum need for added chemical preservatives. The technology has been used by companies in Japan and the United States to produce food products such as ready-to-eat meats and meat products, food spreads (jams), fresh juices and beverages (sake), processed fruits and vegetables, fresh salads, and dips. High-pressure treatments are also used to shuck and retrieve meats from shellfish such as oysters, clams, and lobster.

13.6 Future prospects

The use of food enzymes has grown steadily for several decades now, and it is expected that this trend would continue into the future because of the relative advantages of enzyme-assisted processing methods over existing technologies. Enzymes discriminate in the molecules they select to act on as substrates; thus food products produced by enzyme treatments are more uniform and more consistent from batch to batch. Enzymatic approaches entail milder treatments under mild reaction conditions, and thus are more environmental friendly and would protect the environment better compared with traditional methods based on the use of chemicals and other drastic methods. Consumers perceive food enzymes as nontoxic natural components in foods and are more receptive their use as processing aids. They can be used to selectively remove substances whose presence in foods pose hazards to consumers such as glucosinolates, acrylamide, or phytates.

Recent developments in genetic engineering are creating new and improved enzymes to carry out specific food applications more efficiently. They are enabling useful enzymes from plant and animal sources, and even from harmful or potentially dangerous microorganisms to be produced safely, rapidly, and in high yields on a more consistent basis for food and related uses. These new enzymes can be designed and synthesized to endow them with greater stabilities and higher activities under extreme temperature and pH conditions, and also remain active and functional in nonaqueous systems. Thus, newer products with improved or desired properties could be designed and fabricated with these new enzymes for specific food and/or health benefits to consumers. They would enable the fabrication of biosensors with improved stabilities and selectivities for food and other analytical work.

Although traditional food processing and preservation methods do control the deleterious effects of enzymes in foods to some extent, newer and more effective methods are needed to overcome the various limitations with the traditional methods. In this regard, superior food enzymes are expected to enhance the efficacy of nonthermal food processing and preservation methods to control the undesirable effects of enzymes in foods and preserve food quality.

References

Abd-el-Al, A.T., and H.J. Phaff. 1968. Exo-beta-glucanases in yeast. *The Biochemical Journal* 109(3): 347–360.

Adigüzel, A.C., B.O. Bitlisli, I. Ya a, and N.T. Eriksen. 2009. Sequential secretion of collagenolytic, elastolytic, and keratinolytic proteases in peptide-limited cultures of two *Bacillus cereus* strains isolated from wool. *Journal of Applied Microbiology* 107(1): 226–234.

Ashie, I.N.A., and B.K. Simpson. 1995. Effects of hydrostatic pressure on alpha 2-macroglobulin and selected proteases. *Journal of Food Biochemistry* 18: 377–391.

Ashie, I.N.A., B.K. Simpson, and H.S. Ramaswamy. 1996. Control of endogenous enzyme activity in fish muscle by inhibitors and hydrostatic pressure using rsm. *Journal of Food Science* 61(2): 350–356.

Bachman, E.S., and D.R. McClay. 1996. Molecular cloning of the first metazoan beta-1,3 glucanase from eggs of the sea urchin *Strongylocentrotus purpuratus*. *Proceedings of the National Academy of Science of the United States of America* 93: 6808–6813.

Beschin, A., M. Bilej, F. Hanssens, J. Raymakers, E. Van Dyck, H. Revets, L. Brys, J. Gomez, P. De Baetselier, and M. Timmermans. 1998. Identification and cloning of a glucan- and lipopolysaccharide-binding protein from *eisenia foetida* earthworm involved in the activation of prophenoloxidase cascade. *The Journal of Biological Chemistry* 273(38): 24948–24954.

Bills, D.D., and E.A. Day. 1964. Determination of the major free fatty acids of cheddar cheese. *Journal of Dairy Science* 47: 733–738.

Bogs, J., and K. Geider. 2000. Molecular analysis of sucrose metabolism of *erwinia amylovora* and influence on bacterial virulence. *Journal of Bacteriology* 182(19): 5351–5358.

Bonilha, P.R.M., V. Menocci, A.J. Goulart, M.L.T.M. Polizeli, and R. Monti. 2006. Cyclodextrin glycosyltransferase from *Bacillus licheniformis*: Optimization of production and its properties. *Brazilian Journal of Microbiology* 37(3): 317–323.

Burgess, K., and M. Shaw. 1983. Dairy. In *Industrial enzymology: The application of enzymes in industry*, ed. T. Godfrey and J. Reichelt. New York: The Nature Press.

Chaplin, M.F., and C. Bucke. 1990. Enzyme preparation and use. In *Enzyme technology*, ed. M.F. Chaplin and C. Bucke. Cambridge, UK: Cambridge University Press.

Cheetham, P.S.J., A.J. Hacking, and M. Vlitos. 1989. Synthesis of novel disaccharides by a newly isolated fructosyl transferase from *Bacillus subtilis*. *Enzyme and Microbial Technology* 11(4): 212–219.

Cheftel, J.C. 1992. Effects of high hydrostatic pressure on food constitutes: An overview. In *High pressure and biotechnology*, ed. C. Balny, R. Hayashi, K. Heremans, and P. Masson, 195–209. London: Colloque INSERM/Libby Eurotech Ltd.

Coleman, G.S. 1985. The cellulase content of 15 species of entodiniomorphid protozoa, mixed bacteria and plant debris isolated from the ovine rumen. *The Journal of Agricultural Science* 104: 349–360.

Cooper, R.A., and R.N. Greenshields. 1961. Sucrases in *phaseolus vulgaris*. *Nature* 191(4788): 601–602.

Cooper, R.A., and R.N. Greenshields. 1964. The partial purification and some properties of two sucrases of *phaseolus vulgaris*. *The Biochemical Journal* 92(2): 357–364.

Couto, M.A., S.S.L. Harwig, and R.I. Lehrer. 1993. Selective inhibition of microbial serine proteases by enap-2, an antimicrobial peptide from equine neutrophils. *Infection and Immunity* 61(7): 2991–2994.

Cowan, D. 1983. Proteins. In *Industrial enzymology: The application of enzymes in industry*, ed. T. Godfrey and J. Reichelt. New York: The Nature Press.

Dalev, P.G. 1994. Utilisation of waste feathers from poultry slaughter for production of a protein concentrate. *Bioresource Technology* 48(3): 265–267.

Dalev, P.G., and L.S. Simeonova. 1992. An enzyme biotechnology for the total utilization of leather wastes. *Biotechnology Letters* 14(6): 531–534.

De La Motte, R.S., and F.W. Wagner. 1987. *Aspergillus niger* sulfhydryl oxidase. *Biochemistry* 26(23): 7363–7371.

Dinges, J.R., C. Colleoni, M.G. James, and A.M. Myers. 2003. Mutational analysis of the pullulanase-type debranching enzyme of maize indicates multiple functions in starch metabolism. *The Plant Cell* 15(3): 666–680.

Do Nascimento, W.C.A., and M.L. Leal Martins. 2004. Production and properties of an extracellular protease from thermophilic *Bacillus* sp. *Brazilian Journal of Microbiology* 35(1–2): 91–96.

Elliott, K.J., W.O. Butler, C.D. Dickinson, Y. Konno, T.S. Vedvick, L. Fitzmaurice, and T.E. Mirkov. 1993. Isolation and characterization of fruit vacuolar invertase genes from two tomato species and temporal differences in mRNA levels during fruit ripening. *Plant Molecular Biology* 21(3): 515–524.

Elortza, F., T.S. Nühse, L.J. Foster, A. Stensballe, S.C. Peck, and O.N. Jensen. 2003. Proteomic analysis of glycosylphosphatidylinositol-anchored membrane proteins. *Molecular and Cellular Proteomics* 2(12): 1261–1270.

Ferrarese, L., L. Trainotti, P. Moretto, P.P. De Laureto, N. Rascio, and G. Casadoro. 1995. Differential ethylene-inducible expression of cellulase in pepper plants. *Plant Molecular Biology* 29(4): 735–747.

Gildberg, A., B.K. Simpson, and N.F. Haard. 2000. Uses of enzymes from marine organisms. In *Seafood enzymes: Utilization and influence on postharvest seafood quality*, ed. N.F. Haard and B.K. Simpson. New York: Marcel Dekker.

Goda, S.K., O. Eissa, M. Akhtar, and N.P. Minton. 1997. Molecular analysis of a *Clostridium butyricum* ncimb 7423 gene encoding 4-α-glucanotransferase and characterization of the recombinant enzyme produced in *Escherichia coli*. *Microbiology* 143(10): 3287–3294.

Godfrey, T., and S. West. 1996. Introduction to industrial enzymology. In *Industrial enzymology*, ed. T. Godfrey and S. West. Cambridge: Cambridge University Press.

González-Teuber, M., M.J. Pozo, A. Muck, A. Svatos, R.M. Adame-Álvarez, and M. Heil. 2010. Glucanases and chitinases as causal agents in the protection of acacia extrafloral nectar from infestation by phytopathogens. *Plant Physiology* 152(3): 1705–1715.

Gupta, A.K., and I.S. Bhatia. 1980. Glucofructosan biosynthesis in *fusarium oxysporum*. *Phytochemistry* 19(12): 2557–2563.

Gupta, A.K., and I.S. Bhatia. 1982. Glucofructosan biosynthesis in *fusarium oxysporum*: Regulation and substrate specificity of fructosyl transferase and invertase. *Phytochemistry* 21(6): 1249–1253.

Guzmán, F., S. Barberis, and A. Illanes. 2007. Peptide synthesis: Chemical or enzymatic. *Electronic Journal of Biotechnology* 10(2): 279–314.

Ha, J.K., and R.C. Lindsay. 1993. Release of volatile branched-chain and other fatty acids from ruminant milk fats by various lipases. *Journal of Dairy Science* 76(3): 677–690.

Haard, N.F., and B.K. Simpson. 1994. Proteases from aquatic organisms and their uses in the seafood industry. In *Fisheries processing: Biotechnological applications*, ed. A.M. Martin. London: Chapman and Hall.

Heilskov, N.S.C. 1951. Studies on animal lactase. *Acta Physiologica Scandinavica* 24: 84–89.

Homann, A., and J. Seibel. 2009. Towards tailor-made oligosaccharides-chemo-enzymatic approaches by enzyme and substrate engineering. *Applied Microbiology and Biotechnology* 83(2): 209–216.

Hoober, K.L., S.L. Sheasley, H.F. Gilbert, and C. Thorpe. 1999. Sulfhydryl oxidase from egg white: A facile catalyst for disulfide bond formation in proteins and peptides. *The Journal of Biological Chemistry* 274(32): 22147–22150.

Hou, C.T. 2002. Industrial uses of lipase. In *Lipid biotechnology*, ed. T.M. Kuo and H.W. Gardner. New York: Marcel Dekker.

Hrassnigg, N., and K. Crailsheim. 2005. Differences in drone and worker physiology in honeybees (*Apis mellifera*). *Apidologie* 36(2): 255–277.

Kalisz, H.M. 1988. Microbial proteinases. *Advances in Biochemical Engineering/Biotechnology* 36: 1–65.

Kaneda, M., and N. Tominaga. 1975. Isolation and characterization of a proteinase from the sarcocarp of melon fruit. *Journal of Biochemistry* 78(6): 1287–1296.

Katsaros, G.I., M. Tsevdou, T. Panagiotou, and P.S. Taoukis. 2010. Kinetic study of high pressure microbial and enzyme inactivation and selection of pasteurisation conditions for Valencia orange juice. *International Journal of Food Science and Technology* 45(6): 1119–1129.

Kaufman, S.P., and O. Fennema. 1987. Evaluation of sulfhydryl oxidase as a strengthening agent for wheat flour dough. *Cereal Chemistry* 64: 172–176.

Kikuchi, T., H. Shibuya, and J.T. Jones. 2005. Molecular and biochemical characterization of an endo-β-1,3-glucanase from the pinewood nematode *bursaphelenchus xylophilus* acquired by horizontal gene transfer from bacteria. *The Biochemical Journal* 389(1): 117–125.

Kim, S.M., and E.J. Kim. 2009. Development of chicken breast noodles adding rubus *coreanum miquel* and *opuntia ficus-indica* var. Saboten. *Journal of the Korean Society of Food Science and Nutrition* 38(8): 1111–1117.

Kobayashi, M., and S. Shimizu. 1999. Cobalt proteins. *European Journal of Biochemistry* 261(1): 1–9.

Kozhemyako, V.B., D.V. Rebrikov, S.A. Lukyanov, E.A. Bogdanova, A. Marin, A.K. Mazur, S.N. Kovalchuk, E.V. Agafonova, V.V. Sova, L.A. Elyakova, and V.A. Rasskazov. 2004. Molecular cloning and characterization of an endo-1,3-β-d-glucanase from the mollusk *spisula sachalinensis*. *Comparative Biochemistry and Physiology. Part B, Biochemistry and Molecular Biology* 137(2): 169–178.

Layman, P.L. 1986. Industrial enzymes battling to remain specialties. *Chemical and Engineering News* 64(37): 11–14.

Lee, K.J., J.D. Choi, and J.Y. Lim. 1992. Purification and properties of intracellular fructosyl transferase from *aureobasidium pullulans*. *World Journal of Microbiology and Biotechnology* 8(4): 411–415.

Leubner-Metzger, G. 2003. Functions and regulation of β-1,3-glucanases during seed germination, dormancy release and after-ripening. *Seed Science Research* 13(1): 17–34.

Li, D., P. Adhikari, J.A. Shin, J.H. Lee, Y.J. Kim, X.M. Zhu, J.N. Hu, J. Jin, C.C. Akoh, and K.T. Lee. 2010. Lipase-catalyzed interesterification of high oleic sunflower oil and fully hydrogenated soybean oil comparison of batch and continuous reactor for production of zero trans shortening fats. *LWT-Food Science and Technology* 43(3): 458–464.

Lim, D., and W.F. Shipe. 1972. Proposed mechanism for the antioxygenic action of trypsin in milk. *Journal of Dairy Science* 55: 753–758.

Liu, B., Y. Lu, Z. Xin, and Z. Zhang. 2009. Identification and antifungal assay of a wheat β-1, 3-glucanase. *Biotechnology Letters* 31(7): 1005–1010.

Loncin, N., J.J. Bimbenet, and J. Lenges. 1968. Influence of water activity on the spoilage of foodstuffs. *International Journal of Food Science and Technology* 3: 131–142.

Lopez-Amaya, C., and A.G. Marangoni. 2000a. Lipases. In *Seafood enzymes: Utilization and influence on postharvest seafood quality*, ed. N.F. Haard and B.K. Simpson. New York: Marcel Dekker.

Lopez-Amaya, C., and A.G. Marangoni. 2000b. Phospholipases. In *Seafood enzymes: Utilization and influence on postharvest seafood quality*, ed. N.F. Haard and B.K. Simpson. New York: Marcel Dekker.

Lund, L.R., J. Rømer, T.H. Bugge, B.S. Nielsen, T.L. Frandsen, J.L. Degen, R.W. Stephens, and K. Danø. 1999. Functional overlap between two classes of matrix-degrading proteases in wound healing. *The EMBO Journal* 18(17): 4645–4656.

Mathewson, P.R. 1998. Enzymes: Practical guides for the food industry. In *Enzymes*, Handbook series. American chemical society, ed. P.R. Mathewson. St. Paul: Eagan Press.

McWethy, S.J., and P.A. Hartman. 1979. Extracellular maltase of *Bacillus brevis*. *Applied and Environmental Microbiology* 37(6): 1096–1102.

Menocci, V., A.J. Goulart, P.R. Adalberto, O.L. Tavano, D.P. Marques, J. Contiero, and R. Monti. 2008. Cyclodextrin glycosyltransferase production by new *Bacillus* sp. strains isolated from Brazilian soil. *Brazilian Journal of Microbiology* 39(4): 682–688.
Mistry, V.V. 2006. Chymosin in cheese making. In *Food biochemistry and food processing*, ed. Y.H. Hui, W.K. Nip, L.M.L. Nollet, G. Paliyath, and B.K. Simpson. Ames: Blackwell Publishing.
Mochizuki, K., K. Isobe, and Y. Sawada. 1971. Method for producing candied fruit. US 3,615,687.
Mojovic, L., S. Siler-Marinkovic, G. Kukic, and G. Vunjak-Novakovic. 1993. *Rhizopus arrhizus* lipase-catalyzed interesterification of the midfraction of palm oil to a cocoa butter equivalent fat. *Enzyme and Microbial Technology* 15(5): 438–443.
Moore, M.M., T.J. Schober, P. Dockery, and E.K. Arendt. 2004. Textural comparisons of gluten-free and wheat-based doughs, batters, and breads. *Cereal Chemistry* 81(5): 567–575.
Moore, M.M., M. Heinbockel, P. Dockery, H.M. Ulmer, and E.K. Arendt. 2006. Network formation in gluten-free bread with application of transglutaminase. *Cereal Chemistry* 83(1): 28–36.
Morikawa, M., Y. Izawa, N. Rashid, T. Hoaki, and T. Imanaka. 1994. Purification and characterization of a thermostable thiol protease from a newly isolated hyperthermophilic *Pyrococcus* sp. *Applied and Environmental Microbiology* 60(12): 4559–4566.
Motoki, M., and K. Seguro. 1998. Transglutaminase and its use for food processing. *Trends in Food Science and Technology* 9(5): 204–210.
Mukherjee, K.D., and H.J. Hills. 1994. Lipases from plants. In *Lipases: Their structure, biochemistry and application*, ed. P. Wooley and S.B. Petersen. Cambridge: Cambridge University Press.
Muñoz-Gutiérrez, I., M.E. Rodríguez-Alegría, and A. López Munguía. 2009. Kinetic behaviour and specificity of β-fructosidases in the hydrolysis of plant and microbial fructans. *Process Biochemistry* 44(8): 891–898.
Murakami, Y., T. Yoshida, and A. Hirata. 1998. Enzymatic synthesis of n-formyl-l-aspartyl-l-phenylalanine methyl ester (aspartame precursor) utilizing an extractive reaction in aqueous/organic biphasic medium. *Biotechnology Letters* 20(8): 767–769.
Mustranta, A., E. Karvonen, H. Ojamo, and M. Linko. 1981. Production of mold lactase. *Biotechnology Letters* 3(7): 333–338.
Nair, S.U., R.S. Singhal, and M.Y. Kamat. 2006. Enhanced production of thermostable pullulanase type 1 using bacillus cereus fdta 13 and its mutant. *Food Technology and Biotechnology* 44(2): 275–282.
Novo Nordisk. 1993. *Annual report novo nordisk a/s*. Denmark: Bagsaverd.
Oku, T., T. Tokunaga, and N. Hosoya. 1984. Nondigestibility of a new sweetener, 'neosugar', in the rat. *The Journal of Nutrition* 114(9): 1574–1581.
Olempska-Beer, Z.S., R.I. Merker, M.D. Ditto, and M.J. DiNovi. 2006. Food-processing enzymes from recombinant microorganisms: A review. *Regulatory Toxicology and Pharmacology* 45(2): 144–158.
Owusu-Yaw, J., M.R. Marshall, J.A. Koburger, and C.I. Wei. 1988. Low ph inactivation of pectinesterase in single-strength orange juice. *Journal of Food Science* 53(2): 504–507.
Prapulla, S.G., V. Subhaprada, and N.G. Karanth. 2000. Microbial production of oligosaccharides: A review. In *Advances in applied microbiology*, ed. S.L. Neidleman, A.I. Laskin, J.W. Bennet, and G. Gadd. New York: Academic.
Pugsley, A.P., C. Chapon, and M. Schwartz. 1986. Extracellular pullulanase of *klebsiella pneumoniae* is a lipoprotein. *Journal of Bacteriology* 166(3): 1083–1088.
Quiroz-Castañeda, R.E., E. Balcázar-López, E. Dantán-González, A. Martinez, J. Folch-Mallol, and C. Martínez-Anaya. 2009. Characterization of cellulolytic activities of *bjerkandera adusta* and *pycnoporus sanguineus* on solid wheat straw medium. *Electronic Journal of Biotechnology* 12(4): 15.
Raksakulthai, N., and N.F. Haard. 1992. Correlation between the concentration of peptides and amino acids and the flavour of fish sauce. *ASEAN Food Journal* 7: 86–90.
Ramana Rao, M.V., and S.M. Dutta. 1978. Lactase activity of microorganisms. *Folia Microbiologica* 23(3): 210–215.

Robert, J.H., and B. Darbyshire. 1980. Fructan synthesis in onion. *Phytochemistry* 19: 1017–1020.

Rudenskaya, G.N., A.M. Bogacheva, A. Preusser, A.V. Kuznetsova, Y.E. Dunaevsky, B.N. Golovkin, and V.M. Stepanov. 1995. Macluralisin: A serine proteinase from fruits of maclura pomifera (raf.) schneid. *Planta* 196(1): 174–179.

Sareevoravitkul, R., B.K. Simpson, and H.S. Ramaswamy. 1996. Comparative properties of bluefish (pomatomus saltatrix) gels formulated by high hydrostatic pressure and heat. *Journal of Aquatic Food Product Technology* 5(3): 65–79.

Schmid, R.D., and R. Verger. 1998. Lipases: Interfacial enzymes with attractive applications. *Angewandte Chemie* 37(12): 1609–1633.

Shahidi, F., and X.Q. Han. 1993. Encapsulation of food ingredients. *Critical Reviews in Food Science and Nutrition* 33(6): 501–547.

Shiomi, N., J. Yamada, and M. Izawa. 1979. Synthesis of several fructo-oligosaccharides by asparagus fructosyl transferases. *Agricultural and Biological Chemistry* 43: 2233–2244.

Sidhu, M.S., M.K. KalRa, and D.K. Sandhu. 1986. Purification and characterization of cellulolytic enzymes from *trichoderma harzianum*. *Folia Microbiologica* 31(4): 293–302.

Simpson, B.K., and N.F. Haard. 1984. Trypsin from Greenland cod as a food-processing aid. *Journal of Applied Biochemistry* 6: 135–143.

Sliwkowski, M.X., H.E. Swaisgood, D.A. Clare, and H.R. Horton. 1984. Kinetic mechanism and specificity of bovine milk sulphydryl oxidase. *The Biochemical Journal* 220(1): 51–55.

Sritunyalucksana, K., S.Y. Lee, and K. Söderhäll. 2002. A β-1,3-glucan binding protein from the black tiger shrimp, *penaeus monodon*. *Developmental and Comparative Immunology* 26(3): 237–245.

Sträter, N., I. Przylas, W. Saenger, Y. Terada, K. Fujii, and T. Takaha. 2002. Structural basis of the synthesis of large cycloamyloses by amylomaltase. *Biologia: Section Cell and Molecular Biology* 57(Suppl. 11): 93–99.

Tafazoli, S., A.W. Wong, T. Akiyama, H. Kajiura, E. Tomioka, I. Kojima, H. Takata, and T. Kuriki. 2010. Safety evaluation of amylomaltase from *thermus aquaticus*. *Regulatory Toxicology and Pharmacology* 57(1): 62–69.

Takaha, T., M. Yanase, H. Takata, S. Okada, and S.M. Smith. 1996. Potato d-enzyme catalyzes the cyclization of amylose to produce cycloamylose, a novel cyclic glucan. *The Journal of Biological Chemistry* 271(6): 2902–2908.

Tejada, L., M. Vioque, R. Gómez, and J. Fernández-Salguero. 2008. Effect of lyophilisation, refrigerated storage and frozen storage on the coagulant activity and microbiological quality of cynara cardunculus l. Extracts. *Journal of the Science of Food and Agriculture* 88(8): 1301–1306.

Terada, Y., M. Yanase, H. Takata, T. Takaha, and S. Okada. 1997. Cyclodextrins are not the major cyclic α-1,4-glucans produced by the initial action of cyclodextrin glucanotransferase on amylose. *The Journal of Biological Chemistry* 272(25): 15729–15733.

Thakur, B.R., and P.E. Nelson. 1997. Inactivation of lipoxygenase in whole soy flour suspension by ultrasonic cavitation. *Nahrung Food* 41(5): 299–301.

Thomas, B.R., M. Inouhe, C.R. Simmons, and D.J. Nevins. 2000. Endo-1,3;1,4-β-glucanase from coleoptiles of rice and maize: Role in the regulation of plant growth. *International Journal of Biological Macromolecules* 27(2): 145–149.

Thorpe, C., K.L. Hoober, S. Raje, N.M. Glynn, J. Burnside, G.K. Turi, and D.L. Coppock. 2002. Sulfhydryl oxidases: Emerging catalysts of protein disulfide bond formation in eukaryotes. *Archives of Biochemistry and Biophysics* 405(1): 1–12.

Tucker, M.L., R. Sexton, E. del Campillo, and L.N. Lewis. 1988. Bean abscission cellulase. *Plant Physiology* 88: 1257–1262.

Watanabe, H., and G. Tokuda. 2001. Animal cellulases. *Cellular and Molecular Life Sciences* 58(9): 1167–1178.

Weerasinghe, V.C., M.T. Morrissey, and H. An. 1996. Characterization of active components in food-grade proteinase inhibitors for surimi manufacture. *Journal of Agricultural and Food Chemistry* 44(9): 2584–2590.

Wells, G.H. 1966. *Tenderness of freeze-dried chicken with emphasis on enzyme treatments*. East Lansing: Michigan State University.

Wenham, D.G., T.D. Hennessey, and J.A. Cole. 1979. Regulation of glucosyl- and fructosyltransferase synthesis by continuous cultures of *streptococcus mutans*. *Journal of General Microbiology* 114(1): 117–124.
Whitaker, J.R. 1994. The sulfhydryl proteases. In *Principles of enzymology for the food sciences*, 2nd ed, ed. J.R. Whitaker. New York: CRC Press/Marcel Dekker.
Wilkinson, M.G., and K.N. Kilcawley. 2005. Mechanisms of incorporation and release of enzymes into cheese during ripening. *International Dairy Journal* 15(6–9): 817–830.
Yamada, H. 1989. Localization in skin, activation and reaction mechanisms of skin sulfhydryl oxidase. *The Japanese Journal of Dermatology* 99(8): 861–869.
Yamasaki, Y., S. Nakashima, and H. Konno. 2008. Pullulanase from rice endosperm. *Acta Biochimica Polonica* 55(3): 507–510.
Yoshio, N., I. Maeda, H. Taniguchi, and M. Nakamura. 1986. Purification and properties of d-enzyme from malted barley. *Journal of the Japanese Society of Starch Science* 33: 244–252.
Yun, J.-W., S.-C. Kang, and S.-K. Song. 1995. Continuous production of fructooligosaccharides from sucrose by immobilized fructosyltransferase. *Biotechnology Techniques* 9(11): 805–808.

Chapter 14
Emerging technologies for microbial control in food processing

Michael O. Ngadi, Mohsin Bin Latheef, and Lamin Kassama

14.1 Introduction

Food holds a position of unmatched significance in human history, and the preparation of the same is a practice that has influenced human society like none other. This level of importance enjoyed by food preparation (and processing) automatically lays an almost equal emphasis on a pivotal concept: food safety.

The history of food safety is perhaps as old as the history of food preparation itself. Right from the very modest beginnings of agriculture to the modern food industry, we have faced several challenges to the safety of prepared food (Tannahill 1973; Toussaint-Samat 1992). With advances in food microbiology, more food-borne pathogens have been identified during the last two decades than any other period in history. Food safety has thus become the single most important aspect of food processing and consumption. Yet, the demand for top quality food remains increasingly high. Conventional methods and strategies to ensure food safety are being replaced by novel and emerging methods of microbial control that promise to deliver fresher, higher quality products while striving to be sustainable at the same time. This chapter discusses these new methods after a brief review of historical perspectives, followed by the methods used to control microbial activity in foods, both current and emerging.

M.O. Ngadi (✉) • M.B. Latheef
Department of Bioresource Engineering, McGill University, Montreal, Québec, Canada
e-mail: michael.ngadi@mcgill.ca; mohsin.latheef@mail.mcgill.ca

L. Kassama
Department of Food and Animal Sciences, Alabama A&M University, Normal, AL, USA

J.I. Boye and Y. Arcand (eds.), *Green Technologies in Food Production and Processing*, Food Engineering Series, DOI 10.1007/978-1-4614-1587-9_14,

14.2 Historical perspectives

Food safety was conceived the first time when it was understood that consumption of certain foods could be lethal to humans. The beginning of the nineteenth century saw the advent of thermal processing, one of the safest methods of food preparation. The original method proposed by the French confectioner Nicolas Appert relied on the principle of *airtight food preservation.* It involved keeping food in corked glass bottles that are sealed with wax and heated in boiling water, in what is now acknowledged to be the earliest form of *canning*. The success of Appert's method led to the patenting of a similar method employing tin cans by Peter Durand of England in 1810. Although neither Appert nor Durand knew why their methods were able to successfully preserve food, thermal processing became widely adopted by mass producers of food, who modified these methods to be more efficient and cost effective (Day and McNeil 1996).

Pasteurization became widely used in the early twentieth century. The development of refrigeration technology led to acceptance and increased popularity of relatively safe, shelf-stable rapidly frozen foods in the latter half of the century. Yet, despite improvements in preservation technologies, there has been no respite from food poisoning. The numerous and repeated outbreaks of food-borne diseases further emphasize the need for stricter microbial control in food processing. Early control measures adopted include *severe heating of canned foods* (so as to destroy the most thermal-resistant spores) and a *zero tolerance* approach in processing ready-to-eat foods. However, severe heat processing not only inactivates microorganisms; it also destroys important food nutrients as well as degrades some textural and physical characteristics of food products. This is the major motivating factor driving the interest in developing *novel* methods that can retain the freshness of processed foods.

In spite of advances in analytical methods (for the detection, identification, and isolation of food-borne microbes), food safety was traditionally based on an *end-product analysis* approach. The adoption of the Hazard Analysis and Critical Control Points (HACCP) system brought about a monumental change through the use of a new *safety by design* approach in which all aspects of food processing were integrated into a controlled safety assurance system. Though initially made public in 1971, the system became popular in the 1990s, helped to a large extent by government legislation and endorsements by international regulatory bodies that made it mandatory for food processing units to adhere to it (Hartman 2001; Brandriff and National Agricultural Library Food Safety Research Information Office 2003). Today, HACCP forms one of the pillars of food safety assurance in food processing systems.

Thus, although microorganisms continue to present a formidable challenge in foods, quality and safety have become intertwined in assessing the outcome of food processing.

14.3 Current methods of microbial control

To highlight some of the *emerging* techniques of microbial control, it is necessary to briefly review *conventional* and *current* methods used. These methods, which can be classified as physical, chemical, or biological, exploit the various intrinsic and extrinsic factors that affect microbial growth. Current commercial food processing applications employ these conventional methods as *hurdles* (or combinations) with other traditional or emerging methods.

14.3.1 Physical methods of food preservation

Most physical methods control microbial growth in foods by removing or reducing the *available water* (denoted by water activity or a_w) to lower levels that inhibit microbial growth. Food-borne fungi can tolerate a_w in the range of 0.77–0.94. However, some *xerophilic* and *osmophilic* species can grow in food with a_w as low as 0.6. Most bacteria grow at an a_w of 0.9 and cannot tolerate a_w less than 0.86 (Troller and Christian 1978; International Commission on Microbiological Specifications for Foods ICMSF 1980; Farkas 2007).

Physical dehydration and *low temperature preservation* do not kill microorganisms but simply inhibit their growth. *Thermal preservation* on the other hand uses high temperatures to attain a lethal effect on microbes.

14.3.1.1 Physical dehydration

Available water can be removed by various drying methods employing dry or moist (using steam) heat. In addition to reduction of moisture and water activity, direct microbial inactivation can be achieved when high temperatures are used to achieve food dehydration. Heat used for drying could be generated from natural sources (e.g., solar energy) or artificial methods (e.g., electric or electromagnetic energy). Drying times longer than 30 min provide time-temperature combinations that can effectively kill microbes. Dried foods are microbiologically stable because of their low a_w levels that inhibit microbial growth. However, unless the relative humidity of the storage atmosphere is maintained at 70% or less, there is the danger of the growth of xerotolerant molds (Rose 1983).

Water can also be removed by freeze-drying or lyophilization. In this process, water is sublimated from the solid state directly to the gaseous state (without passing through the liquid phase) by the application of a rapid change in temperature and pressure. Fresh food is frozen before lyophilization. It is a very gentle method of dehydration that ensures that solutes and other chemical constituents of food remain in the food. Thus freeze-dried food retains nearly most of its original physical form and it is easier to rehydrate it to almost its original form. Spontaneous rehydration and oxidation (as a result of the larger surface area of freeze-dried

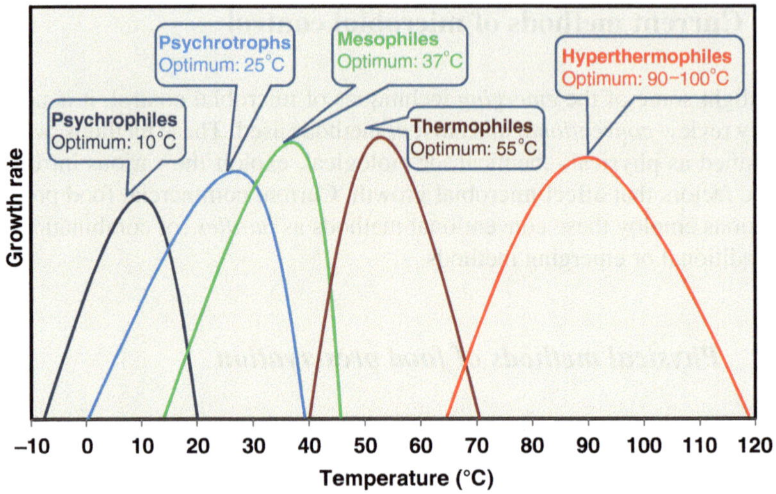

Fig. 14.1 Relationship between growth rate of microorganisms and temperature (Adapted from Montville et al. (2007); Madigan and Brock (2009))

foods) can be prevented through the use of vapor-impermeable packaging in an inert atmosphere. There is sufficient damage caused to microbial cells during both the freezing as well as sublimation stages. However, the antimicrobial action is also influenced greatly by the composition of food, especially the presence of additives (Goldblith et al. 1975; Farkas 2007).

14.3.1.2 Low temperature preservation

Temperature is by far the most important conventional extrinsic factor affecting microbial growth in food. Besides its pivotal role in microbial growth kinetics, temperature has even been found to have a vital role on microbial gene expression. Food microbes not only grow differently, but also express different genes at different temperatures. Thus, the control of processing temperature is one of the most widely used physical methods of microbial control.

The typical growth rates of various groups of microbes at different temperatures are shown in Fig. 14.1. Microbes generally exhibit decreased growth and metabolism at low temperatures; this makes for an effective control strategy. Preservation using low temperatures can also inhibit spoilage enzymes, thus extending shelf life and delaying food spoilage. In general, microbial growth and enzyme reactions are slowed at temperatures below 15ºC. The growth activities of most microorganisms are further reduced significantly or eliminated completely in the temperature range of 0–7ºC except for *psychrophiles,* which are known to thrive at very low temperatures (Madigan and Brock 2009). Almost no microbial growth occurs below –8ºC (International Commission on Microbiological Specifications for Foods ICMSF and Silliker JH 1980; Farkas 2007).

Low temperature preservation could be further classified into *chilling* or *cold storage* (about 7–15°C), *refrigeration* (0–7°C), and *freezing* (0 to −18°C and below) (Fennema et al. 1973).

Chilling and refrigeration involve the processing and/or storage of food in the temperature range of 7–15°C and 0–7°C, respectively. Refrigeration temperatures inhibit the growth of most food-borne microbes. Even psychrotrophs (i.e., microbes that can grow and survive in cold climates) exhibit slow growth at low temperatures. Both these methods could be further combined (or hurdled) with other controlled factors such as pH, a_w, air composition, and pressure. One such method is *controlled-atmosphere (CA) storage*; airtight chilled storage rooms with reduced oxygen content (2–8%) and increased carbon dioxide content (8–10%) are used for processing some fresh fruits and vegetables. Carbon dioxide controls microbial growth by lowering pH and inhibiting respiration, thus leading to a breakdown of microbial metabolism. Although most psychrophilic bacteria are sensitive to carbon dioxide, LAB and yeasts are not (Agar et al. 1997; Farkas 2007). *Modified-atmosphere packaging (MAP)* is quite similar to CA storage, with the difference that food is packaged usually in vacuum with air pressure reduced to 0.3–0.4 bar, hence effectively reducing the oxygen content of air. When packaging material with low gas permeability is used, the oxygen content drops further as a result of respiration and carbon dioxide content increases, thus bringing about a larger antimicrobial effect than CA storage. Widely used for packaging and storage of fruits and vegetables, MAP is also used for packing meat and seafood products. However there is always a concern about the growth of anaerobic microbes such as *C. botulinum* in the latter (Palumbo 1986; Leistner et al. 1994).

Freezing and frozen storage employs temperatures from 0 to −18°C and below to control microbial growth in most commercially available convenience foods. The various methods of freezing include cold air, indirect contact freezing, liquid nitrogen, or by spraying refrigerant onto food (Potter and Hotchkiss 1998). Fruits and vegetables become "completely" frozen at −15°C to −20°C, and meats and seafood at temperatures less than −40°C. Freezing produces antimicrobial action through a variety of mechanisms, including mechanical injury (owing to formation of ice crystals) and osmotic shock. However, the most important factor affecting the antimicrobial effect of freezing is the rate at which the temperature is dropped. Increased freezing rate (which causes lesser damage to food because of the formation of smaller ice crystals), has a reduced antimicrobial effect (Löndahl and Nilsson 1978; Farkas 2007). To maintain the quality of frozen food during long-term storage, it is necessary to maintain a constant storage temperature of −18°C or less and avoid large fluctuations in temperature; this helps prevent recrystallization and iciness (Karel 2003). The composition of food has a major effect on the antimicrobial activity of freezing. For example, salts reduce freezing points, thereby increasing exposure time, whereas saccharose, gelatine, glycerol, and proteins raise the freezing point of food. Although not a good microbicidal process, freezing inhibits several food spoilage microbes, especially gram-negative bacteria. It must also be noted that thawing frozen food exposes it to quicker microbial spoilage and refreezing such foods is not preferred (Farkas 2007).

14.3.1.3 Thermal preservation

Most food-borne microbes are either *psychrotrophs* or *mesophiles*; their optimum growth temperatures fall between room temperature and body temperature (25–35ºC), as can be seen in Fig. 14.1. Some food microbes could grow even at higher temperatures up to 60ºC; these are *thermophiles*. *Hyperthermophiles*, that is, microbes that can grow at extremely high temperatures, are very rarely found in food. In general, microbial growth slows down at temperatures greater than 60ºC; this makes heat one of the most effective ways to control microorganisms (Hui 2006). Microbial inactivation by high heat results from an irreversible denaturation of cellular membranes, proteins, and nucleic acids (Lee and Kaletunç 2002).

Thermal processing of food can be defined as the *heating of food at a specific temperature for a specific period of time often in pressurized and/or airtight containers*. These specific temperature–time combinations are designed to bring about inactivation of both food-borne pathogens as well as spoilage microorganisms. Although ensuring sufficient food safety (complying with established public health standards), there is always some undesirable but unavoidable changes to the nutrient quality of the foods that are thermally processed (Teixeira 2005). Hence, one of the principal aims of food engineering research has been the design of food processing/preservation methods that help retain maximum freshness and nutrient quality while still being as microbiologically safe as possible.

The most basic form of processing food with dry heat is cooking. Whereas the primary aim of cooking is to enhance the culinary quality of food, an equally important objective is to use heat to destroy or inhibit food spoilage microorganisms, food-borne pathogens, their spores, as well as spoilage enzymes. The practice of cooking has evolved over several millennia of human existence, influenced heavily by geography, culture, and even religion. As a result, this process has come to be recognized as one of the defining forces behind human evolution and development (Dell'Amore 2009; Wrangham 2009). There are several different ways to cook food; some of the basic ones include boiling (more of a *wet heat* method), baking, frying, roasting, and smoking. A more recent addition is the use of domestic microwave ovens for cooking food.

When combined with steam (mostly produced from the vaporization of water in the food material by the application of heat), the resulting wet heat is a deadly combination for most viable spoilage microorganisms as well as the equally dangerous spores produced by many of them (Coleman et al. 2007). Consequently, thermal processing is the most widely used antimicrobial method in food processing (Pflug and Gould 2000). Pasteurization takes the distinction of being the first massively commercialized and widely accepted method of thermal preservation of food. This age-old process involves the application of wet heat to kill almost all (99–99.9%) *vegetative non–spore-forming* bacteria contained in unprocessed food; this ensures a higher safety and longer shelf life. However, *conventional* pasteurization methods (with processing temperatures below 100ºC) do not kill *spore-forming bacteria* or their *spores*. Hence, unless complemented with spoilage- and microbial-growth–resistant packaging and storage (refrigeration), they are not considered completely safe (Farkas 2007).

Table 14.1 Different types of pasteurization used in the preservation of milk

Pasteurization type	Typical temperature–time combination(s) used	Microbes destroyed
Vat pasteurization	63°C for 30 min	Vegetative pathogens
High temperature short time (HTST) or flash pasteurization	72°C for 15–30 s	Vegetative pathogens
Higher heat shorter time (HHST) pasteurization	89°C for 1 s, 90°C for 0.5 s, 94°C for 0.1 s, 96°C for 0.05 s, 100°C for 0.01 s	Vegetative pathogens
High temperature pasteurization	120 to 130°C for <1–5 s	Vegetative bacteria and most spores
Ultra-high temperature (UHT) pasteurization	138°C for 2 s	All bacteria and spores

Adapted from Kelly et al. (2006); International Dairy Food Association (2009).

Pasteurization could be differentiated into different types based on the temperature and exposure time to which food is heated. The different types of pasteurization used today are vat (or batch), high temperature short time (HTST), higher heat shorter time (HHST), and ultra-high temperature (UHT) (Table 14.1). With the exception of the first, all remaining methods are *continuous* processes.

One of the most important benefits of using high temperature treatments for shorter times is lesser damage to sensitive nutrients and other important ingredients in foods. The most common method of pasteurization used in the USA is HTST, also called flash pasteurization. However, most pasteurization processes do not use temperatures greater than 100°C, except in high temperature pasteurization and UHT processing. The latter is actually a partial sterilization method; a much more severe process with temperatures always exceeding 100°C, which eliminates *all* food-spoiling microbes. UHT sterilization requires the application of temperatures of up to 140°C and products are always sealed under aseptic conditions (aseptic processing) in airtight packaging such as tin cans. Based on the type of heat exchange employed, UHT processing could be indirect (using heat exchangers of various types such as plate, tubular, or scraped-surface) or direct (using steam infusion or injection). Irrespective of whether the process is direct or indirect, all UHT sterilized products usually have a much longer shelf life compared with pasteurized products; usually from 6 to 9 months, until opened. More importantly, they do not need to be refrigerated until opened. Termed *shelf-stable* or *commercially sterile,* these products may still contain very low quantities of dormant spores but are considered microbiologically safe (Kelly et al. 2006; Farkas 2007; International Dairy Food Association 2009).

Pasteurization has been successfully commercialized for the preservation of milk, other dairy products (e.g., yogurt), fruit juices, and liquid eggs. Pasteurization of intact *in-shell eggs* using a combination of water bath and hot air treatment destroyed *Salmonella enteritidis,* the most significant among egg-borne pathogens (Hou et al. 1996). Since then, several research groups have reported improved microbiological safety and shelf life of pasteurized in-shell eggs (Stadelman et al. 1996; Dawson et al. 2006). Following FDA approval for this process, producers are

Table 14.2 Two classes of antimicrobial preservatives

Chemical preservatives	Naturally occurring food preservatives
Used traditionally for food preservation	Extracted from natural sources
Approved for use as "food additives" in most parts of the world	Do not necessarily require regulatory approval
	Usually organic compounds
Prepared synthetically (even though many of them could be found naturally in certain foods)	Have insignificant or no known negative effect on health and are hence preferred by consumers. For example, salt, sugar, vinegar
Consumers have a rather "negative" opinion about these because of recent and controversial findings about the negative health effects of some of these. For example, benzoates, nitrates, sulfates, etc.	

Adapted from Davidson and Taylor (2007).

required to mark such in-shell pasteurized eggs with a red P to distinguish them from ordinary eggs and are currently available in some regions of the United States (Zeldes 2009).

Although heat processing *before* packaging (as in UHT-processed foods) is effective, the application of heat could be done even *after* packaging. This is primarily employed in canned foods, in which food is sterilized in airtight containers (which are also sterilized separately before filling). In general, aseptic packaging has been very successfully used for processing fruit juices, creams, sauces, and other dairy products. The process has also been commercialized quite efficiently, especially with the application of continuous heat treatment using heat exchangers (Rose 1995).

14.3.2 Chemical antimicrobial preservatives

The use of chemicals is among the oldest methods of microbial control in food preservation that have been used to control microbes in food. Most of these preservatives do not necessarily *kill* microbes but control them by *inhibiting* their growth; that is, they are bacterio*static* rather than baterio*cidal* (Montville and Matthews 2008). Each antimicrobial preservative targets specific products or processes of microbial metabolism to arrest growth. This makes it possible for various preservatives to be used together as hurdles (usually along with other methods of microbial control) to bring about a stronger antimicrobial effect (Leistner and Gorris 1995).

Davidson and Taylor (2007) broadly classified antimicrobials into two groups; chemical (i.e., added to food artificially) and naturally occurring (Davidson and Taylor 2007). Table 14.2 presents these two classes.

14.3.2.1 Organic acids

Organic acids (chemically denoted as RCOOH) or more specifically *short-chain organic acids,* are the largest group of chemical preservatives used in foods. Although they are all found naturally in nature, acids used as food preservatives are

Table 14.3 Commonly used organic acid preservatives

Organic acid	Chemical formula	pK_a	Microbes inhibited	Applications
Acetic acid	CH_3COOH	4.5	Bacteria, yeasts	Bakery, cheeses, gravies, sauces
Benzoic acid	C_6H_5COOH	4.2	Fungi	Beverages, jams, and other similar confectionary
Lactic acid	$CH_3OHCOOH$	4.8	Bacteria	pH control and flavor
Propionic acid	C_2H_5COOH	4.9	Yeasts, moulds, rope-forming bacteria	Bakery, cheeses
Sorbic acid	$C_6H_8O_2$	4.7	Fungi, some bacterial species	Bakery, cheeses, alcoholic beverages (live wine)

Adapted from Montville and Matthews (2008).

usually made chemically (Table 14.3). These are effective mostly in foods having low pH preferably less than 5.5 (Branen and Davidson 1983).

Most organic acids used as preservatives work by reducing the pH of food. The effect could be on the internal pH (the intracellular pH inside microbial cells) and/or external pH (that of the food material or surrounding environment). Only recently was it understood that external pH plays an important role in the expression of certain vital genes in food microbes, some of which regulate important cellular functions such as transmembrane proton gradient, breakdown of amino acids, adjustment to low or high pH conditions, and even virulence (Olson 1993). Food microbes maintain a constant internal pH within their cells and usually counter pH changes through several different mechanisms. These processes cause the microbes to spend valuable cellular energy (in the form of ATP); the lower the internal pH, the more energy spent by the cells to raise it to neutral levels and the lesser energy available for growth, survival, and/or toxin production. Eventually, the microbial cells just run out of energy and are inactivated. For this reason, certain foods with low pH (e.g., cheeses, tomatoes, fruits, juices, yogurt, pickles, alcoholic drinks, carbonated beverages, etc.) require minimal processing and have longer shelf-life than foods with neutral or higher pH (Montville et al. 2007a, b). An effect on membrane permeability has also been suggested (Lambert and Stratford 1999).

14.3.2.2 Parabens or parabenzoic acids

Parabens or parabenzoic acids have a major advantage over short-chain fatty acids in that they are active over a much wider pH range (3.0–8.0). They are usually used in combinations; for example, methyl and propyl parabenzoates are used together in ratios varying from 2:1 to 3:1 to combine the former's higher solubility and the latter's increased antimicrobial activity. Mostly active against yeast and moulds, their exact mode of action is still somewhat unclear. Recent research has suggested that benzoic acids work by disrupting intracellular membrane trafficking pathways and have a synergistic effect with nitrogen starvation (Hazan et al. 2004). Their wide pH range of activity enables the use of parabens as preservatives in a wide range of foods such as wine, cheeses, bakery, salad dressings, syrups, jams, and jellies (Aalto et al. 1953).

14.3.2.3 Nitrites

Nitrites are the most important preservatives used in meat or meat-based products. Besides their antimicrobial activity, nitrites have some interesting beneficial properties, such as antioxidant activity, improved flavor and texture, and even their ability to add a pink color to cured meat (Carlez et al. 1995). The most important target of nitrites is *Clostridium botulinum* and the botulin toxin produced by it, especially in cured meats. They work by inhibiting the outgrowth and swelling of germinated spores, with an enhanced effect at high concentrations and in the presence of certain reducing agents such as ascorbate and isoascorbate (Duncan and Foster 1968; Russell and Gould 1991; Montville and Matthews 2008). Although the use of nitrites as food preservatives (especially in meat products) has been practiced for centuries, their safety has recently been the subject of much debate because of findings that high levels of nitrites could lead to toxic and sometimes fatal effects in humans. In lieu of these concerns, strict regulatory control has been placed on the levels of nitrites that could be used to preserve foods (Cammack et al. 1999).

14.3.2.4 Phosphates

Phosphates such as sodium acid pyrophosphate, tetrasodium pyrophosphate, sodium tripolyphosphate, sodium tetrapolyphosphate, and trisodium phosphate are effective antimicrobial preservatives and work by chelating metal ions (Sofos 1986). They are most effective against gram-positive bacteria like *Clostridium sporogenes, Bacillus subtilis, Clostridium bifermentans,* and *Listeria monocytogenes* (Kelch and Bühlmann 1958; Zaika and Kim 1993). They also effectively inhibit the production of botulin toxin in *Clostridium botulinum* but not the growth of the bacteria itself (Wagner and Busta 1985).

14.3.2.5 Sodium chloride

Sodium chloride or common salt needs no explanation at all for its use as an antimicrobial agent. It is the perhaps the oldest known preservative used. Even today, some foods are still being preserved solely by the addition of salt including raw meats and fish, often along with other processing methods such as canning and pasteurization (Sofos 1984). Salt works by reducing the water activity of food to low levels that inhibit microbial growth through plasmolysis (Davidson and Taylor 2007). *Staphylococcus aureus* and *L. monocytogenes* are two notable salt-tolerant bacteria that are not inhibited by salt (McLean et al. 1968).

14.3.2.6 Other food sanitizers

Sulfites, bisulfates, and metabisulfites are used as antimicrobial agents in wine, fruit juices, and vegetable products. They are most active against fungi and have significant activity against some groups of bacteria (Lund et al. 2000). Chlorine in the form of hypochlorous acid (or its salts) is a commonly used *food sanitizer,* especially used to disinfect water used for washing foods. The most common chlorine-based sanitizer is sodium hypochlorite, commonly called *bleach*. Some of the other potential food sanitizers include quaternary ammonium salts and peroxides. Ozone is another compound that has generated new interest as a novel nonthermal method of microbial control. *Ozonization* is explained in detail in the following sections.

14.3.2.7 Naturally occurring food preservatives

Two of the most common naturally occurring food preservatives are lysozyme and lactoferrin; these are found in milk and eggs. Other examples include avidin (eggs), allicin (onion and garlic), isothiocyanates (mustard family), essential oils (spices), and various phenolic compounds such as flavonoids. However, one of the major drawbacks with their use as direct food additives is their undesirable influence on the sensory (flavor, smell, etc.) and nutritional properties of the foods to which they are added. Another major challenge is the tedious processes of isolation, purification, and stabilization of these compounds before they can be added to foods (Davidson et al. 2005; Davidson and Taylor 2007).

14.3.3 Biological microbial control: biopreservation

This kind of food preservation involves the use of microorganisms (and/or their products) to preserve food and ensure food safety. The most well-studied and widely accepted *biopreservatives* are lactic acid bacteria (LAB) and/or their products. Some LAB species have the ability to produce antimicrobial proteins called *bacteriocins* (Montville and Matthews 2008).

14.3.3.1 Bacteriocins

Bacteriocins are much like narrow-spectrum antibiotics in that they are toxic only to a very narrow range of microorganisms, usually strains that are very closely similar to the producer strain. Though not enzymes in the strict sense, they have similar function. Bacteriocins can be classified into four separate classes based on their nature and mode of action (Table 14.4).

The most notable among bacteriocins is nisin, which is a well-studied food preservative produced by *Lactococcus lactis* subsp. *Lactis*. Nisin has been a favorite among food preservatives because of its qualities such as nontoxicity, *generally-regarded-as-safe* (GRAS) status of its producer strain, absence of cross-resistance in bacteria affecting antibiotic therapeutics and efficient digestion in the human digestive system. The most significant food microbes targeted by nisin include

Table 14.4 Various classes of bacteriocins

Class	Members and examples
I	"Lantibiotics," most notably nisin, lacticin, and carnocin
II	Pediocin PA-1, sakacins A&P, leucocin A, lactococcin G&M, lactacin F
III[a]	Lactacin B, helveticin J&V
IV	Leuconocin S, lactocin 27

[a]Now classified as non-bacteriocin lytic proteins or bacteriolysins (Cotter et al. 2005). Adapted from Klaenhammer (1993); Montville et al. (2007).

Clostridium spp., *Bacillus* spp., and *L. monocytogenes*. Sold under the brand name *Nisaplin* by Danish food-maker Danisco, nisin is the most popular biopreservative and the only commercially available bacteriocin (i.e., approved as a *food additive*) used today (Delves-Broughton 2005). Genetic recombination of LAB and protein engineering of bacteriocins have enabled the production of recombinant heterologous bacteriocins with improved antimicrobial activity, adaptability, and genetic stability (Rodríguez et al. 2003; Montville et al. 2007).

It is much more common to use LAB directly rather than the purified bacteriocins themselves; in fact it is quite difficult to isolate and purify many of the known bacteriocins. Besides, the regulatory status of bacteriocins is still unclear and there is a lack of validated research findings on the development of bacteriocin-resistance by other food microbes (Fields 1996; Cotter et al. 2005).

14.3.3.2 Organic acids naturally produced by lactic acid bacteria

One of the best-known examples of microbial control through the controlled acidification of food with LAB (e.g., *Pediococcus acidilactici*) is the inhibition of *C. botulinum* by decreasing the pH of food to below 4.8, at which levels it is inactivated and hence is unable to produce the toxin (Hutton et al. 1991). LAB have found application as food preservatives in a wide range of foods such as fermented products (cheeses, yogurt, alcoholic beverages), milk (and other pasteurized dairy products), egg products, canned foods, sausages, mayonnaise, salad dressings, and even baby foods (Vandenbergh 1993; Delves-Broughton 2005).

14.3.3.3 Probiotics

Many LAB have been proved to have *probiotic effects*; that is, these bacteria are known to provide various health benefits to humans when administered in their live form. This property has prompted food engineers to try to use these probiotic LAB as food additives rather than their nonprobiotic counterparts. Some examples include *Lactobacillus acidophilus* and *Bifidobacterium* spp. (Montville and Matthews 2008).

14.4 Novel and emerging methods of microbial control

Although *conventional* or *current* methods of microbial control are widely used, and their capacity to assure microbiological safety of foods are well established, significant sacrifices are made in terms of undesirable changes in the flavor, texture, and nutrient composition of food. Conventionally processed foods may be inferior to their fresh forms in terms of nutrients (e.g., vitamins) and other health-promoting bioactive compounds (such as natural antioxidants and essential fatty acids). This is especially true in the case of fruit- and vegetable-based food products for which freshness is an important consumer demand. Since the 1980s, there has been a marked shift in consumer preference from basic safe foods with long shelf lives to safe and high-quality products with comparably reasonable shelf lives. It is this increased emphasis on high quality that drives new developments in food processing. The often conflicting requirements for quality and shelf life present interesting challenges to food processing and product development professionals. *Novel* and *emerging techniques are being developed and used to address these challenges.* Although the safety, antimicrobial effectiveness, and economic feasibility of most of these techniques are still being evaluated fully by food engineers and scientists, some of them have already progressed to the stage of commercialization.

For the purpose of classification, novel methods have been split into two based on their *primary* antimicrobial effect, namely *thermal* and *nonthermal.* Both types of techniques have generated considerable interest in the food industry. In another development, rather than focusing solely on one method of microbial control, new processing systems that combine different techniques to apply cascade of mild and effective treatments are being designed to produce high-quality and safe products. This approach is known as *hurdle technology.* Several processing methods employing different antimicrobial methods employed together have already been incorporated into HACCP systems in many food processing applications (Leistner et al. 1994; Leistner and Gorris 1995; Raso and Barbosa-Cánovas 2003; McNamee et al. 2010).

14.4.1 Novel thermal technologies

Whereas traditional thermal processing methods rely more or less on heat transfer from an external source to food typically by *conduction, convection,* or *radiation mechanisms,* newer and more recent advances in thermal processing have focused more exclusively on *radiation* as a heat transfer mechanism, especially by the use of electromagnetic waves (e.g., microwaves and radiofrequency waves) or generation of heat using electromagnetic fields (Delgado et al. 2006; Sun 2006). The greatest benefit of thermal processing technologies is their ability to effectively destroy most food spoilage microorganisms. When combined with steam (which could be produced from the vaporization of water in the food material by the application of heat), the resulting *wet heat* is deadly for most spoilage microorganisms and the spores produced by them (Coleman et al. 2007). However, as noted, conventional thermal processing causes significant losses to the organoleptic properties of food. Thus, the main focus of novel thermal methods is minimization of such detrimental effects of heat on the freshness and nutritional quality of food (Barbosa-Cánovas and Gould 2000; Sun 2006). Further, newer thermal processing techniques typically use electrical sources of heat (that can be more easily modulated and controlled) rather than conventional fossil fuels, permitting improved energy transfer and conservation with reduced greenhouse gas emissions.

14.4.1.1 Microwave and radiofrequency heating

Radiofrequency (RF) waves are nonionizing waves having longer wavelength than infrared light and falling within the frequency range of 3 kHz to 300 MHz. Microwave (MW) frequencies are higher than RF waves and they fall between 300 MHz and 300 GHz. When food is exposed to the radiation, water molecules (which are electric dipoles) re-orient continuously with the changing direction of the rapidly oscillating electromagnetic fields resulting in intermolecular friction and hence generating heat (Metaxas and Meredith 1983; Mudgett 1985; Farkas 2007).

Microwave technology

Microwave (MW) heating has had remarkable success as a quick and easy household method of heating and/or cooking food. Today, domestic microwave ovens have become one of the most common kitchen appliances. The most important advantage associated with MW/RF heating is the rapidity at which food can be heated to a desired temperature. This means that processing times are much shorter and there are lesser undesirable effects on the nutritional and organoleptic qualities of food. Further, the technology can be easily automated and incorporated into continuous-flow processing (Ramaswamy and Tang 2008).

Zhang and Mittal (2008) reviewed the benefits of MW-assisted drying as an excellent alternative to conventional drying methods. The authors also reviewed some recent novel techniques that involve the combination (or hurdling) of MW heating along with other methods, such as hot air, vacuum, lyophilization, and osmotic rehydration. Besides its application in drying of foods, MW heating has also been successfully commercialized for tempering and precooking meat and meat-based products. From the antimicrobial point of view, the rapid heating property of MW holds immense promise as an effective method of pasteurization and sterilization of food (Zhao 2006; Ramaswamy and Tang 2008).

Household microwave ovens and some industrial processes use MW with a frequency of 2.45 GHz. Although domestic ovens are not designed to be used for the sterilization of food, industrial-scale 2.45 GHz MW sterilization systems have been described in the scientific literature. These industrial MW-based sterilization systems (using 2.45 GHz MW) have achieved successful commercialization in Europe and Japan. However, this technology is yet to be approved for industrial pasteurization

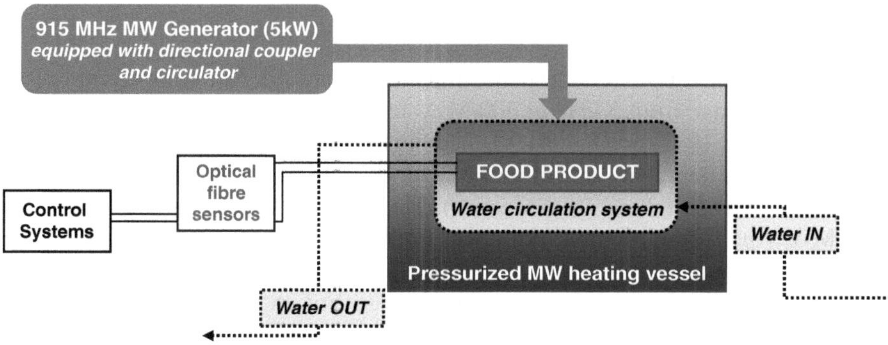

Fig. 14.2 Schematic design of the microwave-circulated water combination (MCWC) heating system developed at Washington State University (Adapted from Guan et al. (2002))

and sterilization by the FDA which has delayed commercialization in North America (Harlfinger 1992; Schlegel 1992; Tang et al. 2002; Ramaswamy and Tang 2008).

Most industrial operations use microwaves of frequency 915 MHz. These systems are not used for domestic purposes. The MW from these systems penetrate food products deeper than 2.45 GHz MW (because of their shorter frequency). Hence they are much more energy efficient, especially from an industrial point of view. There is a lot of information in the scientific literature that suggests that 915 MHz MW systems offer superior processing and antimicrobial efficiency in a wide range of applications. Guan et al. (2003) successfully verified the antimicrobial and processing effectiveness of a 915 MHz microwave-circulated water combination (MCWC) heating system for a macaroni and cheese product using inoculated pack studies. This novel system was developed at Washington State University, Pullman, WA. A schematic depiction of their system is shown in Fig. 14.2. The results clearly indicated that a combination of MW heating using 915 MHz waves and water immersion provided a more efficient sterilization system compared with conventional methods. The authors suggested the use of fiberoptic sensors as an effective solution to problems faced with temperature measurement in MW fields (Guan et al. 2002, 2003).

With the use of packaging materials such as glass, plastic, and paper that are transparent to microwaves, MW heating can even be used to pasteurize sealed and packaged foods. The biggest advantage here is the avoidance of any posttreatment microbial contamination. As with regular thermal treatments, MW pasteurization has the ability to inhibit a wide range of food microbes such as molds, yeast, and many thermolabile bacterial strains (Giese 1992; Kozempel et al. 1997). Foods that have been proved to be pasteurized by MW heating include packaged bakery, fresh pasta, soft cheeses, milk, fruit juices, eggs, and refrigerated ready-to-eat foods; many of these in continuous systems (Nikdel and Mackellar 1992; Nikdel et al. 1993; Kim et al. 1999b; Valero et al. 2000; Sierra and Vidal-Valverde 2001; Tang et al. 2002; Fito et al. 2005; Gentry and Roberts 2005; Dev et al. 2008b; Cinquanta et al. 2010).

Radiofrequency technology

Nonmetallic materials can be heated using electromagnetic waves in the radiofrequency (RF) range from 3 kHz to 300 MHz in a process usually referred to as RF heating (Orfeuil 1987) or RF dielectric heating (Zhao 2006). Radiofrequency waves have more penetrating power than microwaves because of their lower frequency and longer wavelengths (about 20–200 times). This implies that they heat foods better than MW. Apart from rotation of dipolar molecules, RF heating also relies on the additional phenomenon of *ionic migration* to generate heat (Marra et al. 2007). Hence, RF heating may hold much more promise for an industrial application in food processing than MW heating (Wang et al. 2003).

Commercially, RF heating has been used for post-baking drying of freshly baked products (Zhao 2006; Ramaswamy and Tang 2008). The improved penetrating power and heating abilities of RF heating makes it a much better candidate for industrial-scale sterilization and pasteurization than MW heating. Research conducted at Washington State University (WSU) has concluded that RF heating has excellent antimicrobial activity with the ability to inactivate even heat-resistant spores in packaged foods. Further, RF heating was found to have no detrimental effects on the quality of certain foods, besides shortening process times to a third of that required by conventional retort methods. The pilot-scale RF sterilization facility developed at WSU used RF waves of frequency 27 MHz and was co-funded by US Army Natick Soldier Center, US Army Combat Ration Network and Strayfield, United Kingdom (Wang et al. 2003; Luechapattanaporn et al. 2004; Tang et al. 2005). Other reports have tested the use of RF sterilization as an effective antimicrobial processing method for a wide range of foods such as fresh meat, caviar, and even liquid and particulate foods (Starkweather 2000; Lacroix 2001; Zhong et al. 2003, 2004). However, there are no reports on the commercial applications of RF heating as an antimicrobial method. In spite of its superior effectiveness over MW heating, it appears than RF heating has not generated as much research interest as MW heating among food engineers (Tang et al. 2005; Ramaswamy and Tang 2008).

Challenges

Despite reports of the effective antimicrobial properties of MW/RF heating, numerous obstacles prevent these technologies from being used for commercial pasteurization and sterilization. One of the main challenges in MW processing of foods is the nonuniform nature of heating; a result of variation in the dielectric properties within food products and with their surrounding medium (Ramaswamy and Pillet-Will 1992; Ryynanen and Ohlsson 1996). The use of water immersion has been tested successfully as a way to minimize nonuniform heating as in the case of the microwave-circulated water combination heating system developed at WSU mentioned previously (Guan et al. 2002).

The high cost of industrial MW heating-based equipment is another barrier to their commercialization. Furthermore, microwave generators k (e.g., magnetrons and Klystrons) have a shorter life compared with their conventional and traditional counterparts, although the recent development in magnetron and circulator technology has mitigated this problem to some extent (Tang et al. 2002). This has prompted research on combining MW heating with other thermal or nonthermal processes, so as to improve the economics of the processing applications. For example, the application of MW heating at the end of a conventional drying process has been proved to be an improved as well as energy-efficient strategy (DeCareau 1994; Funebo and Ohlsson 1998).

Radiofrequency heating is faced with much similar obstacles as MW heating. Although RF waves penetrate food products deeper than MW, there is still the problem of nonuniform heating. There is need for more efficient RF applicators that can provide uniform pattern of heating as well as more reliable techniques to measure temperature (in the RF field). Computer simulation techniques have already helped in the improvement of existing MW/RF technology and newer programming/software is likely to add further credibility to this method. Additionally, there must be more studies done on the dielectric properties of various foods and how they are influenced by the frequency of the electromagnetic radiation used, the temperature they generate as well, as the other food components (Zhao et al. 2000). Most importantly, the lack of absolute FDA approval has been the biggest obstacle to the commercialization of RF/MW technology. Hence, there is a need for more convincing research on the microbiological safety of RF/MW-treated food (Tang et al. 2005; Ramaswamy and Tang 2008).

14.4.1.2 Electric and magnetic fields

Ohmic heating

Ohmic heating (OH), also called *electrical resistance heating*, *electroconductive heating*, *Joule heating,* or simply *electro-heating* relies on generation of heat in food material (as a result of electrical resistance) by application of electric current (De Alwis and Fryer 1990). The dissipation of electrical energy in the form of heat during OH can be described by the following equation:

$$\dot{\mathbf{Q}} = |\nabla \mathbf{V}|^2 \cdot \boldsymbol{\sigma}$$

where $\dot{\mathbf{Q}}$ is the heat generated and $\nabla \mathbf{V}$ is the applied electric field and $\boldsymbol{\sigma}$ is the electrical conductivity of the product. The equation clearly indicates that heat generated is directly proportional to the square of the electric field strength and electrical conductivity of the material. The latter is a function of temperature, mode of heating, and type of food material. Breakdown of cellular membranes (or walls) and release of cellular cytoplasmic materials occurs at low temperatures when an electric field is applied. This invariably results in a sharp increase in electrical conductivity over a wide range of temperatures. However, for each type of food material, there is a maximum value of field strength beyond which the conductivity does not rise any further (so-called saturation conductivity) (Sastry and Li 1996; Sastry and Barach 2000). Usually, sinusoidal (alternating) or pulse waveform current is used for OH applications. The most popular frequencies used are those of domestic power supply, that is, 60 Hz in North America or 50 Hz in Europe (Sastry 2008).

OH systems can be differentiated from other thermal technologies (e.g., MW/RF systems) by their characteristic electrodes, which are typically in contact with foods and their ability to heat foods rapidly and more uniformly. This makes OH an excellent method of drying, cooking, sterilization, and pasteurization, especially for low-acid and particulate foods (Sarang 2007). The technology was briefly used for pasteurization of milk in the 1930s before being revived again in the 1980s. Current applications commercially developed using OH include processing of seafood (Yongsawatdigul et al. 1995), pasteurization of eggs products (Sastry 2005), pasteurization of fruit juice (Leizerson and Shimoni 2005a, b), pasteurization of milk (Fillaudeau et al. 2006; Cheng 2007), and cooking of meat (Zell et al. 2009).

OH kills food microbes and their spores primarily through thermal effects (hence qualifying as a thermal preservation method), and this has been extensively studied. Initial studies claimed that the lethal effects of OH were solely thermal or that the presence of electrical effects was insignificant (Palaniappan et al. 1992). More recent studies involving microbial inactivation kinetics conducted on *Lactobacillus acidophilus* (Cho et al. 1996), *Bacillus subtilis* spores (Cho et al. 1999), and *Byssochlamys fulva* – a highly thermotolerant fungus (Castro 2007) have shown that the microbial inactivation of OH could also be due to a significant amount of nonthermal effects such as *electroporation* (Imai, Uemura et al. 1995; Wang 1995; Yoon et al. 2002; Vicente et al. 2006). However, this is a very recent development and the effects of electrical processes in microbial inactivation are still not understood very well (Vicente et al. 2006; Sastry 2008). According to Sastry (2005), electroporation may be the result of the accumulation of charges (as a result of the low frequencies used) across cell membranes, eventually leading to the formation of pores. In MW/RF heating, this effect is less pronounced because the much higher frequency of the electromagnetic waves does not allow enough time for electric charges to build up; hence, the membranes do not break down as a result of electroporation (Sastry 2005).

Moderate electric fields

The discovery of significant nonthermal effects in OH (as mentioned) has led to the development of a slightly different process, namely *moderate electric field* (*MEF*) processing. This technology makes use of electric fields less than 1 kV/cm and is hence not included under the same group as pulsed electric field processing (as discussed later). MEF processing is very similar to OH. Both methods are usually grouped together and often combined into a single process. However, MEF has been studied separately for drying of food and extraction of juices. The combined OH-MEF process has been applied to the peeling of fruits and vegetables (Sastry 2008).

Recent research on *Lactobacillus acidophilus* (Loghavi et al. 2009) and *E. coli* (Machado et al. 2010) has confirmed that applications of MEF at either 45 or 50 Hz, even at low temperatures are strong enough to cause microbial inactivation through electroporation. This introduces the immense possibility of using MEF for pasteurization of foods with heat-sensitive components. However, this method is still in its nascent stage and more detailed research on the effects of food matrices on electric field efficiency, microbial death kinetics of electric effects, industrial scale-up, and economic feasibility is required before antimicrobial applications can be developed (Sastry 2008; Machado et al. 2010).

Inductive heating

Inductive heating makes use of strong and rapidly oscillating magnetic fields generated within an electric coil (through which large alternating currents are passed) to rapidly heat up a material placed within the coil or near it. As opposed to a magnetron used to generate electromagnetic fields in MW heating, inductive heating makes use of magnets and metallic coils. Industrial-scale inductive heating systems differ greatly from domestic *induction cookers* in that the former are much more powerful (in the range of several kW) in terms of the magnetic fields and heat generated. Information on industrial-scale inductive heating of food is very scanty. Although used for several non–food-based industrial processes, the application of inductive heating to food processing is a very recent research topic. With the exception of a project on the *thermal treatment of liquids by inductive heating* carried out at the Technical University of Munich and a recent patent filed for *the demolding of ice cream using inductive heating,* there is no other research data available for the use of inductive heating as an industrial-scale antimicrobial method (IFT and FDA 2000; Sastry and Barach 2000; Ramesh 2003; Mezquida et al. 2009). This technology holds immense promise in this area of application owing to its ability to generate rapid heat in food materials; however, more research is required before any further discussion of the technology comes up among food process research groups (Rosenbauer et al. 1996; Sastry et al. 2000; Bengtson et al. 2006).

Challenges

In spite of their immense potential for microbial control in the food industry, OH and MEF processing face numerous issues that need to be addressed before the technology can be commercialized. The most significant obstacle is the high initial operational cost of running an OH system. Recent developments in power supply technology and the availability of low-cost equipment have considerably brought down the costs of running this technology. Yet, more research on the economic feasibility of OH is needed before the process can be fully accepted by food producers. Other challenges include the identification and testing of cold spots, elimination of *electrolytic effects* at the food-electrode interface and more reliable methods of temperature measurement. Further, more research is needed on the understanding of the nonthermal (mainly electric) effects of OH/MEF on food matrices, foodborne pathogens, food spoilage microorganisms, their spores and toxins. One way to offset the disadvantages with OH/MEF technology is through hurdling with other processes such as high pressure processing (Sastry 2005, 2008; Vicente et al. 2006).

14.4.2 Nonthermal technologies

The search for alternatives to thermal food processing techniques has produced several innovative new methods of processing to produce high-quality products that retain most their natural freshness and nutritive values. These technologies range from the application of *electric fields* and *ultrasonic waves* to the use of *cold gas plasma* and *supercritical carbon dioxide*. Although most nonthermal processes can inhibit food spoilage bacteria, molds, and yeast, they are not very effective in inactivating bacterial spores. Thus, their application for sterilization of food is limited, especially when used individually. Industrial-scale applications are usually derived by the hurdling of several nonthermal and/or thermal processes (Leistner et al. 1994; Leistner and Gorris 1995; Raso and Barbosa-Cánovas 2003; McNamee et al. 2010).

Although most of these new methods are termed as *nonthermal*, heat could be generated as a *secondary* effect. The antimicrobial effect produced by such preservation techniques may well be a combined result of the main physical agent along with the action of heat. In some cases, the generation of heat could be responsible for some unavoidable and undesirable changes in the food, similar to or less severe than those produced during thermal processing of foods. However, current research is focused on making these techniques gentler on the food while being uncompromising on microbiological safety. The important distinction is that these methods are *not primarily based on the use of heat* to inactivate, inhibit or kill microbes (Koutchma 2009).

14.4.2.1 Pulsed electric fields

Pulsed electric fields (PEF) (or high intensity pulsed electric fields, HIPEF) technology involves the application of high voltage electric fields to food products held between a set of electrodes in very short pulses. The electric field intensities applied for microbial inactivation for liquids are usually in the range between 20 and 80 kV/cm. In solids the range of electric fields is much small, in the range of 0.5–5 kV/cm. The type of pulse applied could be of exponentially decaying, square wave, bipolar or oscillatory type and the temperature of application could be ambient, subambient, or above-ambient. The application of such short, high voltage pulses has the ability to inactivate microorganisms through the phenomenon of *electroporation* (Ramaswamy et al. 2007).

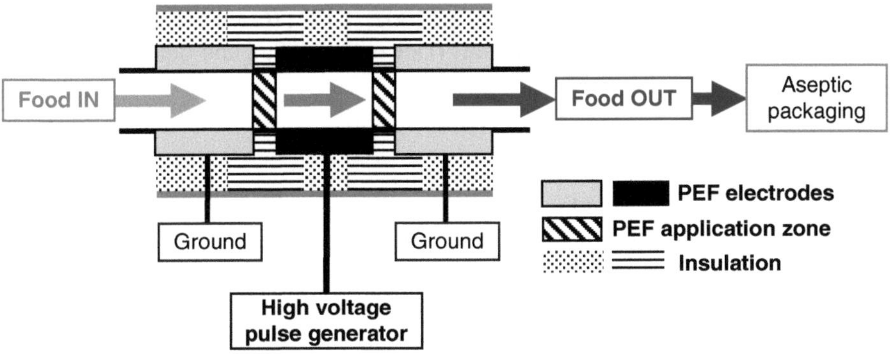

Fig. 14.3 Schematic set-up of an integrated continuous PEF processing system (Adapted from Ramaswamy et al. (2007))

PEF technology is best applicable to liquid and semiliquid foods because of the presence of ions that are electric charge carriers (Zhang et al. 1995; Ramaswamy et al. 2007). Early applications and most lab-scale studies used *static* or *batch* processing systems. However, most industrial applications involving PEF treatment of liquid foods use an *integrated continuous processing system* (e.g., as shown in Fig. 14.3). PEF consists of a fluid handling unit (that can be *sterilized-in-place* and *cleaned-in-place*), a high voltage pulse generator, treatment chambers, and an aseptic packaging system. The treatment chamber consists of at least two electrodes (for the application of high voltage) and specified PEF application zones where the actual treatment occurs. Electrodes used for PEF processing are usually made from inert nonconductive material (e.g., titanium) to prevent their decomposition and contamination of food products during processing. Aseptic packaging and refrigerated storage usually follow PEF processing to avoid the possibility of posttreatment contamination (Ramaswamy et al. 2007). It is possible to use *plastic conductive film electrodes* for PEF treatment, which suggests a possible application for treating prepackaged foods directly (Roodenburg et al. 2010).

Although temperature rise during PEF applications is much lower than during conventional thermal processing, the technique has been used for drying food products. Reports on PEF-assisted drying and dehydration of vegetables and fruits demonstrate higher drying rates and comparable or improved nutrient quality retention compared with conventional drying methods. Among three pretreatment methods (PEF, MW heating, and blanching) investigated for the drying of okra, PEF was able to produce the highest drying rate (Adedeji et al. 2008). PEF treatment also improved mass transfer during osmotic dehydration of apple (Amami et al. 2005). High electric field (HEF)–dried spinach retained the original color and almost a three times higher content of ascorbic acid compared with oven-dried spinach, even after 6 weeks of storage (Bajgai and Hashinaga 2001). PEF-drying has also been found to be effective (both in terms of drying efficiency as well as retention of nutritional qualities) in several other foods, such as bell peppers (Ade-Omowaye et al. 2003), potato (Arevalo et al. 2004; Lebovka et al. 2007), carrot (Gachovska et al. 2008a), beetroots (Shynkaryk et al. 2008), raisins (Dev et al. 2008a), and others.

Antimicrobial applications

The mechanism of microbial inactivation by PEF has been attributed to the process of electroporation in which microbial membranes break down as a result of the high transmembrane potential generated (Ravishankar et al. 2008). Pore formation may be reversible depending on several factors including electric field intensity, pulse duration, number of pulses, processing medium, and microbial species (Wouters and Smelt 1997; Aronsson et al. 2005; Ramaswamy et al. 2007).

The antimicrobial and food preservation efficiency of PEF processing against various foodborne pathogens and spoilage microbes (most notably aerobic mesophilic bacteria, molds, yeasts, enterobacteria, *E. coli*, *E. coli* O157:H7, *Staphylococcus aureus, Listeria,* and *Salmonella*) has been studied (Mosqueda-Melgar et al. 2008a; Saldana et al. 2010) and proved for a variety of fruit juices and nectars, such as orange (Min et al. 2003; McNamee et al. 2010), apple (Schilling et al. 2008), tomato (Aguilo-Aguayo et al. 2008), cranberry (Sen Gupta et al. 2005), carrot (Akin and Evrendilek 2009), cherry, apricot, and peach (Evrendilek et al. 2009). The antimicrobial effects of PEF treatment has been proved in the processing of other liquid foods such as milk (Riener et al. 2009a; Walkling-Ribeiro et al. 2009a), yogurt (Sampedro et al. 2005), liquid egg (Calderón-Miranda et al. 1999; Sampedro et al. 2006; Monfort et al. 2010), and even vegetable soups (Vega-Mercado et al. 1996; Elez-Martinez and Martin-Belloso 2007).

Pulsed electric field hurdle technology

Although the antimicrobial action of PEF as a standalone food processing method has been demonstrated, current research is focused more on bundling it with other thermal or nonthermal processing methods in hurdle approaches. There is some amount of heat generated in the application of PEF. Thus, the generated heat could be made beneficial by hurdling heat treatment with PEF. Combined PEF and thermal processing methods have been successfully evaluated for some low-acid foods, most notably milk (Walkling-Ribeiro et al. 2009a; Guerrero-Beltrán et al. 2010).

One of the most important advantages of PEF processing is the increased microbial membrane permeability produced as a result of mild electroporation; this provides PEF processing with an opportunity to enhance the uptake of certain antimicrobial compounds (e.g., nisin and other bacteriocins, essential oils, spices, etc.) added to the processing medium (Tsong 1991; Calderón-Miranda et al. 1999; Mosqueda-Melgar et al. 2008b; McNamee et al. 2010). Other processing methods hurdled with PEF include the use of sodium chloride (Terebiznik et al. 2002), high pressure methods (Sanchez-Moreno et al. 2009), thermosonication (Walkling-Ribeiro et al. 2009b), UV radiation (Noci et al. 2008), and ozonization (Mahapatra et al. 2005).

Challenges

Although PEF processing has already been commercially exploited for processing of fruit juices, the technology has some significant limitations which prevent application to other foods. These include higher initial costs (compared with conventional pasteurization equipment), nonuniform treatment and varying dielectric properties of foods (e.g., vegetable juices and salted foods conduct electricity better than fruit juices and unsalted foods, respectively). One of the biggest problems faced with PEF treatment systems is the presence of air bubbles in some products (hence requiring preprocessing de-aeration) as well as the formation of bubbles by electrolysis. Further optimization of the technology for application to other types of foods such as particulate foods and a more thorough understanding of the effects of PEF-processed foods on human health is needed before the technology can achieve full producer and consumer acceptance (Barsotti et al. 1999; Espachs-Barroso et al. 2003; Rastogi 2003; Ravishankar et al. 2008).

14.4.2.2 High pressure processing

High pressure processing (HPP) also known as hydrostatic pressure (HHP) or ultra-high-pressure (UHP), involves the processing of foods at elevated pressures, often with the addition of heat.

Current technology

Foods are packaged in flexible containers before immersing them in a pressure-transmitting fluid (usually water) contained in a *high pressure chamber*. After the chamber is sealed, pressure is applied hydrostatically (fluid is pumped into the chamber usually along with mineral or vegetable oil for better lubrication) until the desired pressure (100–700 MPa) is reached. The pressure is maintained at this level for a specific time, usually pulses of 1 ms to periods of up to 20 min (Koutchma 2009). Commercially, processing is carried out over several pressure cycles of specific duration. Most commercial applications for processing solid and liquid foods make use of batch HPP systems. However, semi-continuous HPP systems are also available for liquid foods (Palou et al. 2002; Balasubramaniam and Farkas 2008). HPP has been successfully used for starch gelation and enzyme inactivation without the need for heat treatment (Farkas 2007; Tewari 2007).

The greatest advantage with HPP is that foods can be processed at ambient temperatures (although combination with heat is preferred for more severe antimicrobial action) usually at temperatures ranging between 4°C and 40°C. This ensures very minimal changes in the nutrient quality and organoleptic properties of the processed food. However, commercial systems that can operate at up to 130°C are also available. HPP provides rapid and uniform treatment of food hence processing times are much reduced. Further, it is possible to treat both packaged as well as unpackaged foods (Koutchma 2009).

Antimicrobial applications

High pressure (between 200 and 600 MPa) has been proven to inactivate vegetative cells of most pathogenic and spoilage microbes in food at mild temperatures (Balasubramaniam and Farkas 2008). Noncovalent bonds in proteins and carbohydrates are altered at high pressures, leading to undesirable changes in their molecular structure (Farkas 2007). Microbial inactivation is eventually accomplished through several mechanisms such as removal of membrane lipids, increase in membrane permeability (eventually leading to breakdown), enlargement of cells, collapse of cellular vacuoles, reduction in number of ribosomes and inhibition of microbial enzymes required for survival and growth (Hoover et al. 1989). Notable food microbes inactivated by HPP include various strains of *Salmonella, Clostridium, Listeria, Staphylococcus, Yersinia, Bacillus,* and *E. coli* (Hoover et al. 1989; Welti-Chanes et al. 2005).

Whereas HPP can inactivate microbes at mild temperatures, there is a close correlation between microbial heat-resistance and pressure-resistance. As a result, microbial pressure-resistance peaks at ambient temperatures (that are suitable for the growth of most microbes). Thus, it is preferable to perform HPP at higher temperatures or to hurdle it with other thermal methods of microbial control (Farkas 2007).

HPP (with or without other hurdles) has been used to pasteurize or sterilize a wide variety of foods such as minced meat, skimmed milk (and other dairy products), fish, cereals (especially rice-based products), vegetables (e.g., potato, tomato, radish, carrot), herbs (e.g., thyme, mint, parsley), fruit juices, gel-based products (e.g., jams, jellies; HPP has been found to provide more stability to such products), guacamole (a Mexican delicacy), seafood (especially squid and shellfish such as mussels, oysters, and even lobsters), and eggs among others (Cheftel 1992, 1995; Tauscher 1995; Welti-Chanes et al. 2005; Patterson et al. 2007). One particular market that has seen a resounding success for HPP foods is Japan; high pressure-sterilized foods such as fruit desserts, jams, jellies, juices, and even seafood have been quite successful there (Hayashi 1995).

High pressure processing hurdle technology

In spite of the various benefits associated with the use of HPP, one of the biggest disadvantages with this method is its inefficiency in inactivating microbial spores and toxins at ambient temperatures. Although vegetative microbes are sensitive to high pressure, bacterial spores are more resistant (Tewari 2007). Combining HPP with other hurdles such as heat (Scurrah et al. 2006; Bull et al. 2009), ultrasonic waves (Raso et al. 1998; Demirdöven and Baysal 2009), pulsed electric fields, low pH, antimicrobial agents, and irradiation has been shown to inactivate both vegetative microbes as well as their spores with minimal processing time and at significantly lower costs (Raso and Barbosa-Cánovas 2003; Doona and Feeherry 2007; Zhang and Mittal 2008). Research has suggested that modifications of HPP such as *oscillatory* processing (involving *reciprocal*, *pulsed,* or *cycling* pressure treatments) could also help solve this problem (Hendrickx and Knorr 2002).

Challenges

HPP has been proved to reduce the total energy requirement for sterilization of food by at least 20% compared with conventional processes (Toepfl et al. 2006; Pereira and Vicente 2010). Despite this proven energy efficiency, commercialization has been plagued by problems associated with high costs of operation. High pressure equipment, especially those utilizing pressures greater than 600 MPa, can be extremely costly. Although hurdling HPP with other methods of microbial control have brought down costs significantly, further studies and research on making HPP more economically feasible is necessary. Further, there needs to be more research on the effects of high pressure on carbohydrates and the unfolding mechanisms of food proteins under the influence of other significant factors such as temperature. There are also insufficient data on the effect of HPP on different packaging materials (Farkas 2007). Solutions to these problems may speed up the acceptance of HPP as an effective method of microbial control among food processors as well as consumers. According to Kelly and Zeece (2009), HPP is the most scientifically sound and commercialization-ready process among other emerging novel methods of microbial control in food processing.

14.4.2.3 Ionizing irradiation

Even though irradiation (especially ionizing radiation) as a method for microbial control in foods has been studied and researched right from the beginning of the past century, this versatile technology was commercialized only in the past two decades and still suffers from a major negative public perception. Consumers seem to have carried on the general public skepticism toward nuclear power to the safety of irradiated foods. Yet, in spite of these obstacles, food irradiation has been repeatedly proved to be versatile, safe and highly efficient (Moy 2005; O'Bryan et al. 2008).

The frontrunner among irradiation technologies is ionizing radiation. These are electromagnetic waves that have enough energy to *ionize* atoms through the removal of electrons. The *ionizing power* of the radiation is dependent on the energy of the individual waves rather than the number of waves in it. Ionization requires waves to possess energy higher than a few electron volts (eV). Taking up the left side of the electromagnetic spectrum, ionizing rays have lesser wavelength than UV, visible, and infrared radiation.

Current technology

Three forms of ionizing radiation are currently available for irradiating food; *converted x-rays*, *gamma rays*, and *high-energy electrons* (or *electron beams*). Gamma rays are generated from decaying isotopes of Cobalt-60 or Casium-137. Electrons beams are generated when heated filament emit electrons which are then accelerated (by the application of high voltage) and ejected through a thin window (so as to create a beam) in linear accelerators such as Magnetron or Dynamitron.

Both gamma rays and electron beams have similar penetration depths at 80–100 cm. To generate x-rays, electrons from linear accelerator are collided with dense, heat-resistant plates of tungsten of titanium alloy. The penetration of x-rays is about 10× less than that for gamma rays. The maximum radiation dosage approved by the FDA is 4.5 kGy for nonfrozen and nondry foodstuff and 10 kGy for frozen or dry foods (FDA and HHS 2008).

Antimicrobial applications

Ionizing radiation works through DNA damage caused by ·OH free radicals that are created by the ionization. Microbial and insect DNA at any stage of development subjected to this kind of DNA damage fail to reproduce (Moy 2005). Some damage to cell membranes has also been proposed (Josephson and Peterson 1983). Although different forms of radiation have been suggested to have varying means of antimicrobial action (owing to the different mechanisms of free radical formation), exact details are still unclear and subject to ongoing research (Koutchma 2009).

E. coli O157:H7, *Listeria, Salmonella,* and *Vibrio* are among some of the common food-borne pathogens that can be killed by ionizing radiation (Borsa 2006). Ionizing radiation can rid foods of insect eggs and larvae, thus showing resounding potential for use as a disinfectant of agricultural produce.

Food irradiation can be classified into three processes based on the level of antimicrobial activity obtained. *Radurization* uses relatively high doses ranging from 0.5 to 10 kGy (commonly 5 kGy) of ionizing radiation to substantially eliminate common and normal food spoilage microorganisms (except viruses). However, metabolically less active microbes, like yeasts and lactic acid bacteria, are not eliminated by this process; these along with non–spore-forming pathogenic microbes (including enteropathogenic strains like *Salmonella*) and parasites are removed through *radicidation*–this process make use of radiation up to 8 kGy. *Sterilization irradiation* (or *radappertization*) renders prepackaged, enzyme-inactivated food sterile, that is, devoid of any microbial presence (excluding viruses and including the most resistant spores, such as those of *C. botulinum*); radiation doses of up to 50 kGy are used for this purpose. Radappertized foods are shelf-stable even without refrigeration (Goresline et al. 1964; World Health Organization 1999; Farkas 2007; Koutchma 2009).

One of the biggest advantages with the use of ionizing radiation is that it could be applied to packaged foods, thus ensuring inactivation of any postprocessing and prepackaging microbial contamination. Most of the commonly used packaging materials are radiation safe (Molins 2001). Combining ionizing radiation with other hurdles (most commonly heat, controlled atmospheres or low temperatures) could be further effective in the case of some products (Farkas 2007).

Challenges

Apart from the proven safety of food irradiation, one important benefit is the lack of a need for refrigerated storage; that is, irradiated food can be stored at ambient temperatures: This translates into considerable energy savings (Woods 1994). Further, energy transfer is much more efficient in food irradiation (when compared with conventional thermal processing methods) and has also been proved to be remarkably environmentally friendly in terms of lower greenhouse gas emissions (Berejka 1995).

In spite of all these advantages, commercialization of ionizing radiation has been quite slow. In 1997, the United States Food and Drug Administration (FDA) approved the use of ionizing radiation for controlling microorganisms in red meat. This was followed by endorsement of the technology by several public health agencies such as the American Medical Association and the American Dietetic Association (Wood and Bruhn 2000). At that time, the future of food irradiation technology looked very promising. However, consumer acceptance of irradiated products has remained low. The surprising declaration of bankruptcy in 2004 by SureBeam Corp., the California-based leading provider of food irradiation systems and services brought a major setback to further commercialization of the technology (Goldstein 2004). Though the company renewed operations in 2006 under a new name, Sadex Inc., and despite a minor resurgence of irradiated food products in several markets worldwide, the technology is yet to be completely accepted by consumers as "safe" (Sapp and Downing-Matibag 2009).

Several roadblocks remain before food irradiation (especially with ionizing radiation) can become a commercial and consumer favorite. Although the technology has been approved for use in more than 50 countries worldwide, regulatory approval is pending in many others. In the United States, food irradiation (at specified radiation doses) has already been approved for microbial control in preprocessed meat and poultry and fresh as well as processed fruits and vegetables (FDA and HHS 2008). Regulatory approval is still pending for many other common foods such as ready-to-eat meals and seafood. Additionally, labeling of irradiated foods is still a major controversial issue (MDS Nordion 2010). Undesirable changes in the organoleptic properties of some irradiated foods and recent findings about radiation resistance seen in certain foodborne microorganisms have complicated things further (Farkas 2007). However, the biggest and most significant roadblock is the continuing unfavorable response from consumers. This may call for more assurance, awareness, and education to consumers about the safety of irradiated foods, supported by stronger research findings on the technology.

14.4.2.4 Ultraviolet radiation and high-intensity pulsed light

UV radiation

UV radiation is one of the *gentler* methods of microbial control, especially for certain liquids such as alcoholic beverages, that do not require extensive antimicrobial treatment (Bintsis et al. 2000; Farkas 2007). Short wave ultraviolet rays or UVC light have lesser wavelength than visible light and possess maximum antimicrobial activity between 250 and 260 nm (Bintsis et al. 2000). Traditionally, mercury vapor lamps are used to generate UV light. However, since mercury is a toxic substance, novel nontoxic and more environmentally friendly sources of UV light such as light emitting diodes (LED) and xenon for disinfection of air have been tested and found to be effective, although further research is required to make them more efficient (Ryan et al. 2010).

Microbial inactivation is caused primarily by DNA damage (specifically by removal of thymine dimers); hence inhibiting DNA repair and subsequent microbial reproduction (Boyce and Howard-Flanders 1964). Whereas most gram-negative bacteria are killed by UV light, bacterial spores, viruses and moulds are much harder to kill. Among common foodborne microbes inactivated by UV light are *E. coli, Salmonella, Bacillus, Streptococcus, Staphylococcus* strains, and even some viruses and spores, although these require much higher doses of UV light compared with what is necessary to kill vegetative bacteria. This limited spectrum of antimicrobial activity, coupled with low penetrability and nonuniform treatment make UV radiation a much less-preferred method of microbial control. However this method is commonly used for ancillary processes in the food processing industry such as disinfection of air, packaging, and surfaces of processing equipment and water (when usage of chlorine is undesirable) (Chang et al. 1985; Guerrero-Beltrán and Barbosa-Cánovas 2004; Farkas 2007).

UV hurdle technology

Combining UV radiation with hydrogen peroxide treatment produced a method that was several times more effective (than standalone UV radiation) in eliminating bacterial spores (Bayliss and Waites 1979). Combining UVC light with heat treatment was shown to have good antifungal activity in postharvest fruits and vegetables such as strawberries and cherries (Marquenie et al. 2002). More recent work suggest that combinations of UV radiation with other novel non-thermal methods of microbial control such as pulsed white light (Marquenie et al. 2003a, b), pulsed electric fields (Gachovska et al. 2008b), and RF electric fields (Ukuku and Geveke 2010) are effective methods of microbial control in raw and processed fruit products such as juices. Ozonization is another method that has been successfully combined with UV radiation for the control of *Salmonella* in in-shell eggs (Rodriguez-Romo and Yousef 2005). Recent research has even suggested that exposure to low doses of UV radiation improves the function of some antipathogenic and health-promoting compounds in certain fruits and vegetables (Oms-Oliu et al. 2010).

High-intensity pulsed light technology

The high-intensity pulsed light (HIPL) method involves the use of intense, short pulses (or flashes) of broad-spectrum light. It is analogous to pulsed electric fields in that instead of high-voltage electric fields, food is subjected to very short pulses of high intensity light. Usually a broad spectrum of light is used with wavelengths ranging from 180 nm (UV rays) to 1,100 nm (near-infrared rays). Although primarily a nonthermal method, the antimicrobial effects of HIPL have been attributed both to the effects of UV-assisted DNA damage as well as the thermal effects of visible and infrared light such as protein denaturation and cell rupture owing to heat. Antimicrobial effect is directly proportional to the intensity (energy) of the light pulses. Foodborne microbes that have been successfully inactivated using HIPL include *E. coli* O157:H7, *Listeria, Salmonella,* fungi, yeasts, and even spores of *Clostridium* and *Bacillus* strains (MacGregor et al. 1998; Takeshita et al. 2002, 2003; Guillou et al. 2007). Roberts and Hope (2003) have reported that HIPL technology could effectively inactivate viruses as well (Roberts and Hope 2003).

As with regular UV application, one of the main disadvantages of HIPL is the very low penetration of light and hence is applicable mostly to foods with smooth surfaces or those packaged in UV-transparent material. Besides, HIPL is not preferred for use on protein- and oil-rich foods because of the absorption of some portion of the radiation by these compounds. Another problem is that longer pulses cause the temperature of the foods to rise. This heating up of food is difficult to control and leads to undesirable changes in the color, flavor, and other properties of food (Guillou et al. 2007; Oms-Oliu et al. 2010). These disadvantages have prompted researchers to combine HIPL technology with other thermal and nonthermal methods of microbial control. HIPL (with or without other hurdles) has been tested for antimicrobial activity in foods such as vegetables (Gómez-López et al. 2005) and fruits (Marquenie et al. 2003b). There is much incoherence among the results of these applications, with some of them reporting successful microbial control and elongation of shelf life, and others reporting insufficient to average success, as compared with other conventional and novel methods (Guillou et al. 2007).

High-intensity pulsed uv light technology

A more successful modification of HIPL treatment is high-intensity pulsed UV light (PUVL); here UV rays are the sole sources of light used. This technology has been suggested as an effective alternative to conventional continuous UV water treatment (Huffman et al. 2000; Guillou et al. 2007). From a food processing point of view, PUVL has been shown to be quite effective in the pasteurization and inactivation of pathogens in milk (Smith et al. 2002; Krishnamurthy et al. 2007). PUVL has also been used for the successful inactivation of several food-borne pathogens (e.g., *E. coli* O157:H7, *Salmonella,* and *Listeria monocytogenes*) in seafood (Ozer and Demirci 2006), in-shell eggs (Keklik et al. 2009), fruits (Bialka and Demirci 2007, 2008) and alfalfa seeds (Sharma and Demirci 2003). *Salmonella enteritidis, Pseudomonas aeruginosa, Bacillus cereus,* and *Staphylococcus aureus* are other foodborne pathogens inhibited by PUVL (Rowan et al. 1999; Krishnamurthy et al. 2004). Most recently, Rajkovic et al. (2010) showed that PUVL could be used to decontaminate stainless steel surfaces of meat processing equipment. They were able to successfully eliminate *L. monocytogenes and E. coli* O157:H7 from surfaces that meat could come in contact with during processing. These results (along with necessary further research) suggest that PUVL could be part of HACCP systems in meat processing and could be an energy-efficient alternative to continuous UV treatment (Rajkovic et al. 2010).

14.4.2.5 Ultrasound

One of the simplest and safest non-thermal methods of microbial control developed so far, ultrasound (US), is fast catching up as a favorite among food producers, researchers, and consumers alike. This method employs sound waves in the ultrasonic range; that is, those with frequencies above the human audibility range (20 kHz or more). It is also referred to as *sonication* or *ultrasonication*.

Current technology

US can be divided into two categories based on its intensity; *power US* or *high intensity US* with low frequency (20–100 kHz) and *high frequency US* or *low intensity US* with frequencies in the range of 2–10 MHz. Both power US as well as high frequency US have been used for a wide variety of applications in food processing, such as extraction and separation of specific bioactive compounds, cell disruption, homogenization, emulsification, crystallization of fats and sugars, meat tenderization, cleaning and effluent treatment. However, from an antimicrobial point of view, power ultrasound is more efficient than high frequency US (Piyasena et al. 2003; Zenker et al. 2003; Knorr et al. 2004; Condón et al. 2005; Feng et al. 2008; Demirdöven and Baysal 2009).

When a high-frequency electric field is applied to ferroelectric materials, they undergo *elastic deformation* leading to the generation of ultrasonic waves. US waves are mostly generated using an *electrostrictive transformer* and are transmitted via an amplifier to a *sonotrode,* which eventually transmits them to the processing medium (Raichel 2006).

Antimicrobial applications

When US propagates through liquid media, alternating compression and expansion cycles of the medium particles result. The latter creates a negative pressure that on becoming low enough to overcome the intermolecular forces of the medium leads to the creation of small gas-filled bubbles, or *cavities*. These cavities contract, expand, and implode with the alternating expansion/compression cycle of the sound waves. This phenomenon is called *cavitation* and forms the primary mechanism for the antimicrobial activity of US. There are two types of cavitation caused by US; *stable cavitation* is produced by low-intensity (or high-frequency) US and produces bubbles that oscillate (and stay) in the medium for several thousand cycles and, *transient cavitation,* which is produced by power US results in the formation of bubbles that increase in size during each cycle and eventually implodes.

Whereas the former produces strong *microcurrents* that act as shock waves and create high hydrodynamic shear stress, the latter causes a sudden increase in temperature and pressure in the immediate surrounding region of the processing medium. Both type of cavitations cause cell envelopes (of live vegetative microorganisms) to rupture through the various mechanical actions produced. The rupture of microbial cells by US cavitation is a very well-studied phenomenon and is of much academic interest (Condón et al. 2005; Feng et al. 2008).

Several factors determine the power output of US; frequency and amplitude of the sound waves as well as the hydrostatic pressure and temperature of the processing medium. Although US amplitude and medium temperature are the most significant among these, it is possible to attain high levels of power output by controlling and using different combinations of these factors. The antimicrobial activity of US in food is additionally influenced by factors such as the type of microbes, the volume and composition of the food material being processed. Whereas US has been used successfully for microbial control in several foods like juices (Gómez-López et al. 2010) and milk (Cameron et al. 2009), the lethal effects of the process on microbes are not sufficient for it to be used as a standalone antimicrobial method. For this reason, US is rarely used independently in food processing; it is rather combined with other hurdles like heat, low pressure, chemical antimicrobials, and even pulsed electric fields for better effectiveness (Earnshaw et al. 1995; Raso et al. 1999; Condón et al. 2005; Demirdöven and Baysal 2009; Lee et al. 2009; Walkling-Ribeiro et al. 2009b).

US Hurdle technology

Thermosonication (TS) or ultrasound-assisted thermal (UST) treatment is the combination of US and heat. Besides lesser processing times, TS treats food at temperatures much lower than those required by conventional processes (e.g., pasteurization) to achieve comparable levels of antimicrobial activity (Milly et al. 2007). TS has been tested on foods such as milk (Earnshaw et al. 1995; Villamiel and de Jong 2000; Bermúdez-Aguirre and Barbosa-Cánovas 2008; Bermudez-Aguirre et al. 2009; Riener et al. 2009b) and fruit juices (D'Amico et al. 2006; Valero et al. 2007; Wu et al. 2008; Adekunte et al. 2010). Among food-borne microbes inactivated by TS are fungi (*Aspergillus* and *Penicillium*), *Listeria, Shigella, Lactobacillus, E. coli*, and several *Bacillus* strains (Knorr et al. 2004; Lopez-Malo et al. 2005; Ugarte-Romero et al. 2007; Lee et al. 2009). Studies have also been conducted on combinations of TS along with other hurdles such as pulsed electric fields (Noci et al. 2009; Walkling-Ribeiro et al. 2009a, b), chemical sanitizers (Broda 2007), chemical preservatives such as potassium sorbate and natural antimicrobials such as vanillin and citral (Lopez-Malo et al. 2005; Ferrante et al. 2007)

Manosonication (MS) is the application of US under pressure (at nonlethal temperatures). Usually, pressures of up to 600 kPa are applied to an US process to improve the antimicrobial effect of the combined process (Earnshaw et al. 1995). Some studies have reported increased lethality toward yeasts and an enhanced sensitivity of bacterial spores to lysozyme with MS treatment (Raso et al. 1998).

Manothermosonication (MTS) is the combined application of US and pressure at lethal temperatures. Among all modified US methods, MTS has been shown to possess maximum antimicrobial effect especially toward more thermotolerant microbes; this may be due to the synergistic effect between US and pressure and the additive effect between MS and heat (Ross et al. 2003; Demirdöven and Baysal 2009). MS and MTS have been tested and shown to be effective against yeasts, *Listeria* (Manas et al. 2000), several strains of *Salmonella* (Manas et al. 2000; Alvarez et al. 2003).

Challenges

Although the antimicrobial effects of US have been well understood, the combined effects of US along with pressure, heat, and other hurdles are only beginning to be clearly understood. In spite of its inherent simplicity and safety (compared with high power technologies such as MW/RF and PEF), further research into the antimicrobial mechanisms, processing efficiency, and industrial scale-up is necessary before US can be considered for commercialization.

14.4.2.6 Ozonization

Ozone (O_3) is an unstable allotrope of oxygen produced naturally as a result of UV radiation reactions or lighting, and artificially using UV light or corona discharge generators. The creation of free radicals during the production of ozone and the subsequent unstable nature of the same (the gas dissolves in water at refrigeration temperature and acidic pH but decomposes back to free radicals with increase in temperature and/or pH) make ozone a very effective oxidizing agent and a potent antimicrobial sanitizer (or disinfectant) for use in the food processing industry (Gordon 1995).

Ever since the first commercial ozone-based water treatment plant was built in France in 1906, ozonization has become a very popular method of water treatment, especially in Europe. However, the use of ozone as a disinfectant or sanitizer in the food processing industry was quite limited. This changed in 1997 when ozone was declared as a GRAS substance, following which the USDA and FDA approved its use in food processing (Graham 1997; Karaca and Velioglu 2007).

Among its various advantages are high penetrability in food and spontaneous decomposition to a nontoxic product – oxygen. The latter makes ozone a very environmentally friendly food preservation technology (Guzel-Seydim et al. 2004; Karaca and Velioglu 2007). Microbes inactivated by ozone include various bacteria, molds, yeasts, parasites, and even viruses. The degree of antimicrobial action is directly proportional to the concentration of oxidizable organic compounds in the food or processing medium and indirectly proportional to the "ozone demand" (Kim et al. 1999a; Kim et al. 2003; Novak and Yuan 2007).

Hurdling ozonization with other treatments such as chemical disinfectants (e.g., hydrogen peroxide), antimicrobial preservatives (e.g., bacteriocins), irradiation

(Mahapatra et al. 2005), pulsed UV light (Bialka and Demirci 2007), and pulsed electric fields (Unal et al. 2001) has been shown to improve the antimicrobial activity of ozone (Kim et al. 2003). Ideal for use on foods with solid, smooth surfaces (e.g., fruits, vegetables, eggs, meat, and some dry foods), ozonization could also be used to sanitize food processing equipment and packaging materials.

Although some ozone-disinfected products recently have been introduced into the US market, success has been limited to certain fruits and vegetables only. The most notable problem with the use of ozone is that it cannot be stored or transported; this means that the ozone gas has to be produced on-site or at a facility close to the processing unit. Further, each separate food processing application requires appropriate modification and incorporation of the ozonization process. The high costs of the electrical power and specialized ozone-compatible materials and equipment necessary to generate ozone are other disadvantages. The process also needs to be well controlled. Extreme ozonization of food leads to discoloration and loss of some organoleptic properties. Additionally, excessive inhalation of ozone could be toxic to humans. This calls of extensive safety measures to be put in place at processing units (Karaca and Velioglu 2007; Novak and Yuan 2007; Pascual et al. 2007).

A better understanding of ozone's antimicrobial mechanisms and more research into efficient system design and monitoring may improve the prospects of ozone being used as an effective and environmentally friendly disinfectant and sanitizer in food processing (Kim et al. 1999a; Kim et al. 2003; Novak and Yuan 2007).

14.4.2.7 Other potential nonthermal methods

Pulsed magnetic fields (PMF) is another method that has only recently been shown to have antimicrobial effects. Whereas MW technology uses magnetic fields, PMF can be differentiated from MW in that the latter relies on thermal effects to achieve microbial inactivation. Work done by Pothakamury et al. (1993) suggests that short pulses (ranging from 25 μs to 10 ms) of oscillating magnetic fields of intensity ranging from 5 to 50 T at frequencies ranging from 5 to 500 kHz could be used for pasteurization of packaged foods (Pothakamury et al. 1993). In another more recent work, it was shown that pulsed magnetic fields of intensity 160 mT at a frequency of 62 kHz was able to sterilize *E. coli* (Li et al. 2004). The exact mechanism of microbial inactivation by magnetic fields is still largely unclear. It has been suggested that these fields interfere with the magnetic dipoles of large biomolecules such as nucleic acids and proteins, rendering them unstable and susceptible to undesirable structural and functional changes. Detrimental effects on fluidity of biological membranes have also been suggested as a possible mechanism (Farkas 2007).

Cold plasma processing involves the use of gases in their *nonthermal* or *cold plasma* (i.e., partially ionized) form to decontaminate surfaces of foods, packaging material, and processing equipment. Work done at USDA has shown that cold plasma could effectively inactivate EHEC O157:H7 and Salmonella on the surfaces of apples (Niemira and Sites 2008). *Escherichia coli* was also inactivated in several other foods treated with pulsed cold plasma including apple juice (Montenegro et al.

2002), almonds (Deng et al. 2007) and cut fruits (Perni et al. 2008a,b). Netherlands' Wageningen University Food Technology Centre has also been investigating the use of this technology for disinfecting the surfaces of fresh meat and foods with fragile surfaces like cut vegetables and fruits (Wageningen 2006).

Supercritical carbon dioxide (i.e., liquid carbon dioxide at high pressure) has been found to effectively pasteurize liquid food products such as fruit juices (Del Pozo-Insfran et al. 2006; Tahiri et al. 2006; Chen et al. 2009; Ferrentino et al. 2009; Yuk et al. 2010), whole egg (Garcia-Gonzalez et al. 2009), and dairy products (Werner and Hotchkiss 2006). The gas is solubilized at high pressure and mixed with the food product while maintaining conditions that would prevent the liquefied carbon dioxide from freezing. Following treatment, the food product is de-aerated to remove the carbon dioxide (Brennan 2006).

The mentioned methods are just three of the several emerging techniques being currently investigated by researchers. Although many of these methods show promise, research data are extremely limited because of their very recent nature and results are inconsistent. As with other recently developed and poorly understood methods, much detailed information on their mode of antimicrobial activity and safety of processed food is required before any commercial application can be considered.

14.5 Future directions and conclusions

As mentioned, the distinction among *conventional*, *current,* and *emerging* methods is slowly disappearing with the application of *hurdle* concepts. Most modern food processing applications do not rely on any single method (conventional, current, or emerging) to achieve microbial control. These are rather composed of several different methods (or *hurdles*) incorporated into the processing system at different stages of HACCP programs. The many different combinations of all of the mentioned methods result in an endless number of hurdle technologies for microbial control. However, the surface has only just been scratched. As research into novel methods progresses, we are likely to see a torrent of research data on hundreds of feasible hurdle technologies customized for very specific applications. Not only does this mean more work for the FDA, this also means that food producers now have a choice of several different technologies available to ensure the microbiological safety of their food products. Two research areas that are likely to shape the future of microbial control in food processing are *predictive microbiological modeling* and *rapid automated methods of microbial detection.*

Mathematical models can be used to explain and *predict* how microbes behave under the influence of different parameters such as time, environment, and other factors that affect microbial growth and function (Whiting 1995; Betts and Everis 2005). Although these models are based on actual experimental results; that is, based on microbial responses under specific growth conditions in vitro, they could be used to predict responses for conditions that have not been laboratory tested. This ability to predict microbial behavior enables food producers to test how microbes would grow (or behave) in new food recipes in silico; and hence their profound applications in new product development (NPD). Since the 1980s and early 1990s, several mathematical modeling systems were developed and became available for use in food safety design systems. Some of the most prominent of such modeling programs and tools currently available include the Refrigeration Index calculator, Pathogen Modeling Program (PMP), Seafood Spoilage and Safety Predictor (SSSP), ComBase Predictor, and Growth Predictor. These modeling concepts are likely to influence the design and evaluation of new and novel hurdle concepts for microbial control in food processing (Tamplin 2009).

Another major research direction will most likely be into more advanced rapid methods of microbial detection. It must be understood that a *rapid* test does not necessarily mean one that produces instant (within seconds or minutes) results; in fact such a test is yet to be created! Feng (2007) defines a rapid method as one that brings any substantial reduction to the analysis time taken by a conventional method or assay (Feng 2007). Hence one can imagine the vast variety of methods that could fit this definition. Although microbial detection is in itself an important aspect of HACCP approaches, this topic has not been discussed in this chapter because it encompasses another major field of microbial control in food processing that is simply too large to include here. The field of rapid methods (for microbial detection) has been accelerated by the monumental advances in molecular biology and computing technologies over the last quarter of the twentieth century. Progress has been so revolutionary that it is now possible to detect even *single* cells of food microbes. However major challenges such as problems with sampling and high costs of maintaining equipment and trained personnel have slowed down the full-fledged application of these methods to commercial food processing systems. In any case, this is probably one field that will continue to grow exponentially and shape the landscape of food microbiology research in the coming years (Feng and Heredia 2009).

The development of most of the novel methods discussed in this chapter was mainly a result of the need for foods that are minimally processed and hence retain much of their natural organoleptic properties. However, over the last few decades, there has been an equally significant emphasis placed on the development of more "green" or environmentally friendly technologies in food processing. With advances in novel methods, researchers have already proved that several of these methods (notably MW/RF, OH, PEF, HPP, and irradiation) provide both higher energy savings as well as reduced greenhouse gas emissions in comparison with conventional methods (Pereira and Vicente 2010). Additionally, many of the recently developed methods incorporate designs for minimal or no use of chemical preservatives and recycling of processing water (e.g., in ozonization).

Water and energy are two of the most pivotal driving resources for the food processing industry and conservation of both has become a significant objective for researchers and producers alike. Hence, energy management and water efficiency constitute two factors that will continue to influence the "greenness" of any novel method that has been developed at the laboratory scale (Masanet et al. 2008). This also means that there needs to be increased interaction and understanding between environment engineers and food engineers. In conclusion, the future of food processing looks determined to be dominated by novel hurdle technologies that not only produce food products of high quality but are also environmentally friendly and energy efficient.

Another interesting concluding point to note is that most of the technologies that are currently commercialized have had their origins in academic research that has been jointly funded by government organizations and private sector companies. Pilot systems built with such industry-academia collaboration have been the starting point for the commercialization of almost all major current and emerging techniques used in food processing today such as UHT, MW, PEF, and many others. This adds credibility to the suggestion that governments, environment agencies, and private sectors companies must provide increased funding to academic institutions to further improve their research infrastructure. More importantly, such linkages help build an understanding of mutual cooperation between these two sectors that could eventually transform into successful technology that is both beneficial to human well-being and considerate of the environment.

References

Aalto, T.R., M.C. Firman, and N.E. Rigler. 1953. p-Hydroxybenzoic acid esters as preservatives. I. Uses, antibacterial and antifungal studies, properties and determination. *Journal of the American Pharmacists Association* 42(8): 449–457.

Adedeji, A.A., T.K. Gachovska, M.O. Ngadi, and G.S.V. Raghavan. 2008. Effect of pretreatments on drying characteristics of okra. *Drying Technology* 26(10): 1251–1256.

Adekunte, A.O., B.K. Tiwari, P.J. Cullen, A.G.M. Scannell, and C.P. O'Donnell. 2010. Effect of sonication on colour, ascorbic acid and yeast inactivation in tomato juice. *Food Chemistry* 122(3): 500–507.

Ade-Omowaye, B.I., N.K. Rastogi, A. Angersbach, and D. Knorr. 2003. Combined effects of pulsed electric field pre-treatment and partial osmotic dehydration on air drying behaviour of red bell pepper. *Journal of Food Engineering* 60(1): 89–98.

Agar, I.T., J. Streif, and F. Bangerth. 1997. Effect of high CO_2 and controlled atmosphere (CA) on the ascorbic and dehydroascorbic acid content of some berry fruits. *Postharvest Biology and Technology* 11(1): 47–56.

Aguilo-Aguayo, I., I. Odriozola-Serrano, L.J. Quintao-Teixeira, and O. Martin-Belloso. 2008. Inactivation of tomato juice peroxidase by high-intensity pulsed electric fields as affected by process conditions. *Food Chemistry* 107(2): 949–955.

Akin, E., and G.A. Evrendilek. 2009. Effect of pulsed electric fields on physical, chemical, and microbiological properties of formulated carrot juice. *Food Science and Technology International* 15(3): 275–282.

Alvarez, I., P. Manas, F.J. Sala, and S. Condon. 2003. Inactivation of *Salmonella enterica serovar* enteritidis by ultrasonic waves under pressure at different water activities. *Applied and Environmental Microbiology* 69(1): 668–672.

Amami, E., E. Vorobiev, and N. Kechaou. 2005. Effect of pulsed electric field on the osmotic dehydration and mass transfer kinetics of apple tissue. *Drying Technology* 23(3): 581–595.

Arevalo, P., M.O. Ngadi, M.I. Bazhal, and G.S.V. Raghavan. 2004. Impact of pulsed electric fields on the dehydration and physical properties of apple and potato slices. *Drying Technology* 22(5): 1233–1246.

Aronsson, K., U. Ronner, and E. Borch. 2005. Inactivation of *Escherichia coli, Listeria innocua* and *Saccharomyces cerevisiae* in relation to membrane permeabilization and subsequent leakage of intracellular compounds due to pulsed electric field processing. *International Journal of Food Microbiology* 99(1): 19–32.

Bajgai, T.R., and F. Hashinaga. 2001. Drying of spinach with a high electric field. *Drying Technology* 19(9): 2331–2341.

Balasubramaniam, V.M., and D. Farkas. 2008. High-pressure food processing. *Food Science and Technology International* 14(5): 413–418.

Barbosa-Cánovas, G.V., and G.W. Gould. 2000. *Innovations in food processing*. Boca Raton, FL: CRC.

Barsotti, L., P. Merle, and J.C. Cheftel. 1999. Food processing by pulsed electric fields. I. Physical aspects. *Food Reviews International* 15(2): 163–180.

Bayliss, C.E., and W.M. Waites. 1979. The combined effect of hydrogen peroxide and ultraviolet irradiation on bacterial spores. *Journal of Applied Bacteriology* 47(2): 263–269.

Berejka, A.J. 1995. Irradiation processing in the '90's: energy savings and environmental benefits. *Radiation Physics and Chemistry* 46(4-6, Part 1): 429–437.

Bengtson, R., E. Birdsall, S. Feilden, S. Bhattiprolu, S. Bhale and M. Lima. 2006. Ohmic and inductive heating. *Handbook of food science, technology, and engineering*. Y. H. Hui. Boca Raton, Taylor & Francis.

Bermúdez-Aguirre, D., and G.V. Barbosa-Cánovas. 2008. Study of butter fat content in milk on the inactivation of *Listeria innocua* ATCC 51742 by thermo-sonication. *Innovative Food Science and Emerging Technologies* 9(2): 176–185.

Bermudez-Aguirre, D., R. Mawson, K. Versteeg, and G.V. Barbosa-Canovas. 2009. Composition properties, physicochemical characteristics and shelf life of whole milk after thermal and thermo-sonication treatments. *Journal of Food Quality* 32(3): 283–302.

Betts, G., and L. Everis. 2005. Modeling systems and impact on food microbiology. In *Novel food processing technologies*, ed. G.V. Barbosa-Cánovas, M.P. Cano, and M.S. Tapia. Boca Raton, FL: CRC.

Bialka, K.L., and A. Demirci. 2007. Decontamination of *Escherichia coli* O157:H7 and *Salmonella enterica* on blueberries using ozone and pulsed UV-light. *Journal of Food Science* 72(9): M391–M396.

Bialka, K.L., and A. Demirci. 2008. Efficacy of pulsed UV-light for the decontamination of *Escherichia coli* O157:H7 and *Salmonella* spp. on raspberries and strawberries. *Journal of Food Science* 73(5): M201–M207.

Bintsis, T., E. Litopoulou-Tzanetaki, and R.K. Robinson. 2000. Existing and potential applications of ultraviolet light in the food industry: a critical review. *Journal of the Science of Food and Agriculture* 80: 637–645.

Borsa, J. 2006. Introduction: food irradiation moving on. In *Food irradiation research and technology*, ed. C.H. Sommers and X. Fan. Ames, IA: Blackwell.

Boyce, R.P., and P. Howard-Flanders. 1964. Release of ultraviolet light-induced thymine dimers from DNA in *E. coli* K-12. *Proceedings of the National Academy of Sciences of the United States of America* 51(2): 293–300.

Brandriff, K.M., and National Agricultural Library Food Safety Research Information Office. 2003. A focus on Hazard Analysis and Critical Control Points. http://hdl.handle.net/1030.28/NAL00000027.

Branen, A.L., and P.M. Davidson. 1983. *Antimicrobials in foods*. New York: Dekker.

Brennan, J.G. 2006. *Food processing handbook*. Weinheim, Germany: Wiley-VCH.

Broda, D.M. 2007. The effect of peroxyacetic acid-based sanitizer, heat and ultrasonic waves on the survival of *Clostridium estertheticum* spores in vitro. *Letters in Applied Microbiology* 45(3): 336–341.

Bull, M.K., S.A. Olivier, R.J. van Diepenbeek, F. Kormelink, and B. Chapman. 2009. Synergistic inactivation of spores of proteolytic *Clostridium botulinum* strains by high pressure and heat is strain and product dependent. *Applied and Environmental Microbiology* 75(2): 434–445.

Calderón-Miranda, M.L., G.V. Barbosa-Cánovas, and B.G. Swanson. 1999. Inactivation of *Listeria innocua* in liquid whole egg by pulsed electric fields and nisin. *International Journal of Food Microbiology* 51(1): 7–17.
Cameron, M., L.D. McMaster, and T.J. Britz. 2009. Impact of ultrasound on dairy spoilage microbes and milk components. *Dairy Science and Technology* 89(1): 83–98.
Cammack, R., C.L. Joannou, X.-Y. Cui, C. Torres Martinez, S.R. Maraj, and M.N. Hughes. 1999. Nitrite and nitrosyl compounds in food preservation. *Biochimica et Biophysica Acta: Bioenergetics* 1411(2–3): 475–488.
Carlez, A., T. Veciana-Nogues, and J.-C. Cheftel. 1995. Changes in colour and myoglobin of minced beef meat due to high pressure processing. *Food Science and Technology* 28(5): 528–538.
Castro, I. 2007. Ohmic heating as an alternative to conventional thermal treatment. Dissertation, University of Minho, Portugal.
Chang, J.C., S.F. Ossoff, D.C. Lobe, M.H. Dorfman, C.M. Dumais, R.G. Qualls, and J.D. Johnson. 1985. UV inactivation of pathogenic and indicator microorganisms. *Applied and Environmental Microbiology* 49(6): 1361–1365.
Cheftel, J.C. 1992. Effects of high hydrostatic pressure on food constituents: an overview. In *High pressure and biotechnology*, ed. C. Balny, R. Hayashi, K. Heremans, and P. Masson. Paris, France: INSERM.
Cheftel, J.C. 1995. High pressure, microbial inactivation and food preservation. *Food Science and Technology International* 1(2): 75–90.
Chen, J.L., J. Zhang, Z.S. Feng, L.J. Song, J.H. Wu, and X.S. Hu. 2009. Influence of thermal and dense-phase carbon dioxide pasteurization on physicochemical properties and flavor compounds in hami melon juice. *Journal of Agricultural and Food Chemistry* 57(13): 5805–5808.
Cheng, X.D. 2007. Apparatus for ohmic heating e.g. to pasteurize or sterilize product such as milk comprises electrode clearing device having scrapper that contacts surface of electrodes and moves continuously over a cylindrical surface during ohmic heating, Auckland Uniservices Ltd. (AUCK-Non-standard): 30.
Cho, H.Y., A.E. Yousef, and S.K. Sastry. 1996. Growth kinetics of *Lactobacillus acidophilus* under ohmic heating. *Biotechnology and Bioengineering* 49(3): 334–340.
Cho, H.Y., A.E. Yousef, and S.K. Sastry. 1999. Kinetics of inactivation of Bacillus subtilis spores by continuous or intermittent ohmic and conventional heating. *Biotechnology and Bioengineering* 62(3): 368–372.
Cinquanta, L., D. Albanese, G. Cuccurullo, and M.D. Matteo. 2010. Effect on orange juice of batch pasteurization in an improved pilot-scale microwave oven. *Journal of Food Science* 75(1): E46–E50.
Coleman, W.H., D. Chen, Y-q Li, A.E. Cowan, and P. Setlow. 2007. How moist heat kills spores of *Bacillus subtilis*. *Journal of Bacteriology* 189(23): 8458–8466.
Condón, S., J. Raso, and R. Pagán. 2005. Microbial inactivation by ultrasound. In *Novel food processing technologies*, ed. G.V. Barbosa-Cánovas, M.P. Cano, and M.S. Tapia. Boca Raton, FL: CRC.
Cotter, P.D., C. Hill, and R.P. Ross. 2005. Bacteriocins: developing innate immunity for food. *Nature Reviews Microbiology* 3(10): 777–788.
D'Amico, D.J., T.M. Silk, J.R. Wu, and M.R. Guo. 2006. Inactivation of microorganisms in milk and apple cider treated with ultrasound. *Journal of Food Protection* 69(3): 556–563.
Davidson, P.M., and T.M. Taylor. 2007. Chemical preservatives and natural antimicrobial compounds. In *Food microbiology: fundamentals and frontiers*, ed. L.R. Beuchat and M.P. Doyle. Washington, DC: ASM.
Davidson, P.M., J.N. Sofos, and A.L. Branen. 2005. *Antimicrobials in food*. Boca Raton, FL: Taylor & Francis.
Dawson, P.L., S. Mangalassary, and B.W. Sheldon. 2006. Thermal processing of poultry products. In *Thermal food processing: new technologies and quality issues*, ed. D.-W. Sun. Boca Raton, FL: CRC/Taylor & Francis.

Day, L., and I. McNeil. 1996. *Biographical dictionary of the history of technology*. London/New York: Routledge.

De Alwis, A.A.P., and P.J. Fryer. 1990. A finite-element analysis of heat generation and transfer during ohmic heating of food. *Chemical Engineering Science* 45(6): 1547–1559.

DeCareau, R.V. 1994. The microwave sterilization process. *Microwave World* 15(2): 12.

Del Pozo-Insfran, D., M.O. Balaban, and S.T. Talcott. 2006. Microbial stability, phytochemical retention, and organoleptic attributes of dense phase CO_2 processed muscadine grape juice. *Journal of Agricultural and Food Chemistry* 54(15): 5468–5473.

Delgado, A.E., D.W. Sun, and A.C. Rubiolo. 2006. Thermal physical properties of foods. In *Thermal food processing: new technologies and quality issues*, ed. D.-W. Sun. Boca Raton, FL: CRC/Taylor & Francis.

Dell'Amore, C. 2009. Cooking gave humans edge over apes? Chigago National Geographic News. http://www.nationalgeographic.com.

Delves-Broughton, J. 2005. Nisin as a food preservative. *Food Australia* 57(12): 525–527.

Demirdöven, A., and T. Baysal. 2009. The use of ultrasound and combined technologies in food preservation. *Food Reviews International* 25(1): 1–11.

Deng, S., R. Ruan, C.K. Mok, G. Huang, X. Lin, and P. Chen. 2007. Inactivation of *Escherichia coli* on almonds using nonthermal plasma. *Journal of Food Science* 72(2): M62–M66.

Dev, S.R.S., T. Padmini, A. Adedeji, Y. Gariepy, and G.S.V. Raghavan. 2008a. A comparative study on the effect of chemical, microwave, and pulsed electric pretreatments on convective drying and quality of raisins. *Drying Technology* 26(10): 1238–1243.

Dev, S.R.S., G.S.V. Raghavan, and Y. Gariepy. 2008b. Dielectric properties of egg components and microwave heating for in-shell pasteurization of eggs. *Journal of Food Engineering* 86(2): 207–214.

Doona, C.J., and F.E. Feeherry. 2007. *High pressure processing of foods*. Ames, IA; Chicago, IL: Blackwell, IFT.

Duncan, C.L., and E.M. Foster. 1968. Effect of sodium nitrite, sodium chloride, and sodium nitrate on germination and outgrowth of anaerobic spores. *Applied Microbiology* 16(2): 406–411.

Earnshaw, R.G., J. Appleyard, and R.M. Hurst. 1995. Understanding physical inactivation processes: combined preservation opportunities using heat, ultrasound and pressure. *International Journal of Food Microbiology* 28(2): 197–219.

Elez-Martinez, P., and O. Martin-Belloso. 2007. Effects of high intensity pulsed electric field processing conditions on vitamin C and antioxidant capacity of orange juice and gazpacho, a cold vegetable soup. *Food Chemistry* 102(1): 201–209.

Espachs-Barroso, A., G.V. Barbosa-Canovas, and O. Martin-Belloso. 2003. Microbial and enzymatic changes in fruit juice induced by high-intensity pulsed electric fields. *Food Reviews International* 19: 253–274.

Evrendilek, G.A., F.M. Tok, E.M. Soylu, and S. Soylu. 2009. Effect of pulsed electric fields on germination tube elongation and spore germination of *Botrytis cinerea* inoculated into sour cherry juice, apricot and peach nectars. *Italian Journal of Food Science* 21(2): 171–182.

Farkas, J. 2007. Physical methods of food preservation. In *Food microbiology: fundamentals and frontiers*, ed. M.P. Doyle and L.R. Beuchat. Washington, DC: ASM.

FDA (Food and Drug Administration) and HHS (Health and Human Services). 2008. Irradiation in the production, processing and handling of food. Final rule. Food and Drug Administration and Health and Human Services. *Federal Register* 73: 49593–49603.

Feng, P. 2007. Rapid methods for the detection of foodborne pathogens: current and next-generation technologies. In *Food microbiology: fundamentals and frontiers*, ed. L.R. Beuchat and M.P. Doyle. Washington, DC: ASM.

Feng, P., and N. Heredia. 2009. Rapid methods for foodborne bacterial enumeration and pathogen detection. In *Microbiologically safe foods*, ed. N. Heredia, I. Wesley, and S. García. Hoboken, NJ: Wiley.

Feng, H., W. Yang, and T. Hielscher. 2008. Power ultrasound. *Food Science and Technology International* 14(5): 433–436.

Fennema, O.R., W.D. Powrie, and E.H. Marth. 1973. *Low-temperature preservation of foods and living matter*. New York: Marcel Dekker.

Ferrante, S., S. Guerrero, and S.M. Alzamora. 2007. Combined use of ultrasound and natural antimicrobials to inactivate *Listeria monocytogenes* in orange juice. *Journal of Food Protection* 70(8): 1850–1856.

Ferrentino, G., M.L. Plaza, M. Ramirez-Rodrigues, G. Ferrari, and M.O. Balaban. 2009. Effects of dense phase carbon dioxide pasteurization on the physical and quality attributes of a red grapefruit juice. *Journal of Food Science* 74(6): E333–E341.

Fields, F.O. 1996. Use of bacteriocins in food: regulatory considerations. Journal of Food Protection 59(3)Suppl: 72–77.

Fillaudeau, L., P. Winterton, J.C. Leuliet, J.P. Tissier, V. Maury, F. Semet, P. Debreyne, M. Berthou, and F. Chopard. 2006. Heat treatment of whole milk by the direct joule effect—experimental and numerical approaches to fouling mechanisms. *Journal of Dairy Science* 89(12): 4475–4489.

Fito, P., A. Chiralt, and M.E. Martin. 2005. Current state of microwave applications to food processing. In *Novel food processing technologies*, ed. G.V. Barbosa-Cánovas, M.P. Cano, and M.S. Tapia. Boca Raton, FL: CRC.

Funebo, T., and T. Ohlsson. 1998. Microwave-assisted air dehydration of apple and mushroom. *Journal of Food Engineering* 38(3): 353–367.

Gachovska, T.K., A.A. Adedeji, M. Ngadi, and G.V. Raghavan. 2008a. Drying characteristics of pulsed electric field-treated carrot. *Drying Technology* 26(10): 1244–1250.

Gachovska, T.K., S. Kumar, H. Thippareddi, J. Subbiah, and F. Williams. 2008b. Ultraviolet and pulsed electric field treatments have additive effect on inactivation of *E. coli* in apple juice. *Journal of Food Science* 73(9): 412–417.

Garcia-Gonzalez, L., A.H. Geeraerd, K. Elst, L. Van Ginneken, J.F. Van Impe, and F. Devlieghere. 2009. Inactivation of naturally occurring microorganisms in liquid whole egg using high pressure carbon dioxide processing as an alternative to heat pasteurization. *Journal of Supercritical Fluids* 51(1): 74–82.

Gentry, T.S., and J.S. Roberts. 2005. Design and evaluation of a continuous flow microwave pasteurization system for apple cider. *Lebensmittel-Wissenschaft und - Technologie* 38(3): 227–238.

Giese, J. 1992. Advances in microwave food processing. *Food Technology* 46(9): 118–123.

Goldblith, S.A., L. Rey, and W.W. Rothmayr. 1975. *Freeze drying and advanced food technology*. London/New York: Academic.

Goldstein, H. 2004. Surebeam Goes Under. IEEE Spectrum.

Gómez-López, V.M., F. Devlieghere, V. Bonduelle, and J. Debevere. 2005. Intense light pulses decontamination of minimally processed vegetables and their shelf-life. *International Journal of Food Microbiology* 103(1): 79–89.

Gómez-López, V.M., L. Orsolani, A. Martínez-Yépez, and M.S. Tapia. 2010. Microbiological and sensory quality of sonicated calcium-added orange juice. *Lebensmittel-Wissenschaft und - Technologie* 43(5): 808–813.

Gordon, G. 1995. The chemistry and reactions of ozone in our environment. *Progress in Nuclear Energy* 29(Suppl 1): 89–96.

Goresline, H.E., M. Ingram, P. Macuch, G. Mocquot, D.A.A. Mossel, C.F. Niven, and F.S. Thatcher. 1964. Tentative classification of food irradiation processes with microbiological objectives. *Nature* 204(4955): 237–238.

Graham, D. 1997. Use of ozone for food processing. *Food Technology* 51(6): 72–73.

Guan, D.S., V.C.F. Plotka, S. Clark, and J.M. Tang. 2002. Sensory evaluation of microwave treated macaroni and cheese. *Journal of Food Processing and Preservation* 26(5): 307–322.

Guan, D., P. Gray, D.-H. Kang, J. Tang, B. Shafer, K. Ito, F. Younce, and T.C.S. Yang. 2003. Microbiological validation of microwave-circulated water combination heating technology by inoculated pack studies. *Journal of Food Science* 68(4): 1428–1432.

Guerrero-Beltrán, J., and G. Barbosa-Cánovas. 2004. Advantages and limitations on processing foods by UV light. *Food Science and Technology International* 10(3): 137–147.

Guerrero-Beltrán, J.A., D.R. Sepulveda, M.M. Góngora-Nieto, B. Swanson, and G.V. Barbosa-Cánovas. 2010. Milk thermization by pulsed electric fields (PEF) and electrically induced heat. *Journal of Food Engineering* 100(1): 56–60.

Guillou, S., F. Leroi, N. Orange, A. Bakhrouf, M. Federighi, and N. Elmnasser. 2007. Pulsed-light system as a novel food decontamination technology: a review. *Canadian Journal of Microbiology* 53: 813–821.

Guzel-Seydim, Z.B., A.K. Greene, and A.C. Seydim. 2004. Use of ozone in the food industry. *Lebensmittel-Wissenschaft und - Technologie* 37(4): 453–460.

Harlfinger, L. 1992. Microwave sterilization. *Food Technology* 46(12): 57–60.

Hartman, P.A. 2001. The evolution of food microbiology. In *Food microbiology: fundamentals and frontiers*, ed. L.R. Beuchat, M.P. Doyle, and T.J. Montville. Washington, DC: ASM.

Hayashi, R. 1995. Advances in high pressure processing technology in Japan. In *Food processing: recent developments*, ed. A.G. Gaonkar. Amsterdam/Lausanne/New York: Elsevier.

Hazan, R., A. Levine, and H. Abeliovich. 2004. Benzoic acid, a weak organic acid food preservative, exerts specific effects on intracellular membrane trafficking pathways in *Saccharomyces cerevisiae*. *Applied and Environmental Microbiology* 70(8): 4449–4457.

Hendrickx, M.E.G., and D.W. Knorr. 2002. *Ultra high pressure treatments of foods*. New York: Kluwer Academic/Plenum Publishers.

Hoover, D.G., C. Metrick, A.M. Papineau, D.F. Farkas, and D. Knorr. 1989. Biological effects of high hydrostatic pressure on food microorganisms. *Food Technology* 43(3): 99–107.

Hou, H., R.K. Singh, P.M. Muriana, and W.J. Stadelman. 1996. Pasteurization of intact shell eggs. *Food Microbiology* 13(2): 93–101.

Huffman, D.E., T.R. Slifko, K. Salisbury, and J.B. Rose. 2000. Inactivation of bacteria, virus and *Cryptosporidium* by a point-of-use device using pulsed broad spectrum white light. *Water Research* 34(9): 2491–2498.

Hui, Y.H. 2006. *Handbook of food science, technology, and engineering*, vol. 1. Boca Raton, FL: Taylor & Francis.

Hutton, M.T., P.A. Chehak, and J.H. Hanlin. 1991. Inhibition of botulinum toxin production by *Pediococcus acidilactici* in temperature abused refrigerated foods. *Journal of Food Safety* 11(4): 255–267.

IFT and FDA. 2000. Kinetics of microbial inactivation for alternative food processing technologies—ohmic and inductive heating. Retrieved 20 March 2011, from http://www.fda.gov/Food/ScienceResearch/ResearchAreas/SafePracticesforFoodProcesses/ucm101246.htm.

Imai, T., K. Uemura, N. Ishida, and S. Yoshizaki. 1995. Ohmic heating of Japanese white radish Rhaphanus sativus L. *International Journal of Food Science and Technology* 30(4): 461.

International Commission on Microbiological Specifications for Foods (ICMSF). 1980. *Microbial ecology of foods: Factors affecting life and death of microorganims*, vol. 1. New York: Academic.

International Commission on Microbiological Specifications for Foods (ICMSF), and J.H. Silliker. 1980. *Factors affecting life and death of microorganisms*. New York: Academic.

International Dairy Food Association. 2009. Pasteurization: definition and methods. Retrieved 13 June 2010, from http://www.idfa.org/files/249_Pasteurization%20Definition%20and%20Methods.pdf.

Josephson, E.S., and M.S. Peterson. 1983. *Preservation of food by ionizing radiation*. Boca Raton FL: CRC.

Karaca, H., and Y.S. Velioglu. 2007. Ozone applications in fruit and vegetable processing. *Food Reviews International* 23(1): 91–106.

Karel, M. 2003. Freezing. In *Physical principles of food preservation*, ed. M. Karel and D.B. Lund. New York: M. Dekker.

Keklik, N.M., A. Demirci, P.H. Patterson, and V.M. Puri. 2009. Decontamination of shell-eggs with pulsed UV-light. ASABE Paper No. 095974. American Society of Agricultural Engineers. St. Joseph, MI.

Kelch, F., and X. Bühlmann. 1958. Effect of commercial phosphates on the growth of microorganisms. *Fleischwirtschaft* 10: 325–328.

Kelly, A.L., N. Datta, and H.C. Deeth. 2006. Thermal processing of dairy products. In *Thermal food processing: new technologies and quality issues*, ed. D.-W. Sun. Boca Raton, FL: CRC/Taylor & Francis.

Kelly, A.L., and M. Zeece. 2009. Applications of novel technologies in processing of functional foods. *Australian Journal of Dairy Technology* 64(1): 12–15.

Kim, J.-G., A.E. Yousef, and S. Dave. 1999a. Application of ozone for enhancing the microbiological safety and quality of foods: a review. *Journal of Food Protection* 62(9): 1071.

Kim, S.S., J.H. Lee, and S.Y. Kim. 1999b. Pasteurization efficiency of a continuous microwave HTST system for milk. *Korean Journal of Food Science and Technology* 31: 1392–1396.

Kim, J.-G., A.E. Yousef, and M.A. Khadre. 2003. Ozone and its current and future application in the food industry. In *Advances in food and nutrition research*, vol 45. San Diego: Academic, pp. 167–218.

Klaenhammer, T.R. 1993. Genetics of bacteriocins produced by lactic acid bacteria. *FEMS Microbiology Reviews* 12(1–3): 1–3.

Knorr, D., M. Zenker, V. Heinz, and D.-U. Lee. 2004. Applications and potential of ultrasonics in food processing. *Trends in Food Science and Technology* 15(5): 261–266.

Koutchma, T. 2009. Traditional and high-technology approaches to microbial safety in foods. In *Microbiologically safe foods*, ed. N. Heredia, I. Wesley, and S. García. Hoboken: Wiley.

Kozempel, M., O.J. Scullen, R. Cook, and R. Whiting. 1997. Preliminary investigation using a batch flow process to determine bacteria destruction by microwave energy at low temperature. *Lebensmittel-Wissenschaft und - Technologie* 30(7): 691–696.

Krishnamurthy, K., A. Demirci, and J. Irudayaraj. 2004. Inactivation of *Staphylococcus aureus* by pulsed UV-light sterilization. *Journal of Food Protection* 67(5): 1027–1030.

Krishnamurthy, K., A. Demirci, and J.M. Irudayaraj. 2007. Inactivation of *Staphylococcus aureus* in milk using flow-through pulsed UV-light treatment system. *Journal of Food Science* 72(7): M233–M239.

Lacroix. K. 2001. Dielectric heating for antimicrobial treatment of fresh meats. Dissertation, McGill University, Montreal, QC.

Lambert, R.J., and M. Stratford. 1999. Weak-acid preservatives: modelling microbial inhibition and response. *Journal of Applied Microbiology* 86(1): 157–164.

Lebovka, N.I., N.V. Shynkaryk, and E. Vorobiev. 2007. Pulsed electric field enhanced drying of potato tissue. *Journal of Food Engineering* 78(2): 606–613.

Lee, J., and G. Kaletunç. 2002. Evaluation of the heat inactivation of *Escherichia coli* and *Lactobacillus plantarum* by differential scanning calorimetry. *Applied and Environmental Microbiology* 68(11): 5379–5386.

Lee, H., B. Zhou, W. Liang, H. Feng, and S.E. Martin. 2009. Inactivation of *Escherichia coli* cells with sonication, manosonication, thermosonication, and manothermosonication: microbial responses and kinetics modeling. *Journal of Food Engineering* 93(3): 354–364.

Leistner, L., and L.G.M. Gorris. 1995. Food preservation by hurdle technology. *Trends in Food Science and Technology* 6(2): 41–46.

Leistner, L., and L.G.M. Gorris, Eds. 1994. Food Preservation by Combined Processes. Final Report of FLAIR Concerted Action No. 7, Subgroup B, EUR 1577 EN. Brussels, European Commission, Directorate-General XII, Brussels, Belgium, p. 100.

Leizerson, S., and E. Shimoni. 2005a. Effect of ultrahigh-temperature continuous ohmic heating treatment on fresh orange juice. *Journal of Agricultural and Food Chemistry* 53(9): 3519–3524.

Leizerson, S., and E. Shimoni. 2005b. Stability and sensory shelf life of orange juice pasteurized by continuous ohmic heating. *Journal of Agricultural and Food Chemistry* 53(10): 4012–4018.

Li, M., J.H. Qu, and Y.Z. Peng. 2004. Sterilization of *Escherichia coli* cells by the application of pulsed magnetic field. *Journal of Environmental Sciences (Beijing)* 16(2): 348–352.

Loghavi, L., S.K. Sastry, and A.E. Yousef. 2009. Effect of moderate electric field frequency and growth stage on the cell membrane permeability of *Lactobacillus acidophilus*. *Biotechnology Progress* 25(1): 85–94.

Löndahl, G., and T.E. Nilsson. 1978. Microbiological aspects of the freezing of meat and prepared foods. *International Journal of Refrigeration* 1(1): 47–52.

Lopez-Malo, A., E. Palou, M. Jimenez-Fernandez, S.M. Alzamora, and S. Guerrero. 2005. Multifactorial fungal inactivation combining thermosonication and antimicrobials. *Journal of Food Engineering* 67(1–2): 87–93.

Luechapattanaporn, K., Y. Wang, J. Wang, M. Al-Holy, D.H. Kang, J. Tang, and L.M. Hallberg. 2004. Microbial safety in radio-frequency processing of packaged foods. *Journal of Food Science* 69(7): M201–M206.

Lund, B.M., T.C. Baird-Parker, and G.W. Gould. 2000. *The microbiological safety and quality of food*. Gaithersburg, MD: Aspen Publishers.

MacGregor, S.J., N.J. Rowan, L. McIlvaney, J.G. Anderson, R.A. Fouracre, and O. Farish. 1998. Light inactivation of food-related pathogenic bacteria using a pulsed power source. *Letters in Applied Microbiology* 27(2): 67–70.

Machado, L.F., R.N. Pereira, R.C. Martins, J.A. Teixeira, and A.A. Vicente. 2010. Moderate electric fields can inactivate *Escherichia coli* at room temperature. *Journal of Food Engineering* 96(4): 520–527.

Madigan, M.T., and T.D. Brock. 2009. *Brock biology of microorganisms*. San Francisco, CA: Pearson/Benjamin Cummings.

Mahapatra, A.K., K. Muthukumarappan, and J.L. Julson. 2005. Applications of ozone, bacteriocins and irradiation in food processing: a review. *Critical Reviews in Food Science and Nutrition* 45(6): 447–461.

Manas, P., R. Pagan, and J. Raso. 2000a. Predicting lethal effect of ultrasonic waves under pressure treatments on *Listeria monocytogenes* ATCC 15313 by power measurements. *Journal of Food Science* 65(4): 663–667.

Manas, P., R. Pagan, J. Raso, F.J. Sala, and S. Condon. 2000b. Inactivation of *Salmonella enteritidis, Salmonella Typhimurium,* and *Salmonella Senftenberg* by ultrasonic waves under pressure. *Journal of Food Protection* 63(4): 451–456.

Marquenie, D., J. Lammertyn, A.H. Geeraerd, C. Soontjens, J.F. Van Impe, B.M. Nicolaï, and C.W. Michiels. 2002. Inactivation of conidia of *Botrytis cinerea* and *Monilinia fructigena* using UV-C and heat treatment. *International Journal of Food Microbiology* 74(1–2): 27–35.

Marquenie, D., A.H. Geeraerd, J. Lammertyn, C. Soontjens, J.F. Van Impe, C.W. Michiels, and B.M. Nicolaï. 2003a. Combinations of pulsed white light and UV-C or mild heat treatment to inactivate conidia of *Botrytis cinerea* and *Monilia fructigena*. *International Journal of Food Microbiology* 85(1–2): 185–196.

Marquenie, D., C.W. Michiels, J.F. Van Impe, E. Schrevens, and B.N. Nicolaï. 2003b. Pulsed white light in combination with UV-C and heat to reduce storage rot of strawberry. *Postharvest Biology and Technology* 28(3): 455–461.

Marra, F., J. Lyng, V. Romano, and B. McKenna. 2007. Radio-frequency heating of foodstuff: solution and validation of a mathematical model. *Journal of Food Engineering* 79(3): 998–1006.

Masanet, E., E. Worrell, W. Graus, and C. Galitsky. 2008. *Energy efficiency improvement and cost saving opportunities for the fruit and vegetable processing industry—an energy star guide for energy and plant managers*. Berkeley, CA: Ernest Orlando Lawrence Berkeley National Laboratory.

McLean, R.A., H.D. Lilly, and J.A. Alford. 1968. Effects of meat-curing salts and temperature on production of *Staphylococcal enterotoxin* B. *Journal of Bacteriology* 95(4): 1207–1211.

McNamee, C., F. Noci, D.A. Cronin, J.G. Lyng, D.J. Morgan, and A.G.M. Scannell. 2010. PEF based hurdle strategy to control *Pichia fermentans, Listeria innocua* and *Escherichia coli* k12 in orange juice. *International Journal of Food Microbiology* 138(1–2): 13–18.

MDS Nordion. 2010. The history of food irradiation. Retrieved 15 June 2010, from http://www.nordion.com/documents/The-History-of-Food-Irradiation.pdf.

Metaxas, A.C., and R.J. Meredith. 1983. *Industrial microwave heating*. London, UK: P. Peregrinus on behalf of the Institution of Electrical Engineers.

Mezquida, G.M., G.E. Dede, M.J. Jordan, and M.P. Moratalla. 2009. Induction heating system for demolding e.g. ice creams, has induction coils that are moved closer to and/or away from molds to couple and uncouple with respect to molds or to fix with the molds, MEZQUIDA G M (MEZQ-Individual) DEDE G E (DEDE-Individual): 6.

Milly, P.J., R.T. Toledo, M.A. Harrison, and D. Armstead. 2007. Inactivation of food spoilage microorganisms by hydrodynamic cavitation to achieve pasteurization and sterilization of fluid foods. *Journal of Food Science* 72(9): M414–M422.

Min, S., Z.T. Jin, S.X. Min, H. Yeom, and Q.H. Zhang. 2003. Commercial-scale pulsed electric field processing of orange juice. *Journal of Food Science* 68(4): 1265–1271.

Molins, R.A. 2001. Global status of food irradiation in 2000. In *Food irradiation: principles and applications*, ed. R.A. Molins. New York: Wiley.

Monfort, S., E. Gayan, G. Saldana, E. Puertolas, S. Condon, J. Raso, and I. Alvarez. 2010. Inactivation of *Salmonella typhimurium* and *Staphylococcus aureus* by pulsed electric fields in liquid whole egg. *Innovative Food Science and Emerging Technologies* 11(2): 306–313.

Montenegro, J., R. Ruan, H. Ma, and P. Chen. 2002. Inactivation of *E. coli* O157:H7 using a pulsed nonthermal plasma system. *Journal of Food Science* 67(2): 646–648.

Montville, T.J., and M.L. Chikundas. 2007. Biopreservation of foods. In *Food microbiology: fundamentals and frontiers*, ed. L.R. Beuchat and M.P. Doyle. Washington, DC: ASM.

Montville, T.J., and K.R. Mathews. 2007. Growth, survival, and death of microbes in foods. In *Food microbiology: fundamentals and frontiers*, ed. L.R. Beuchat and M.P. Doyle. Washington, DC: ASM.

Montville, T.J., and K.R. Matthews. 2008. *Food microbiology: an introduction*. Washington, DC: ASM.

Mosqueda-Melgar, J., P. Elez-Martínez, R.M. Raybaudi-Massilia, and O. Martín-Belloso. 2008a. Effects of pulsed electric fields on pathogenic microorganisms of major concern in fluid foods: a review. *Critical Reviews in Food Science and Nutrition* 48(8): 747–759.

Mosqueda-Melgar, J., R.M. Raybaudi-Massilia, and O. Martin-Belloso. 2008b. Non-thermal pasteurization of fruit juices by combining high-intensity pulsed electric fields with natural antimicrobials. *Innovative Food Science and Emerging Technologies* 9(3): 328–340.

Moy, J.H. 2005. Food irradiation—an emerging technology. In *Novel food processing technologies*, ed. G.V. Barbosa-Cánovas, M.P. Cano, and M.S. Tapia. Boca Raton, FL: CRC.

Mudgett, R.E. 1985. Dielectrical properties of foods. In *Microwaves in the food processing industry*, ed. R.V. DeCareau and R.E. Mudgett. Orlando, FL: Academic.

Niemira, B.A., and J. Sites. 2008. Cold plasma inactivates *Salmonella stanley* and *Escherichia coli* O157:H7 inoculated on golden delicious apples. *Journal of Food Protection* 71(7): 1357–1365.

Nikdel, S., and D.G. Mackellar. 1992. A microwave system for continuous pasteurization of orange juice. *Proceedings of the Florida State Horticultural Society* 105: 108–110.

Nikdel, S., C.S. Chen, M.E. Parish, D.G. Mackellar, and L.M. Friedrich. 1993. Pasteurization of citrus juice with microwave energy in a continuous-flow unit. *Journal of Agricultural and Food Chemistry* 41(11): 2116–2119.

Noci, F., J. Riener, M. Walkling-Ribeiro, D.A. Cronin, D.J. Morgan, and J.G. Lyng. 2008. Ultraviolet irradiation and pulsed electric fields (PEF) in a hurdle strategy for the preservation of fresh apple Juice. *Journal of Food Engineering* 85(1): 141–146.

Noci, F., M. Walkling-Ribeiro, D.A. Cronin, D.J. Morgan, and J.G. Lyng. 2009. Effect of thermosonication, pulsed electric field and their combination on inactivation of *Listeria innocua* in milk. *International Dairy Journal* 19(1): 30–35.

Novak, J.S., and J.T.C. Yuan. 2007. The ozonization concept: advantages of ozone treatment and commercial developments. In *Advances in thermal and non-thermal food preservation*, ed. G. Tewari and V.K. Juneja. Ames, IA: Blackwell.

O'Bryan, C.A., P.G. Crandall, S.C. Ricke, and D.G. Olson. 2008. Impact of irradiation on the safety and quality of poultry and meat products: a review. *Critical Reviews in Food Science and Nutrition* 48(5): 442–457.

Olson, E.R. 1993. Influence of pH on bacterial gene expression. *Molecular Microbiology* 8(1): 5–14.

Oms-Oliu, G., O. Martín-Belloso, and R. Soliva-Fortuny. 2010. Pulsed light treatments for food preservation: a review. *Food and Bioprocess Technology* 3(1): 13–23.

Orfeuil, M. 1987. *Electric process heating: technologies, equipment, applications*. Columbus, OH: Battelle.

Ozer, N.P., and A. Demirci. 2006. Inactivation of *Escherichia coli* O157:H7 and *Listeria monocytogenes* inoculated on raw salmon fillets by pulsed UV-light treatment. *International Journal of Food Science and Technology* 41(4): 354–360.
Palaniappan, S., S.K. Sastry, and E.R. Richter. 1992. Effects of electroconductive heat treatment and electrical pretreatment on thermal death kinetics of selected microorganisms. *Biotechnology and Bioengineering* 39(2): 225–232.
Palou, E., A. López-Malo, and J. Welti-Chanes. 2002. Innovative fruit preservation methods using high pressure. In *Engineering and food for the 21st century*, ed. J. Welti-Chanes, G.V. Barbosa-Cánovas, and J.M. Aguilera. Boca Raton, FL: CRC.
Palumbo, S.A. 1986. Is refrigeration enough to restrain foodborne pathogens? *Journal of Food Protection* 49(12): 1003–1009.
Pascual, A., I. Llorca, and A. Canut. 2007. Use of ozone in food industries for reducing the environmental impact of cleaning and disinfection activities. *Trends in Food Science and Technology* 18(Suppl 1): S29–S35.
Patterson, M.F., M. Linton, and C.J. Doona. 2007. Introduction to high pressure processing of foods. In *High pressure processing of foods*, ed. C.J. Doona and F.E. Feeherry. Ames, IA/Chicago: Blackwell/IFT.
Pereira, R.N., and A.A. Vicente. 2010. Environmental impact of novel thermal and non-thermal technologies in food processing. *Food Research International* 43(7): 1936–1943.
Perni, S., D.W. Liu, G. Shama, and M.G. Kong. 2008a. Cold atmospheric plasma decontamination of the pericarps of fruit. *Journal of Food Protection* 71(2): 302–308.
Perni, S., G. Shama, and M.G. Kong. 2008b. Cold atmospheric plasma disinfection of cut fruit surfaces contaminated with migrating microorganisms. *Journal of Food Protection* 71(8): 1619–1625.
Pflug, I.J., and G.W. Gould. 2000. Heat treatment. In *The microbiological safety and quality of food*, ed. B.M. Lund, T.C. Baird-Parker, and G.W. Gould. Gaithersburg, MD: Aspen.
Piyasena, P., E. Mohareb, and R.C. McKellar. 2003. Inactivation of microbes using ultrasound: a review. *International Journal of Food Microbiology* 87(3): 207–216.
Pothakamury, U.R., G.V. Barbosa-Cánovas, and B.G. Swanson. 1993. Magnetic-field inactivation of microorganisms and generation of biological changes. *Food Technology* 47(12): 85–93.
Potter, N.N., and J.H. Hotchkiss. 1998. *Food science*. Gaithersburg, MD: Aspen.
Raichel, D.R. 2006. *The science and applications of acoustics*. New York: Springer Science+Business Media.
Rajkovic, A., I. Tomasevic, N. Smigic, M. Uyttendaele, R. Radovanovic, and F. Devlieghere. 2010. Pulsed UV light as an intervention strategy against *Listeria monocytogenes* and *Escherichia coli* O157:H7 on the surface of a meat slicing knife. *Journal of Food Engineering* 100(3): 446–451.
Ramaswamy, H.S., and T. Pillet-Will. 1992. Temperature distribution in microwave-heated food models. *Journal of Food Quality* 15(6): 435–448.
Ramaswamy, H., and J. Tang. 2008. Microwave and radio frequency heating. *Food Science and Technology International* 14(5): 423–427.
Ramaswamy, R., T. Jin, V.M. Balasubramaniam, and H. Zhang. 2007. Pulsed electric field processing: fact sheet for food processors. Retrieved 13 June 2010, from http://ohioline.osu.edu/fse-fact/pdf/0002.pdf.
Ramesh, M.N. 2003. Sterilization of foods. In *Encyclopedia of food sciences and nutrition*, ed. B. Cabellero. Oxford, UK: Academic.
Raso, J., and G.V. Barbosa-Cánovas. 2003. Nonthermal preservation of foods using combined processing techniques. *Critical Reviews in Food Science and Nutrition* 43(3): 265–285.
Raso, J., A. Palop, R. Pagán, and S. Condón. 1998. Inactivation of *Bacillus subtilis* spores by combining ultrasonic waves under pressure and mild heat treatment. *Journal of Applied Microbiology* 85(5): 849–854.
Raso, J., P. Mañas, R. Pagán, and F.J. Sala. 1999. Influence of different factors on the output power transferred into medium by ultrasound. *Ultrasonics Sonochemistry* 5(4): 157–162.
Rastogi, N.K. 2003. Application of high-intensity pulsed electrical fields in food processing. *Food Reviews International* 19(3): 229–251.

Ravishankar, S., H. Zhang, and M.L. Kempkes. 2008. Pulsed electric fields. *Food Science and Technology International* 14(5): 429–432.

Riener, J., F. Noci, D.A. Cronin, D.J. Morgan, and J.G. Lyng. 2009a. Effect of high intensity pulsed electric fields on enzymes and vitamins in bovine raw milk. *International Journal of Dairy Technology* 62(1): 1–6.

Riener, J., F. Noci, D.A. Cronin, D.J. Morgan, and J.G. Lyng. 2009b. The effect of thermosonication of milk on selected physicochemical and microstructural properties of yoghurt gels during fermentation. *Food Chemistry* 114(3): 905–911.

Roberts, P., and A. Hope. 2003. Virus inactivation by high intensity broad spectrum pulsed light. *Journal of Virological Methods* 110(1): 61–65.

Rodríguez, J.M., M.I. Martínez, N. Horn, and H.M. Dodd. 2003. Heterologous production of bacteriocins by lactic acid bacteria. *International Journal of Food Microbiology* 80(2): 101–116.

Rodriguez-Romo, L.A., and A.E. Yousef. 2005. Inactivation of *Salmonella enterica serovar Enteritidis* on shell eggs by ozone and UV radiation. *Journal of Food Protection* 68(4): 711–717.

Roodenburg, B., S.W.H. de Haan, L.B.J. van Boxtel, V. Hatt, P.C. Wouters, P. Coronel, and J.A. Ferreira. 2010. Conductive plastic film electrodes for Pulsed Electric Field (PEF) treatment: a proof of principle. *Innovative Food Science and Emerging Technologies* 11(2): 274–282.

Rose, A.H. 1983. *Food Microbiology*. London/New York: Academic.

Rose, D. 1995. Advances and potential for aseptic processing. In *New methods of food preservation*, ed. G.W. Gould. London; New York: Blackie Academic & Professional.

Rosenbauer, G., M. Rudolph, and F. Traub. 1996. Gentle thermal treatment of liquids by induction heating. *Chemie-Ingenieur-Technik* 68(3): 287–289.

Ross, A.I.V., M.W. Griffiths, G.S. Mittal, and H.C. Deeth. 2003. Combining nonthermal technologies to control foodborne microorganisms. *International Journal of Food Microbiology* 89(2–3): 125–138.

Rowan, N.J., S.J. MacGregor, J.G. Anderson, R.A. Fouracre, L. McIlvaney, and O. Farish. 1999. Pulsed-light inactivation of food-related microorganisms. *Applied and Environmental Microbiology* 65(3): 1312–1315.

Russell, N.J., and G.W. Gould. 1991. *Food preservatives*. Glasgow/New York: Blackie/AVI.

Ryan, K., K. McCabe, N. Clements, M. Hernandez, and S. Miller. 2010. Inactivation of airborne microorganisms using novel ultraviolet radiation sources in reflective flow-through control devices. *Aerosol Science and Technology* 44(7): 541–550.

Ryynanen, S., and T. Ohlsson. 1996. Microwave heating uniformity of ready meals as affected by placement, composition, and geometry. *Journal of Food Science* 61(3): 620–624.

Saldana, G., E. Puertolas, S. Condon, I. Alvarez, and J. Raso. 2010. Inactivation kinetics of pulsed electric field-resistant strains of *Listeria monocytogenes* and *Staphylococcus aureus* in media of different pH. *Food Microbiology* 27(4): 550–558.

Sampedro, F., M. Rodrigo, A. Martinez, D. Rodrigo, and G.V. Barbosa-Cánovas. 2005. Quality and safety aspects of PEF application in milk and milk products. *Critical Reviews in Food Science and Nutrition* 45(1): 25–47.

Sampedro, F., D. Rodrigo, A. Martinez, G.V. Barbosa-Canovas, and M. Rodrigo. 2006. Review: application of pulsed electric fields in egg and egg derivatives. *Food Science and Technology International* 12(5): 397–405.

Sanchez-Moreno, C., B. De Ancos, L. Plaza, P. Elez-Martinez, and M.P. Cano. 2009. Nutritional approaches and health-related properties of plant foods processed by high pressure and pulsed electric fields. *Critical Reviews in Food Science and Nutrition* 49(6): 552–576.

Sapp, S.G., and T. Downing-Matibag. 2009. Consumer acceptance of food irradiation: a test of the recreancy theorem. *International Journal of Consumer Studies* 33(4): 417–424.

Sarang, S.S. 2007. *Ohmic heating for thermal processing of low-acid foods containing solid particulates*. Columbus, OH: Ohio State University.

Sastry, S.K. 2005. Advances in ohmic heating and moderate electric field (MEF) processing. In *Novel food processing technologies*, ed. G.V. Barbosa-Cánovas, M.P. Cano, and M.S. Tapia. Boca Raton, FL: CRC.

Sastry, S. 2008. Ohmic heating and moderate electric field processing. *Food Science and Technology International* 14(5): 419–422.
Sastry, S.K., and J.T. Barach. 2000. Ohmic and inductive heating. *Journal of Food Science* 65(4): 46.
Sastry, S.K., and Q. Li. 1996. Modeling the ohmic heating of foods. *Food technology* 50(5): 246–248.
Schilling, S., S. Schmid, H. Jaeger, M. Ludwig, H. Dietrich, S. Toepfl, D. Knorr, S. Neidhart, A. Schieber, and R. Carlet. 2008. Comparative study of pulsed electric field and thermal processing of apple juice with particular consideration of juice quality and enzyme deactivation. *Journal of Agricultural and Food Chemistry* 56(12): 4545–4554.
Schlegel, W. 1992. Commercial pasteurization and sterilization of food products using microwave technology. *Food Technology* 46(12): 62–63.
Scurrah, K.J., R.E. Robertson, H.M. Craven, L.E. Pearce, and E.A. Szabo. 2006. Inactivation of *Bacillus* spores in reconstituted skim milk by combined high pressure and heat treatment. *Journal of Applied Microbiology* 101(1): 172–180.
Sen Gupta, B., F. Masterson, and T.R.A. Magee. 2005. Inactivation of *E-coli* in cranberry juice by a high voltage pulsed electric field. *Engineering in Life Sciences* 5(2): 148–151.
Sharma, R.R., and A. Demirci. 2003. Inactivation of *Escherichia coli* O157:H7 on inoculated alfalfa seeds with pulsed ultraviolet light and response surface modeling. *Journal of Food Science* 68(4): 1448–1453.
Shynkaryk, M.V., N.I. Lebovka, and E. Vorobiev. 2008. Pulsed electric fields and temperature effects on drying and rehydration of red beetroots. *Drying Technology* 26(6): 695–704.
Sierra, I., and C. Vidal-Valverde. 2001. Vitamin B1 and B6 retention in milk after continuous-flow microwave and conventional heating at high temperatures. *Journal of Food Protection* 64(6): 890–894.
Smith, W.L., M.C. Lagunas-Solar, and J.S. Cullor. 2002. Use of pulsed ultraviolet laser light for the cold pasteurization of bovine milk. *Journal of Food Protection* 65: 1480–1482.
Sofos, J.N. 1984. Antimicrobial effects of sodium and other ions in foods: a review. *Journal of Food Safety* 6(1): 45–78.
Sofos, J.N. 1986. Use of phosphates in low-sodium meat products. *Food Technology* 40(9): 52–68.
Stadelman, W.J., R.K. Singh, P.M. Muriana, and H. Hou. 1996. Pasteurization of eggs in the shell. *Poultry Science* 75(9): 1122–1125.
Starkweather, K.A. 2000. *Radio frequency thermal inactivation of Listeria innocua in salmon caviar*. Pullmann, WA: Washington State University.
Sun, D.-W. 2006. *Thermal food processing: new technologies and quality issues*. Boca Raton, FL: CRC/Taylor & Francis.
Tahiri, I., J. Makhlouf, P. Paquin, and I. Fliss. 2006. Inactivation of food spoilage bacteria and *Escherichia coli* O157: H7 in phosphate buffer and orange juice using dynamic high pressure. *Food Research International* 39(1): 98–105.
Takeshita, K., H. Yamanaka, T. Sameshima, S. Fukunaga, S. Isobe, K. Arihara, and M. Itoh. 2002. Sterilization effect of pulsed light on various microorganisms. *Journal of Antibacterial and Antifungal Agents, Japan* 30(5): 277–284.
Takeshita, K., J. Shibato, T. Sameshima, S. Fukunaga, S. Isobe, K. Arihara, and M. Itoh. 2003. Damage of yeast cells induced by pulsed light irradiation. *International Journal of Food Microbiology* 85(1–2): 151–158.
Tamplin, M.L. 2009. Predictive microbiology: growth *in silico*. In *Microbiologically safe foods*, ed. N. Heredia, I. Wesley, and S. García. Hoboken, NJ: Wiley.
Tang, J., F. Hao, and M. Lau. 2002. Microwave heating in food processing. In *Advances in bioprocessing engineering*, ed. X.H. Yang and J. Tang. River Edge, NJ: World Scientific.
Tang, J., Y. Wang, and T.V.C.T. Chan. 2005. Radio-frequency heating in food processing. In *Novel food processing technologies*, ed. G.V. Barbosa-Cánovas, M.P. Cano, and M.S. Tapia. Boca Raton, FL: CRC.
Tannahill, R. 1973. *Food in history*. New York: Stein and Day.
Tauscher, B. 1995. Pasteurization of food by hydrostatic high pressure: chemical aspects. *Zeitschrift für Lebensmittel-Untersuchung und -Forschung* 200(1): 3–13.

Teixeira, A.A. 2005. Simulating thermal food processes using deterministic models. In *Novel food processing technologies*, ed. G.V. Barbosa-Cánovas, M.P. Cano, and M.S. Tapia. Boca Raton, FL: CRC.

Terebiznik, M., R. Jagus, P. Cerrutti, M.S. de Huergo, and A.M. Pilosof. 2002. Inactivation of *Escherichia coli* by a combination of nisin, pulsed electric fields, and water activity reduction by sodium chloride. *Journal of Food Protection* 65(8): 1253–1258.

Tewari, G. 2007. High-pressure processing of foods. In *Advances in thermal and non-thermal food preservation*, ed. G. Tewari and V.K. Juneja. Ames, IA: Blackwell.

Toepfl, S., A. Mathys, V. Heinz, and D. Knorr. 2006. Potential of high hydrostatic pressure and pulsed electric fields for energy efficient and environmentally friendly food processing. *Food Reviews International* 22(4): 405–423.

Toussaint-Samat, M. 1992. *A history of food*. Oxford, UK: Blackwell Reference.

Troller, J.A., and J.H.B. Christian. 1978. *Water activity and food*. New York: Academic.

Tsong, T.Y. 1991. Electroporation of cell membranes. *Biophysical Journal* 60(2): 297–306.

Ugarte-Romero, E., H. Feng, and S.E. Martin. 2007. Inactivation of *Shigella boydii* 18 IDPH and *Listeria monocytogenes* Scott A with power ultrasound at different acoustic energy densities and temperatures. *Journal of Food Science* 72(4): M103–M107.

Ukuku, D.O., and D.J. Geveke. 2010. A combined treatment of UV-light and radio frequency electric field for the inactivation of *Escherichia coli* K-12 in apple juice. *International Journal of Food Microbiology* 138(1–2): 1–2.

Unal, R., J.G. Kim, and A.E. Yousef. 2001. Inactivation of *Escherichia coli* O1 57:H7, Listeria monocytogenes, and *Lactobacillus leichmannii* by combinations of ozone and pulsed electric field. *Journal of Food Protection* 64(6): 777–782.

Valero, E., M. Villamiel, J. Sanz, and I. Martinez-Castro. 2000. Chemical and sensorial changes in milk pasteurised by microwave and conventional systems during cold storage. *Food Chemistry* 70(1): 77–81.

Valero, M., N. Recrosio, D. Saura, N. Muñoz, N. Martí, and V. Lizama. 2007. Effects of ultrasonic treatments in orange juice processing. *Journal of Food Engineering* 80(2): 509–516.

Vandenbergh, P.A. 1993. Lactic acid bacteria, their metabolic products and interference with microbial growth. *FEMS Microbiology Reviews* 12(1/3): 221–238.

Vega-Mercado, H., O. Martin-Belloso, F.J. Chang, G.V. Barbosa-Canovas, and B.G. Swanson. 1996. Inactivation of *Escherichia coli* and *Bacillus subtilis* suspended in pea soup using pulsed electric fields. *Journal of Food Processing and Preservation* 20(6): 501–510.

Vicente AA, Castro I, Teixeira JA (2006) Ohmic heating for food processing. In: Sun D-W (ed) Thermal food processing: new technologies and quality issues. CRC/Taylor & Francis: Boca Raton, FL

Villamiel, M., and P. de Jong. 2000. Influence of high-intensity ultrasound and heat treatment in continuous flow on fat, proteins, and native enzymes of milk. *Journal of Agricultural and Food Chemistry* 48(2): 472–478.

Wageningen, U.R. 2006. Cold Plasma. Retrieved 27 July 2010, from http://www.ftc.wur.nl/EN/PullDown3/KO_2_12.asp.

Wagner, M.K., and F.F. Busta. 1985. Inhibition of *Clostridium botulinum* 52A toxicity and protease activity by sodium acid pyrophosphate in media systems. *Applied and Environmental Microbiology* 50(1): 16–20.

Walkling-Ribeiro, M., F. Noci, D.A. Cronin, J.G. Lyng, and D.J. Morgan. 2009a. Antimicrobial effect and shelf-life extension by combined thermal and pulsed electric field treatment of milk. *Journal of Applied Microbiology* 106(1): 241–248.

Walkling-Ribeiro, M., F. Noci, D.A. Cronin, J.G. Lyng, and D.J. Morgan. 2009b. Shelf life and sensory evaluation of orange juice after exposure to thermosonication and pulsed electric fields. *Food and Bioproducts Processing* 87(2): 102–107.

Wang, W.-C. 1995. Ohmic heating of foods: Physical properties and applications, Ohio State University. PhD.

Wang, Y., T.D. Wig, J. Tang, and L.M. Hallberg. 2003. Sterilization of foodstuffs using radio frequency heating. *Journal of Food Science* 68(2): 539–544.

Welti-Chanes, J., A. López-Malo, E. Palou, D. Bermúdez, J.A. Guerrero-Beltrán, and G.V. Barbosa-Cánovas. 2005. Fundamentals and applications of high pressure processing of foods. In *Novel food processing technologies*, ed. G.V. Barbosa-Cánovas, M.P. Cano, and M.S. Tapia. Boca Raton, FL: CRC.

Werner, B.G., and J.H. Hotchkiss. 2006. Continuous flow nonthermal CO_2 processing: the lethal effects of subcritical and supercritical CO_2 on total microbial populations and bacterial spores in raw milk. *Journal of Dairy Science* 89(3): 872–881.

Whiting, R.C. 1995. Microbial modeling in foods. *Critical Reviews in Food Science and Nutrition* 35(6): 464–494.

Wood, O.B., and C.M. Bruhn. 2000. Position of the American Dietetic Association: food irradiation. *Journal of American Dietetic Association* 100(2): 246–253.

Woods, R.J. 1994. Food irradiation. *Endeavour* 18(3): 104–108.

World Health Organization. 1999. High-dose irradiation: wholesomeness of food irradiated with doses above 10 kGy. Technical report series; 890.

Wouters, P.C., and J.P.P.M. Smelt. 1997. Inactivation of microorganisms with pulsed electric fields: potential for food preservation. *Food Biotechnology* 11(3): 193–229.

Wrangham, R.W. 2009. *Catching fire: how cooking made us human*. New York: Basic Books.

Wu, J., T.V. Gamage, K.S. Vilkhu, L.K. Simons, and R. Mawson. 2008. Effect of thermosonication on quality improvement of tomato juice. *Innovative Food Science and Emerging Technologies* 9(2): 186–195.

Yoon, S.W., C.Y.J. Lee, K.M. Kim, and C.H. Lee. 2002. Leakage of Cellular Materials from Saccharomyces cerevisiae by Ohmic Heating. *Journal of Microbiology and Biotechnology* 12: 183–188.

Yongsawatdigul, J., J.W. Park, and E. Kolbe. 1995. Electrical conductivity of pacific whiting surimi paste during ohmic heating. *Journal of Food Science* 60(5): 922–925.

Yuk, H.G., D.J. Geveke, and H.Q. Zhang. 2010. Efficacy of supercritical carbon dioxide for non-thermal inactivation of *Escherichia coli* K12 in apple cider. *International Journal of Food Microbiology* 138(1–2): 91–99.

Zaika, L.L., and A.H. Kim. 1993. Effect of sodium polyphosphates on growth of *Listeria* monocytogenes. *Journal of Food Protection* 56(7): 577–580.

Zeldes, L.A. 2009. Eat this! Old-fashioned eggnog, made safer, thanks to Chicago-area eggs. Dining Chicago. Chicago, IL. www.diningchicago.com.

Zell, M., J.G. Lyng, D.A. Cronin, and D.J. Morgan. 2009. Ohmic cooking of whole beef muscle: optimisation of meat preparation. *Meat Science* 81(4): 693–698.

Zenker, M., V. Heinz, and D. Knorr. 2003. Application of ultrasound-assisted thermal processing for preservation and quality retention of liquid foods. *Journal of Food Protection* 66(9): 1642–1649.

Zhang, H., and G. Mittal. 2008. Effects of high-pressure processing (HPP) on bacterial spores: an overview. *Food Reviews International* 24(3): 330–351.

Zhang, Q., G.V. Barbosa-Cánovas, and B.G. Swanson. 1995. Engineering aspects of pulsed electric field pasteurization. *Journal of Food Engineering* 25(2): 261–281.

Zhao, Y. 2006. Radio frequency dielectric heating. In *Thermal food processing: new technologies and quality issues*, ed. D.-W. Sun. Boca Raton, FL: CRC/Taylor & Francis.

Zhao, Y., B. Flugstad, E. Kolbe, J.W. Park, and J.H. Wells. 2000. Using capacitive (radio frequency) dielectric heating in food processing and preservation—a review. *Journal of Food Process Engineering* 23: 25–56.

Zhong, Q., K.P. Sandeep, and K.R. Swartzel. 2003. Continuous flow radio frequency heating of water and carboxymethylcellulose solutions. *Journal of Food Science* 68(1): 217–223.

Zhong, Q., K.P. Sandeep, and K.R. Swartzel. 2004. Continuous flow radio frequency heating of particulate foods. *Innovative Food Science and Emerging Technologies* 5(4): 475–483.

Chapter 15
Green technologies in food dehydration

Stefan Grabowski and Joyce I. Boye

15.1 Introduction

Drying of foods, as well as other wet materials, is a high energy consuming industrial operation. As much as possible it should be avoided or replaced by other less energy demanding processes. High energy consumption and greenhouse gas emissions (GHG) are the main concerns related to industrial drying. In general, the energy required for drying food and the load of evaporated moisture should be reduced as much as possible during food processing to reduce environmental impacts. In addition to the high amount of energy consumed during drying, other environmental issues that need to be addressed include preventing or reducing pollution from exhaust gases and proper utilization of the different streams of wastes or other materials in the wet feed pretreatment and preforming, etc.

In this chapter, each step of a typical food dehydration process, that is (1) wet feed pretreatment, (2) preparing (mostly heating) of the drying agent, (3) drying process, (4) maximal retrieval of the dry product, and (5) heat recovery from the exhaust gases, is reviewed from the point of view of conserving energy and minimizing environmental impacts. Detailed information about the theory and practice of food drying, however, are not presented. For this basic information on food drying, the reader is referred to several books, book chapters, research publications, patents, equipment folders, etc., in which the subject is extensively covered. Examples of such sources include *Advances in food dehydration* (2009) by Ratti, *Food drying science and technology* (2007) by Hui et al. *Handbook of industrial drying* (2007) by Mujumdar, "Energy aspects in drying" (2004) by Kudra, "Drying of fruits, vegetables and spices" (2003) by Grabowski et al., *Dehydration of foods* (1996) by Barbosa-Canovas and Vega-Mercado, etc.

S. Grabowski (✉) • J.I. Boye
Food Research and Development Centre, Agriculture and Agri-Food Canada,
Saint Hyacinthe, QC, Canada
e-mail: Stefan.Grabowski@agr.gc.ca; Joyce.Boye@agr.gc.ca

J.I. Boye and Y. Arcand (eds.), *Green Technologies in Food Production and Processing*,
Food Engineering Series, DOI 10.1007/978-1-4614-1587-9_15,

15.2 "Green" drying scheme

The general purpose of food drying is to extend the shelf life of food products by lowering water activity, reduce the final mass and volume for less-costly transportation, and in some instances modify the food matrix. The diversity of drying processes and equipment used in food processing reflects the difficulty of handling and processing of solid materials, and the special requirements for various food products. In addition, economics (investment and operation) is an important factor, especially for large-tonnage low-cost products, such as dried skim milk. The equipment for drying ranges from crude environmentally friendly sun and solar dryers to sophisticated multistage industrial dryers. Typically, batch dryers are used for solid throughput below 50 kg/h, whereas continuous operation is preferred above 1,000 kg/h.

The main elements of industrial drying operations do not vary for "green" drying technologies (Fig. 15.1) versus other regular drying processes; however, in the case of the former, each element in the drying process has to be analyzed from the point of view of its environmental impact. In addition to the basic process engineering requirements, food dryers must meet the strict standards for food quality and food hygiene and safety.

The basic elements of "green" drying include:

1. Wet feed preparation: thickening, preconcentration, pretreatment, preforming, preheating, etc.
2. Heating of the drying air or agent (for convection dryers) or drying surfaces (for conduction dryers).
3. Drying in energy efficient dryers (i.e., selection of new dryers or improvement of existing equipment to improve energy efficiency).
4. Retrieval and separation of the final product from the exhaust gases.
5. Heat recovery from the exhaust gases.

Each of the preceding elements is described in more details in the next section.

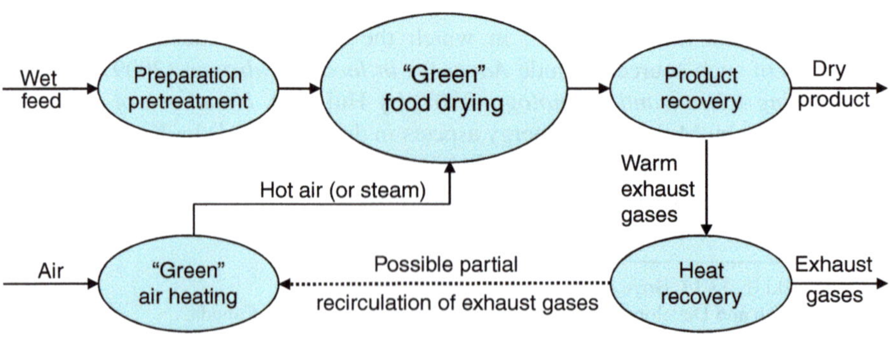

Fig. 15.1 Typical elements of a "green" drying installation scheme

15.3 Wet feed preparation before drying

15.3.1 Thickening, preconcentration, and preforming

In general, nonthermal water removal is 10 to almost 100% less energy intensive than drying (Table 15.1). Therefore, whenever possible, other water removal processes (e.g., pressing, evaporation, membrane separation, filtration, etc.) must be considered so as to reduce the water load before drying. In the food processing industry, reduction of the evaporation load, such as thickening, preforming, or other methods of preparation of the wet feed are used to reduce energy consumption. Thus, before selecting the type of dryer to use a basic question that has to be answered is whether the moisture content in the ready-to-dry food can or cannot be reduced by other means prior to drying. Thickening or preconcentration of initial wet feed can be typically performed by: (1) filtration, (2) centrifugation, (3) coagulation, (4) evaporation, (5) sedimentation, etc. (Tutova and Kutz 1987; Orsat and Raghavan 2008). Each of these methods is more energy efficient than thermal drying. Therefore, application of these processes before drying can be a positive step in reducing energy consumption.

By changing initial moisture content, many parameters of the wet food feed are changed as well. The binding energy of colloidal type of particles rises markedly as the moisture content falls. The use of mechanical means to squeeze out moisture may render subsequent thermal drying more difficult if the particulate mass becomes more consolidated. For example, powders produced by spray drying are made from feedstock that can be pumped and atomized, whereas feeds of relatively high solid content cannot be normally handled. However, fairly dry pastes and filter cakes can be treated in a spray dryer using special nozzles and feeders or in other types of dryers, for example jet-spouted bed dryer with inert bodies (Fig. 15.2) or spin-flash dryer (Fig. 15.3). For the later type of dryer, Kraglund (1983) achieved fuel savings in the range of 60% and one third less power consumption by changing the moisture content from 75% to 65%. Cook and DuMont (1991) also found that a decrease in

Table 15.1 Examples of specific energy consumption for industrial water removal operations

Method or equipment	Specific heat of water removal (kJ/kg of water removed)
Membrane filtration (UF, RO)	50–150
Osmotic dehydration	200–500
Evaporation (concentration)	
Single effect	2,600
Double effect	1,300
Spray dryer	4,000–6,000
Drum dryer	5,000
Tunnel dryer	4,000
Freeze dryer	Up to 100,000

Source: Ramaswamy and Marcotte (2006) and author's own unpublished data.

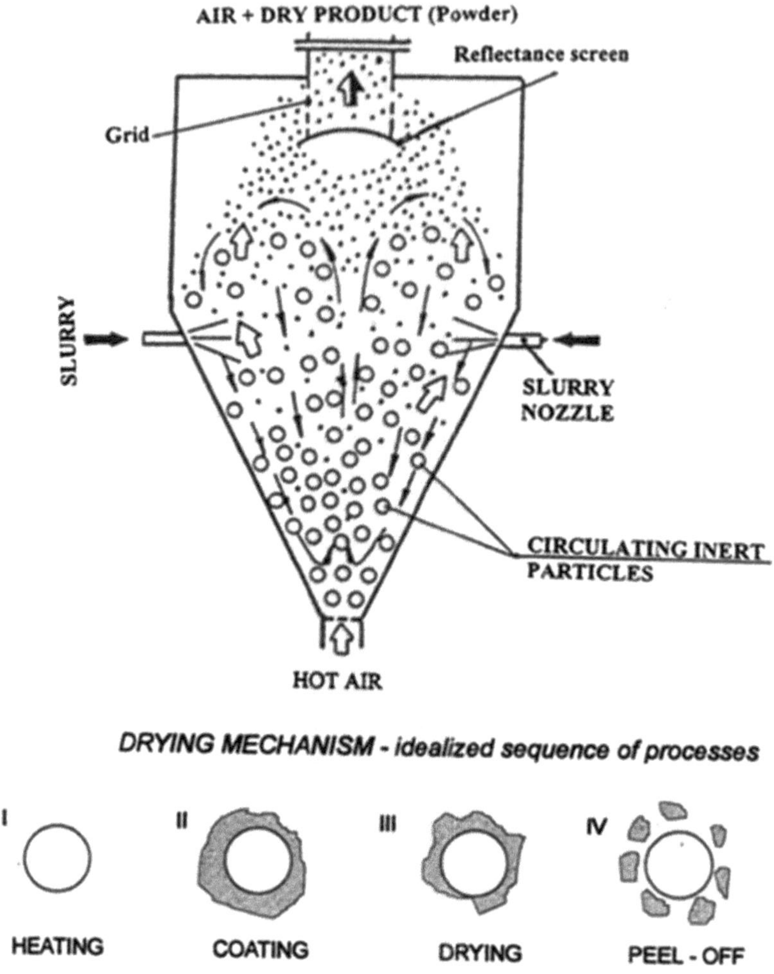

Fig. 15.2 Principles of drying of a slurry in a spouted bed of inert particles (Adapted from Kudra and Mujumdar 2009)

moisture content of wet feed from 32% to 30%, that is, by just 2%, resulted in some 9% reduction in energy consumption and 12–15% in electric power use. Similar or slightly lower values are reported elsewhere (Tutova and Kutz 1987); however, this effect depends on the type and characteristics of the food product and the range of moisture content. Contact-sorption drying or adding carrier materials, etc., can sometimes be optional solutions for more energetically effective dehydration (Kudra and Mujumdar 2009). To accommodate these new upstream operations, new feeders as well as new dryers and/or drying conditions may be required. For example, for spray dryers, several new feeders and nozzles for concentrated pastelike materials have been developed (Mujumdar 2007).

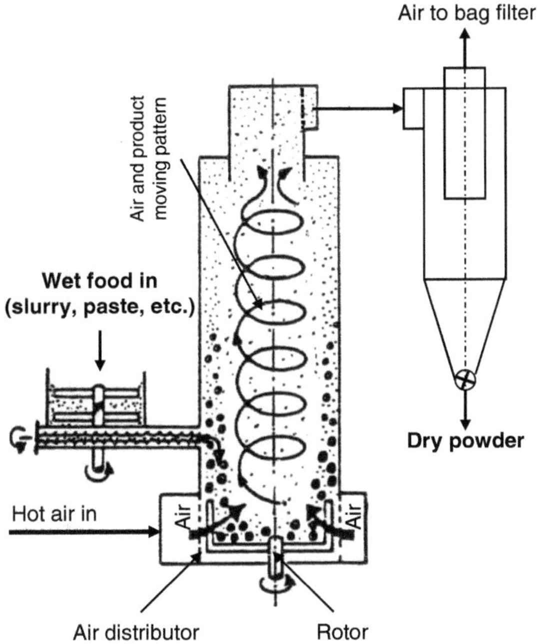

Fig. 15.3 Spin-flash dryer for paste-like foods as a possible replacement of a spray dryer (Adapted from Kraglund 1983)

15.3.2 Preheating of wet feed

Heating of the wet feed inside the dryer is much less effective than heating of this feed in much more effective heat exchangers before entering the dryer. Thus, preheating of the wet feed to as high a temperature as the product and process will allow before introduction into the dryer can be another good approach to enhance energy savings, especially for solutions, emulsions, slurries, and other pastelike materials. Heating of solid feeds can be more difficult, but benefits are generally even greater on a percentage basis. Calculations show that preheating a 30% solid feed to 56°C will reduce the heat requirement and the electric power in the range of 9–15%, dropping off slightly at higher air temperatures (Cook and DuMont 1991).

15.3.3 Osmotic pretreatment or dehydration

Biological type cellular foods can be subjected to low-energy consuming osmotic dehydration. This process can be considered as a pretreatment as well as a final dehydration step for the production of intermediate-moisture foods, for example, fruits for yogurts (Grabowski and Marcotte 2003). It is, however, usually used as a first step in hybrid drying technologies (Kudra and Mujumdar 2009) because it allows significant reduction of moisture content and water activity in foods before the final thermal drying treatment.

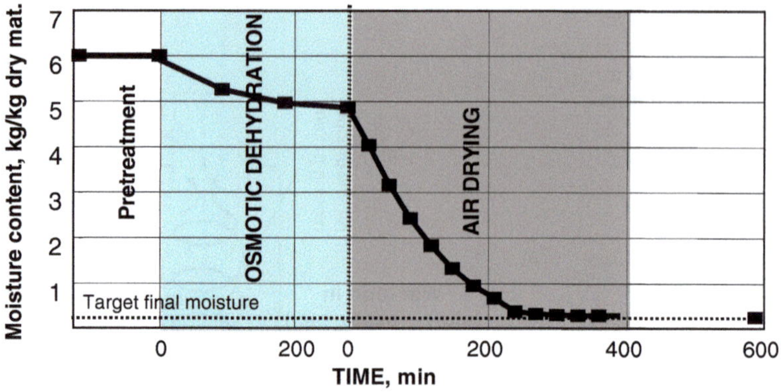

Fig. 15.4 Two-steps dehydration of small fruits with osmotic dehydration in the first step (Adapted from Grabowski and Marcotte 2003)

The mechanism of osmotic dehydration is based on a tendency to reach equilibrium between the osmotic pressure inside biological cells (fruits, meats, fishes, vegetables, etc.) and the surrounding osmotic solution, which has an increased osmotic pressure because of the high concentration of soluble osmotic agents. Molecular diffusion of water through semipermeable cell membrane takes place and the product loses its water. The osmotic dehydration process is stopped when the osmotic pressure inside the cells reaches the osmotic pressure of the surrounding syrup. Theoretically, molecules of the solute, being large in size, do not migrate en masse from the osmotic syrup into the product. Practically, however, some migration (infusion) of the solute occurs. A net mass loss is therefore a difference between water removal and solids gain. Osmotic dehydration can provide reduction of water activity to some extent, and is limited mainly by the water activity of the osmotic syrup itself. For example, a typical osmotic syrup, such as a sucrose solution of 60° Brix, has a water activity of about 0.85. This means that foods osmotically dehydrated in this solution would be equilibrated to a water activity of 0.85 and would therefore not necessarily be safe from a microbial point of view, thus making a second stage of drying necessary. This second stage could be a typical hot air drying, solar drying as well as vacuum, freeze, microwave, or any other type of drying. An example of the hybrid technology with osmotic dehydration in the first stage is dehydration of small fruits, such as cranberries and blueberries (Grabowski and Marcotte 2003). Figure 15.4 presents a schematic of such a process, which is already being applied on an industrial scale. High quality dry carrot and potato products (Mazza 1989; Raoult-Wack et al. 1992) have been produced; however, dehydrated celery has not given satisfactory results (Jen 1989). Special attention has to be given to the application of the spent syrup after the osmotic process. Reuse of such syrups as a possible ingredient in soft drinks, bakery, confectionary products, etc. is a potential solution. Re-concentration in evaporators can be also considered as evaporators have typically much higher thermal efficiency than dryers, and the net energy savings will be still considerable (Marcotte and Grabowski 2008).

Typical batch-type osmotic dehydration equipment consists of a mixer tank for the syrup with heating elements to maintain the temperature around 50°C; however, higher or lower temperatures are possible depending on the type of dehydrated product. Continuous osmotic dehydration units have been also developed (Marcotte and Tanguay 2005). More detailed data on practical and theoretical aspects of fruit and vegetable osmotic dehydration can be found elsewhere (Grabowski and Marcotte 2003; Lewicki and Lenart 2007). Osmotic infusion of some additives, for example, caffeine, vitamins, etc., before regular drying, is a technique used to fortify some novel dry foods (Cohen and Yang 1995).

15.4 Heat sources for drying

15.4.1 Natural solar and wind energy

The most appropriate "green" energy source for the drying of foods is solar and, to a less extent, wind or possibly, geothermal energy. Generally, solar and wind energy sources are natural, abundant, and environmentally friendly and create no, or minimal, greenhouse gases. The simplest case of application of these types of energy is field sun drying, whereas the most complicated is solar-assisted convection drying.

15.4.1.1 Field sun and wind drying

Field sun and wind drying, used for large tonnages of vegetables or fruits, can be very inexpensive in areas of adequate climatic conditions and additionally (but not necessarily, as for example, in California) where labor costs are low. Generally, no costs are incurred to heat or circulate the air (by natural convection). Typically, fruits, herbs, spices, and to a lesser extent vegetables are spread out in the sun and wind on the field, special mats, concretes, etc. The radiant energy of the sun provides the heat to evaporate the water while the wind helps to move the moisture and accelerates the process. This type of drying is ideal in warm and dry conditions on the field or other locations (e.g., shadowy areas for light-sensitive products such as herbs, spices, etc.), although at night and during rainy periods, this type of drying cannot be used. The temperature of the food during sun drying is usually 5–15°C above ambient temperature. Sun drying techniques are based mainly on experience and tradition. Thickness of the wet material layer, drying time, and material handling before, during, and after drying are conditions that are normally standardized based on practice and can differ from one location to another. Some approximate prediction of sun drying performance could be based on energy balance from available data on the distribution of solar energy radiation throughout the world, absorbency/reflectance characteristics of the wet and dry material (Sokhansanj and Jayas 2007), local climatic conditions such as ambient air temperature, speed and humidity, and rainfall

Table 15.2 Solar cabinet drying throughput[a]

Product	Amount of fresh matter dried per unit time[b]	Maximum allowable temperature (°C)
Garlic	2.6 kg/per 2 days	60
Grapes	5.7 kg/per 4 days	88
Okra	3.0 kg/per 2 days	66
Onions	3.0 kg/per 2 days	71
Apricots	4.0 kg/per 2 days	66

[a]Source: Lawand (1975) (with permission of Brace Research Institute, McGill University, Montreal, Canada).
[b]Cabinet dimensions: 1.93 × 0.6 m; location: Syria.

frequency and intensity. For example, the thermal efficiency of sun drying of Colombian coffee beans, calculated as a ratio of heat necessary to remove the moisture to the total solar radiation heat, was reported to be in the range of 13.7–22.6% (Szulmayer 1976).

The time of sun drying depends on the product characteristics and drying conditions and typically ranges from 3 to 4 days, but can be longer (Sokhansanj and Jayas 2007). During this time, the product has to be protected from rain, insects, birds, animals, etc. Several practical methods have been developed to reduce the length of time of drying. One of the most important, especially for production at larger scales, is the application of different product pretreatment before sun drying. Several pretreatment methods and materials are used for this purpose, for example, steaming, immersion in boiling water or sugar and/or salt solutions, the use of ethyl-, methyl- or sodium oleate, etc.

Another way to reduce sun-drying time is to use solar energy concentrators with or without natural or forced airflow inside the dryer. This technique is normally called "solar drying." More detailed description of solar drying principles, equipment, and conditions may be found in selected publications (Bansal and Garg 1987; Parker 1991). Some practical values of the efficiency of solar drying collectors are given by Hall (1979) and Imre (2007). Examples of solar dryer developments for fruits, vegetables, herbs, and spices are detailed in the following.

15.4.1.2 Solar natural dryers (cabinet, tent, greenhouse, etc.)

Solar cabinet is the simplest type of solar natural dryer and is very popular in many locations around the world. It is well suited for small production (family) scale enterprises. The wet product is placed in an enclosure and the solar heat, generated through a conversion of solar radiation into low-grade heat, accelerates the evaporation of moisture from the product. The airflow inside the cabinet is driven by natural convection. The dryer throughput and the drying time are estimated from practice (Table 15.2); however, considerable effort has been made to model solar collector performance (Parker 1991). The thermal efficiencies of solar drying were reported to be from 11% to 13% (Puigalli and Tiguert 1986) with the highest value at the beginning of the drying run. These are not the general values for all solar drying applications, and specific parameters may need to be estimated in each particular drying process.

The quality of the dried product is the most important parameter in drying. Patil (1989) compared the quality of peppers after sun drying, polyethylene solar drying, and solar cabinet drying. Solar cabinet drying offered some advantages over direct sun drying in terms of better quality and faster rates of drying. Jayaraman et al. (1992) compared ascorbic acid retention in cauliflower, cabbage, and bitter gourd dried by three different methods: (1) directly under the sun, (2) under the sun in a black polyethylene tent, and (3) in a drying cabinet with three solar collectors. Direct exposure of these products to the sun decreased quality factors such as vitamin retention. The use of solar energy in an indirect way could, therefore, improve the quality of the final product. In some cases, this was not practical because of the large cost of solar collectors. Less expensive solar tents or covers are often used for bigger production scale. Also solar greenhouses, terraces, or room dryers are very practical for larger productions because they are simple in construction and relatively low in investment costs. Figure 15.5 presents two examples of low-cost solar dryers for relatively larger production scale.

15.4.1.3 Dryers with solar collectors and natural or artificial airflow

Some vegetable, fruit, and spice dryers utilize both direct and indirect solar radiation. In these dryers, radiant energy from the sun falls directly onto the product being dried. In addition, an air-preheater (solar collector) is also used to raise the drying air temperature. The circulation of air in the solar pre-heater is either done by free convection or a fan. In both situations, this air stream accelerates significantly the drying rate. Figure 15.6 presents two examples of very simple and inexpensive dryers of this type. Family-scale dryers (Fig. 15.6a) work well for drying mushrooms, apricots, apples, etc., usually in 1 day. Similarly, by adding a wind-driven fan for more intensive airflow (Fig. 15.6b) a higher dryer throughput can be achieved. The fan provides air circulation in the dryer and pulls the air up through the product. Dampers can be installed in the stack to control airflow rate, which also helps to control the amount of heat build-up. Several other modifications of this concept are in use for drying purposes in tropical and hot climate areas. Further extension of the dryer throughput can be achieved by applying forced convection airflow. Another solution in this type of installations is the use of very long metal air-preheater pipes (Westeco Drying Inc. 1987) or large surface solar collectors (Roa and Macedo 1976). The flow of air is then artificial with the use of powerful fans. In the construction design presented by Westeco Drying Inc (1987), the length of the solar air-preheating pipe was about 7 miles. The heating rate of this preheater was reported to be sufficient for proper operation of a solar-assisted sonic dryer for several fruits and vegetables.

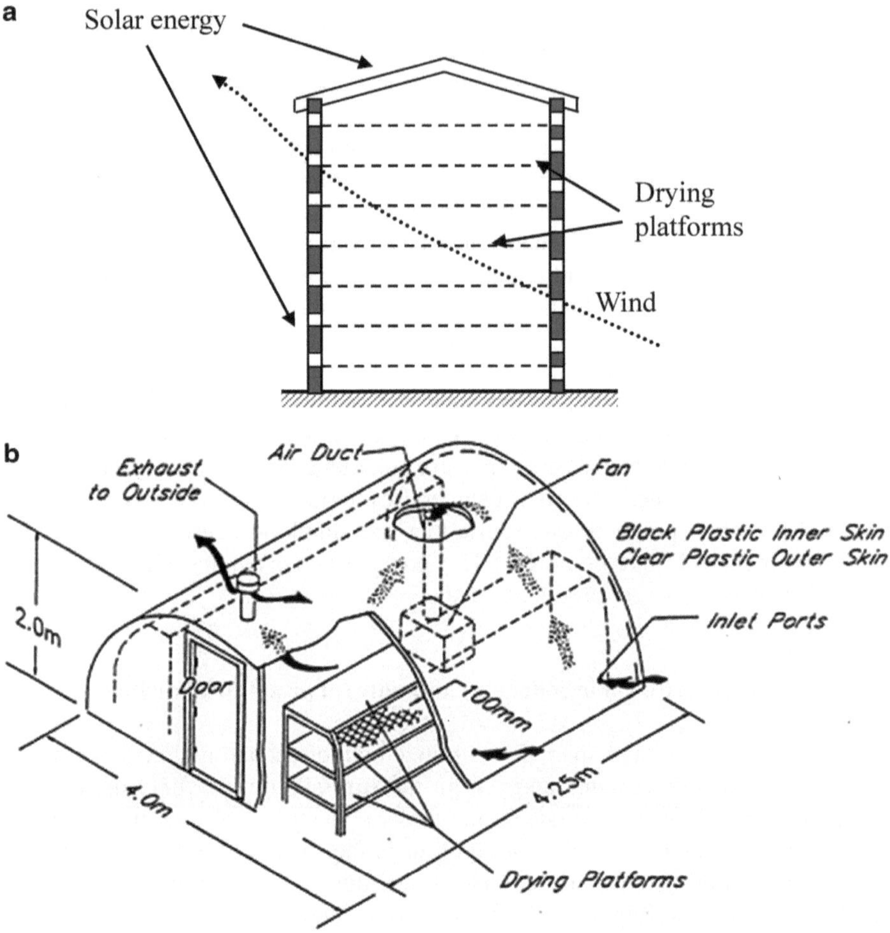

Fig. 15.5 Examples of large-scale solar dryers: (**a**) Australian type dryer, (**b**) plastic-tent solar dryer (Modified from Parker (1991) and Bansal and Garg (1987))

15.4.1.4 Solar-assisted artificial dryers

For commercial applications, in which a large tonnage of product needs to be dried, some source of auxiliary energy is needed to initiate forced air movement and/or to provide the supplemental energy. In this scenario, the solar drying equipment is part of a whole drying system and generally assists operation of a typical air dryer, supplying it with some amount of heat. Construction of a solar-assisted artificial drying system could be simple or combined with other elements, such as a source of stored heat, heat pumps, etc. In the simplest system the solar energy is directly used to provide a part of the heat required for moisture evaporation in a conventional air dryer. The drying of the food is normally intensive during sunshine hours, whereas at night

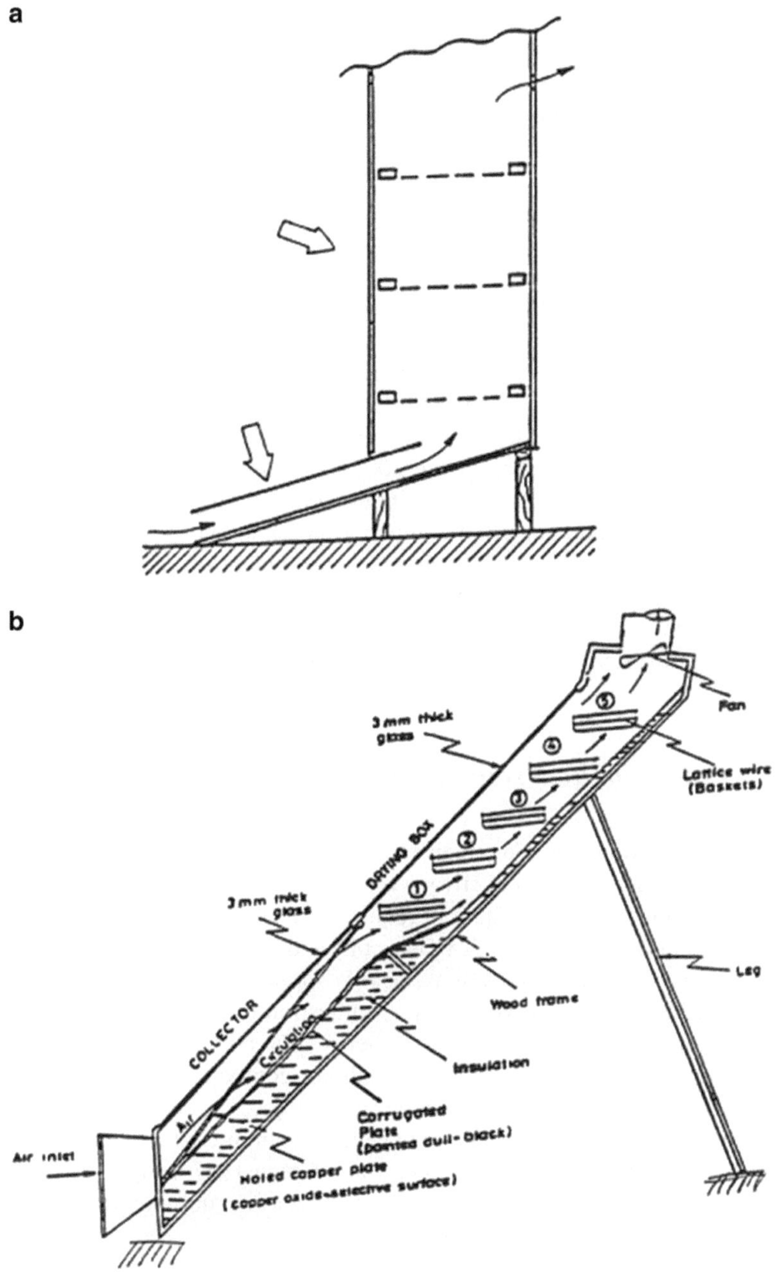

Fig. 15.6 Family-scale low-cost solar dryers for fruits and vegetables (Reproduced with permission of Brace Research Institute, McGill University, Montreal, Canada)

or during low solar radiation periods an auxiliary heat source supplements the heat energy. Several conventional types of dryers could be combined with solar energy collectors. For example, Smith et al. (1977) describe commercial convective air-drying installation for the final dehydration of potato cubes from 11% down to 4% moisture (wet basis). This system utilizes solar double-glazed collectors and an auxiliary gas burner to heat the drying air. The solar collectors operate at about 40°C above the ambient temperature. In another example, considerable reduction of the fuel-gas was reported by West Growers and Packers (cited by Bolin and Salunkhe 1982), when a tunnel dryer was modified for utilization of solar radiation and by using exhaust air to preheat inlet air to the dryer. Another technique that has not been fully explored is the use of a desiccant to remove moisture from the drying agent (used for dehydration of heat-sensitive foods), and then solar energy to regenerate the desiccant.

To reduce the effect of periodicity of solar radiation several physical heat storage systems have been developed for drying applications. Water-type and rock-bed type heat storage are the most typical (Imre 2007). Proposed schemes of solar energy application should also be easy to integrate into other drying systems. More generally, solar energy should be applied as much as possible in energy-consuming processes such as drying. Active research and development of various methods and components of solar drying systems have been taking place internationally, especially for the reduction of the investment costs associated with solar collectors and adequate control systems (Carnegie 1991; Imre 2007).

15.4.2 Industrial heat sources for food drying

Preparing, i.e., heating and sometimes dehumidifying of the drying agent (mostly air) or drying surfaces (for conductive dryers), is a very important step to consider in a greener drying processes. The heat required for food drying is provided mostly by direct heating (firing) with fuels (gas or oil) and saturated or overheated steam. In some rather small-size applications, heat is supplied in the form of hot water, infrared radiation, or microwave energy. Combustion of fuel (natural gas, fuel oil, etc.) is the simplest and most economical energy source for industrial drying applications; however, direct contact of foods with hot combustion gases is sometimes prohibited from a sanitary point of view. In such cases, saturated steam and hot water are used, although they are more expensive than direct fuel combustion. The energy efficiency of such equipment (Perry and Green 1997; Mujumdar 2007) has to be one of the most important criteria for selection of the most appropriate heating method of the drying agent (mostly air).

Natural gas, used for direct firing of the dryer, is suitable for food and agricultural drying (Kudra 2009). High temperatures are obtainable, which substantially increases drying rate and shortens drying time. Natural gas contains about 98% methane and ethane, and its heating value is 37.3 MJ/m^3. The heating value of propane is 50.4 MJ/kg, and that of fuel oils is 41.7 MJ/kg.

Except for sanitary conditions, the major limiting factor of using direct heating by natural gas and fuel oil in drying is the presence of significant amounts of water vapor in the fuel gases, produced by combustion of the hydrocarbons. Stoichiometric calculations indicate that combustion gases contain up to 19% by volume of water vapor, but in practice lower concentrations are observed because of the use of excess air. As a result, the capacity of the drying agent to remove water in the dryer is reduced, and low moisture contents may not be reached easily for some foods, which may, therefore, require a finishing drying step using lower humidity air, for example, in bin dryers.

Another group of energy-saving methods are the use of internal heat exchangers, often referred to as coils, which can be located inside the drying chamber. This is mostly beneficial for drying free-moving and particulate foods. These heat exchangers are mostly used for fluidized and spouted bed dryers. Because of the strong movement of solid particles in fluidized or spouted bed, the heat transfer coefficient between the heated surface and the solid-gas mixture is very high, providing high efficiency for heat exchange. According to Cook and DuMont (1991), energy costs can be reduced and productivity increased by one third or more. Unfortunately, fouling of exchanger surfaces precludes their use with sticky materials. Moreover, foods of low thermal resistance cannot be in contact with internal hot exchanger surfaces; additionally, cleaning and sanitizing of dryers equipped with internal heat exchangers are more difficult.

Utilization of renewable energy resources, such as solar, wind, and geothermal energy, can reduce excessive use of fossil energy for industrial food drying. Furthermore, it will reduce pollution (chemical and thermal) of the environment, and preserve fuel reserves.

15.5 Energy-efficient dryers

The dryer is the hearth of the drying installation and has to fulfill the conditions of high heat and mass transfer efficiency. For air drying, the forced convective heat transfer coefficient (*h*) depends on many factors such as velocity, physical properties of the food and drying agent, temperature, etc. Normally *h* values vary from 25–100 $W/(m^2K)$ (Ratti and Crapiste 1995; Simal et al. 2000) and has a significant influence on the thermal efficiency of the drying process.

15.5.1 Energy efficiency parameters

For typical industrial drying systems, several energy intensity parameters have been developed (Pakowski and Mujumdar 2007; Strumillo et al. 2007; Marcotte and Grabowski 2008):

1. Energy efficiency coefficient (η), defined as:

$$\eta = \frac{E_1}{E_2}$$

where E_1 is the energy required for the moisture evaporation, that is ($E_1 = m_w \cdot r$), m_w is the stream of evaporated moisture (kg/s) and r latent heat of evaporation (J/kg), and E_2 is the total energy supplied to the dryer (from fossil fuel and electricity, etc.).

2. Thermal efficiency of convective drying (η_T), measured by the air temperature profile, is defined as:

$$\eta_T = \frac{(T_1 - T_2)}{(T_1 - T_0)}$$

where T_1 and T_2 are the temperatures of drying agent on the inlet and outlet from the dryer, respectively, and T_0 is an ambient temperature. This coefficient is applicable to convection dryers only where streams of drying agent (mostly air) are responsible for the drying kinetics. The temperature difference ($T_1 - T_2$) reflects the stream of heat released in the dryer not only for moisture evaporation but also for dried material heating and heat losses in the dryer, etc. The difference ($T_1 - T_0$) reflects the stream of heat provided to the drying agent in the heater of the dryer. If heating occurs internally in the dryer, this parameter is not applicable. Because drying is a heat and mass transfer process, the temperature of the drying agent in contact with a wet material theoretically cannot be lower than the wet bulb (wb) temperature. Thus, the maximum value of η_T is related to ($T_2 = T_{wb}$) when conditions of saturation of the gas on the outlet from the dryer exist. Coefficient η_T should be used with some precautions because it reflects a combination of the effect of energy consumption for moisture evaporation, heat losses, and material heating in the dryer.

3. Drying efficiency (η_D) is defined by Pakowski and Mujumdar (2007) as:

$$\eta_D = \frac{(E_1 + E_M)}{E_2}$$

where E_1 is the energy required for moisture evaporation, E_M energy used for the food material heating, and E_2 total energy supplied to the dryer. This parameter is not as useful as the thermal efficiency parameter. However, in some cases it is proposed as an additional reference value in the literature.

4. Energy consumption per 1 kg of evaporated moisture. This value is a complex parameter defining global energy efficiency and consumption in the drying process. Theoretically, the amount of heat required for evaporation of 1 kg of water under typical conditions is 2,200–2,700 kJ/kg. The upper limit of this value refers to the removal of bound water. The only drying regime in which such a result could be obtained is an ideal adiabatic equilibrium in which there is no heating of the food product, that is, it enters at the water evaporation temperature. However, in practice, the drying process has to be carried out at some realistic conditions, sometimes far from such an adiabatic environment. Realistic conditions of this ratio can be as high as 8,000 kJ/kg of moisture, for example, for drying of fine particulate materials. Higher energy consumption per kg of

Table 15.3 Energy and thermal efficiency of selected industrial dryers

Method or dryer type	Energy or thermal efficiency (%)
Tray, batch	85
Tunnel	35–40
Spray	50
Tower	20–40
Flash	50–75
Conveyor	40–60
Fluidized bed, standard	40–80
Vibrated fluidized bed	56–80
Pulsed fluidized bed	65–80
Sheeting	50–90
Drum, indirect heating	85
Rotary, indirect heating	75–90
Rotary, direct heated	40–70
Cylinder dryer	90–92
Vacuum rotary	Up to 70
Infrared	30–60
Dielectric	60
Freeze	10 or lower

Source: Combined data of Strumillo et al. (2007), Ramaswamy and Marcotte (2006) and author's own unpublished results.

moisture evaporated occurs in industrial convective drying of pastelike materials (Strumillo et al. 2007).

5. Energy consumption per 1 kg of dry product (kJ/kg). This value is a very practical parameter that is used mainly for the comparison of energy consumption of the same material dried using several dryers and/or various drying conditions. Careful precautions must be taken to ensure that the quality of the final products is the same and not compromised while generating data for such comparisons.
6. Specific electric power consumption per 1 kg of product (kJ/kg) is used for electrically driven dryers therefore very limited in terms of its application. However, in relation to the unit mass of final product, this parameter has a very practical importance.

The energy efficiency coefficient is the most global parameter for drying and it is calculated for the evaluation of all drying processes. Statistical data obtained from energy balances of convective dryers show that 20–60% of the energy supplied to the dryer is used for moisture evaporation (E_1), 5–25% for product heating, 15–40% for heat losses with exhaust gases, 3–10% for heat losses from dryer walls to the atmosphere, and 5–20% for other losses (Danilov and Leontchik 1986). For convective type of dryers, the energy efficiency coefficient is in the range of 20–60%. More detailed data for specific dryer types is given in Table 15.3. In some cases, the actual energy cost of energy sources (natural gas, oil, and/or electricity) per 1 kg of evaporated moisture is the best parameter to consider as drying efficiency. Among all typical food drying technologies freeze drying requires up to 100,000 kJ/kg of evaporated water, which is 10–30 times higher than the energy

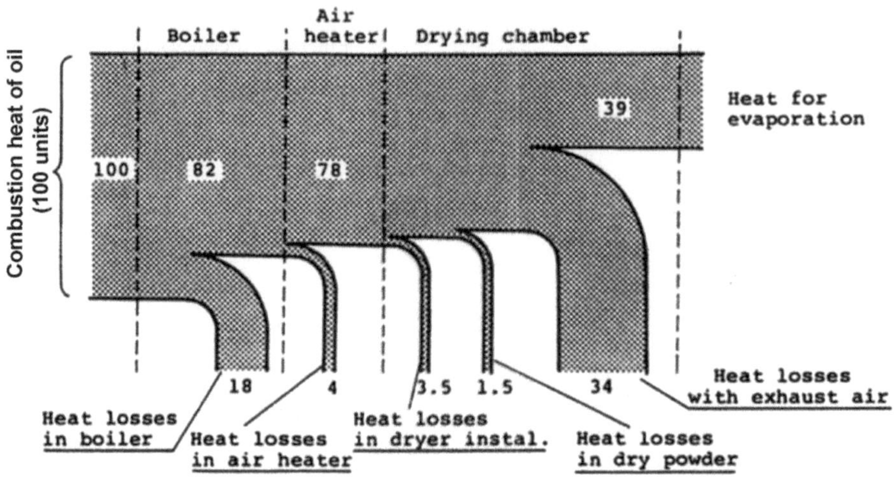

Fig. 15.7 Thermal efficiency of steam-heated spray-dryer for inlet and outlet temperatures of 180°C and 90°C, respectively (From Strumillo (1983), courtesy of WNT, Warsaw)

required using other drying methods (e.g., spray, conveyor, or tunnel drying). Very long drying times are involved and the need for energy to freeze first and then dry is responsible for these exceptional values. Superior quality of the final product is, however, a big advantage of this drying technology, which explains why it is applied usually to expensive products.

Figure 15.7 presents an example of heat consumption of a spray dryer (Strumillo 1983). In this steam-heated direct dryer, with inlet air temperature of 180°C, a thermal efficiency of 39% was achieved, whereas when the drying air was heated to 500°C an efficiency of 83% was possible. For food drying processes, inlet air temperatures are normally in the lower range, however drying efficiency can be improved by means other than increasing the inlet air temperature as described later.

Generally, it is possible to use other drying efficiency parameters but mostly for special applications (Flink 1977). A dynamic analysis of drying energy consumption was proposed for nonfood material by Menshutina et al. (2004) and Kudra (2004). A process integration study of a milk powder plant was also conducted by Robertson and Baldwin (1993) and offers another example of a possible way of analyzing drying process performance.

15.5.2 New drying installation

15.5.2.1 General recommendations

Planning a new drying installation requires a complex analysis of several factors such as (1) quality of the end-product (2) assumed throughput, (3) energy consumption characteristics, and (4) available space for installation, energy sources, and work-forces, etc. If it is decided that there has to be a drying step, various types of installations are available for successful drying of food products. Raw food material as well as final product must be characterized for their drying properties. Using the food drying characteristics data will allow the selection of a suitable dryer type and design. Further information may, however, be required on the handling characteristics of the product and particular constraints such as temperature sensitivity.

There are many recommendations in the literature on how to select the right type of dryer as well as information on how to conduct appropriate testing for the drying of food products in small or pilot scale experiments (Barbosa-Canovas and Vega-Mercado 1996; Mujumdar 2007; Kudra and Mujumdar 2009). Most textbooks and handbooks cover conventional types of dryers, which can be helpful in the decision-making process. Technical catalogues published by various manufacturers of drying and ancillary equipments are also very valuable. To facilitate the selection of greener drying equipment, manufacturers should provide specific data on energy efficiency and consumption of their drying installations for the dehydration of different food products. Today, more than 200 types of dryers are used in the food and industrial sectors.

With this knowledge, energy consumption should be considered as one of the most important criteria for dryer selection. For example, it is well known that indirect dryers are more energy efficient than direct dryers. Direct dryers depend on convection, whereas those depending on conduction are referred to as indirect. Direct dryers usually rely on heated air or combustion gases but in some cases may use hot neutral gases or superheated steam. Conduction dryers are indirect and the heating medium may be condensing steam, electrical heating elements, combustion gases, and hot liquid heating media such as high pressure water. There is always a separation (mostly metal surface) between the heating medium and the food. Indirect dryers require little or no hot air for drying, which significantly reduces heat loses with the exhaust air. This type of dryer is generally recommended, but is not always possible to use because of specific product characteristics that may limit application.

15.5.2.2 Selection of new energy-efficient dryers

In the early stages of a new drying project, product quality is often of more concern than energy consumption. The selection of a dryer at this stage frequently includes, but is not limited to:

1. The production capacity
2. Initial moisture content of the product
3. Particle size distribution or form of products
4. Drying characteristics of the product
5. Maximum allowable product temperature
6. Explosion characteristics (for spray or fluid bed drying, etc.)
7. Moisture isotherm
8. Physical data of the material

In the later stages of the project development, the cost of operation becomes important, and it is at this stage that the cost of the operation of the dryer and energy consumption are considered. Options in efficient dryer selection include whether a single dryer or a combination of several dryers in different configurations (hybrid technologies, etc.) will be used. For a single dryer, energy efficiency can be estimated and compared relatively simply, for example, using Table 15.3, dryer manufacturer information, experimental tests, etc. For example, spray drying of liquids or slurries can be replaced by alternative spouted-bed drying with inert bodies (see Fig. 15.2) or spin-flash dryer (see Fig. 15.3). Also, mostly from an energy consumption point of view, fluidized bed (FB) dryers for particulate materials can be modified into spouted bed fluidized bed dryer (Fig. 15.8b), vibrated fluidized bed (VFB) (Fig. 15.8c), or pulse fluidized bed (PFB) dryers (Fig. 15.8d). All of these generally require less hot drying air for food particles fluidization, giving better thermal efficiency than the original FB dryer.

For combined hybrid-type installations, the prediction of energy efficiency is much more difficult and is based mostly on experiment, experience or step-wise analysis. Combined drying installations typically have advantages over the single dryer because each step of such drying installations can be optimized separately to give the desired kinetics for the entire drying process. For example, drying of milk powder can be done using a spray dryer alone, but this operation is not efficient and most recently drying of milk powder is performed using a combination of two (spray dryer plus fluidized bed dryer, including sometimes granulation process) or three-stage drying (spray, integrated fluidized bed and external fluidized bed). Figure 15.9 presents a schematic of a three-stage drying installation for drying of milk powder. The approximate energy savings using different drying operations is provided in Table 15.4.

The multistage drying configuration is worth considering for other types of food products and for other types of dryers as well (Kudra and Mujumdar 2009). It is well known that for drying of some fruits, vegetables, and spices, a two-stage drying is often applied (Grabowski et al. 2003), with the final step done in stationary bins using partially dehumidified air. This allows not only improved quality of the finished product, but also significant reduction in the energy required for drying.

Another example of combined energy-efficient drying is the two-step dehydration of small fruits developed for many fruits, including blueberries and cranberries (Grabowski and Marcotte 2003). The process consists of an osmotic dehydration as the first step and a convection drying in the final step (see Fig. 15.4). Approximately 50% of the moisture content of berries is removed in the osmotic step, which is a low-energy consuming operation; this, therefore, provides a definite advantage from an energy consumption point of view. Pretreatment is recommended in the case of cranberries and, in general, pre-treatment of many other raw food materials can be considered as a way to improve drying efficiency (Tarhan et al. 2006; Grabowski and Marcotte 2003). Fernandes and Rodrigues (2007), as an example, achieved an 11% reduction in energy consumption in a banana dehydration process using ultrasound as a pretreatment operation.

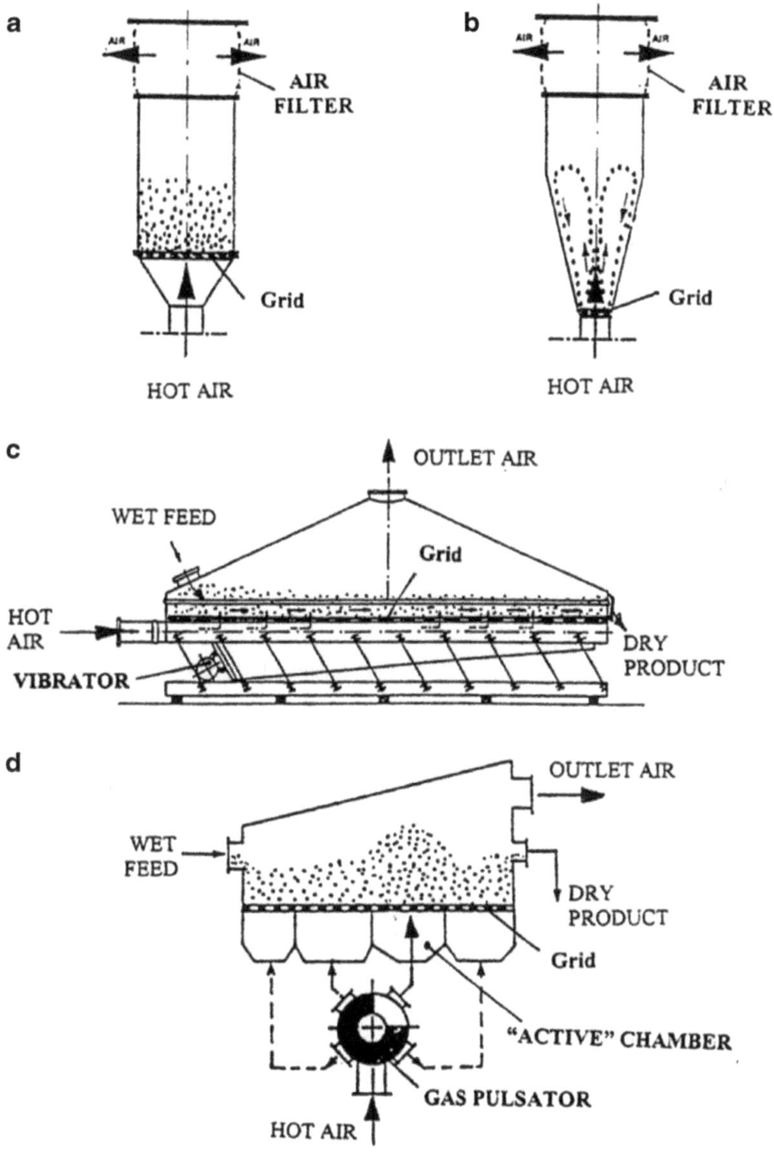

Fig. 15.8 Modifications of fluidized bed dryers for better thermal efficiency: (**a**) standard, (**b**) spouted bed, (**c**) vibrated fluidized bed, and (**d**) pulse fluidized bed (Modified from Mujumdar (2007) and Grabowski et al. (2003))

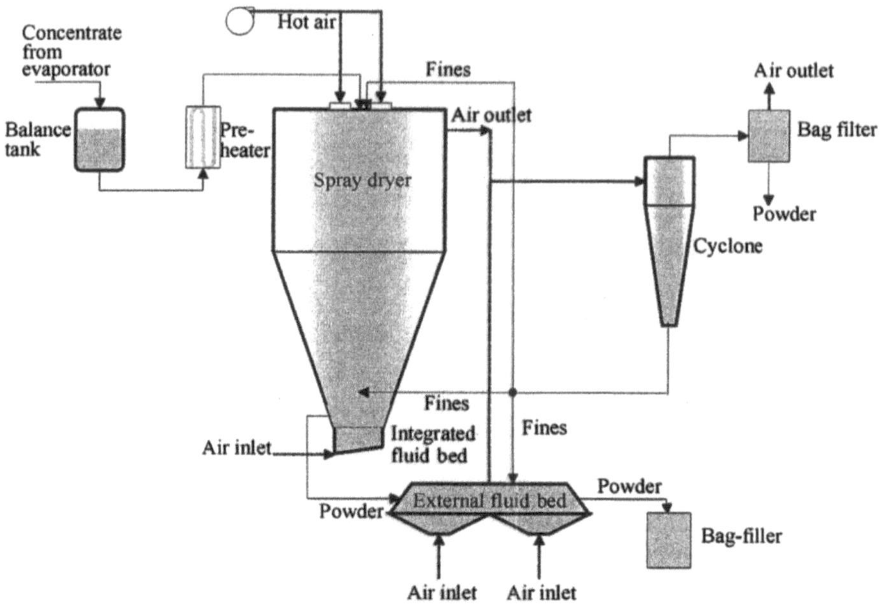

Fig. 15.9 Schematic of an energy efficient three-stage industrial dryer for milk (From Verdurmen and de Jong (2003), with permission of Woodhead Publishing Ltd and CRC)

Table 15.4 Example of energy consumption and savings in a multi-stage drying installation

Installation	Energy consumption[a] (kJ/kg of evaporated water)	Energy savings (%)
Single stage (spray)	5,265.1	0
Two-stage (spray + VFB)	4,454.4	15.4
Three-stage (spray + FB + VFB)	3,985.9	24.3

[a] Average energy consumption for dairy feed drying from 50% to 4% w.b.
Source: Modified from Tang et al. (1999).

15.5.2.3 Superheated steam drying

As mentioned, improved energy efficiency, enhanced quality, reduced emissions, and safer operation are some of the essential advantages of superheated steam drying (SSD) systems. For the drying of slurries, pulps, brewer's spend grain, biomass, fruit pomaces, fish to produce fish meal, and so on, current steam-drying technologies exist that are already cost and energy effective. The basic principle of this method of drying is to use superheated steam as a source of energy for evaporation. One of the obvious advantages of SSD is that the dryer exhaust is also steam, albeit at lower specific enthalpy. In air drying, the latent heat of the exhausted steam is generally difficult and expensive to recover. Indeed, at current high prices of energy sources and for most direct dryers with low-to-medium temperature exhaust, it is often more expensive to recover energy in the exhaust stream than to evacuate the gases through the smokestacks (Kumar and Mujumdar 1990). In principle, any direct dryer can be converted to a superheated steam operation (e.g., spray, conveyor, flash, fluidized bed dryers, etc.). Mujumdar (2007) provides an example of SSD operation for drying of beet pulp in a sugar factory. Apparently, Niro A/S of Denmark has also successfully commercialized a pressurized steam-fluidized bed dryer for particulate and sludge-like or pulpy materials. Capacities of 2–40 t/h water evaporation are available. Compared with conventional rotary dryers, energy savings of up to 90% are feasible. The only important limitation in case of SSD for foods is possible heat sensitivity of the wet and final products.

15.5.2.4 Other modern methods

Contact-sorption drying (Kudra and Mujumdar 2009) is another technique of food drying based on the use of a hot desiccant directly in contact with the food to remove the moisture. It is often referred to as contact-sorption drying, desiccant drying, or adsorption drying. Different types of sorbents can be used, for example, molecular sieves, activated carbon, silica gel, zeolite, etc. Solar or secondary sources energy can be used for desiccant regeneration (removal of moisture).

Other novel drying technologies for foods involve the use of an electromagnetic field as source of heat (Hui et al. 2007; Ratti 2009). Examples of such methods are microwave, radiofrequency, and infrared drying, which have become increasingly applicable for food dehydration. For specific applications (thin layer foods, chips, post-baked biscuits, cereals, etc.), these methods give high levels of energy efficiency with good quality products. For microwave drying, an average ratio of the energy absorbed by the food over the energy being supplied to the microwave oven is roughly 50–60%. In infrared drying, similar values in the range of 40–46% for radiation sources from combustion of natural gas, or approximately 78–85% for electrically powered infrared lamps (Ramaswamy and Marcotte 2006) have been reported. These novel drying methods could be considered as potential alternatives to conventional drying techniques in the initial steps of the drying operation. Other dehydration methods, typical for biological materials (foam, foam-mat, explosion puffing, etc.) generally require more energy; therefore, specific cost analysis, including energy consumption, should be performed before a decision is made on its application. Further details of these technologies are given in Ratti (2009), Hui et al. (2007), Mujumdar (2007), Grabowski et al. (2003), etc. In regard to freeze-drying techniques, limited use is recommended because they are very energy consuming; therefore, they should be applied only to relatively expensive products in which quality is an important parameter that is not achievable by other drying methods.

15.5.3 Improvement of energy performance of existing dryers

15.5.3.1 Change of the heat source

The combustion of gas or fuel by direct burners located directly in the dryer generates water vapor to the air drying agent. According to Cook and DuMont (1991) of the common fuels, natural gas adds the most moisture, heating oil about 43% less, and no moisture is added using direct heating. If a dryer is switched from natural gas to heating oil, the reduced amount of water vapor formed lowers the total heat consumption by about 4%. This change also lowers the outlet gas humidity which allows the reduction of the gas temperature and the increase of thermal effectiveness. Electric heating of drying gases is rather effective, but generally the cost of this type of energy is higher than that of natural gas or fuel oil. Natural biomass can be considered as an environmentally friendly source of heat for food drying, especially for small-scale operations.

15.5.3.2 Heat pump

Recently, heat pumps have become very important and popular tools for significant energy savings. Drying is one of the potential application areas for these equipment in terms of the reduction of energy consumption to evaporate moisture. Heat pumps extract heat energy from a source at low temperature (e.g., on the air outlet from the dryer) and makes it available as useful heat energy at a higher temperature (e.g., on the drying air inlet). The use of heat pumps in drying food products is an efficient and controllable method. Further information on heat pump applications in drying can be found in Kudra and Mujumdar (2009), Raghavan and Orsat (2009), Strumillo et al. (2007), and Kiang and Jon (2007).

Compared with conventional oil or gas-fired dryers, the application of heat pumps can improve energy efficiency and reduce external heat consumption by as much as 60–80% (Strommen 1994) or 40–65% (Lopez-Cacicedo 1986). From the evaporator, the heat pump recuperates sensible and latent heat from the dryer exhaust, hence the energy is recovered. Condensation occurring at the evaporator reduces the humidity of the drying agent, thus increasing the effectiveness of drying (Fig. 15.10).

15.5.3.3 Optimization of the air temperature and moisture content at the inlet and outlet

It is well known that raising the inlet air temperature will generally improve thermal efficiency. A similar effect is expected by reducing the outlet air temperature. However, there is a serious limitation for the increase of the inlet air temperature because of the heat sensitivity of most food materials. For a reasonably efficient commercial drying operation, the air inlet temperature should be within 6–11°C

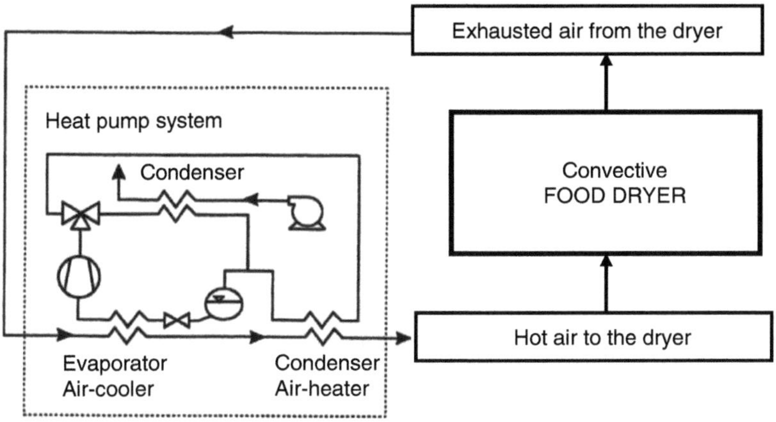

Fig. 15.10 Typical scheme of a heat pump application for food drying (Adapted from Kudra and Mujumdar (2009))

below its maximum. The moisture condensation at the outlet part of the drying installation gives the limit for the application of low temperatures at the outlet. Cook and DuMont (1991) concluded that each degree of increase of the air inlet temperature is less beneficial than 1° decrease of the outlet temperature. The inlet temperature can usually be changed more easily than the outlet temperature. However, this advantage diminishes at higher temperatures. A decrease of the outlet air temperature by 5.6°C resulted in the reduction of energy consumption by greater than 8% at low temperature, but less than 3% at high temperature. In some cases, for better efficiency (e.g., drying of onion), a two-step drying was used with dehumidified air of relatively low temperature (50°C) at the final drying step (Grabowski et al. 2003).

15.5.3.4 Automatic control of the dryer

Optimization of drying conditions coupled with automatic control of operations is a standard task routinely performed for new installation as well as existing ones. Adjustment of all drying parameters to actual conditions of the wet food inlet stream and outlet air parameters is a minimum requirement for an effective drying process. This is very important during the start-up and shutdown periods as well as during the continuous operation of the dryer. The initial and final drying periods usually represent the least energy efficient part of drying operation and these have to be minimized (Gardiner 1997). Basic principles as well as examples of practical applications of automatic control applied to food dryers are presented by Bhuyan (2007), Mujumdar (2007), etc. Benefits can be significant. For example, according to information published on the Internet by ABB Group from Finland, an automatic control of the fans supplying the heating air to large-scale industrial dryer of 105,000 t of malt per year allows a 9% energy savings with payoff time of the investment of 3 years (ABB Group 2005).

Table 15.5 Examples of effects of simple changes of operational parameters on approximate energy consumption in drying[a]

Drying parameter	Change of value[b]	Fuel savings (%)	Electric power savings (%)
Initial moisture of food	From 32% to 30%	9	12–15
Inlet air temperature, °C	+5	12.6	12.9
Outlet air temperature, °C	−2.5	12.6	12.6
Wet feed temperature, °C	+25	9	9
Air leakage, %	−5	4	10
Insulation thickness, cm	+5	5	5

[a]Data for air temperatures on inlet and outlet: 149°C and 77°C, respectively.
[b](+) increase, (−) decrease in value of specific parameter.
Modified from Cook and DuMont (1991).

15.5.3.5 Insulation against thermal loss

Heat losses to the surroundings can account for significant energy requirements. Proper thermal insulation as well as reduction of the size of the dryer will significantly reduce energy losses. For cylindrically shaped dryers, a critical thickness of insulation has to be calculated to prevent an increase of heat loss as the surface area in contact with the environment is increased. Good-quality insulation materials, characterized by low thermal conductivity, must be used. In typical dryer energy balances, a conservative value of 15% of the evaporation energy is usually assumed to be used to compensate for the thermal losses of the dryer. A more precise calculation of the heat loss is recommended to improve the insulation effectiveness of older dryers (Mujumdar 2007).

15.5.3.6 Simple changes to improve energy savings

Application of even small changes in the drying installation or drying parameters can result in substantial energy savings. Table 15.5 presents examples of the effects of such simple changes in operational parameters of drying on energy consumption.

15.6 Exhaust gases: prevention from environment pollution

Maximum recovery of the finished dry product from the drying installation is an important task from an economical as well as environmental point of view (Pell and Dunson 1997). Solid/air separators are used to recover dry food particles from exhaust air in various convective drying installations. Large scale spray, pneumatic, conveyors, spouted-bed, and fluidized-bed dryers are examples of installations fitted with such separators. The most typical design for finished product recovery from dryers is based on a two-stage separation: cyclone or multi-cyclone separators in the first stage and bag filters in the second one as described further in the following. Sometimes, but rather rarely, electrical filters or wet scrubbers are applied.

15.6.1 Cyclone separators

Cyclones are simple and inexpensive equipment that are effective in removing solid food particles larger than 10 μm from air streams. The particle/air mixture enters the cyclone tangentially, following a spiral flow pattern from the top to bottom, and a vortex flow from bottom to top, with an exit from the collector. The solid food particles are subjected to self-generated centrifugal forces which throws them to the cyclone wall from which they fall and are collected at the bottom.

The self-generated centrifugal force, developed by the cyclone spiral flow, is very large, because of the small cyclone diameter, reaching up to 1,000 times the gravitational force. The efficiency of a cyclone collector can be estimated from the Lapple diagram (Perry and Green 1997). Saravacos and Kostaroupoulos (2002), for example, have shown that for dry milk powder separated in cyclones of 20 cm diameter and inlet air velocity of 15 m/s, 90% of particles smaller than 3.9 μm in diameter will pass through the cyclone; that is, the collector efficiency will be 10% for these small particles only. To increase the collecting efficiency, the diameter of the cyclone should be smaller, and a number of collectors, operated in parallel, will be required to handle the large volumes of exhaust gases. Such collectors, built in compact structures, are installed as multicyclones. The ideal case of cyclone separation is full retrieval of the dry product in a single step. In practice, however, a second step (bag filters, wet scrubbers, etc.) is necessary.

15.6.2 Bag filters

Small food particles, passing through cyclone collectors, can be separated from exhaust air by bag filtration, which is similar in principle to solid/liquid filtration. Depending on their size (d), particles are collected mainly by inertia ($d > 1$ μm) or electrostatic forces ($d < 0.5$ μm). As with other filtration processes, solid food particles in bag filters form a mat on the surface, which acts as a filter medium, increasing the efficiency of filtration but, unfortunately, the pressure drops as well. Filtration is interrupted when the pressure drop exceeds a specified limit, at which time the filter is then cleaned.

Bag filters are usually made of woven cloth or felt. Typically, they are made of cotton, wool, nylon, Teflon, and Dacron. Bag filters have diameters of 12–20 cm and a length of 2.5–5 m, and they are often assembled in compartments, called "bag houses" of 100–200 m^2 cloth surface. The diameter of large cylindrical bag filter compartments of 120 m^2 can be 3 m and their height 6–7 m. Their weight is about 2 t and the power consumption is 1 kW. Volumetrically it is one of the largest parts of the drying installation. Typical air filtration rate is 0.5–2 m/min for woven filters and 2–5 m/min for felt filters (Saravacos and Kostaroupoulos 2002). This gives an approximate surface area of the bag filter for known exhaust air stream from the dryer. More detailed information on dimensioning and scale-up of cyclone and bag filters can be found in Pell and Dunson (1997), Purchas and Wakeman (1986), and manufacture flyers, etc. As previously mentioned, solid/air separation is associated with relatively high pressure losses in the drying installation and has to be optimized by selection of the most effective separators, fans, and other auxiliary equipment. Energy savings as well as maximum purity of exhausting air have to be important selection criteria.

15.7 Heat recovery from the exhaust air

Recirculation of the exhaust air/gases and the recovery of heat losses from the dryer surfaces are strongly recommended because it limits the amount of heat rejected to the environment. Almost all convective dryers use some form of recirculation, recovering only sensible heat, because it is difficult to recover the latent heat of evaporation of water in the exhaust gases. The heat pump application described in the preceding is an interesting solution (see Fig. 15.10), although it is a rather costly alternative. Therefore, it is used, for example, with large batch dryers or rotary-tray units. Exhaust gases are often used indirectly to preheat the dryer supply air. As well, various proportions of exhaust air can be combined with the supply of fresh inlet air so that the energy is recuperated. Precise monitoring and control of the air humidity must be in place to avoid unwanted moisture saturation of the air which would retard drying. Thermal wheels, pipes, or plate heat exchangers can be applied for this purpose (Strumillo et al. 2007). For spray drying of some milk products and other foodstuffs, Reay (2008) presented data of heat recovery in the range of 10–30%. Moraitis and Akritidis (1996) also earlier proposed a complex system of heat exchangers between the exhaust gases and the inlet air to recover part of the latent heat of evaporation of food moisture. Exhaust gases can also be used to preheat the moist feed before drying, etc.

15.8 Conclusion

Drying is an important unit operation in many food processing installations. Because of its operation at very large scale in many food processing industries, it represents a significant cost investment and a major source of energy consumption. However, drying techniques and equipment used by many industries today are characterized by relatively low thermal efficiency, making this an exemplary process for consideration in terms of the development of greener more environmentally friendly food production and processing practices. Complete elimination of this process from the production line or replacing it by other, more environmentally friendly operations, is the primary recommendation to be considered. If this is not possible, each step of the drying operation should be analyzed separately to identify areas to reduce energy consumption and improve efficiency. The use of natural energy sources (sun and wind, biomass, etc.), some advanced technologies (heat pumps, microwaves, radiofrequency, etc.) and full automatic regulation, is strongly recommended. Nonthermal dewatering, such as osmotic dehydration, contact-sorption, etc., also provide good alternative solutions. Finally, energy-efficient dryers which include heat recovery systems from exhaust gases are recommended for adoption as part of a "greener" drying operation. Additionally, special attention needs to be given to the full recovery of finished dry food product to prevent pollution of the environment.

References

ABB Group. 2005. Saving energy at a malting plant in Finland. www.abb.com. Accessed 15 Dec 2010.

Bansal, N.K., and H.P. Garg. 1987. Solar crop drying. In *Advances in drying*, vol. 4, ed. A.S. Mujumdar. Washington: Hemisphere.

Barbosa-Canovas, G.V., and H. Vega-Mercado. 1996. *Dehydration of foods*. New York: Chapman & Hall.

Bhuyan, M. 2007. *Measurement and control in food processing*. Boca Raton: CRC Press, Taylor & Francis Group.

Bolin, H.R., and D.K. Salunkhe. 1982. Food dehydration by solar energy. *Critical Reviews in Food Science and Nutrition* 16: 327–353.

Carnegie, E.J. 1991. Solar fruit drying. In *Solar energy in agriculture*, ed. B.F. Parker. Amsterdam, The Netherlands: Elsevier.

Cohen, J., and T.C.S. Yang. 1995. Osmotic dehydration and its application in nutrient infusion of various foods. Report No. TR-95/034, U.S. Army Research, Development and Engineering Center, Natick, MA.

Cook, E.M., and H.D. DuMont. 1991. *Process drying practice*. New York: McGraw-Hill.

Danilov, O.L., and B.I. Leontchik. 1986. *Energeticzna ekonomija w termicznoj suszkie (Energy economics in thermal drying)*. Moscow, Russia: Energoatomizdat.

Fernandes, F.A.N., and S. Rodrigues. 2007. Ultrasound as pre-treatment for drying of fruits: Dehydration of banana. *Journal of Food Engineering* 82: 261–267.

Flink, J.M. 1977. Energy analysis in dehydration processes. *Food Technology* 1977: 77–84.

Gardiner, S.P. 1997. Dryer operation and control. In *Industrial drying of foods*, ed. C.G.J. Baker. London, UK: Blackie Academic & Professional.

Grabowski, S., and M. Marcotte. 2003. Pretreatment efficiency in osmotic dehydration of cranberries. In *Transport phenomena in food processing*, ed. J.W. Chanes, J.F. Velez-Ruiz, and G.V. Barbosa-Canovas. Boca Raton, FL: CRC.

Grabowski, S., M. Marcotte, and H.S. Ramaswamy. 2003. Drying of fruits, vegetables and spices. In *Handbook of postharvest technology*, ed. A. Chakraverty, A.S. Mujumdar, G.S. Raghavan, and H.S. Ramasawmy. New York: Marcel Dekker.

Grabowski, S., and M. Marcotte. 2003. Kinetics and quality aspects of Canadian blueberries and cranberries dried by osmo-convective method. *Drying Technology* 25: 367–374.

Hall, C.W. 1979. *Dictionary of drying*. New York: Marcel Dekker.

Hui, Y.-H., C. Clary, M.M. Farid, O.O. Fasina, A. Noomhorn, and J. Welti-Chanes (eds.). 2007. *Food drying science and technology*. Lancaster, PA: Destech Publications Inc.

Imre, L. 2007. Solar drying. In *Handbook of industrial drying*, 3rd ed, ed. A.S. Mujumdar. Boca Raton, FL: CRC Press, Taylor & Francis Group.

Jayaraman, K.S., D.K. Das Gupta, and N. Babu Rao. 1992. Solar drying of vegetables-Quality improvement using cabinet with multiple flat plate collectors and pretreatment. In *Drying of solids*, ed. A.S. Mujumdar. New York, New Delhi, India: International Science Publ. and Oxford & IBH Publ. Co.

Jen, J.J. 1989. Post harvest handling and processing of selected fruits and vegetables. In: Ghee, A.H. (ed) Trends in food processing II. Proceedings of the 7th World Congress of Food Science and Technology. Singapore, October 1987. Singapore Inst. Food Science and Technology, Singapore.

Kiang, C.S., and C.K. Jon. 2007. Heat pump drying systems. In *Handbook of industrial drying*, 3rd ed, ed. A.S. Mujumdar. Boca Raton, FL: CRC Press, Taylor & Francis Group.

Kraglund, A. 1983. Energy savings in the production of powders. *Journal of Separation Process Technology* 4: 1–9.

Kudra, T. 2004. Energy aspects in drying. *Drying Technology* 22: 917–932.

Kudra, T. 2009. Energy aspects in food dehydration. In *Advances in food dehydration*, ed. C. Ratti. Boca Raton, FL: CRC Press, Taylor & Francis Group.

Kudra, T., and A.S. Mujumdar. 2009. *Advanced drying technology*. Boca Raton, FL: CRC Press, Taylor & Francis Group.

Kumar, P., and A.S. Mujumdar. 1990. Superheated steam drying. A review. In *Drying of solids*, ed. A.S. Mujumdar. Meerut, India: Sarita.

Lawand, T.A. 1975. A survey of solar agricultural dryers. Report No. R-99. Brace Research Institute, Macdonald Campus of McGill University, Canada.

Lewicki, P., and A. Lenart. 2007. Osmotic dehydration of fruits and vegetables. In *Handbook of industrial drying*, 3rd ed, ed. A.S. Mujumdar. Boca Raton, FL: CRC Press, Taylor & Francis Group.

Lopez-Cacicedo, C.L. 1986. Electrical methods for drying. In *Drying '86*, vol. 1, ed. A.S. Mujumdar. Boston, MA: Hemisphere.

Marcotte, M., and S. Grabowski. 2008. Minimising energy consumption associated with drying, baking and evaporation. In *Handbook of water and energy management in food processing*, ed. J. Klemes, R. Smith, and J.-K. Kim. Cambridge, UK: Woodhead.

Marcotte, M., and M. Tanguay. 2005. Design and scaling-up a continuous osmotic dehydration contractor for small fruits. In: Proceedings of 3rd Inter-American drying conference, Paper I-4. Montreal, Canada, 21–23 August 2005.

Mazza, G. 1989. Carrot. In *Quality and preservation of vegetables*, ed. N.A.M. Eskin. Boca Raton, FL: CRC Press.

Menshutina, N.V., M.G. Gordienko, A.A. Voynovskiy, and T. Kudra. 2004. Dynamic analysis of drying energy consumption. *Drying Technology* 22: 2281–2290.

Moraitis, C.S., and C.B. Akritidis. 1996. Energy savings in industrial drying plants by partial recovery of the latent heat of the exhaust air. In: Stumillo, C., and Z. Pakowski (eds.) Proceedings IDS 96. Lodz Technical University, Lodz, Poland.

Mujumdar, A.S. (ed.). 2007. *Handbook of industrial drying*, 3rd ed. Boca Raton, FL: CRC Press, Taylor & Francis Group.

Orsat, V., and G.S.V. Raghavan. 2008. Dewatering for food waste. In *Handbook of water and energy management in food processing*, ed. J. Klemes, R. Smith, and J.-K. Kim. Cambridge, UK: Woodhead.

Pakowski, Z., and A.S. Mujumdar. 2007. Basic process calculations in drying. In *Handbook of industrial drying*, 3rd ed, ed. A.S. Mujumdar. Boca Raton, FL: CRC Press, Taylor & Francis Group.

Parker, B.F. 1991. *Solar energy in agriculture*. Amsterdam: Elsevier.

Patil, R.T. 1989. Drying studies on black pepper. *Journal of Food Science and Technology* 26: 230–231.

Pell, M., and J.B. Dunson. 1997. Gas-solid operations and equipment. In *Perry's chemical engineering handbook*, ed. R.H. Perry and D.W. Green. New York: McGraw Hill.

Perry, R.H., and D.W. Green. 1997. *Perry's chemical engineering handbook*. New York: McGraw Hill.

Puigalli, J.R., and A. Tiguert. 1986. The building and use of a performance model for a solar dryer. *Drying Technology* 4: 555–581.

Purchas, D.B., and R.J. Wakeman. 1986. *Solid/liquid equipment scale-up*. London, UK: Uplands Press.

Raghavan, V.G.S., and V. Orsat. 2009. Nonconventional heating sources during drying. In *Advances in food dehydration*, ed. C. Ratti. Boca Raton, FL: CRC Press, Taylor & Francis Group.

Ramaswamy, H.S., and M. Marcotte. 2006. *Food processing principles and applications*. Boca Raton, FL: CRC Press.

Raoult-Wack, A.L., A. Lenart, and S. Guilbert. 1992. Recent advances in dewatering through immersion in concentrated solutions. In *Drying of solids*. ed. A.S. Mujumdar. New York, New Delhi, India: International Science Publ. and Oxford & IBH Publ. Co.

Ratti, C. 2009. *Advances in food dehydration*. Boca Raton, FL: CRC Press, Taylor & Francis Group.

Ratti, C., and G.H. Crapiste. 1995. Determination of heat transfer coefficients during drying of foodstuffs. *Journal of Food Process Engineering* 18: 41–53.

Reay, D. 2008. Heat recovery in the food industry. In *Handbook of water and energy management in food processing*, ed. J. Klemes, R. Smith, and J.-K. Kim. Cambridge, UK: Woodhead.

Roa, G., and I.C. Macedo. 1976. Stationary bin and solar collector module. *Solar Energy* 18: 445.

Robertson, L.J., and A.J. Baldwin. 1993. Process integration study of a milk powder plant. *The Journal of Dairy Research* 60: 327–338.

Saravacos, G.D., and A.E. Kostaroupoulos. 2002. *Handbook of food processing equipment*. New York: Kluwer Academic.

Simal, S., A. Femenia, P. Llull, and C. Rossello. 2000. Dehydration of Aloe Vera: simulation of drying curves and evaluation of functional properties. *Journal of Food Engineering* 43: 109–114.

Smith, C.C., J.A. Maga, and J.C. Chapman. 1977. Solar process drying of potato products. Paper No. 77-6521 presented at ASAE Meeting, Chicago, IL.

Sokhansanj, S., and D.S. Jayas. 2007. Drying of foodstuffs. In *Handbook of industrial drying*, 3rd ed, ed. A.S. Mujumdar. Boca Raton, FL: CRC Press, Taylor & Francis Group.

Strommen, I. 1994. New applications of heat pumps in drying processes. *Drying Technology* 12: 889–901.

Strumillo, C. 1983. *Podstawy teorii i techniki suszenia (Principles of the drying theory and technique)*. Warsaw, Poland: Wydawnictwo Naukowo-Techniczne.

Strumillo, C., P.L. Jones, and R. Zylla. 2007. Energy aspects in drying. In *Handbook of industrial drying*, 3rd ed, ed. A.S. Mujumdar. Boca Raton, FL: CRC Press, Taylor & Francis Group.

Szulmayer, W. 1976. Drying principles and thermodynamics of sun drying. Technical Report No. T-124. Brace Research Institute, Macdonald Campus of McGill University, Canada.

Tang, J.X., Z.G. Wang, and L.X. Huang. 1999. Recent progress of spray drying in China. *Drying Technology* 17: 1747–1757.

Tarhan, S., G. Ergunes, and F. Taser. 2006. Selection of chemical and thermal pretreatment combination to reduce the dehydration time of sour cherry (*Prunus cerasus L.*). *Journal of Food Process Engineering* 29: 651–663.

Tutova, E.G., and P.S. Kutz. 1987. *Sushka produktow microbiologiczeskogo proizwodstwa (Drying of materials of microbiological origin)*. Moscow, Russia: Agropromizdat.

Verdurmen, R.E.M., and P. de Jong. 2003. Optimising product quality and process control for powdered dairy products. In *Dairy processing*, ed. G. Smit. Cambridge, UK: Woodhead.

Westeco Drying Inc. 1987. First commercial sonic-assisted drying plant starts up in California. *Chilton's Food Engineering* 59(12): 120.

Chapter 16
Green packaging

Susan E.M. Selke

16.1 Introduction

No matter how "green" the technology is for food production and processing, all is lost if the food does not get to the ultimate consumer in an acceptable condition. Condition in this context includes not only the obvious factors such as safety, quality, and nutritional value, but also factors such as appearance that factor into consumer acceptance. We depend on packaging to enable the food to get from where it is produced to where it is consumed. Even items that consumers may think of as "not packaged," such as bulk produce, must be packaged for transport. Consequently, packaging is an essential part of overall "green" food production. It should not be surprising that efforts to improve sustainability or reduce the environmental footprint of packaging are getting an increasing amount of attention across the industry. Of course, such efforts are not limited to food packaging, but with food and beverage packaging accounting for roughly half of all packaging, this sector is obviously extremely important in these efforts.

This chapter starts with a discussion of packaging functions. Efforts to make packaging "greener" fail completely if the package does not perform its required functions, as that will result in waste of not only the package, but also the product. The chapter also discusses the importance of considering the entire packaging system, not just the consumer package. Also, remember that the entire package (and sometimes product) life cycle must be considered in any efforts to make packaging more "green." This consideration will lead to examinations of definitions of sustainable packaging, and of tools for measuring environmental impacts of packaging systems. Finally, the chapter takes a special look at end-of-life considerations. All of this leads up to a discussion of guidelines for green packaging design.

S.E.M. Selke (✉)
School of Packaging, Michigan State University, East Lansing, MI, USA
e-mail: sselke@msu.edu

J.I. Boye and Y. Arcand (eds.), *Green Technologies in Food Production and Processing*, Food Engineering Series, DOI 10.1007/978-1-4614-1587-9_16,

16.2 Packaging functions

As mentioned, we depend on packaging to enable the delivery of products from where they are produced to where they are consumed. Along the way, packaging performs a variety of functions. Experts differ in how they categorize these functions, with some having three basic categories, whereas others have six or more. This discussion keeps it simple and opts for three categories: containment and protection, communication, and utility.

16.2.1 Containment and protection

Historically, the very first function of packaging was simply to contain products to enable them to be moved. Early containers included animal skins, baskets, even leaves from trees. For free-flowing products such as liquids, powders, grains, etc., the necessity for containment is obvious. Virtually all containers also provide some level of protection for the products they contain, which is the reason we are grouping these two functions together. The animal skin full of water not only contains the water, but also keeps out dust and other debris. The plastic bottle of milk in addition helps keep out microorganisms in the environment and limits evaporation, among other functions.

Providing adequate containment and protection is fundamental to package performance. The types and degree of protection required depend on two major factors: the sensitivity of the product, and the severity of the environment. For example, hamburger requires a different level of protection against microbes than does a whole uncooked potato. A carbonated soft drink in a plastic bottle will lose carbon dioxide at a faster rate if it is in the back of a truck on a hot summer day in Arizona than if it is on the supermarket shelf in Chicago in November.

Therefore, the first key to optimal package design is to understand the product requirements, and the second key is to understand the environments through which the product will travel. Certain product requirements may be quite specific to the individual product, but there are a variety of generalizations that can typically be made, to inform required design. One of these, in fact, has already been mentioned: the need for adequate containment. Similarly, although the environments that products must survive differ not only for different products, but importantly, vary quite a bit for each individual product, we can—and must—generalize. Virtually always, we design package systems to perform adequately in environments that can be realistically expected; we do not try to design to protect against all possible hazards.

For foods and beverages, required protection often includes protection from contamination by microorganisms, as well as dust, dirt, etc. Often protection is needed against excessive gain or loss of water or water vapor. Many products are sensitive to oxidation and need limits on oxygen exposure. Some require protection against light exposure. Products may be sensitive to breakage or crumbling. They may be sensitive to loss of volatile components such as carbon dioxide, odors, or flavors.

Protection may be required against temperatures that are too high, promoting degradation, or too low, causing freezing. And the list can go on.

Environmental influences encountered by products and potentially requiring protection include commonly recognized factors such as shock from being dropped, vibration from transport vehicles, compression from stacking, etc. A variety of atmospheric factors have been alluded to in the preceding, including light, oxygen, water vapor, temperature, etc. Perhaps less obvious, products may be exposed to lower than atmospheric pressure, such as in a cargo plane or a truck going over a high mountain pass. The levels of all of these hazards depend on the typical conditions in the physical surroundings that will be encountered, as well as on the transport mode.

One area of interaction between products and environments that is often overlooked in thinking about protection is the human environment. If toxic substances are being transported, an important packaging function is to protect humans from undesired exposure to that substance. For example, iron-containing vitamins must be packaged in child-resistant packaging to help protect that vulnerable section of the population from ingesting a toxic dose. Similarly, shipment of hazardous materials must satisfy a set of regulations intended to provide protection for both the human and natural environment against accidental spills, etc. Although this is not typically a major issue for food packaging, it is an important aspect of protection for certain categories of products.

On the flip side, packages may also need to provide protection for the products from the human environment. For example, a variety of small expensive items are packaged in oversize, hard-to-open packages in large part to make shoplifting more difficult. Again, this is more common outside the area of food packaging.

The required level of protection cannot always be obtained (or be economically obtained) from packaging alone. Especially for fresh products, temperature is a major influence on product degradation. It is seldom feasible to control temperature solely through packaging; rather, the supply chain depends on air conditioning, refrigeration, freezing, or in certain cases heating, to keep temperatures within reasonable limits. Packaging can be used to provide an additional level of protection, however. Insulated packages can slow down temperature changes. Ice (water or dry ice) can be incorporated in distribution packages. For certain high value products, gel packs can be employed, although this is rare for foods.

Because food products are by their nature perishable to at least some extent, and often are highly perishable, providing adequate protection is extremely important in design of food packaging. In many parts of the world, lack of adequate packaging is a major factor contributing to both inadequate overall food supply and poor nutritional quality of the food that is available. Kader (2005) estimates that postharvest losses of fresh fruits and vegetables in the United States average about 12% between production and consumption locations. He further reports that estimates for developing countries sometimes exceed 50%. In total, Kader estimates that about a third of worldwide production of fruits and vegetables is not consumed by humans. Of course, not all of these losses can be attributed to inadequate packaging, but proper packaging can certainly reduce loss.

16.2.2 Communication

The second major function of packaging is communication. This aspect of packaging functionality includes all the "messages" that the package sends to a prospective consumer. Although it certainly encompasses all the printed text on the package, it is not limited to written information. The communication function of packaging also encompasses attributes such as package shape and color, and even simple transparency of the packaging material to allow consumers to view the product inside. In short, any package attribute that assists in providing the user information about the product can be regarded as part of the communication function of packaging.

Many aspects of communication are legally required. These include product name, manufacturer, amount, nutritional information, etc. Specific requirements vary for different categories of products. Details about these requirements are beyond the scope of this chapter. For food packaging, in the United States, most regulations fall under the authority of the U.S. Food & Drug Administration, although in some product categories the U.S. Department of Agriculture is the primary regulatory authority.

Other aspects of the communication function are designed primarily to entice the consumer to purchase the product. Examples include appetizing pictures of the product on the label, instructions, recipes, etc. For many products, a unifying look for a product line, a brand identity, is an important part of the communication function. These nonverbal aspects of communication are one reason manufacturers may be reluctant to make major changes in package design—they are concerned about whether their loyal customers will still be able to recognize the product on the shelf at a glance. If these customers have a hard time finding their usual product, they may opt to purchase something different instead. At the same time, companies want to entice new purchasers by offering packages that look new, fresh, and different so they catch the eye of the prospective customer. Finding the right balance between these competing influences can be tricky.

Further, in the area of "green packaging," there is a strong desire for the package to convey its environmental benefits. Manufacturers may make significant changes in packaging primarily for the resulting environmental benefits, but they certainly would like to be rewarded by increased sales!

16.2.3 Utility

The final category of packaging function is utility, sometimes termed "convenience." This category encompasses all the packaging attributes that provide added value to users of the product or package. It is important to remember that these users are not necessarily the final consumer; they may be anywhere along the product supply chain.

Examples of packaging attributes that provide convenience abound in the area of food packaging. Microwavable entrees, fresh cut carrots, diced pineapple, boil-in-bag frozen vegetables in cheese sauce, steam-in-pouch broccoli, and many other products depend on specialized packaging to provide convenience for consumers in a rush to put food on the table. But there are also a wide variety of other packaging attributes that provide utility: zippers on pouches of luncheon meat or grated cheese, tuck tabs on cereal cartons, shaker tops on spice jars, molded-in grips on bottles of fruit punch, etc.

As mentioned, utility need not be intended for the ultimate consumer. The hole in the cheese pouch that allows it to hang suspended for display provides utility for the retailer, as does the bar code. A radiofrequency identification (RFID) tag on a pallet load of products can signal to the distribution center as it moves in—and later out—of the warehouse, simplifying the product tracking and reordering system. The plastic identification code on the bottom of a plastic bottle of vegetable oil is intended to provide utility to those involved in the collection and recycling of the bottles at the end of their life. Again, many other examples could be provided.

16.2.4 Package attributes with multiple functions

It should be obvious, but nonetheless should be stated explicitly, that a single package attribute can provide multiple functions. The cap on a bottle of milk has as its primary function containment and protection of the milk, but the color of the cap typically provides an important communication function by signaling the type of milk in the bottle—whole, 2%, skim, etc. The recipe on the pasta carton provides both utility and communication. The aseptic brick-pack for shelf-stable milk provides both containment and protection for the milk, but also utility by allowing distribution and storage without refrigeration.

16.3 Packaging systems

The discussion to this point has focused mostly on consumer packages. It is essential to realize that distribution of products actually relies on systems of packages or packaging components, working together. Some terminology follows.

16.3.1 Primary package

The package that contains the product and associated components of that package is referred to as the "primary package." For example, in a jar of mayonnaise, the primary package consists of the jar, the cap, the label, and often a shrink band around the cap for tamper evidence. The primary package is usually in direct contact with the product.

16.3.2 Secondary package

In the classic definition, a secondary package contains a number of primary packages; it provides a bundling function. For example, if that mayonnaise jar is sold in a club store as a two-pack, the shrink wrap or plastic ring connector that bundles those two jars together is the secondary package. The products bundled together need not be identical. A multipack of assorted flavors of juice boxes is an example, as is a pack of mini salt and pepper shakers.

16.3.3 Tertiary package or distribution package

The tertiary package is the shipping container, and typically contains a number of secondary packages. For the mayonnaise jar, the tertiary or distribution package likely consists of a corrugated tray containing a number of two-pack jars, a pallet containing a number of trays, and stretch wrap to unitize the load. There may also be a label and/or an RFID tag on the stretch wrap. All of these would make up the distribution package system.

16.3.4 Multiple categories and blurry distinctions

Although the categorizations in the preceding sound clear, in practice these categories are not always easy to apply. Some people find it more useful to distinguish between retail packages and distribution packages, for example, rather than the primary/secondary/tertiary distinctions. For example, what is the folding carton that cereal comes in? If the cereal is in a pouch, as is typically the case, the pouch is clearly the primary package. The carton does not fit the definition of secondary package, because it contains only one primary package, but neither does it have direct product contact, which is part of the definition of primary package; and clearly it is not a tertiary package. In the retail/distribution package distinction, clearly it is a retail package, so the categorization becomes clear.

However, adopting the retail/distribution package distinction does not totally solve the categorization problem, either. If that cereal is sold in a club store in a multipack, it is likely packaged in a corrugated box, which in turn is palletized and stretch wrapped. Typically, corrugated boxes are categorized as distribution packages, but in this case, the box is also a retail package. Here, incidentally, it might be usefully classified as a secondary package!

In sum, it is important to remember that packages do not function alone. The classifications of primary, secondary, and tertiary, or retail and distribution, can usefully remind one to include the whole packaging system in the analysis, even though it may be difficult to reach consensus on the appropriate category for a particular package (or one may simply have to accept that a single package can fit in more than one category, even for a specific use). It is especially crucial to remember the need to consider all parts of the system when attempting to compare the environmental impact of different packaging systems. A change in one part of the system may necessitate changes in other parts of the system. For example, if tuna is packaged in retort pouches rather than steel cans, the distribution package will also have to be modified. It may be that an increase in box strength will be required to make up for the decrease in stacking strength of the pouches compared with the cans (or perhaps the reduced weight will mean that less stacking strength is required). Certainly changes in filling and handling machinery are required when the package type and configuration are dramatically changed. All of these factors must be taken into account when package design changes are considered and implemented.

Similarly, as is discussed in other parts of this book, in determining whether one package system is more or less "green" than another, it is essential to consider the whole life cycle of the package system. This includes not only production and eventual disposal or reuse of the packaging materials, but also the various process steps associated with manufacture, transport, and use of the package system. Life cycle analysis is the tool that formalizes consideration of the full life of packaging systems.

One of the aspects that is very difficult to include in a typical life cycle analysis, but that is perhaps even more important for perishable products such as food than for other products, is the impact of the packaging system on product loss and waste. As mentioned, a significant proportion of foods never reaches human consumption. To the extent that investment in packaging systems can reduce this loss, the avoided impacts associated with production of "replacement" food may more than compensate for the added impacts associated with the packaging system. It should be remembered that this consideration of loss and waste should not be limited to losses during distribution. Losses that occur after the products are purchased by consumers are also important, and can be substantial.

A related consideration is that the storage and distribution requirements for products, including requirements at the consumer level, are often strongly influenced by the choice of packaging and processing system. Shelf-stable products do not require refrigeration or freezing. Therefore, they reduce requirements for temperature control at the manufacturer (once the food is packaged), during distribution, at the retailer, and in the home. However, such products typically also require enhanced barrier performance from the package. Therefore, the added impacts associated with packaging must be balanced against the reduced impacts associated with the enhanced stability of the product. (And, of course, the added processing requirements to achieve that stability must also enter into the calculation).

16.4 Green packaging and sustainability

The goal, then, behind "green" food and beverage packaging is to assist in minimizing the overall impacts associated with production and delivery of the food or beverage. As has been mentioned, the interaction between the packaging food system and the product is very important. Not only must the package fulfill its required functions, but also the way in which the package fulfills these functions may contribute toward reducing the environmental impacts associated with the product itself. With that said, the remainder of this chapter concentrates on package-specific considerations, as the product-related factors are covered in detail in other chapters.

Rather than talking about "green" packaging in a broad context, it is useful to focus on sustainability of packaging systems. It seems obvious that a packaging system that can "go on" indefinitely over the years has a lower environmental impact, at least in some sense, than a system that is not sustainable. In essence, the reason that a system is not sustainable is that it is "using up" natural resources, either by actually depleting resources required for the system, or by causing impacts such as pollution that either deteriorate a specific resource or that impact broader categories of resources, for example, affecting air quality.

Various organizations around the world have promulgated definitions of sustainable packaging. This section examines a sampling of them only, listed in order of increasing complexity, as measured by the number of characteristics listed. It discusses this further in Sect. 16.7, which looks at guidelines for green packaging design.

16.4.1 Wikipedia

Even Wikipedia now has a definition of sustainable packaging: "Sustainable packaging is the development and use of packaging which results in improved sustainability." The entry goes on to state "Sustainable packaging must meet the functional and economic needs of the present without compromising the ability of future generations to meet their own needs" (Wikipedia 2010).

16.4.2 Sustainable packaging alliance

The Sustainable Packaging Alliance (SPA) (2007) in Australia has identified four principles for sustainable packaging:

- Effective—provide social and economic benefits.
- Efficient—provide benefits by using materials, energy and water as efficiently as possible.
- Cyclic—be recoverable through industrial or natural systems.
- Safe—nonpolluting and nontoxic.

16.4.3 Environmental protection agency

The U.S. Environmental Protection Agency (2010) states “Packaging can be made more sustainable by applying the principles of product stewardship” and adds that this involves the following five considerations:

- Eliminating toxic constituents
- Using less material
- Making packaging more reusable
- Using more recycled content
- Making it more readily recyclable

16.4.4 Wal-Mart

Wal-Mart has instituted a Packaging Scorecard that is intended “to rate suppliers on the sustainability of their packaging.” The company states, “we have seen our suppliers making noticeable changes to their product packaging” with products being “packaged in more sustainable materials,” “transported using less fuel,” or “created using less energy and fewer natural resources” (Wal-Mart 2010). The scorecard, which was launched in the United States in 2008, is now expanding to Canada and Mexico, with implementation planned for the end of 2011. Wal-Mart’s definition of sustainable packaging focuses on what it refers to as the “7 R’s of Sustainable Packaging” (Wal-Mart 2008):

- Remove: Eliminate unnecessary packaging, boxes or layers, and harmful materials.
- Reduce: Right-size packages, optimize material strength, and design packages appropriately for contents and merchandising requirements.
- Reuse: Wal-Mart has a goal that all transport packaging will be reused or recycled by 2011 through improved pallets and reusable plastic containers (RPCs).
- Renew (able): Use materials made of renewable resources as measured using ASTM D6866, or select biodegradable materials that meet ASTM D6400 or ASTM D6868.
- Recycle (able): Use materials made of the highest recycled content without compromising quality, including post-consumer recycled material (PCR) where appropriate. Components should be chosen based on recycle-ability post-use, with a goal of increasing the municipal recycling rate to 35% by 2011.
- Revenue: Achieve all principles at cost parity or cost savings, which requires a supply chain approach.
- Read: Get educated on sustainability and how suppliers play a part.

16.4.5 *Sustainable packaging coalition*

Within the United States, the Sustainable Packaging Coalition (SPC 2010) is arguably the most influential in defining sustainable packaging and working toward its implementation. SPC defines eight characteristics of sustainable packaging:

- Is beneficial, safe, and healthy for individuals and communities throughout its life cycle.
- Meets market criteria for both performance and cost.
- Is sourced, manufactured, transported, and recycled using renewable energy.
- Optimizes the use of renewable or recycled source materials.
- Is manufactured using clean production technologies and best practices.
- Is made from materials healthy in all probable end of life scenarios.
- Is physically designed to optimize materials and energy.
- Is effectively recovered and utilized in biological and/or industrial closed loop cycles.

16.5 Tools for quantifying environmental impacts of packaging systems

Just as there are different definitions for sustainable packaging, there are a number of tools that are available to provide quantitative information about certain categories of environmental impacts, and thus certain aspects of sustainability. Various life cycle inventory databases and life cycle analysis systems are available, intended for use for both products and packages. Discussion of these systems is beyond the scope of this chapter and is addressed elsewhere. This section examines a few systems that are designed specifically for packaging. It is worthwhile to note, however, that of LCA systems (other than those belonging to and exclusively used by consulting organizations such as Franklin Associates), SimaPro (Pre Consultants 2010) appears to be the most widely used for evaluation of packaging systems.

16.5.1 *Packaging impact quick evaluation tool*

The Sustainable Packaging Alliance has developed the Packaging Impact Quick Evaluation Tool (PIQET). It is described as "a streamlined life cycle assessment (LCA) tool" whose purpose "is to provide the packaging supply chain a quick and credible environmental assessment tool which will assist them in making decisions on packaging development and innovation strategies" (SPA 2010). Release 2, which came out in 2009, contains datasets for Western Europe, New Zealand, the United States, China, and Japan. It is designed to be easy to use for businesses, does not require expertise in LCA, and is available as a web-based system by subscription.

The user is required to define the components of the packaging system, identifying the materials and the processes used; provide information on the supply chain; and define the end of life profile for these components. PIQET then generates a range of reports, including both graphs and tables that outline the environmental impact of the packaging system for indicators that include water, energy, and land use; solid waste generation; and global warming. The system methodology is based on packaging LCA models developed at RMIT University in Melbourne, using SimaPro and a variety of data sources. One of its common uses is "to identify and prioritise investment in full life cycle assessment studies."

16.5.2 Wal-Mart scorecard

The Wal-Mart packaging scorecard has already been mentioned. Wal-Mart requires its major U.S. suppliers to enter information about packaging into a database, which calculates scores for the packages entered. The idea is that buyers for the company can then use the packaging scores as part of the process of making a decision about which products to buy. Users enter information about the package system, including materials and amounts for both primary and secondary packages, cube utilization, transport distances, etc. This information is then combined with a database of metrics for various materials, using formulas to calculate scores for the package system. Separate scores are available for the "selling unit materials" and for the "transport materials," and these are combined into an overall score. Scores are calculated for package systems for greenhouse gas emissions from package production, sustainability, average transport distance, product to package ratio, cube utilization, recycled content, recovery, renewable energy and "innovation different from energy standard." The last category includes credits for ISO 14001 certification of manufacturing facilities, certification of renewable packaging materials by selected organizations, use of additional postconsumer recycled content, manufacturing innovations to decrease overall energy use, and manufacturing innovations to use a greater proportion of renewable energy. These scores are calculated based on information about the package system provided by the user plus metrics for individual packaging materials or types of packages included in the built-in database. These metrics are a high/low ranking of sustainability, average postconsumer content, recovery value, CO_2 emissions, nonrenewable energy use, and average number of uses. The system is designed to facilitate comparisons of various options for package systems. The objective is to achieve a low score.

16.5.3 Comparative packaging assessment

Comparative Packaging Assessment, or COMPASS, is a web-based LCA system for packaging offered by the Sustainable Packaging Coalition. It is designed as a "tool for packaging designers and engineers to assess the human and environmental impacts of up to four package designs side-by-side" (COMPASS 2008). It facilitates comparison of various packaging system options within projects, and allows creation of a database of user-specific components that can then be used for various packaging systems. As in the other systems, the user inputs information about the specific packaging systems, and the program then calculates metrics using a built-in database. Lifecycle metrics reported include fossil fuel consumption, greenhouse gas emissions, water consumption, human impacts, biotic resource consumption, aquatic toxicity, mineral consumption, and eutrophication. In addition, a set of packaging attributes is reported that includes measures of virgin/recycled content, percent certified/noncertified sourcing, and percent disposal/recovery for solid waste. Finally, there is also a tally of the number of times carcinogens; reproductive toxicants; and persistent, bioaccumulative and toxic substances are used in the final stage of manufacture of the materials in the packaging system. One of the unique features of COMPASS is a database of packaging components that are available to all users. These can be incorporated as is into packaging systems being assembled, or can be copied and modified to meet specific characteristics. For example, available components at the time of this writing include compact disc (CD) snap trays, both clear and black, polypropylene (PP) labels, pump assemblies, spring assemblies, as well as a number of others.

At the time of this writing, version 2 of COMPASS is scheduled to be released within a few weeks, with a number of enhanced features, including the ability to compare reusable with single-use packaging systems.

16.6 End-of-life considerations

Since the mid-1980s, municipal solid waste and litter have often been the centers of publicly expressed concern about the environmental impacts of packaging. The source of packaging materials is also a frequent concern—especially as it relates to their renewable or nonrenewable nature.

16.6.1 Solid waste

In the mid-1980s, beginning when the garbage barge Mobro sailed down the east coast of the United States, to the Caribbean, and back again, trying to find a place that would accept the load of New York City garbage it carried, the common public perception in the United States was that most municipal solid waste was packaging material. This perception was fueled by what ordinary people saw themselves putting in their own trash cans each day. Plastics were a special focus, as they were perceived as "filling up" landfills with non-degradable trash that would stay around for centuries. This perception led to a number of attempts to legislate packaging reduction of various types. Some of these efforts were successful, but most were not.

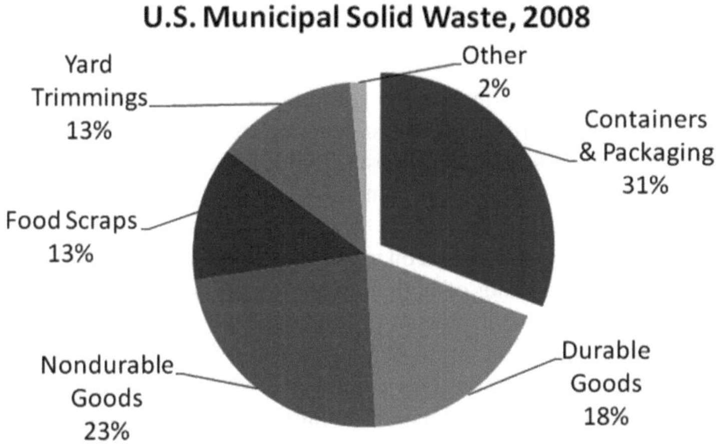

Fig. 16.1 U.S. Municipal Solid Waste, 2008 (EPA 2009)

Similar concerns in many other parts of the world led to implementation of various types of producer responsibility systems, to make the product manufacturers responsible for the ultimate fate of the packaging materials they chose to use. These are generally coupled with specific requirements for recovery and recycling of specified percentages of materials.

Packaging is a significant component of municipal solid waste (MSW), which can be loosely defined as the "ordinary" waste generated by individuals and businesses. In many parts of the world, dealing with municipal solid waste is problematic for a variety of reasons, including general unsightliness, sanitation concerns, odor, attraction of vermin, air and water pollution, etc.

International comparisons of the amount of MSW generated are often difficult because the precise definitions vary from one country to another. In fact, even within the United States, definitions vary in different states. However, the most generally accepted definition is that promulgated by the EPA under which all packaging materials entering the waste stream are included, regardless of where they originate (EPA 2009). With this definition, nearly one third of MSW generated in the United States is packaging materials (Fig. 16.1), a proportion that has changed little over the last several decades, although total amounts have risen substantially during this period. In other industrialized countries, packaging is also a major component of the MSW stream. Because of the strong association between lack of proper packaging and food waste, one might expect some degree of inverse correlation between the amount of packaging entering the waste stream and the amount of food waste.

Options for dealing with municipal waste are in essence limited to land deposition, combustion, and recovery. The first two of these can be combined under the term "disposal." It should be noted, however, that use of these terms is not always consistent. For example, in the European Union, recovery generally includes waste to energy systems, whereas in the United States it does not, and instead is limited to recycling and composting.

16.6.2 Disposal

Land deposition varies from uncontrolled dumping (including littering) all the way to high-tech sanitary landfills equipped to generate energy from the methane generated as the waste anaerobically decomposes. In the United States and other industrialized countries, "open" dumps are generally illegal. Regulations in the United States, for example, require that in any operating landfill, a daily cover meeting certain requirements must be applied on top of the deposited waste each day. Further, barrier layers, often consisting of both clay and a polymer membrane, must be in place under the waste to help prevent groundwater pollution. Completed landfill cells have a similar barrier on top to help prevent ingress of water. The whole site must be maintained and monitored for a number of years after it is closed. A properly designed, operated, and maintained landfill is designed to sequester wastes indefinitely. What it is not designed to do is to enable the occupied space to become available for future use, with the exception of limited use of the space above the landfill, for example, as a park or ski hill.

Combustion does not eliminate the need for land deposition of waste, but can cut down very substantially on the volume of waste requiring land disposal. At the time when open dumps were common in the United States, it was also common to periodically set fire to the waste, thus cutting down on problems such as odor and vermin, as well as creating more space for the next load of garbage. Just as open dumping is no longer permitted in much of the developed world, uncontrolled combustion has also disappeared in many places, owing primarily to air pollution issues. Modern incineration not only burns material much more cleanly, employing emission control systems similar to those used in power plants, but also is designed to recover energy from the combustion. A waste-to-energy plant typically transfers the heat from the combustion to steam, which drives electrical generators. Thus, an MSW waste-to-energy (WtE) plant reduces the need for combustion of fossil fuels such as coal. There are also systems to process MSW to remove noncombustibles and even to compact the resulting refuse-derived-fuel (RDF) into a form that allows it to be co-fired with coal in municipal power plants. However, most MSW incineration facilities are of the mass-burn design where little pre-processing of the MSW is done; rather the collected waste, combustibles and noncombustibles alike, is fed into the combustion chamber. The output of an MSW incineration facility, in addition to energy, is ash from the combustion chamber (bottom ash) and ash from the pollution control equipment (flyash). Although some systems for use of ash have been developed, for the most part the bottom ash and flyash are combined and disposed of in dedicated landfills.

16.6.3 Composting and biodegradability

Composting systems are designed to promote biodegradation of materials that can be consumed (assimilated) by microorganisms, yielding a combination of water, carbon dioxide, and a humus-type material. Some refer to it as "nature's recycling" because it returns the carbon in carbon-based materials to the earth's natural carbon cycles. Although in some locations, organized collection of organic materials for composting is common, in the United States it is quite rare except for specialized waste streams such as yard trimmings. There are international as well as national standards for determining whether a material is compostable in commercial facilities. There are as yet no generally accepted standards for determining whether or not materials are compatible with individual composting facilities, generally referred to as backyard composting. One of the issues is the tremendous variability in the operation of such systems. A similar issue arises with the term "biodegradable." In essence, a material is biodegradable if it can be consumed—serve as a food source—for biological organisms, especially microorganisms. The difficulty with operationalizing this decision is that the extent and rate of biodegradation are heavily dependent on conditions such as temperature, availability of water, and the population of organisms. Also, there is no general consensus on what rate of degradation is sufficient to categorize the material as biodegradable. For example, most plastics can eventually be consumed by microorganism, but only in time frames of decades or even centuries. Most experts agree that if a particular plastic article will take centuries to biodegrade, categorizing it as a biodegradable material is misleading at best. However, how long for total breakdown is too long? What if there is mechanical breakdown within a reasonably short period, but the molecules persist in the soil for decades or more? These are some of the unresolved issues, and the reason that many authorities avoid the use of the term biodegradable.

For the most part, paper-based packaging materials are compostable. However, packaging components such as coatings, added layers of plastic or aluminum, etc., can significantly interfere with the biodegradability of such materials, rendering them not compostable. Metal and glass packaging are not biodegradable at all. They can break or corrode, and can lose their visible identity, but they cannot be consumed by biological organisms. Most plastics are not compostable, and have exceedingly slow rates of biodegradation. There are a few exceptions, such as biobased polylactides (PLA), polyhydroxyalkanoates (PHBV), and starch-based plastics. Certain synthetic polymers, for example polyvinyl alcohol and polybutylene adipate co-terephthalate (PBAT), are also compostable. In addition, cellophane is a transparent film but not a plastic, and is compostable unless coated with materials that make it incompatible with composting systems. For some of these materials, the mechanism depends on hydrolysis at compost temperatures to reduce the molecular weight to a point that enzyme systems from microorganisms can effectively attack the remaining molecular bonds.

Biodegradability of packaging materials is an asset if it allows materials to be diverted from disposal (landfill or incineration) to a compost facility. It is not an advantage if materials are destined for incineration, because carbon-based structures are generally combustible regardless of whether or not they are compostable. Similarly, although perhaps less obviously, biodegradability is not an asset in a modern sanitary landfills. The dry anaerobic environment in a modern sanitary landfill is not conducive to microbial activity. It is well established that biodegradation of even such readily assimilable materials as food and yard waste is typically extremely slow. Investigators have dug up carrots after 10 years in a landfill and found them still intact, and found chicken meat still clinging to the bone. It is certainly unrealistic to expect that food packaging materials would degrade more quickly in a landfill than the foods they contain.

16.6.4 Litter

There is one area in which biodegradability of packaging materials, even if not on a cycle or to an extent that is compatible with composting, can be a significant advantage—litter. When wastes are simply dispersed into the environment rather than entering a proper disposal system, they can cause a variety of problems, from simply being unsightly to clogging sewer systems to actually killing animals. Littered packaging materials can entrap birds or animals. When the physical form of the package is the main issue, materials that degrade on exposure to water, microorganisms, or simply sunlight can decrease the strength and embrittle the materials, so that they are no longer entrapment hazards. For example, the rings that are used to connect beverage cans are required in the United States to be photodegradable. On exposure to ultraviolet (UV) light, such as from sunlight, the plastic gets brittle and loses its tensile strength. Then mechanical stress can readily cause the material to break into small pieces that cannot trap birds or other animals. Such changes do not require the total breakdown of the molecular structure of the material to be effective. However, consideration must be given to whether the resulting fragments pose their own set of environmental hazards.

A litter issue that is likely even more serious than litter on land is the effects of packaging materials, especially plastics, that get into bodies of water, either through "direct" littering into bodies of water or through runoff from land, overflows from sewer systems, etc. In such environments, floating packaging materials can be ingested by marine organisms that mistake them for food. Many people have probably heard of the issue of endangered sea turtles mistaking floating plastic bags for jellyfish, for example. These animals are sometimes not able to eliminate the plastics that they ingest. The build-up of plastics in the digestive system can eventually result in death. Plastics are a particular problem because they often do not biodegrade, and in many cases they float rather than sinking to the bottom. Metals and glass, although also not biodegradable, generally sink and become covered by sediment, so they are not as much of a problem. Paper-based materials eventually biodegrade, and if ingested they are more readily cleared by the digestive system. One solution is to change human behavior to prevent these materials from entering water systems in the first place. There has been some success in cutting down such inputs into water through regulations, such as those that prohibit discharge of garbage from boats. However, the problem is certainly not yet solved. Therefore, for types of packages that are often associated with litter, especially litter that enters water systems, using biodegradable materials can provide real environmental benefits. One caution: Materials that are compostable do not always biodegrade quickly in bodies of water.

16.6.5 Recycling

Recycling refers to converting materials that would otherwise enter the waste stream into raw materials for manufacture of new products or packages. It is differentiated from reuse, which is discussed later, because the materials are not directly used again, but rather are remanufactured. The recycling rate for packaging materials varies considerably by material category, and also varies for different types of packages within a material category. There are also, of course, large differences in recycling rates among countries. The information presented here will, for the most part, be limited to the United States, in which a primary source of information is the annual reports on the U.S. municipal waste stream that are put out by the EPA (2009).

16.6.5.1 Collection of recyclables

Before examining recycling rates for selected packaging materials, it is useful to take a general look at the structure of recycling systems. Over the years, a number of people have tried to develop systems to take MSW and "harvest" recyclable materials from the mixed waste. Such systems would have the considerable advantage of not requiring the cooperation of individual people to divert the target materials from the waste they generate, and therefore could potentially achieve very high recovery rates. However, for the most part these efforts have been unsuccessful. Either they have not been economical, or the quality of the recovered materials has been unacceptable, primarily owing to high levels of contaminants. Therefore, attention has turned to how to achieve high levels of participation in recycling, measured both by the proportion of desired materials recovered, and by the proportion of the population that engages in recycling behavior. Two essential factors in getting people to engage in desired recycling-related behaviors are motivation and convenience.

As for many other human behaviors, if an individual has high motivation to participate in recycling, they are likely to do so. A variety of systems to increase motivation for recycling can be employed to increase recycling behavior. Motivation can come from a variety of sources. Monetary motivation in the form of deposits is largely responsible for the significant differences in recovery rates for carbonated beverage containers between deposit and non-deposit states in the United States. In many communities, there is a per-container fee for garbage collection (pay-as-you-throw) while recyclables are collected at no cost, which is another way of providing a monetary incentive for recycling. Businesses often have an economic motivation for recycling, as they must directly pay the costs of waste disposal, while in some cases they can actually sell separated recyclables.

Motivation can also be internal; many people are motivated to "do the right thing" so if they see recycling as environmentally and socially beneficial, they are motivated to participate. And, peer pressure is not limited to children—adults, too, can be motivated to recycle by not wanting their friends and neighbors to think less of them.

Recycling can also be increased by making it convenient. If something is easy to do, people are more likely to do it. If there is a recycling bin located right by the garbage container in the hall, it does not take as high a level of motivation to get people to put the aluminum can into the recycling bin as would be needed if instead they had to take the can outside to a special recycling area. Most major communities in the United States (and many in other places around the world) provide curbside collection of recyclables, primarily because this makes household recycling much more convenient than systems that require people to deliver the recyclables somewhere. The majority of recycling collection systems require relatively little preprocessing of the recyclable materials. Most require that materials be reasonably clean, but if requirements are too high, convenience decreases and people more often choose to put the recyclable into the trash rather than go to the effort to get it "ready" for recycling. For example, in the early days of plastic milk bottle recycling, participants in Grand Rapids, Michigan, were asked not only to clean the bottles, remove the caps, and deliver them to drop-off locations, but they were also instructed to remove the paper labels from the bottles! Needless to say, few people complied.

The relationship between motivation and convenience is obvious—the more convenient the action, the less motivation is required; the more inconvenient, the stronger the motivation must be. Another important determinant of recycling behavior is communication and education. If you want people to participate in recycling, they need to know what to recycle, and how and where to do so. They also need to be informed, on an ongoing basis, of the benefits that ensue from recycling. Many communities have found that the local school systems are an important vehicle for disseminating this information. One of the hopes is that if the habit of recycling is built early, it will persist throughout life.

Of course, these factors must be balanced against the cost of recycling. Often, systems that make recycling easier for people are also more costly. A community can set up a drop-off site with a bunch of bins for different materials for much less cost than it can implement a curbside collection program. Education and other forms of communication are not free. Providing a rolling container for recyclables to be placed in makes it easy to store them and move them out to the curb for collection, but the containers are certainly not free. Collecting recyclables in commingled form (mixed together) increases convenience, but adds to processing costs.

16.6.5.2 Packaging recycling rates

Recycling rates for materials of various types differ substantially. Rates also differ within the same material category for different types of products. In general recycling rates for packaging compare favorably with recycling rates for many other categories of MSW, as shown in Fig. 16.2. Using the standard EPA categories, only yard trimmings have a higher recovery rate than containers and packaging. Nearly all the recovery of yard trimmings is through composting. There is very little composting of packaging materials in the United States, so the recovery rate for packaging essentially equals the recycling rate. The exception is wood packaging, in which options such as composting are important. The overall recovery of wood packaging in 2008 was 1.6 million tons, for a recovery rate of 14.8% (US EPA 2009). The primary

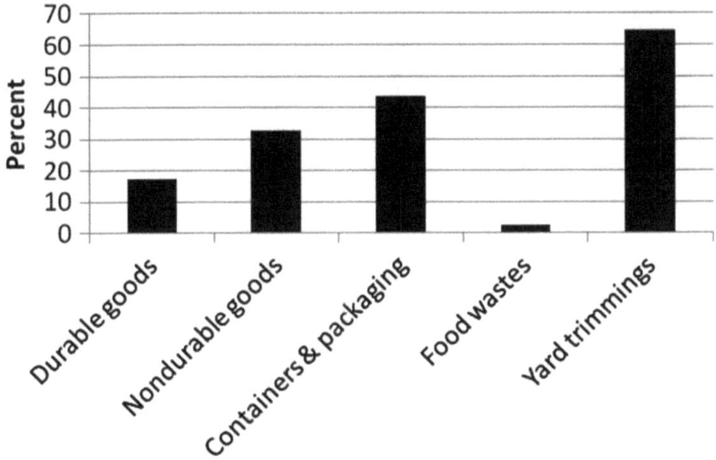

Fig. 16.2 Recycling rates for U.S. MSW, 2008 (EPA 2009)

source of this material was wood pallets. Recovery includes use of recovered material in paper making, as mulch, animal bedding, and other applications.

16.6.5.3 Recycling of paper-based packaging

Paper-based packaging materials make up the majority of all packaging as well as the majority of recycled packaging (Fig. 16.3). In the paper and paperboard packaging category, the majority of packaging produced and a sizable majority of that recycled is corrugated boxes. In 2008, corrugated boxes accounted for 29.7 million tons of the 38.3 million tons of paper-based packaging waste generated, and for 22.8 million tons of the 25.1 million tons recycled, resulting in a recycling rate of 76.6%. The overall recycling rate for paper and paperboard packaging was 65.5%. The rate for folding cartons was 35.2%, for bags and sacks was 37.9%, and for other categories of packaging was listed as negligible (US EPA 2009).

One issue in recycling paper is that paper properties depend on the fiber structure, which is damaged somewhat with each pass through the paper manufacturing system. Therefore, there are limitations on the number of times a paper fiber can be used in making new paper, and paper properties are affected somewhat by the proportion of recycled fiber they contain. For direct food contact applications, potential migration of contaminants is also a concern. Therefore, use of recycled fiber in direct food contact is limited. However, in many applications the paper is not in direct contact with the food, so migration is of little or no concern.

The main reason for the high recycling rate for corrugated boxes is that the primary generator of OCC (old corrugated cartons or old corrugated containers) is business and industry, rather than individual consumers. For example, retailers often generate large quantities of corrugated boxes because their merchandise arrives in

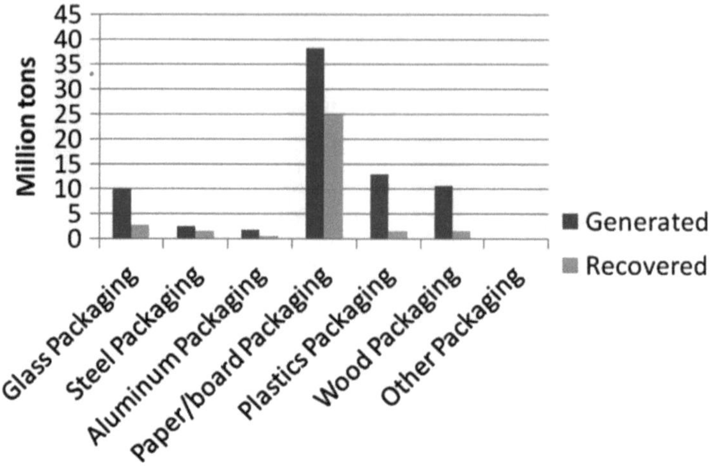

Fig. 16.3 Generation and recovery of packaging materials in the United States (EPA 2009)

the boxes but leaves in bags. These businesses, as mentioned, must pay for disposal of their wastes, typically by volume. If, rather than placing the boxes in the dumpster, they separate the boxes and bale them, usually they can find a buyer for the baled corrugated. The value of the material plus the avoided disposal cost usually easily covers the cost of the baler and the added labor required.

16.6.5.4 Recycling of glass packaging

In terms of mass, the next most important recycled packaging material in 2008 was glass, at 2.8 million tons recycled. Most of this, 2.3 million tons, was beer and soft drink bottles. Generation of glass packaging in 2008 totaled 10.1 million tons, of which 6.4 million tons was beer and soft drink bottles. The recycling rate for glass packaging was 28.0%. Beer and soft drink bottles were recycled at a rate of 35.6%, whereas the rate for wine and liquor bottles was 14.9% and for food and other bottles and jars was 14.8% (US EPA 2009).

In contrast to paper-based materials, the properties of glass are not affected by previous uses. In principle, glass can be recycled indefinitely. Of course, nonglass contaminants must be removed, and to make new container glass, the recycled glass must also be container glass rather than glass of some differing chemical composition. The majority of glass container recycling is closed loop, in which old glass containers are recycled into new glass containers.

Contamination with colored glass is also a concern, especially in manufacture of flint (colorless) glass. Although especially in the past, glassmakers have expressed concern about their ability to use high concentrations of cullet (recycled glass), in practice, use of recycled content in glass containers is influenced more by price and availability of cullet than by issues related to quality of the finished product.

The primary source of recycled glass is deposit systems and curbside and drop-off collection programs. In a number of communities, green glass is not accepted in collection programs, as there are no markets for green cullet within an acceptable transport distance. Cullet is heavy and therefore expensive to ship, and competes with relatively low-value mineral resources such as sand and limestone. Research has been directed to increasing markets for green and mixed-color cullet. Although some markets exist, such as aggregate in construction materials and asphalt, use in decorative tiles, and sandblasting, economical recycling of green glass and mixed-color cullet remains problematic.

16.6.5.5 Recycling of steel packaging

Steel packaging is next in terms of tonnage of recycled material, at 1.6 million tons recycled in 2008 out of an amount generated of 2.5 million tons. Most of this was food and other steel cans, 2.3 million tons of the total generated and 1.5 million tons of the total recycled, for a recycling rate of 63.1% overall for steel packaging, and 62.8% for food and other cans (US EPA 2009).

Like glass, steel packaging can be recycled indefinitely without loss of performance properties. However, unlike glass recycling, most recycling of packaging steel is open loop. A recycled steel can may become part of a new steel can, or it may end up in a variety of other steel products. This is related to the steel production process. The two main processes for producing steel are the basic oxygen furnace and the electric arc furnace. The basic oxygen furnace is limited to about 25% scrap steel, whereas the electric arc furnace requires essentially 100% scrap steel. Recycled steel is categorized in a variety of grades, and typically is directed to the nearest facility, where various grades are combined to achieve the desired set of properties. Because there are only a handful of basic oxygen furnaces and many electric arc furnaces, most recycled steel goes to the latter. Can sheet is produced exclusively in basic oxygen furnaces. Therefore, the recycled content of steel cans is limited to about 25%. Increases in the recycling rate for steel cans would not be expected to result in greater recycled content in the cans, owing to process limitations. However, recycling cans is still effective in reducing the need to draw on natural resources to produce new steel. Thus steel recycling is a prime example of how closed loop recycling is not necessarily superior to open loop recycling. It would not be environmentally beneficial to truck collected cans long distances to a basic oxygen furnace, simply to allow the recycled content in cans to be from old cans rather than from refrigerators, automobiles, or any of the many other sources of recycled steel.

Collection of steel for recycling also differs somewhat from collection of other materials. Although steel cans are collected in most curbside and drop-off collection programs, these alone cannot account for the relatively high recycling rate for steel packaging. The explanation is that steel is an exception to the general rule that separation of recyclables from mixed waste is not viable. Because steel is magnetic, electromagnet systems are able to pull steel out of waste streams and enable it to be recycled. In fact, steel can even be recovered from incinerator ash after the combustion process. Many landfills and even a higher proportion of MSW incinerators are set up to recover steel for recycling. Therefore, steel recycling is less dependent on individual participation than is recycling of most other packaging materials.

16.6.5.6 Recycling of aluminum packaging

In 2008, 720,000 t (0.72 million tons) of aluminum packaging were recycled in the United States, out of total generation of 1.88 million tons. Beer and soft drink cans accounted for 1.39 million tons of generation, and 670,000 t of the recycled aluminum packaging. Recovery of food and other cans totaled 10,000 t, and 40,000 t of foil and closures were recycled. This amounts to an overall recycling rate of 38.3% for aluminum packaging, a rate of 48.2% for beer and soft drink cans, 14.3% for other cans, and 9.5% for foil and closures (US EPA 2009). Collection of aluminum for recycling is through a combination of deposit, curbside, and drop-off programs.

As for steel and glass, recycling does not decrease the performance of the recycled metal. Like glass but unlike steel, most recycling of aluminum cans is closed loop. A recycled beverage can will probably be back on the shelf as a beverage can within just a few weeks.

A characteristic of recycled glass and metal that has not yet been discussed is the issue of organic contaminants. The very high temperatures involved in the melting and reforming processes ensure that any organic contaminants, whether toxic or not, will be destroyed. This includes not only microorganisms, but also organic chemicals. Therefore, potential toxicity associated with the use of recycled content is limited to nonorganics such as heavy metals. Proper control over the source of the recycled feedstocks can, therefore, allow these materials to be as safe for use in direct contact with food as packaging made from virgin materials.

16.6.5.7 Recycling of plastics packaging

Recycling of plastics packaging in the United States in 2008 totaled 1.7 million tons, out of generation of 13.0 million tons, for an overall recycling rate of 13.3% (US EPA 2009). As can be seen, this is the lowest rate for the major categories of packaging materials. Recycling rates for some types of plastics were significantly higher than for others, as shown in Table 16.1. In general, recycling rates for bottles and jars are higher than those for other plastics packaging. Collection is through a combination of deposit programs, curbside collection, and drop-off systems. Recycling of low density polyethylene includes pallet stretch wrap collected from businesses in a similar manner to corrugated. Collection of plastic film—mostly HDPE and L/LDPE merchandise sacks—also occurs through a network of drop-off bins located at various retailers. Flexible plastics such as film and bags are rarely included in curbside recycling programs.

The performance of plastics, like that of paper, is generally affected somewhat by its previous uses. Although for paper the cause of the changed performance was

Table 16.1 Recycling of packaging plastics in the United States, 2008 (US EPA 2009)

Resin	Generation (million tons)	Recycling (million tons)	Recycling rate (%)
High density polyethylene (HDPE)	3.89	0.57	14.7
Low and linear low density polyethylene (LDPE and LLDPE)	2.84	0.33	11.6
Polyethylene terephthalate (PET)	2.89	0.73	25.3
Polypropylene (PP)	1.98	0.07	3.5
Polystyrene (PS)	0.36	0.02	5.6
Polyvinyl chloride (PVC)	0.37	—	—
Other resins	0.68	0.01	1.5

mostly changes in the physical structure of the fiber, for plastics the change is mostly owing to changes in the molecular structure of the polymers. Just as fibers are subject to damage and breakage, polymer molecules are subject to chemical changes, typically involving some combination of chain scission, cross-linking, oxidation, and hydrolysis. These changes in chemical structure, including in molecular weight distribution, can affect performance. In addition, again similar to the case with paper, migration of contaminants to food is a concern. Approaches that have been deemed by the U.S. FDA to be acceptable for use of recycled plastic in food packaging include arrangements in which the recycled plastic is separated from the food by a "barrier" layer of virgin plastic; systems in which the plastic is broken down to monomers, purified, and repolymerized; and systems employing mechanical reprocessing that have been proved to be capable of removing contaminants to an acceptable level. Some of these are approved for all types of food packaging; others have significant use limitations. A related issue is that because in the recycling process food-grade and non–food-grade plastics might be mixed, substances not approved for food contact might be found in the recycled plastic. A complete discussion of these issues is beyond the scope of this chapter, but a good starting point is FDA's web page on "Recycled Plastics in Food Packaging" (US FDA 2009).

16.7 Sourcing of materials

Although end-of-life considerations are often a public concern when it comes to the environmental impacts of packaging materials, life cycle analysis studies typically show that it is the manufacture of the materials, not their disposal, which has the greatest impacts. Therefore, consideration of the source and amount of packaging materials is a major consideration in design of "green" food packaging systems.

One point of consideration is whether the materials being used are renewable or nonrenewable. For packaging materials, wood and paper are made from renewable materials, whereas glass and metal are nonrenewable. Plastics are, for the most part, made from petrochemicals so are nonrenewable. However, there has been increasing interest in bio-based plastics. It is important to differentiate between plastics that are bio-based—are made from biological materials—and those that are biodegradable, as these terms are not synonymous. Although some plastics, such as PLA, are both bio-based and biodegradable, others are not. Polyvinyl alcohol (PVOH), for example, is made from petrochemicals, but is biodegradable. On the other hand, Braskem in Brazil is making polyethylene from sugar cane, so the PE is bio-based, but in performance it is identical to petro-based polyethylene, and so it is not biodegradable.

Long term sustainability of renewable resources requires, of course, that the resource base be managed appropriately. For wood and materials derived from wood, organizations exist that certify the forest management operation. One of the most respected of these is the Forest Stewardship Council (FSC 2010), which operates internationally. A number of other certification systems also exist, some with government backing, some organized by industry, and some independent.

For nonrenewable resources, recycling is a key to maintaining the long-term availability of the feedstock materials. Choosing packaging materials that are recycled at an acceptable rate and using packaging materials with post-consumer recycled (PCR) content are both important in enhancing the sustainability of the packaging system.

It is important to remember that it is not only the material resources that are involved in manufacture of packaging systems. Energy is always a requirement, and water is virtually always used at least somewhere in the process. Therefore, the source of energy should also be a consideration. Is the energy from a renewable source, or is it fossil-fuel-based? Is water reused internally in closed-loop systems, or is it discharged? If it is discharged, is this done in a manner that minimizes potential harm to ecosystems? All these considerations are also part of "greening" packaging design.

Finally, but certainly not least important, source reduction and reuse must be considered. Two types of considerations are typically included in the concept of source reduction. The first is "using less"—downgauging films, lightweighting containers, etc. If less material is used to deliver the same function, with everything else remaining the same, then the savings are obvious. Further, these savings percolate throughout the supply chain, as less material needs to be stored and transported, as well as manufactured. Most segments of the packaging industry have made strides over the past few decades in reducing the amount of material required for a given package. Packages from glass bottles to plastic grocery sacks deliver the same (or at least acceptable) performance with less mass. Of course, not all packaging reduction involves simply using less of the same material. One of the major changes in packaging over the last couple of decades is the ongoing switch from rigid packaging, typically glass and metal, to flexible packaging, typically plastic or combinations of paper and plastic. In these cases, the evaluation of overall environmental costs and benefits becomes much more complex, because the replacement materials are often better in some impact categories and worse in others. Overall, however, both the environmental and economic impacts are often positive—especially when the whole product/package system, including distribution, is taken into account.

Reuse systems have also enjoyed some resurgence. Although reusable packages typically use more materials and energy in their manufacture and distribution than one-way packaging systems, under appropriate circumstances they can provide substantial savings both economically and environmentally. Two of the main requirements for this to happen are that the packages must remain in the system—they must not be "lost" at an unacceptable rate—and the reverse distribution system to return the packages for reuse must function efficiently. If transport distances are very long and many "middlemen" are involved in the distribution chain, the system is likely to fail. For this reason, most successful reusable packaging systems involve distribution packaging—transport packaging—rather than retail containers.

16.8 Guidelines for green packaging design and selected examples

The following general guidelines are presented as guidance for the development of "green" food packaging systems. Although not all encompassing, they are designed to provide a starting point.

First, the packaging must perform its intended functions. It is especially critical that the package provide the required containment and protection for the product. However, packages that do not meet other supply chain needs, including being reasonably attractive to the target market, will also result in wasted food in addition to wasted packaging. In many (probably most) cases, impacts associated with food production significantly outweigh impacts associated with packaging systems.

Second, minimize the amount of packaging material used, to the extent compatible with maintaining needed performance. Some caution is needed when using less material because it may also involve change in the type of material, but usually less material translates into reduced environmental impact. Be sure to consider the entire packaging system, not just the primary package. Pay special attention to minimizing potentially toxic components from packaging materials, even if they do not pose any migration hazards.

Third, when feasible, use materials with the maximum viable amount of post-consumer recycled content, and design package systems to facilitate recycling. When multiple materials are used in a single package, they should, if possible, be either compatible with each other when recycled or be readily separable from each other. Especially for distribution packaging, consider opportunities for reusable packaging systems where efficient closed loop systems can be designed to recover the packages.

Use life cycle assessment methodology to evaluate the tradeoffs between differing packaging systems. Because LCA always requires drawing of boundaries and making assumptions, be sure these are designed to present fair comparisons among the alternatives being considered. Recognize that there will likely be tradeoffs among alternatives, with each being better in certain indicators and worse in others, so value judgments will be required.

Finally, be sure that any "green" claims made about products or packaging are both factual and verifiable. In general, claims should be specific. It should be clear whether the claims are for the product or for the package. Third party certification and eco-labels can be useful, as long as the certifying organizations themselves have strong standards. Green marketing is of lasting value only when it does not become "greenwashing." Informed consumers want accurate and dependable information about the environmental impact of the products and packages they may purchase.

References

COMPASS. 2008. https://design-compass.org/about.gsp. Accessed 17 June 2011.

Forest Stewardship Council. 2010. http://www.fsc.org/. Accessed 21 June 2010.

Kader, A.A. 2005. Increasing food availability by reducing postharvest losses of fresh produce. *Acta Horticulturae* 682: 2169–2176.

Pre Consultants, SimaPro. http://www.pre.nl/simapro/. Accessed 14 June 2010.

Sustainable Packaging Alliance. 2007. Defining sustainable packaging. http://www.sustainablepack.org/research/subpage.aspx?PageID=10&id=7. Accessed 11 June 2010.

Sustainable Packaging Alliance. 2010. http://www.sustainablepack.org. Accessed 11 June 2010.

Sustainable Packaging Coalition. 2010. http://www.sustainablepackaging.org/. Accessed 11 June 2010.

United States. Environmental Protection Agency. 2009. Municipal solid waste generation, recycling, and disposal in the United States: Facts and Figures for 2008, http://www.epa.gov/waste/nonhaz/municipal/pubs/msw2008rpt.pdf. Accessed 17 June 2010.

United States. Environmental Protection Agency. 2010. Packaging. http://www.epa.gov/osw/partnerships/stewardship/products/packaging.htm. Accessed 11 June 2010.

United States. Food and Drug Administration. 2009. Recycled plastics in food packaging, http://www.fda.gov/food/foodingredientspackaging/foodcontactsubstancesfcs/ucm093435.htm. Accessed 21 June 2010.

Wal-Mart. 2008 Packaging factsheet. Available through link at http://walmartstores.com/pressroom/FactSheets/. Accessed 11 June 2010.

Wal-Mart. 2010. Global sustainability report, 2010 progress update. http://walmartstores.com/sites/sustainabilityreport/2010/environment_waste.aspx. Accessed 11 June 2010.

Wikipedia. 2010. Sustainable packaging. http://en.wikipedia.org/wiki/Sustainable_packaging. Accessed 11 June 2010

Part V
Environmentally friendly approaches to R&D and QA

Chapter 17
Microtechnology and nanotechnology in food science

Filomena Nazzaro, Florinda Fratianni, and Raffaele Coppola

17.1 Introduction

For many years, microtechnology and nanotechnology have taken a fundamental role in different fields of social life, including electronics, communications, energy production, computer, cosmetics, medicine, and health. More recently, they have also revolutionized agriculture and food, not only in areas concerning food quality and safety, but also in food packaging.

The terms "nanotechnology" and "microtechnology" were originally introduced to indicate new technologies that went beyond controlling materials and engineering on the micrometer scale.

Today, the terms correspond to molecule and atom-by-atom manipulative hard-tech processing methodology. The definitions of "microtechnology" and "nanotechnology" are based on the Greek prefixes *micro* and *nano,* respectively. In the case of microtechnology, the information is related to a special scale known as micro (10^{-6} m = 1 μm). Nanotechnology, on the other hand, refers to the scale known as nano (10^{-9} m = 1 nm). Therefore, microtechnologies and nanotechnologies focus on design, characterization, production, and application of microscale and nanoscale systems and components, which can also be food or health components.

Their relationship with other basic or innovative technologies is also leading to more advanced innovations, expected to make a stronger impact along the entire food chain, from production through processing, storage, and transportation, to improving the traceability, tracking, and safety of foods (Fig. 17.1).

F. Nazzaro (✉) • F. Fratianni • R. Coppola
Istituto di Scienze dell'Alimentazione, Avellino, Italy
e-mail: mena@isa.cnr.it; Fratianni@isa.cnr.it; direttore@isa.cnr.it

J.I. Boye and Y. Arcand (eds.), *Green Technologies in Food Production and Processing*, Food Engineering Series, DOI 10.1007/978-1-4614-1587-9_17,

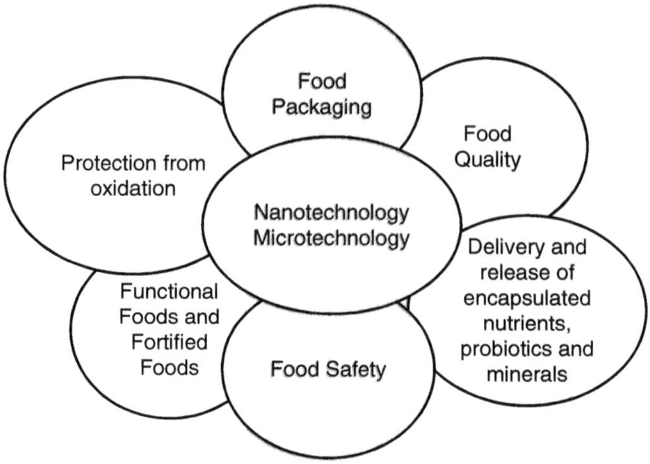

Fig. 17.1 Schematic of the principal application areas of nanotechnology and microtechnology in food science

For example, the integration of nanotechnology with biotechnology allowed the development of nano-biosensors, which are useful for the detection of pathogens and contaminants in food. The manipulation of substances closer to their molecular level opened up ways to develop and optimize new food textures, tastes and sensations, which all form part of food quality.

17.2 Nanotechnology

Nanotechnology can be considered as the manipulation or self-assembly of individual atoms, molecules, or molecular clusters, to form devices or materials with different properties.

The application of nanotechnology in the food field, although emergent, is projected to see dramatic growth. Nanotechnology opens up new markets and scientific opportunities and is considered as one of the essential technologies of the twenty-first century. The importance of this sector is so great that many of the world's biggest food companies are exploring its potential use in food quality, safety, and packaging (Cientifica 2006).

For food applications, nanotechnology may be generally used in two different ways, the so-called bottom-up or top-down approaches. Top-down implies the reduction of size of the smallest structures. In food science, for example, such an approach can be realized by means of physical processes that involve the food matrix, that is, grinding or milling. This is similar to the approach usually applied to wheat flour with the aim of obtaining a finer size with high water-binding capacity (dry-milling technology) (Yaghmur and Glatter 2009), or improving the antioxidant activity in green tea powder, by enhancing dismutase activity and providing a higher range of active oxygen-eliminating potency (Shibata 2002). In this last case, the reduction of size to 0.5–1 μ by using dry-milling technology permits a higher ratio of nutrient polyphenols digestion and subsequent absorption.

On the contrary, the bottom-up technology moves from the biological concepts of self-assembly and self-organization. This concept involves the manipulation of individual atoms or molecules to form nanostructures. A typical example of the bottom-up concept is given by the organization of casein micelles or starch, as well as by the folding of globular protein, which create more stable biological entities (Morris and E. Dickinson 2003).

Because of a more widespread surface area of nanoparticles/mass unit compared with larger-sized particles, nanoparticles are more biologically active given the same chemical composition.

This becomes of particular importance in the case of functional foods, in which nanoparticles have demonstrated greater activity, with better global effectiveness on human health (Chen et al. 2006), by decreasing, for example, the incidence of certain types of serious pathologies such as cancer (Rafter 2002). In addition to improving some of their chemical properties, such as solubility, and providing greater resistance against gastrointestinal enzymes and microflora, the reduction in size of nanoparticles renders them more easily absorbable. Today, prebiotics, omega 3 and omega 6 fatty acids, vitamins, and minerals seem to be more effective when administered as nanoparticles (Sozer and Kokini 2009). From the health and nutritional perspectives, however, some concerns have been expressed about accumulation of nanoparticles in the body and potential negative impacts that make continued research in this field important in order to maximize the potential positive benefits.

The food industry could benefit from nanotechnologies, because they offer amazing opportunities for the miniaturization of micelles, emulsions, and liposomes, transforming them to nanomicelles, nanoemulsion, and cubosomes, respectively (Esposito et al. 2005; Yih and Al-Fandi 2006; Gutiérrez et al. 2008). In the same way, nanotechnology could help improve the sector of food safety through the development of nanosensors, which, for example, can be used in the detection of dangerous toxins (Ligler et al. 2003), pesticides, or pathogenic microorganisms in water as well as along the entire food chain. Other examples of nanotechnology include the use of edible oil-containing nutraceuticals encased within nanocapsules to improve their application as nutraceuticals, or the building of nanocapsules capable of selectively binding and removing dangerous chemicals from food matrices (www.nanoforum.org).

17.2.1 Nanotechnology and food packaging

For many years, the development of smart packaging systems capable of enhancing the shelf life of foods has been a subject of interest for various companies and academic institutions. These systems would be capable of responding to different environmental conditions, such as temperature and moisture, and alert consumers or industry about the possibility that a certain food could be contaminated by toxin, microorganisms, or other biotic and abiotic pollutants.

Nanotechnology provides a solution, by modifying the packaging material to improve the quality and safety of packaged foods.

Currently, this can be considered the widest appliance of nanotechnology in the food field. The main applications for food contact materials (FCMs) can be summarized as follows:

- Nanomaterials incorporated to FCMs to improve their packaging properties, in terms of flexibility, stability to temperature or moisture, specific gas barrier forming properties, etc.
- Nanoparticles with antimicrobial or oxygen-scavenging properties that transform FCMs into active materials.
- Nanosensors that form part of FCMs as "intelligent" packaging matrices, making continuous monitoring of the food condition possible.
- Nanoparticles that make the packaging matrix biodegradable through use of specific polymer–nanomaterial composites.

In all cases, a relatively small amount of nanoparticles can be satisfactory to change the qualities of packaging materials, improving their properties without changing their chemical characteristics, such as transparency, density, and the steps used in manufacturing (Lei et al. 2006). Indeed, the addition of some nanoparticles, in certain cases, can improve the mechanical and thermal performance of packaging, providing them with greater stability and enhancing their durability (Wang et al. 2003), increasing their resistance to flame (Ray et al. 2002) and temperature (Alexandre and Dubois 2000), and improving their optical, viscosity, and recycling properties (Schartel et al. 2005). This has led to the development of nanocomposites containing up to 5% (w/w) nanoparticles (Schartel et al. 2005). Nanocomposites can also contain nanominerals, such as bentonite (or montmorillonite), a cheap and widely available material resulting from volcanic activity, that can improve the gas barrier properties of nanocomposites and limit gas permeation (Ke and Yong Ping 2005), making them useful in a wide variety of applications, such as packaging cheeses, cereals, processed meats, bakery, or other uses, such as in the extrusion–coating applications for fruit juices and dairy products, or in the coextrusion process during the manufacturing of beer and carbonated drinks (Akbari et al. 2006).

Several nano-materials of natural origin such as starch, cellulose, polylactic acid, gelatine, collagen, and chitosan, are used as nano-encapsulation agents in the food industry (Fig. 17.2). Such matrices provide several advantages in food packaging applications; for example, they can be used to encapsulate bioactive compounds, such as prebiotics, vitamins, flavors, flavonoids, and polyphenols, allowing them to be released in a controlled manner (Lopez-Rubio et al. 2006).

Among the biopolymers generally used in food packaging, starch and its derivatives can be further improved through nanotechnology approaches. Nanotechnology can extend the shelf life of foods and enhance quality because it can serve not only as barrier to moisture, gases, and solutes, but also as a carrier of different active substances, such as antimicrobials and antioxidants (Zhao et al. 2008). These bio-nanocomposites

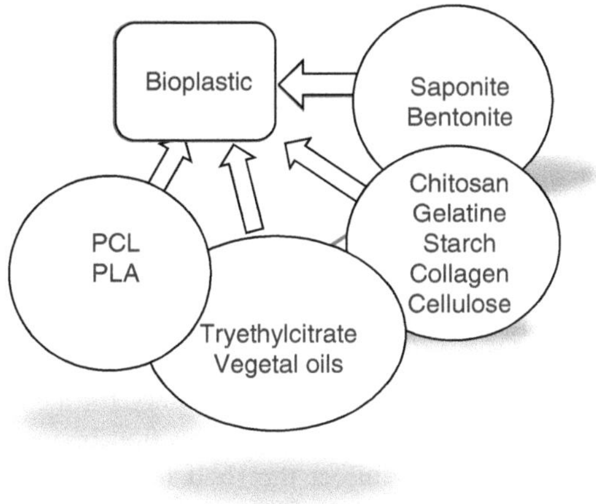

Fig. 17.2 Some materials used as nano-encapsulation agents in the production of bioplastic in the food industry. *PCL* Polycaprolacton, *PLA* polylactic acid

are significant because of their nanoscale dispersion with size less than 1,000 nm (a nanometer is a billionth of a meter) (Sanguansri and Augustin 2006).

Other biopolymers, such as biopolyesters formed by biological monomers of polylactic acid, polyhydroxybutyrate, or polycaprolacton can be improved through nanotechnology to provide better gas-barrier properties and brittleness. The physical properties of such materials can be improved also through the support of nanoclay, used as supportive filling agents (Chen et al. 2003).

At present, the world packaging market stands at much more than $US 3.5 billion. Several food companies are investing significant amounts of money to develop and optimize smart packaging systems using nanotechnologies. Kraft has developed an electronic tongue to be used as a nanosensor in the packaging of different foods. The nanosensor is extremely sensitive to gases released as a result of spoilage, which, when present, can cause a colorimetric variation of the sensor, giving a clear signal as to the state of freshness of the food. Bayer, a global company, is developing a packaging material that is lighter, stronger, and more resistant compared with the materials currently available on the market. This is a typical hybrid system, enriched with silicate nanoparticles, which is capable of diminishing the entrance of oxygen and other gases, and therefore the presence of moisture and subsequent food spoilage (Durethan, Bayer, Leverkusen, Germany). Food safety is also a major topic of concern for other companies, who are developing nanosystems containing engineered luminescent proteins capable of binding to the surface of some pathogens, such as *Escherichia coli* and *Salmonella* spp., so that the presence of these microorganism can be visualized by the formation of a visible glow, which varies in intensity depending on the amount of contamination (NanoBio Luminescence Detection Spray, BioMark, AgroMicron, Singapore). In other cases, nanotechnologies are supporting the development of new materials, useful for more accurate detection of toxins, dangerous chemicals, and pathogens, thus making the systems cheaper and the products safer. Current systems are able to detect the presence of pathogens or unwanted microorganisms only after several days. Nanosystems are capable of immediate detection, through the use of DNA device biochips that can be used to continuously monitor the safety of different food systems, such as meat, fish products, or fruit contaminated by fungi and yeasts, or by using microarray sensors to monitor and detect xenobiotics in fruits and vegetables and in the general environment (www.goodfood-project.org) (Spielbauer and Stahl 2005). In many cases, nanotechnology combined with other technologies, such as radiofrequency identification and microprocessors, has been developed to monitor an item from the manufacturer to the warehouse to the consumer. The fusion of nanotechnology and electronics created nanotronics, making such tags easier, cheaper, and more efficient. In the sector of beverages, nanotechnologies are looked on with increasing interest by food companies. One of the main goals is the formulation of new materials that could substitute glass, mainly for alcoholic beverages, without deleterious secondary effects, such as the reaction between alcohol and conventional plastic, which negatively affect the shelf life and quality of the final product. In the last several years, a new nanocomposite was developed, called Inoperm, containing clay nanoparticles, which is lighter and stronger than glass and is less likely to shatter. It decreased the loss of carbon dioxide from alcoholic carbonated beverages such as beer and limited the entrance of oxygen into the bottle, keeping the beer fresher and of superior quality even after 6 months.

The development of new food packaging systems is one of the topics in the EU 7th Framework Programme and could constitute an important interdisciplinary link among several research and development areas (i.e., food safety, nutrigenomics, technology development, biological renewal in food sector and biological production, consumers' needs, innovations in food and feed, and systems biology in food research).

Nanotechnologies are giving new impulse to the use of biodegradable polymers in the food industry. One of the most used compounds in this regard is zein, a prolamin and a major component of corn protein, which dissolves in ethanol or acetone, and can give rise to biodegradable zein films with good tensile and water barrier properties. Using nanotechnology, zein can be organized into nanobeads and nanoparticles that can be used as biodegradable materials to enhance the strength of plastic and bioactive food packages, as well as edible carriers for flavor biomolecules, or materials used to encapsulate nutraceuticals (Guo et al. 2005). The barrier property of zein polymer or of other biodegradable materials can be increased by using silicate complexes (bentonite, saponite, etc.) (Lee et al. 2005), or polysaccharides and lipids, that improve the physical properties of these films (Weiss et al. 2006).

At present, many natural sources of materials have been tested for use in nano form. Obtaining nanotubes from milk proteins such as alpha-lactalbumin through a partial hydrolysis of the molecule results in alpha-lactalbumin nanotubes capable of increasing viscosity and stiffness. These nanotubes can bind other food components because of the size of their cavities (8 nm), and their cavities can be also used to encapsulate and protect important nutraceuticals or mask unwanted flavors or aromas. Because of their source, these nanotubes can be considered food-grade materials; therefore, they could have a relatively easy introduction in the international market as well as wider applications in nanoencapsulation of nutrients, supplements, or pharmaceuticals (Sozer and Kokini 2009) if proved to be safe.

17.2.2 Nanotechnology, food quality, and processing

Nanotechnology is making an impressive impact on other fields of food science as well. The principal goal of the food industry seems to be the optimization and improvement of methods that permit a safer food supply and provide longer shelf life, without altering qualities such as color, flavor, texture, consistency, or nutritional characteristics. Other important considerations taken into account, in the case of functional foods, are the presence and programmed release of nutritional molecules within the body and the possibility of enhanced absorption. In all of these cases, nanotechnologies, through the use of nanosensors or nanocapsules can support the food industry to extend the shelf life of food, improve stability, remove unwanted molecules (e.g., by nanofiltration), and allow consumers to modify their nutritional practice and make them more relevant to their own needs.

For example, nanotechnologies can increase the absorption of nano-sized nutrients and supplements, ameliorating the nutritional value of food, developing new tastes and creamier textures through nanostructuring of food ingredients, with less (or no additional) fat (Chaudhry et al. 2008).

In response to increasing demand by consumers for functional and healthier foods, nanotechnologies could aid food companies and scientists to develop new "on demand" foods, containing specific biomolecules that can remain dormant in the food formulation until they are carried into the host body, becoming reactive and delivering nutrients to the cells only when needed. In fact, research has allowed the development of nanocapsules that are incorporated into food and that release nutrients in a programmed way (Srinivas et al. 2010). In other cases, for example, fortified foods, nanocapsules containing micronutrients, vitamins, or polyphenols are incorporated into existing food so as to increase their absorption into the organism. In certain cases, nanocapsules are developed to be resistant to different physical and thermal stresses (Weiss et al. 2006). Nanocapsules containing tuna fish oil, which is rich in omega-3 fatty acids, have been developed and incorporated in Tip-Top bread (Farhang 2007). The tuna fish oil is protected by nanocapsules and is degraded only when in the body to release the active contents (www.foodscience.afisc.csiro.au/foodfacts/foodfacts11-fishoil.htm). In Israel, the company Nutralease has developed nanosize particles to be included into cells by using nano-sized self-assembled liquid structures technology. Particles are represented by expanded micelles, made from fat, with an aqueous inner layer, and a diameter no more than 30 nm. Different nutraceuticals can be incorporated within this structure, including some polyphenols and other antioxidants such as lycopene, beta-carotene, lutein, and phytosterols. The patented system allows these compounds to easily penetrate at the level of the gut, where they are released, metabolized by gut microflora, and more easily absorbed into the blood. Similarly, it is reported that nanocapsules capable of competing for bile solubilization have been developed, and can be useful to decrease blood cholesterol levels by up to 15% (www.nanoforum.org). Clearly further research on efficacy, safety, and toxicity are required in this emerging area.

Another important sector of increasing interest for food industries is the use of nanotechnologies to improve the shelf life of biocomponents such as vitamins, lycopene, and omega fatty acids. In this case nanoparticles of about 50 nm can coat such biomolecules and deliver them more efficiently to the cells (Farhang 2007).

Nanotechnology can also be used to reduce the level of nutritionally problematic components, such as fat, in food. In fact, Unilever developed low-fat ice creams by decreasing the size of particles in the emulsion so as to mimic ice cream texture. By doing so, the technology permits the use of up to 90% less of the emulsion, reducing fat content from 16% to about 1% ("How super-cows and nanotechnology will make ice cream healthy," Daily Telegraph, August, 21, 2005).

Food quality can also be improved indirectly by using other types of nanoparticles. For example, the use of silver nanoparticles allows the inhibition of bacterial growth and odors in refrigerators, thereby preserving the sensory as well as the microbial quality of foods when refrigerated.

17.2.3 Nanotechnology and food safety

Food spoilages can be detected through the use of so-called nanosensors, formed by an array of thousands of particles designed to fluoresce in different colors on contact with specific concentration of one or more pathogens. Considering the fundamental importance of time in food microbiology and food safety, it is crucial to diminish the time for pathogen detection from many hours or days to a few hours or minutes (Bhattacharya et al. 2007). "Nanocantilevers" are another innovative class of nanosensors, whose principle is essentially based on their capability to detect specific biological–binding interactions, such as antibody–antigens, enzyme–substrate, cofactor–receptors, and so on through physical or mechanical signals. Recently, they have been used to also detect toxins as well as pathogens, antibiotic residues, or chemical contaminants in foods (Hall 2002). In a "green" approach to science, nanotechnologies can be also helpful to detect the potential hazard arising from the exposure of foods to micropesticides and nanopesticides, as well as veterinary products, nanoantibiotics, herbicides, and phytochemicals. This aspect is of particular relevance, because many companies throughout the world are developing, in the agricultural field, different types of pesticide and chemical nanoforms, whose presence and accumulation in all sectors of the agro-food and health chain can be detected by nanotechnology as well.

The two major challenges for the rapid detection of bacteria are the capacity to achieve short to real-time detection and ultrasensitivity in bioanalysis. The reduction of detection time raises the issue of the requirement of a minimal amount of sample for manipulation. The ultrasensitivity of the analysis should be high, to allow for an accurate identification (even to a single microorganism) in a short time. Many novel technologies have been developed to amplify the analytical signals from bio-recognition events, aimed to improve the sensitivity of different bioassays for bacteria detection (Yao et al. 2006).

The discovery of the polymerase chain reaction (PCR) opened the way for a new era of biological research. The impact of this technology constitutes a milestone not only in the field of molecular biology, but also in other allied fields of science involving food quality and safety.

However, other fast and ultrasensitive immunological methods for bacterial detection that use antibody-conjugated nanoparticles have been developed in the last few years, allowing for quick and accurate single bacterium detection without any amplification or enrichment. Such methods are fast (less than 20 min with a spectrofluorometer), convenient, and highly sensitive. Indeed, because multiple samples can be analyzed concurrently, the test allows for high-throughput analysis of multiple pathogens, in food, the environment, and clinical samples. Using fluorescent-bioconjugated silica particles, Zhao et al. (2004) developed ultrasensitive methods for microbial bioassays. Each nanoparticle encapsulated thousands of fluorescent dye molecules in a protective silica matrix, providing a highly reproducible signal for fluorescence-based bioanalysis. Compared with conventional immunoassays, in which only few dye molecules are coupled to an antibody molecule and used to signal a specific antibody–antigen binding event, the bioconjugated nanoparticles can enable significant amplification of the analytical signal because many dye molecules are present inside each nanoparticle that is attached to the antibody molecule. As is well known, for each bacterium many surface antigens are present and are available for specific recognition through the use of antibody-conjugated nanoparticles. At the same time, thousands of nanoparticles can bind to a single bacterium, producing a greatly amplified signal. As indicated, through the development of these nanoparticles, it is possible to detect a single bacterium per given sample in minutes. Such nanoparticles have been successfully used to detect *Escherichia coli* O157:H7 bacterial cells in ground beef (Alocilja 2008).

Novel classes of semisynthetic DNA–protein conjugates have been developed. These are self-assembled oligomeric networks consisting of streptavidin and double-stranded DNA, which can be converted into well-defined supramolecular nanocircles (Niemeyer 2002).

The DNA–streptavidin conjugates can be used and applied in the field of nanotechnology for the production of new immunological reagents for the ultrasensitive trace analysis of xenoproteins and other antigens by means of micro-immuno-PCR methodology, a combination of the specificity of an antibody-based immunoassay with the exponential power of the amplification of PCR, that results in a 1,000-fold degree of sensitivity as compared with standard enzyme-linked immunosorbent assay (ELISA) methods. Covalent conjugates of single-stranded DNA and streptavidin are used as biomolecular adapters for the immobilization of biotinylated macromolecules to solid substrates through nucleic acid hybridization. This DNA-directed immobilization allows for reversible and site-selective functionalization of solid substrates with metal and semiconductor nanoparticles, or vice versa, for the DNA directed functionalization of gold nanoparticles with proteins, such as immunoglobulins and enzymes, suitable as diagnostic tools in bioanalytics, and applicable also to the food and environment fields.

17.2.4 Nanotechnology, health, and nutrition

The three principal constituents of foods—proteins, carbohydrates, and lipids—although subjected to different digestive processes, are transformed into nanomolecules in the gastrointestinal (GI) tract. Thus, the processing of foods at the nanoscale level could be used to improve their speed or efficiency, as well as their bioavailability and metabolism. On the contrary, it is not very well known how such processes can affect the relationship between the food components and the gastrointestinal tract (Chaudhry et al. 2008) and whether there are any negative side effects. The passage of manufactured nanoparticles through the GI epithelium is strictly dependent not only on their size, but also on their physical and chemical features, that is, hydrophobicity, and the presence or absence of ligands, as well as the status of the host's intestinal tract. Thus, all of these different aspects must be taken into account in the development of new or functional foods using nanotechnologies. In addition, it should be noted that the use of nanoparticles such as titanium dioxide, silicon, aluminium, and titanium, which are commonly added to granular and nano-powdered foods, are stable after passage through the GI tract, and can accumulate various compartments in the body, such as Peyer's patches, and be passed on to macrophages. Because these latter cannot digest nanoparticles, often their accumulation can give rise to lymphoid aggregates. This aspect, which could be negative, could also be surprisingly useful in the development of nanocomposites containing calcium ions and lipopolysaccharides, which could target peripheral blood mononuclear cells or sites in the gastrointestinal tract and be used as adjuvants in some immune responsiveness applications (Powell et al. 2010). More extensive research is required to ascertain safety and determine appropriate conditions of use to minimize negative outcomes. Recently, nanotechnology is also being used as a link between food technology and cosmetics, because of the formulation of nanocomposites or nanocapsules containing nutritionally important molecules such as antioxidants, which enhance their absorption by the skin (nutricosmetics) when added to cosmetics (cream, powder, etc.).

17.3 Microtechnologies

Microtechnologies are used to apply microsystems, techniques, and components to many fields of science. They are giving an extraordinary boost to food science along all the steps of the food chain, including manufacturing, processing, transport, and storage. Additionally they are helping to enhance traceability, health, safety, and food analysis. The development of new analytical techniques in food science appears extremely important and, recently has seen a huge rise because of consumers' increased interest in food quality and safety. As a result, the use of more powerful, cleaner, faster, and cheaper analytical techniques that can be also miniaturized are increasingly required both at the academic and industrial levels. The same requirements apply to domestic and international regulatory agencies and quality control labs, for whom it is essential to work, generate, and give information about food quality and safety in a short time, as well as to detect—in this era of globalization—rapid information about adulteration, contamination, or product tampering (Ibanez and Cifuentes 2001; Garcia-Canas et al. 2004). For these reasons, laboratories are abandoning the use of classical procedures and adopting new methodologies, which are more suited to attaining these requirements. Clearly, we do not put forward claims in regard to all analytical procedures and the use of new microtechnologies in all fields of food science. Thus, this chapter is limited to a few important examples of how microtechnologies, individually or combined, can play a crucial role in food science and analysis.

17.3.1 Microfluidics

Microfluidics is a branch of technology developed in the 1990s. Microfluidic devices came into existence in the beginning of the integrated circuit (IC) industry, and were essentially based on silicon channels for species separation in a carrier fluid.

With recent scientific and technological advances in this field, the use of microfluidics is coming to the fore with several applications in biology, pharmaceutics, medicine, food, and green environment. Micrototal analysis systems (microTAS), also called labs-on-a-chip (LOC), are becoming important. Chemical analyses in microvolumes are performed in shorter time spans and with high or higher sensitivity. Today, microfluidic devices provide new capabilities at the microscale and can facilitate commercialization of micro-platforms, creating new, unimaginable applications as well as functions in food science.

The field of microfluidics has received tremendous benefits from the concurrent participation of biologists, chemists, material scientists, physicists, and engineers. Microfluidic electrophoresis devices have been applied successfully for the separation and analysis of amino acids, proteins, nucleic acids, lipids, and carbohydrates (El Rassi 2010).

17.3.2 Lab-on-a-chip concept

Lab-on-a-chip (LOC) is a generic term used to describe all devices that integrate different lab functions on a single chip, capable of handling extremely small fluid volumes, even down to a few picoliters. Such technology allows chemical and biological processes to be performed on a small glass plate with fluid channels also called microfluidic capillaries. Chips, initially created from glass, quartz, or silicone, today are built with different polymers, depending on their use, sensitivity, and the type of molecules to be analyzed. The LOC separation technique is based on conventional capillary electrophoresis, which is itself a recent and developing technology.

Although the application of LOC is still novel, it is attracting increasing interest from different companies and academies and different groups are using it for chemical analysis, environmental monitoring, food quality and safety, and medical diagnostics. At present, LOC is often used in combination with other techniques, such as molecular techniques, which creates an impressive analytical method that combines the selectivity and sensitivity provided by any molecular technique with the speed, resolving power, and low sample requirements of the LOC techniques. Thus, on combining PCR-based techniques with LOC, for example, the opportunity is created for a terrific analytical alternative that can be used, in a multidisciplinary way, for:

- Identification of species.
- Detection of starter, probiotic, and pathogen microorganisms.
- Detection of genetically modified organisms.

17.3.3 Species identification

Many factors (e.g., economic impact, food safety, or consumer health) account for the necessity for species identification and availability of essential information on food quality, food species, and fraudulent substitution, contamination, or improper addition of materials to foodstuffs. For instance, the increasing number of people allergic to different molecules present in foods makes necessary the use of analytical methods capable of identifying in a sensitive, reliable way, species that can be responsible for allergens present as a result of inappropriate practices in manufacturing, such as dairy products or infant formulas. Currently, terminal-RFLP (restrictive fragment length polymorphism, T-RFLP), PCR-RFLP, and microchip-electrophoresis approaches are successfully applied, which can detect goat, sheep, buffalo, and cow's milk in several dairy products (Abdel-Rahman and Ahmed 2007; Pirondini et al. 2010), and the detection of low-quality milk resulting from mastitis (Lee et al. 2008). This may constitute an additional powerful tool that can be employed for meat speciation (Fajardo et al. 2009) or the detection of adulteration, such as fraudulent replacement of more expensive meats for poor quality meats, or the risk of transmission of certain zoonotic diseases to humans (Garcia-Canas et al. 2004).

17.3.4 Food microbiology

Traditionally, microbiological analysis is based on growth, isolation, and morphological as well as physiological characterization of microorganisms. All these methods, although essential in food microbiology, have limitations, mainly because they cannot give the rapid responses required by the pace of modern research. The application of DNA-based detection and quantification overcame some of these limitations, making avoidable the growth and isolation of specific microorganisms before detection and monitoring, and speeding up the process of species identification even to the level of strain. DNA-LOC has been successfully applied to detect food poisoning bacteria using bead assay with a microfluidic chip-based system (Ikeda et al. 2006; Jing et al. 2010; Saleh-Lakha and Trevors 2010). In the case of *Salmonella,* its genotyping is a powerful tool to differentiate strains, helping to trace a potential origin of contamination. Use of DNA-LOC combined with amplified fragment length polymorphism (AFLP) can provide a discriminatory power comparable with that of pulsed-field electrophoresis, but obtainable in a much shorter time when compared with conventional molecular methods. Similar results might be observed for one of the emergent food-borne pathogens, *Listeria monocytogenes,* whose detection and identification through conventional methods can be time consuming and expensive. T-RFLP and LOC have overcome the problem, and the combined method has been effectively used for other pathogens, such as *Vibrio, Aeromonas,* and *Flavobacterium* (Garcia-Canas et al. 2004). Some species of microorganisms in foods can be particularly hazardous when they are toxin producers, and thus can directly or indirectly affect human health. *Clostridium botulinum,* a neurotoxigenic and anaerobic spore-forming bacteria, can be identified by traditional bacteriological approaches. However, today the more sophisticated and rapid microtechnologies allow not only their detection, but also the detection of the toxin gene, in only a few hours (Vinayaka and Thakur 2010). The same methodological approaches are effective in the detection of the entero-haemorrhagic *E. coli* O157:H7 or of its entero-toxin (Law et al. 2009), with excellent results in terms of speed and reproducibility of analysis. The fast technological evolution taking place in the last two decades in molecular biosciences, combined with the possibility of using microtechnologies such as LOC, has resulted in an increased use of finer molecular techniques in many research areas, among which are food microbiology and food biotechnology. Thanks to the shift from conventional toward molecular microbiology, genome- and proteome-related knowledge has become available for many microorganisms, including important food-related bacteria, such as lactic acid bacteria (LAB) and yeasts (Hierro et al. 2004; Zagorec et al. 2008; Lonigro et al. 2009), which are of great industrial importance in the production of fermented foods (e.g., dairy products, fermented sausages, wine, beer, and sourdoughs), and contributing considerably to their safety and organoleptic properties (Leroy and De Vuyst 2004).

17.3.5 Detection of transgenic foods

Genetic engineering is used in agriculture and the food industry to improve the performance of plant varieties (e.g., to make them more resistant against herbicides or environmental stresses), enhance the technological properties of products during storage and processing, and/or improve their sensory and nutritional properties (Garcia-Canas et al. 2004). As a result of intense debate at the international level, several food rules regulate the threshold of the presence of transgenic plants and other genetically modified organisms in food, fixing a maximum threshold of 1%, which is considered to be accidental contamination and therefore not subjected to any labeling requirement. This has, of course, gained demand for more specific, faster, and new analytical methods capable of detecting and quantifying the presence and amount of GMOs in foods. Use of genetically modified organisms (GMOs) in food and food products is becoming more widespread. Polymerase chain reaction technology is extensively used for the detection of GMOs in food products to verify compliance with labeling requirements. A combination of different biochemical and molecular methods seems to be a valid alternative to the conventional singular methods (PCR for direct detection of the transgenic DNA, immunological methods, etc.). Recently, a novel close-loop ferric fluid-driven PCR microchip for rapid amplification of GMOs was developed, that combines advantages of cycling flexibility and quick temperature transitions associated with two existing microchip PCR techniques, providing a cost saving and less time-consuming way to conduct preliminary screening of GMOs. Polymerase chain reaction solution was contained in a circular closed microchannel and was driven by magnetic force generated by an external magnet through a small oil-based ferric fluid plug. Successful amplification of genetically modified soya and maize were achieved in less than 13 min (Sun et al. 2009). Use of microfluidic systems offer improvements in quantitative accuracy, objectivity, and ease of use compared with traditional agarose electrophoresis.

17.3.6 Microtechnologies applied to food proteins

The protein profile, also in food, is quite complex as a result of various internal and external factors that can continuously affect the proteome. The internal factors influencing the proteome are genes and their relationships, gene mutations, and post-translational processes taking place within the cell. External factors can be related to the effect of certain compounds, changes in the status of the food (raw, processed, stored, contaminated, etc.). Separation and quantification of individual food proteins are an important consideration in all branches of food research. In most cases, the identification of proteins is based on their separation by electrophoretic techniques. SDS-PAGE (sodium dodecyl sulfate polyacrylamide gel electrophoresis) under reducing conditions is still one of the most important techniques used in practically all laboratories for the separation, identification, and quantification of food proteins. It is often used with high-performance liquid chromatography (HPLC) techniques, capillary electrophoresis, or immuno-based assays. However, one disadvantage of almost all separation techniques remains the time required for the preparation of samples, the physical separation of the proteins, and the final integration and quantification of the individual protein components in the sample. Many of the separation techniques can also use high volumes of materials such as buffers and solvents, many of which are expensive and toxic for both humans and the environment. In recent years microtechnology, using the microfluidic LOC technique, has offered noticeable support to the study of nucleic acids and the separation and quantification of proteins (Wu et al. 2008), allowing the processing of 10 samples in 30 min, and using very small quantities of materials and samples, typically less that 0.5 ml total volume per chip (10 samples), including samples and all reagents. This aspect is attractive in that very low levels of toxic chemicals are consumed by the analysis technique, and minimal sample volumes are required. Depending on the chips selected, the technique is reported to separate proteins of up to 260 kDa, and also claims to have a linear dynamic range of 2.5–1,000 mg/ml. In recent years, LOC technology has been successfully used to study different food matrices, such as milk and dairy products (Anema 2009; Nazzaro et al. 2009). In the meat sector, LOC could provide important information about species identification (Fratianni et al. 2008), and the correctness of manufacturing, storage, and the safety of the final product (Nazzaro et al. 2008a; Fratianni et al. 2010a). Labs-on-a-chip can also be useful to study the protein pattern of vegetable species. Detailed knowledge of cereal products is essential, especially because of the diversity of foods and feeds made from cereal grains. Varietal identity and ability to predict dough properties of wheat consignments are largely dependent on protein composition, which have been studied for many species of cereals (Bhandari et al. 2004; Uthayakumaran et al. 2005, 2006). Such analysis can be complementary for providing complete information about both variety and likely processing properties, as well as reflecting the contributions of growth and storage conditions (Wrigley et al. 2006).

17.3.7 Proteomic-on-a-chip

The technique of proteome analysis can allow the monitoring of global changes occurring in the protein complement of tissues and subcellular compartments, and the verification of the mechanism of protein formation and catabolism. Very promising applications of proteomics in the field of food science, technology, and biotechnology have been reported (Bendixen et al. 2005). In view of the emerging importance of nutrition, often closely linked to the development of chronic diseases, there is an increasing need to develop new strategies and products that are not only safe but also effective in maintaining good health or preventing specific diseases. The wealth of proteomic information, proteomic-based technologies, and model systems provide a wide spectrum of tools that can be used in the field of food science and technology. Proteomic techniques offer a promising approach for the identification of proteins in a food matrix, and the ability to study the interactions that can occur in raw and processed foods, or the interactions taking place between proteins and other food components. At the same time, it can contribute to the evaluation of safety, body distribution, and metabolism of food ingredients, as well as the detection and control of food spoilage and the presence of beneficial or pathogen microorganisms (Liu and Zhu 2005). Furthermore, proteomes of certain foods (i.e., wine, wheat, fish) can be used in their identification and assessment of their quality. Proteome and/or metabolome of starter cultures in fermentation processes of cheese, sausage, beer, and so on, can be used to predict the quality of the fermented

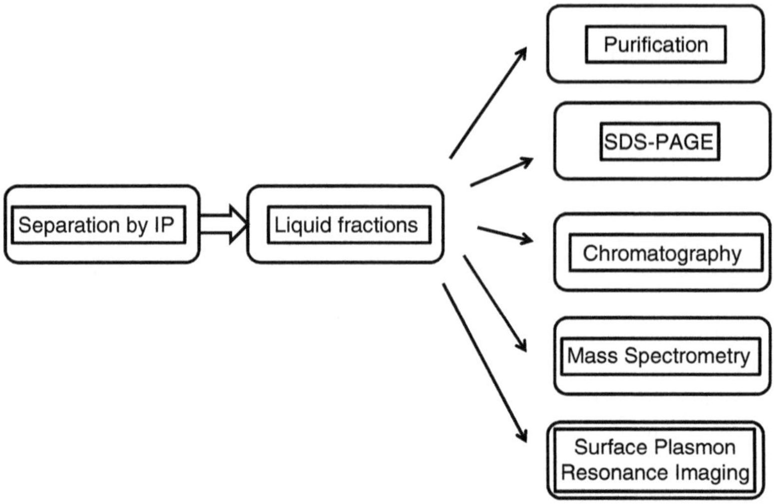

Fig. 17.3 Application of fractionation by isoelectric point (IP). Liquid fractions from a run can be collected, pooled, and even refractionated for further purification. The protein fractions might be analyzed by their molecular weight (2D) or processed by chromatography, surface plasmon resonance imaging (3D), and/or mass spectrometry (4D). *PAGE* polyacrylamide gel electrophoresis

end-product (Lin and Lee 2010). Despite this progress, there continues to be the need to improve and optimize existing proteomics technologies, enhance their analytical capabilities, diminish cost and consumption of materials and reagents, and shorten processing times. Also in this case, microtechnology can offer a good support, in terms of devices and analysis of data. Proteomic-on-a-chip is essentially a miniaturized device suitable to separate and detect a protein mixture, enabling low reagent consumption, easy manipulation, and speed of analysis. Indeed, the data obtained by the microfabricated device are comparable to those obtained by conventional 2D-PAGE. Proteomics-on-a-chip has become available only recently. A microfabricated device replacing 2D-PAGE comprises a microisoelectric focusing (1D) unit that allows the separation of proteins in liquid fractions based on their isoelectric point (IP). Whether used alone or as part of a broader separation or purification scheme, such microsystems offer rapid, simple, and effective fractionation and enrichment of proteins from complex protein mixtures. Thus, liquid fractions from a run can be easily collected, pooled, and even re-fractionated for further purification, or when present in their native state, they can be subjected to native electrophoresis, in case of the detection of biological activity, and then subjected to a usual or microelectrophoresis. In this case, reference is made to the so-called 3D (micro) electrophoresis. The protein fractions might be analyzed by their molecular weight, on a microchip (2D) or, processed by chromatography, surface plasmon resonance imaging (3D) and/or mass spectrometry (4D) (Fig. 17.3). In the case of the LOC mass spectrometry (MS) double system, the direct online connection of microfluidic separation with MS allows for the identification of a broad range of proteins.

Electrospray ionization (ESI) or nano-ESI interfaces have been used as compatible devices with the fluidics in the microchannel. MS was successfully coupled to micro-capillary electrophoresis with different electrospray microdevices (El Rassi 2010). In some cases, microchip CE systems that contained an integrated protein-digested bed filled with trypsin coated beads have been described (Kustos et al. 2007).

For the most used methodology that combines 1D and 2D, the proteomic-on-a-chip system can constitute an interesting application, because each protein present in each pI fraction can be analyzed by the following microchip on the basis of both its molecular weight and pI, and expressed as amount of protein present in each fraction, in terms of ng/μ l. In some micro-proteomic systems the separation of protein mixtures by IP and molecular weight is performed in just 3–4 h, comprising analysis and evaluation of data. This system has been recently applied for the characterization of the whole protein profile of different durum and soft wheat varieties, allowing for the detection and identification of some typical molecular markers and wheat cultivars present in southern Italy (Nazzaro 2011, unpublished data).

17.3.8 Microencapuslation

Microencapsulation (ME) can be defined as "the technology of packaging solid, liquid and gaseous materials in small capsules that release their contents at controlled rates over prolonged periods of time." Microencapsulation is the envelopment of small solid particles, liquid droplets, or gases in a 1–1,000-μm coating. In general, one can distinguish between mononuclear capsules, which have one core enveloped by a shell, and aggregates, which have many cores embedded in a matrix. The shell or matrix materials are usually polymers. The most available technologies for microencapsulation can be divided into two categories, one that uses a liquid as a suspending medium (complex coacervation, interfacial, and in situ polymerization or solvent evaporation from emulsions) and another that uses a gas as a suspending medium into which a liquid phase is sprayed (spray-drying or spray-cooling, fluidized-bed coating, or co-extrusion) (Schrooyen et al. 2001).

The technology is of particular interest not only in the medical and pharmaceutical fields (e.g., for drug and vaccine delivery); but it also has an evident impact on the food industry, mainly in the development of fortified or functional foods. In food science, microencapsulation involves the incorporation of food ingredients, enzymes, bacteria, or other materials in small capsules. Microcapsules offer to food processors a means with which they can protect sensitive food components, ensuring them against nutritional loss, utilize otherwise sensitive ingredients, incorporate uncommon or time-release mechanisms into the formulation, mask or preserve flavors and aromas, and transform liquids into easily handled solid ingredients. Controlled release of food ingredients at the right place and right time is a vital function provided by microencapsulation. Timely and targeted release improves the effectiveness of food additives and broadens the application range of food ingredients. It ensures optimal dosage, thereby improving the cost and effectiveness for the food manufacturer. Reactive, sensitive, or volatile additives (vitamins, cultures, flavors, etc.) can be turned into stable ingredients through microencapsulation. With carefully fine-tuned controlled-release properties, microencapsulation is no longer just a value-added technique, but can be defined as a real source of totally new ingredients with wide properties. Microencapsulation can effectively constitute a support for food packaging, and different systems of ME have been developed to mask unpleasant flavors and odors, or to supply barriers between the sensitive bioactive materials and the environment (which can be food or oxygen). Food manufacturing requires the addition of some ingredients that can improve quality while extending the shelf life of the end product. Recently, ingredients with potential health benefits have also been included. In the case of functional foods, these processes represent an important step, particularly with respect to the stability of bioactive compounds during processing and storage, and when it is necessary to prevent unwanted reactions with the carrier food matrix. Microencapsulation can help to resolve some of these requirements. Essentially, powerful support is required for functional foods, to ensure the stability of bioactive compounds and beneficial bacteria during their transit through the gastrointestinal system, and facilitate their controlled release.

Different materials are used for microencapsulating bioactive components in emulsion systems. The most used materials are water and oil. A typical water-oil–water emulsion consists of small water droplets, dispersed in large oil droplets, that in turn are themselves dispersed in an outer aqueous phase. The functional component can be encapsulated within the inner phase, the oil phase, or the outer water phase after drying; thus, a single delivery system can contain multiple functional components. Nanoemulsions and microemulsions, with droplets sizes between 100 and 500 nm, are produced by microfluidization or micelle formation techniques and are gaining interest for both pharmaceutical and food applications. Microencapsulation can serve to fortify existing foods with important nutritional and healthy biocomponents in which these nutrients are inadequate. Indeed, it can promote the delivery of vitamins and minerals to foods mainly by preventing their interaction with other food components, also increasing their bioavailability in certain cases. Water-soluble (ascorbic acid) and lipid-soluble vitamins (A, β-carotene, D, E, and K) can be encapsulated using various technologies. The most common reason for encapsulating these ingredients is to extend the shelf life, either by protecting them against oxidation or preventing reactions with components in the food system in which they are present. Generally, water soluble vitamins are encapsulated by spray cooling and spray chilling, or by fluidized bed coating, mainly when they must be added to solid foods, such as cereal bars, biscuits, or bread. For application in liquid food systems, water-soluble vitamins are generally microencapsulated in liposomes, single or multilayered vesicles of phospholipids containing either aqueous-based or lipophilic compounds. A good example of this is ascorbic acid (vitamin C), which is added extensively to a variety of food products as either an antioxidant or vitamin supplement (Kirby et al. 1991). Its application as a vitamin supplement is impaired by its high reactivity and, hence, poor stability in solution, resulting in a series of reactive intermediates. Lipid-soluble vitamins such as vitamin A, β-carotene, and D, E, or K are much easier to encapsulate than water-soluble ingredients. Spray-drying is the industrial method of choice for encapsulation of oil-based compounds such as flavors, vitamins, and fatty acids (Ré 1998). In a typical process all matrix materials, such as arabic gum and maltodextrin, are hydrated in water and the oil-based substance is added while stirring. The mixture is subsequently homogenized to form an emulsion. The emulsion can then be spray-dried to yield a powder that can contain up to 50% (w/w) oil. These microencapsulated vitamins are often used to prepare tablets. Omega-3 fatty acids are well known to have huge health benefits, and modern nutritional guidelines recommend the presence of these fatty acids as fundamental dietary requirements. Thus, enrichment of food with fish oil, a well-known source of these components, is considered an acceptable and effective means of increasing their level in the diet. However, their propensity to rapidly oxidize because of the high number of unsaturated double bonds in the fatty acyl chains is known. This factor limits their shelf life to 6 months, when stored at 4°C in closed containers under N_2. Different attempts have been made to fortify food with such important biomolecules; however, many problems have been encountered with regard to the taste and smell of these oils. Encapsulation by microemulsion plus spray drying has been used successfully to increase the shelf life, availability, and daily intake of this type of ingredients, allowing their use in a large variety of foods such as infant formulas and bread mixes (Andersen 1995). Microencapsulation is also applied to enrich foods, in particular with iron and calcium. From a nutritional point of view, iron is one of the most important elements, and its deficiency affects about one third of the world's population. The best way to prevent this problem is through iron fortification of food. However, iron bioavailability is generally affected by interactions with food ingredients, such as tannins, phytates, and polyphenols. In turn, iron itself catalyzes the oxidative degradation of fatty acids and vitamins. In this case, iron can be microentrapped through liposome technology, so that its capability to react with food components can be reduced while enhancing its bioavailability for human health. For people with cholesterol problems who cannot consume milk and dairy products, the low level of calcium in their diet, which is normally provided by these foods, can negatively affect the structure of bones, mainly in children, women, and elderly people. The fortification of soya milk with Ca salts (calcium triphosphate and calcium citrate) initially gave unsuccessful results because of coagulation and precipitation of soya proteins by calcium. It is more effective to encapsulate calcium lactate in lecithin liposomes. With this microtechnology, it is possible to fortify dairy-free milk, such as soymilk, which naturally contains much less Ca (12 mg/100 g soya milk), with levels of calcium equivalent to those found in cow's milk (120 mg/100 g cow's milk), while preventing undesirable calcium–protein reactions (Schrooyen et al. 2001).

Microencapsulation has been investigated for the protection of lactic acid bacteria (starters and/or probiotics) in food products and the gastrointestinal tract (Fávaro-Trindade and Grosso 2002); advantages include prevention of interfacial inactivation, stimulation of production and excretion of secondary metabolites, and continuous utilization. Additionally, microencapsulation may enhance microbial survival and operating efficiency during fermentation Champagne et al. (1994). Chandramouli et al. (2004) and De Giulio et al. (2005) found that microencapsulation of LAB helped overcome inactivation during drying or exposure to artificial gastric conditions. Different polymer matrices are used for the microencapsulation of lactic acid bacteria (starters and/or probiotics). The most commonly used, all from natural sources, are alginate (a polymer extracted from seaweed), carrageenan, chitosan from arthropods, starch, and whey proteins. All of these are nontoxic, biocompatible, and inexpensive. LAB present two sets of problems: their size (between 1 and 5 mm in diameter), which immediately excludes nanotechnologies, and the fact that they must be kept alive. This latter is the key aspect in selecting the appropriate ME technology. Today, at least five ME methods are commonly applied to LAB: spray coating, spray drying, extrusion, emulsion, and gel-particle technologies (including spray chilling). Among them, spray coating and gel-particle technologies are the most often used, the latter being considered the least expensive encapsulation technology. Spray chilling deserves greater attention, because it could expand the range of matrices used. Moreover, this technology could be used to generate smaller beads, which might be desirable in food processing. Spray-drying offers the most potential, supporting the cells that can be stress adapted to the high processing temperatures. In the last years, lactic acid bacteria have been used to produce new fermented functional foods containing no dairy products, especially targeting people allergic to milk and dairy components. However, LAB are less stable in nondairy compared with dairy products. However, if microencapsulated, and also in presence of other healthy important biomolecules such as prebiotics, the fermentation process can proceed correctly (Fratianni et al. 2010b). Indeed, the presence of the envelope protects LAB from adverse environments. Thus, these new products retain all of their nutritional characteristics and show the high viability of probiotics that are protected from the acidic environment and other stress conditions. The coencapsulation of microorganisms and other important components such as prebiotics has been demonstrated to protect probiotics better in food systems and the gastrointestinal tract owing to symbiosis (Chen and Chen 2008; Nazzaro et al. 2008b). Coencapsulation has also been found to be beneficial in regard to vitamins and minerals. The presence of calcium in a hydrophilic phase and vitamins A or D in a hydrophobic phase could promote calcium absorption in the GI tract. Encapsulation of ascorbic acid in a liposome together with vitamin E creates a synergistic antioxidant effect (Reineccius 1995; Champagne and Fustier 2007). In this case, vitamin E is incorporated into the liposome wall, while the ascorbic acid is entrapped in the aqueous inner layer. As pointed out, ME could be useful in helping to deliver bioactive ingredients both to the food matrix itself and the GI tract. To date, the literature shows that ME has primarily served for the delivery of bioactives into the matrix and it has not been fully explored for more efficient delivery in the GI tract. The idea of not wanting added bioactive compounds that could alter food sensory properties or their appearance is weak (Siro et al. 2008). Commercial products are already available (a yogurt and a breakfast cereal) in which particles containing biomolecules (polyphenols, vitamins, or antioxidant) or probiotics are clearly seen in the foods and even advertised on the labels. Therefore, ME might be used to create particles that clearly indicate to consumers that the bioactive ingredients are present in the functional foods, thus promoting marketing strategies for natural, healthy, and green food product differentiation.

17.4 Conclusion

The use of microtechnologies and nanotechnologies is currently increasing as a result of its selectivity, sensitivity, and speed of analysis in all areas of food science.

Thus, it is expected that their application will become routine in the not-too-distant future. Promising results and applications are already being developed in the areas of food packaging, quality, and safety. Natural biopolymers of micro-size or nano-size, such as polysaccharides, can be used for the encapsulation of vitamins, prebiotics, and probiotics; drug delivery systems; nutraceuticals; or the development of so-called "new" foods. Further research on positive and negative potential outcomes is required.

References

Abdel-Rahman, S.M., and M.M.M. Ahmed. 2007. Rapid and sensitive identification of buffalo's, cattle's and sheep's milk using species-specific PCR and PCR-RFLP techniques. *Food Control* 18: 1246–1249.

Akbari, Z., T. Ghomashchi, and A. Aroujalian. 2006. Potential of nanotechnology for food packaging industry. In *Proceedings of "Nano and Micro Technologies in the Food and HealthFood Industries,"* Amsterdam, The Netherlands, October 25–26, 2006.

Alexandre, M., and P. Dubois. 2000. Polymer-layered silicate nanocomposites: Preparation, properties and uses of a new class of materials. *Materials Science and Engineering R-Reports* 28(1–2): 1–63.

Alocilja, E.C. 2008. Biosensors for detecting pathogenic bacteria in the meat industry. In *Meat biotechnology*, ed. F. Toldra. New York, NY: Springer Science+Business Media.

Andersen, S. 1995. Microencapsulated marine omega-3 fatty acids for use in the food industry. *Science and Technology in Europe* 1995: 104–106.

Anema, S.G. 2009. The use of "lab-on-a-chip" microfluidic SDS electrophoresis technology for the separation and quantification of milk proteins. *International Dairy Journal* 19: 198–204.

Bendixen, E., R. Taylor, K. Hollung, K.I. Hildrum, B. Picard, and B. Bouley. 2005. Proteomics, an approach towards understanding the biology of meat quality. In *Indicators of milk and beef quality*, ed. J.F. Hocquette and S. Gigli. Wageningen, The Netherlands: EAAP publication No.112, Wageningen Academic Publishers.

Bhandari, D.G., S. Church, A. Borthwick, and M.A. Jensen. 2004. Automated varietal identification using lab-on-a-chip technology. In *Proceedings of the 12th ICC Cereal and Bread Congress*, Harrogate, UK, May 24–26, 2004.

Bhattacharya S., J. Jangy, L. Yangz, D. Akin, and R. Bashiryy. 2007. Biomems and nanotechnology-based approaches for rapid detection of biological entities. Birck Nanotechnology Center, Birck and NCN Publications:1–33.

Champagne, C.P., and P. Fustier. 2007. Microencapsulation for the improved delivery of bioactive compounds into foods. *Current Opinion in Biotechnology* 18: 184–190.

Champagne, C.P., C. Lacroix, and I. Sodini-Gallot. 1994. Immobilized cell technologies for the dairy industry. *Critical Reviews in Biotechnology* 14: 109–134.

Chandramouli, V., K. Kailasapathy, P. Peiris, and M. Jones. 2004. An improved method of microencapsulation and its evaluation to protect lactobacillus spp. In simulated gastric conditions. *Journal of Microbiological Methods* 56: 27–35.

Chaudhry, Q., M. Scotter, J. Blackburn, B. Ross, A. Boxall, L. Castle, R. Aitken, and R. Watkins. 2008. Applications and implications of nanotechnologies for the food sector. *Food Additives and Contaminants: Part A* 25: 241–258.

Chen, M.J., and K.N. Chen. 2008. Encapsulation of probiotics in alginate systems. *CRS Delivering Bioactives Newsletter* 25(2): 6–9.
Chen, C.C., J.Y. Chueh, H. Tseng, H.M. Huang, and S.Y. Lee. 2003. Preparation and characterization of biodegradable PLA polymeric blends. *Biomaterials* 24: 1167–1173.
Chen, H., J. Weiss, and F. Shahidi. 2006. Nanotechnology in nutraceuticals and functional foods. *Food and Technology* 60(3): 30–36.
Cientifica. 2006. Quietly nanotechnology joins the food chain. The Sunday Telegraph. October 22, 2006.
Cientifica. 2007. Half way to the trillion dollars market. A critical review of the diffusion of nanotechnologies. The nanotechnology market in 2007. http://cientifica.eu/blog/category/finance/ april 16, 2007.
De Giulio, B., P. Orlando, G. Barba, R. Coppola, M. De Rosa, A. Sada, P.P. De Prisco, and F. Nazzaro. 2005. Use of alginate and cryo-protective sugars to improve the viability of lactic acid bacteria after freezing and freeze-drying. *World Journal of Microbiology and Biotechnology* 21: 739–746.
El Rassi, Z. 2010. Electrophoretic and electrochromatographic separation of proteins in capillaries: An update covering 2007–2009. *Electrophoresis* 31: 174–191.
E. Dickinson, T. Van Vliet Eds Food Colloids Biopolymers and Materials, Royal Society of Chemistry (2003).
Esposito, E., R. Cortesi, M. Drechsler, L. Paccamiccio, P. Mariani, C. Contado, E. Stellin, E. Menegatti, F. Bonina, and C. Puglia. 2005. Cubosome dispersions as delivery systems for percutaneous administration of indomethacin. *Pharmacological Research* 22: 2163–2173.
Fajardo, V., I. González, J. Dooley, S. Garret, H.M. Brown, T. García, and R. Martín. 2009. Application of polymerase chain reaction-restriction fragment length polymorphism analysis and lab-on-a-chip capillary electrophoresis for the specific identification of game and domestic meats. *Journal of the Science of Food and Agriculture* 89: 843–847.
Farhang, B. 2007. Nanotechnology and lipids. *Lipid Technology* 19: 132–135.
Fávaro-Trindade, C.S., and C.R.F. Grosso. 2002. Microencapsulation of *L. Acidophilus* (La-05) and *B. Lactis* (Bb-12) and evaluation of their survival at the pH values of the stomach and in bile. *Journal of Microencapsulation* 19: 485–494.
Fratianni, F., A. Sada, P. Orlando, and F. Nazzaro. 2008. Micro-electrophoretic study of the sarcoplasmic fraction in the dry-cured goat raw ham. *The Open Food Science Journal* 2: 89–94.
Fratianni, F., L. De Martino, A. Melone, V. De Feo, R. Coppola, and F. Nazzaro. 2010a. Preservation of chicken breast meat treated with thyme and balm essential oils. *Journal of Food Science* 75: M528–M535.
Fratianni, F., R. Coppola, A. Sada, J. Mendiola, E. Ibañez, and F. Nazzaro. 2010b. A novel functional prebiotic product containing phenolics and anthocyanins. *International Journal of Probiotics and Prebiotics* 5: 85–90.
Garcia-Canas, V., R. Gonzales, and A. Cifuentes. 2004. The combined use of molecular techniques and capillary electrophoresis in food analysis. *Trends in Analytical Chemistry* 23: 637–643.
Guo, Y., Z. Liu, H. An, M. Li, and N. Hua. 2005. Nano-structure and properties of maize zein studied by atomic force microscopy. *Journal of Cereal Science* 41: 277–281.
Gutiérrez, J.M., C. González, A. Maestro, I. Solè, C.M. Pey, and J. Nolla. 2008. Nano-emulsions: New applications and optimization of their preparation. *Current Opinion in Colloid and Interface Science* 13: 245–251.
Hall, R.H. 2002. Biosensor technologies for detecting microbiological food borne hazards. *Microbiology and Infection* 4: 425–432.
Hierro, N., A. Gonzalez, A. Mas, and J.M. Guillamon. 2004. New PCR-based methods for yeast identification. *Journal of Applied Microbiology* 97: 792–801.
Ibanez, E., and A. Cifuentes. 2001. New analytical techniques in food science. *Critical Reviews in Food Science and Nutrition* 41: 413–450.
Ikeda, M., N. Yamaguchi, K. Tani, and M. Nasu. 2006. Rapid and simple detection of food poisoning bacteria by bead assay with a microfluidic chip-based system. *Journal of Microbiological Methods* 67: 241–247.

Jing, J.Z., P. Zhao, and B. Wang. 2010. Application of Agilent 2100 bioanalyzer in detection of foodborne pathogens Shigella. In *Bioinformatics and Biomedical Engineering (iCBBE), 4th International Conference 18–20 June, 2010 Chengdu, China.*

Ke, Z., and B. Yong Ping. 2005. Improve the gas barrier property of PET film with montmorillonite by in situ interlayer polymerisation. *Materials Letters* 59: 3348–3351.

Kirby, C.J., C.J. Whittle, N. Rigby, D.T. Coxon, and B.A. Law. 1991. Stabilization of ascorbic acid by microencapsulation in liposomes. *International Journal of Food Science and Technology* 26: 437–449.

Kustos, I., B. Kocsis, and F. Kilár. 2007. Bacterial outer membrane protein analysis by electrophoresis and microchip technology. *Expert Review of Proteomics* 4: 91–106.

Law, W.S., S.F. Li, and L.J. Kricka. 2009. Detection of enteropathogenic *Escherichia coli* by microchip capillary electrophoresis. *Methods in Molecular Biology* 509: 169–179.

Lee, C.J., D.A. Scheufele, and B.V. Lewenstein. 2005. Public attitudes toward emerging technologies: Examining the interactive effects of cognitions and effect on public attitudes toward nanotechnology. *Science Communication* 27: 240–267.

Lee, K.H., J.W. Lee, S.W. Wang, L.Y. Liu, M.F. Lee, S.T. Chuang, Y.-M. Shy, and C.H. Chi. 2008. Development of a novel biochip for rapid multiplex detection of seven mastitis-causing pathogens in bovine milk samples. *Journal of Veterinary Diagnostic Investigation* 20: 463–471.

Lei, S.G., S.V. Hoa, and M.T. Ton-That. 2006. Effect of clay types on the processing properties of polypropylene nanocomposites. *Computer Science and Technology* 66: 1274–1279.

Leroy, F., and L. De Vuyst. 2004. Lactic acid bacteria as functional starter cultures for the food fermentation industry. *Trends in Food Science and Technology* 15: 67–78.

Ligler, F.S., C.R. Taitt, L.C. Shriver-Lake, K.E. Sapsford, Y. Shubin, and J.P. Golden. 2003. Array biosensor for detection of toxins. *Analytical and Bioanalytical Chemistry* 377: 469–477.

Lin, G., and A.P. Lee. 2010. Microfluidics: An emerging technology for food and health science. *Annals of the New York Academy of Sciences* 1190: 186–192.

Liu, W.T., and L. Zhu. 2005. Environmental microbiology-on-a-chip and its future impacts. *Trends in Biotechnology* 23: 174–179.

Lonigro, S.L., F. Valerio, M. De Angelis, P. De Bellis, and P. Lavermicocca. 2009. Microfluidic technology applied to cell-wall protein analysis of olive related lactic acid bacteria. *International Journal of Food Microbiology* 130: 6–11.

Lopez-Rubio, A., R. Gavara, and J.M. Lagaron. 2006. Bioactive packaging: Turning foods into healthier foods through biomaterials. *Trends in Food Science and Technology* 17: 567–575.

Morris, E. 2003. Food colloids, biopolymers and materials.

Nazzaro, F., A. Sada, F. Scognamiglio, and F. Fratianni. 2008a. Monitoraggio della frazione proteica sarcoplasmatica in salami tipo Napoli prodotti con estratti vegetali come additivi naturali. In *Proceedings of the Conference "Qualicibi, cibi di ieri e di domani:qualità e sicurezza tra tradizione e innovazione."* May, 2008 Positano, Italy 27–30.

Nazzaro, F., F. Fratianni, A. Sada, and P. Orlando. 2008b. Synbiotic potential of carrot juice supplemented with *lactobacillus* spp. And inulin or fructooligosaccharides. *Journal of the Science of Food and Agriculture* 88: 2271–2276.

Nazzaro, F., F. Fratianni, A. Sada, and R. Coppola. 2009. Microfluidic lab-on-a-chip methodology as alternative method to study milk proteins. In *Proceedings of the Conference MICheese*. Avellino, September, 2009 Italy.

Niemeyer, C.M. 2002. The developments of semisynthetic DNA–protein conjugates. *Trends in Biotechnology* 20: 395–401.

Pirondini, A., U. Bonas, E. Maestri, G. Visioli, M. Marmiroli, and N. Marmiroli. 2010. Yield and amplificability of different DNA extraction procedures for traceability in the dairy food chain. *Food Control* 21: 663–668.

Powell, J.J., N. Faria, E. Thomas-McKay, and L.C. Peleaù. 2010. Origin and fate of dietary nanoparticles and microparticles in the gastrointestinal tract. *Journal of Autoimmunity* 34: J226–J233.

Rafter, J.H. 2002. Scientific basis of biomarkers and benefits of functional foods for reduction of disease risk: Cancer. *British Journal of Nutrition* 88: S219–S224.

Ray, S.S., P. Maiti, M. Okamoto, K. Yamada, and K. Ueda. 2002. New polylactide/layered silicate nanocomposites. 1. Preparation characterization properties. *Macromolecules* 35: 3104–3110.

Ré, M.I. 1998. Microencapsulation by spray drying. *Drying Technology* 16: 1195–1236.
Reineccius, G.A. 1995. Liposomes for controlled release in the food industries. In: Encapsulation and controlled release of food ingredients. *ACS Symposium Series* 590: 113–131.
Saleh-Lakha, S., and J.T. Trevors. 2010. Microfluidic applications in microbiology. *Journal of Microbiological Methods* 82: 108–111.
Sanguansri, P., and M.A. Augustin. 2006. Nanoscale materials development: A food industry perspective. *Trends in Food Science and Technology* 17: 547–556.
Schartel, B., P. Potschke, U. Knoll, and M. Abdel-Goad. 2005. Fire behaviour of polyamide 6/multiwall carbon nanotube nanocomposites. *European Polymer Journal* 415: 1061–1070.
Schrooyen, P.M.M., R. van der Meer, and C.G. De Kruif. 2001. Microencapsulation: Its application in nutrition. *Proceedings of the Nutrition Society* 60: 475–479.
Shibata, T. 2002. Method for producing green tea in microfine powder. US Patent 6,416,803.
Siro, I., E. Kapoln, and A. Lugasi. 2008. Functional food: Product development, marketing and consumer acceptance—A review. *Appetite* 51: 456–467.
Sozer, N., and J.L. Kokini. 2009. Nanotechnology and its applications in the food sector. *Trends in Biotechnology* 27: 82–89.
Spielbauer, B., and F. Stahl. 2005. Impact of microarray technology in nutrition and food research. *Molecular Nutrition and Food Research* 49: 908–917.
Srinivas, P.R., M. Philbert, T.Q. Vu, Q. Huang, J.L. Kokini, E. Saos, et al. 2010. Nanotechnology research: Applications in nutritional sciences. *Journal of Nutrition* 140: 119–124.
Sun, Y., Y.C. Kwok, P. Foo-Peng Lee, and N.T. Nguyen. 2009. Rapid amplification of genetically modified organisms using a circular ferrofluid-driven PCR microchip. *Analytical and Bioanalytical Chemistry* 394: 1505–1508.
Uthayakumaran, S., I.L. Batey, and C.W. Wrigley. 2005. On-the-spot identification of grain variety and wheat-quality type by lab-on-a-chip capillary electrophoresis. *Journal of Cereal Science* 41: 371–374.
Uthayakumaran, S., Y. Listiohadi, M. Baratta, I.L. Batey, and C.W. Wrigley. 2006. Rapid identification and quantitation of high-molecular-weight glutenin subunits. *Journal of Cereal Science* 44: 34–39.
Vinayaka, C., and M.S. Thakur. 2010. Focus on quantum dots as potential fluorescent probes for monitoring food toxicants and foodborne pathogens. *Analytical and Bioanalytical Chemistry* 397: 1445–1455.
Wang, K.K., C.M. Koo, and I.J. Chung. 2003. Physical properties of polyethylene/silicate nanocomposite blown films. *Journal of Applied Polymer Science* 89: 2131–2136.
Weiss, J., P. Takhistov, and J. McClements. 2006. Functional materials in food nanotechnology. *Journal of Food Science* 71: 107–116.
Wrigley, C.W., I.L. Batey, S. Uthayakumaran, and W.G. Rathmell. 2006. Modern approaches to food diagnostics for grain quality assurance. *Food Australia* 58: 538–542.
Wu, D., J. Qin, and B. Lin. 2008. Electrophoretic separations on microfluidic chips. *Journal of Chromatography A* 1184: 542–559.
Yaghmur, A., and O. Glatter. 2009. Characterization and potential applications of nanostructured aqueous dispersions. *Advances in Colloid and Interface Science* 147–148: 333–342.
Yao, G., L. Wang, Y. Wu, J. Smith, J. Xu, W. Zhao, E. Lee, and W. Tan. 2006. FloDots: Luminescent nanoparticles. *Jornal of Analytical and Bioanalytical Chemistry* 385: 518–524.
Yih, T.C., and M. Al-Fandi. 2006. Engineered nanoparticles as precise drug delivery systems. *Journal of Cellular Biochemistry* 97: 114–119.
Zagorec, M., S. Chaillou, M.C. Champomier-Vergès, and A.M. Crutz-Le Coq. 2008. Role of bacterial 'Omics' in food fermentation. In *Molecular techniques in the microbial ecology of fermented foods*, ed. L. Cocolin and D. Ercolini. New York: Springer.
Zhao, X., L.R. Hilliard, S.J. Mechery, Y. Wang, R. Bagwe, S. Jin, and W.A. Tan. 2004. A rapid bioassay for single bacterial cell quantitation using bioconjugated nanoparticles. *PNAS* 101: 15027–15032.
Zhao, R.X., P.E. Torley, and J. Halley. 2008. Emerging biodegradable materials: Starch- and protein-based bio-nanocomposites. *Journal of Materials Science* 43: 3058–3071.

Chapter 18
Bead-based arrays: multiplex analyses

Anne Maltais and Joyce I. Boye

18.1 Introduction

Foodborne diseases cause approximately 76 million illnesses, 325,000, hospitalizations, and 5,000 deaths in the United States each year (Mead et al. 1999). More than 2,000 known diseases are transmitted through food poisoned by bacteria, viruses, parasites, toxins, pesticides, antibiotics, metals, and prions, which contaminate food at different points during the production and preparation process (Bryan 1982). With growing international trade, travel, and globalization of food marketing and distribution, pathogens and contaminants have spread faster in the last decade. We now face the threat of affecting the health of people in numerous countries at the same time if no rigorous and efficient control of food contamination is put in place at key steps of food production.

Food contamination by potential allergens represents another growing challenge for the quality assurance of food industry. From 1997 to 2007, the prevalence of food allergy among children under the age of 18 increased by 18% in the United States (Branum and Lukacs 2009). There are some indications that the occurrence of food allergy may be increasing in the United States as well as other countries (Sicherer 2002). Food allergy is an adverse immune response to a food protein mainly from eight sources: milk, eggs, peanuts, tree nuts, fish, shellfish, soy, and wheat (Sampson 2004). Allergic reactions can range from mild (dermatitis, tingling sensation around the mouth and lips, gastrointestinal disorders) to severe and life-threatening reactions, especially when respiratory distress occurs as a result of anaphylaxis. For this reason, allergenic components must be clearly indicated on food labels to allow avoidance by allergic individuals. Because cross-contamination can

A. Maltais (✉) • J.I. Boye
Food Research and Development Centre, Agriculture and Agri-Food Canada,
Saint Hyacinthe, QC, Canada
e-mail: Anne.Maltais@agr.gc.ca; Joyce.Boye@agr.gc.ca

J.I. Boye and Y. Arcand (eds.), *Green Technologies in Food Production and Processing*, Food Engineering Series, DOI 10.1007/978-1-4614-1587-9_18,

occur in manufacturing lines even when allergen-free ingredients are used and decontamination is performed, the food industry continues to require tools to monitor for the presence of multiple allergens.

Several methods have been developed and refined over the last several decades to detect and quantify undesired contaminants in food. Traditional methods to detect foodborne bacteria often rely on time-consuming growth in culture media, followed by isolation, biochemical identification, and sometimes serology. It usually takes many days before results can be analyzed. In addition to being time consuming, microbiological methods often require the use of high volumes of materials and reagents and a wide array of growth media specific to each kind of microorganism. Enzyme-linked immunoabsorbent assays (ELISA) and polymerase chain reaction (PCR) methods have been developed for the detection of pathogens, toxins, allergens, and microbial analytes in foods. Traditionally, food products have been challenging matrices for conducting immunoassays and PCR-based tests because proteins, carbohydrates, and other macromolecules present in food can interfere with the tests. Removal of particulate matter may also be a time-consuming and challenging process before analyses. However, one of the greatest limitations of these assays is their inability to test for the presence of multiple contaminants or pathogens in a given sample. This represents a major drawback because a different test has to be performed for each pathogen of interest, which lengthens the analysis time. Additionally, more assays also mean more use of reagents, increased requirements for energy, human, financial, and material resources and overall material waste.

A greener and more efficient approach would be to utilize a multiplexed assay in which several different pathogens, allergens, or contaminants can be detected in one single procedure. Microsphere-based flow cytometric assays represent a promising avenue in multiplexed analysis. They are sensitive, specific, and can be multiplexed to a high level for the detection of analytes (Vignali 2000). Compared with bacterial culture and immunological methods, microsphere array assays are faster, may require fewer reagents, and demand less technical expertise. With the emergence of new pathogens and the growing number of people allergic to food proteins, food industry and regulatory agencies increasingly need to have fast reproducible results to assure consumers of the safety of foods sold on the market. This chapter reviews the principles of bead-based flow cytometric assay, their existing applications, and future trends in the detection of pathogens, contaminants, toxins, allergens, and other bioactive molecules.

18.2 Background and history

18.2.1 Microspheres and cytometry

The association of flow cytometry and microspheres occurred in the 1970s with the commercialization of the first flow cytometer. This instrument was first designed for counting and sorting microscopic particles, such as cells and chromosomes, by suspending them in a stream of fluid and passing them by an electronic detection apparatus. This allowed simultaneous multiparametric analysis of the physical and/or chemical characteristics of up to thousands of particles per second. Calibration was an important aspect of the instrument setup and small, cell-size polystyrene/latex microspheres (approximately 3–5 μm in diameter) proved to be ideal for alignment, compensation, and calibration. This heralded the connection between these microspheres and the flow cytometer. Not long after, it was recognized that microparticles coupled to antigens could serve as solid support for the capture of antibodies from blood (Horan and Wheeless 1977). In this study, differently sized microspheres were distinguished according to their optical properties using a flow cytometer, resulting in multiplexing of the method. A few years later, fluorescein isothiocyanate-labeled latex particles were used for the measurements of phagocytosis by neutrophils and macrophages (Dunn and Tyrer 1981; Steinkamp et al. 1982; Parod and Brain 1983). Not long after, Stewart and Steinkamp (1982) used fluorescent beads and a flow cytometer to count cell populations in a blood sample. With the discovery of monoclonal antibodies, microbead-based assays were proposed as an alternative to ELISA for the detection of antigens or antibodies in a sample. Lisi et al. (1982) reported the first immunoassay using fluorescent bead-based and flow cytometric detection to quantify human Immunoglobins (IgG). Since then, numerous immunoassays have been successfully established using microsphere-based assays.

18.2.2 Medical applications

The medical field has found many applications for multiplexed analysis, especially for the quantitation of cytokines in biological fluids and tissue culture supernatants. Cytokines are small proteins that are secreted by specific cells of the immune system and that carry signals locally between cells, and thus can exert effects on neighboring cells. Different cytokines possess biologically overlapping functions and may be present in the same microenvironments such as at an inflammation site to provoke a cascade of reactions. A large number of cytokines may, therefore, be expressed in a specific pathogenic condition. For this reason, multiplexed bead based assays are well suited for cytokine detection and quantitation in biological fluids. In vitro and in vivo studies have been successfully conducted on cell cultures and animal sera, respectively (Fulton et al. 1997; Gordon and McDade 1997; Olivier et al. 1998; Carson and Vignali 1999). Comparisons between bead-based assays and standard ELISA techniques, as to their ability to quantitate cytokines, have also been reported (Carson and Vignali 1999; Prabhakar et al. 2002; De Jager et al. 2003; Elshal and McCoy 2006). Equivalent dosages were found with both techniques, suggesting that bead-based assays could be used as an alternative to ELISA. Accuracy, sensitivity, and reproducibility using this novel technique have been superior to the standard technique. However, the major advantage of using multiplexed analysis remains its time effectiveness.

Bead-based assays have now found applications in immunology, infectiology, cancerology, and genetics. Because the technology can be applied to almost any protocol requiring analysis of molecular interactions, numerous biological fields will eventually find applications for it. Food analysis represents one of many new avenues for this promising technique and some successful applications are discussed later in this chapter.

18.3 Theory

18.3.1 Materials and instrumentation

18.3.1.1 Beads and dyes

As mentioned, the use of microspheres as solid supports for molecular analysis using flow cytometer dates back three decades, but a renewed interest in this technique in the area of multiplexing, has led to the development of reagents and instruments specifically designed to support multiplexed microparticle-based analyses. The development of fluorescence-encoded microspheres exhibiting discrete intensities of fluorescence from two different fluorescence dyes by the Luminex Corporation (xMAP Luminex Technology, Austin, TX) was the key element for the resurgence of bead-based assays. These 5.6-μm polystyrene microspheres set are dyed with orange and red fluorophores (Fl 2 and Fl 3, respectively). Ten intensities of the two colors enable the encoding of 100 different microspheres that can be distinguished by a flow cytometer. Hence, these color codes can be thought of as forming a microarray, showing the same function as the more common flat microarray, but with single beads representing array elements in a two-dimensional fluorescence space, replacing spots on a flat surface. In theory, the addition of ten intensities or one or two additional dyes would enable 1,000-plex and 10,000-plex bead sets, but this exponential expansion of multiplexing capability is limited by several technical challenges, including emission spectral overlap among encoding dyes (Keij and Steinkamp 1998). Quantum dots have been proposed as alternatives to organic fluorophores for encoding, owing to the narrower emission spectra compared with organic fluorophores (Gao and Nie 2004).

For low-scale multiplex applications, microspheres encoded with different intensities of a single fluorophore in one or more sizes are also available from several commercial sources. Because particle-based assays are recognized to have a number of advantages over flat-surface microarrays and new particle-encoding schemes using colloidal silica (Battersby et al. 2002; Zhao et al. 2008), microfabricated nanorods (Nicewarner-Peña et al. 2001), and small radiofrequency traducers (Mandecki et al. 2000) are being developed to expand the multiplex levels of microparticle assays.

Promoting interactions between microbeads and biomolecular targets require functionalization of their surface with a chemical reactive group. Surface carboxyl and amino groups are the most popular for many assay applications. Conjugation capacity (targeting sites) on functionalized microspheres can vary from different commercialized beads because of different factors (e.g., size of the bead). The 5.6-μm microspheres provide a large surface area, which can accommodate up to 100,000 capture molecules per bead (McBride et al. 2003). Because polystyrene beads are hydrophobic, they exhibit nonspecific binding to variable degrees. The use of blocking agents such as albumin, Tween 20, or serum allows control of undesired binding reactions such as microsphere aggregation.

18.3.1.2 Reagents

Reagents play two main functions in a multiplexed bead-based assay. The first type of reagent is chosen to react with the functions covering the surface of the bead. Such reagents are called "capture reagents" because their role is to bind with the specific analyte targeted for detection or quantification. Capture reagents can be antibodies for the detection of antigens, antigens for the detection of antibodies, or probes for the detection of a nucleic acid-specific sequence. The second type of reagent is a labeled reporter molecule that can specifically bind to the analyte once captured at the bead surface. Once conjugated with the analyte, the reporter molecule is usually coupled to a green emitting fluorophore—normally phycoerythrin (Alexa 532, Alexa 488, and Cy3 are also used), owing to its high quantitative yield, low light sensitivity, and stability—which allows detection of the interaction between the capture reagent and analyte occurring at the surface of the bead. Reporter molecules traditionally used are antiimmunoglobins for detection of an antibody–antigen reaction and streptavidin for nucleic acid sequence detection. The intensity of the emitted signal on each bead using flow cytometry can be correlated to the quantity of analyte present in the sample analyzed.

Even though multiplexed assay reagents and materials can be prepared from scratch, it requires time and experience. For this reason, read-to-use assay kits are available for many common assays and targets (e.g., cytokines, chemokines, signaling molecules, immunoglobulins, infectious-disease agents, autoantibodies, allergens, and DNA sequences from infectious-disease organisms) (Becton Dickinson, Bender Medsystems-BioRad, Biosource, OneLambda, R&D Systems, Upstate, and others) (Nolan and Mandy 2006). For immunoassays, these kits include capture antibodies immobilized on encoded microspheres, labeled reporter antibodies, standards, and buffers. For nucleic acid–based assays, encoded microspheres provided are functionalized with oligonucleotides. For both kinds of analysis, kit providers also supply custom assay–development services.

18.3.1.3 Flow cytometer

A flow cytometer is an instrument that allows illumination of cells or other particles as they flow individually in front of a light source, and then detects and correlates the signals from those cells or particles resulting from the illumination (Givan 2004). Data from two specific measurements are thus required for microsphere array analyses. First, the encoding signature on each set of beads has to be measured

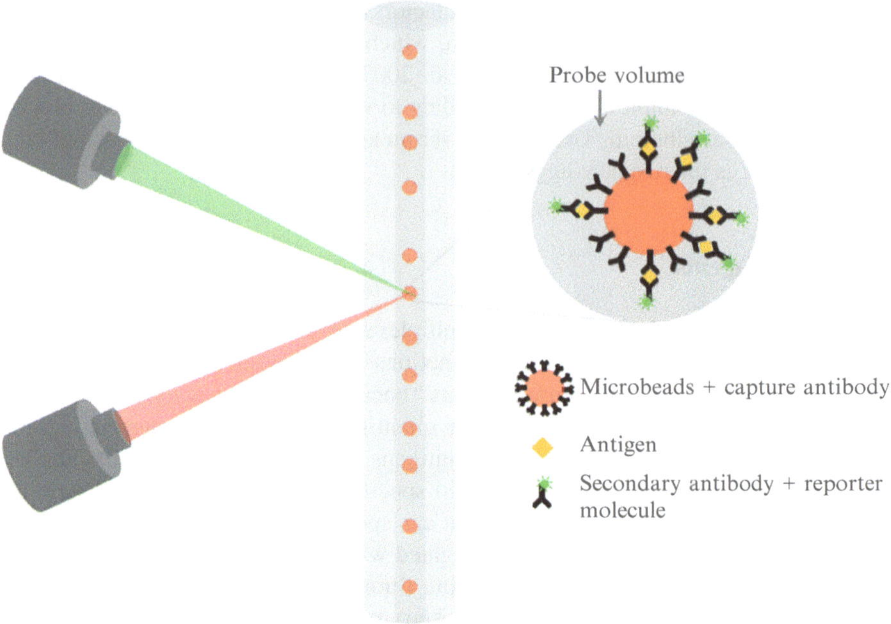

Fig. 18.1 Schematic representation of analyte detection using Luminex technology. The beads are arranged in a thin streamline before being introduced one at a time in the flow cytometer. A red laser (*dotted beam*) classifies the bead, identifying the bead type. A green laser (*full beam*) quantifies the assay on the bead surfaces. Only beads having an antigen–antibody complex fluoresce in the green, and the signal is a function of the antigen concentration

according to their dual-fluorescent intensities to sort them. A second measurement detects reporter signal intensity, which allows quantitation of analytes. Flow cytometers are well designed for these purposes.

It should be mentioned here that whereas flow cytometry has become widely used in many areas of biomedical science for the study of cells, later applications have been successfully conducted on single fluorescent molecules (Kellar et al. 2002). Although based on the principles of flow cytometry, for this type of measurements, the more apt term "flow fluorometry" should be used.

In a typical multiplexed array analysis, the encoded microbeads with varied reporter fluorescence intensities are hydrodynamically focused to a very thin sample stream, typically on the order of 10 μm in diameter. This focused sample stream passes through a focused laser beam on the order of 10 μm in height. The intersection of sample stream and laser beam, the probe volume (average dimension of 10 μm^3), is where cells or particles of up to 10 μm in diameter are lined up in a single file and pass one at a time through the laser beams, which excites the fluorescent dyes, as shown in Fig. 18.1 (Nolan and Yang 2007). In the illustration, the first diode red laser beam (635 nm) allows excitation of orange and red internal dyes of the beads. In response to the degree of excitation, emission spectra at extreme wavelengths are collected and associated to a specific bead set. Once sorted, detection and quantitation of the analyte can be carried out using a second laser beam (532 nm), which excites the green fluorescent dye (usually phycoerythrin) conjugated at the surface of the bead according to the degree of analyte reaction with the capture reagent on the bead. The intensity of the fluorescence emitted is finally detected through a photomultiplier tube. Measurement of side scatter light can also be simultaneously performed on the flowing particle stream to detect and reject oversized microspheres resulting from bead aggregation. Typical transit times through the probe volume are 10 μs or less for many commercial flow cytometers, enabling sample analysis rates of thousands of cells or particles per second. High-speed cell sorters can analyze up to tens of thousands of cells per second (Ibrahim and van den Engh 2003).

18.3.1.4 Microplate reader and software

Multiplexed flow cytometers can be automated by the use of a 96-well microtiter plate (ELISA type). In such sampling systems, a needle sucks up liquid sample from each well and brings it to the sample stream to expose the beads to the dual laser beam in the probe volume. The beads are sorted and counted until each type of bead in a set is detected at least 100 times. Specific software, providing communication between the hardware and the flow cytometer, processes the data and shows progress of the data acquisition. The software also detects aggregated beads, which is not taken into account in data acquisition. The real-time data acquisition allows comparison of the bead cloud positioning to a preestablished gate, corresponding to a set of microspheres with coded fluorescent red and orange dye intensities. A deviation of the cloud in relation to the theoretical scheme is usually caused by a defective internal marking of beads. Fluorophore stability of beads is affected by photobleaching. Carson and Vignali (1999) found that beads were irreversibly photobleached by prolonged exposure to light, causing them to fall outside their gates. Care must be taken to reduce light exposure to a minimum because only 2 h of direct white light exposure are sufficient to cause a significant number of beads to move out of their gates. Bad calibration of the flow cytometer according to the temperature at which the assay is performed is another issue causing deviation of the bead cloud. Lasers are sensitive to temperature variations as low as 3°C (Moalic et al. 2004). Therefore, it is recommended that bead-based arrays be performed in a temperature stable room to avoid temperature variation.

18.3.2 Bead-based assay procedure

In general, performing a multiplexed bead-based array assay consists of five main steps as follows:

1. *Preparation of coupled microbeads.* This step involves the conjugation of individual sets of microspheres with the target molecules required for each reaction

between analyte and capture molecule. Target molecule may be antigens, antibodies, oligonucleotides, receptors, peptides, enzyme substrates, etc.

2. *Preparation of fluorescent labeled reactants.* Reactants, which may be antibodies, antigens, oligonucleotides, and receptors, should have the ability to bind the target molecule. The labeling of reactants (i.e., reporter reactant) is fundamental for detecting analyte–target molecule reaction. A biotin molecule is often used to conjugate the reporter molecule, which allows subsequent binding of a reporter green fluorochrome (e.g., streptavidin-phycoerythrin).
3. *Optimization of the parameters for each assay.* In a non-multiplexed format, each analyte is detected separately using different reagent concentrations and incubation periods to verify positive reaction and set reagent concentration needed in multiplexed format analyses.
4. *Multiplexing of the assay.* Multiplexing of the assay is carried out by simply mixing the different sets of microbeads. The fluorescent reagents are also mixed to form a cocktail for the multiplexed reactions.
5. *Reaction with sample analytes.* The microspheres are reacted with the mixture of analytes, such as serum or blood samples, followed by the addition of the cocktail of tracer molecules. Samples are incubated for a given period to allow the mixture of microspheres and reagents to react. Various amount of green fluorescence appear on the beads' surface.
6. *Data acquisition.* Data acquisition, analysis, and reporting are performed in real time on all microsphere sets included in the multiplexed assay. As the beads are analyzed through the flow cytometer, they are immediately classified into their respective sets on the basis of orange and red fluorescence. The green fluorescence value of each bead is recorded as well. At least 100 individual microspheres of each set are analyzed and the mean value of their green fluorescence intensity is reported on the screen. The concentration of analytes in the sample can be determined by extrapolation from an internal standard.

Because the individual bead sets are separated by the fluid in the flow cytometer, many assays can be performed simultaneously to detect multiple analytes in a single sample. Multiplexing capability is only one of the benefits that this new technology provides.

18.3.3 Characteristics: benefits

Although some biological fields of application of bead-based assay are now well documented, the full potential of the technique has not yet been reached. There were eight key considerations for developing the perfect multiplexed assay as described by Vignali (2000). These considerations are detailed below and also highlight the benefits of the technique.

18.3.3.1 Specificity

Bead-based assays can be very specific, as methods can be developed that rely on the exclusive specificity of antibodies. The quality of the antibodies used can guarantee extremely high specificity. Monoclonal antibodies technology that allowed the production of a unique antibody in large quantities first reported by Köhler and Milstein (1975) is widely used for immunoassay purposes. Today, it is possible to produce pure and highly specific antibodies against almost any type of antigen.

18.3.3.2 Sensitivity

Sensitivity greatly depends on the quality of the fluorescent marker and the stability of the dye. Precautions have to be taken to prevent excessive exposure of the beads to light owing to the photosensitivity of the dyes. Sensitivity can be enhanced by the use of the chromophore phycoerythrin, an exceptionally bright reporter dye (Krishhan et al. 2009). Effective and optimal excitation of the label has also been achieved with a green yttrium aluminum garnet (YAG) laser in systems such as Luminex (Mandy et al. 2001).

Washing steps also influence sensitivity of multiplexed assay, especially if microbeads are lost through problems encountered by the use of filter plates (e.g., clogging, leaking, or unspecific adsorption of analyte onto the large surface of filters). A no-wash assay format is feasible because the reporter molecule is measured only at the surface of the bead. Residual or unreacted reporter molecules will remain in solution and will not contribute to the assay value, which can simplify the assay protocol relative to typical ELISA procedures (Haasnoot and Du Pré 2007). Reducing overall processing time and using washing reagents contributes to the greenness of the technique.

Sensitivity can be enhanced further by reducing the number of beads used per test. This increases the ratio of analyte to capture molecule for each test without reducing potential signal strength of the assay (the number of capture molecules per bead).

Background signal can be a factor that limits sensitivity. The measurement probe volume required in flow cytometry is rather small (approximately 10 μm^3), making this instrument especially sensitive for fluorescence detection. Although custom instruments have reported single molecule sensitivity (Kellar et al. 2002; Habbersett and Jett 2004), most commercial cytometers have detection limits of a few hundred molecules of a small organic fluorophore.

18.3.3.3 Simplicity

The simpler an assay is, the less there is that can go wrong. Although simplifying assays can lead to reduced sensitivity and specificity, it is an important consideration for routine analysis such as multiplexed assays. Elimination of the washing step in some protocols (Olivier et al. 1998; Tripp et al. 2000) and automation are advances that enhance the simplicity of bead-based arrays.

Moreover, the fluorescent readout of the multiplexed assay is more direct and sensitive than other techniques, such as ELISA, which requires enzyme amplification before colorimetric readout can be achieved. This additional step is time consuming and prone to variability and errors.

18.3.3.4 Reliability

Reliability and reproducibility as well as standardization between assays performed at different times and in different locations are of high importance for development of routine techniques. Previous studies demonstrated that reproducibility of bead-based assays was superior to conventional ELISA procedures for cytokine quantitation (Carson and Vignali 1999; Vignali 2000). Replication of multiplexed standards from several separate experiments exhibit very little variation in fluorescence in comparison with cytokine standards measured by ELISA. The high reliability and accuracy of multiplexed assays is also enhanced by the high number of measures (i.e., data are calculated from the mean of at least a 100 beads).

18.3.3.5 Multiplexing capabilities

Multiplicity is known here as the ability of an assay to measure multiple analytes at the same time, referred to as multiplexing capacity. Bead-based arrays are fundamentally meant for carrying multiplexed assays. Theoretically, taking the Luminex technology as an example, up to 100 different analytes can be detected in a single sample. Although there may be challenges in working with that many analytes, it potentially represents a huge benefit, which can be explored in several fields of applications. Additionally, only very low sample volumes are required for a wide range of analyses. Because high sample volumes can represent a challenge, especially in medical applications, multiplexed analysis may prove to be a very suitable technology to replace standard immunoassays for diagnostic assay. Reagent usage for microsphere-based assays is 10- to100-fold less than for microtiter-based assays because all assays are done at the same time with the same amount of reagents. Multiplexing ability of bead-based assay hence represents an advantage in terms of greenness.

18.3.3.6 Internal controls

Internal controls are needed during experimentation to avoid false-negatives or positives. With the multiplexing capabilities of bead-based assays, the addition of true internal controls for every individual test samples is possible and was demonstrated by Martins (2002). Using this feature, quality control parameters can be established to ensure that each individual analyte has been in contact with the appropriate reagent and that the reaction is free from any interaction effects.

18.3.3.7 Cost

Although the costs of instrumentation and reagents used for multiplexed assays are not negligible, bead-based assays are less expensive than classic immunoassays when multiplexing capability is used. A high degree of multiplexing will considerably reduce the cost of usage in comparison with individual analyses. In general, multiplexed assays become cheaper than ELISA when six or more analytes are measured simultaneously (Vignali 2000). The reduced costs include less reagents, energy, and material use and shorter manipulation time by technicians.

18.3.3.8 Time

Rapid analysis is always desirable, particularly in diagnostic settings. Simultaneous analyses on the same allow for a considerably decreased working time. Even though the measurement step takes longer for multiplexed analysis than for ELISA (10–90 s for sufficient measurements by flow cytometer compared with 3 s for photometric measure of 96 wells), time can be saved by using protocols that do not involve washing steps. For high-throughput studies, bead-based assays can be automated to deliver results rapidly. With high-affinity antibodies, such assays can be completed in less than 3 h.

18.3.4 Limitations

The multiplexed flow cytometric immunoassay, as an emerging technology, has some limitations. As with other immunoassays, multiplexed assays are prone to cross-reaction effects among reagents, sample analytes (antibodies, oligonucleotides, peptides, etc.), and capture molecules (Balsari and Caruso 1997; Vignali 2000; de Jager et al. 2003; Phillips et al. 2006). Such interference directly affects results, causing false-positive findings and a loss of quantitative response at high multiplexed levels. For this reason, pretesting can be carried out to ensure single reactions between analytes and capture molecules. A full bead mixture with the detection molecules for each kind of bead is incubated in the presence of a single analyte standard. If a positive reading is only found for the microspheres labeled with the specific capture molecule, no cross-reaction will be expected in the multiplexed assay. This pretesting step can be time consuming. The use of high-quality specific capture molecules that can only bind to one type of molecule remains the best option for successful multiplexed assays. The invention of monoclonal antibodies was a major step forward in the generation of unlimited resources of defined capture molecules. However, the identification and production of monoclonal antibodies can be laborious and could become very expensive, especially when rare antibodies that require screening of a large number of colonies are needed. In the quest for more economic options, novel approaches based on aptameters, which are synthetic oligonucleotide ligands able to recognize virtually any class of target molecules with high affinity and specificity and that could be directly immobilized on microspheres, are being developed (Jayasena 1999; Porschewski et al. 2006). This approach could be useful in a large number of applications because specific aptamers can be rapidly generated for almost any protein target.

In addition to multitarget capture molecules, proteins, soluble receptors, immune complexes, lectins, and other sources present in samples can cause interference or inhibition (Weber et al. 1990; Selby 1999). These occur especially in complex matrices such as blood and food products. Therefore, adequate measures have to be taken to minimize these effects (e.g., proper methods of sample preparation and the inclusion of adequate control steps).

18.4 Applications

Bead-based array assay relies on two general approaches: immunoassays (antibody–antigen reaction) and nucleic acid hybridization (probe–amplicon reaction). The first approach is used either for the detection of specific antigens (allergens, cytokines, toxins, etc.) or antibodies (immunoglobulins), whereas the latter approach can detect DNA or RNA sequences specific to viruses or pathogens. A brief summary of these techniques is presented, after which other known and emerging applications in food quality analysis are discussed.

18.4.1 Immunoassays

Antibodies or immunoglobulins (abbreviated Ig) are gammaglobulin proteins that are found in blood or other bodily fluids of vertebrates. They are used by the immune system to identify and neutralize foreign objects (antigens), such as bacteria and viruses. They are composed of basic structural units, each with two large heavy chains and two small light chains, forming a Y-shaped glycosylated proteins. There are several different kinds of antibodies, which are grouped into five isotypes (IgA, IgD, IgE, IgG, and IgM) based on the type of heavy chain they possess. Ig isotypes perform different roles, helping to direct the appropriate immune response for each different type of foreign object they encounter.

The general structure of all antibodies is very similar, except for a small region at the tip of the protein, known as the hypervariable region, which is extremely variable. This specific variability allows millions of antibodies with slightly different structures at the tip to exist, known as antigen binding sites. Each of these variants can bind to a different target (i.e., the antigen). This huge diversity of antibodies allows the immune system to recognize an equally wide variety of antigens. The epitope is the unique part of the antigen recognized by an antibody. These specific sites bind with their antibody in a highly specific interaction, called induced fit, which allows antibodies to identify and bind only their unique antigen among the millions of different molecules present in an organism.

Specific antibodies can be produced by injecting an antigen into a mammal, such as a mouse, rat, or rabbit. Blood isolated from these animals contains polyclonal antibodies in the serum (the antiserum). Polyclonal antibodies are multiple antibodies that bind to the same antigen, whereas monoclonal antibodies are all clones of a unique cell. These antibodies are widely produced for immunology testing, including multiplexed bead-based array assays.

Classic bead-based immunoassays generally employ the sandwich assay design. An antibody showing a high affinity for a specific antigen is conjugated to a set of microbeads to capture the antigen through its epitope. A fluorescent-labeled secondary antibody with a fluorophore is then used as a reporter for measurement (Fig. 18.1). Multiplexing this process, it is possible to measure simultaneously the levels of multiple analytes. The use of standards with known concentrations of analytes allows a relationship between the fluorescence signal and concentration to be established. Nonfood applications of this method are mainly in immunology with the detection of multiples cytokines specific to certain tissues or diseases such as cancers (Carson and Vignali 1999; De Jager et al. 2003; Keyes et al. 2003; Weber et al. 2003).

In the reverse situation, when the aim is to detect and characterize antibodies in fluids, beads are coupled with purified antigens. In a typical assay configuration, purified antigen is immobilized on beads and used to capture antibodies present in sera such as blood and plasma. Detection is carried out by a labeled class- or isotype-specific secondary antibody. This approach finds applications in infectiology by detecting exposure to infectious diseases (Faucher et al. 2004; Moss et al. 2004) and to monitor antigen-specific antibody responses to infection or vaccination (Emeny et al. 2002; Opalka et al. 2003).

Microsphere-based competitive inhibition immunoassay protocols similar to classic ELISA competitive immunoassays have also been developed (Kelly et al. 2005). Competitive inhibition assays are often used to measure small analytes because they only require the binding of one antibody rather than two, as is used in the standard sandwich immunoassay format. Because of the high probability of steric hindrance when two antibodies attempt to bind to a small molecule at the same time, a sandwich assay format may not be feasible; therefore, a competitive inhibition assay is preferable. This technique is based on the principle that the antigen in the unknown sample competes with labeled antigen to bind with antibodies. The competitive antigen is covalently attached to the binding groups of a distinct fluorescent bead set. For the coating with low molecular weight compounds, the compound is usually conjugated to a carrier protein such as biotin or ovalbumin. After the addition of the sample containing the unknown antigens, specific antibodies compete between the antigen-coupled beads and free antigen. The bead–antigen–antibody complex is then incubated with a fluorophore conjugated reporter antibody and analyzed by multiparametric flow cytometry. The amount of labeled antigen bound to the antibody site is subsequently measured. In this method, the response is inversely related to the concentration of antigen in the unknown sample (i.e., the greater the response, the less antigen in the unknown was available to compete with the labeled antigen). Therefore, this procedure is well suited for low concentration analytes for which response and sensitivity can be enhanced.

In the same way, inhibition immunoassays can be carried out by coating the beads with antibodies instead of antigens (Bellisario et al. 2000; Young Song et al. 2005). The sample containing the antigen to be analyzed and free biotinylated antibodies are then added to the beads to create a competitive reaction. The inhibition degree is revealed through detection of the fluorophore, which is added after the reaction is completed.

18.4.2 DNA hybridization

Hybridization is the process of establishing a noncovalent, sequence-specific interaction between two complementary strands of nucleic acids to form a single hybrid, referred to as a "duplex." In this process, DNA (or oligonucleotide or RNA) is heated to a temperature at which its backbone separates (also called the "melting temperature"). The DNA is allowed to mingle with the backbone of DNA from another species (i.e., DNA from a different species) to form hybrid DNA with one backbone from one species and a second backbone from another. This process works because hydrogen bonds form only between complementary base pairs. Oligonucleotides, DNA or RNA, will bind to their complement under normal conditions, so two perfectly complementary strands will bind to each other readily. In the absence of external negative factors, the processes of hybridization and melting may be repeated in an indefinitely long succession, which is the basis of the polymerase chain reaction (PCR), a technique used in molecular biology to amplify a sequence of DNA and generating thousands to millions of copies of a particular DNA sequence. The method consists of cycles of repeated heating and cooling of the reaction for DNA melting and enzymatic replication of the DNA. Primers consisting of short sequence of oligonucleotides, or amplicons (short DNA sequence) containing sequences complementary to the target region along with a DNA polymerase are key components to enable selective and repeated amplification.

Multiplexed nucleic acid analyses provide a new platform for high-throughput nucleic acid detection and are being utilized with increasing frequency. Specific applications of xMAP technology for nucleic acid detection are in the areas of single nucleotide polymorphism (SNP) genotyping, genetic disease screening, gene expression profiling, human leukocyte antigen (HLA) DNA typing, and microbial detection (Dunbar 2006). This technique requires functionalization of the microsphere surface with nucleic acid. Synthetic oligonucleotide is used for this purpose, even though natural DNA or RNA can also be used depending on the application. The most popular approach is to couple amino-modified nucleic acids to carboxylated surfaces using carbodiimide chemistry (Fulton et al. 1997). Target DNA or RNA is amplified using PCR and biotinylated-specific primers. The biotinylated PCR products obtained are melted to produce single biotinylated oligonucleotide sequences. These biotinylated amplicons are then put in contact with the microbeads. Fixation of amplicon to a specific probe at the surface of the beads is revealed by addition of streptavidin conjugated to phycoerythrin (Fig. 18.2).

Competitive hybridization assay format can also be useful when direct labeling of the target nucleic acid by PCR or other methods is not possible or undesirable.

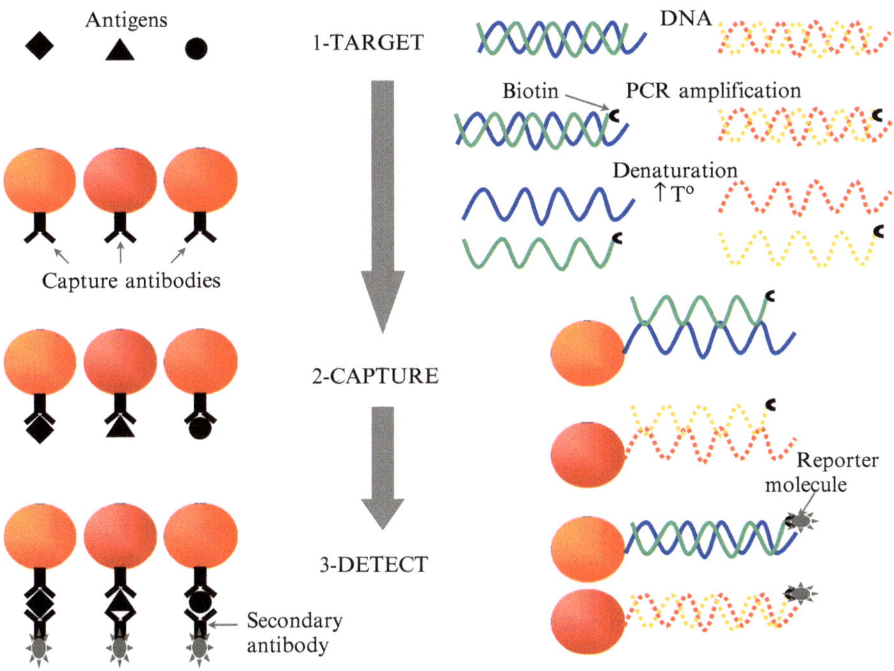

Fig. 18.2 Schematic representation of the xMAP assay formats. *Left* capture-sandwich immunoassay. *Right* nucleic acid hybridization

The protocol used is similar to the direct hybridization technique. In the absence of target DNA, the biotinylated competitor oligonucleotides hybridize to the allele-specific probe-coupled microsphere sets. The hybridized microsphere sets are labeled with streptavidin-phycoerythrin, resulting in 100% signal. When target DNA is present, the biotinylated competitor oligonucleotides hybridize to the target DNA instead of the allele-specific probe-coupled microsphere sets. The target DNA/competitor oligonucleotide hybrids are labeled with streptavidin-phycoerythrin, resulting in a reduction of signal on the allele-specific probe-coupled microsphere sets.

These techniques have found many applications in the medical fields such as for leukemia detection (Wallace et al. 2003) and bacteria and viruses detection through their specific sequence of nucleic acids (Cowan et al. 2004; Wallace et al. 2005).

18.4.3 Applications in food quality assessment

18.4.3.1 Detection of pathogens

The presence of pathogens (bacteria, viruses, and parasites) is a major concern in the assurance of food safety and quality. Rapid identification of pathogens prevents foodborne illness as it allows better control of processed foods. For this reason, many attempts have been made to replace isolation and identification methods requiring time-consuming and labor-intensive plating and biochemical testing procedures. Among these techniques, bioluminescence, cell counting, ELISA, and nucleic acid amplification are some of the most popular and effective (de Boer and Beumer 1999). However, these techniques are costly and can be affected by interfering substances in the sample. It is also desirable to have a test that can detect a wide range of pathogenic species using a single sample.

Dunbar et al. (2003) reported the first use of multiplexed assays for the detection of bacterial pathogens found in foods. The technique is rapid, cost effective with high multiplexing ability, and is well suited for food analysis. Nucleic acid detection of *Salmonella, L. monocytogenes,* and *C. jejuni*-specific DNA was carried out via hybridization. The assay was completed in 30–40 min postamplification and was shown to be sensitive. All of the bacterial species were correctly identified with a detection sensitivity of 2.5–500 organisms per ml depending on the species. Capture-sandwich immunoassays were carried out as well, with organism-specific antibodies coupled to biotinylated detection antibodies and streptavidin-R-phycoerythrin. Detection of a few to several hundreds of organisms per milliliter was obtained within 3 h with no apparent cross-reactivity. This first approach of multiplexed assays in food analysis demonstrated that it was a powerful technique that could be used as a routine analysis of food microbiological quality.

Further development in bead-based assay detection of pathogens led to the analysis of complex food matrices. A major issue is the presence of particulates found in such complex matrices. Because washing steps for microsphere-based assays rely on size exclusion filters or centrifugation to collect microspheres and remove excess reagents, food components are not separated. This problem, previously encountered for ELISA analysis of food sample, was resolved by Malkova et al. (1998), who developed an immunomagnetic separation technique to enrich samples before detection. In the same way, MagPlex microspheres have recently been developed by Luminex to assist streamlining of the sample preparation step. These microspheres are fluorescently coded for use in Luminex assays, but also contain superparamagnetic material for collection by a permanent magnet. The concept is to use magnetic force to selectively separate microspheres, leaving the particulate matter and interfering macromolecules generated by the food sample.

MagPlex microspheres were used by Bergervoet et al. (2008) to perform a detection of potato viruses from infected potato leaves, and recently by Kim et al. (2010) for the detection and identification of multiple pathogens in food matrices (apple juice, green pepper, tomato, ground beef, alfalfa sprouts, milk, lettuce, spinach, and chicken washes). The procedure proposed by the latter consists of multiple steps, allowing matrix food separation immunological reactions to happen:

1. Dilute homogenized food matrix.
2. Incubate MagPlex microspheres with sample food matrix dilution.
3. Capture pathogen-bound microspheres with magnet.
4. Remove debris and interfering macromolecules (proteins, carbohydrates, fibers, etc.) using magnet.
5. Resuspend and add biotinylated tracer antibodies.

6. Remove excess tracer antibodies with magnet.
7. Resuspend and add streptavidin-phycoerythrin.
8. Remove excess fluorophore with magnet.
9. Resuspend complexed immunoassay complex.
10. Analyze using a flow cytometer.

It should be mentioned, however, that although MagPlex microspheres facilitated recovery of the microspheres and targets from the food matrices using this protocol, analytes detection was somewhat inhibited. Inhibition can be attributed to the release of interfering proteins, carbohydrates, and small molecules from the food matrix during sample preparation. For example, high levels of polyphenols found in many fruits and vegetables have been shown to interfere with immunoassays (Henning et al. 2006). Milk matrices have also been found to decrease detection performance of some immunoassays (Homola et al. 2002; Son et al. 2007).

MagPlex microspheres can be used in immunoassays in complex food matrices, but higher limits of detection than those obtained in standard buffers can be found depending on the type of food matrix analyzed. Considerable interference occurs when assays are conducted in nonliquid homogenized foods. The impact on the limit for each assay needs to be ascertained before implementation of a testing protocol. For many food products, performing a simple wash instead of homogenization could provide better pathogen detection. Assay inhibition could be minimized and sample processing simplified by the elimination of the matrix removal steps, which would result in a more sensitive and overall more rapid assay. To enhance detection signals of the food fluids recuperated, an enrichment step could be carried out to increase the number of target pathogen cells present. Another approach using MagPlex-based assay to detect contaminants in the presence of food products would include the direct addition of growth media to the food sample followed by direct analysis of cultures immediately after enrichment. In the event of positive detection, such a procedure could significantly decrease the complexity of materials needed and even allow on-site testing.

18.4.3.2 Detection of toxins

Food consumers are susceptible to many illnesses that can occur from unintended contamination and that may occur during the manufacturing, handling, or distribution of food. Microbial pathogens represent an important source of contamination; however, other sources exist such as pesticides, antibiotics, pathogenic toxins, and undesired genetically modified organisms. Adaptations of bead-based assays have already been reported for detection of multiple contaminants in foodstuffs, which may replace time-consuming single-plexed methods in a field in which rapid methods are increasingly required. Examples of some recent food applications are described in the following.

Simultaneous detection of proteinaceous toxins in complex heterogeneous food mixture was recently reported by Garber et al. (2010). Phytotoxins (abrin and ricin), botulinum toxins, and *Staphylococcus* enterotoxins were blended in various diluted beverages and condiments and in bread and lettuce. Polyclonal and monoclonal antibodies were used against each toxin in a multiplexed bead-based assay. Paramagnetic microspheres were used to facilitate the detection of analytes in foods containing particulate matter. Similar studies were also conducted by Pauly et al. (2009) and Wang et al. (2009). Sensitivity of the method varies depending on the number of capture beads, sample volume, and incubation time. The right balance between sensitivity and throughput time of analysis has to be set depending on the purpose of the analysis. For example, an analysis aiming at an initial screening for food quality is meant to be rapid and sensitivity is not a major issue. Hence, high numbers of capture beads, low volume samples, and short incubation times are included in such screening protocols.

Antibiotics—compounds with antimicrobial activity—are used in veterinary medicine to treat and prevent disease and for other purposes, including growth promotion in animals. Consumption of antibiotics still raises questions in terms of potential risks of developing bacterial resistance to antibiotics. As a result, some regulatory agencies have banned their presence in meat products. Therefore, antibiotic detection methods have been developed, including microbial inhibition assays, immunoassays, chromatography techniques, and surface plasmon resonance techniques, some of which specifically focus on the detection of sulfonamides (Mellgren et al. 1996; Wang et al. 2006). Multiplexed screening analysis can favorably replace these techniques in terms of cost, rapidity, and throughput, as demonstrated by de Keizer et al. (2008) in a multiplexed immunoassay for the detection of multiple sulfonamides in raw milk. A single step of sample preparation consisted of filtration of the raw milk to remove milk particles that could interfere with the reactions owing to their diameter being in the same range as the microbeads. Multi-sulfonamide mutant antibodies were used against 11 different sulfonamide molecules in an inhibition competitive assay. Beads were coated with a sulfathiazole derivate, which captured the biotinylated antibodies. In the presence of free sulfonamide molecules in the sample, however, a certain proportion of the sulfathiazole derivate were not bound by the antibodies. The percentage of inhibition was quantified via the expression of the reporter molecule after only 1 h of analysis (after milk pretreatment). A similar procedure was used by Zou et al. (2008) for the detection of clenbuterol, a β-adrenergic agonist growth promoter associated with a series of severe food poisoning outbreaks worldwide (Prezelj et al. 2003). Although analysis was not performed on samples from real food matrices, it demonstrated high efficiency, biocompatibility, and selectivity, as well as ultra-low trace sample consumption and low cost.

Multiplexed assays for the detection of other food contaminants, including undesired genetically modified organisms (Fantozzi et al. 2007) and carcinogenic polycyclic aromatic hydrocarbons (Meimaridou et al. 2010) were successfully reported, showing a promising future for this technique in these food safety areas. Similarly, a triplex competitive inhibition immunoassay for simultaneous detection of cheaper plant proteins (soy, pea, and soluble wheat proteins) as potential fraudulent adulterants in milk powder has been reported using a rapid no-wash procedure (Haasnoot and Du Pré 2007). The examples presented here show the wide range of possibilities for bead-based assays and the economy of time and money they could bring to the food industry.

18.5 Future trends

Allergen detection is another important field in which multiplexed assays will no doubt be developed in the near future. Several immunoassay methods have been developed for the detection of food allergenic proteins such as peanut, the most common allergen found in foods (Mills et al. 1997; Blais and Phillippe 2000), hazelnut (Blais and Phillippe 2001), soy (Yeung and Collins 1997), and milk (Mariager et al. 1994). However, all of these assays present a major drawback, which is that they can only detect one allergenic protein at a time. The use of multiplexed assay protocols would represent a much more time- and labor-efficient approach, considering that a wide range of allergens can be detected in a single sample assay. A planar-type method using macroporous hydrophobic polyester cloth for immobilization of multiple allergen, developed by Blais et al. (2003), was successfully used for allergen detection from nuts. In a similar approach, microbeads could serve as solid support for specific antibodies against common allergens in foods.

Functional foods are health foods claimed to have health-promoting or disease-preventing properties beyond the basic function of supplying nutrients. With the growth in this market requirements for labeling have become increasingly complex. Analyses must be carried out to detect and quantify the presence of beneficial molecules or microorganisms and their bioactivity. Multiplexed analysis may provide a potential application in high throughput multianalyte detection of nutraceutical molecules in food products.

Detection of pathogens, contaminants, toxins, allergens, and bioactive molecules represent great future trends for multiplexed analysis. To improve the technology and its capabilities further, future efforts will be required in the expansion of bead labelling. Spectral analysis–based approaches to single particle analysis, coupled with new types of nanoparticle labels, offer a potential for significant increase in the levels of multiparametric analysis (Nolan and Yang 2007). A high throughput multiparametric analysis calls for a wide range of capture molecules such as antibodies. Although antibodies possess high specificity and sensitivity, they have been shown to cross-react with serum proteins or assay immunoglobulins (Andersson et al. 1989; Balsari and Caruso 1997; Vignali 2000). The development of specific ligand molecules (i.e., aptameters), offers an interesting avenue for highly multiplexed analysis (Porschewski et al. 2006). Future trends also include further automation of the multiplexed assays from sample preparation to analyte detection. Along with miniaturization and portability, this could provide a unique tool for on-site assays that could help to reduce time and cost.

18.6 Conclusion

Bead-based assays have found applications in a wide range of fields, including food safety and quality. In addition to contaminants and residues, which can be of concern when present in foods even at very low levels, rapid methods are also required for the detection of food allergens for which detection methods are still laborious. The growing functional foods market also presents challenges as the concentration of bioactive molecules needs to be accurately assessed for labeling. Additionally, fast and accurate analytical methods are required for nutritional labeling of regular foods, novel foods, and fortified and dietetic foods to ensure that appropriate claims are made on packaging. Mutliplexed bead-based assays have proved to be simple and robust techniques for the analysis of a wide range of analytes. Adaptation of current methods for the detection of a wider range of analytes can be made through the use of specific capture molecules coupled to labeled microbeads. Other advantages of bead-based assays as described in the preceding include speed, economy, versatility, and advanced multiplexing analytical capabilities. Furthermore, reduced labor and energy costs resulting from simultaneous as opposed to sequential detection of analytes provides economy of time and money. Furthermore, significant environmental benefits can be accrued because of lower reagent usage especially when the multiplexing capability of the method is used.

References

Andersson, G., H.P.T. Ekre, G. Alm, and P. Perlmann. 1989. Monoclonal antibody two-site ELISA for human IFN-γ. Adaptation for determinations in human serum or plasma. *Journal of Immunological Methods* 125: 89–96.

Balsari, A., and A. Caruso. 1997. Natural antibodies to IL-2. *Biotherapy* 10: 25–28.

Battersby, B.J., G.A. Lawrie, A.P.R. Johnston, and M. Trau. 2002. Optical barcoding of colloidal suspensions: Applications in genomics, proteomics and drug discovery. *Chemical Communications* 14: 1435–1441.

Bellisario, R., R.J. Colinas, and K.A. Pass. 2000. Simultaneous measurement of thyroxine and thyrotropin from newborn dried blood-spot specimens using a multiplexed fluorescent microsphere immunoassay. *Clinical Chemistry* 46: 1422–1424.

Bergervoet, J.H.W., J. Peters, J.R.C.M. van Beckhoven, G.W. van den Bovenkamp, J.W. Jacobson, and J.M. van der Wolf. 2008. Multiplex microsphere immuno-detection of potato virus Y, X and PLRV. *Journal of Virological Methods* 149: 63–68.

Blais, B.W., and L.M. Phillippe. 2000. A cloth-based enzyme immunoassay for detection of peanut proteins in foods. *Food and Agricultural Immunology* 12: 243–248.

Blais, B.W., and L.M. Phillippe. 2001. Detection of hazelnut protein in foods by enzyme immunoassay using egg yolk antibodies. *Journal of Food Protection* 64: 895–898.

Blais, B.W., M. Gaudreault, and L.M. Phillippe. 2003. Multiplex enzyme immunoassay system for the simultaneous detection of multiple allergens in foods. *Food Control* 14: 43–47.

Branum, A.M., and S.L. Lukacs. 2009. Food allergy among children in the United States. *Pediatrics* 124: 1549–1555.

Bryan, F.L. 1982. *Diseases transmitted by foods*. Atlanta: CDC.

Carson, R.T., and D.A.A. Vignali. 1999. Simultaneous quantitation of 15 cytokines using a multiplexed flow cytometric assay. *Journal of Immunological Methods* 227: 41–52.

Cowan, L.S., L. Diem, M.C. Brake, and J.T. Crawford. 2004. Transfer of a *mycobacterium tuberculosis* genotyping method, oligotyping, from a reverse line-blot hybridization, membrane-based assay to the Luminex multianalyte profiling system. *Journal of Clinical Microbiology* 42: 474–477.

de Boer, E., and R.R. Beumer. 1999. Methodology for detection and typing of foodborne microorganisms. *International Journal of Food Microbiology* 50: 119–130.

de Jager, W., H. te Velthuis, B.J. Prakken, W. Kuis, and G.T. Rijkers. 2003. Simultaneous detection of 15 human cytokines in a single sample of simulated peripheral blood mononuclear cells. *Clinical and Diagnostic Laboratory Immunology* 10: 133–139.

de Keizer, W., M.E. Bienenmann-Ploum, A.A. Bergwerff, and W. Haasnoot. 2008. Flow cytometric immunoassay for sulfonamides in raw milk. *Analytica Chimica Acta* 620: 142–149.

Dunbar, S.A. 2006. Applications of Luminex® xMAP™ technology for rapid, high-throughput multiplexed nucleic acid detection. *Clinica Chimica Acta* 363: 71–82.

Dunbar, S.A., C.A. Vander Zee, K.G. Olivier, K.L. Karem, and J.W. Jacobson. 2003. Quantitative, multiplexed detection of bacterial pathogens: DNA and protein applications of the Luminex LabMAP™ system. *Journal of Microbiological Methods* 53: 245–252.

Dunn, P.A., and H.W. Tyrer. 1981. Quantitation of neutrophils phagocytosis, using fluorescent latex beads. Correlation of microscopy and flow cytometry. *The Journal of Laboratory and Clinical Medicine* 98: 374–381.

Elshal, M.F., and J.P. McCoy. 2006. Multiplex bead array assays: Performance evaluation and comparison of sensitivity to ELISA. *Methods* 38: 317–323.

Emeny, R.T., C.M. Wheeler, K.U. Jansen, W.C. Hunt, T.M. Fu, J.F. Smith, S. MacMullen, M.T. Esser, and X. Paliard. 2002. Priming of human papillomavirus type 11—Specific humoral and cellular immune responses in college-aged women with a virus-like particle vaccine. *Journal of Virology* 76: 7832–7842.

Fantozzi, A., M. Ermolli, M. Marini, D. Scotti, B. Balla, M. Querci, S.R.H. Langrell, and G. van den Eede. 2007. First application of a microsphere-based immunoassay to the detection of genetically modified organisms (GMOs): Quantification of Cry1Ab protein in genetically modified maize. *Journal of Agricultural and Food Chemistry* 55: 1071–1076.

Faucher, S., A. Martel, A. Sherring, T. Ding, L. Malloch, J.E. Kim, M. Bergeron, P. Sandstrom, and F.F. Mandy. 2004. Protein bead array for the detection of HIV-1 antibodies from fresh plasma and dried-blood-spot specimens. *Clinical Chemistry* 50: 1250–1253.

Fulton, R.J., R.L. McDade, P.L. Smith, L.J. Kienker, and J.R. Kettman Jr. 1997. Advanced multiplexed analysis with FlowMetrix system. *Clinical Chemistry* 43: 1749–1756.

Gao, X., and S. Nie. 2004. Quantum dot-encoded mesoporous beads with high brightness and uniformity: Rapid readout using flow cytometry. *Analytical Chemistry* 76: 2406–2410.

Garber, E.A.E., K.V. Venkateswaran, and T.W. O'Brien. 2010. Simultaneous multiplex detection and confirmation of the proteinaceous toxins abrin, ricin, botulinum toxins, and *Staphylococcus* enterotoxins A, B, and C in food. *Journal of Agricultural and Food Chemistry* 58: 6600–6607.

Givan, A.L. 2004. Flow cytometry: An introduction. In *Methods in molecular biology: Flow cytometry protocols*, 2nd ed, ed. T.S. Hawley and R.G. Hawley. Totowa, NJ: Humana Press.

Gordon, R.F., and R.L. McDade. 1997. Multiplexed quantification of human IgG, IgA and IgM with the FlowMetrix™ system. *Clinical Chemistry* 43: 1799–1801.

Haasnoot, W., and J.G. du Pré. 2007. Luminex-based triplex immunoassay for the simultaneous detection of soy, pea, and soluble wheat proteins in milk powder. *Journal of Agricultural and Food Chemistry* 55: 3771–3777.

Habbersett, R.C., and J.H. Jett. 2004. An analytical system based on a compact flow cytometer for DNA fragment sizing and single-molecule detection. *Cytometry Part A* 60A: 125–134.

Henning, S.M., W. Aronson, Y. Niu, F. Conde, N.H. Lee, N.P. Seeram, R.P. Lee, J. Lu, D.M. Harris, A. Moro, J. Hong, L. Pak-Shan, R.J. Barnard, H.G. Ziaee, G. Csathy, V.L.W. Go, H. Wang, and D. Heber. 2006. Tea polyphenols and theaflavins are present in prostate tissue of humans and mice after green and black tea consumption. *Journal of Nutrition* 136: 1839–1843.

Homola, J., J. Dostálek, S. Chen, A. Rasooly, S. Jiang, and S.S. Yee. 2002. Spectral surface plasmon resonance biosensor for detection of staphylococcal enterotoxin B in milk. *International Journal of Food Microbiology* 75: 61–69.

Horan, P.K., and L.L. Wheeless Jr. 1977. Quantitative single cell analysis and sorting. *Science* 198: 149–157.

Ibrahim, S.F., and G. van den Engh. 2003. High-speed cell sorting: Fundamentals and recent advances. *Current Opinion in Biotechnology* 14: 5–12.

Jayasena, S.D. 1999. Aptameters: An emerging class of molecules that rival antibodies in diagnostics. *Clinical Chemistry* 45: 1628–1650.

Keij, J.F., and J.A. Steinkamp. 1998. Flow cytometric characterization and classification of multiple dual-color fluorescent microspheres using fluorescence lifetime. *Cytometry* 33: 318–323.

Kellar, R.A., W.P. Ambrose, A.A. Arias, H. Cai, S.R. Emory, P.M. Goodwin, and J.H. Jett. 2002. Analytical applications of single molecule detection. *Analytical Chemistry* 74: 316A–324A.

Kelly, S., T. Green, J. Van Doren, D. Puchalski, P. Boerckel, N. Chirmule, and M.T. Esser. 2005. Automation of a multiplexed competitive luminex immunoassay for measuring antibodies to human papillomavirus types 6, 11, 16, and 18. *Journal of the Association for Laboratory Automation* 10: 301–309.

Keyes, K.A., L. Mann, K. Cox, P. Treadway, P. Iverson, Y.F. Chen, and B.A. Teicher. 2003. Circulating angiogenic growth factor levels in mice bearing human tumors using Luminex multiplex technology. *Cancer Chemotherapy and Pharmacology* 51: 321–327.

Kim, J.S., C.R. Taitt, F.S. Ligler, and G.P. Anderson. 2010. Multiplexed magnetic microsphere immunoassays for detection of pathogens in foods. *Sensing and Instrumentation for Food Quality and Safety* 4: 73–81.

Köhler, G., and C. Milstein. 1975. Continuous cultures of fused cells secreting antibody of predefined specificity. *Nature* 256: 495–497.

Krishhan, V.V., I.H. Khan, and P.A. Luciw. 2009. Multiplexed microbead immunoassays by flow cytometry for molecular profiling: Basic concepts and proteomics applications. *Critical Reviews in Biotechnology* 29: 29–43.

Lisi, P.J., C.W. Huang, R.A. Hoffman, and J.W. Teipel. 1982. A fluorescence immunoassay for soluble antigens employing flow cytometric detection. *Clinica Chimica Acta* 120: 171–179.

Malkova, K., P. Rauch, G.M. Wyatt, and M.R.A. Morgan. 1998. Combined immunomagnetic separation and detection of *Salmonella enteritidis* in food samples. *Food and Agricultural Immunology* 10: 271–280.

Mandecki, W., E. Ernst, and N. Kogan. 2000. Light-powered microtransponders for high multiplex-level analyses of nucleic acids. *Abstracts of Papers of the American Chemical Society* 219: U755–U755.

Mandy, F.F., T. Nakamura, M. Bergeron, and K. Sekiguchi. 2001. Overview and application of suspension array technology. *Clinics in Laboratory Medicine* 21: 713–729.

Mariager, B., M. Solve, H. Eriksen, and C.H. Brogren. 1994. Bovine β-lactoglobulin in hypoallergenic and ordinary infant formulas measured by indirect competitive ELISA using monoclonal and polyclonal antibodies. *Food and Agricultural Immunology* 6: 73–83.

Martins, T.B. 2002. Development of internal controls for the Luminex instrument as part of a multiplex seven-analyte viral respiratory antibody profile. *Clinical and Diagnostic Laboratory Immunology* 9: 41–45.

McBride, M.T., S. Gammon, M. Pitesky, T.W. O'Brien, T. Smith, J. Aldrich, R.G. Langlois, B. Colston, and K.S. Venkateswaran. 2003. Multiplexed liquid arrays for simultaneous detection of simulants of biological warfare agents. *Analytical Chemistry* 75: 1924–1930.

Mead, P.S., L. Slutsker, V. Dietz, L.F. McCaig, J.S. Bresee, C. Shapiro, P.M. Griffin, and R.V. Tauxe. 1999. Food-related illness and death in the United States. *Emerging Infectious Diseases* 5: 607–625.

Meimaridou, A., W. Haasnoot, L. Noteboom, D. Mintzas, J. Pulkrabova, J. Hajslová, and M.W.F. Nielen. 2010. Color encoded microbeads-based flow cytometric immunoassay for polycyclic aromatic hydrocarbons in food. *Analytica Chimica Acta* 672: 9–14.

Mellgren, C., Å. Sternesjö, P. Hammer, G. Suhren, L. Björck, and W. Heeschen. 1996. Comparison of biosensor, microbiological, immunochemical, and physical methods for detection of sulfamethazine residues in raw milk. *Journal of Food Protection* 59: 1223–1226.
Mills, E.N.C., A. Potts, G.W. Plumb, N. Lambert, and M.R.A. Morgan. 1997. Development of rapid dipstick immunoassay for the detection of peanut contamination of food. *Food and Agricultural Immunology* 9: 37–50.
Moalic, V., B. Mercier, and C. Ferec. 2004. Technologie Luminex™: Principe, applications, et perspectives. *Immuno-analyse and Biologie Spécialisée* 19: 181–187.
Moss, D.M., J.M. Montgomery, S.V. Newland, J.W. Priest, and P.J. Lammie. 2004. Detection of cryptosporidium antibodies in sera and oral fluids using multiplex bead assay. *The Journal of Parasitology* 90: 397–404.
Nicewarner-Peña, S.R., R.G. Freeman, B.D. Reiss, L. He, D.J. Peña, I.D. Walton, R. Cromer, C.D. Keating, and M.J. Natan. 2001. Submicrometer metallic barcodes. *Science* 294: 137–141.
Nolan, J.P., and F. Mandy. 2006. Multiplexed and microparticle-based analyses: Quantitative tools for a large-scale analysis of biological systems. *Cytometry Part A* 69: 318–325.
Nolan, J.P., and L. Yang. 2007. The flow of cytometry into systems biology. *Briefings Functional Genomics and Proteomics* 6: 81–90.
Olivier, K.G., J.R. Kettman, and R.J. Fulton. 1998. Multiplexed analysis of human cytokines by use of the Flowmetrix system. *Clinical Chemistry* 44: 2057–2060.
Opalka, D., C.E. Lachman, S.A. MacMullen, K.U. Jansen, J.F. Smith, N. Chirmule, and M.T. Esser. 2003. Simultaneous quantitation of antibodies to neutralizing epitopes on virus-like particles for human papillomavirus types 6, 11, 16, and 18 by multiplexed Luminex assay. *Clinical and Diagnostic Laboratory Immunology* 10: 108–115.
Parod, R.J., and J.D. Brain. 1983. Uptake of latex particles by macrophages: Characterization using flow cytometry. *The American Journal of Physiology* 245: C220–C226.
Pauly, D., S. Kirchner, B. Stoermann, T. Schreiber, S. Kaulfuss, R. Schade, R. Zbinden, M.A. Avondet, M.B. Dorner, and B.G. Dorner. 2009. Simultaneous quantification of five bacterial and plant toxins from complex matrices using a multiplexed fluorescent magnetic suspension assay. *The Analyst* 134: 2028–2039.
Phillips, D.J., S.C. League, P. Weinstein, and W.C. Hooper. 2006. Interference in microsphere flow cytometric multiplexed immunoassays for human cytokine estimation. *Cytokine* 36: 180–188.
Porschewski, P., M.A.M. Grättinger, K. Klenzke, A. Erpenbach, M.R. Blind, and F. Schäfer. 2006. Using aptameters as capture reagents in bead-based assay systems for diagnostic and hit identification. *Journal of Biomolecular Screening* 11: 773–781.
Prabhakar, U., E. Eirikis, and H.M. Davis. 2002. Simultaneous quantification of proinflammatory cytokines in human plasma using the LabMAP™ assay. *Journal of Immunological Methods* 260: 207–218.
Prezelj, A., A. Obreza, and S. Pecar. 2003. Abuse of clenbuterol and its detection. *Current Medicinal Chemistry* 10: 281–290.
Sampson, H.A. 2004. Update on food allergy. *The Journal of Allergy and Clinical Immunology* 113: 805–819.
Selby, C. 1999. Interference in immunoassay. *Annals of Clinical Biochemistry* 36: 704–721.
Sicherer, S.H. 2002. Food allergy. *The Lancet* 360: 701–710.
Son, J.R., G. Kim, A. Kothapalli, M.T. Morgan, and D. Ess. 2007. Detection of *Salmonella enteritidis* using a miniature optical surface plasmon resonance biosensor. *Journal of Physics* 61: 1086–1090.
Steinkamp, J.A., J.S. Wilson, G.C. Saunders, and C.C. Stewart. 1982. Phagocytosis: Flow cytometric quantitation with fluorescent microspheres. *Science* 215: 64–66.
Stewart, C.C., and J.A. Steinkamp. 1982. Quantitation of cell concentration using the flow cytometer. *Cytometry* 2: 238–243.
Tripp, R.A., L. Jones, L.J. Anderson, and M.P. Brown. 2000. CD40 ligand (CD154) enhances the Th1 and antibody responses to respiratory syncytial virus in the BALB/c mouse. *Journal of Immunology* 164: 5913–5921.

Vignali, D.A.A. 2000. Multiplexed particle-based flow cytometry assays. *Journal of Immunological Methods* 243: 243–255.
Wallace, J., Y. Zhou, G.N. Usmani, M. Reardon, P. Newburger, B. Woda, and G. Pihan. 2003. Barecode-all: Accelerated and cost-effective genetic risk stratification in acute leukemia using spectrally addressable liquid bead microarray. *Leukemia* 17: 1411–1413.
Wallace, J., B.A. Woda, and G. Pihan. 2005. Facile, comprehensive, high-throughput genotyping of human genital papillomaviruses using spectrally addressable liquid bead microarrays. *The Journal of Molecular Diagnostics* 7: 72–80.
Wang, S., H.Y. Zhang, L. Wang, Z.J. Duan, and I. Kennedy. 2006. Analysis of sulphonamide residues in edible animal products: A review. *Food Additives and Contaminants* 23: 362–384.
Wang, J., Y. Yang, L. Zhou, J. Wang, Y. Jiang, K. Hu, X. Sun, Y. Hou, Z. Zhu, Z. Guo, Y. Ding, and R. Yang. 2009. Simultaneous detection of five biothreat agents in powder samples by a multiplexed suspension array. *Immunopharmacology and Immunotoxicology* 31: 417–427.
Weber, T.H., K. Käpyaho, and P. Tanner. 1990. Endogenous interference in immunoassays in clinical chemistry. A review. *Scandinavian Journal of Clinical and Laboratory Investigation* 201(Suppl): 77–82.
Weber, J., V.K. Sondak, R. Scotland, R. Phillip, F. Wang, V. Rubio, T.B. Stuge, S.G. Groshen, C. Gee, G.G. Jeffery, S. Sian, and P.P. Lee. 2003. Granulocyte-macrophage-colony-stimulating factor added to a multipeptide vaccine for resected stage II melanoma. *Cancer* 97: 186–200.
Yeung, J.M., and P.G. Collins. 1997. Determination of soy proteins in food products by enzyme immunoassay. *Food Technology and Biotechnology* 35: 209–214.
Young Song, E., C. VanDunk, T. Kuddo, and P.G. Nelson. 2005. Measurement of vasoactive intestinal peptide using a competitive fluorescent microsphere immunoassay or ELISA in human blood samples. *Journal of Immunological Methods* 300: 63–73.
Zhao, Y., X. Zhao, C. Sun, J. Li, R. Zhu, and Z. Gu. 2008. Encoded silica colloidal crystal beads as supports for potential multiplex immunoassay. *Analytical Chemistry* 80: 1598–1605.
Zou, M., H. Gao, J. Li, F. Xu, L. Wang, and J. Jiang. 2008. Rapid determination of hazardous compounds in food based on a competitive fluorescence microsphere immunoassay. *Analytical Biochemistry* 374: 318–324.

Chapter 19
Greening of research and development

Joyce I. Boye, Anne Maltais, Steve Bittner, and Yves Arcand

19.1 Introduction

A typical laboratory building uses 5–10 times as much energy and water per square foot as a typical office building (USEPA 2008). Research and development buildings need extensive computer and networking systems and expensive equipment that generate heat, consume energy, and require a constant energy supply and back-up emergency power. In addition, research facilitates use large amounts of hazardous chemicals and solvents that, if not managed properly, result in significant waste. To protect the health of workers, many of these buildings must meet stringent requirements for ventilation and material handling, which further contribute to energy use.

Maintaining a pollution-free environment, reducing costs of R&D, and reducing chemical wastes are subjects of increasing concern to all scientists, educators, and the general public (www.microscale.org). As an example, the recent increase in the cost of acetonitrile because of a worldwide shortage has forced the research community to rethink the usage of this important solvent. Whereas much effort has been devoted in recent times to identify ways to green the industrial sector, much less emphasis has been placed on the greening of R&D as a whole, and more particularly in the agricultural and agri-food processing sectors, in which there continues to be growing concerns about the environmental impacts of current practices.

There is some information in the literature on the greening of technology and ecotechnology; improving systems of land utilization, social and industrial, and energy supply; models of decision making of various supply chain structures

J.I. Boye (✉) • A. Maltais • S. Bittner • Y. Arcand
Food Research and Development Centre, Agriculture and Agri-Food Canada, Saint Hyacinthe, QC, Canada
e-mail: Joyce.Boye@agr.gc.ca; Anne.Maltais@agr.gc.ca; Steve.Bittner@agr.gc.ca; Yves.Arcand@agr.gc.ca

J.I. Boye and Y. Arcand (eds.), *Green Technologies in Food Production and Processing*, Food Engineering Series, DOI 10.1007/978-1-4614-1587-9_19,

and analyzing their impact on players; and the supply chain and technology-economy-ecology linkages that may help to define and accomplish environmentally sustainable development, to name a few examples (Kemp and Soete 1992; Matsuda et al. 1995; Jorgensen 2001; Lukman et al. 2009; Ghosh and Shah 2011).

As early as 20 years ago, Kemp and Soete (1992) discussed concerns about the focus of the industrial sector on short-term benefits instead of longer-term optimality and the need for institutional adaptation, and talks about the relationship between economic growth and particular trajectories of technological change. They found that although it was clear that some of the technological trajectories being used at the time had reached their environmental limits and needed to be replaced with environmentally friendlier trajectories, change was hindered by inertia, risk aversion, technical, economic, and institutional barriers.

In another relevant study conducted at the Engineering Campus of the University of Maribor in Slovenia, researchers found that heating and construction of buildings were the "hot spots" for most environmental impacts (Lukman et al. 2009). In comparing different waste management options for plastic and paper, including recycling, incineration, and landfill, combining 70% recycling, 29% incineration, and 1% landfill was found to be the most economically and environmentally sustainable approach for managing waste generated from the facility.

Greening of R&D in the agriculture and agri-food sectors goes further than waste management, however. This chapter proposes a greener philosophy and framework for the conception, management, and execution of R&D that looks at the greening of R&D from a much broader perspective. In addition to identifying ways to decrease energy use, greenhouse gas (GHG) emissions, waste, and operational costs, a greener R&D management framework focuses efforts on sustainable management of all resources, including human, physical, material, and financial resources required in the conduct of research. As research organizations increasingly have limited access to such resources, the greening of R&D must include reflections on how to foster partnerships and collaboration so as to avoid duplication while maximizing the potential for success. Therefore, this chapter further provides information on how to improve the management of intellectual property so as to enhance partnerships.

19.2 Greener approaches to research and development

19.2.1 Chemical use

Research activities require the use of expensive chemical products and create great amounts of waste. In addition to the high cost of chemicals and the attendant economic losses when these chemicals are not used judiciously, disposal of hazardous wastes created through research activities presents a danger to people as well as the environment. In many jurisdictions, regulatory agencies determine which chemicals are considered to be hazardous. In Canada the Workplace Hazardous and Materials

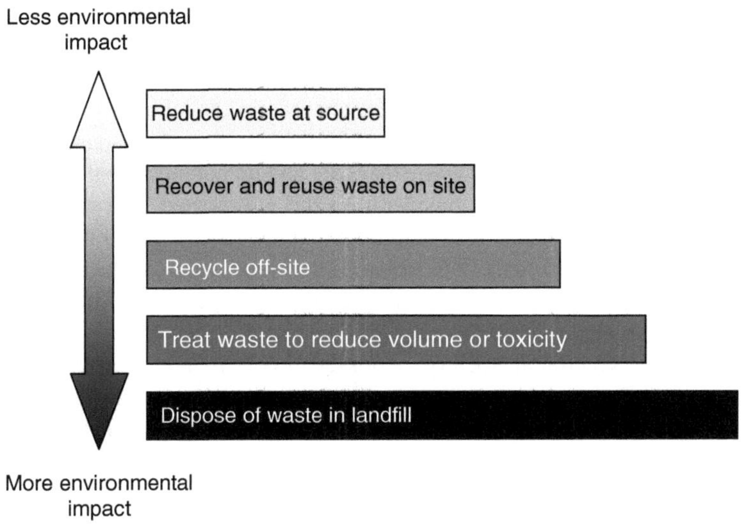

Fig. 19.1 The waste management hierarchy

Information System (WHMIS) classifies hazardous substances into the following six groups: compressed gases, flammable and combustive materials, oxidizing materials, poisonous and infectious materials, corrosive materials, and dangerously reactive materials. The waste management hierarchy, an internationally recognized waste management guide, classifies waste management strategies according to their desirability (Fig. 19.1) to improve environmental outcomes. Because hazardous wastes are highly undesirable, they should be minimized at the source through different management strategies, starting with avoiding use, reducing use, reusing, recycling, recovering, treatment, and disposal, the latter being the least preferable option. Cost of disposal of wastes is also a good incentive to minimize use. The cost of methanol disposal, for example, can double when diluted to a 50% concentration.

19.2.2 Purchasing chemicals

The question of which chemical and how much to purchase is one of the most important decisions in waste minimization and a critical step in the greening of R&D. Because most chemicals have an expiration date, buying more chemicals than the amount needed for specific R&D activities can result in significant waste if these products are not used on time. It is estimated that unused chemicals constitute up to 40% of the waste generated by a laboratory (ACS 1994). In addition to chemicals becoming useless or unstable after long periods of storage, the integrity of the containers may also become compromised because of poor conditions of storage, increasing hazardous risks. In many laboratories, these challenges are further compounded by poor labeling practices.

The desire to reduce costs through bulk purchasing can be very tempting; however, unused chemicals can present a safety hazard in the laboratory and are likely to be difficult and expensive to dispose of. For this reason, bulk purchasing is not recommended. When the cost of waste disposal and its potential health and environmental impacts are taken into consideration the anticipated cost savings of bulk purchase becomes less attractive. Additionally, the risk of generating inaccurate data increases when chemicals with expired shelf lives are used during experimentation. The economic consequences of such an outcome can be dire and are frequently not considered.

The American Chemical Society's booklet *Less is better: laboratory chemical management for waste reduction* (ACS 1993) gives several reasons for ordering chemicals in smaller containers, even if that might be more expensive and may mean using several containers of specific chemicals for a single experiment. Some of these reasons are provided in the following:

- The risk of breakage is substantially reduced for small package sizes.
- The risk of accident and exposure to the hazardous chemical is less when handling smaller containers.
- Storeroom space needs are reduced when only a single size is inventoried.
- Containers are emptied faster, resulting in less chance for decomposition of reactive compounds.
- The large "economy size" often dictates a need for other equipment, such as a smaller transfer container, funnels, pumps, and labels. Added labor to subdivide the larger quantities into smaller containers, as well as additional personal protective equipment for the hazards involved also may be needed.
- If unused hazardous material must be disposed of, the disposal cost per container is less for smaller containers.

Thus, in addition to determining the simple costs of purchasing chemicals, it is important to assess costs associated with the life cycle of each chemical (i.e., all costs associated with the use, storage, and disposal of each chemical through its lifetime at a research center) (National Research Council 1995). The initial purchasing cost only represents a part of the costs associated with a product's life cycle (i.e., handling cost, human, financial, and disposal costs must be taken into account as well). Paying closer attention to the management of chemicals in a laboratory at the early stages of purchasing allow a shift to the upper sections of the management hierarchy presented in Fig. 19.1, thereby lowering chemical wastage at the source.

19.2.3 Experimentation

In most research laboratories, the design of experiments focuses on ensuring that the right types of treatments are administered to the experimental units to be tested and that accurate and reliable data are generated to answer research questions posed or to test stated hypotheses. To reduce risks and avoid waste of energy, as well as human and chemical resources, reflections at the design stage of experimentation should also include an evaluation of all potential sources of hazardous waste expected from manipulation and equipment usage. Selection of methods, chemicals, and type of equipment should take into consideration their environmental impacts and waste output. Knowing at which stage of the experiment hazardous wastes will be accumulated makes it easier to incorporate strategies in the experimental design to minimize waste. Examples of some strategies that can be considered include the following:

- Use literature reviews and computer modeling to get as much information as possible on possible experimental outcomes. Use all available information to the fullest before designing experiments.
- Carefully plan and revise experimental protocols to avoid the production of unnecessary and/or unusable data.
- Use adequate statistical experimental designs to maximize results (and/or confidence in results) while minimizing the number of samples to analyze. This may include stepwise experiments of smaller size instead of one huge experiment only to discard unimportant variables; estimating the right number of replicates, etc.
- Carry out chemical reactions and other laboratory procedures on a smaller scale.
- Consider minimization of material used at each step of an experiment.
- Use less solvent to rinse equipment, by carrying out several rinses with small volumes of solvent instead of few rinses with large amounts of solvent.
- Use more sensitive analytical equipment.
- Recycle and reuse materials when possible.
- Substitute less hazardous chemicals when possible.
- Isolate nonhazardous waste from hazardous waste.
- Include follow-up reaction steps in the experiment plan that deactivate hazardous chemicals or reduce toxicity.

Some of these strategies can have huge positive environment impacts. Among these, the reduction of chemical use to minimal levels at which experiments can be effectively performed, an approach known as microscale chemistry, seems to be a promising trend.

Microscale chemistry is an environmentally safe pollution prevention method of performing chemical processes using small quantities of chemicals without compromising the quality and standard of chemical applications in education and industry (www.microscale.org). The approach focuses on markedly reducing the amounts of chemicals used during experimentation. Although traditional chemistry often requires grams of materials in the conduct of experiment, only milligram quantities are required for microscale chemistry. The approach further promotes the use of safe and easy-to-manipulate techniques, miniature laboratory ware, and high-quality skills, allowing a decrease of chemical use by as many as three orders of magnitude. Microscale chemistry is increasingly being used as a teaching method in schools and universities to promote environmental consciousness in the conduct of research. Benefits include:

- Reduced chemical use promoting waste reduction at the source
- Improved laboratory safety owing to better laboratory air quality
- Reduced exposure to toxic chemicals
- Absence of fire and explosion hazards
- Reduced risks of spills and accidents
- Reduced laboratory costs
- Shorter experimentation times
- Implementation of excellent laboratory manipulative techniques
- Reduction in glass breakage cost
- Fewer requirements for storage space
- Cleaner, more productive work environment
- Promotion of the principle of the 3Rs: Reduce, Recover, and Recycle
- Promotion of "green chemistry"
- Change in the psychology of people in regard to the use of chemicals
- User-friendly for people with physical disabilities

A wide range of experiments can be scaled down to follow microscale protocols (Skinner 1997). A typical Kjeldahl reaction is a good example of an experiment that calls for potentially harmful chemicals that should be reduced to a minimum, as indicated in Fig. 19.1. In fact, the Kjeldahl reaction protocol uses mercury as a catalyst, which causes disposal problems at the end of the experiment. Converting to a micro-Kjeldahl equipment reduces the quantities of chemicals needed, which can reduce waste by up to 90%. This could translate into a reduction of several liters of waste per day in laboratories that routinely run Kjeldahl reactions. Furthermore, Campins-Falco et al. (2008) have demonstrated the speed and accuracy of using the microscale Kjeldahl reaction protocol for nitrogen determination.

19.3 Greener buildings

19.3.1 Green construction

In addition to the potential impacts of R&D, energy conservation and reduction of greenhouse gases and the carbon footprint should form part of the initial considerations in the design and construction of chemical laboratories. The availability of some nonrenewable raw materials has decreased significantly over the past decades. Growing consumption of renewable and nonrenewable materials at a rate higher than that at which these materials can be replaced is greatly responsible for this situation. This trend could be reduced, at least partially, through better utilization of materials and the creation of appropriate opportunities for renewable resources to replenish themselves.

This has led to the development of green constructions (e.g., also called sustainable construction, green building) which brings life cycle thinking into the construction of building. It involves the practice of creating structures and using processes that are environmentally responsible and resource efficient throughout a building's complete life cycle from site selection to design, construction, operation, maintenance, renovation, and deconstruction. This practice expands and complements the classical building design concerns of economy, utility, durability, and comfort (http://www.epa.gov/greenbuilding/pubs/about.htm). Green buildings are designed to reduce the overall impact of the built environment on human health and the natural environment by: (1) efficiently using energy, water, and other resources; (2) protecting occupant health and improving employee productivity; and (3) reducing waste, pollution, and environmental degradation. For example, it may involve choosing the site and the building structure to minimize excavation, using solar passive, geothermal, and "vegetalized" roofs as heating/cooling systems, using windows with shades and optical channels to bring natural light in all rooms, using low flow toilets and rainwater as gray water, etc.

Green building codes and standards has appeared in numerous countries. They are sets of rules created by standards development organizations that establish minimum requirements for elements of green building such as materials or heating and cooling designs (http://en.wikipedia.org/wiki/Green_building#cite_note-epa.gov-0) and are all based on ISO/TS 21931:2006 sustainability in building construction, which is not an assessment system in itself but is intended to be used in conjunction with, and following, the principles set out in the ISO 14000 series of standards. Most standards have different levels of achievement, such as the "certified," "silver," "gold," and "platinum" levels of the LEED system.

Whenever possible, greener alternatives should be considered when choosing materials for construction. A product's environmental impact is the sum of several factors. To help in making the best choice from a safety, economic as well as an environmental perspective, tools such as product life cycle assessment (PLCA) are increasingly being used. Product life cycle assessment is a framework that can be employed to identify environmental inputs, outputs and impacts within the lifetime of a product. The framework considers material types, water, and energy use. All manufacturing processes are inventoried for the evaluation. In general, the life cycle of a product can be grouped into four main stages: raw material acquisition, manufacturing, use (including reuse and maintenance) and recycling and waste management (Public Works and Government Services Canada 2000). These are explained in greater detail in the following.

19.3.1.1 Phase 1: raw material acquisition

The predominant environmental factor that should be considered in the first stage of PCLA is to select a renewable raw material over a nonrenewable resource. Nonrenewable resources such as petroleum or tropical rain forests are created over such a long period of time that replenishment is not taken into consideration for sustainable development. Renewable resources include products obtained in much shorter periods of time and for which the length of time needed for renewal can be monitored. This makes it much easier to ensure the sustainability of renewable resources. Today, certification programs are increasingly being established that provide an assurance that harvesting and extraction of resources (e.g., wood) are being conducted in a sustainable manner. A green choice of raw materials should, in addition to originating from a renewable resource, not be listed on the Convention in Trade in Endangered Species (CITIES) list. Products are considered renewable when they can be verified through a certification mechanism that they are made of materials harvested under controlled conditions, providing the necessary regeneration time required, unless their supply is so abundant that self-regeneration is good enough to minimize environmental impacts.

Recycling is another answer to problems of resource depletion. Recycling materials significantly reduces the amount of solid waste ending up in landfills. The use of materials with recycled content and remanufactured products are, therefore, another great green choice for building construction because they reduce the necessity to extract and process raw materials, thereby reducing the environmental footprint of the product. When put together, these smart choices have lower environmental impacts and contribute toward sustainable development. Furthermore, they help to supply the needs of the present without compromising the ability of future generations to meet their needs.

19.3.1.2 Phase 2: manufacturing

To avoid health hazards and negative environmental impacts, the reduction and/or elimination of the use of toxic materials should be a central priority during manufacturing. Whenever possible, these should be replaced with greener alternatives. Vapors of solvent and particulates produced from toxic materials during construction can cause adverse effects on workers, but also on building occupants, which can continue many years after the end of the construction. The disposal of materials fabricated or treated with toxic substances represents another issue because few options exist. Proper disposal of these undesired wastes requires shipment to facilities designed to handle hazardous waste. This is important because burying toxic materials in landfill sites contributes to land, water, and air pollution. With increased awareness of the need for greener behaviors and policies, numerous international eco-labeling programs are emerging that establish acceptable toxicity levels as part of the product guidelines.

Greenhouse gases released during the life cycle of a product have to be taken into consideration in the manufacturing process. Carbon dioxide and methane are two of the most significant greenhouse gases, and both contribute to global warming. Reducing energy requirements through the use of energy-efficient products and greener manufacturing processes can reduce the generation of carbon dioxide. Compared with carbon dioxide (CO_2), however, methane is more than 20 times more effective in trapping heat in the atmosphere (www.epa.gov). Landfill sites produce up to 30% of the methane gas produced by artificial sources and other activities; thus, the less waste put in landfills, the lower will be the rate of methane release.

In addition to carbon dioxide and methane, chlorofluorocarbons (CFCs) also contribute to global warming through thinning of the ozone layer. Because these gases have global warming potential (GWP) of 500–15,000 (for most popular commercial substances) and can remain in the atmosphere for many decades, controlling their release into the environment is of primary importance. Chlorofluorocarbons are frequently used in cooling appliances and as propellants in different building products and materials. Although presently banned in many countries, CFCs and hydrochlorofluorocarbons (HCFCs) still represent a threat to the planet, and care should be taken to ensure that they are eliminated when establishing building material specifications.

19.3.1.3 Phase 3: use, reuse, and maintenance

Energy conservation, efficiency, and economy are all important criteria that should be taken into consideration in the selection of materials and equipment for the construction of greener research and chemical laboratories. Energy conservation calls for behavioral changes, such as turning off lights in empty rooms, and installation and use of operable windows that allow for natural cooling and ventilation. Energy efficiency, on the other hand, focuses on the use of eco-labeled appliances and fixtures that reduce overall energy demand. Energy-efficient products reduce environmental impact by reducing the environmental cost associated with the generation of electricity. For example, carbon dioxide emissions resulting from the burning of fossil fuels and flooding of lands for the building of hydroelectric dams are major drawbacks of high electricity demand. Thus, appliances that reduce energy use contribute positively to a greener environment. Energy economy applies to all material or equipment that provides effective energy use through the regulation of electricity consumption (e.g., use of thermal insulation or high performance lighting systems).

Construction materials can also have an impact on air quality. Depending on the construction materials selected and their composition (e.g., adhesives, paint, particle boards, and fabrics), variable concentrations of volatile organic compounds (VOCs) and formaldehyde can be released into the air, which can affect the health of workers. Symptoms of VOC exposure can vary and may include headaches, fatigue, dizziness, and blurred vision, which can affect employee morale and productivity. The occurrence of these chemicals in the workplace environment can be minimized through the use of proper ventilation methods and the selection of greener alternative materials.

The durability of materials selected for construction is another important criterion that should be taken into consideration because it provides reduced environmental impact by minimizing the maintenance or replacement requirements of a product. The warranty provided by manufacturers can give a good estimation of product durability. Product testimonials can also provide another source of information to verify durability claims.

19.3.1.4 Phase 4: recycling and waste reduction

Products can continue to have huge environmental impacts even at the end of their usefulness. Selecting materials and products that can be recycled and returned for reprocessing into new material is a greener choice for the environment. Ideally, products fabricated from many different materials should be designed in such a way as to facilitate recycling. This can be facilitated by easy disassembly and identification of material types by icons and codes. Moreover, degradable products that break down into materials that can be used for building construction or the development of other products can be sent to specific commercial composting facilities at the end of their life.

Green construction also calls for source reduction, which can be easily incorporated in the planning stages of construction projects. For example, suppliers can be required to minimize packaging, liquid products such as paint can be purchased in bulk, and careful demolition can allow the reuse of items such as doors and windows. As clients increasingly request for products manufactured with minimal waste generation, including packaging, industry will be encouraged to respond by finding new and effective ways to reduce waste.

19.4 Greener laboratory design and operation

The design of chemical laboratories has greatly evolved in the last decades. There is currently an increased demand for construction favoring social interaction and team-based research. Examples of this trend include open laboratory design concepts and flexible engineering, allowing easy reorganization of the work space. In addition to facilitating collaboration among researchers, the new model of laboratory design has to ensure sustainability and the reduction of negative potential environmental impacts.

19.4.1 Social buildings for team-based research

The benefits of collaboration and tapping into mass ideas to enhance creativity and innovation are increasingly being recognized in the scientific and research communities. As a result, scientific research is increasingly characterized by social activities and engagements such as team work, conferences and meetings, interlaboratory collaboration, and the use of subcontracting services. The most productive and successful scientists and scientific research centers are those who remain intimately familiar with the work going on in other laboratories and who actively seek collaboration and partnerships. These scientists and laboratories develop a strong capacity to adopt new research approaches and tools as quickly as they become available.

In response to this trend, scientific and research activities function best when they are supported by architecture that facilitates both structured and informal interactions, flexible use of space and sharing of resources (http://www.wbdg.org/resources/labtrends.php). A key consideration in designing such an environment is to establish places such as meeting rooms and even break rooms in which people can gather together outside their labs to meet and exchange ideas.

Effective collaboration within a research team also calls for scientists with varying expertise to form interdisciplinary research units. To connect people and organizations as well as to share data and protocols within a research team, laboratories can be specifically designed to facilitate such interaction. Creating flexible engineering systems that encourage research teams to adjust their spaces to meet their needs, designing offices as places where people can work in teams, and creating all the space necessary for research team members to operate properly in close proximity when needed are examples of essential elements favoring collaborative work.

In order to respond to the demands of team-based research, open laboratory designs have become increasingly popular. The open laboratory concept is very different from the closed laboratory concept typical of older laboratory buildings, which was based on accommodating the needs of an individual researcher. In open laboratories, researchers share not only the space itself, but also equipment, bench space, and support staff. The concept of open laboratories also includes the use of shared software and data analysis tools to improve speed and efficiency. Networking software that can be connected to several computers rather than having one for each equipment or team saves money and increases speed and efficiency. Overall, open laboratory concepts can facilitate communication among scientists and make the laboratory space more adaptable for future needs.

However, closed laboratories will continue to be essential for some specific research or equipment. Electron microscopes, gas chromatographs coupled to sniffing detectors, darkrooms, tissue culture labs, allergen testing labs, and glass washing are all examples of equipment and activities that must be housed in separate spaces.

19.4.2 Increasing efficiency through collaboration and contracting-out of research

Beyond construction, chemical management, and waste minimization, the greening of R&D should also include efforts toward enhanced inter-laboratory collaboration and the use of specialized analytical laboratory services when appropriate. Scientific research calls for the use of very expensive materials, research equipment, and technologies. Increased partnership and collaboration among scientists and use of professional analytical services can help to lower the overall cost of research, enhance efficiency, and reduce waste. Moreover, sharing of resources is eco-friendly, facilitates social interaction in many regards, and can help to avoid the duplication of resources.

When there is a need for a specific analysis that is not routinely done in a given laboratory, scientific personnel should consider using external laboratories that provide such services routinely (contracting-out of research), rather than purchase reagents and equipment that will only be used minimally. This not only saves money, but also saves the time and energy required to train personnel and manage materials needed to conduct these types of analyses. Analytical laboratories that run specialized experiments can manage better the optimization of chemical and equipment usage. Contracting-out research also applies to animal trials and other clinical trials in which specialized laboratories may be the most efficient and logical approach to use.

Laboratories providing analytical services may themselves have to subcontract out part of the obligations of their contract. One of the most important factors in subcontracting analytical services is the need to certify the quality of the results obtained from the service provider. In this regard, testing laboratories are governed by the international standard ISO/IEC 17025 General Requirements for the Competence of Testing and Calibration Laboratories. ISO/IEC 17025 provides standards and guidelines to follow for all aspects of a laboratory's quality system, from management requirements and document control, to sampling of products, creating test reports, and subcontracting (www.iso.org).

Collaboration with partners who conduct specific activities not routinely carried out in a given laboratory is another approach that may be considered to increase efficiency. In this case, the terms of the collaboration need to be clearly discussed, defined, and documented before beginning any experiments so as to avoid issues surrounding the management of intellectual property (IP). The next few sections provide some insights about IP management.

19.4.2.1 Management of intellectual property

Intellectual property takes many forms, such as copyrights, confidential information, patents, trade secrets, trademarks, designs, plant breeder's rights, etc. It is essentially the property emerging from creative undertakings. The challenge of collaborative research is not merely the difficulties posed by the research question or technological hurdle for which a solution must be found. There are other challenges related to managing resources, accountability, administration, accounting, finances, interim and final reporting, confidentiality, protection of intellectual property, valuation of intellectual property, and marketing of the results that need to be managed effectively for the collaboration to be truly successful.

Collaborative research has many advantages because it benefits from access to additional teams from different organizations, including the complementarity of the skills of specialists, access to laboratory equipment and pilot facilities, and even to additional funding from different sources. This makes it possible to develop more ambitious research protocols with greater scientific risks and potential innovation outcomes, and there is also a higher probability for successful transfer of knowledge and technology with social and commercial benefits.

Successful collaborative partnerships must start with well-written research proposals and protocols, having clear general objectives, scientific subobjectives, activities, and subactivities, which reflect the capacities and potential competencies of each team and each organization. Evidently, such proposals must respect the mission of each organization and its statutes, regulations, directives, and policies.

Once the research protocol is completed and each team's contribution is clearly defined, a relatively accurate picture develops of the scientific process and the timelines and schedule for completion. This step is already a significant achievement in the establishment of a multidisciplinary and multipartite research partnership. Research activities should, however, not be initiated until an agreement has been drafted, reviewed, revised, finalized, and approved by the duly authorized signatories for each organization.

In addition to clear understanding of the roles of the partners actively involved in the collaboration, parts of the research work may need to be contracted to a service provider. That provider of such specialized services will require a specific mandate that clearly outlines what he or she is expected to produce and deliver at the end of the contract. This, however, does not make the provider a collaborator if there is no scientific or financial risk assumed by the provider. Such a specialized service provider will deliver the services provided and will be paid for these services and can therefore not be considered as a collaborator. The services of such a provider may be critical to the success of the collaboration because of their recognized scientific and technological expertise and because that phase of the project is best handled within a service contract. That is to say that, in some instances even though the signatories to a collaborative endeavor may have the skills and equipment themselves to conduct the phase in question, it might be more appropriate to contract it out as a means of optimizing the resources allocated to the collaborative research project.

What is meant by collaborative research, then? It is a sharing of the financial and scientific risks, human and material resources, and the joint management of a research activity, project, or program among institutions, organizations, or people. In some cases, results generated fall within the public domain and can be handled by a joint publication agreement. Other collaborative research results are protected as intellectual property by patents, whereas others are protected as industrial secrets.

Intellectual property–related issues and research collaboration among government, universities, and small and medium enterprises (SMEs) are a fascinating world. Licensing and collaborative research agreements between institutions collaborating on a project are essentially business agreements. Despite the numerous types of agreements that exist, each new collaborative agreement has unique attributes and may present unique challenges. This is particularly the case when partnerships involve federal research departments that have a dual economic and scientific mission (public good research).

Supporting scientific creativity is the pith and marrow of research. As such, researchers increasingly have to find novel ways to create partnerships by stimulating creativity, while creating the right environment for IP right and investment protection. The key question in this regard is how to sustain and grow the business of R&D (i.e., create economic value) while expanding the boundaries of knowledge and the generation of commercial applications and outcomes. This question is of critical importance because once results become public knowledge, they become the property of the whole world, which is why the control of confidential and strategic research information is an important cornerstone of science and innovation, and a key driver of competitiveness.

In developing collaborative agreements, it is of the utmost importance to understand the respective research environments of each party. Some IP terms and provisions can be very error-prone and even contentious, because of the unwillingness of partners to understand the respective research environment of each party. This frequently occurs in situations in which there are both academic and commercial-oriented projects and outcomes. Government research institutions often have greater flexibility to develop alternative research collaborative agreements with partners than the private sector.

These challenges have resulted in the general perception that agreements on the handling of IP are deal breakers or that they can significantly slow down the negotiation of agreements. However, this perception is false. Although it is true, in some cases, that these negotiations take too long, IP terms are not the most problematic issue in collaborative agreements.

Other issues that slow down the licensing, drafting of research agreements, and negotiation processes are secrecy, confidentiality and security obligations, liability, indemnification, infringement, litigations, disputes, warranties, termination provisions, publication terms, overhead calculation, sources of funding, budget and payment terms, different fiscal year end dates for each party, and human resource issues related to the people involved in the research projects (e.g., students, post doc, technical staff) and the agreement approval process (i.e., financial officers, accountants, and legal advisors of all parties). In cases of multi-partite agreements, care must be taken to ensure that discussions are held with all key stakeholders, which may be time consuming. Early planning and engagement is therefore necessary to avoid unnecessary delays. Furthermore, honest, direct, and open communication is a central tenet of successful collaboration; unfortunately, this is often neglected and reduced to a prolonged exchange of e-mails.

Collaboration is vital in today's competitive economic environment and is no longer an option for companies and institutions that want to remain competitive. Research institutions need to learn the value and dynamics of collaboration to get the most value from invested resources. The challenge is to build on the value of established relationships in a timely manner, and to use a win–win effective collaborative or partnership approach for commercialization and/or technology transfer of outcomes. Thus, there is a shared responsibility to find new ways of winning together.

A list of some of the key steps that should be taken into consideration in the handling and management of IP are provided in the following.

1. Respect the scientific and financial contributions of all collaborators.
2. Identify background intellectual property rights.
3. Recognize joint ideas.
4. Recognize foreground intellectual property rights.
5. Recognize mutual and complementary scientific and technological skills of the parties.
6. Recognize co-inventorship and deal with co-ownership rights related to invention disclosures developed jointly.
7. Ensure equity in the joint development of protocols and experimental designs.

8. Recognize contributions of co-inventors/co-innovators during the conduct of research and in obtaining joint results.
9. Respond to expectations of partners in regard to demands for exclusive rights or "not-to-exploit" IP.
10. Maximize the chances of success of collaborators in finding financial and other partners to scale up and commercialize useful, usable results.
11. Maximize spinoffs from the return on R&D investment with partners who co-develop and use the results.
12. For government institutions, move from a government subsidy or sponsorship model to a research collaborative or partnership paradigm.

19.4.2.2 A refined understanding of intellectual property

A refined understanding of IP can infuse new life and change the perceived social and economic benefits of partnerships and collaboration. Protection of IP fosters innovation in all sectors, including the agri-food sector, and provides benefits to all innovators and the future of the industry facing a competitive world and society as a whole. This also refers to governments, which are often signatory members of many world trade agreements and treaties.

In recent years we have witnessed a world legislative boom in the IP sector and its applications in the fields of health and agriculture. Not only is IP protection important, it is also vitally and critically a major piece of the puzzle for all those involved in creativity, innovation, and business. Intellectual property protection is important because it gives companies and investors the necessary confidence to invest more in research, development, and innovation that will benefit subsectors, sectors, and society as a whole.

Intellectual property protection is a shared responsibility among many parties, entities, institutions, and individuals and plays a key role in successful partnerships and projects, which is why it is important to educate all players in the innovation chain (including users) about its benefits.

Technology is a key driver of the economy and productivity of many countries and institutions, and it is increasingly important to fully understand the role of IP and IP protection in this dynamic. Intellectual property attracts investors and is an integral part of value-added creation and plays a fundamental role in venture-capital funding. As an example, small emerging technology firms often have a challenge in accessing funds for start-up. Venture capitalists will frequently only support companies who can clearly demonstrate that they control their IP and that it offers competitive and sustainable advantages. Furthermore, very few companies will be able to afford law suits and judicial procedures owing to IP infringement, which can sometimes cost up to half a million dollars, which is another reason why IP control is of the utmost importance.

All personnel engaging in research should, therefore, be educated and encouraged to develop, protect, and exploit IP arising out of research activities that result in potential commercial IP in accordance with the IP policy and terms of their respective establishments. Intellectual property–related R&D projects can be translated into improvements in cereals and crops, yields, pest tolerance, disease resistance, agronomic efficiency, new cultivars, increase in farm income, healthier foods, specialty food ingredients (e.g., low saturated fat, high omega oils, high antioxidants), new vaccines, new renewable resources, larger choice of adapted plant cultivars, improved competitiveness, improved productivity, and so on.

For the funding organization, these inventions represent significant investments in R&D. For example, it takes on average 12 years to develop a cherry tree variety and 15 years to develop a wheat or cereal cultivar, representing millions of dollars in research investment. For IP rights owners and holders, economic benefits may be translated into millions of dollars a year for a single commodity such as a new wheat variety. A small fraction of these benefits, of the order of 2% or 3%, is captured in the form of royalties and returned to licensors. The remaining benefits goes to farmers, growers, industry, and the society as a whole.

Intellectual property is an asset, and entrepreneurs and business firms have to pay attention and care during all initial stages of technology emergence and startup. Intellectual property rights can differentiate a business from its competitors and can be very appealing to consumers and the public at large. Intellectual property is critical to all corporate growth strategies just like HR. Companies and licensees can, therefore, not expect to effectively commercialize their IP related products and services if they do not pay attention to the chain of IP ownership. This is why early recognition of IP protection is critical and should be considered as an important asset that contributes to a company's commercial success.

Increasingly, the value of IP is seen as being equal to, if not more important than, other assets. Some corporations and firms today recognize IP as the most valuable asset of the business, which needs to be protected. Intellectual property is a multi-million dollar business to copyright holders (publishing business, film business, CDs, etc.), trademark holders, patent holders, trade secret holders, etc. The questions of due diligence, freedom-to-operate (FTO), IP risks, business risks, financial liabilities, legal issues, and infringement are all concerns to IP developers that should, therefore, be effectively managed.

As an anecdotal example, protection of IP is really not new. In a different "world" soldiers were instructed to chop off the hand of the craftsmen who built the Taj Mahal so that another one could never be created. Today, in a more civilized world, IP rights rest on access to a pool of intellectual resources, which is why they matter: They also suggest free authorized access or controlled and regulated usage or access so as to capture innovation returns. It is for this reason that many organizations have Technology Transfer and Intellectual Property Offices that dedicate efforts to assess invention disclosures for potential IP protection needs and identify commercialization strategies.

In summary, the simple objective behind IP management is to encourage investment in innovation. Intellectual property rights are not necessarily a hurdle to collaboration. Intellectual property management and use, however, may present challenges that need to be properly handled. Therefore, serious thought needs to be given to the cost of developing IP, protection costs, access costs, innovations gains and investments, and the net social cost or gain. For those who are against the notion of IP, it may be important to note that even if IP protection (copyright or patent) was abolished, firms, universities, and governments would probably come up with contractual restrictions enforceable against unauthorized end-users. It is, therefore, important to foster a positive environment of debate on the subject of IP handling and management and impress upon stakeholders the importance of adopting flexible, creative policies that respect intellectual property rights while enhancing partnership and innovation.

19.5 Greener administration

In addition to the issues discussed in the preceding, the greening of R&D should include other peripheral activities such as office administration–related activities, which can contribute positively to the environment and result in eco-friendly behaviors beyond the office space. Ten good behaviors that could be adopted by office administration to decrease energy demand and consumption loads identified on the website of the "journaldunet" (www.journaldunet.com/management/0602/0602121gestes-ecolo.shtml) are summarized in the sections that follow.

19.5.1 Print as little as possible

It may not always be necessary to print a document, especially if it can be read and reviewed on a computer screen. Saving web pages, documents, and articles as PDF files and classifying them into folders to ease reading and accessibility are good options for consideration. Using e-mail instead of faxes or sending faxes directly from your computer also eliminates the need for hard copies. However, when printing is absolutely needed, using both sides of a sheet of paper and selecting only the pages you really need are simple ways to reduce the amount of paper used. Furthermore, documents can be reformatted by reducing the size of the margins and the images, which not only saves paper, but also ink cartridges. Paper printed on only one side can also be used to make notepads by cutting each sheet in quarters and making a stack held together by a staple.

19.5.2 Recycle ink cartridge and paper

Many people are unaware that ink cartridges can be recycled. More than 13 cartridges are discarded in the United States every second, according to *CartridgeFundraising.com*. This represents hundreds of millions of cartridges (made with plastics) that end up in landfills every year, and that can take up to

1,000 years to decompose. Solutions to cut down on this waste include the following: first, some inkjet cartridges can be used more than once by buying a refill kit; second, many companies, such as Hewlett-Packard, take back used cartridges for free and they even provide mailers to ship them back to be recycled or refurbished; third, if a company is too busy to return used cartridges, many charities pick them up and recycle them to fund their activities.

In regard to paper recycling, this has been increasingly encouraged by advertising campaigns and the growing availability of recycling bins. Paper shredders also should be provided in offices for confidential documents and the shredded documents recycled rather than thrown in the garbage.

19.5.3 Turn the lights off when leaving any empty room

When offices do not use automatic lighting system, employees should be encouraged to turn off lights when leaving a room. This can significantly lower energy costs and prolong the lifetime of lamps. Many offices keep their lights on both during the day and at night, which is a huge and unnecessary waste.

19.5.4 Chose sustainable lighting designs and low energy light bulbs

Lighting can be the greatest load for electricity in laboratories, and yearly costs can run as high as that for heating (i.e., for cold climates). With recent advances in light technology, the energy associated with lighting can be reduced significantly while still providing required levels of illumination. Sustainable lighting strategies include the use of linear and compact fluorescents lamps (CFLs) rather than incandescent lamps and employing various photosensing technologies to conserve energy.

Incandescent lamps are extremely energy inefficient, using only 10% of the energy they consume to produce light (the rest is given off as heat). Fluorescent lamps should be used instead, as they are about three to five times as efficient as standard incandescent lamps and can last about 10–20 times longer. Regular dust removal on light bulbs is also a good way to improve light efficiency.

Other than using efficient sources of light, maximizing the availability of natural daylight is an important principle of sustainable design (www.wbdg.org). Not only does it reduce energy use, but it also increases comfort and enhances productivity. In the planning and design of chemical laboratories and research facilities, every effort should be made to include in the design concepts the direction of natural light into as many laboratory spaces and public areas as possible such that employees can have the opportunity to look outdoors and be cognizant of weather conditions and time of day. A general principle should be to use day lighting as the primary source of illumination and resort to artificial lighting only as a supplement to day lighting.

Although daylight may provide many benefits in regard to energy saving and a sense of well-being, it may also present some challenges to chemical laboratories (e.g., wide variations in temperature, which is sometimes undesired in laboratories; difficulties in controlling lighting when tests are carried out on photosensitive materials; etc.). Hence, the art and science of proper day lighting design is not so much the provision of adequate daylight to an occupied space, but how to do so without any undesirable side effects (www.wbdg.org). In addition to adding windows or skylights to a given space, attention also has to be paid to the careful balancing of heat gain and loss, glare control, and variations in daylight availability. For example, successful day lighting designs invariably pay close attention to the use of shading devices to reduce glare and excess contrast in the workspace (www.wbdg.org).

There are day lighting control systems available today that can determine the amount of light available in a given space and switch off one or more banks of lights whenever there is enough sunlight. Photosensing devices, for example, can be used to control off-on switches for exterior lights, triggering fixtures to add light to a particular area when light levels decline. Also, a number of new fluorescent and metal halide fixtures are available and employ daylight harvesting by storing solar energy in the fixture during daylight hours and then using that energy to run the lamp when daylight diminishes (www.wbdg.org).

19.5.5 Reduce air conditioning

In addition to being costly in terms of energy use, air conditioners contribute to the emission of greenhouse gases. During the summer, the outdoor temperature may be tolerable enough (i.e., not too high) making air conditioning unnecessary. At other times, alternatives such as the use of fans can provide enough fresh air to working spaces. In the same way as air conditioning is overused during summer time, some offices are overheated during the winter. In addition to high costs related to heating, broad variations of temperature between outdoors and indoor can be potentially harmful to human health. The use of monitored thermostats, which maintain a more uniform temperature and can be set at lower temperatures at night and during weekends for example, can help to save a lot of money by lowering energy demand and reducing temperature variations.

19.5.6 Use reusable cups

Plastic cups take up to 100–1,000 years to decompose when not recycled. Although their use may appear to have insignificant impacts on the environment, an office of 100 workers uses anywhere from 25,000 to 75,000 nonreusable cups yearly. In many Western countries, this represents billions of cups thrown in the garbage every year. An easy and cheap way to avoid such waste is to provide individual cups or encourage employees to use their own personal cups.

19.5.7 Favor virtual meetings to gathering

Current developments in web-based, tele-conferencing, and video-conferencing technologies provide modern facilities with many alternatives to actual face-to-face meetings. Although face-to-face gatherings and meetings are important and should not be ignored, the use of conference calls, videoconferences, and webinars should be considered as the first options whenever possible. Virtual meetings, especially with distant collaborators, can be a great way to cut down cost while maintaining frequent exchange of information and can be beneficial for the environment as well as contributing to better work–life balance.

19.5.8 Work with energy star–labeled equipment

Working with Energy Star–labeled computers and instruments can help to save costs and energy, and reduce GHG emissions. The Energy Star system in North America was jointly developed by the United States (US) Environmental Protection Agency and the US Department of Energy and is aimed at protecting the environment through the identification and labeling of energy-efficient products and practices (www.energystar.gov). The label guarantees that a piece of equipment or an instrument uses lower energy than the average for other similar equipment in the same category, and it also includes all computers and computer-related equipment such as printers, scanners, and photocopiers. Another easy way to reduce energy use is to turn off computer devices and laboratory equipment during the night and on weekends. This also helps to decrease energy costs. For example, a French study carried out in 2002 estimated that employers who did not encourage their employees to adopt such energy saving behaviors were responsible for a national loss of 128 million euros (http://www.journaldunet.com/management/0602/0602121gestes-ecolo.shtml). Furthermore, this excess energy used causes supplementary CO_2 emissions.

19.5.9 Encourage carpooling

Carpooling is an environmentally friendly and sustainable way to travel that reduces carbon emissions, road traffic, and the need for parking space. Authorities often encourage carpooling, especially during high pollution periods and after fuel costs rise. Employers can facilitate and promote carpooling by advertising and creating a website with the schedules of employee to make matching easier. In addition to being an environmentally friendly way to travel, carpooling can bring other benefits, such as creating new friendships and decreasing driving stress as each driver gets a break from being at the wheel.

19.5.10 Reduce use of water

Although the proper washing of hands is of critical importance in chemical and research laboratories for health and safety reasons, a significant amount of water wastage can be reduced by not leaving water running when washing hands. Furthermore, it may not always be necessary to wash hands with hot water, which uses not only heat but also wastes cold water before it gets hot. Employees can be reminded about these simple behavioral changes through the use of eye-catching posters strategically placed in bathrooms and offices. Water use can also be greatly reduced by providing toilets with dual flow devices, which can cut down water use by 3–5 L per flush.

Laboratory techniques and equipment often need a temperature-controlled environment. It was common practice in the past to run cold tap water as a coolant. This practice should be avoided because it results in substantial waste of water, and in some localities can result in water bills that can rapidly add up to several thousand dollars per year per equipment. Long-term sustainable thinking suggests the purchase and use of recirculating chiller baths dedicated to an equipment or moved from bench to bench as needed.

19.6 Conclusion

Innovation and creativity, underpinned by R&D, will continue to be key drivers of the economy as we move into the future. With the increasing costs of chemicals, energy and human resources, government institutions and employers in both the private and not-for-profit sectors must begin to embrace a culture change in which the greening of R&D is not simply seen as a burdensome environmental sustainability endeavor, but as a modified way to improve efficiency, increase productivity, and establish and maintain economically viable partnerships. For example, pharmaceutical laboratories have to follow strict guidelines on methods used for analysis, which limits flexibility in miniaturization as well as the replacement of solvents and other chemicals with greener alternatives. Efforts to simplify this process need to be considered. Another important factor to mention is that the "silo mentality" in R&D, which continues to be pervasive in some fields of research, creates unhealthy duplication and competition and is increasingly recognized as unproductive. In addition to reducing waste at the source and selecting the right materials to use in the design and construction of R&D buildings and in the conduct of research, efforts should be made to minimize waste and duplication by fostering greater partnerships and collaboration. In this regard, greater understanding and simplification of IP agreements and management will contribute to more effective collaboration and knowledge generation. From the instrumentation perspective, the development and purchase of highly sensitive weighing scales are needed to facilitate the measurements of smaller amounts of materials for preparing solutions to minimize waste. Until then, a good practice may be to buy pre-weighed chemicals or prepared solutions that have the required accuracy in weight and concentration. Although this may appear more expensive in the short term, when the cost of wastage and manipulation time are factored in the benefits become more evident. Finally, employee education and the promotion of the benefits of a greener approach to R&D should be maintained as part of a greener R&D management framework because it will likely have benefits far beyond the present time and far beyond the workplace environment.

Disclaimer: The views expressed by the authors are meant only to provide general information and do not represent the official position of Agriculture & Agri-Food Canada. Final advice on any of the matters discussed should be sought from recognized qualified professionals.

References

ACS (American Chemical Society). 1993. *Less is better: Laboratory chemical waste management for waste reduction*, 2nd ed. Washington: ACS (American Chemical Society).

ACS (American Chemical Society). 1994. *Laboratory waste management: A guidebook.* Washington, D.C: ACS (American Chemical Society).

Campins-Falco, P., T. Meseguer-Lloreta, T. Climent-Santamariaa, and C. Molins-Legua. 2008. A microscale Kjeldahl nitrogen determination for environmental waters. *Talanta* 75: 1123–1126.

Ghosh, D., and J. Shah. 2011. A comparative analysis of greening policies across supply chain structures. *International Journal of Production Economics*, In Press, Corrected proof, Available online 30 May 2011, ISSN 0925-5273, doi: 10.1016/j.ijpe.2011.05.027.

Jorgensen, U. 2001. Greening of technology and ecotechnology. In *International encyclopedia of the social & behavioral sciences*, 6393–6396. Oxford: Pergamon. 1016/B0-08-043076-7/04181-4. ISBN 978-0-08-043076-8.

Kemp, R., and L. Soete. 1992. The greening of technological progress: An evolutionary perspective. *Futures* 24: 437–457.

Lukman, R., A. Tiwary, and A. Azapagic. 2009. Towards greening a university campus: The case of the University of Maribor, Slovenia, Resources. *Conservation and Recycling* 53(11): 639–644.

Matsuda, S., T. Kojima, and H. Komiyama. 1995. Technological problems in systems approach of greening. *Energy Conversion and Management.* In *Proceedings of the second international conference on carbon dioxide removal*, vol. 36 Oxford/New York, pp 927–930.

National Research Council. 1995. *Prudent practices in the laboratory: Handling and disposal of chemicals*. Washington, D.C: National Academy Press.

Public Works and Government Services Canada. 2000. *The environmentally responsible construction and renovation handbook*, 2nd ed.

Skinner, J. 1997. *Microscale chemistry*. London: The Royal Society of Chemistry.

USEPA. 2008. Laboratories for the 21st century: An introduction to low-energy design. A US Environmental Protection Agency Publication. DOE/GO-102OOO-1161 (http://www.labs21century.gov/pdf/lowenergy_508.pdf).

Chapter 20
Massidea.org: a greener way to innovate

Teemu Santonen

20.1 Introduction

According to Wikipedia—the free encyclopedia that anyone can edit—"green" or "environmental" technology stands for that which conserves nature and limits the negative impacts of human involvement. "Green" or "environmental friendly" is a synonym for products, services, and processes that have minimal or no effects on the environment. The definition of "sustainable development" takes this a step further: meeting human needs, but preserving the environment for present and future generations. As a result, a greener innovation process itself should meet all these goals and produce outcomes that also fulfil the defined requirements.

This chapter focuses on the very early phase of the innovation process (Cooper 1988) and challenge our current understanding of innovation practices at the individual but also at the system level. Typically the first phase is named as the fuzzy front end (later FFE) of innovation (Smith and Reinertsen 1991). Interestingly, there are some variations among the activities included in the FFE (e.g., Khurana and Rosenthal 1998; Nobelius and Trygg 2002), even though existing models seems to have somewhat similar activities (Jetter 2003). In simple terms, FFE includes stages from the idea generation to decisions on further development (Murphy and Kumer 1997). According to Wagner and Ehrenmann (2010), a great majority of whole life cycle costs and features are defined at the FFE stage. Basically, in this stage the main principles of the forthcoming product or service are formulated without actually developing or manufacturing the product or service, which is done in the subsequent new product development (NPD) process (e.g., Cooper and Kleinschmidt 1986). Significantly, even if the FFE stage sets the course for the entire development

---

T. Santonen (✉)
Laurea University of Applied Sciences, Espoo, Finland
e-mail: teemu.santonen@laurea.fi

J.I. Boye and Y. Arcand (eds.), *Green Technologies in Food Production and Processing*, Food Engineering Series, DOI 10.1007/978-1-4614-1587-9_20,

project, only minor costs are actually generated during this stage. Typically in this stage individuals or small core teams conduct “desk” research to define the guidelines for further development. This can consume a substantial amount of development time but does not require a significant amount of financial resources. Because of this leveraging effect, FFE can support green movements with modest resource inputs. Fortunately this cost structure enables risk taking at the FFE stage, which evidently is more uncertain than other innovation process stages (Koen et al. 2001; Zhang and Doll 2001).

Social media and online social networks (OSNs) such as Wikipedia, Facebook, Twitter, and YouTube have revolutionized the way we collaborate among each other. Communication with others has never been as easy as it is today. online social networks generally refer to communities and hosted services facilitating collaboration and sharing among users (Cachia et al. 2007). In principle, the OSNs facilitate interaction among members by providing a dynamic platform that enables versatile services such as discussions and sharing of multimedia content and information. These networks comprise millions of active members from all continents and all age groups. Some consider this technological change process to be a ubiquitous revolution, in which all technologies will be integrated to other technologies and technology is a seamless part of the everyday life of ordinary people. Rapidly updating technology such as iPhones—seamlessly integrating Internet access, camera, multimedia files, compass, maps, and voice control—is taking us closer to this vision of the future.

Online social networks and open networks are also permanently changing how working life is operating and innovating. Therefore, some authors, including us, argue that business success will depend more and more on the ability to utilize external resources (Chesbrough 2003). As a result, not only must individuals change their current working habits to be more open, but also innovation systems and strategies must evolve and provide tools to individuals to effectively operate in this openly networked environment. We expect that OSN-based innovation systems can enhance creativity through the unexpected linkage between insights and people behind these thoughts. We believe that OSNs will be critical and significant parts of forthcoming innovation systems, especially when greener approaches are emphasized.

This chapter introduces a novel OSN based on a national open innovation system (NOIS) approach named *Massidea.org*. Massidea.org is founded on series of innovation theories and is a free-of-charge open innovation community in which people can share their ideas and discuss today’s challenges as well as visions of the future, which are key factors when creating new innovations. By intelligently connecting people and their insights with the help of content recommendation, a creative space that can boost individual and communal creativity is constructed. In Massidea.org public, private and educational sector organizations and nations can collaborate in a green way with a wide range of people.

The chapter is structured as follows: The first section presents some theoretical foundations, followed by a presentation of the Massidea.org concept in detail, and finally presents some conclusions.

20.2 Theoretical foundations of online social network based open innovation systems

20.2.1 What is the difference between an idea and an innovation?

> An idea is a novel representation in an individual's mind relating to conception or notion of something to be done or carried out. At the first stage, novel thinking is an intention or plan in an individual's mind, which arises from the individual's creative thinking process. In the second stage after the individual's thinking process, the individual shares his or her idea with other people by verbal or written communication (Santonen et al. 2007).

Innovation literature has identified numerous definitions for innovation, but there are some things that remain common to most of them. The term "innovation" is typically used as a synonym for something new (Huiban and Boushina 1988), which has been put into practice (Ståhle et al. 2004) and is bringing added value to companies and customers (Haho 2002; Urabe 1988). Various types of economic innovations exist, such as product, process, market, organizational, new source of supply of raw materials, or half-manufactured goods-related innovations (Schumpeter 1934). However, to qualify as financially and commercially successful innovation, the implemented innovation must produce more financial value than has been invested in creating them (Stevens and Burley 1997; Taatila et al. 2006). To simplify the difference between idea and innovation the following summary can be made: An idea is always the starting point, plan or intention for potential innovation. Idea changes to innovation during the successful execution process. Without the successful execution, the idea will not change to innovation (Santonen et al. 2007).

20.2.2 Expected and unexpected findings in innovation

In the case of expected findings, the phenomenon fits with human expectations as relates to the future, whereas in the case of unexpected findings, the phenomenon is not coherent with the individual's cognitive and belief system and it breaks the conventional habit of a people (Santonen et al. 2007). If an innovator is consciously trying to create a solution to a defined problem based on previous knowledge and systematic development efforts, he or she is looking for an expected finding. For example, Thomas Alva Edison and his scientific team experimented with many different materials and numerous versions before they were able to create a long-glowing light bulb. Moreover, their work was strongly based on previous patents and discoveries made by other researchers. They had a clear goal and after serious development work, they succeeded.

By nature, unexpected finding–based innovations are discovered totally differently compared with expected finding–based innovations. Researchers at Pfizer were trying to find a treatment for high blood pressure. Interestingly, during clinical trials, it was observed that the emerging medicine actually had a strong effect on penile erection. Viagra was invented. The pharmaceutical industry provides another classical example of an unexpected finding. In the 1920s penicillin was accidently discovered as a result of a careless error. A petri dish of *Staphylococcus* bacteria was left uncovered in the laboratory for a few days. As a result, excluding a small area, the tray was filled with bacterial growth. When the uninhabited area was investigated, it was discovered to be crowded by *penicillin notatum*, which prevented bacterial growth. Penicillin was invented. In innovation literature these phenomena can be called surprise (Suomala et al. 2006) or serendipity—lucky insight. Serendipity is a process in which one accidentally discovers something fortunate, especially when looking for something else (Thagard and Croft 1999).

As described, the expected and unexpected findings are opposite by nature. Therefore, innovation systems in general should support both approaches. Most importantly, the system should create an environment that consciously supports the birth of unexpected findings.

20.2.3 Open innovation: combing internal and external resources

Recently, growing attention has been devoted to the concept of "open innovation." Chesbrough, who coined the term, describes in his book, *Open Innovation: The New Imperative for Creating and Profiting from Technology* (2003) how organizations have shifted from so-called closed innovation processes toward a more open way of innovating. According to Chesbrough (2003), open innovation can be described as combining internal and external ideas as well as internal and external paths to market to advance the development of new technologies. Basically, open innovation strategy suggests that instead of doing everything by yourself, you should also look for help from external resources. Most importantly, an effective open innovation strategy includes inbound and outbound processes. One should not only search for new technologies and ideas outside (i.e., inbound) of the firm, but also export (i.e., outbound) those ideas and technologies that do not fit the firm's current strategy.

20.2.4 Open source and free software

When software is freely and publicly available for use and modification it is referred to as open source (www.opensource.org) or free software (www.gnu.org). The definitions behind these terms are similar, but not fully identical because there are some fundamental philosophical differences between these two initiatives. Basically nearly all free software is open source, and nearly all open source software is free.

Open source or free software does not only mean access to the source code, but also for example, users' freedom to run, copy, distribute, study, change, and improve the software. In practice there are dozens of licenses such as the GNU general public license (GPL), which define the terms of use in more detail.

The open source concept—a free sharing of technology information—is not a new phenomenon. The earliest examples date back to the beginning of the nineteenth century when the automobile industry in the United States made a cross-licensing agreement. As a result all US auto manufacturers developed technology and filed patents, but shared these patents among other manufacturers openly and without the exchange of money (Flink 1976). Another classic example from the 1960s is the development of telecommunication network protocols. This collaborative process eventually led to the birth of the Internet. If all the key players in a specific market are able to come to a sharing agreement, there is a great possibility of speeding up the evolution of the market.

20.2.5 Business success depends on the ability to operate in larger innovation networks

The network economy believers associate business success with the ability to cooperate with external resources and the circulation of know-how (Pöyhönen and Smedlund 2004). Therefore, instead of focusing on individual companies (Bontis 2004), it is also important to look closely at the innovation power of larger networks, including regions and nations. Innovations have major impacts on national and regional economies, and they are major sources of competitive advantages among different nations (Porter 1998; Tuomi 2002). Because of this, the most competitive countries in the world typically have extensive and sophisticated National Innovation Systems (later NISs) (Lundvall 2007). National Innovation Systems are a set of institutions whose interaction determines the innovative performance of a nation (adapted from Nelson 1993). Theoretical foundations for NISs were seeded in the late 1980s (Freeman 1987) when our understanding of communication and networking possibilities were totally different compared with the current network-based global economy (Shapiro and Varian 1998). Instead of focusing purely on the national scope, a country-specific NIS today should have an open interface to the NIS of other countries. Together these NISs then construct a global open innovation network

20.2.6 Users as innovators

Users as innovators theories are grounded on the basic belief that a company that satisfies its customers' individual wants and needs better will eventually have greater sales (Pine 1993). A customer orientation strategy, which commonly is linked to the market orientation strategy (Kohli and Jaworski 1990), can be defined as a strong desire to identify customer needs and the ability to answer these recognized needs, or as a firm's sufficient understanding of its target buyers, which allows it to create superior value for them continuously (Narver and Slater 1990). Recently, in some contexts customer and market orientation terms have been replaced by user-driven innovation terms, which both basically having closely related definitions. User-driven innovation strategy promotes systematic engagement of users at all stages of the innovation process. In most organizations, development processes are grounded on stage-gates in which the continuation of the process is decided by a steering committee (Cooper 2001). In traditional development processes customers' roles during different stages or gates have been limited. A user-driven innovation strategy instead typically emphasizes users' involvement in all development phases, from initial idea generation to final launch and production.

However, user driven innovation methods also have recognizable characteristics that some might see as limitations. It has been identified that users are not all equally capable of identifying their needs. "Normal average users" (if there is such a thing) are typically producing incremental innovations from practice, which involve small evolutionary improvements that add value to them. This is not necessarily a bad thing, but will most likely direct the development outcome to smaller improvements.

On the other hand, the lead user method has proved to be a very successful source of commercially attractive innovations. Lead users are defined by two characteristics (von Hippel 1986; Urban and von Hippel 1988). *First*, they experience needs ahead of the majority of a target market and *second*, they benefit greatly if they obtain a solution to these needs. Compared with average users, lead users are more likely to be able to break old habits and discover novel solution for their needs, which are beyond incremental improvement.

20.2.7 Users as content creators

The term "open innovation" is typically associated with companies and organizations executing innovation processes. In the case of users and online social networks, the literature refers to *user-generated content* and *sharing*. User-generated or user-created content is typically associated with publically available contents in which the consumer is the creator, consumer, and distributor of content (OECD 2007; Le Borgne-Bachschmidt et al. 2009). The quick rise of this amateur content creation has taken media and publications industries by surprise, which have been traditionally founded on professional and centralized publication process.

The most well-known examples, *Wikipedia*—a multilingual, web-based, free-content encyclopedia project that is written collaboratively by volunteers, allowing most of its articles to be edited by nearly anyone with access to the website—and *YouTube*—allowing users from all over the world to easily upload and share their original video clips on the Internet—have challenged our media usage habits. Instead of relying on a centrally controlled publication process, both examples of

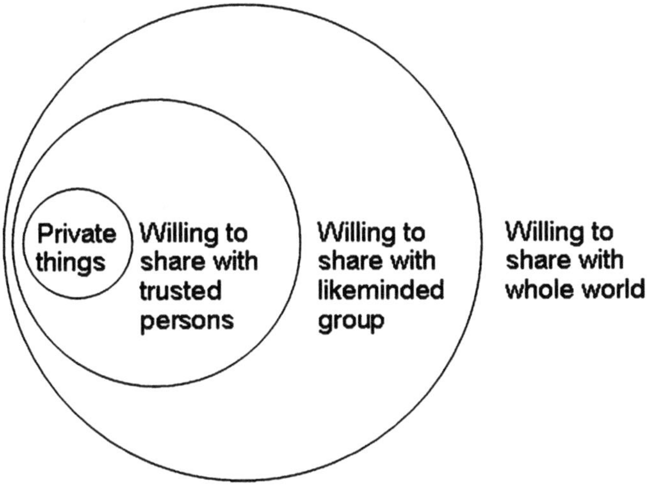

Fig. 20.1 Multilayered sharing strategy

these online services are using crowdsourcing. In crowdsourcing a task is delegated to a large group of people, which then suggests their own solutions for the defined task (Howe 2006). If successful, this kind of process provides significant cost savings compared with traditional publication processes. However, according to Nielsen (1997, 2006) OSN are affected by participation inequality. In most communities, 90% of users are lurkers who never contribute, 9% of users contribute a little, and 1% of users account for almost all the action. A sharing request is simple to demand, but in practice it appears to be very difficult to execute at the personal level.

20.2.8 Successful sharing strategy is based on multiple layers

We all have thoughts and insights that we want to keep private. Therefore, open innovation networks and OSN service providers should understand this and enable a multilayered sharing strategy (Fig. 20.1) in which users and organizations are able to share only the contents that they are willing to, but at the same time encourage and tempt them to share more than they originally thought.

Typically, these strategies include tools to share insights easily with trusted people such as friends, family members, and coworkers. However, if the sharing strategy is limited only to trusted people, the coverage and number of interaction will stay modest and results will most likely follow prior expectations. Thus in some cases we are willing to open the door to like-minded people who are interested in solving the same issues as we are. When individual interests are combined with those of other like-minded people, the possibility of getting the mission accomplished increases. Actions that are led by mission-based interest become stronger when individual voices are combined in user communities. In online communities, groups and campaigns are gathering a wide range of people together who share the same mission and offer a good platform to channel common interests.

The more one follows an open strategy, the more likely it is to have new unexpected connections. Sharing insights openly to the whole world increases the chances to find and be contacted by a valuable partner. Most importantly skillful and capable players are constructing their sharing strategy in a way that also supports their secret missions, yet without exposing these private interests.

20.2.9 Mass innovation: innovating with masses

When combining a wide range of people and their different but complementary insights and creative interaction, a novel thinking outside the box is possible and mass innovations emerge (adapted from Leadbeater 2008). Some authors call this mass collaboration, which occurs when a large group of people work independently to achieve shared outcomes through communication technologies and loose voluntary networks (adapted from Tapscott and Williams 2006). Without OSN and supporting technology this kind of mass cooperation would be impossible.

The term "mass" in mass communication indicates great volume, range, or extent (of people or production) and reception of messages (McQuail 2005). Accordingly, to qualify as a mass idea/innovation, large segments of the population (even at the same time) must be involved in open creative processes. A genuine mass idea or innovation process is not possible without the contribution of many people; therefore, one might call it a mass collaboration. In the case of mass idea, the initial idea is also first born in the individual's mind, but as a result of the (even simultaneous) interaction with a great quantity of other users and shred contents. Furthermore, afterward the newborn mass idea is openly shared back to the community. As in the typical idea-to-innovation processes, mass idea transfers to mass innovation as a result of a successful implementation process which again should involve masses of people.

20.2.10 Taxonomy for an online social network–based open innovation system

Based on the theory provided in the Introduction, the following taxonomy is defined to describe a scalable open innovation system

- *First*, the system should be grounded on open innovation, which suggests that instead of doing everything by yourself, you should look also for external resources and not forget inbound and outbound strategies.
- *Second*, supporting technology should be based on an open source solution in which the software is freely and publicly available for use and modifications.

- *Third*, the system should be scalable and include interfaces and tools to individual, group, organizational, regional, national, and global cooperation.
- *Fourth*, users and customers should be involved by utilizing user-generated content and user-driven innovation models.
- *Fifth*, cooperation and communication in the system should be based on mass collaboration and crowdsourcing.
- *Sixth*, the system should seamlessly combine online and offline structures.
- *Seventh*, the system should consciously support expected and unexpected innovation approaches.
- *Finally*, the system must be a personal tool that helps to boost personal creativity.

20.3 Introducing the foundations for massidea.org: combing complementary innovation sources with the help of an innovation triangle and recommendations

20.3.1 Origins of massidea.org: smashing ideas

The original idea for Massidea.org was created by Teemu Santonen in the autumn of 2006. He was stimulated by a new academic working environment and realized that his previously identified practical need—how masses of people could effectively innovate without causing significant extra costs and spending plenty of resources—might be solved. Santonen started to openly spread the idea to others. As a result, the development process included numerous people right from the beginning.

The Massidea.org name and related *smashing idea* slogan was coined by Santonen in the autumn of 2009. Each word in the *Massidea.org—Smashing ideas* name and slogan combination has an important message to deliver.

1. *Mass* indicating masses of people and global reach
2. *Idea* meaning ideas and other ingredients including challenges and vision, which help generate novel insights
3. *Org* reflecting open innovation and open source environment
4. *Smashing ideas* indicating that generated ideas and idea ingredients can be "great, superb, spectacular, magnificient" or if the idea (or other insight) is not sustainable it might be "smashed by other users"

Theoretical foundations for Massidea.org and related innovation triangle were initially introduced by Santonen et al. (2007) and later on refined in a series of publications including:

1. The integration of OSNs to the Triple Helix model (Santonen et al. 2008b).
2. Increase of individual creativity with the help of content recommendation (Santonen et al. 2008a).
3. A definition of a solid reward model (Santonen et al. 2011).
4. Implementation as a part of higher education (Santonen 2009).

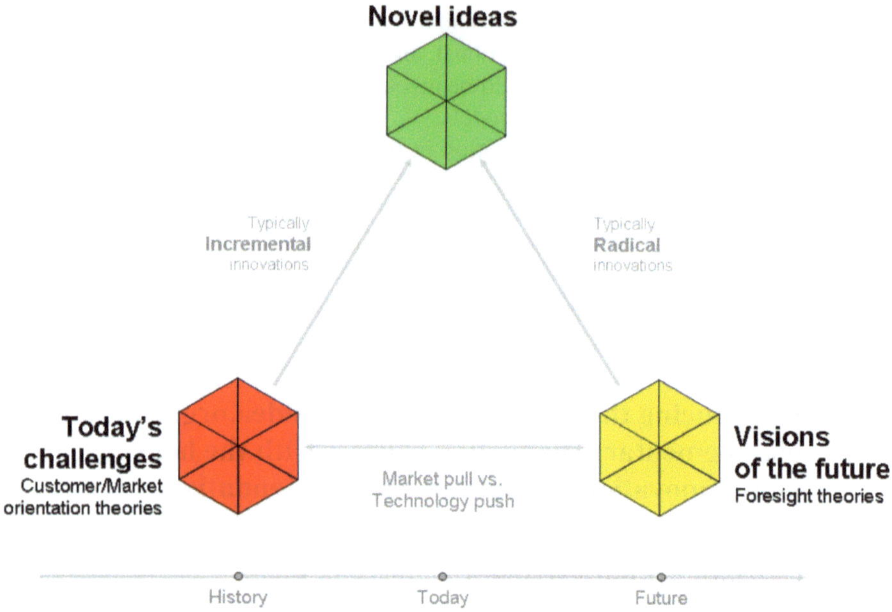

Fig. 20.2 The innovation triangle

5. Higher education students' motivation to participate in online mass innovation (Santonen and Lehtelä 2010).
6. Analyzing massinnovation process from Systemic Perspectives (Kaivo-oja and Santonen 2010).
7. Defining the digital business ecosystem (DBE) for Massidea.org (Santonen and Karhu 2010).
8. Defining Massidea.org as a business model innovation from Stage-Gate Process point (Santonen and Schallmo 2011).
9. Describing how intercultural virtual student teams open innovate via online social networks (Santonen 2011).

The preceding studies are suggested as additional reading for those who are interested in more detailed theoretical information.

20.3.2 Three repositories of innovation triangle

Figure 20.2 presents the innovation triangle, which integrates various innovation theories as a compact construct that can be implemented as an OSN such as Massidea.org (Santonen et al. 2007).

With the aim of generating new ideas (i.e., the top box) innovation triangle framework includes two different yet complementary innovation sources: *first,* current market environment information, presenting today's challenges derived from history (i.e., the left box); and *second,* future market environment information, presenting visions of the future (i.e., the right box). The following focuses on explaining the dynamics between these repositories.

To fully understand today's challenges, one should systematically collect and analyze people's needs and wants, people's problems and challenges, and the occurrences and situations that people are currently facing in their everyday life. Summing these up one is able to understand the current market needs, which sometimes are also called the "market pull." Today's challenges-based innovation process is producing novel ideas from practice, which typically generates small incremental improvements to current offerings. This approach is certainly important, but it is not enough. Thus, typically customers and average users are not able to see or imagine the forthcoming future as well as the top experts in the field. The famous Henry Ford quote—*If I had asked people what they wanted, they would have said faster horses*—is the classic example.

Humankind needs developers and researchers who are able expand our current understanding and knowledge into new fields. These frontierspeople must see the following: (1) trends and antitrends, (2) expected future scenarios, and especially (3) emerging weak signals and seeds of change to create new technology and sciences (Santonen et al. 2007). In contrast with today's challenges-based small improvements—incremental innovations—this foresight-driven approach is more likely to lead to real novelties. These radical or disruptive innovations and technologies are innovations that eventually overturn the existing dominant technologies and innovations in the market (Bower and Christensen 1995). The list of examples is endless: cars overturning horses or mobile phones overturning wired telephones, which overturned telegraphs, which overturned letters, and so on. Genuinely radical innovation usually needs a new concept or word (Thagard 1999).

Even if an idea is possible to construct and implement as a concrete entity, it does not necessarily mean that there is a market need for it. According to *technology-push* theory, research leads to inventions, which then leads to the development, production, marketing, and finally introduction of innovations to the market (Santonen et al. 2007). However, inventions transfer to innovations only if there is a balance between market pull and technology push. In practice, the world is full of needless solutions that no one wants. If the bulletproof recipe for creating innovation existed, then we would not witness the numerous misses in the market, but only smash hits.

20.3.3 Using recommendations to increase the likelihood of unexpected findings

When a wide range of people and their different but complementary insights are brought together, novel ideas generated by thinking outside the box are possible, especially if people and their insights can be intelligently linked together. This creative space systematically boosting individual and communal creativity is presented in Fig. 20.3. By integrating recommendation tools to the innovation triangle (i.e., the

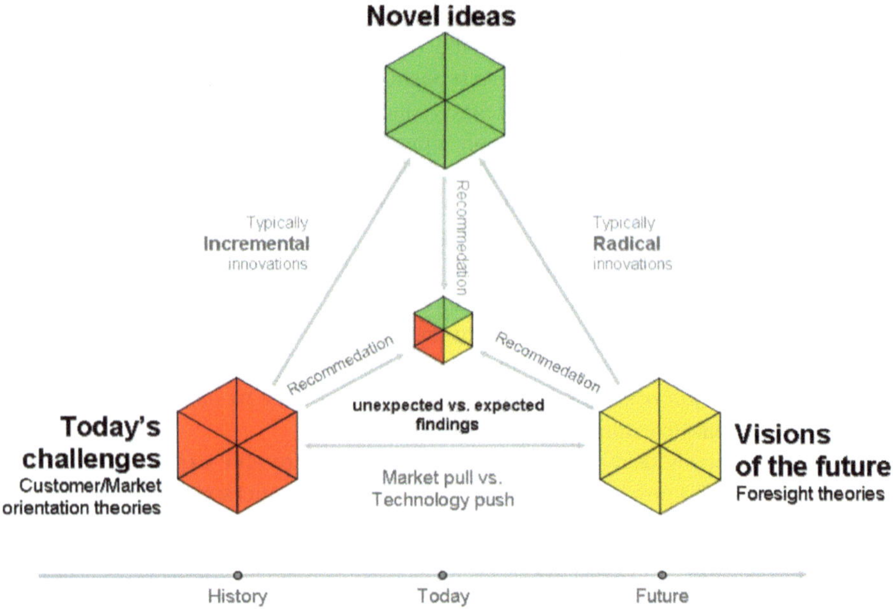

Fig. 20.3 Increasing the likelihood of unexpected findings with content recommendation

arrows in the middle), we can increase the dynamics of the individual's creativity and increase the likelihood of unexpected findings from expected findings.

When the amount of content increases in a website, one must provide intelligent services to end-users in order to create a solid user experience. Site-specific search functions, content classifications, and keywords have typically been the fastest and easiest way to help users find what they want. However, this approach mainly supports expected finding events as a typical source of novel idea (i.e., the user needs to find something specific and can complete the task with the help of the search and classification functionality). Obviously, the likelihood of unexpected findings naturally increases, when the number of interacting users and content increases even with the basic search functions. Without advanced recommendation systems, the unexpected findings potential might remain modest.

Content recommendation on a mass production scale is a kind of mass customization management system, whose history goes back 40 years (Toffler 1970; Davis 1987; Pine 1993). In the online environment, the term "personalization" often replaces "customization" or more specifically "mass customization," although the definitions of these terms are similar in our opinion. Personalization generally refers to making a site more responsive to the unique and individual needs of each user (Cingil et al. 2000), whereas in a mass customization management system, the goal is to develop, produce, market, and deliver affordable goods and services with enough variety and customization that nearly everyone will find exactly what they want (Pine 1993). In practice, mass customization means that customers can select, order, and receive a specially configured product—often choosing from among hundreds of product options—to meet their specific needs (Bourke and Kempfer 1999).

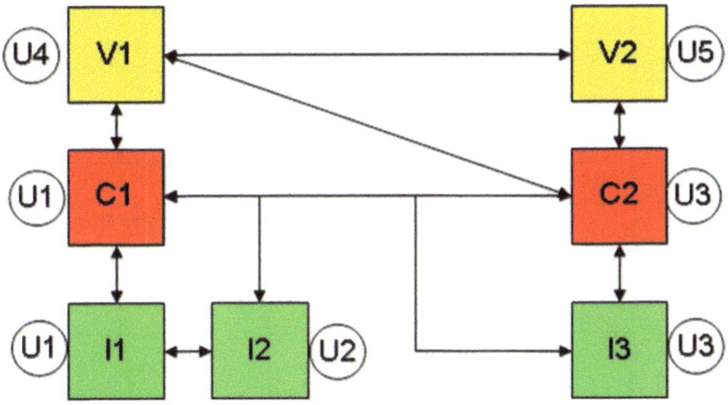

Fig. 20.4 Intelligently connecting people and their insights

Most importantly, in many cases heterogeneous customer needs mean that a true desire and willingness to listen to customer needs (i.e., customer orientation) should lead to mass-customized products and services (Santonen 2007). In principle at the extreme level of customization, a company can produce and market unique products for all customers. Pepper and Rogers (1993) defined this extreme customer orientation strategy approach as one-to-one marketing, while making a difference between individual customers and customer segments, which are more commonly related to mass customization management systems. The authors behind the ideas of mass customization and one-to-one marketing (Pine et al. 1995) later joined forces and argued that a company hoping to deliver customers exactly what they want (i.e., implement the extreme customer orientation strategy) must utilize both mass customization and one-to-one marketing management systems. When these suggestions are implemented in an open innovation environment, one should provide a personal experience to each user because each user is different.

According to Santonen (2003), websites' recommendations can be based on user preferences, content, or user similarity (i.e., collaboration). Manual decision rule-based systems allow site administrators to specify rules based on end-user preferences, demographics, or static profiles, which are collected through a registration process or session history (Mobasher et al. 2000). In a pure content-based recommendation system, recommendations are made on the basis of a profile generated by analyzing content, while a pure collaborative recommendation system does not analyze content at all, but recommends items that similar end-users have liked or used (Balabanović and Shoham 1997).

20.3.4 Practical example of creating unexpected finding through content recommendation

In Fig. 20.4 we have presented an illustrative example of linking multiple insights and users.

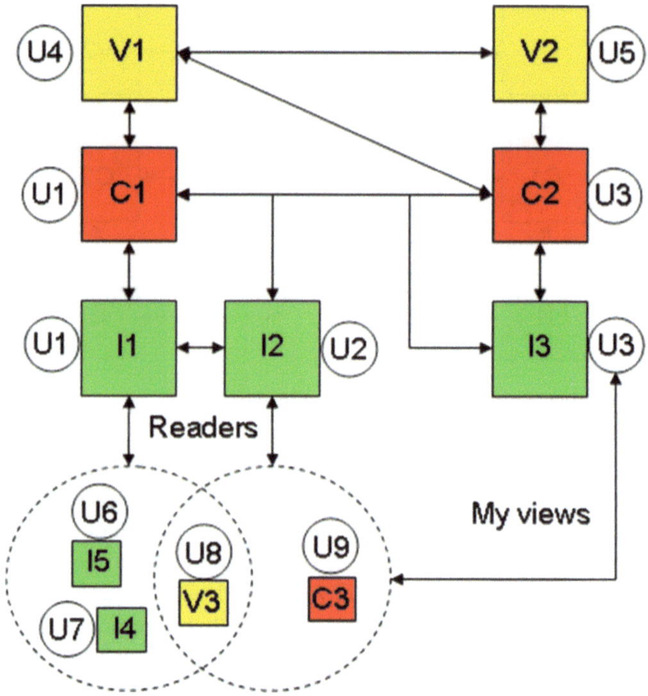

Fig. 20.5 Expanding the possibilities to generate unexpected findings

In Fig. 20.5, C1 stands for one of today's challenges, which has been posted by user U1. User U1 has also posted idea I1, which solves challenge C1. Once user 2 finds out about challenge C1 or idea I1 with the help of Massidea.org functionalities supporting (1) expected findings (e.g., search functions) or (2) unexpected findings (e.g., related keywords–based recommendation), he or she might post an alternative idea (I2) and link it to one or both C1/I1 insights. As a result these three contents and two users will be linked together. They then form a network.

Meanwhile, user U3 might be working on a completely different challenge (C2) and idea (I3). However, later on, he or she might similarly realize that I3 solution might also solve challenge C1. Now the five insights and three users are linked together. If challenge C1 was related to vision V1 posted by user U4 and challenge C2 to vision V2 posted by user U5, all seven contents and five users are linked together. Now the whole group of people (U1 to U5) and their insights (C1, C2, I1, I2, I3, V1, and V2) are linked together, even though they might not seem related at first sight.

In online social network sites such as Massidea.org, contents include groups of keywords (i.e., words that capture the essence of the topic of the content). Therefore, the preceding network example of contents and users can easily be expanded by doing automated searches based on these user-defined keywords. The possibility of unexpected findings increases when showing automated search results in one content page based on the group of keywords, which the human linked contents construct. Moreover, in Fig. 20.5, we have expanded the possibilities to generate unexpected findings.

As argued in the theoretical foundations section, OSN including Massidea.org, is affected by participation inequality, meaning most users are lurkers who never contribute. To make contribution as easy as possible, Massidea.org is following logged-in users' usage patterns. This information is then utilized to enhance the possibility of unexpected findings. In practice this means that each time a user reads an insight, he or she will be included into insight's *readers' list*. The list of readers show who else has been interested in this particular topic. In the author's personal profile page, a summary list of all his or her readers and his or her own insight reads are kept. Moreover, instead of just keeping track of individual insight reads, it is possible to analyze longer usage paths and recommend content based on these paths. This combination of content recommendation approaches is seamlessly uniting content-based and collaborative recommendation systems and is offering an increased chance for generation of novel thoughts.

20.4 Defining a common content structure to enable effective insight sharing

20.4.1 *Without clear guidelines, the content production and consumption become blurry*

If masses of people collaborate and openly share their ideas on today's challenges as well as visions of the future, eventually a large cumulative database of contents in each repository will be created. To make the communication and interaction between repositories and people easy, a common content structure and format should be defined. Contents for OSNs are written and read by numerous individuals, who might have significantly different backgrounds (e.g., cultural, educational, age, and working history to name a few). Our background and previous life experiences affect our ability to produce and consume information. As in the case of traffic, we need simple guidelines and content structure to make OSN usage smooth and effective. Imagine for example academic literature without shared understanding on how to write a quality article. A common structure and way of writing acts as a universal language, helping readers to determine if the content is of interest quickly and effectively.

20.4.2 *A good headline is sometimes enough to trigger a novel thought*

In the Internet and mobile communication era, the need for short and fast messaging has emerged. Therefore, OSNs such as Twitter have gained great popularity. Twitter is a real-time information network powered by people all around the world that lets users share and discover what is happening now (source: Twitter.com). Twitter lets users write and read messages (i.e., tweets) of up to 140 characters. For this reason tweets are sometimes called as short message service (later SMS) of the Internet. Interestingly, in the case of generating new ideas, visions, and challenges—key factors for new innovations—the stimulus for novel thought might be triggered also from very sort stimuli. Therefore, users must have a channel to share their thought also in a short SMS format. This means that content headlines should be considered as the most valuable content. A good headline grabs the reader's attention, summarizes the whole thought, and attracts reader to read the rest of the story.

20.4.3 *Defining a compact way to distribute innovation-related information: 5WA4*

In practice it is often very difficult to understand the whole meaning of the author's thoughts if the information is restricted only to a headline. Idea generation methods, such as brainstorming, often produce short descriptions or headlines that introduce the author's initial insight. After these brainstorming sessions the insights that result are summarized together by one of the participants of the brainstorming session or an external facilitator (i.e., the one organizing the session). It is typical that during the idea generation session, the participants understand each other rather well, but when these short written thoughts are later passed on to the external group, they are not able to discover the original thoughts behind these short lines. Therefore, especially in the case of OSN-based innovation systems, in which there is not necessarily a direct communication between reader and writer, a longer and more comprehensive format is needed to ensure accurate interpretation.

How long is enough and what is comprehensive enough? In a fast-paced world readers must quickly find the interesting and valuable information. Readers do not want to spend more than a few seconds in the information-gathering process, yet they do want to collect all the required information that is needed to understand the author's whole thought. For this reason a somewhat adjusted press release and news format is suggested as a good tool to share innovation-related information.

Generally speaking, a press release is typically kept to one page (ISO 216, A4 format) or roughly 300–500 words. In practice this means no more than 1,800–3,000 characters (because one word on average includes six characters). Effectively summarizing a message in only one page is certainly a challenge, but it allows readers to scan through the content in a few seconds and find out if it interests them. In press releases and news journalism, the Five Ws concept (who, what, when, where, and why) is a popular way to deliver the whole story in a compact format. In the case of distributing innovation-related information, content should give basic answers to the following questions depending on whether it is a challenge, a vision, or an idea: (1) what is the thought? (2) why is the thought important and valuable? (3) who is the target group? (4) when (temporal dimension) is the thought topical? and (5) where (geographical or physical location or circumstances) is the thought happening? Most importantly, the writer must attract the readers to read the whole story, not just answer the suggested questions.

The following suggests our basic structure (5WA4) for writing quality content for the innovation triangle.

- *Headline.* The headline should fit in SMS format (minus some additional information such as a user ID) and explain in one sentence the whole thought. Basically this results in a maximum length of 140 characters.
- *Lead paragraph.* A headline is followed by a lead paragraph, which together with the headline, sum up the entire content. The lead paragraph should give short answers to the adapted Five Ws concept questions. This does not specifically require five individual sentences, but two to three sentences. The lead paragraph is also important because it can be used as a tool to give information to search engines and automated recommendation systems.
- *Body content:* The body content elaborates the headline and lead paragraph and justifies the author's thought in more detail. The body content should not exceed 4,000 characters.
- *References:* It is important to provide reference to existing knowledge and show readers what these thoughts are based on.
- *Keywords:* Keywords are important because they will determine how and whether readers and search engines find the content.
- *Related organizations and companies:* Author should also be able to identify a potential organization or company who might be interested in the author's thought.

By following the guidelines indicated in the preceding, easy-to-read and -link cumulative content repositories can be created. Appendix 20.1 presents an example of an article, "Massidea.org is a good tool to stimulate collaboration before, during and after conferences and other events," which follows our 5WA4 format and describes a greener way to organize conferences and events.

20.5 Massinnovation requires multiple players with different but complementary roles

20.5.1 Triple helix players are not enough

In order to identify key players, we base our suggestions on the enhanced Triple Helix model. The Triple Helix is the most well-known framework to describe collaboration among universities, policy institutions, and industry (Etzkowitz and Leydesdorff 1999, 2000). In the Triple Helix model each actor has its own task: Universities produce research, industries manufacture, and governments secure stability for maintaining exchange and interaction. The Triple Helix regime operates on these complex dynamics of innovation as a recursive overlay of interactions and negotiations among the three institutional spheres. The different partners

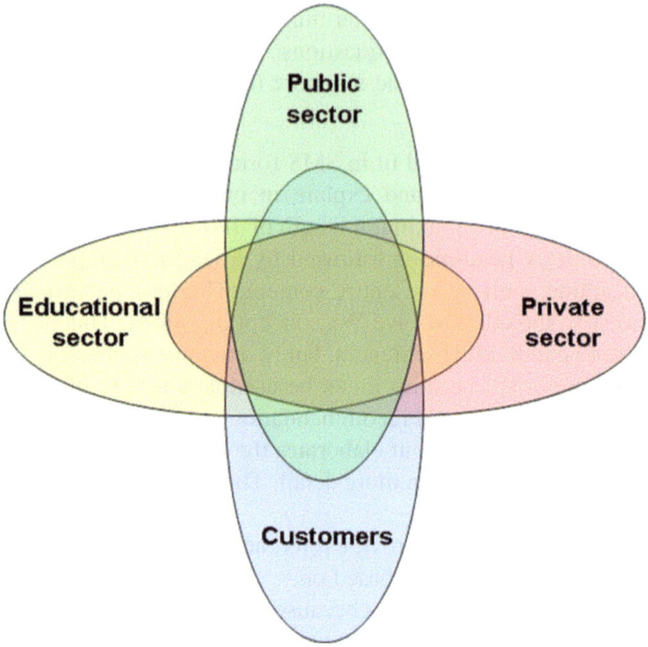

Fig. 20.6 Key players in the Massidea.org innovation system

engage in collaborations and competitions as they calibrate their strategic direction and niche positions.

In the past this national innovation system model has been very successful. However, in our opinion Triple Helix is lacking a genuine market orientation and viewpoint, which are currently emphasized in innovation literature. Critical thinkers might say that the voice of user, consumer, and people is totally missing in the Triple Helix. Moreover, the Triple Helix does not recognize the innovation potential of other educational sectors, such as basic education and upper secondary education, which cover children and young people. Compared with people in working life, youth are typically more open-minded and not as restricted by the historical burden of life and work experience. Therefore, they are able to look at things differently compared with those in working life and they might offer valuable second opinions. It is for this reason that broad-minded companies are using students as shadow groups for example. In contrast with the traditional Triple Helix model, the taxonomy for OSN-based open innovation systems such as Massidea.org requires strong end-user interaction and does not exclude other educational sectors besides universities (Fig. 20.6).

20.5.2 *Educational sector role*

We can identify two main user groups in the educational sector: students and teachers.

A significant share of the content for the OSN-based innovation system could be produced by numerous students performing their everyday studies. Students spend thousands of hours learning new competences on the way to their work life or the next level of education. While doing everyday course assignments through OSN-based innovation systems student could learn new insights, expand their contact network, and accumulate their personal knowledge portfolio. All this can be done with no or only a little extra effort to studies, but with the possibility to capitalize on personal resource investments. OSN's knowledge portfolio promotes students' capabilities and know-how, which later on increases the likelihood of opening the door to a desired job. A tool that reveals a person's ability to produce novel thought is a far greater indicator than certificate of graduation or traditional CV explaining one's working history.

The supervision and management of the student's work could be integrated as part of everyday teaching tasks, whereas the strategic overall resource allocation could be conducted with the help of school specific curriculums. Teachers set course-specific study tasks for their students. These tasks can have different viewpoints depending on the course objectives. A study task's outcome—the critical mass of content—would be shared and linked by students with other users and their contributions. A global learning network can be built by sharing and combining study tasks with other teachers and courses around the world. The traditional learning model based on individual and group interaction would become an up-to-date network communication model.

To fully integrate something to educational processes, there should also be integration to school-specific curricula and strategies. In fact the curriculum integration takes easily years, because curricula are not changed annually. When OSN-based learning is not a part of the curriculum's objectives, it must be implemented by one teacher and one course at a time. Basically, this means winning hearts one by one and convincing each teacher personally about the rationality and importance of OSN-based learning. Teacher autonomy in schools is very strong. Typically, individual teachers have almost total freedom to plan course activities as he or she will. Therefore, only if teachers change their current behavior and provide studying tasks for students that instead of individual performance and student–teacher interaction are based on networked learning models, a critical mass of students who exploit the social network can be achieved. Especially the teachers responsible for basic courses (i.e., having many students) can play a key role. According to preliminary results there is a genuine will among students to participate in content production and development of OSN-based innovation systems.

In the educational sector, there are also good possibilities to content production in thesis processes. Essential parts of the educational sector studying outcomes are bachelor, master, or doctorate theses. However, regardless of the small group-based thesis seminars and one-to-one conversations with the thesis supervisor, the process of constructing the final thesis is rather lonely, not only from the student's point of view, but also from the supervisor's. This might sound strange because typically there are often multiple students and researchers working on similar topics at the same time. If thesis processes would genuinely exploit open network-based working models such as Massidea.org, then the research plans would be published and shared, for example as Massidea.org challenges. Later when the thesis is ready and results are discovered, the outcome of the thesis could be published, for example as a Massidea.org idea, and the full study could be linked to this content and the user's knowledge profile. As a result of the thesis outcome, the supervising teacher and student would be integrated into the network of content and users working on similar issues. More importantly, this network would include players from outside the educational sector who would also provide practical points of view to the theoretical work.

Similarly, the research and development (R&D) projects conducted in educational sector would significantly benefit from transformation to open network-based working models. Many university-based projects are government funded or funded by other similar public operators such as European Union (EU) or local governments. Therefore, educational sector project outcomes and processers are typically publicly shared with the rest of the world without extra costs. However, it is commonly argued that the educational sector has had problems in effectively sharing these results with others such as industry especially during the middle of the project. Thus, new tools are needed, which will help the public to find out and easily understand what kind of new technologies and science is on the way in the near and distant future. Traditional scientific literature format is not a good tool for the general public, who are not familiar with scientific terminology and way of writing. By publishing educational sector R&D project challenges, ideas and visions in a format similar to that proposed for the Massidea.org 5WA4-format, the range of people who could exploit the results would be expanded. In this transformation process, the project funders have the key role, because they can define conditions for funding, which then automatically guide projects to operate in a more open way.

20.5.3 Public sector role

The national innovation policy gives guidelines as to where a particular nation is heading, whereas the national innovation system (NIS) defines the tools for implementing this policy (Lundvall 2007). Typically this policy is managed throughout several ministries and other public sector operators, including towns and municipalities and their numerous service operators. To commit a public sector player to participate in OSN, the player must include open innovation and sharing as a part of their strategy. Today, collaboration between different public sector players is often limited; as a result, the public sector is blamed that the left hand does not know what the right hand is doing. Furthermore, citizens are alienated from processes that influence their lives. Therefore, many people see the traditional way of influencing through politics to be ineffective.

Current OSN-based communication tools challenge our understanding of democracy. If desired, citizens can be called on to participate directly in decision-making processes that affect their lives. The term "direct democracy" refers to citizens making decisions without using representatives. In practice, the direct access to most public processes has been limited because of difficulties of effectively coping with large masses of people. With the help of OSN-based innovation systems, genuine collaboration between citizens and various public sector operators becomes possible. Free of charge open source-based software and ready-made models for easy collaboration offer a fast track to direct democracy, if public sector players are willing to open their processes to direct interaction. This interaction does not have to cost much, and actually is more likely to reduce costs in the long run. Because of laws and regulations, many of the public processes are identical in different geographical regions. Thus great possibilities for cooperation are offered when the number of public players in OSN-based innovation networks increases.

20.5.4 Private sector role

In the connected world, business success greatly depends on the ability to utilize external resources. Moreover, human resources are always limited no matter who you ask in the industry. Pure customer listening is not enough anymore. Instead, companies must genuinely co-create with their customers if they desire success. The search for novel solutions and opportunities becomes possible when individual business challenges are linked to the surrounding business environment. Industry will have an access to both the extensive repositories of information and also people who might be valuable to the company. The concept of open innovation must be included as a part of the strategy to convince companies to participate. Management has to implement and operationalize this strategy to the everyday life of their business units. Finally, because novel ideas are first created in the individual's mind, developers in the companies must accept massinnovation as a personal tool for their employees.

20.5.5 Role of customers

Customer actions are led by personal interest. When individual interests are combined with other like-minded people, the possibility of accomplishing a mission increases. Therefore, participating in nonprofit organizations and user communities offers a good platform for channeling interests. Online social network-based innovation systems provide effective tools for these groups and help like-minded people to join forces. Especially in Western countries, forecasts relating to the amount of available workforce have seen an unhealthy trend and in general there is a growing need to activate aged and retired people (Katajisto and Kimari 2005). This voluntary workforce—senior citizens—could be engaged as content providers alongside with more organized educational sector. In principal the active members of aging people will share and communicate their experiences with youth. Marketing and resourcing this possibility will be conducted throughout the network of voluntary organizations.

20.6 Conclusion

Massidea.org has been used in this chapter as a good example of innovation theories, which suggests that changing the way we work as well as the work environment can give birth to new ideas. Moreover, it is also a serious attempt to implement a group of innovation theories in practice. The concept has gained a significant amount of interest from its inception and in 2008 was awarded as the best school-related innovation in Finland by the Support Association for Finnish Inventors. Some funders seem to also believe in the concept, because presently Massidea.org-related projects have gathered more than EU€1.5 million funding, and the official project consortium includes 15 universities. More projects are on the drawing board, and suggestions for cooperation are welcome. Evidently a great number of people and organizations are seeing the assumed benefits of Massidea.org and other similar initiatives.

Yet its practical implementation process faced a number of challenges (Santonen 2009), which cannot be considered a big surprise. The biggest challenges were the recruitment of human resources to content production and development, and the ability to change the current practices of organizations and people. Even when many people and organizations are convinced of the goodness and benefits of an idea, it can be amazingly hard to recruit them to contribute. Recruitment to content production appears to be even more difficult, if technical solutions do not fully answer the needs and wants of the users. Therefore, Massidea.org is in continuous development.

This observation is in line with previous studies, which have identified participation inequality in the case of OSNs in general. The diffusion of innovations theory—a specific theory explaining the rate at which new ideas or technologies spread on the market—can also offer some explanations (Rogers 1962). It appears that only innovators and early adapters—the first and second individuals to adopt an innovation—join OSN-based innovation networks and are willing to explore new ways of innovating. Contribution to innovation-related OSNs evidently requires changes to the current way of thinking and acting. Necessarily these changes are not big from the workload point of view, but are demanding from the state-of-mind viewpoint. The old habits of working individually or with a small team, but not openly with masses of people, have been printed hard in our minds. Changing this is a slow process, but if successful might deliver a substantial competitive advantage to participating individuals, organizations, and nations. Especially the vision of the greener future developed with greener innovation processes should challenge our current understanding of innovating.

As a result, it has been suggested that learning to operate and innovate in open networks should be started early on, even before youth start their university studies and enter into working life (Santonen and Lehtelä 2010). The process should also ensure positive experiences relating to motivation factors to participate in OSNs (Gaston-Breton et al. 2009), such as presenting individual expertise to employers, sharing information and getting feedback from other users, and reward or recognition. It is suggested, however, that the development of an OSN-based open innovation system such as Massidea.org be initiated with experienced lead users (von

Hippel 1986; Urban and von Hippel 1988). This combination of inexperience and experience offers the best combination to transform from closed to open innovation systems and to greener massinnovation systems. Because this change does not happen only by reading books, the author of this chapter challenges readers to start learning to massinnovate, post green insights at Massdea.org, and start discovering unexpected findings. Further information can be obtained at *Massidea.org—a greener way to smashing ideas.*

20.7 5WA4 format example

Massidea.org is a good tool to stimulate collaboration before, during, and after conferences and other events. Massidea.org brings complimentary insights together and boosts creativity by intelligently connecting people and their insights. Especially scientific conferences, which are typically limited to face-to-face interaction and linear content format could expand collaboration possibilities with the help of Massidea.org.

20.7.1 What is the idea?

Massinnovation occurs when a wide range of people and their different but complimentary insights are brought together and novel ideas generated by thinking outside the box are generated. Massidea.org (www.massidea.org) is an open innovation community for sharing challenges, ideas and visions. It boosts individual and communal creativity by intelligently connecting people and their insights.

Because in Massidea.org people's posts (i.e., insights) are cumulative of users' knowledge portfolio, it helps them to build their collaboration network based on their insights while promoting their expertise to others. Interaction before, during, and after the conference is possible when combing these individual persons and their insights as groups such as scientific conferences.

20.7.2 Why might the idea work?

Scientific and other academic-oriented conferences in particular mainly rely on face-to-face interaction. This limits collaboration because the interaction is restricted to conference members and small groups during the conference days. This traditional process is also founded on linear content production and usage process (i.e., the process does not include interaction). Following linear process most likely results in expected findings. However, the possibility to discover unexpected findings increases when following nonlinear and interactive content reading processes. Unexpected findings have been identified as an important tool to generate novel innovations.

20.7.3 At whom is the idea aimed?

This idea should interest especially scientific conference organizers (or other written document–based events) and the participants for these events. Because the majority of the required content is already created, there is limited extra work required to exploit this possibility. Massidea.org is free of charge so there is no extra cost to run it.

20.7.4 Where can the idea be used?

Implementing this idea in practice expands the events from the physical world to the digital online environment. Thus, even a very local event can have a global audience. The automated translation tools in Massidea.org also enable the usage of one's mother tongue. This lowers the barrier to interact with a global audience.

20.7.5 When can the idea be used?

Before the event: Using Massidea.org should start well in advance before the actual conference. The main idea is to start collaboration and group building preferably weeks before the event. As a result, when the conference members enter into the conference venue, they will already be familiar with those members who are interested in the same issues. Because the collaboration is based on people's insights and linkages (even unexpected) between insights, this helps building up the right network.

During the event: Massidea.org will also act as a tool to collect insights during the conference presentations. Typical academic format includes 15–25 min' presentation time and then a few minutes for a couple of questions. With the help of Massidea.org participants are able to give comments during the presentation. Now not only the session participants, but also other conference members and the rest of the world are able to "hear" what questions the topic raised. Also, if questions are raised before a session, the person giving the presentation can note them during the presentation. This process also enables a kind of simultaneous participation to multiple sessions, which is impossible in the face-to-face–based operating model.

After the event: All the insights, comments, and user profiles can be used as for networking interface after the event. Because Massidea.org collects different topics, including both academic and practical insights, this also links one event to the rest of the world.

References

Balabanović, M., and Y. Shoham. 1997. Fab: Content-based, collaborative recommendation. *Communications of the ACM* 40(3): 66–72.

Bontis, N. 2004. National intellectual capital index: A United Nations initiative for the Arab region. *Journal of Intellectual Capital* 5(1): 13–39.

Bourke, R., and L. Kempfer. 1999. *Achieving success with mass customisation: The vital contribution of engineering*. Cleveland: Computer-Aided Engineering.

Bower, J. L., and C. M. Christensen. Disruptive technologies: Catching the wave. *Harvard Business Review* 73(1): 43–53.

Cachia, R., R. Compañó, and O. Da Costa. 2007. Grasping the potential of online social networks for foresight. *Technological Forecasting and Social Change* 74(8): 1179–1203.

Chesbrough, H. 2003. *Open innovation: The new imperative for creating and profiting from technology*. Harvard: Harvard Business School Press.

Cingil, I., A. Dogac, and A. Azgin. 2000. A broader approach to personalisation. *Communications of the ACM* 43(8): 136–141.

Cooper, R.G. 1988. Predevelopment activities determine new product success. *Industrial Market Management* 17(2): 237–248.

Cooper, R.G. 2001. *Winning at new products: Accelerating the process from idea to launch*. New York: Basic Books.

Cooper, R.G., and E.J. Kleinschmidt. 1986. An investigation into the new product process: Steps, deficiencies, and impact on performance. *Journal of Product Innovation Management* 3: 71–85.

Davis, S.M. 1987. *Future perfect*. Reading: Addison-Wesley.

Etzkowitz, H., and L. Leydesdorff. 1999. The future location of research and technology transfer. *The Journal of Technological Transfer* 24(2–3): 111–123.

Etzkowitz, H., and L. Leydesdorff. 2000. The dynamics of innovation: From national systems and 'Mode 2' to a Triple Helix of university-industry-government relations. *Research Policy* 29: 109–123.

Flink James, J. 1976. *The car culture*. Cambridge: MIT Press.

Freeman, C. 1987. *Technology policy and economic performance: Lessons from Japan*. London: Pinter.

Gaston-Breton, C., L. Duque, and N. Lado. 2009. What's keeping people in virtual communities? A survey study of the determinants of participant's satisfaction. In *Proceedings from the 8th international marketing trends conference*, Paris, France.

Haho, P. 2002. Strategic and operational process innovations in NPD processes applying evolutionary change management. In *Continuous innovation in business processes and networks, 4th International CINet 2002 conference*, Espoo, Finland, 15–18 Sep 2002.

Howe, J. 2006. The rise of crowdsourcing. *Wired* 14(16). http://www.wired.com/wired/archive/14.06/crowds.html.

Huiban, J.P., and Z. Boushina. 1988. Innovation and the quality of labour factor. *Small Business Economics* 10: 389–400.

Jetter, A. 2003. Educating the Guess: Strategies, Concepts and Tools for the Fuzzy Front End of Product Development. In *PICMET 03 Portland International Conference on Management of Engineering and Technology Technology Management for Reshaping the World 2003*, Publisher: Portland State Univ, Pages: 261–273.

Kaivo-oja, J., and T. Santonen. 2010. Open Innovation in a Systemic Innovation Context: Analyzing Online Mass Innovation Process from Systemic Perspectives. In *Proc. of The Open Innovation Research Seminar*, Kouvola, Finland, 23 August 2010.

Katajisto J., and M. Kimari. 2005. Education, Training and Demand for Labour in Finland by 2015. Finnish Helsinki, National Board of Education.

Khurana, A., and S.R. Rosenthal. 1998. Towards holistic "front ends" in new product development. *The Journal of Product Innovation Management* 15(1): 57–74.

Koen, P., G. Ajamian, R. Burkart, A. Clamen, E. Fisher, S. Fountoulakis, A. Johnson, P. Puri, and R. Seibert. 2001. Providing clarity and a common language to the "fuzzy front end". *Research Technology Management* 44(2): 46–55.

Kohli, A.K., and B.J. Jaworski. 1990. Market orientation: The construct, research propositions, and managerial implications. *The Journal of Marketing* 54(2): 1–18.

Le Borgne-Bachschmidt, F., S. Girieud, M. Leiba, S. de Munck, S. Limonard, M. Poel, L. Kool, N. Helberger, L. Guibault, E. Janssen, N. Van Eijk, C. Angelopoulos, J. Van Hoboken, and E. Swart. 2009. *User generated content: Supporting a participative information,* Final Report, Society Study for the European Commission DG Information Society.

Leadbeater, C. 2008. *We-think: mass innovation, not mass production: The power of mass creativity*. London: Profile Books.

Lundvall, B.Å. 2007. National innovation systems—Analytical concept and development tool. *Industry and Innovation* 14(1): 95.

McQuail, D. 2005. *McQuail's mass communication theory*. Thousand Oaks: Sage.

Mobasher, B., R. Cooley, and J. Srivastava. 2000. Automatic personalisation based on Web usage mining. *Communications of the ACM* 43(8): 142–151.

Murphy, S.A., and V. Kumer. 1997. The front end of new product development: A Canadian survey. *R&D Management* 27: 5–15.

Narver, J., and S. Slater. 1990. The effect of a market orientation on business profitability. *The Journal of Marketing* 54(4): 20–35.

Nelson, R. (ed.). 1993. *National innovation systems: A comparative analysis*. New York/Oxford: Oxford University Press.

Nielsen, J. 1997. Community is dead; long live mega-collaboration. *Jakob Nielsen's Alertbox* for 15 Aug 1997.

Nielsen, J. 2006. Participation inequality: Encouraging more users to contribute, *Jakob Nielsen's Alertbox*, 9 Oct 2006.

Nobelius, D., and L. Trygg. 2002. Stop chasing the front end process: Management of the early phases in product development projects. *International Journal of Project Management* 20(5): 331–340.

OECD. 2007. *Participative Web: User-created content, DSTI/ICCP/IE(2006)7/FINAL,* prepared by Sacha Wunsch-Vincent and Graham Vickery, Directorate for science, technology and industry; Committee for information, computer and communications policy.

Pepper, D., and M. Rogers. 1993. *The one to one future*. New York: Currency-Doubleday.

Pine, J. 1993. *Mass customisation: The new frontier in business competition.* Boston: Harvard Business School Press.

Pine, J., D. Peppers, and M. Rogers. 1995. Do you want to keep your customers forever? *Harvard Business Review* 73(2): 103–114.

Porter, M. E. 1998. *The Competitive Advantage of Nations*. New York: Free Press (First published 1990).

Pöyhönen, A., and A. Smedlund. 2004. Assessing intellectual capital creation in regional clusters. *Journal of Intellectual Capital* 5(3): 351.

Rogers, E.M. 1962. *Diffusion of innovations*. Glencoe: Free Press.

Santonen, T. 2003. The effect of mass customization on demand turbulence. *IADIS International Journal on WWW/Internet* 1(2): 64–75.

Santonen, T. 2007. Defining a mass customization strategy for online banking. *International Journal of Mass Customisation* 2(1/2): 95–113.

Santonen, T. 2009. Creating the foundations for mass innovation implementing national open innovation system (NOIS) as a part of higher education. In *Proceedings of the 2nd ISPIM innovation symposium—Stimulating recovery—The role of innovation management*, eds. K.R.E. Huizingh, S. Conn, M. Torkkeli, and I. Bitran. New York, 6–9 Dec 2009.

Santonen, T. 2011. Intercultural virtual student teams open innovating via online social networks, in The 4th ISPIM Innovation Symposium - Managing Innovation for Sustained Productivity: Creating Advantage and Resilience - New Zealand on 29 November–2 December 2011.

Santonen, T., and M. Lehtelä. 2010. Higher education student's motivation to participate in online mass innovation. In The XXI ISPIM Conference - The Dynamics of Innovation - Bilbao, Spain on 6–9 June 2010.

Santonen, T., and K. Karhu. 2010. Massidea.org: Defining a digital business ecosystem (DBE) for massinnovation: in The 3rd ISPIM Innovation Symposium - Managing the Art of Innovation: Turning Concepts into Reality - Quebec City, Canada on 12–15 December 2010.

Santonen, T., and D. Schallmo. 2011. Evaluating Industry Business Model Innovation Stage-Gate Process: Case Massidea.org: in the The XXII ISPIM Conference – Sustainability in Innovation: Innovation Management Challenges - Hamburg, Germany on 12–15 June 2011.

Santonen, T., J. Kaivo-oja, and J. Suomala. 2007. *Introduction to National Open Innovation System (NOIS) paradigm: A preliminary concept for interchange*. Turku: Finland Futures Research Centre, Turku School of Economics.

Santonen, T., J. Kaivo-Oja, J. Suomala. 2008a. Brief Intorduction to National Open Innovation System (NOIS) Paradigm: Integrating Online Social Networks and the Triple Helix Model, In: "*Proceedings of the 2nd International Multi-Conference on Society, Cybernetics and Informatics: The 6th International Conference on Politics and Information Systems, Technologies and Applications*: PISTA 2008, Volume III", Edited by Friedrich Welsch, Jose Vicente Carrasquero, Angel Oropeza, Chie Bein Chen, ISBN-10: 1-934272-48-5 (Volume III), Orlando, Florida, USA, June 29th – July 2nd 2008, pp.126–131.

Santonen, T., J. Suomala, J. Kaivo-Oja. 2008b. Brief Introduction to the National Open Innovation System (NOIS) Paradigm: Supporting individual creativity in an online social network with content recommendation, In: K.R.E. Huizingh, M. Torkkeli, S. Conn and I. Bitran (eds). Proc. of the XIX ISPIM Annual Conference. Open Innovation: Creating Products and Services through Collaboration. Tours, France, 15–18 June 2008.

Santonen, T., J. Kaivo-Oja, M. Antikainen. 2011. National Open, Innovation System (NOIS): Defining a Solid Reward Model for NOIS, International Journal of Innovation and Regional Development, International Journal of Innovation and Regional Development (IJIRD), Vol 3., No. 1, 2011.

Schumpeter, J. 1934. *The theory of economic development: An inquiry into profits, capital, credit, interest and the business cycle*. Cambridge: Harvard University Press.

Shapiro, C., R. Hal, Varian. 1998. Information Rules: A Strategic Guide to the Network Economy, Harvard Business School Press.

Smith, P.G., and D.G. Reinertsen. 1991. *Developing products in half the time*. New York: van Nostrand Reinhold.

Ståhle, P., A. Poyhonen, and K. Kylakeiko. 2004. Towards operationalization of dynamic capabilities. In *Internet economy: Opportunities and challenges for the developed and developing regions of the world*, ed. Y. Hosni. Amsterdam: Elsevier.

Stevens, G.A., and J. Burley. 1997. 3000 raw ideas equals 1 commercial success! *Research Technology Management* 40(3): 16–27.

Suomala, J., V. Taatila, R. Siltala, and S. Keskinen. 2006. Chance discovery as a first step to economic innovation. In *Proceedings of the twenty-eighth annual conference of the cognitive science society*, ed. R. Sun and N. Miyake. Mahwah: LEA.

Taatila, V., J. Suomala, R. Siltala, and S. Keskinen. 2006. Framework to study the social innovation net-works. *European Journal of Innovation Management* 9(3): 312–326.

Tapscott, D., and A.D. Williams. 2006. *Wikinomics: How mass collaboration changes everything*. New York: Portfolio Hardcover.

Thagard, P. 1999. *How scientists explain disease*. Princeton: Princeton University Press.

Thagard, P., and D. Croft. 1999. Scientific discovery and technological innovation: Ulcers, dinosaur extinction, and the programming language JAVA. In *Model-based reasoning in scientific discovery*, ed. L. Magnani, N. Nersessian, and P. Thagard. New York: Plenum.

Toffler, A. 1970. *Future shock*. New York: Bantam Books.

Tuomi, I. 2002. *Networks of innovation. Change and meaning in the age of Internet*. New York: Oxford University Press.

Urabe, K. 1988. Innovation and the Japanese management system. In *Innovation and management: International comparisons*, ed. K. Urabe, J. Child, and T. Kagono. Berlin: De Gruyter.

Urban, G., and E. von Hippel. 1988. Lead user analyses for the development of new industrial products. *Management Science* 34(5): 569–582.

von Hippel, E. 1986. Lead users: A source of novel product concepts. *Management Science* 32(7): 791–805.

Wagner, L., and S. Ehrenmann. 2010. How diversity management supports the fuzzy front end of innovation. In *Proceedings of the XXI ISPIM conference. the dynamics of innovation*, eds. K.R.E. Huizingh, S. Conn, M. Torkkeli, and I. Bitran, Bilbao, Spain, 6–9 June 2010.

Zhang, Q., and W.J. Doll. 2001. The fuzzy front end and success of new product development: A causal model. *European Journal of Innovation Management* 4(2): 95–112.

Part VI
Health and social perspectives

Chapter 21
Reducing process-induced toxins in foods

Humayoun Akhtar

21.1 Introduction

Food is a complex mixture of ingredients, and is consumed by humans to provide the essentials for sustenance. Over many millennia, human diets have undergone several evolutions, including the use of heat (fire) for food preparation to enhance taste, flavor, digestibility, etc. It is conceivable that modern human race (50,000–100,000 years ago, after the extinction of *Homo erectus, Homo neanderthalensis,* and archaic *Homo sapiens*) initially ate grains, fruits, vegetables, and meat, mostly raw. It is needless to say there has been serious evolution of human diets, especially after the innovation of cooking, an integral part of all human societies. The importance of cooking on the development of human race and societal behavior is best summed up by the statement, "cooking food, especially starchy food, as an innovation has, perhaps more than any other in human history, enabled people to extend the ranges into habitats that were impossible to live in before" (Ungar 2007).

The industrial revolution allowed the food industry to make extensive use of a variety of processing techniques, and added steps that include long storage, and improved texture, flavor, and safety through the destruction of microorganisms, enzymes, and toxins.

Human's modern day diet consists of carbohydrates, proteins, fats, and minor nutrients, vitamins, and minerals that provide energy for healthy living. Water, is not a nutrient but is an integral and essential part of the digestive process. These components have well-defined chemical structures with reactive functional groups such as –COOH, $–NH_2$, –OH, and are obtained from a variety of sources, including grains, fruits, vegetables, milk, meat, and eggs. It is not unexpected that a process that heats carbohydrates, fats, proteins, and nitrite (preservative) at moderate to

H. Akhtar (✉)
Guelph Food Research Centre, Agriculture and Agri-Food Canada, Guelph, ON, Canada
e-mail: Humayoun.Akhtar@agr.gc.ca

J.I. Boye and Y. Arcand (eds.), *Green Technologies in Food Production and Processing*, Food Engineering Series, DOI 10.1007/978-1-4614-1587-9_21,

high temperatures produces new product(s)—volatile and nonvolatile—with different characteristics—both desirable (taste) and undesirable (toxicants, e.g., acrylamide).

Before early 2002, there were only three major groups of potential dietary carcinogens, classified as nitrosamines, polyaromatic hydrocarbons (PAHs), and heterocyclic aromatic amines (HAAs) (Gray and Morton 1981). Considerable scientific literature existed on their occurrences, nature, analysis, and mitigation steps. However, on April 22, 2002, the food industries, as well as national and international food regulatory agencies, were surprised and jolted by the announcement made by the Swedish National Food Administration and University of Stockholm researchers declaring their discovery of high levels of acrylamide in various fried and baked products. A new highly toxic category of acrylamide was added because of its potential toxicity. This announcement, as well as the heightened health concerns (potential human neurotoxin), initiated unprecedented extensive independent and collaborative research projects to provide a better understanding of the processes leading to the formation of acrylamide, and to define steps for its reduction while maintaining the highly desirable organoleptic (sensory aroma, color, feel, and taste) characteristics of the food. Several books and reviews have been published on the nature, occurrences, and mitigation of heat-induced toxicants (Pegg and Shahidi 2000; Skog and Alexander 2006; Stadler and Lineback 2009). This chapter surveys the most recent research activities that tackled these issues and proposed solutions.

21.2 Important processing steps

Food processing can be defined as a step or a combination of steps that make food digestive, appealing (aromatic, tasty, color, etc.) and safe. Some common processing steps include rinsing, soaking, washing, treatment with acid or base solution, blanching, warming, heating (roasting, boiling, pan broiling, pan frying, grilling), steaming, and/or treatment with substances to destroy pathogens and increase shelf life. Processing is a trade-off because each of these steps has the potential to remove some of the nutrients from food, but will also soften the tissues and cells for easier chewing (reducing the size and shape), consumption, and digestion.

21.3 Heat-induced toxicants

Heating raw materials while softening tissues and killing pathogens also initiates chemical reactions between functional groups in endogenous and/or exogenous materials. Such chemical reactions produce both desirable and undesirable products. In this chapter efforts are concentrated on compounds that are not only undesirable, but are also carcinogenic and mutagenic.

$$(CH_3)_2NH + \overset{+}{N}{=}\overset{-}{O} \longrightarrow [(CH_3)_2\overset{+}{N}H\text{-}N{=}O] \xrightarrow{-H^+} (CH_3)_2N\text{-}N{=}O$$

dimethyl amine nitrosyl cation

N-nitroso-N,N-dimethyl amine (NDMA)

Fig. 21.1 Reaction pathways for the formation of N-nitroso-N,N-dimethyl amine

21.3.1 Animal products

Meat is an excellent source of protein compared with fruits and the major source of vitamin B_{12} in a typical Western diet. In 2000, average meat consumption was considerably higher (around 88 kg per capita) in industrialized countries. The top five meat consuming countries are the United States, Demark, Spain, Australia, and New Zealand, where on average 120 kg meat are consumed per capita (NCSU). Meat consumption is increasing steadily in developing countries from around 10 kg per capita in the early 1990s to 26 kg per capita in 2000, and is expected to reach 37 kg per capita in 2030 (WHO). Meat is cooked (from rare to well done) at higher temperatures with or without other ingredients so as to soften the tissues for digestibility and bioavailability of essential elements, as well as for taste.

21.3.1.1 Nitrosamines

For centuries foods have been treated with rock salts to prevent the growth of *Clostridium botulism* to control botulism. It was later established that the potent ingredient was in fact sodium nitrite. It was further documented that curing meats such as pork with sodium nitrite also retained the original reddish color. However, sodium nitrite was seen to be dangerous after a herring meal preserved with large amounts of sodium nitrite caused serious liver disorders and cancers when fed to farm animals in Norway. The compound responsible for the toxic symptoms was identified as N-nitroso-N,N-dimethylamine (NDMA). Nitrite ion under appropriate conditions form nitrosyl cation, (http://chemistry2.csudh.edu/rpendarvis/NOcation.GIF), which then reacts with dimethylamine, as shown in Fig. 21.1, to produce NDMA.

This initiated discussion, and the food industry started questioning the unabated use of sodium nitrite in human foods as a processing agent in meat curing. Scientific communities were tasked with developing reliable analytical methods to determine NDMA and other nitrosamines in fried bacon, sausages, and various cured meats. Although bacon strips are fried at very high temperature for crispiness and flavor (owing mainly to aldehydes and ketones) (Xie et al. 2008), the process is also conducive to the formation of nitrosamines, including NDMA. Incidentally, there was no regulation for the use of sodium nitrite in food preservation, but by mid-1980 numerous analytical techniques allowed various national and international regulatory agencies to set limits of salt in products at 200 ppm, except for bacon, which was restricted to 150 ppm (Anonymous 1984).The Canadian Federal-Provincial-Territorial Committee on Drinking Water has established a guideline of 45 ppm (i.e., 45 mg/L) for nitrate and less than 3.2 ppm (i.e., <3.2 mg/L) for nitrite (Anonymous 2010). The average levels of nitrite in Canadian cured meat over 25 years (1972–1997) have ranged between 28 and 44 ppm (Sen and Baddoo 1997).

In 1981 the US National Academy of Sciences estimated that sources of nitrite intake were cured meat (39%), baked goods and cereal (34%), vegetables including leaves and roots (16%), and drinking water (USNAS 1981). The source of nitrite via nitrate in vegetables and drinking water is linked to overuse of nitrogen fertilizer.

Factors affecting occurrence/formation

The formation of N-nitrosamines is greatly dependent on the method, temperature, and duration of cooking. Conventional heating always produces more nitroso compounds than microwave heating. N-nitrosopyroline (NPYR) and N-nitroso-N,N-dimethylamine (NDMA) are the major volatiles produced when bacon is cooked in a skillet or microwave, but the amounts produced are much higher with conventional methods compared with microwave heating. Oesterdahl and Elriksson (1990) recorded 1.3 μg/kg NDMA, 4.5 μg/kg NPYR in skillet-fried bacon compared to 0.3 and 1.2 μg/kg, respectively, in a microwave. A 1997 study on the volatile nitrosamines in bacon reported a considerably lower value for NDMA at 0.95 μg/kg. This confirms a lowering trend in the amount of carcinogenic nitrosamines since the early days before microwave ovens were available (Gloria et al. 1997; Pegg and Shahidi 2000). Short duration cooking (45 s) in a microwave does not produce nitroso compounds (Miller et al. 1989).

Methods for analysis

Nitrosamines are found in human diets as both volatiles and non-volatiles requiring a variety of extraction and detection techniques. Extraction of nitrosamines including NDMA and other volatiles are performed using AOAC methods (AOAC 1995). Although solid phase microextraction techniques (SPME) have been employed, issues exist with extraction efficiencies (Andrade et al. 2005; Ventanas et al. 2006). The detection and quantification of N-nitrosamines are performed by GC on both packed and/or capillary columns using nitrogen-phosphorus detector (NPD) in nitrogen mode. The use of SPME-MS (mass spectrometry) to measure NDMA in beer was reported (Perez et al. 2008). A very recent review outlines the progress of analytical techniques for the determination of N-nitrosamines in food (Colin 2010).

Toxicological consequences

Foods are the major source of human exposure to highly carcinogenic N-nitrosamines (IARC 1978). Humans are mainly exposed to NDMA by consuming cured meats cooked at high temperatures, eating vegetables, drinking beer and water, and smoking.

New research from the University of Hawaii that followed nearly 200,000 men and women for 7 years reported that consuming processed meats increases the risk of pancreatic cancer. According to the study, people who consumed the most processed meats (hot dogs and sausage) showed a 67% increased risk of pancreatic cancer over those who consumed little or no meat products. Although authors did not confirm the source for the higher incidences, they opined that the source was neither the fat nor the meat. They speculated that the sources came from the processing steps that might have produced nitrosamines and/or heterocyclic amines (Nothlings et al. 2008).

The continued use of nitrite in cured and processed ready-to-eat meat is still problematic, especially for unborn children. Residual nitrite when consumed in bacon, bologna, hot dog/frankfurters, pepperoni, smoked ham, sausages, salami, and others products may form traces of N-nitrosamines with endogenous amines in the stomach. Several epidemiological studies have linked a higher than normal incidence of childhood leukemia and paediatric brain tumors to consumption of cured meats by pregnant mothers of low socioeconomic status (Preston-Martin et al. 1996). However, there appears to be disagreement on this conclusion because the level of nitrite in cured meats has decreased over the years (Pagoda and Preston-Martin 2001; Klurfeld 2001). Also, there appears to be some concerns over the presence of high amounts of nitrate in ground water. Recent studies suggest more occurrence rates of Alzheimer's disease and diabetes may be linked to nitrate in water, and recommend reduction in the use of nitrate fertilizers (De la Monte et al. 2009; Derave and Taes 2009).

Approaches for reduction

The amount of nitrite in processed foods has declined considerably owing to adherence of new guidelines and continued efforts by the food industry; and as such formation of nitroso compounds including the levels of NDMA in food have also declined below the acceptable level, and does no longer require outline surveillance (Colin 2010). However, it would be a worthwhile research endeavor to investigate the use of naturally occurring substances to replace nitrite in meat curing processing. The use of gamma irradiation has been studied to destroy N-nitrosamines in salted and fermented anchovy sauce and pork sausages. NDMA and NPYR are significantly decreased from anchovy at 5 kGy, but required greater than 30 kGy for pork sausages (Ahn et al. 2003, 2006). NDMA is removed from water by absorbents, UV radiation, and sunlight photolysis (Mhlongo et al. 2009). Other major sources of nitrite include industrial and human-derived contamination, microbial action, disinfectants, and the natural degradation of precursors.

Heterocyclic aromatic amines

Unlike nitrosamines, which are produced via a thermal reaction between exogenous nitrite ions and endogenous amines in processed animal products, heterocyclic aromatic amines (HAAs) are formed in animal food products by a thermal reaction between endogenous materials—amino acids, creatine/creatinine found in muscle

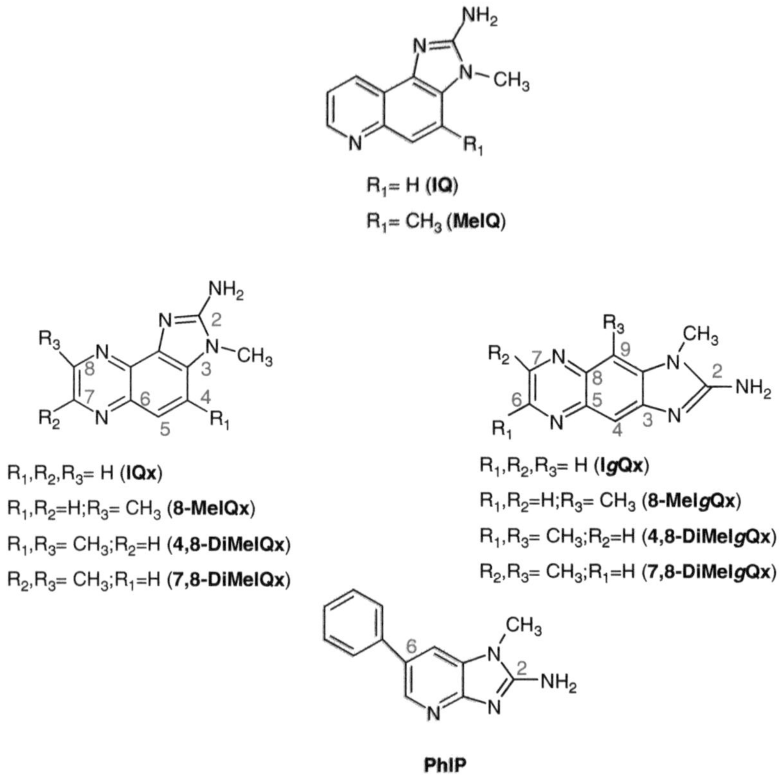

Fig. 21.2 Structure of major heteroaromatic amines

meat (beef, pork, fowl, and fish) and sugar. A reaction between amino acids and sugars at high temperature is known as Maillard reaction, a highly desired process used in the preparation of cereal, bread, potato chips, and French fries for appetizing appeal (Mottram et al. 2006).

Factors affecting occurrence/formation

Since the first reported case of HAAs formation (Sugimura et al. 1977), more than 20 such compounds have been identified in cooked beef, pork, poultry, and fish. Two major classes of HAAs are formed when meat is cooked at elevated temperatures (>150°C). The main group consists of compounds in which the 2-iminoimidazo group is fused to quinoline (e.g., IQ, MeIQ), quinoxaline (e.g., MeIQx, MeIgQx) or a pyridine ring (PhIP) (Fig. 21.2). The non-IQ type products formed at more than 250°C are referred to as carbolines (structures not shown).

The formation and concentrations, which can vary by more than 100-fold, of HAAs in cooked meat products are highly dependent on the source and type of meat (beef, chicken, pork), and the temperature and duration of cooking and doneness (a combination of temperature and cooking duration produces just done, well done, and very well done meat products). Many studies have reported the effect of temperature on the amounts of MeIQ and PhIP produced. One recent study reported that total amounts of HAAs were higher when beef patties were cooked at 400°F (204°C) than 375°F (191°C) for 5 min each side (11.82 ng/g versus 8.10 ng/g), and the amounts of both MeIQx and PhIP increased with increasing temperature, with the amounts of MeIQx produced being slightly higher (Puangsombat and Smith 2010). In general, pork products produced less PhIP than beef and chicken (Knize et al. 1994; Sinha et al. 1994, 1995, 1998b, c; Turesky 2007). In the majority of just done pork products HAAs were not detected, whereas very well done products contained almost three times more MeIQx than the well done pan-fried pork chop (1.3 ng/g versus 3.8 ng/g) (Sinha et al. 1998b). A recent study quantified 13 HAAs in cooked beef, pork, and chicken using liquid-chromatography-electron spray ionization/tandem mass spectrometry and observed PhIP was the most abundant HAAs in barbecued chicken (up to 305 μg/kg i.e., 305 ng/g), broiled bacon (16 μg/kg, i.e., 16 ng/g), and pan-fried bacon (4.9 μg/kg, i.e., 4.9 ng/g). The most abundant HAAs in beef were 7-MeIgQx in well-done pan-fried beef and steak and beef gravy (about 30 μg/kg, i.e., 30 ng/g) (Ni et al. 2008).

Consumption of fish is encouraged because of its omega-3 fatty acid content and related proven health benefits (Leaf 2008). As with other meats, fish is also barbecued, broiled, fried, and grilled at temperatures between 150°C and 250°C, depending on the type of fish and level of doneness (personal preferences). It is expected that fish should also produce mutagenic and carcinogenic HAAs in various amount depending on the cooking choices. A recent study (Costa et al. 2009) confirms that HAAs were indeed formed in various amounts when sardines and Atlantic salmon were barbecued between 280°C and 300°C for varying length of times: rare (5 min on each side), medium (6 min on each side), and well done (7 min on each side). No HAAs were detected in "rare" samples, but the amounts of MeIQx and PhIP were 4.4 μg/kg (i.e., 4.4 ng/g) and 3.3 μg/kg (i.e., 3.3 ng/g) for medium, and 2.2 μg/kg (i.e., 2.2 ng/g), and 6.5 μg/kg (i.e., 6.5 ng/g) for well done, respectively. Salmon when grilled under the same conditions produced considerably larger amounts of PhIP than MeIQx. Broiling of sardines on charcoal at 180–200°C did not produce detectable amounts of HAAs, but detectable and quantifiable amounts of PhIP were formed with salmon under these conditions (Costa et al. 2009).

In general, PhIP is the most abundant HAAs followed by MIQx in cooked meats. It is formed when muscle meat is heated between 180°C and 225°C following complex reaction pathways (Murkovic 2006). In a model system study, the formation of PhIP was accelerated when glucose was present, but was produced in very small amounts when no sugar was added (Skog and Jagerstad 1991). Other model studies indicated following steps: (i) thermal degradation of phenylalanine [1] to phenylacetaldehyde [2], (ii) condensation of [2] with creatinine [3] to yield aldol addition product [4], (iii) a rapid dehydration of [4] to aldol condensation product [5]. The condensation product [5], but not the addition product [4], was detected in both the model system and fried meat by LS-MS. The final step is the reaction of the condensation product [5] with any number of amino acids, including phenylalanine [1] and sugar to yield PhIP [6] (Zochling and Murkovic 2002) (Fig. 21.3).

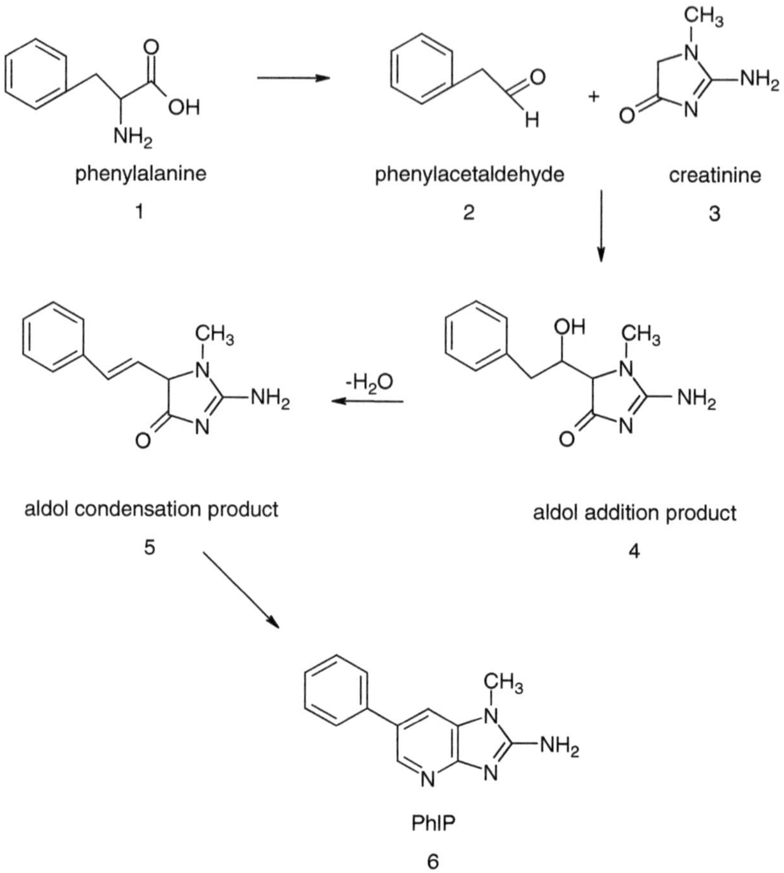

Fig. 21.3 Chemical reactions involved in the production of PhIP during high temperature cooking (Adapted with permission from Elsevier Zochling and Murkovic 2002)

Smoked cheese is a delicacy that can be produced using both natural methods (smoking on an open pit in which combustion is not controlled) and commercial methods. A number of HAAs have been identified in smoked Provola cheese from Calabria (Naccari et al. 2009), and it has been suggested that uncontrolled combustion of the fire materials (oat straw, shrubs, poplar shaving, olive, and orange wood) were their sources (Kataoka et al. 1998).

Toxicological consequences

The International Agency for Research on Cancer (IARC 1993; see also Turesky 2007) lists MeIQ, MeIQx, and PhIP as potential human carcinogens. In rats, these compounds exhibit mutagenic and carcinogenic characteristics (Abdulkarim and Smith 1998; Chen and Chiu 1998). Several studies have provided evidence that linked the consumption of grilled and fried meat to higher risks of various forms of cancers in human. For example, PhIP has been associated with colorectal adenomas (Sinha et al. 1999), stomach cancer (Ward et al. 1997), lung cancer (Sinha et al. 1998a), breast cancer (Sinha et al. 2000; Sinha and Snyderwine 2001), and prostate cancer (Cross et al. 2005). Two recent review articles on the roles of HAAs concluded that high frequencies of consumption of red meat and very dark browned meat (heavily broiled) contributed to carcinogenicity, but the authors could not link the cancerous effects solely to HAAs in cooked meat (Alaejos et al. 2008a, b). Tang et al. (2007) showed that prostate cancer in men is directly linked to the PhIP-DNA adduct formed on consumption of grilled meat products (steak or pork chop, hamburger, hot dog, chicken with or without skin, fish). In a separate study strong correlations between the consumption of fried and processed meat and DNA-adduct were observed in breast tissues of 44 women undergoing mammary reduction surgery. Because the DNA-adduct measurement was nonspecific for HAAs, these effects could not be directly linked to HAAs. However, authors suggested that in this incidence HAAs could be surrogate to other genotoxic substances, such as polyaromatic hydrocarbons (Rohrmann et al. 2009).

Approaches to reduction

The link between HAAs and various forms of cancers along with the IARC's classification as potential and probable human carcinogenic and mutagenic compounds forced the national and international scientific communities to define ways to reduce consumer exposure through foods. A simple solution would be to avoid cooked meat consumption, but this is not a practical solution for a majority of the population. Numerous scientific studies have established the physical and chemical factors as well as the reaction intermediates that influence HAAs formation. The physical factors include cooking temperatures, cooking time (just done to very well done), and water-holding capacity (Persson et al. 2003). The chemical factors, on the other hand, include the endogenous carbohydrate and free amino acids (their water-solubility), and creatine and creatinine in muscle meat (Food mutagens ; Sinha et al. 1998a, b). As previously indicated, thermal reactions among amino acid, carbohydrates, and creatine/creatinine are an extension of simple reaction between an amino acid and carbohydrate (monosaccharide) commonly referred to as a Maillard reaction—a highly desirable step in the production of appetizing and tasty ready-to-eat food. Thus, a single intervention or one that combines the reduction of temperature, time, and availability of endogenous precursors leading to the formation of reactive intermediates, or the ability to trap or nullify the reactive intermediates would eliminate or considerably reduce the HAAs content in cooked meat, and which in turn would result in less exposure to the public.

Today, a microwave oven is as common a household appliance as a conventional oven, toaster, or refrigerator. Most often a microwave oven complements a conventional oven by reducing the cooking time and rapidly heating already cooked foods at home as well as at work to make them more appealing and appetizing. The effect of a combination of microwave heating followed by conventional heating was investigated. Beef patties (15% fat content) were heated in a petri dish at 80% power for various lengths of time in a household oven to remove liquid. The liquid collected contained water, fat, and amino acids along with creatine, which accounted for up to 40% of the liquid content, and increased with increasing time of microwaving. After microwave treatment, the meat was fried at 200°C or 250°C for 6 min on each side and analyzed for HAAs. The microwaved patties contained a considerably lower level of HAAs than the fresh patties (pre-microwaved). Again, the main HAAs were MeIQx and PhIP (Felton et al. 1992). Thus, adding microwave heating before frying meat could reduce HAAs considerably, as long as dripped liquid is not reused in frying or cooking steps. Further, reducing conventional heating time should not affect taste, aroma, or appeal of cooked meat products.

Natural products, especially herbs and fruits, are known to possess antioxidants that can act as scavengers of free radicals. Over the years researchers have investigated their uses for protection against lipid oxidation. Several naturally occurring products, including tea (Weisburger et al. 1994), fruits (Cheng et al. 2007), herbs, culinary ingredients, and vitamin E (Balogh et al. 2000) have been tested for their ability to reduce the levels of HAAs in cooked animal products. In one of the earliest studies, rosemary, thyme, sage, garlic, and brine alone were investigated for their effects on the reduction of heterocyclic amines in fried beef. After heating a piece of beef at 180°C for 20 min on both sides, it was shown that all the spices studied reduced the HAAs (MeIQx + PhIP) content by almost 60% compared with the control. Rosemary, thyme, and sage were better inhibitors than garlic and brine; and were much more effective at eliminating PhIP than MeIQx (Murkovic et al. 1998).

Rosemary is a culinary and medicinal herb (helps improve memory), which also exhibits high antioxidant properties. Dried rosemary leaves have been shown to reduce the formation of HAAs in beef patties (Murkovic et al. 1998; University of Arkansas 2009). Recently, the effects of many rosemary preparations, such as commercially available rosmarinic acid (97%), rosemary antioxidant powder (15–18% carnosic acid), and dried rosemary extracts (with 10, 20, 30, and 40% ethanol in water) and water (100%) at various concentrations were evaluated in beef patties cooked at the routinely used temperatures 375°F (191°C) and 400°F (204°C) for 5, 6, and 7.5 min on each side. All rosemary preparations reduced the formation of major HAAs (MeIQx and PhIP) at both temperatures and at all concentrations. Rosemarinic acid and rosemary powder were most effective in reducing HAAs content at 0.02% level. The overall reduction was between 57% and 66% for MeIQx, and 48–77% for PhIP, and was slightly lower at the lower temperature (Tsen et al. 2006). Several studies from the same laboratory showed that ethanolic extracts of rosemary also performed very well in reducing HAAs in beef patties. The highest reduction up to 91.7% for MeIQx and up to 85.3% for PhIP was achieved with 20% rosemary extracts (Puangsombat and Smith 2010). In a conflicting study rosemary extracts (2–15% dried leaf in water) increased the total amount of HAAs produced in beef Shawarma (a Middle Eastern shaved meat for pita filling), but reduced sevenfold in grilled chickens compared with the control. Amounts of PhIP in Shawarma and chicken were reduced to an almost undetectable level on treatment of meat with 2% rosemary extract (Awney and Sindi 2010).

It is a common practice to marinate meat for some time before frying, grilling, or barbecuing. Today a variety of marinades are commercially available in several flavors to suit the taste preferences of consumers, although some still prefer their own recipes for the tastiest and juiciest hamburger, steaks, etc. The effectiveness of commercially available steak marinades made up of a mixture of spices, instead of an individual spice, for the reduction of heterocyclic amines in grilled beef steaks was evaluated. Eye of round steaks were treated individually with one of the three marinades composed of a combination of spices, and were: Caribbean (thyme, red pepper, black pepper, allspice, rosemary, and chives), Southwest (paprika, red pepper, oregano, thyme, black pepper, garlic, and onion) and herb (oregano, basil, onion, jalapeno pepper, parsley, and red pepper) and grilled at 204°C (400°F) for 5 min on each side. All the test steaks contained fewer HAAs, mainly MeIQx and PhIP, than the untreated steak. The overall reduction in HAAs followed the order Caribbean (88%) > the Herb (72%) > Southwest (57%) (Smith et al. 2008). In certain countries, especially in France and Spain, marinating with wine is a common practice. Frying marinated chicken breast with red wine, which is composed of a large number of amino acids (proline being in the range of 72–85% of all amino acids) at 220°C for 5 min on each side reduced the HAAs content, more specifically PhIP, by almost 85% (from 72 ng/g in uncooked meat versus 8–12 ng/g in marinated cooked pieces). The role of the high content of proline and some trace metals are not fully clear (Busquets et al. 2006). Also, steaks marinated with pilsner beer (5.4% alcohol) and red wine (13% alcohol) for 1, 2, 4, and 6 h at 180°C, and then fried (180–200°C) in a Teflon-coated pan for 4 min each side produced less MeIQx and PhIP than untreated sample. The highest levels of PhIP reduction (88%) were observed in samples marinated in beer and wine for 6 h, but there was no significant difference between the two modes of marination. Reduction of MeIQx was also observed, but it was not as pronounced as the reduction observed for PhIP (Melo et al. 2008).

These observations were verified with meat cooked at home. Analysis of 42 Singaporean Chinese home-cooked grilled chickens cooked at around 190°C revealed that the total HAAs in marinated (ingredients consisted of ground white pepper 74%, soya sauce 67%, garlic, ginger, onion, and wine in addition to sugar and salt) grilled chicken were slightly lower than nonmarinated pieces (2.46 versus 2.75 ng/g)—a serious deviation from those observed under controlled conditions. However, the amount of PhIP was reduced considerably from 2.37 to 1.35 ng/g, and increased for MeIQx from 0.18 to 0.75 ng/g (Salmon et al. 2006).

Generally, the reduction of HAAs with natural products is linked to their antioxidant properties, that is, their scavenging capacity of free radicals during food processing. However, this may not always be the only mechanism. Epigallocatechin gallate (EGCG), a polyphenol in tea, and its fully acetylated derivative (peracetate) reduced the formation of PhIP when heated with phenylacetaldehyde, a major intermediate in a model system (see Fig. 21.3). The reduction was slightly lower for peracetate. Further, the isotope labeling experiments also identified adducts, which showed that the role of natural products could be a combination of radical scavenger as well as trapper of intermediates such as phenyl acetaldehyde (Cheng et al. 2009).

Methods of analysis

The success of an analytical method is highly dependent on the simplicity of sample preparation (extraction), purification and clean up before analysis, and detection and quantification. For universal acceptability, the accessibility of reliable, reproducible, and affordable detection systems is also very important. The high mutagenicity of a number of HAAs in cooked meat makes it almost mandatory that methods are accurate, robust, and as simple as possible so that results can be reproduced even in less-developed countries. In a thorough review of the literature, Galceran and Puignou (2006) traced the history of methods development for HAAs in cooked meat. They have identified the best options for extraction, purification, and analysis. In addition, they have also provided a large table of methods, including extraction, separation, and detection of HAAs in various types of cooked meat. Table 21.1 lists methods recently used for the detection and quantification of HAAs in various food matrixes.

21.3.2 *Plant-derived carbohydrate-rich products*

Cereal grains and potatoes are the main staple foods that provide the basic ingredients for life sustenance. Their use and consumption patterns vary considerably around the globe based on their availability and cultural preferences. These foods, and other products high in carbohydrates, are most prone to the development of acrylamide owing to food processing.

21.3.2.1 Acrylamide

April 22, 2002 will be remembered as the date that forced the food processing industry to rethink applications of high temperature and cooking duration in food preparation. It is the date when the Swedish National Food Administration and University of Stockholm issued an International Press Release “Fried, oven-baked and deep-fried potato and cereal products may contain high levels of acrylamide, which is a probable human carcinogen” (http://192.71.90.8/enginformationakryl.htm), which was followed by a publication by Tareke et al. (2002). Since then there has been extraordinary independent and international collaborative research, which in addition to being published in scientific journals has also been reported at various symposia and conferences. For example, presentations at some of these symposia have appeared in special issues: (1) Conference Chemical Reactions in Foods V (2004); (2) a 3-day American Chemical Society Symposium on the Chemistry and Toxicology of Acrylamide held in Boston in 2007 (Symposium 2008); (3) an updated account of the progress of acrylamide research by the Institute of Food Science and Technology (IFT 2010); (4) The Heatox Project established by the European Commission on Food Quality and Safety (ECFQ&S 2005), which lasted 40 months, involved 24 groups in 14 countries, and issued a FINAL report (Heatox 2007); and (5) several

Table 21.1 Selected methods of analysis of HAAs in various food matrices

Matrix	Column and operating conditions	Detector	References
Cheese	C-18 (3 μm), gradient: (0.3 mM $CH_3CO_2H+NH_4OAC$ pH 3.7), 0.3 mL/min	(i) Fluorescence $\lambda_{ex\text{-}em}$ 263–380 (ii) HPLC-ESI-MS	Naccari et al. (2009)
Beef, pork, chicken	C-18 (3 μm), gradient (1 mM NH_4OAc, 0.1% HCO_2H-10% CH_3CN-89.9% H_2O+0.1% HCO_2H-4.9% H_2O-CH_3CN) 40 μL/min	LC-ESI-MS/MS	Ni et al. (2008)
Beef, bacon, pork, chicken	RX-C-8 (5 μm), gradient (CH_3CN+30 mM NH_4CO_2H pH 3.7), 0.3 mL/min	HPLC-MS	Back et al. (2009)
Beef	TSK gel ODS80 TM (5 μm), Temperature at 40°C Linear gradient (0.01 M trimethylamine pH 3.6+CH_3CN), 1 mL/min.	DAD UV–vis/Fluorescence (programmable)	Puangsombat and Smith (2010)
Beef	TSK gel ODS80 TM gradient (0.01 M trimethyl (amine phosphate, pH 3.2)+0.01 M trimethylamine phosphate pH 3.6)+CH_3CN; 1 mL/min	HPLC UV-258 nm Fluorescence: $\lambda_{ex\text{-}em}$ 300–440	Jautz et al. (2008)
Beef	LiChropsher Si-60 Pre-condition followed by methanol: chloroform (1:9)	HPTLC/UV-FLD	Jautz et al. (2008)
Salmon	TSK gel ODS80 (5 μm) linear gradient (0.01 M trimethyl phosphate pH 3.2+phosphate pH 3.2+0.01 M trimethylamine phosphate pH 3.6+CH_3CN)	HPLC-DAD/FLD DAD λ 263 nm Fluorescence: $\lambda_{ex\text{-}em}$ 307–370	Costa et al. (2009)

others (Studer et al. 2004; Wedzicha et al. 2005; Ames 2009). Based on the Heatox report, the EU issued recommendations on how companies can change their processes to lower the formation of acrylamide (El Amin 2007).

Factors affecting formation/occurrence

It is coincidental that several international research groups within a year independently reported the formation of acrylamide from natural ingredients present in raw materials, and proposed a variety of reaction pathways, including a reaction between asparagine and a sugar (fructose, glucose, sucrose) (Mottram et al. 2002; Becalski et al. 2003). Subsequently, several researchers proposed a thermal reaction between asparagine and a reducing sugar (glucose, fructose) as the predominant route for acrylamide production. For example, Stadler et al. (2002) heated 20 amino acids with different sugars (glucose, fructose, galactose, lactose, sucrose) and noted that only asparagine with a reducing sugar provided the highest yield of acrylamide. At the same time results of several experiments carried out at the Procter & Gamble Company laboratories in Ohio were reported. Zyzak et al. (2003) provided insight on the routes leading to the formations of acrylamide, and observed that when asparagine was heated with glucose, larger amounts of acrylamide were produced when compared with a reaction among glutamine, alanine, arginine, and so on (9,270 μg/kg with asparagine versus between 156 and 590 μg/kg for other amines). Further, these workers followed the progress of the reaction between U-$^{13}C_4$, U-$^{15}N_2$-asparagine [1] with glucose [2] using LC-MS. They detected the formation of the Schiff base [3] with labeled carbon and nitrogen atoms (MW 300 versus 294 for nonlabeled product), which then proceeded via azomethine ylide [4], following different pathways to yield acrylamide [6] in which all carbon and nitrogen atoms were labeled with ^{13}C and ^{15}N (MW 75 versus 71 for nonlabeled compound). Further, these workers also showed that removal of asparagine with asparaginase considerably reduced the amounts of acrylamide formed. Similarly, a reaction between ^{15}N (amido)asparagine with glucose yielded corresponding ^{15}N-labeled acrylamide, whereas a reaction between $^{13}C_6$ glucose and asparagine yielded acrylamide with no measurable amount of ^{13}C (Becalski et al. 2003). Hence the importance of asparagine in the formation of acrylamide was established, but not that of glucose. Formation of smaller amounts of acrylamide in the absence of asparagine provide evidence that carbonyl compounds other than asparagine can also lead to acrylamide, but to a smaller extent (Perez-Locas 2008). Reaction paths leading to the formation of acrylamide with doubly labeled asparagine (^{13}C and ^{15}N) are presented in Fig. 21.4. Other mechanistic studies have been reported (Yaylayan and Stadler 2005; Zhang and Zhang 2008).

Several studies on the effects of time and temperature showed that the amount of acrylamide formed during frying is highly dependent on the frying temperature and frying time. At temperatures between 150°C and 175°C, relatively small amounts of acrylamide were produced, with levels increasing dramatically at temperatures greater than 175°C. Also, the amount of acrylamide increased linearly with time at a constant temperature. This linear pattern was not seen when there was no increase

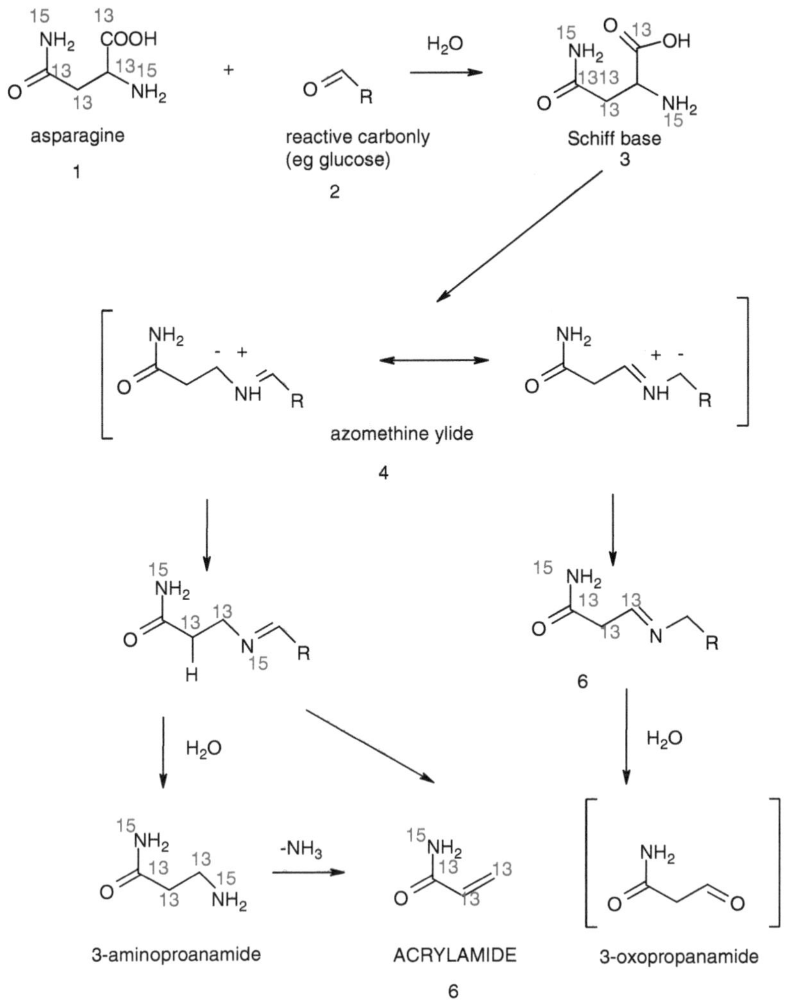

Fig. 21.4 Chemical reactions leading to formation of labeled acrylamide (Adapted with permission from American Chemical Society Becalski et al. 2003; Zyzak et al. 2003). (Note that 13 and 15 refer to the isotope enrichment for carbon atom and nitrogen atom, respectively)

in temperature for a fixed time (Matthaus et al. 2004; Mestdagh et al. 2004). Studies have shown that the addition of inorganic salt in a cereal matrix reduces the amount of acrylamide formed during processing (Kukurova et al. 2009). Similarly, Vass et al. (2004) reported an 80% reduction in acrylamide content in wheat crackers when sodium hydrogen carbonate (sodium bicarbonate) instead of ammonium hydrogen carbonate was used as the baking agent. Further, it was observed that fermentation of bread dough reduces formation of free asparagines, which in turn has less acrylamide than in bread (Fredriksson et al. 2005).

Published data clearly establish fried potato products as the major dietary source of acrylamide for adults and children/adolescents alike in Western countries in which French fries and chips are consumed regularly as part of a meal or snack. Coffee is the most common hot drink in the developed countries, and is the second major source of acrylamide for adults. Other common foods responsible for large amounts of dietary acrylamide are bread, cake, cookies, cereals, and crackers (Friedman and Levin 2008). For example, in one survey, amounts of acrylamide were: potato chips (700–3,700 ng/g), French fries (200–1,900 ng/g), cereal (100–170 ng/g), bread (<100 ng/g, except toasted French bread 260 ng/g), roasted almonds (260 ng/g) (Becalski et al. 2003). Analyses of cereals and potato chip products sold in Canada between 2002 and 2008 showed consistently high levels of acrylamide (with great fluctuation) in French fries and potato chips, followed by considerably lower amounts (<100 ppb) in cereals, and almost negligible (<10 ppb) in several baby foods. The variations in the amount of acrylamide present between these years was attributed to various factors, including sampling design, changes in the amounts of precursors (notably asparagine and sugars in raw materials), as well as to the processing conditions (Health Canada 2008a).

Methods of analysis

The mutagenic and carcinogenic properties of acrylamide in model animals, and the classification by several national and international regulatory agencies as a suspect human carcinogen makes it mandatory that foods be monitored on a regular basis for its presence. This requires access to state-of-the-art methodologies that are accurate, robust, and reproducible for detecting and quantifying various foods for acrylamide content.

Serious concerns around the presence of acrylamide in food were evident by the number of scientific papers in print soon after its discovery in 2002. First among these was a highly informative paper by Canadian researchers at Health Canada, the regulatory arm of food and drug regulations, who proposed various sources of acrylamide produced during food processing. They also developed a liquid chromatography/tandem mass spectrometry (LC-MS/MS) method to detect and quantify, and determined acrylamide levels in several carbohydrate-rich foods that are consumed regularly by Canadians (Becalski et al. 2003).

To develop a workable, reproducible, and inexpensive method of detection, sound knowledge of both the physical and chemical properties of acrylamide is essential for its extraction from a variety of foods and to develop accurate analytical techniques for the determination of its presence and content. Acrylamide is a low molecular weight compound that is more soluble in water than in most commonly used extracting organic solvents (215, 155, 86, 63, 13, and 3 g in 100 mL of water, methanol, ethanol, acetone, ethyl acetate, and chloroform, respectively, at 30°C) (IARC 1994). It lacks spectroscopic properties for detection in minute amounts by UV, and GC without derivation. Further acrylamide produced during cooking is deposited mostly on the surface of products. A separation system (LC and GC) equipped with a mass detection unit would be an option for detection and quantification, even though such a system may not be affordable for a number of developing countries. Thus, an analytical method that employs water for extraction and uses LC or GC equipped with a mass detector would be highly advantageous.

The analysis of acrylamide in foodstuff is a multistep process: (1) sample collection and subsampling; (2) homogenization with or without an internal standard; (3) extraction with water or organic solvent alone or in a mixture; (4) clean up (minimum clean up with water extract); followed by (5) direct analysis and or derivation (bromination) and subsequent clean up; and finally (6) detection and quantification using LC-MS/MS, GC-MS (Castle and Eriksson 2005; Zhang et al. 2005).

Acrylamide in food matrices has been extracted with water at room temperature, organic solvents (methanol, ethanol), and a mixture of water and miscible organic solvents. However, published data do not show any significant benefits for the use of organic solvents over water. Often, extraction with water is complete within minutes. However, for fatty samples, hot water (to disperse the fat molecules) and/or a mixture of water and organic solvent is required for completeness (Zyzak et al. 2003). Liquid pressurized fluid extraction (Yusa et al. 2006), matrix solid-phase dispersion, and solid phase extraction procedures have also been employed to extract acrylamide from a variety of food matrices (bread, toast, cereals, snack, cookies biscuits, coffee) (Soares et al. 2006; Soares and Fernandes 2009). All procedures provided excellent recoveries (93–101%), and the limit of detection and quantification was about 5 and 15 μg/kg, respectively.

Globally, both GC-MS and LC-MS/MS have been used extensively following extraction to identify and quantify acrylamide in foods. The extracts can be analyzed by GC-MS with or without derivation (bromination) and clean up before analysis. A number of researchers have effectively used this technique for monitoring acrylamide levels in various food matrices (Castle and Eriksson 2005; Zhang et al. 2005; Heatox 2007; Soares and Fernandes 2009). The LC-MS/MS technique has also been used extensively, including several national regulatory agencies (FDA 2003; Health Canada 2003; Granby and Fagt 2004; Yusa et al. 2006; Bagdonaite et al. 2008; El-Ziney et al. 2009).

Toxicological significance on human health

Acrylamide is defined as a human neurotoxicant and a rat mutagen (IARC 1994). The Joint Food and Agriculture Organization and World Health Organization Expert Committee on Food Additives and Contaminants determined in 2006 that acrylamide in food may be of health concern (FAO/WHO 2006). In 2009, Health Canada placed acrylamide on the dangerous substances list and initiated a monitoring program (Health Canada 2008b, 2009a). The European Chemical Agency proposed inclusion of acrylamide on the list of hazardous chemicals of substances of Very High Concern after receiving input from EU member states and the European Environment Agency (Harington 2009). To date, epidemiological studies have failed to link acrylamide content in food to cancer. However, there are considerable pressures on national and international regulatory bodies to set a maximum daily intake tolerance for acrylamide. Very recently the use of state-of-the-art physiologically based pharmacokinetics (or toxicokinetics) model (PBPK or PBTK) allowed the tolerable daily intake of acrylamide to be estimated as 40 μg/kg body weight per day for neurotoxicity, and 2.6 μg/kg body weight per day for cancer. The cancer potency of glycinamide (acrylamide conjugate) is decreased considerably and lies at 16 μg/kg (body weight) per day (Tardiff et al. 2010). The daily intake of acrylamide for adults has been estimated by Health Canada, Sweden, and the USFDA, and lies between 0.3 and 0.5 μg/kg. For children and adolescents the daily intake lies between 0.3 and 1.4 μg/kg of body weight per day (i.e., children have a higher exposure than adult) (Health Canada 2008b). The mean highest dietary intake of acrylamide at 3.82 μg/kg body weight (considerably higher) was reported in children between 3 and 6 years of age as compared with 0.49 μg/kg body weight for adults 50 years and older in Gizza, Egypt (Saleh and El-Okazy 2007). A study involving 58,279 men and 62,573 women did not find any link between dietary intake of acrylamide and brain cancer or lung cancer (Hogervorst et al. 2009a, b).

In December 2009 Health Canada sought input from Canadians through a Notice of Intent for the use of L-asparaginase (an enzyme that breaks down asparagine) as a food additive. Currently asparaginase is approved as an injectable drug to treat some forms of cancer. The use of asparaginase in food has been approved in the United States, Australia, New Zealand, and Denmark with support from the Joint Food and Agriculture Organization/World Health Organization Expert Committee on Food Additives (Health Canada 2009b).

Approaches to reduction of acrylamide

The Swedish study (Tareke et al. 2002) identified acrylamide, a compound of great human health concern, in several baked products that are consumed routinely by young and old alike in the developed countries. Concerted efforts have been made worldwide to determine the sources (precursors), processing conditions, mechanism of formation, and steps that would minimize acrylamide levels in heat-treated foods.

Research clearly indicates the reaction between asparagine and reducing sugars as the main cause of acrylamide formation. Potato, cereal grains, and green coffee beans all have sufficient amounts of asparagine and reducing sugars, which when cooked at higher temperature produce acrylamide (Skog and Alexander 2006). Hence, it would be extremely difficult to prevent the formation of acrylamide during cooking unless one of the precursors is removed in significant quantities from plants, and/or cooking is done at a considerably lower temperature and/or for very short duration. This is an unlikely solution because asparagine and reducing sugars are important for plant growth and development, and the organoleptic properties sought by consumers. However, there may be ways to remove a good portion of one or both raw ingredients using processing techniques or genetic modification. Similarly, lowering the processing temperature and time could also reduce the rate of formation of acrylamide. However, in all cases, organoleptic characteristics of food must be maintained. Food will not be accepted by consumers if it does not have the correct attributes, especially taste.

The food processing industry is massive, complex, and highly mechanized, and will require extreme ingenuity and the proper use of resources and protocols to produce foods that not only contain less acrylamide, but also retain all the desirable organoleptic characteristics. The CIAA (Confederation des Industries Agro-Alimentaires de L'Eu-Confederation of the Food and Drink Industries), the voice of the European food and drink industry, provides legislative frameworks for innovation and competitiveness of industry. In an effort to help its clients the CIAA was very proactive and issued 11 versions of "Toolbox" before 2007 (CIAA 2007), which has been referred to as a living document. It is a result of serious collaboration among industries within the European Union (EU). The latest "Toolbox" version 12 issued (CIAA 2009) includes for the first time information from the US Grocery Manufacturers Association (GMA). Further, CIAA and the European Commission have produced pamphlets in more than 20 languages for five important sectors: biscuits, crackers and crisp breads, breads, breakfast cereals, and fried potato products including potato crisps and French fries. These efforts are good for addressing this issue globally.

- *Cereal grain products*: There are 12 grass seeds that are classified as cereal grains. Maize (corn), rice, wheat, barley, oat, and rye, in that order, constitute more than 85% of the world's cereal grain production. Cereals are a major source of nonanimal protein for humans globally, and are the main ingredient for biscuits, breads, cereals, and snacks. Other cereal grains produced in smaller amounts are sorghum, millet, triticale, fonio, buckwheat, and quinoa—the last two are pseudocereals that are cultivated in specific regions. Today it is not uncommon to find bread, biscuits, and cereals carrying wholegrain, 7 grain, and 12 grains labels signifying the health benefits of grains, and a value-added factor in marketing. A wholegrain product provides better nutritional value to consumers, and is made from milled flour from which bran, endoplasm, and germ have not been removed through fractionation. Milled grain flour is routinely refined to separate bran, endoplasm, and germ for use in other bakery goods to meet consumer demands for taste, aroma, texture, digestibility, and shelf life. White flour consists mainly of endocarp and has less nutritional value compared with wholegrain flour.

  Acrylamide is formed when raw food materials containing asparagine and reducing sugars are cooked at temperatures above 120°C (Claus et al. 2008). Both precursors are needed for acrylamide formation, but research has shown that the importance of each ingredient is product dependent. It has been shown that free asparagine is the predominant precursor in cereals, and the amount of acrylamide is increased if asparagine is added to the cereal raw materials before cooking. The same is not true when glucose and sucrose are added (Surdyk et al. 2004). The amount of asparagine in major cereal grain commercial flours follow the order rye > oat > wheat > barley > rice (Hamlet et al. 2008). Asparagine is concentrated mainly in the outer parts of the grain (bran and germ). For example, asparagine in wheat germ averaged 4.9 g/kg compared with sifted flour

0.17 g/kg; in rice germ 0.23 g/kg compared with 0.07 g/kg in flour (Fredriksson et al. 2005). Thus, cereals derived from wheat should have more acrylamide than those from maize (corn) and rice. This was validated with higher acrylamide content in wheat cereals than in maize (corn) cereals when processed under the same conditions (Stadler 2006). Although acrylamide content in cereals is considerably lower compared with fried potato products, its presence is a cause of concern because bread and breakfast cereals are consumed as part of a regular diet by children and adults alike.

Studies have shown that fermented dough produced low amounts of acrylamide in the processed food and stored processed rye crisp bread under moist conditions (Mustafa et al. 2008).

Unpublished studies have shown that asparaginase performs well (70–90%) in reducing acrylamide in most cereal grain products in which high moisture content is required (biscuits, crisp breads, snacks, etc.) except from breakfast cereals, in which the distribution of enzyme is limited because of penetration in coarse and large pieces of grains under low moisture operation conditions (CIAA 2009; Ciesarova et al. 2009).

These details clearly establish the factors influencing the formation of acrylamide in routinely consumed foods derived from cereal grains (bread, biscuits, breakfast cereal, cereal snacks, and crisp bread). They are: (1) the amount of precursors (asparagine and reducing sugar), (2) processing techniques employed prior to processing, (3) cooking temperature and time. The question one needs to ask is: What is the best way to reduce acrylamide in foods without changing the desirable qualities of that product? It is well accepted that total elimination of acrylamide from cooked food is unlikely because regular cereal grains are the main sources of such products. Each end product has its own special requirements. Any effort to remove and/or reduce these ingredients in the end product (diet) must consider its impact on the organoleptic qualities desired by consumers. The statement in the CIAA "Toolbox" sums up the situation: "there is in most cases no single solution to reduce acrylamide in foods even in a given product category" (CIAA 2009).

- *Potato products*: Potatoes are the fourth largest food crop worldwide behind cereal grains (rice > wheat > maize [corn]). Asia and Europe accounted for more than 80% of total world potato production in 2007, but Europe led in its consumption at 87.8 kg/per capita, followed by North America at 60 kg/per capita. Africa has the lowest potato consumption rate at 13.9 kg/capita, but it is gaining acceptance rapidly. In general, potatoes consist of more than 75% water, and more than 20% of starch and carbohydrates; and contain vitamins, minerals, and several amino acids, asparagine being the most abundant and varying considerably in different cultivars (Synge 1977). A recent study employing nine varieties of potatoes sold in Italy and 22 varieties in the United Sates found a significant variation between the lowest and highest values in asparagine (49.3-fold), glucose (31.3-fold), sucrose (35-fold), and fructose (19.4-fold). Highest amounts of asparagine were found in Russet potatoes in the United States at 57.65 mmol/kg fresh weight (Vivanti et al. 2006). Potatoes are not eaten raw, but are baked, steamed, and fried between 150°C and 220°C depending on the desirable attributes of the finished product. Color and aroma/taste are the main consumer requirements for potato products such as baked potatoes, French fries, and potato chips, with the desirable flavors being achieved through the Maillard reaction. Processing of potato materials above 120°C under low moisture contents produces desirable characteristics along with nondesirable toxicant acrylamide in various amounts depending on the processing conditions. To date there has been no report of acrylamide formation in boiled potatoes.

Currently, there is no internationally acceptable average daily intake (ADI) level for acrylamide. Because average daily intake of potato products are much higher in Europe followed by North America, consequently higher acrylamide exposure is experienced by Europeans. In Canada it is estimated that almost 70% of acrylamide intake is from French fries and potato chips (Health Canada 2008a). Thus, the reduction of acrylamide in French fries and potato chip would go a long way in reducing its overall human exposure in Canada.

Several breeding studies have produced potatoes with low sugar levels. It has also been shown that appropriate storage conditions (with a storage temperature greater than 8°C, high humidity, high carbon dioxide concentration) also reduces sugar content, the controlling ingredient in the formation of acrylamide in potato products (Amrein et al. 2003; De Wilde et al. 2004; Hasse 2006; Wicklund et al. 2006; Heatox 2007; Muttucumaru et al. 2008). Using cultivars with low sugar levels seems to be the easiest solution, but other quality criteria, such as organoleptic, growing, or processing attributes, must also be met if the cultivar is to be acceptable to processors and consumers. This area of research needs to be pursued further.

The foregoing clearly establishes that the total removal of acrylamide from fried potato products (French fries, potato chips/crisp) is not feasible without considerable loss of the organoleptic attributes of these products. A study on the risk and benefit of mitigation measures on acrylamide content was reported recently (Seal et al. 2008). In summary, acrylamide reduction could be achieved through the proper use of: (1) raw materials: reduced amounts of precursors through agronomic and genetic approaches; (2) pre-frying treatments: blanching and soaking with water, sodium chloride, calcium chloride, and citric acid solution, adding amino acids (lysine, glycine, cysteine), microwaving precooked fries before frying or cooking (Brathen et al. 2005; Erdoglu et al. 2007; Pedreshi et al. 2007; El-Saied et al. 2008); (3) storage of potato tuber at more than 8°C (Burch et al. 2008); (4) use of asparaginase; (5) reduce cooking/baking time and temperature (avoid browning and/or overcooking); and (6) employing temperature program frying (Palazoglu and Gokmen 2008).

There have been many patents issued on the reduction of acrylamide in processed foods. A recent US patent issued to Zyzak and Stojanovic (2009) for the reduction of acrylamide in food products proposes the reduction of asparagine in food material before cooking. The method adds an enzyme (not limited to asparaginase) capable of hydrolyzing the amide group of asparagine to the food material before cooking at high temperatures. Further, Health Canada (2009b) is also proposing an amendment to Food and Drug Regulations to allow addition of asparaginase in certain food products.

- *Coffee*: Coffee is the fastest growing drink globally. It is a morning ritual in most Western countries because it is viewed as "feel good drink" because of its caffeine content. Coffee consumption is highest at 10 kg/capita per year in the Nordic countries (Denmark, Finland, Iceland, Norway, and Sweden), whereas in Canada and the United States coffee consumption is 6 and 4 kg, respectively, but increasing steadily. Currently more than 57% of Americans use coffee regularly. Recent studies have shown some health benefits for moderate coffee drinkers compared with their non-coffee drinking colleagues. These benefits include considerable reduction in risk of Parkinson's disease (Ascherio et al. 2001), type 2 diabetes (van Dam and Feskens 2002), and the incidence of gallstones in women (Leitzmann et al. 2002). However, detection of acrylamide in coffee is a cause of concern, especially in Nordic countries where the average coffee consumption is the highest in the world.

  Coffee beans are roasted at a very high temperature (240–300°C) for different lengths of time to obtain a desired color, aroma, and taste. During roasting the color of the coffee beans changes from green to dark brown (American preference) to almost dark black (French), with an increase in pleasant and appealing aroma (Cirilo et al. 2003). The bean is then ground, and the coffee drink is prepared by filtering hot water through the bean grinds. Upon brewing, all water soluble components in the coffee grinds are then transferred into the drinkable portion. Instant coffee is produced by dehydrating brewed coffee, followed by rehydration that produces a coffee beverage. Acrylamide being highly water soluble is transferred totally in the drink (Andrzejewski et al. 2004). Acrylamide is known to degrade at temperatures above 170°C, and its amount in the drink is highly dependent on the processing conditions of beans (i.e., roasting time and temperature). Thus, it is not surprising that the highest amounts of acrylamide (>7,000 ng/g) are formed during the first minute of roasting and then they slowly decrease as the roasting time increases, degrading acrylamide. Dark coffee beans (roasted at higher temperatures for a longer duration) have less acrylamide than those derived from light brown beans (roasted at lower temperatures for a shorter duration) (Bagdonaite and Murkovic 2004). The amounts of acrylamide in pre-brewed roasted ground coffee ranged between 45 and 539 μg/kg. On the other hand, a 300-mL brewed coffee from Folgers Classic Roast ground (medium) contained 3.98 μg acrylamide compared to mid-dark variety, which had 2.24 μg. Similarly, a 300-mL instant Folgers Classic Roast coffee made of 2.5 g dry crystals contained 1.14 μg acrylamide. There was no significant difference between brewed and instant coffee when evaluated for the same factors (Andrzejewski et al. 2004). A recent study (Bagdonaite et al. 2008) compared the amounts of acrylamide in two major varieties of coffee beans that constitute greater than 70% of the global market, and found considerably higher amounts of acrylamide in Robusta coffee than in Arabica coffee (mean 708 versus 374 ng/g).

  As with fried potato products, raw coffee beans containing high amounts of asparagine are processed at high temperatures that also produce acrylamide via the Maillard reaction. The preceding details suggest that roasting beans at higher temperatures for a longer duration would reduce or destroy acrylamide totally.

However, this process may also impact negatively on the aroma, taste, and other qualities in the coffee drink. Collaborative research between the European Commission and Nestle Product Technology Centre concluded that roasting coffee at high temperatures reduces radical scavenging capacity (reduced antioxidant capacity), impacts on organoleptic properties, and introduces other undesirable products that may result in poor consumer acceptability of such products. The organoleptic characteristics are very important to coffee drinkers (Summa et al. 2007). A study using ^{14}C-acrylamide revealed: (1) loss of ^{14}C during storage of roasted and ground coffee; (2) nondetectable ^{14}C in volatile(s); (3) a good portion of ^{14}C in filter residue was un-extractable even with polar solvents. These observations are good news in that all acrylamide in roasted and ground coffee may not be bioavailable owing to its covalent binding to insoluble matrix of coffee thus preventing total extractability in brew (Baum et al. 2008), and less human exposure. Hence these factors must be taken into account while devising methods for the reduction of acrylamide in coffee beans. One US patent holder claims to have developed a method that would reduce the amount of acrylamide in roasted coffee beans through the use of an asparagine-reducing enzyme before roasting. This would remove a portion of asparagine, thus reducing the Maillard reaction, and in turn, the amount of acrylamide produced in roasted beans (US Patent 2007).

21.3.3 *Miscellaneous*

In addition to the toxicants produced during food processing, other compounds are also produced during cooking processes. A few important ones among them are evaluated in the following.

21.3.3.1 Furan

Furan has been found in various packaged and roasted foods (Persson and von Sydow 1973; Maga 1979). Before 2004, furan in food was not an issue. It was the discovery of acrylamide in many heat processed foods with rich carbohydrate components that opened the exploratory doors for other reaction products of health consequences. In 2004, researchers at the US Food and Drug Administration (USFDA) found furan in several canned and jarred foods as levels as high as 170 ppb and determined that this was not resulting from interaction between food and can (Hasnip et al. 2006). The IARC (1995), based on animal model studies, had identified furan as a possible human carcinogen, and the United States listed it in the Department of Health and Human Services Report under Carcinogens. Further, the European Food Safety Authority recently concluded that there is a very narrow safety margin between possible human exposure and the no observable adverse effect level (EFSA 2004). Hence, it is necessary to collect additional furan data in routinely consumed food.

Two recent article reviews the literature for the precursor, formation, detection, and analysis of furan (Yaylayan 2006; Vranova and Ciesarova 2009).

Because furan is volatile, it is generally detected by gas chromatography–mass spectrometry using head-space technology (FDA 2004; Becalski and Seaman 2005; Morehouse et al. 2007). As with acrylamide, the formation of furan may follow several pathways (Fan et al. 2008). Perez-Lucas and Yaylayan (2004), and Becalski and Seamen (2005), Yaylayan (2006) have postulated several pathways that involve oxidation of PUFA to 4-hydroxy-2-butanol followed by dehydration, deamination/decarboxylation of amino acids (serine, alanine, and aspartic acid) to produce one or both precursors (acetaldehyde and glycoaldehyde), which undergo aldol condensation to produce furan in multi-steps.

A recent survey of furan in foods and coffee from five EU countries showed that all jarred baby food contained detectable amounts (on average 37 ppb) of furan. Coffee is the primary source of furan in the diet. It is interesting to note that the amount of furan in coffee depends on its preparation. Instant coffee generally has lower furan, with the exception of two—Instant Expresso and Instant Mocha contained around 150 ppb furan. Further, the survey also showed that coffee in Italy and Portugal contained the highest levels of furan. Four of five balsamic vinegars from Spain contained 159–662 ppb of furan, but considerably less in the other samples in Spain and two from Italy were very low (Crews et al. 2009).

21.3.3.2 Polyaromatic hydrocarbon

Polyaromatic hydrocarbons (PAHs), commonly referred to as PAHs, are the byproducts of incomplete combustion of fossil fuels (wood, coal, oil) used in several settings, including food production and processing at home. Polyaromatic hydrocarbons have been known for their toxic effects on chimney sweepers in Britain, especially for lung diseases. More than 100 PAHs have been identified, but 16 have been listed as priority pollutants by the European Food Safety Authority (EFSA) and the US EPA, based on their carcinogenicity and relative potency. These are Benzo[a]pyrene, Dibenz[*a,h*]anthracene and Benz[a]anthracene, ranked 1, 2, and 3, respectively, for carcinogenicity in humans (EFSA 2008) (Fig. 21.5). There are several sources of PAH contamination in foods, but the main routes of PAHs in food are through preparation and processing such as smoking, roasting, and grilling, frying, drying, and steaming food products, and also the environmental contamination of raw food materials before processing (grains, fruits, vegetables, fish, invertebrates—mussels, clams, and oysters). The content of PAHs greatly depends on the type of food and the duration and type of processes employed. For example, smoked and charred fish and meat have higher amounts of PAHs as compared with poached or boiled fish. Also, maximum amounts of PAHs are found in fish and meat that are cooked on wood fire, ranging in temperature between 500°C and 700°C. Hence, avoidance of cooking meat/fish on an open firewood pit would greatly reduce the amount of PAHs in these foods. Precooking meat before barbecuing/grilling/smoking/roasting would also reduce PAHs in meat and fish (Akpambang et al. 2009; Park and Penning 2009).

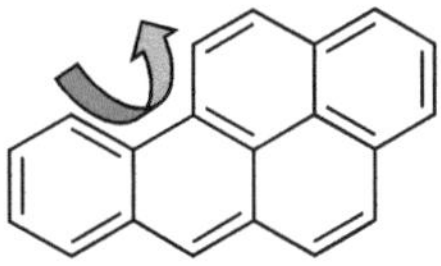

Benzol[*a*]pyrene

1

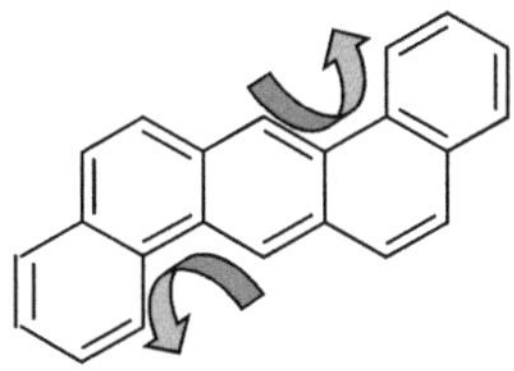

Dibenz[*a,h*]anthracene

2

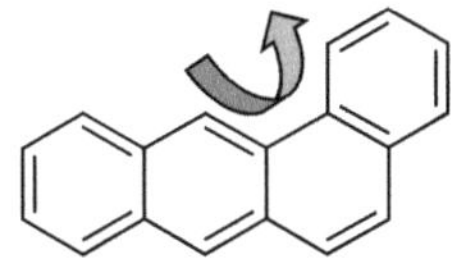

Benz[*a*]anthracene

3

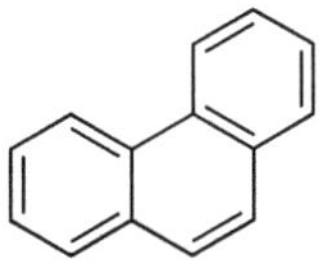

Phenanthrene

4

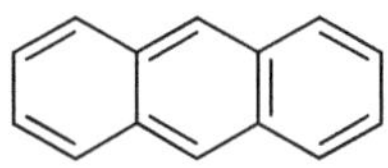

Anthracene

4

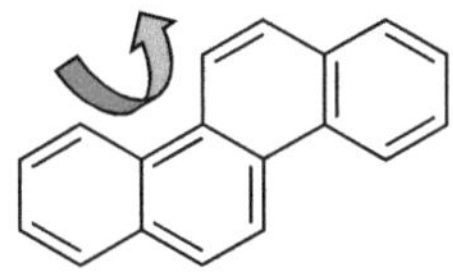

Chrysene

4

Fig. 21.5 Chemical structures of important polycyclic aromatic hydrocarbons. Arrows refer to active regions (Bay)

21.3.3.3 Bisphenol

Bisphenol, commonly referred to as BPA, is a synthetic organic compound and is an important ingredient in plastic making. It is a key monomer for polycarbonate plastic, which is routinely used in the production of household food containers, baby bottles, inside coating of metal food cans, dental fillings, and sealants because of its transparency and shatterproof properties. Bisphenol is not produced during cooking of food ingredients, but it may be introduced, leached, or transferred to food from the mentioned BPA-containing products during heating.

Bisphenol is classified as an endocrine disruptor, and can mimic natural body hormones, which in turn may lead to serious health issues. It can also cause other health problems such as cancers, heart disease, diabetes, reproductive issues. Because of serious potential health issues national and international regulatory agencies are monitoring BPA's indirect exposure to humans, mainly through foods. In 2007, a panel consisting of 38 experts released a consensus report that stated that average human BPA exposure levels were already above the level that cause harms to model animals (vom Saal et al. 2007). In 2008, Canada became the first country to formally declare BPA hazardous to human health and especially to newborns and infants up to 18 months of age (Canada Gazette 2009). Health Canada also terminated the use of BPA in baby bottles and other baby products (varying levels of concern because of BPA exposure). For example, the FDA viewed BPA exposure with mid-level concern for potential effects on brain, behavior, and prostate in fetuses, infants, and young children (FDA 2010). At the same time, the United States Department of Health & Human Services (HHS) issued advice to parents to reduce children's exposure to BPA (HHS 2010). The preferred method to reduce BPA in baby formula and food for regular consumption is to avoid or minimize use of bottles and containers made of polycarbonated plastics. Further, one should reduce food contact times with BPA containers, especially after heating, as this would significantly lower BPA residues in foods.

21.4 Future perspectives

The described toxicants are chemical food safety issues of great concern among researchers, processors, regulators, and health care personnel, as well as consumers (Jackson 2009). Future advances in processing technologies for meals and snacks will be dictated by consumers' demands for healthy, tasty, aromatic, appetizing, and nutritious products that are free of process-induced toxicants. Heat and preservatives will continue to be important processing techniques to produce safe, nutritious, and appetizing foods and snacks until alternative technologies are developed. A recent meta-analysis study on the consumption of meats (lamb, beef, or pork, but not chicken) and deli meats that contain preservatives showed 42% higher incidence of coronary heart disease (CHD) and 19% higher risk of diabetes mellitus (type 2), but not with unprocessed red meats (Micha et al. 2010).This area certainly deserves investigation for better understanding of potential mechanism for observed effects. There will be considerable demands from national and international regulatory agencies for toxicological data that can be used to improve our understanding of health effects as well as to satisfy trading requirements. Research in the development of new crop varieties and processing technologies (a combination of conventional and microwave energy, for example) would continue to reduce toxicant levels in ready-to-eat meals and snacks. There are already some efforts in these areas. For example, three major UK research groups—potato producers, processors, and marketers—are working together to identify varieties and breeding lines of potato with low levels of asparagine and sugars. The aim is to identify potatoes that can retain color and flavor but have low acrylamide risk (Rothamsted Research). Some potato varieties with a lowered sugar contents have been produced, but appetizing capacities of products derived from such varieties have not been fully investigated. The current collaboration among regulatory agencies, the food industry, and researchers is solid and will continue to flourish. There is a strong emphasis to use ingredients with antioxidative properties in food, but information is lacking on the interactions and reactions with naturally occurring amino acids and sugars in foods during processing. Communication, including education and information dissemination to consumers for home cooking, should be expanded. Recent literature and regulatory actions in Canada, the United States, and the European Union provide ample examples of future directions and trends. There is a need for example to clearly establish a link between acrylamide in food and serious health effects. There is also a need to do risk–benefit analysis of the mitigation steps.

Transfer of toxic chemicals from food processing utensils and containers is an emerging issue of great concern. An article in the popular weekly Time magazine under the heading "The perils of plastics" explores the harmful effects of various ingredients, including BPA and phthalates used in modern plastics. There are growing concerns in the United States; and the EPA has launched a new investigation into BPA (Walsh 2010). There is, thus, a greater need for closer collaboration among the scientific communities, regulatory agencies, and plastics manufacturers to resolve the issue of leaching BPA into foods, especially baby foods.

Acknowledgments The author gratefully acknowledges timely assistance of Sheridan Alder, Mike Bryan, and Francesco Lai; and offers special thanks to Kyly Whitfield for reviewing the manuscript.

References

Abdulkarim, B.G., and J.S. Smith. 1998. Heterocyclic amines in fresh and processed meat products. *Journal of Agricultural and Food Chemistry* 46: 4680–4687.

Ahn, H.J., J.H. Kim, C. Jo, et al. 2003. N-nitrosamine reduction in salted and fermented anchovy sauce by ionization energy. *Food Control* 14: 553–557.

Ahn, H.J., J.H. Kim, C. Jo, et al. 2006. Reduction of carcinogenic N-nitrosamines and residual nitrite in model system sausage irradiation. *Journal of Food Science* 67: 1370–1373.

Akpambang, V.O.E., G. Purcaro, L. Lajide, et al. 2009. Determination of polycyclic aromatic hydrocarbons (PAHs) in commonly consumed smoked/grilled fish and meat. *Food Additives and Contaminants Part A* 26: 1096–1103.

Alaejos, M.S., V. Gonzalez, and A.M. Afonso. 2008a. Exposure to heterocyclic aromatic amines from the consumption of cooked red meat and its effect on human cancer risk: A review. *Food Additives and Contaminants Part A* 25: 2–24.

Alaejos, M.A., V. Pinto, and A.M. Afonso. 2008b. Metabolism and toxicology of heterocyclic aromatic amines when consumed in diet: Influence of the genetic susceptibility to develop human cancer: A review. *Food Research International* 41: 327–340.

Ames, J.M. 2009. Dietary Maillard reaction products: Implications for human health and disease. *Czech Journal of Food Sciences* 27(special issue): S66–S69.

Amrein, T.M., S. Bachmann, A. Noti, et al. 2003. Potential of acrylamide formation, sugars, and free asparagines in potatoes: A comparison of cultivars and farming system. *Journal of Agricultural and Food Chemistry* 51: 555–5560.

Andrade, R., F.G.R. Reyes, and S. Rath. 2005. A method for the determination of volatile N-nitrosamines in food by HS-SPME-GC-TEA. *Food Chemistry* 91: 173–179.

Andrzejewski, D., J.A.G. Roach, M.L. Gay, et al. 2004. Analysis of coffee for the presence of acrylamide by LC-MS/MS. *Journal of Agricultural and Food Chemistry* 52: 1996–2002.

Anonymous. 1984. The food and drugs act and regulations. Table XI with amendments to January 1982 Issued by the Department of Health and Welfare (now Department of Health or commonly referred to as Health Canada).

Anonymous. 2010. Guidelines for canadian drinking water quality: Summary Table, Health Canada. http://www.hc-sc.gc.ca/ewh-semt/pubs/water-eau/2010-sum_guide-res_recom/index-eng.php.

AOAC. 1995. *Official methods of analysis of the association of official analytical chemists*, 16th ed. Gaithersburg: Association of Official Analytical Chemists.

Ascherio, A., S.M. Zhang, I. Hernan, et al. 2001. Prospective study of caffeine consumption and risk of Parkinson's disease in men and women. *Annals of Neurology* 50: 56–63.

Awney, H.A., and H. Sindi. 2010. The effect of rosemary on the mutagenicity of heterocyclic amines extracted from common food consumed in Saudi Arabia. *International Journal of Food Sciences and Nutrition* 61: 192–203.

Back, Y.M., J.H. Lee, H.S. Shin, et al. 2009. Analysis of heterocyclic amines and ß-carbolines by liquid chromatography-mass spectrometry in cooked meats commonly consumed in Korea. *Food Additives and Contaminants Part A* 26: 298–305.

Bagdonaite, K., and M. Murkovic. 2004. Factors affecting the formation of acrylamide in coffee. *Czech Journal of Food Sciences* 22(special issue): 22–24.

Bagdonaite, K., K. Derler, and M. Murkovic. 2008. Determination of acrylamide during roasting of coffee. *Journal of Agricultural and Food Chemistry* 56: 6081–6086.

Balogh, Z., J.I. Gray, E.A. Gomma, et al. 2000. Formation and inhibition of heterocyclic amines in fried ground patties. *Food and Chemical Toxicology* 38: 395–401.

Baum, M., N. Bohm, J. Gorlitz, et al. 2008. Fate of ^{14}C-acrylamide in roasted and ground coffee during storage. *Molecular Nutrition and Food Research* 52: 600–608.

Becalski, A., B.P.-Y. Lau, D. Lewis, et al. 2003. Acrylamide in foods: Occurrence, sources and modelling. *Journal of Agricultural and Food Chemistry* 51: 802–808.

Becalski, A., and S. Seaman. 2005. Furan precursors in food: A model study and development of simple headspace method for determination of furan. *Journal of AOAC International* 88: 102–106.

Brathen, E., A. Kita, S.H. Knutsen, et al. 2005. Addition of glycine reduces the content of acrylamide in cereal and potato products. *Journal of Agricultural and Food Chemistry* 43: 3259–3264.

Burch, R., A. Trzesicks, M. Clarke, et al. 2008. The effects of low-temperature potato storage and washing and soaking pre-treatment on the acrylamide content of French fries. *Journal of the Science of Food and Agriculture* 88: 989–995.

Busquets, R., L. Ouignou, M.T. Galceran, et al. 2006. Effect of red wine marinades on the formation of heterocyclic amines in fried chicken breast. *Journal of Agricultural and Food Chemistry* 54: 8376–8384.

Canada Gazette. 2009. Order amending Schedule I to the hazardous product act (bisphenol A). 143(26). http://www.gazette.gc.ca/rp-pr/p1/2009/2009-06-27/html/reg5-eng.html. Accessed 27 June 2009.

Castle, L., and S. Eriksson. 2005. Analytical methods used to measure acrylamide concentrations in foods. *Journal of AOAC International* 88: 274–283.

Chen, B.H., and C.P. Chiu. 1998. Analysis, formation and inhibition of heterocyclic amines in foods: An overview. *Journal of Food and Drug Analysis* 6: 625–636.

Cheng, K.-W., C.C. Wong, J. Chao, et al. 2009. Inhibition of mutagenic PhIP formation by epigallocatechin gallate via scavenging of phenylacetaldehyde. *Molecular Nutrition and Food Research* 53: 716–725.

Cheng, K.-W., Q. Wu, Z.P. Zheng, et al. 2007. Inhibitory effect of fruits extracts on the formation of heterocyclic amines. *Journal of Agricultural and Food Chemistry* 55: 10359–10365.

CIAA. 2007. Acrylamide toolbox: A "toolbox" for the reduction of acrylamide in breakfast cereals. http://www.ciaa.be/asp/documents/l1.asp?doc_id=822.

CIAA. 2009. Acrylamide toolbox-rev. http://www.ciaa.eu/asp/documents/brochures_form.asp?doc_id=65. Accessed 12 Feb 2009.

Ciesarova, Z., K. Kukurova, A. Bednarikova, et al. 2009. Improvement of cereal product safety by enzymatic way of acrylamide mitigation. *Czech Journal of Food Sciences* 27(special issue): S96–S98.

Cirilo, M.P.G., S. Coelho, C.M. Araujo, et al. 2003. Profile and levels of bioactive amines in green and roasted coffee. *Food Chemistry* 82: 397–402.

Claus, A., R. Carle, and A. Schieber. 2008. Acrylamide in cereal products: A review. *Journal of Cereal Science* 47: 118–133.

Colin, C. 2010. The determination of N-Nitrosamines in food. *Quality Assurance and Safety of Crops & Foods* 2: 2–12.

Conference on Chemical Reaction in Foods V. 2004. *Czech Journal of Food Science*. Special issue.

Costa, M., O. Viegas, A. Melo, et al. 2009. Heterocyclic aromatic amine formation in barbecued sardines (*sardine pilchardus*) and Atlantic Salmon (*Salmo salar*). *Journal of Agricultural and Food Chemistry* 57: 3173–3179.

Crews, C., D. Roberts, S. Lauryssen, et al. 2009. Survey of furan in foods and coffees from five European Union countries. *Food Additives and Contaminants Part B* 2: 95–98.

Cross, A.J., U. Peters, V.A. Kirsh, et al. 2005. A perspective study of meat and meat mutagens and prostate cancer risk. *Cancer Research* 65: 11779–11784.

Derave, W., and Y. Taes. 2009. Beware of the pickle: Health effects of nitrate intake. *Journal of Applied Physiology* 107: 1677.

De la Monte, S.M., A. Neusner, J. Chu, et al. 2009. Epidemiological trends strongly suggest exposures as etiologic agents in the pathogenesis of sporadic Alzheimer's disease, diabetes mellitus, and non-alcoholic steatohepatitis. *Journal of Alzheimer's Disease* 17: 519–529.

De Wilde, T., B. De Meulenaer, F. Mestdagh, et al. 2004. Acrylamide formation during frying of potatoes: Thorough investigation on the influence of crop and process variables. *Czech Journal of Food Sciences* 22(special issue): 15–18.

ElAmin A. 2007. EU reveals acrylamide recommendations. http://www.confectionarynews.com/Publications/Food-Beverage-Nutrition/Food.

El-Saied, M.H., A.M. Sharaf, M.M. Abul-Fadl, et al. 2008. Reduction of acrylamide formation in fried potato strips by different pre-frying treatments. *World Journal of Dairy & Food Sciences* 3: 17–24.

Erdoglu, E., T.K. Palazoglu, V. Gokmen, et al. 2007. Reduction of acrylamide formation in French fries by microwave pre-cooking of potato strips. *Journal of the Science of Food and Agriculture* 87: 133–137.

ECFQ&S. 2005. (European Commission Food Quality and Safety) Heat-generated toxicants: Identification, characterization and risk minimisation (The HEATOX Project 506820).

EFSA. 2004. (European Food Safety Authority) Report of the scientific panel on contaminants in the food chain on provisional findings on furan in food. *EFSA Journal* 137: 1–20. http://www.efsa.europa.eu/EFSA/efsa_locat_1178620753812_1178620772979.htm.

EFSA. 2008. (European Food Safety Authority) EFSA opinion on suitable indicators for both the occurrence and toxicity of polycyclic aromatic hydrocarbons (PAHs) in food. http://www.efsa.europa.eu/EFSA/efsa_locale-1178620753812_1211902034860.htm.

El-Ziney, M.G., A.A. Al-Turki, and M.S. Tawfik. 2009. Acrylamide status in selected traditional Saudi foods and infant estimation of daily exposure. *American Journal of Food Technology* 4: 177–191. doi:DOI=ajft.2009.177.191.

Fan, X., L. Huang, and K.J.B. Sokorai. 2008. Factors affecting thermally induced furan formation. *Journal of Agricultural and Food Chemistry* 56: 9490–9494.

FAO/WHO. 2006. Evaluation of certain food contaminants: Sixty-fourth report of the joint FAO/WHO Expert Committee on Food Additive. WHO Technical Report Series 930.

FDA. 2003. Detection and quantitation of acrylamide in foods. http://www.fda.gov/Food/FoodSafety/FoodContaminantsAdultration/ChemicalContami.

FDA. 2004. Exploratory data on furan in food. http://www.fda.gov. Determination of furan in foods, http://www.cfsan.fda.gov.

FDA. 2010. Update on bisphenol A for use in food contact applications: January 2010. http://www.fda.gov/NewsEvents/PublicHealthFocus/ucm197739.htm.

Felton, J.S., E. Fultz, F.A. Dolbeare, et al. 1992. Reduction of heterocyclic aromatic amine mutagens/carcinogens in fried beef patties by microwave treatment. *Mutatation Research* 259: 205–218.

Food mutagens: The cooking makes a difference. https://www.llnl.govt/str/FoodSections3.html.

Friedman, M., and C.E. Levin. 2008. Review of methods for the reduction of dietary content and toxicity of acrylamide. *Journal of Agricultural and Food Chemistry* 56: 6113–6140.

Fredriksson, H., J. Tallving, J. Rosen, et al. 2005. Fermentation reduces free asparagines in dough acrylamide content in bread. *Journal of Agricultural and Food Chemistry* 53: 3259–3264.

Gloria, M.B.A., J.F. Barbour, and R.A. Scanlan. 1997. Volatile nitrosamines in fried bacon. *Journal of Agricultural and Food Chemistry* 45: 1816–1818.

Granby, K., and S. Fagt. 2004. Analysis of acrylamide in coffee and exposure to acrylamide from coffee. *Analytica Chimica Acta* 520: 177–182.

Gulceran, M.T., and L. Puignou. 2006. Latest developments in the analysis of heterocyclic amines in cooked foods. In *Acrylamide and other hazardous compounds in heat-treated foods*, ed. K. Skog and J. Alexander. Cambridge, UK: Woodhead.

Gray, J.I., and J.D. Morton. 1981. Some toxic compounds produced in food by cooking and processing. *International Journal of Food Sciences and Nutrition* 35: 5–23.

Hamlet, C.G., P. Saad, and L. Liang. 2008. Correlation between the amount of free asparagine and saccharides present in commercial cereal flours in the United Kingdom and the generation of acrylamide during cooking. *Journal of Agricultural and Food Chemistry* 56: 6145–6153.

Harington, R. 2009. EU proposal to add acrylamide to hazardous chemical list. http://www.foodproductiondaily.com/content/view/print/258349.

Hasse, N.U. 2006. The formation of acrylamide in potato products. In *Acrylamide and other hazardous compounds in heat-treated foods*, ed. K. Skog and J. Alexander. Cambridge, UK: Woodhead.

Hasnip, S., C. Crews, and L. Castle. 2006. Some factors affecting the formation of furan in heated foods. *Food Additives and Contaminants Part A* 23: 219–227.

Health Canada. 2003. The determination of acrylamide in foods by LC-ESI-MS-MS. http://www.hc-sc.gc.ca/fn-an/res-rech/analy-meth/chem/lps_003-eng.php.

Health Canada. 2008a. Acrylamide levels in selected Canadian foods. http://www.hc-sc.gc.ca/fn-an/securit/chem-chim/food-aliment/acrylamide/acrylamide_level-acrylamide_niveau-eng.php.

Health Canada. 2008b. Canadian exposure assessment for acrylamide in food. http://www.hc-sc.gc.ca/fn-an/securit/chem-chim/food-aliment/acrylamide/can_exp_acryl_food-alim-eng.php.

Health Canada. 2009a. Health Canada's acrylamide monitoring program. http://www.hc-sc.gc.ca/fn-an/securit/chem-chim/food-aliment/acrylamide/monitoring-prog-surveillance-eng.php.

Health Canada. 2009b. Health Canada's proposal to amend the *food and drug regulations* to permit the use of a second source (*Aspergillus niger* ASP72) of the enzyme asparaginase in certain food products. http://www.hc-sc.gc.ca/fn-an/consultation/init/_dec2009-asparaginase/asp-niger-draft-ebauche-eng.php. Accessed December 2009.

HEATOX. 2007. Heat-generated food toxicants: Identification, characterization and risk minimisation. Final report, April

HHS. 2010. United States Department of Health and Human Service Bisphenol A (BPA). Information for Parents. 15 January 2010. http://www.hhs.gov/safety/bpa/.

Hogervorst, J.G.E., L.J. Schouten, E.J.M. Konings, et al. 2009a. Dietary acrylamide intake and brain cancer risk. *Cancer Epidemiology Biomarkers and Prevention* 18: 1663–1666.

Hogervorst, J.G.F., L.J. Schouten, E.J.M. Konings, et al. 2009b. Lung cancer risk in relation to dietary acrylamide intake. *Journal of the National Cancer Institute* 101: 651–662.

IARC. 1978. *Evaluation of the carcinogenic risk of chemicals to humans (Some N-Nitroso compounds) International Agency for Research on Cancer*, vol. 17, 125. Lyon, France: IARC.

IARC. 1993. *Monographs on the evaluation of carcinogenic risks to human (Heterocyclic aromatic amines)*, vol. 56, 165–195. Lyon, France: International Agency for Research on Cancer.

IARC. 1994. *Monographs on the evaluation of carcinogenic risks to human (Acrylamide)*. Lyon, France: International Agency for Research on Cancer.

IARC. 1995. *Monographs on the evaluation of carcinogenic risks to human (Furan)*. Lyon, France: International Agency for Research on Cancer.

IFT. 2010. Acrylamide. http://www.ifst.org/science_technology_resources/for_food_professionals/information_and_news/acrylamide/.

Jackson, L.S. 2009. Chemical food safety issues in the United States: Past, present, and future. *Journal of Agricultural and Food Chemistry* 57: 8161–8170.

Jautz, U., M. Gibis, and G.E. Morlock. 2008. Quantification of heterocyclic aromatic amines in fried meat by HPTLC/UV-FLD and HPLC/UV-FLD: A comparison of two methods. *Journal of Agricultural and Food Chemistry* 56: 4311–4319.

Kataoka, H., K. Kijima, and G. Maruo. 1998. Determination of mutagenic heterocyclic amines in combustion smoke samples. *Bulletin of Environmental Contamination and Toxicology* 60: 60–67.

Klurfeld, D.M. 2001. Maternal cured meat consumption during pregnancy and risk of paediatric brain tumour in offspring: Potential harmful levels of intake (Letters to the Editor). *Public Health Nutrition* 4: 1303–1305. doi:10.1079/PHN2001270; 10.1079/PHN2001271.

Knize, M.G., F.A. Dolbeare, K.L. Carroll, et al. 1994. Effect of cooking time and temperature on the heterocyclic amine content of fried beef patties. *Food and Chemical Toxicology* 32: 595–603.

Kukurova, K., Z. Ciesarova, A. Bednarikova, et al. 2009. Effect of inorganic salts on acrylamide formation in cereal matrices. *Czech Journal of Food Sciences* 27(special issue): S425–S428.

Leaf, A. 2008. Historical overview on n-3 fatty acids and coronary heart disease. *The American Journal of Clinical Nutrition* 78: 237–245.

Leitzmann, M.F., M.J. Stampler, W.C. Willet, et al. 2002. Coffee intake in associated with lower risk of symptomatic gallstone disease in women. *Gastroenterology* 1232: 1823–1830.

Maga, J.A. 1979. Furans in foods. *CRC Critical Reviews in Foods Science and Nutrition* 11: 355–400.

Matthaus, B., N.U. Haase, and K. Vosmann. 2004. Factors affecting the concentration of acrylamide during deep-fat frying of potatoes. *European Journal of Lipid Science and Technology* 106: 793–801.

Mestdagh, F., B. Meulenaer, C. Van Peteghem, et al. 2004. Towards a better understanding in acrylamide formation, degradation and reduction in model systems (and foodstuffs). *Czech Journal of Food Sciences* 22(special issue): 11–14.

Melo, A., O. Viegas, C. Petisca, et al. 2008. Effect of beer/red wine marinades on the formation of heterocyclic aromatic amines in pan-fried beef. *Journal of Agricultural and Food Chemistry* 56: 10625–10632.

Mhlongo, S.H., B.B. Mamba, and R.W. Krause. 2009. Nitrosamines: A review on their prevalence as emerging pollutants and potential remedial options. *Water SA* 35: 735–739.

Micha, R., S.K. Wallace, and D. Mozaffarian. 2010. Red and processed meat consumption and risk of incident coronary heart disease, stroke, and diabetes mellitus. A systematic review and meta analysis. *Circulation* 121: 2271–2283. doi:10.1161/CIRCULATIONAHA.109.924977.

Miller, B.J., S.M. Billedeau, and D.W. Miller. 1989. Formation of N-nitrosamines in microwaved versus skillet-fried bacon containing nitrite. *Food and Chemical Toxicology* 27: 295–299.

Morehouse, K.M., P.J. Nyman, T.P. McNeal, et al. 2007. Summary of furan in heat processed foods by headspace gas chromatography/mass spectrometry and estimated adult exposure. *Food Additives and Contaminants Part A* 25: 259–264.

Mottram, D.S., M.Y. Low, and J.S. Elmore. 2006. The Maillard reaction and its role in the formation of acrylamide and other potentially hazardous compounds in foods. In *Acrylamide and other hazardous compounds in heat-treated foods*, ed. K. Skog and J. Alexander. Cambridge, UK: Woodhead, CRC Press.

Mottram, D.S., B.L. Wedzicha, and A.T. Dodson. 2002. Acrylamide is formed in the Maillard reaction. *Nature* 419: 448–449. doi:10.1038/419448a.

Murkovic, M. 2006. Mechanism for the formation of PhIP in foods. In *Acrylamide and other hazardous compounds in heat-treated foods*, ed. K. Skog and J. Alexander. Cambridge, UK: Woodhead.

Murkovic, M., D. Steinberger, and W. Pfannhauser. 1998. Antioxidant spices reduce the formation of heterocyclic amines in fried meat. *Zeitschrift fur Lebensmittel Untersuchung und Forschung A* 207: 477–480.

Mustafa, A., R. Andersson, K.-E. HellEnas, et al. 2008. Moisture enhances acrylamide reduction during storage in model studies of rye crisp bread. *Journal of Agricultural and Food Chemistry* 56: 11234–11237.

Muttucumaru, N., J.S. Elmore, T. Curtis, et al. 2008. Reducing acrylamide precursors in raw materials derived from wheat and potato. *Journal of Agricultural and Food Chemistry* 56: 6167–6172.

Naccari, C., M.T. Galceran, E. Moyano, et al. 2009. Presence of heterocyclic aromatic amines (HAs) in smoked "Provola" cheese from Calabria (Italy). *Food and Chemical Toxicology* 47: 321–327.

Ni, W., L. McNaughton, D.M. LeMaster, et al. 2008. Quantitation of 13-heterocyclic aromatic amines in cooked beef, pork, and chicken by liquid chromatography-electron spray ionization/tandem mass spectrometry. *Journal of Agricultural and Food Chemistry* 56: 68–78.

Nothlings, U., L. Wilkens, S. Murphy, et al. 2008. Meat intake increases the risk for pancreatic cancer: The multiethnic cohort. Amer Assoc Cancer Res, Annual Meeting, San Diego, April 12–16, Abstract #5821.

NCSU: FS Muscle Foods & Eggs: Meat consumption trends & Nutritional value. www.cals.ncsu.edu/course/ans020/Trends.ppt.

Oesterdahl, B.-G., and E. Elriksson. 1990. Volatile nitrosamines in microwave-cooked bacon. *Food Additives and Contaminants Part A* 7: 51–54.

Pagoda, J.M., and S. Preston-Martin. 2001. Maternal cured meat consumption during pregnancy and risk of paediatric brain tumour in offspring: Potentially harmful levels of intake. *public Health Nutrition* 4(183–189): 1304–1305.

Palazoglu, T.K., and V. Gokmen. 2008. Reduction of acrylamide level in French Fries by employing a temperature program during frying. *Journal of Agricultural and Food Chemistry* 56: 6162–6166.

Park, J.H., and T.M. Penning. 2009. Polyaromatic hydrocarbons. In *Process-induced food toxicants occurrence, formation, mitigation, and health risks*, ed. R.H. Stadler and D.H. Lineback. Hoboken: Wiley.

Pedreshi, F., K. Kaack, K. Granby, et al. 2007. Acrylamide reduction under different pre-treatments in French Fries. *Journal of Agricultural and Food Chemistry* 79: 1287–1294.

Pegg, R.B., and F. Shahidi (eds.). 2000. *Nitrite curing of meat: The nitrosamine problem and nitrite alternatives*. Hoboken: John Wiley InterScience. DOI:10.1002/9780470385081.

Perez, D.M., G.G. Alatorre, E.B. Alvarez, et al. 2008. Solid-phase microextraction of N-nitrosomethylamine in beer. *Food Chemistry* 107: 1348–1352.

Perez-Locas, C. 2008. Mechanism of formation of thermally generated potential toxicant in food related model system. Dissertation, McGill University, Montreal, Canada.

Perez-Lucas, C., and V.A. Yaylayan. 2004. Origin and mechanistic pathways of formation of the parent furan- a food toxicant. *Journal of Agricultural and Food Chemistry* 52: 6830–6836.

Persson, E., I. Sjoholm, and K. Skog. 2003. Effect of high water-holding capacity on the formation of heterocyclic amines in fried beefburgers. *Journal of Agricultural and Food Chemistry* 51: 4472–4477.

Persson, T., and E. von Sydow. 1973. Aroma of canned beef: Gas chromatographic and mass spectrometric analysis of volatiles. *Journal of Food Sciences* 38: 377–385.

Preston-Martin, S., J.M. Pagoda, B.A. Mueller, et al. 1996. Maternal consumption of cured meats and vitamin in relation to pediatric brain tumors. *Epidemiology Biomarkers & Prevention* 5: 599–605.

Puangsombat, K., and J.S. Smith. 2010. Inhibition of heterocyclic amine formation in beef patties by ethanol extracts of Rosemary. *Journal of Food Sciences* 75: T40–T47. doi:10.1111/j.1750-3841.2009.01491.x.

Rohrmann, S., S.-U.L. Jung, J. Linseisen, et al. 2009. Dietary intake of meat and meat-derived heterocyclic aromatic amine and their correlation with DNA-adduct in female breast. *Mutagenesis* 24: 127–132.

Rothamsted Research "Low Acrylamide Potatoes". http://www.acrylamide-potato.org.uk.

Saleh, S.I., and M. El-Okazy. 2007. Assessment of the mean daily dietary intake of acrylamide in Alexandria. *The Journal of the Egyptian Public Health Association* 82: 331–345.

Salmon, C.P., M.G. Knize, J.S. Felton, et al. 2006. Heterocyclic aromatic amines in domestically prepared chicken and fish from Singapore Chinese household. *Food and Chemical Toxicology* 44: 484–489.

Seal, C.J., A. De Mula, G. Eisenbrand, et al. 2008. Risk-benefit considerations of mitigation measures on acrylamide content of foods: A case study on potatoes, cereals and coffee. *British Journal of Nutrition* 99: S1–S46.

Sen, N.P., and P.A. Baddoo. 1997. Trends in the levels of residual nitrite in Canadian cured meat product over the past 25 years. *Journal of Agricultural and Food Chemistry* 45: 4714–4718.

Sinha, R., W.H. Chow, M. Kulldorff, et al. 1999. Well-done, grilled red meat increases the risk of colorectal adenomas. *Cancer Research* 59: 4320–4324.

Sinha, R., D.R. Gustafson, M. Kulldorff, et al. 2000. 2-Amino-1-methyl-6-phenyl- imidazole[4,5-b]pyridine, a carcinogen in high temperature cooked meat, and breast cancer risk. *Journal of the National Cancer Institute* 92: 1352–1354.

Sinha, R., M. Kulldorf, J. Curtin, et al. 1998a. Fried, well-done red meat and risk of lung cancer in women (United States). *Cancer Causes Control* 9: 621–630.

Sinha, R., N. Rothman, C.P. Salmon, et al. 1998b. Heterocyclic amine content in beef cooked by different methods and to varying degrees of doneness. *Food and Chemical Toxicology* 36: 279–287.

Sinha, R., M.G. Knize, C.P. Salmon, et al. 1998c. Heterocyclic amine content of pork products cooked by different methods and to varying degrees of doneness. *Food and Chemical Toxicology* 36: 289–297.

Sinha, R., N. Rothman, E.D. Brown, et al. 1994. Pan-fried meat containing high levels of heterocyclic aromatic amines but low levels of polycyclic aromatic hydrocarbons induces cytochrome P4501A2 activity in human. *Cancer Research* 54: 6154–6159.

Sinha, R., N. Rothman, E.D. Brown, et al. 1995. High concentrations of carcinogen 2-amino-1-methyl-6-phenylimidazo[4,5-b]pyridine (PhIP) occur in chicken but are dependent on the cooking method. *Cancer Research* 55: 4516–4519.

Sinha, R., and E.G. Snyderwine. 2001. Heterocyclic amines (HCAS) and risk of breast cancer. *Breast Cancer Research* 3(Suppl 1): A60. doi:10.1186/bcr389.

Skog, K., and J. Alexander (eds.). 2006. *Acrylamide and other hazardous compounds in heat-treated foods*. Cambridge, UK: Woodhead, CRC Press.

Skog, K., and M. Jagerstad. 1991. Effects of glucose on the formation of PhIP in model system. *Carcinogenesis* 12: 2297–2300.

Smith, J.S., F. Ameri, and P. Gadgil. 2008. Effect of marinades on the formation of heterocyclic amines in grilled beef steaks. *Journal of Food Sciences* 73: T100–T105.

Soares, C., S. Cunha, and J.O. Fernandes. 2006. Determination of acrylamide in coffee products by GC-MS using an improved SPE clean up. *Food Additives and Contaminants Part A* 23: 1276–1282.

Soares, C., and J.O. Fernandes. 2009. MSPD method to determine acrylamide in food. *Food Analytical Methods* 2: 197–203.

Stadler, R.H. 2006. The formation of acrylamide in cereal products and coffee. In *Acrylamide and other hazardous compounds in heat-treated foods*, ed. K. Skog and J. Alexander. Cambridge, UK: Woodhead.

Stadler, R.H., I. Blank, N. Varga, et al. 2002. Acrylamide from Maillard reaction products. *Nature* 419: 449.

Stadler, R.H., and D.H. Lineback (eds.). 2009. *Process-induced food toxicants: occurrence, formation, mitigation, and health risk.* New York: Wiley.

Studer, A., I. Blank, and R.H. Stadler. 2004. Thermal processing contaminants in foodstuffs and potential strategies of control. *Czech Journal of Food Sciences* 22(special issue): 1–10.

Sugimura, T., N. Nagao, T. Kawachi, et al. 1977. Mutagen-carcinogens in food, with special reference to highly mutagenic pyrolysis products in broiled foods. In *Origin of human cancer*, ed. H.H. Hiatt, J.D. Watson, and J.A. Winstein, 1561–1577. Cold Spring Harbor, NY: Book C. Cold Spring Harbor Laboratory.

Summa, C.A., B. de la Calle, M. Brohee, et al. 2007. Impact of the roasting degree of coffee on the in vivo radical scavenging capacity and content of acrylamide. *LWT- Food Science and Technology* 40: 1849–1854.

Surdyk, M., J. Rosen, R. Anderson, et al. 2004. Effects of asparagines, fructose, and baking conditions on acrylamide content in yeast-leavened wheat bread. *Journal of Agricultural and Food Chemistry* 52: 2047–2051.

Symposium on the chemistry and toxicology of acrylamide. 2008. *Journal of Agricultural and Food Chemistry* 56:5983–6776 (special issue).

Synge, R.L.M. 1977. Free amino acids of potato tubers: A survey of published results set out according to potato variety. *Potato Research* 20: 1–7.

Tang, D., J.J. Liu, A. Rundle, et al. 2007. Grilled meat consumption and PhIP-DNA adducts in prostate carcinogenesis. *Cancer Epidemiology Biomarkers and Prevention* 16: 803–808.

Tardiff, R.G., M.L. Gargas, C.R. Kirkman, et al. 2010. Estimation of safe dietary intake levels of acrylamide for human. *Food and Chemical Toxicology* 48: 658–667. doi:10.1016/j.fct.2009.11.048.

Tareke, E., P. Ryrberg, P. Karlsson, et al. 2002. Analysis of acrylamide, a carcinogen formed in heated foodstuffs. *Journal of Agricultural and Food Chemistry* 50: 4998–5006.

Tsen, S.Y., F. Ameri, and J.S. Smith. 2006. Effects of rosemary extracts on the reduction of heterocyclic amines in beef patties. *Journal of Food Sciences* 71: C469–C473.

Turesky, R.J. 2007. Formation and biochemistry of carcinogenic heterocyclic aromatic amines in cooked meats. *Toxicology Letters* 168: 219–227.

Ungar, P.S. (ed.). 2007. *Evolution of the human diet: The known, unknown, and the unknowable.* New York: Oxford University Press.

University of Arkansas, Food Safety Consortium. 2009. Try Thai or Rosemary when spicing the meat to curb carcinogenesis. ScienceDaily. http://www.sciencedaily.com/releasees/2009/05/090527180038. htm, 1 June 2009.

USNAS—US National Academy of Sciences. 1981. *The health effects of nitrate, nitrite, and N-nitrosocompounds*. Washington, D.C: National Academy.

US Patent. 2007. Patent # 7220440-Method for reduction of acrylamide in roasted coffee beans, roasted coffee beans having reduced level of acrylamide, an article of commerce.

van Dam, R.M., and E.J. Feskens. 2002. Coffee consumption and risk of type 2 diabetes mellitus. *Lancet* 360: 1477–1478.

Vass, M., T.M. Amrein, B. Schonbachler, et al. 2004. Ways to reduce the acrylamide formation in cracker products. *Czech Journal of Food Sciences* 22: 19–21.

Ventanas, S., D. Martin, M. Estevez, et al. 2006. Analysis of volatile nitrosamines from model system using SPME-DED at different temperatures and times of extraction. *Food Chemistry* 99: 842–850.

Vivanti, V., E. Finotti, and M. Friedman. 2006. Level of acrylamide precursors asparagines, fructose, glucose, and sucrose in potatoes sold at retail in Italy and the United States. *Journal of Food Sciences* 71: C81–85.

vom Saal, F.S., B.T. Akingbemi, S.M. Belcher, et al. 2007. Chapel Hill bisphenol A expert panel consensus statement: Integration of mechanisms, effects in animal and potential to impact human health at current levels of exposure. *Reproductive Toxicology* 24: 13–18.

Vranova, J., and Z. Ciesarova. 2009. Furan in food: A review. *Czech Journal of Food Sciences* 27: 1–10.

Ward, M.H., R. Sinha, E.F. Heinmann, et al. 1997. Risk of adenocarcinoma of the stomach and esophagus with meat cooking method and doneness preferences. *International Journal of Cancer Control* 14: 14–19.

Walsh, B. 2010. The perils of plastic. Time Magazine April 12 issue. http://www.time.com/time/specials/packages/article/0,28804,1976909_1976908,00.html.

Wedzicha, B.L., D.S. Mottram, J.S. Elmore, et al. 2005. *Chemistry and safety of acrylamide in food.* New York: Springer.

Weisburger, J.H., M. Nagao, K. Wakabayashi, et al. 1994. Prevention of heterocyclic amine formation by tea and polyphenol. *Cancer Letters* 83: 143–147.

Wicklund, T., H. Ostile, O. Lothe, et al. 2006. Acrylamide in potato crisp: The effect of raw material and processing. *LWT- Food Science and Technology* 39: 571–575.

WHO (World Health Organization) Global and regional food consumption patterns and trends-Availability and changes in consumption of animal products http://www.who.int/nutrition/topics/3_foodconsumption/en/index8.html.

Xie, J., B. Sun, F. Zhang, et al. 2008. Volatile flavour constituents in roasted pork. *Food Chemistry* 109: 508–514.

Yaylayan, V.A. 2006. Precursors, formation and determination of furan in food. *Journal für Verbraucherschutz und Lebensmittelsicherheit* 1: 5–9.

Yaylayan, V.A., and R.H. Stadler. 2005. Acrylamide formation in food: A mechanistic perspective. *Journal of AOAC International* 88: 262–267.

Yusa, V., G. Quintas, O. Padro, et al. 2006. Determination of acrylamide in foods by pressurized fluid extraction and liquid chromatography-tandem mass spectrometry used for a survey of Spanish cereal-based foods. *Food Additives and Contaminants* 23: 237–244.

Zhang, Y., G. Zhang, and Y. Zhang. 2005. Occurrence and analytical methods for acrylamide in heat-treated foods- Review and recent developments. *Journal of chromatography. A* 1075: 1–21.

Zhang, Y., and Y. Zhang. 2008. Effect of natural antioxidants on kinetic behaviour of acrylamide formation and elimination in low-moisture asparagines-glucose model system. *Journal of Food Engineering* 85: 105–115.

Zochling, S., and M. Murkovic. 2002. Formation of the heterocyclic aromatic amine PhIP: Identification of precursors and intermediates. *Food Chemistry* 79: 125–134.

Zyzak, D.V., R.A. Sanders, M. Stojanovic, et al. 2003. Acrylamide formation mechanism in heated food. *Journal of Agricultural and Food Chemistry* 51: 4782–4787.

Zyzak, D., and M. Stojanovic. 2009. Method for reducing acrylamide in foods, foods having reduced levels of acrylamide and article of commerce. USPTO, US Patent 7524518 B2 issued on April 28.

Chapter 22
The fallacy of bio-based materials and biodegradability

Rita Schenck

22.1 Introduction

Bio-based is a wonderful word. It conjures images of sun-dappled forests and kittens and babies wrapped in organic cotton and amber waves of grain. Human beings are programmed to feel good about healthy living systems. In fact, research shows that workers with windows onto natural scenes are substantially more productive (Kellert et al. 2008). There is even a name for the phenomenon—biophilia (Wilson 1984). The true drivers of human behavior are biological. The need to reproduce, the need for status among our peers, the need for continued quality of life: They are the basis of everything humans do, and hence all that is valued. That is why humans love living systems. They provide food clothing and shelter. Deep in our genes is the knowledge that living systems assure the continuance of life for ourselves and our progeny. This recognition is the basis of the environmental movement. Promoting and protecting healthy living systems means taking care of the deepest biological drives.

There has been a great deal of discussion about how climate change is having a negative effect on ecosystems—and all the research confirms this. However, the flip side is that ecosystems control the climate, too. The biosphere—the thin layer of life that covers the earth—also participates actively in our climate. It controls most of the atmosphere composition, chemical cycling, and even rainfall. The amount of rain in an area determines whether you can get a forest—but the density of trees also controls the amount of rain you get. The rainforest recycles its rain through transpiration and re-precipitation. Once you cut down the trees, you reduce the rainfall, too. The hypothesis that the biosphere controls our environment is called the Gaia hypothesis, after the Greek earth goddess. It was first developed by James Lovelock (1979). The study of biogeochemistry (how things cycle in the environment) is really the study of Mother Nature's physiology.

R. Schenck (✉)
Institute for Environmental Research and Education, Tacoma, WA, USA
e-mail: rita@iere.org

J.I. Boye and Y. Arcand (eds.), *Green Technologies in Food Production and Processing*, Food Engineering Series, DOI 10.1007/978-1-4614-1587-9_22,

The correlate to biophilia is that biomass, the stuff that is produced by healthy living systems, is our most valuable resource. It feeds us, clothes us, and maintains our health and well-being. And it is most valuable when it is alive. Berries are best eaten right from the bush. Fresh-cut flowers smell better than dried ones.

What are human activities doing with the biological wealth of the planet? How much of that biology can be used for human needs? Are we conserving it for future generations? How strong is the web of life on which we all depend? These are the questions at the heart of the environmental movement.

How much human life can the planet support? This is a topic that was first discussed by Malthus (1798). More recently it was discussed by Meadows et al. (1972).

Almost all life is based on photosynthesis, so the answer lies in the following:

1. How much sun shine we receive
2. What proportion of that sunshine gets converted to the goods we need
3. How much space and sunshine we must leave for other organisms to support healthy living systems
4. How much resource we need to support human populations
5. How humans can share space and still support ecosystems

Answering these questions requires knowledge taken from many different disciplines, and some of the answers have large error bars around them. One answer that is very well known is how much energy we get from the sun. On an annual average, the sun gives the earth 1.37 kW of energy per square meter as measured out in space. This figure is called the solar constant. It changes a bit (a fraction of a percent) over periods of decades to millennia, and it changes substantially over periods of billions of years. The atmosphere absorbs much of the sun's radiation, and at the land surface, we get 342 W per square meter.

Of course we know that different parts of the globe get different amounts of light. Climate, latitude, and seasonal changes all have an effect, but on an annual average, almost everywhere gets about the same amount of light. The poles get 6 months of light and 6 months of darkness. The equator gets 12 h of light a day, year round. Of course there is more light at the equator than at the poles, but less there is less difference than you would think, only about a factor of five difference when you look on an annual average.

How much of the light gets converted to usable products? To get an answer to this question, we need to understand how energy flows in ecosystems, including agricultural systems. Plants capture sunlight through photosynthesis and convert it to biomass. This activity is called primary production. Some of the energy they capture is used for the plant metabolism. What is left over is called net primary production. On average, plants capture about 0.05% of the solar radiation and turn it into net primary production (University of Michigan 2008; Pidwirny 2006). A certain fraction of net primary production is available to animals (including people) to eat.

How much of net primary production transfers to animals depends on the plant and the animal. For example, we only eat wheat grain, but the entire stalk can be eaten by cattle, which can also eat grasses people do not eat. Once you include losses in production (not all wheat acres planted get harvested. In the United States some

15% are lost because of droughts and floods) (USDA 2009), we find that only 0.03% of the sunlight energy gets converted into wheat that we can eat, about 8 ounces per square meter. Other crops are a bit different. The harvestable fraction for potatoes, for example, is four times higher: 0.12% of the sunshine gets turned into potato Calories you can eat, for a yield of 9 lb of potatoes per square meter. Corn (maize) is even better, capturing about 0.15% of the energy from the sun (USDA 2009).

Consider then another conversion. A hayfield in the United States produces about 1,100 usable calories per square meter. This can be eaten by cattle and other ruminants. However, the conversion rate is not too good—only about 2.5% of the energy gets turned into beef we can eat, so we only get one ounce of beef from that square meter of land. The conversion of plant to animal biomass is called secondary production. It is this huge loss of energy between net primary production and secondary production that is the basis of the ecological argument for vegetarianism. We can feed ourselves on one tenth the land as vegetarians than we can as carnivores.

According to the FAO (2009), almost 40% of the world's land area is currently used for agriculture: About 1% in permanent crops such as orchards, about 11% in annual crops, and 26% in pasturage. How much food can we grow in the world's croplands? Assuming that you grow only potatoes, and no crops fail, you can get about 44 quadrillion calories per year, or about eight and half times as many calories as the world population needs for a 2,000-calorie-a-day diet.

So we are golden, right? Wrong. If we add up the caloric value of all the fish and seafood and crops and animals harvested in 2004 (except for fruits and vegetables) we get less than three times the total required amount needed to feed the world population.

Why is that? First of all, we cannot grow potatoes everywhere—they do not like hot climates, for one thing. The Irish potato famine taught us that we need more than potatoes to live, and most other crops have a smaller yield: You do not get many calories per square meter from spinach or other greens and vegetables, but the nutrients they provide are essential to good health.

Second, we grow other things besides food on crop lands—fibers for our clothes, and many other useful goods. Perhaps one third of the crop land is used for nonfood crops (USDA 2009). That means we are down to about two times as many calories as we need.

Third, in the wealthy nations, we are feeding our grains to animals to make meat. What does this mean? In the United States, only about 20% of our food calories come from grain—as opposed to about 50% of the world as a whole. About 28% of our diet is based on animal products, in contrast to the world median of 16% (figures from the FAO 2009). As a result, 40% of our arable land is in pasture or dedicated to growing feed for animals, and of course rangeland (almost twice as big as the arable land: four times as big as the direct food cropland) is also dedicated to growing animals.

Finally, not all the crops make it into hungry stomachs. In the United States, about 40% of food is wasted somewhere between harvest and consumption (Bloom 2010). If that number holds worldwide, then we are down to only 1.6 times as many calories as we need. That does not give us much margin for error. Failure in delivery, bad harvests, use of cereals to create alcohol and fuels, and many other things can make a super-sufficiency into not enough—and then we see hunger and starvation. According to the Food and Agriculture Organization of the United Nations, millions of people are starving even as you read these words.

We still have the range land, right? This land is range land because it has inadequate rainfall for crops. It cannot be turned into cropland via irrigation, because there is not enough water to do so. Worldwide, we use 70% of our fresh water resources on agriculture—and essentially all of that goes to crops (FAO 2009). If the net primary production of all rangeland were converted like hay into beef, we would have lots of meat—over 200 lb per person, worldwide. However, how productive is this land? According to the FAO (2009), worldwide production is only about 75 lb of meat per person. Obviously, the average world pastureland is not as productive as an average United States hayfield. The primary limiting factor is water availability.

We are running out of fresh water resources, too. In the United States, people have siphoned off enough water for irrigation that two major river systems, the Colorado and the Rio Grande, no longer reach the sea. The Ogallala aquifer, a huge aquifer in the middle of the continent, has dropped as much as 35 ft in some places (USGS 2000). There is not a good enough inventory of our freshwater resources to say how much water is left, but it is clear that lakes, rivers, and aquifers are all suffering from overuse. We see the same trends in other countries, too. The Aral Sea (a large freshwater lake) in Uzbekistan has lost 90% of its area since its water sources were diverted for irrigation in the 1960s (UCSD 2011).

Irrigation has more effects than just using up our water, as well. Over time, irrigated lands eventually tend to become saline or have excess aluminum. In some cases, selenium from the groundwater gets high enough to poison all the wildlife in an area. California's Central Valley is an example of where some of this has happened. These lands are poisoned and can no longer support crop production. Only decades of rainfall can restore these lands, and many of them are in deserts where rainfall is scant. Centuries may pass before they are suitable for agriculture (Barrett-Lennard 2002).

Another issue is the amount of energy it takes to produce our food, and the greenhouse gas emissions they imply. In conventional farming, we use energy to manufacture fertilizers and pesticides, and we use tractors on-farm to plow, plant, cultivate, and harvest. Organic farming uses at least as much energy as conventional farming, because it requires much heavier fertilizer sources that use much more energy to transport, and it uses the same amount of fuel for tractors, with less yield per acre. In the United States, about 6–7% of the greenhouse gas inventory is in the agricultural sector (EPA 2011).

What is becoming more and more evident is that on-farm activities release greenhouse gases independent of the use of fossil fuels. The act of plowing itself releases greenhouse gases. In fact, a healthy soil has a complex ecology, and substantially more biodiversity lies below ground than above it. When we plow a field, we are destroying many organisms that release greenhouse gases in decay. Furthermore, some of the nitrogen in soil gets turned into nitrous oxide, which is about 300 times more potent than CO_2 as a greenhouse gas (IPCC 2007). Agriculture is the overwhelming source of nitrous oxide emissions in the United States (EPA 2011).

Conventional farming methods often act to damage the inherent ecological values of our farmlands. A typical cornfield in the United States has had two treatments of herbicide, and uses anhydrous ammonia to provide nitrogen fertilizer (USDA 2009). Anhydrous ammonia (made from natural gas) acts a fumigant, killing all the microorganisms in the soils. The field is planted with a single genetic strain of corn, and often it is a gene-modified strain. During its growth it is treated with insecticides, so there are few or no insects in the field, and after harvest, it is treated with additional fumigants to control pests. The net result is that our agricultural systems resemble laboratory petri dishes more than natural ecologies.

Compare a corn field to a native prairie. The fertilizer in a prairie comes from the natural decay of last year's plants and of course from the manure from the animals that ate those plants. There are likely to be more than 100 different species of plants, and thousands of strains of microorganisms in the soil. More than 800 plant species have been identified in Illinois prairies alone (INHS 2011). Insects and other animals are abundant. A native prairie hosts a high diversity of herbivores: the deer, the antelope and the buffalo that are in the lyrics of past years' song do not live in corn fields. Compared with this vibrant ecology, a monoculture corn field is an ecological wasteland.

One of the uses of crop land is to produce biofuels to substitute for petroleum. The total world agricultural net primary productivity is about 57 quadrillion calories per year (FAO 2009). The total petroleum consumption is about 620 quadrillion calories per year (BP 2010). We use about ten times as much petroleum energy as all the energy we capture through plants' primary productivity. Remember that not all of the plant can be converted into useful fuel. We would be very lucky if even 10% of it could be converted—so probably we would need to use at least 100 times the total crop net primary production to meet our petroleum needs. Clearly, we cannot meet our fuel needs with biofuels. There simply is not enough land.

What about biofuels from forests? About 30% of the world's land surface is in forests. They typically have substantially higher primary productivity (University of Michigan 2008) than do agricultural lands—but still, if we captured all of the primary productivity of the world forests, we would only have 40% of the energy we burn every year in petroleum alone. Many studies look at the potential for using wood and wood byproducts to make fuels, but the technology in this area is not well developed. For the most part, we cannot eat forest products. Instead, what we tend to do is cut down forests to grow energy crops or food. The world demand for biodiesel is leading to the destruction of rainforests in Indonesia, converting them to palm oil plantations (Knudsen 2009; The Jakarta Post 2011). The use of US corn to make ethanol is leading to the conversion of forests there and in Brazil into farms growing corn and soybeans (Houghton et al. 1999; Grunewald 2008; Houghton 2008).

Every time we cut down forests, we also cause a loss of stored carbon in the soils. Although the amount of carbon released is under debate, with wide margins of error, there is no question that this is an important source of climate change. This fact was recognized by the Brazilian government in the runup to the recent climate talks in Copenhagen. They proposed that they stop deforestation and that first world countries assist in that effort with monetary support (Bloomberg 2009).

The conversion of farm to agricultural land continues, though at a decreasing rate. Currently about 0.1 percent of forests are being converted annually (FAO 2010). Not only is this an indicator of the loss of these rich ecosystems, but it implies a reduced primary production worldwide. The flip side of reduced primary production is reduced oxygen production, for with every carbon atom taken from the air, an atom of oxygen is produced. The ocean is a source of oxygen too, but cutting down the forest is tantamount to cutting down our lungs. No one knows how long we can keep reducing forests before we can see any effect, but caution is certainly necessary.

Some are recommending that we make our plastics out of corn rather than petroleum or natural gas. These bio-based plastics can in theory be degraded in compost systems and are renewable. However, there are actually not very many commercial composting facilities—and these plastics do not degrade well in home-based compost piles. Composting itself is not without its problems. During composting, there can be substantial release of toxic compounds—including dioxin, which is made during the composting process. Also, it is quite common for composting processes to become anaerobic. When that happens, methane, a potent greenhouse gas, is emitted. The state of California estimates that reactive gas emissions from waste management are dominated by the composting process (California Air Resources Board 2009).

Then there is the problem of capacity. Once again, if we look at the numbers, we see that to substitute bio-based plastics for the plastics currently being made, we would need to apply all of our agricultural lands to plastic production. Add to that all the ecological damage created by growing corn, and bio-based plastics lose their luster.

There are several things we can do to improve our agricultural practices. At Washington State University, there is a program called "Climate friendly farming" that studies agricultural practices to find which ones reduce emissions of greenhouse gases (Center for Sustaining Agriculture 2010). No-till farming is an example of that. Mixed agriculture rather than monoculture is another example. When animals are grown where their food is grown, the fossil energy use goes down, and nutrients are recycled rapidly. Planting many crops rather than a single crop makes a difference, too.

Clearly we need farms to produce the food we need. Human beings do have a right to be here. This book documents the work of many researchers seeking to make better use of our agricultural resources, seeking to feed the world while reducing resource use and pollution. Choosing to have more efficient consumption patterns is more likely to allow a continued enjoyment of the natural world into future generations.

Collectively, we need to recognize that we are very near the breaking point where we could simply not have enough food for us all. Cutting down forests for agriculture puts the health of the planet at risk. Use of the world's limited agricultural resources to grow biofuels and bio-based products is a self-limiting and short-sighted choice. The challenge for us all is to find a way to live with less, recognizing the limits to our bio-based resources. Taking care of Mother Nature truly is taking care of our deepest selves.

References

Barrett-Lennard, E.G. 2002. Restoration of saline land through revegetation. *Agricultural Water Management* 53: 213–226.

Bloom, J. 2010. *American wasteland: How America throws away nearly half of its food (and what we can do about it)*. New York: Da Capo Press.

Bloomberg. 2009. Brazil want limits on tropical trees for CO2 credits. http://www.bloomberg.com/apps/news?pid=20601086&sid=at0BYn3fpSpU.

British Petroleum (BP). 2010. Statistical review of world energy. http://www.bp.com/productlanding.do?categoryId=6929&contentId=7044622.

California Air Resources Board Emissions database (waste). 2009. http://www.arb.ca.gov/app/emsinv/emssumcat_query.php?F_YR=2008&F_DIV=1&F_SEASON=A&SP=2009&F_AREA=CA#1.

Center for Sustaining Agriculture and Natural Resources (CSANR). 2010. http://csanr.wsu.edu/CFF/.

Food and Agriculture Organization of the United Nations (FAO). 2009. Databases. http://faostat.fao.org/site/339/default.aspx.

Food and Agriculture Organization of the United Nations. 2010. Global Forest Resources Assessment 2010. FAO Forestry Paper 163.

Grunewald, M. 2008. The clean energy scam. *Time Magazine*. http://www.time.com/time/magazine/article/0,9171,1725975,00.html.

Houghton, R.A. 2008. Carbon flux to the atmosphere from land use change 1850 to 2005. http://cdiac.ornl.gov/trends/landuse/houghton/houghton.html.

Houghton, R.A., J.L. Hackler, and K.T. Lawrence. 1999. The U.S. carbon budget: Contributions from land-use change. *Science* 285(5427): 574–578.

Illinois Plant Information Network (IPIN). 2011. The Tallgrass prairie in Illinois. http://www.inhs.illinois.edu/~kenr/prairieplants.html.

Kellert, S.R., J. Heerwagen, and M. Mador. 2008. *Biophilic design: The theory, science and practice of bringing buildings to life*. Hoboken: Wiley.

Knudsen, R. 2009. The cost of the biofuel boom: Destroying Indonesia's forests. Environment 360. http://e360.yale.edu/feature/the_cost_of_the_biofuel_boom_destroying_indonesias_forests/2112/.

Lovelock, J. 1979. *Gaia: A new look at life on earth*. Toronto: Cambridge University Press.

Malthus, T. 1798. *An essay on the principle of population*. London: J. Johnson.

McGuire, V.L., M.R. Johnson, R.L. Schieffer, J.S. Stanton, S.K. Sebree, and I.M. Verstraeten. 2005. Water in storage and approaches to ground-water management, High Plains Aquifer, 2000. U.S. Geological Survey Circular 1243.

Meadows, D.H., D.L. Meadows, and J. Randers. 1972. *The limits to growth. Club of Rome*. New York: Universe Books.

Pidwirny, M. 2006. Primary productivity of plants: *Fundamentals of Physical Geography, 2nd Edition*. Date viewed 20 October 2010 http://www.physicalgeography.net/fundamentals/9l.html.

Solomon, S., D. Qin, M. Manning, Z. Chen, M. Marquis, K.B. Averyt, M. Tignor, and H.L. Miller (eds.). 2007. *Contribution of working group I to the fourth assessment report of the intergovernmental panel on climate change*. New York: Cambridge University Press.

The Jakarta Post. 2011. Bright future for biodiesel in Indonesia. http://www.thejakartapost.com/news/2010/02/08/bright-future-biodiesel-indonesia.html.

University of California. 2011. Aral sea. http://visearth.ucsd.edu/VisE_Int/aralsea/index.html.

University of Michigan. 2008. Global change. http://www.globalchange.umich.edu/globalchange1/current/lectures/kling/energyflow/energyflow.html.

United States. Department of Agriculture (USDA). 2009. U.S. Department of Agriculture Agricultural Statistics Services has been the source nearly all statistics related to crops in the United States. http://www.nass.usda.gov/.

United States. Environmental Protection Agency (US EPA). 2011. Draft inventory of the U.S. greenhouse gas emissions and sink: 1990–2009. http://www.epa.gov/climatechange/emissions/downloads11/US-GHG-Inventory-2011-Complete_Report.pdf.

USGS. 2009. Digital map of water-level changes in the High Plains aquifer in parts of Colorado, Kansas, Nebraska, New Mexico, Oklahoma, South Dakota, Texas, and Wyoming, predevelopment (about 1950) to 2007. U.S. Geological Survey Scientific Investigations Report 2009–5019.

Wilson, E.O. 1984. *Biophilia*. Cambridge: Harvard University Press.

Chapter 23
"Green" food processing technologies: factors affecting consumers' acceptance

Elisabeth Gauthier

23.1 Introduction

Scientists and companies are very focused on designing and implementing environmentally friendly technologies. They ultimately hope to meet consumer demand. However, what do consumers really think about green food processing technologies? Little research has, in fact, been done on this question. Consumers generally know little about the technologies used in food processing. However, there is abundant literature in a number of disciplines regarding what consumers think about food technologies in general, green products, and environmental certification labels (ecolabels) used in the food industry.

Before we examine the literature we must discard the notion that there is a single, clear, and universal definition of green technologies and products. Their green nature is a social construct.[1] This means that the nature and objective of green technologies and products may differ significantly depending on the individuals or groups referring to them: business leaders, scientific experts, the media, environmental

[1] In the following, we will be using the terms "green" and "environmentally friendly" as synonyms when qualifying products or technologies which have a low or reduced environmental impact. Although they will be used occasionally in the text, the terms "sustainable" and "ethical" are not equivalent to the first two; in addition to environmental considerations, they also refer to other aspects such as social justice. Also, as we will see, in the case of food, consumers readily associate health and environmental issues when identifying a product as "green." In the present chapter, we limit the scope of this concept to environmental issues.

E. Gauthier (✉)
Agriculture and Agri-Food Canada, Food Research and Development Centre,
Saint Hyacinthe, QC, Canada
e-mail: elisabeth.gauthier@agr.gc.ca

J.I. Boye and Y. Arcand (eds.), *Green Technologies in Food Production and Processing*, Food Engineering Series, DOI 10.1007/978-1-4614-1587-9_23,

activists, consumers, and so forth. Within companies, the meaning of green technologies and products is constructed based on technical, strategic, and marketing criteria, just like the environmental issues that these technologies and products are designed to alleviate (Champion and Gendron 2002).

For a food science expert, the object of green technologies maybe, for example, to reduce the use of raw materials and energy, decrease effluent pollutants, reclaim byproducts, and dispose of waste in an environmentally friendly manner. However, even from the perspective of technical criteria alone, scientists do not always agree on what the best environmental choice is. Understandably, it is difficult for laypeople to tell the chaff from the wheat, even if they adopt the stance of experts, assuming the latter know what this is. Similarly, for food science experts "processing" generally includes the steps in the food production line downstream from primary production and upstream from retail trade. It is not immediately clear that the public makes a distinction among production, processing, and secondary processing, or interprets the concept of processed foods in the same way as do food science professionals. Therefore, this chapter will take a broad approach to the concept of food technologies and include in the sources some of the abundant literature on the acceptability of genetically modified foods and organic production.

The public also constructs its own understanding of food processing technologies, and that understanding either coincides with, or is opposed to, that of scientific experts. From the perspective of experts, two technologies could for example serve as models for reducing environmental impacts. The development of genetically modified plants is one of the avenues proposed by Lillford and Edwards (1997) to reduce at the source the use of raw materials (e.g., high-yield plants), pollution (plants resistant to pests or requiring less fertilizer), and waste (e.g., shorter stalk cereals, leafless peas). Food irradiation could also be thought of as an environmentally friendly alternative to thermal processes to control the food microbial flora, because of its reduced energy cost (Sadat 2004). It is interesting to note that genetically modified organisms (GMOs) and irradiation are precisely among those technologies that cause the most apprehension among consumers (Hwang et al. 2005; Cardello et al. 2007), mainly for health and environmental reasons (Knight 2007; European Commission 2008; Gauthier 2010). Moreover, studies aimed at determining how consumers perceive environmentally friendly food technologies and certain food ecolabels suggest that the very concept of a green food technology is a contradiction in terms for many of them.[2]

---

[2] We will be reporting in this chapter on studies emphasising consumers' reluctance with respect to food technologies, and contrasting consumers' and experts' viewpoints. We do not mean however, to characterize consumers as a homogeneous, antitechnology group nor all food science professionals and other scientific experts as blindly protechnology either.

Lastly, those called citizens, consumers, or the public[3] do not form a homogenous group. To analyze what an individual[4] thinks about a particular product or technology, we have to consider that person's attitudes and beliefs as well as his or her cultural and social environment. What each person thinks is influenced among other things by the characteristics of the product or technology, the person's trust in the people who developed it or supervised its application, the media, and inputs from other actors. The disciplines that address this issue utilize very different theories and methodologies whose interpretation is often limited to a specific context. With this caveat, a general picture emerges from this mosaic of knowledge. It is this general picture that we attempt to bring out, although it is not exhaustive by any means.

So, what do consumers think about green food processing technologies? There are few direct answers to this question, but each of the topics that is addressed sheds light on the factors determining the acceptability of green technologies. We begin with a brief historical review showing how environmental concerns have become widespread among the public, how the industry has adjusted, the role of the media, and the beginnings and first pitfalls of green marketing. We then review the factors that influence consumers' attitudes toward food technologies in general and their environmental impact.

23.2 The progressive "greening" of consumption

23.2.1 The rise of environmental issues

The publication in September 1962 of Rachel Carson's *Silent Spring* led to what some have called a paradigm shift. Denouncing the ravages of synthetic pesticides on flora, fauna, and human health, Carson's book generated broad public awareness of environmental issues, giving considerable impetus to the environmental movement. In the 1960s, the public began worrying about the environmental impact of technological change on agriculture, particularly chemical pesticides. In the 1970s

---

[3] The various studies to which we will be referring use different terms to refer to the individuals and groups they are examining, depending on the discipline to which they belong (consumer science and communications science, sociology, psychology, marketing, etc.), the theoretical approach used and the aspect that interests them. We will try to respect the terminology used in these studies, that is, either "public," "citizen," or "consumer." However, these terms are not functional equivalents. A buyer seeking to satisfy his or her personal needs will best be described by the term "consumer" whereas the term citizen encompasses a larger spectrum of motivations. The concept of "citizen-consumer" has been proposed to describe a buyer who simultaneously seeks to meet his or her needs and to protect the common good, for example, the environment. This concept is widely used in the literature on ethical consumption. It translates the idea that through their choices, consumers can be agents of social change or "vote with their wallets." For a critical review of this concept, see Johnston (2008).

[4] A majority of the studies we will be referring to take individuals as their focus of study. For a detailed review of existing theoretical approaches of consumer behavior, see Jackson (2005).

and 1980s, people became concerned about the intensification of agricultural production and its impact on the ecology of farmland, pollution, and changes in the landscape. Environmentalists lobbied for public policy reforms and encouraged consumers to make environmentally friendly and socially equitable choices[5] (Lowe 1992; Lowe et al. 2008).

A metaanalysis of opinion polls in the United States shows that environmental protection began weighing heavily on the public's mind in the late 1960s in the United States and in the European Community (Dunlap 2002). The global boycott in the late 1980s of CFC-based aerosols, the worldwide success in 1988 of the *Green Consumers Guide,* and reaction to the repercussions of tuna fishing on dolphins—which led to the "dolphin-safe" label (Elkington and Hailes 1988; Teisl et al. 2002; Peattie and Crane 2005)—also illustrate this public interest. In the early 1990s, 79% of Americans considered themselves environmentalists; 82% said that they had engaged in recycling; 83% reported that they had changed their consumption habits to protect the environment; and 67% indicated that they were willing to pay between 5% and 10% more for environmentally friendly products (Coddington 1990). Despite fluctuations in the 1990s (Dunlap 2002), environmental concerns have become a global phenomenon, regardless of the prosperity of the countries in which opinion polls are conducted (Dunlap and York 2008). On the eve of the United Nations Framework Convention on Climate Change Summit in Copenhagen, two thirds of respondents in 23 countries felt that climate change was a "very serious" issue, a level of concern not seen since 1998 (GlobeScan and BBC World Service 2009). According to another survey in 15 countries, pollution and the availability of water were considered "very serious" issues by 72% and 71% of respondents, respectively, exceeding even concerns about global warming (Waterviews and Globe Scan 2009). Citizens were also aware that food production, processing, and distribution have a significant environmental impact (Lea and Worsley 2008).

In view of these data, one might think that corporate environmental initiatives would unquestionably be welcomed by the public and the media. However, various analysts and observers report that several companies in the United States and Europe are reluctant to emphasize their environmental efforts, for fear of being accused by environmental groups of "greenwashing" or alienating the media and the public (Peattie and Crane 2005; Thøgersen 2006; Anonymous 2009; Roth 2009). How has this situation developed? To provide insights into this question, the following sections review how two important categories of actors reacted to the rising environmental concerns of citizens: the agri-food industry and the media.[6] These actors play a major role in influencing citizens' reaction to novel green technologies and products.

---

[5] As observed by Lowe et al. (2008), the environmental movement strongly contributed to the confluence of environmental and ethical concerns in the minds of consumers, an association that is very present nowadays, as we will see in Sect. 23.3.4.

[6] We have left aside the role of states, which, through the establishment of environmental policies and regulations, largely determine the actions of other actors. For reviews and examples, see Hoffman (2001), Mason (2006), Pfeifer and Sullivan (2008), Andrews and DeVault (2009), and Jørgensen et al. (2010).

23.2.2 *The industry: from skeptic to proactive to cautious*

Initially, the agri-food industry's response to the citizens' nascent environmental concerns was not necessarily very positive.[7] This would gradually change following pressure by various stakeholders (citizens, environmental groups, regulatory bodies, competitors, etc.). Initially on the defensive, companies then focused on compliance with new environmental regulations following the creation of the Environmental Protection Agency (EPA), the United Nations Environment Program, and the first European environmental policies in the early 1970s.[8] Industry subsequently adopted a more proactive approach. The beginning of the 1990s saw companies creating vice president positions to manage environmental affairs as well as processes for identifying environmental issues Groups[9] were being formed that promoted collaboration among investors, social groups, and other actors on environmental and social issues (Menon and Menon 1997; Hoffman 2001).

This shift toward a more proactive approach was not homogenous. An analysis of specific cases (Wong et al. 1996; Crane 2000; Hoffman 2001) shows the diversity of strategies used by companies in connection with environmental issues, based on consumers' expectations, regulations (existing or projected), and competitors' strategies. As noted by Hoffman (2001), the shift in social norms on the issue of the environment puts pressure on companies to conform to avoid the consequences of nonconformity: legal penalties, public protests, and loss of market share to competitors. Some responded to such pressure by making fundamental changes in their practices and technologies, whereas others tried to avoid these changes by limiting themselves to public relations offensives or other avoidance tactics (For a review, see Peattie and Crane 2005).

Social groups rose up against this superficial form of "greening." The word "greenwashing," which was added to the *Oxford English Dictionary* in 1999, would then come to mean "disinformation disseminated by an organization so as to present an environmentally responsible public image" (Holcomb 2008). This monitoring, the enactment by several countries of regulations on misleading advertising and the

---

[7] In 1962, Monsanto Chemical Co. responded to Rachel Carson's book in the pages of *Monsanto Magazine* with a parody entitled *The Desolate Year,* which described the ills that a world without pesticides would face (Monsanto Chemical Company 1962). In the same tone, an editorial in the *American Agriculturist* featured a grandfather explaining to his grandson that a book called *Quiet Summer* caused a return to a natural way of life, resulting in famine and malaria, of which the little boy's parents had died (Brooks 1972).

[8] The founding of the EPA in 1970 coincided in the United States with a phenomenal increase in the number of environmental health and safety regulations and the number of lawsuits against industry (Ausubel and Sladovich 1989). At the time, economist Milton Friedman described as "pure and unadulterated socialism" any expense by industry to control pollution beyond the requirements of the law (Friedman 1970, cited in Hoffman 2001).

[9] Examples of such groups include the Council for Environmentally Responsible Economies (http://www.ceres.org) or the World Business Council for Sustainable Development (1992) (http://www.wbcsd.org).

cost associated with technological change, have caused some companies thinking about entering the green market to reconsider (McGrath 1992; Roberts 1996). Greenwashing practices continue to be actively monitored by certain groups (Terrachoice Environmental Marketing 2009; Greenpeace 2010) and as we have seen, the threat of being labeled as a greenwasher is taken seriously by companies, leading some of them to refrain from publicizing their environmental efforts.

As this brief overview shows, the reactions of companies to the rising environmental concerns of citizens were diverse. In general, however, analysts observed an initial defensive reaction, followed by a perhaps overenthusiastic jump into the new green market by several companies, some of which drifted into greenwashing tactics. The latter were cited by several studies among factors which lead to consumers' cynicism toward green claims and the industry in general, as seen in Sect. 23.2.4. Media coverage of environmental issues and the related issue of greener production is likely to be another contributing factor. The following section builds on the literature on media theory as well as on studies of the public perception of environmental, food technology, and food risk issues. It aims to highlight the complex dynamics by which media influence the view of citizens on these issues.

23.2.3 The media, the environment, and food technology

As environmental concerns have become widespread among the public, media coverage of environmental issues has also increased over the years. The number of environment-related news items doubled between 1988 and 1989 (Research Alert 1990, cited in Roberts 1996). At the same time, some journalists began specializing in covering these issues, as is evidenced by the founding in 1990 of the Society of Environmental Journalists, whose membership stood at 800 at the beginning of the 2000s (Hoffman 2001).

The role of the media is particularly significant in the case of environmental issues. When individuals have no direct experience with an issue, information about it comes to them mainly through two sources: the information media and personal networks (Kasperson et al. 1988). Few citizens have any direct experience with issues such as global warming or the shrinking ozone layer, which increases their dependency on the media for information (Zucker 1978; Anderson 1997). The same is true of food technologies and food risks (Frewer et al. 1993; Bauer et al. 2001), although citizens' avowed trust in the media varies from country to country and technology to technology (Lyndhurst 2009) and depending on the media (Frewer et al. 1996).

The media choose the issues they consider to be most important and thus determine to a large extent what will be seen as most important by citizens and public policy makers (e.g., Protess et al. 1987). It is their "agenda-setting" function.[10]

[10] Of course, media may also pick issues that are already widely discussed in public arenas, making these issues even more salient. For this concept, see McCombs and Ghanem (2003).

Empirical studies have not only confirmed the existence of a correlation, but also of a causal link between the coverage of certain issues in the media and their later saliency in public opinion polls, particularly in the case of environmental issues (Ader 1995; Soroka 2002). Media coverage of an issue is also likely to affect citizens' attitudes and subsequently their behavior (e.g., Fishbein and Yzer 2003). In the wake of the "agenda-setting" theory, the "quantity of coverage theory" goes further about news reporting on environment and technological risk: Increased coverage not only makes an issue salient, but also turns public opinion in a negative direction and generates increased opposition to risky technologies and heightened fear of environmental hazards (Mazur 1987; Mazur and Lee 1993). There are empirical demonstrations of this theory in the case of food technologies. Frewer et al. (2002) demonstrated that during periods when media coverage of GMOs increases in the United Kingdom, consumers have a heightened perception of the risks associated with this technology and value its benefits less. Frentzen et al. (2000) analyzed the coverage of food irradiation in newspapers and on television in 18 states in the United States between 1996 and 1999. They observed that the number of articles and reports and declared intentions to purchase irradiated foods showed an inverse trend (Food Marketing Institute 1996–1999).

Other researchers have focused particularly on the influence the media have on public opinion through the way in which they frame questions, that is, the way in which they select, highlight, or exclude certain information and suggest interpretations.[11] Food safety experts sometimes view the media as alarmist in their presentation of the risks associated with new technologies and that they underreport microbial food safety issues (De Boer et al. 2005; Van Kleef et al. 2006).[12] Empirical content analysis of the coverage of specific food technologies or food risk issues sometimes contradicts these accusations. The media coverage of food irradiation in the United States was shown to be generally neutral or positive with regard to the process (Thomsen et al. 2003). In Canada, in their coverage of a public consultation on food irradiation in 2002, the media actually conveyed a balanced representation of the risks associated with irradiating or not irradiating food and they insisted more on the latter than did irradiation supporters in their own public statements (Gauthier 2008).

The media continuously endeavor to refresh their coverage of long-term issues (e.g., most environmental problems) so as not to lose the interest of their readers and viewers, either by addressing specific problems or changing their framing. After a while, people lose interest in certain issues either because they have already made up their minds about them or they have begun to find them boring, unless they affect them directly (Zucker 1978). For example, Thøgersen (2006) notes that the Danish media, which were initially very enthusiastic about NGOs and green companies (Hansen 1993; Linné 1993), between 1996 and 2002 took a more negative and pessimistic

[11] On the concept of framing, see Entman (1993), Tankard (2003), and Gitlin (1980).

[12] Interestingly, citizens' views on the media coverage of food issues may differ from those of experts; studies report that the general public considers that the media are alarmist in their presentation of foodborne microbial hazards (Miles et al. 1999; Miles and Frewer 2001).

turn, putting greater emphasis on the misdeeds of certain green companies and the drop in sales of organic products. The author saw no clear difference in the amount of coverage given to these products, and attributed this to the change in framing a negative impact on consumers' attitudes toward organic foods, which contributed to stagnating sales of these foods in the late 1990s in Denmark.

Lastly, the efforts of journalists to meet a journalistic ideal of objectivity often translate into a practice of balanced reporting, which means presenting both sides of an issue with equal emphasis (Nelkin 1987). This standard of practice tends to result in the coverage of certain food technologies as being polarized, i.e., presented as a conflict between proponents' and opponents' viewpoints on the technology, overshadowing the technical aspects of these processes (Ten Eyck 1999, 2002; Ten Eyck and Deseran 2001). This practice of balanced reporting can also result in marginal viewpoints receiving a treatment similar to that of viewpoints that are the object of a vast consensus, or in the negative aspects or criticisms of steps taken to protect the environment receiving considerable emphasis, as was observed, for example, in the coverage of the climate change issue in the United States (Boykoff and Boykoff 2004; Boykoff 2008).

In summary, the media thus play a critical role in conveying information to citizens about environmental and technology issues. The salience and framing of these issues in the media have been shown to influence both public attitudes and practices.

23.2.4 Beginnings and first pitfalls of green marketing

The late 1980s marked the heyday of green marketing.[13] Citizen-consumers' professed willingness to pay a premium for green consumer products (Coddington 1990; Roper Organization 1990; Worcester 1993; Mintel 1995) caused a small revolution. The proportion of new green consumer products introduced into the United States market rose from 0.5% (24 products) in 1985 to 13.4% (810 products) in 1991; and green claims increased by 430% in print advertising and by 367% in television advertising (Ottman 1994). At the beginning of the 1990s, 92% of European multinationals reported that they had modified their products and 85% their processes in response to consumers' environmental concerns (Vandermerwe and Oliff 1990). Many analysts were of the opinion at the time that the growth of the green market was "significant, continuing and unstoppable" (Crane 2000).

Enthusiasm would soon give way to doubts. Green products were not the success that environment concerns voiced in surveys suggested they would be (e.g., Pearce 1990; Simmons Market Research Bureau 1991, cited in Roberts 1996). The same is true

[13] This discipline studies how the environmental variable can be incorporated into corporate marketing decisions (Crane 2000). In 1971, the *Journal of Marketing* devoted a special issue to *green marketing*. However, it was not until the 1980s that this new speciality took off (Peattie and Crane 2005).

today (European Commission 2008). According to Mintel (2006), despite consumers' favorable attitudes toward environmental protection, their fears about pollution and their willingness to pay for environmentally friendly products, few have translated these attitudes into regular purchasing behaviors, resulting in a decrease in the frequency and prominence of these claims on consumer products (National Consumer Council 1996; Peattie and Crane 2005).

This "green backlash" led to a few realizations. Companies realized that the green attribute can open up markets for a product, but it is alone not enough to generate sales (Ottman 1994; Wong et al. 1996). Other elements of marketing (such as product quality and performance, price, promotion, and distribution) must be chosen so as to give green products a competitive advantage. Some analysts point for example to the price of green products, deemed excessive by consumers (McGrath 1992),[14] or the lack of investment in promotion and public relations to communicate the benefits of green products (Wong et al. 1996). Another finding is that green attributes can be perceived negatively by consumers, for example, associated with a lack of performance or with "weediness" (National Consumer Council 1996, cited in Crane 2000; Wong et al. 1996).

Consumers were found to be cynical and confused about green claims and the companies making them (Pearce 1990; Schot and Fischer 1993; Prothero et al. 1994; Kangun and Polonsky 1995; Mendleson and Polonsky 1995; Crane 2000). Greenwashing was responsible for this cynicism in part. In 1991, two thirds of adults in the United States felt that companies were making exaggerated environmental claims about their products (Stisser 1994), and 47% did not believe the environmental claims being made in advertising (Fierman 1991). The complexity of measuring the environmental impact of comparable products and processes added to the confusion not only among consumers but also companies, which were unsure about which changes to make (Kangun and Polonsky 1995; Peattie 1995).

These findings illustrated the need to refocus green marketing around a better understanding of consumers and to win their trust. A raft of studies has since attempted to profile the typical green consumer and the practices he or she is willing to adopt, including purchasing environmentally friendly products (Roberts 1996; Roberts and Bacon 1997; Diamantopoulos et al. 2003; Hirsh 2010). The need to publicize the environmental attributes of these products in a positive and credible manner gave rise to a multitude of ecolabels (Kangun and Polonsky 1995). Ecolabeling is a system in which a third party (certification agency) guarantees that a product, service, or company has a lesser impact on the environment. This guarantee hinges on the producer of the product or service adhering to a set of specifications. Ecolabels, represented by a logo affixed to products, therefore makes it possible in principle for consumers to consider the environmental impact of products in their

[14] The same observations were made in the case of fair trade, which is concerned with environmental protection and social justice. Analysts observe that consumers do not always want to bear the consequences of the practices that they demand companies to adopt, such as increased costs, lost time, or greater efforts (Carrigan and Attalla 2001; Sen and Bhattacharya 2001; Page and Fearn 2005).

decision to purchase. Ecolabeling was started by environmental and consumer groups, and they continue to be very active in monitoring their use (e.g., Terrachoice Environmental Marketing 2009; Plant 2010). With the proliferation of ecolabels around the world, efforts are currently under way to create an inventory of their use (Ecolabel Index 2010).

In this first part, we reviewed some of the factors that have led companies to market green products, and also factors that influenced market performance of these products. Despite setbacks and challenges in the marketing of green products, consumers continue to be interested in goods that are produced in a sustainable[15] manner, particularly foods (Food Marketing Institute 2009; Robinson and Smith 2002). In 2009, a majority of US consumers reported purchasing more sustainably produced foods than a year earlier, and 45% said that they were continuing to buy green products and services, even though prices were high and the economy in rough shape (Borra 2009, cited in Bruhn 2009). The second part of this chapter focuses on the factors that affect the acceptability of food technologies among consumers. It emphasizes the relative weight of environmental protection among these factors. Lastly, it looks at some of the benchmarks used by consumers to identify environmentally friendly products on grocery store shelves.

23.3 Food technologies and the environment: in the eye of the consumer

Consumers are generally not very aware of processes and technologies that have been used on the foods they buy. Their understanding of the scientific aspects of food technologies is generally limited, even in the case of known technologies (Lyndhurst 2009; Clery and Bailey 2010). This does not mean, however, that they are not interested in these questions, as is shown by their stances in the studies cited in the following.

23.3.1 Information

Dissemination of information is often one of the first initiatives put forward to convince the public of the benefits of a given technology. Studies on the impact of information and knowledge about food technologies on attitudes have yielded conflicting results (Lyndhurst 2009). Hearing positive information about a technology

[15] As discussed, the term "sustainable" does not only refer to environmental protection but encompasses social and economic issues as well. As will be seen in Sect. 23.3.4, consumers do not necessarily distinguish between the various aspects of "sustainable" or "ethical" consumption. The term is used here to respect the terminology used in the study cited.

encourages favourable attitudes towards it (e.g., Bruhn and Noell 1987; Schutz et al. 1989; Delgado-Gutierrez and Bruhn 2008), but negative information has an even greater influence (Sapp et al. 1995; Fox et al. 2001). For example, Hayes et al. (2002) showed that information coming from an environmental group opposed to irradiation had a greater influence on US consumer attitudes than positive information from an organization dedicated to the promotion of technological innovations.

Information about technical or scientific subjects is not transmitted linearly to citizen-consumers, and neither is it passively received. It is interpreted, constructed by individuals based on their particular context, their worldview, their values and prior knowledge as well as their trust in the sources of the information (Wynne 1995; Wagner 2007). We saw earlier that the media as well as friends and relatives are key sources of information about topics outside the individual's direct experience, such as technology and environmental issues. The role of friends and family in the transfer of information and construction of a common understanding is not insignificant: It can prove decisive in the formation of attitudes about a technology (Sapp and Harrod 1990; Sapp et al. 1994; Heller 2003; Sapp and Korsching 2004; Mellman Group 2006, cited in Lyndhurst 2009).

23.3.2 *Individual factors*

Sociodemographic variables (e.g., gender, age, income, education) are easy to obtain and use to identify population segments. Some of them have been reported to influence attitudes toward food technology in general as well as toward specific food technologies, but evidence about their effect is mixed. Values, worldviews, trust in various actors (e.g., the government, the food industry, scientists, etc.), attitudes toward science and the environment are actually considered better variables to predict consumers' attitudes toward new food technologies (Lyndhurst 2009). For example, several studies report that women generally are more fearful of new technologies and associate fewer benefits with them (Cardello et al. 2007; Lyndhurst 2009). However, among people who share strongly proenvironmental attitudes, this gender gap disappears (Frewer et al. 1998). In another example, seniors, also alleged to fear novel technologies, are paradoxically receptive to nanotechnologies used in packaging, which according to Lyndhurst (2009) could be explained by the fact that many of them also have deeply negative attitudes toward food waste (Defra 2008).

A positive attitude toward science and technology is considered a good predictor of a positive attitude toward agricultural biotechnology (Grunert et al. 2004, cited in Gaskell et al. 2006; Traill et al. 2004; Chen and Li 2007; Cormick 2007; Knight 2007; Lyndhurst 2009) and nanotechnologies (Priest 2006). Favorable attitudes toward environmental protection, as described by various constructs, were linked in a number of studies to rejection of GMOs (Frewer et al. 1997a; Siegrist 1998; Rimal et al. 2005; Cormick 2007; Knight 2007; Essoussi and Zahaf 2009). Another study related anti-GMO activism not to general attitudes towards health and the environment, but rather to anticorporate sentiments, fear of globalization, and the belief that

GMOs are fundamentally unnatural (West and Larue 2005). Finally, some studies relate proenvironmental beliefs to a lack of trust in science and technology in general (Steger et al. 1989; Eden 2011). Conversely, those who adhere strongly to a set of prevailing values termed the "Dominant Social Paradigm," which include the belief in abundance, the faith in science and technology, a commitment to a "laissez-faire" economy, limited government planning and private property rights (Dunlap and van Liere 1978) are less likely to hold proenvironmental attitudes (Kilbourne et al. 2001).

Trust is another key factor. Individuals who trust regulatory bodies and industry tend to hold the opinion that genetic modification technologies are less risky, whereas those who tend instead to trust militants in environmental or consumer groups believe the opposite (Traill et al. 2004). The same influence of trust was observed for nanotechnologies (Lee et al. 2005; Siegrist et al. 2007) and food irradiation (Bord and O'Connor 1990; Sapp et al. 1995; Sapp 2003). According to a model developed in the Netherlands, trust that foods in general are safe is mainly influenced by trust in food manufacturers and the government (de Jonge et al. 2008, 2010).

Trust in a source of information depends on the degree to which that source is perceived as honest, accurate, competent, and concerned about the welfare of the public (Frewer et al. 1996). Sources considered trustworthy vary from country to country. Americans are said to place more trust in regulatory bodies than Europeans, who tend instead to trust environmental and consumer groups considered to be independent of the interests of government and industry (Lyndhurst 2009). On the other hand, Cardello et al. (2007) observed that manufacturers and certain environmental groups (among which, Greenpeace) both inspired less confidence in American citizens than the Surgeon General of the United States as a "source of product endorsement." Turkish consumers (Gunes and Tekin 2006) and their Brazilian counterparts (Deliza et al. 2003a) are reported to be more receptive to a new food technology if it is associated with a well-known company or brand. Conversely in the United Kingdom pairing potentially nonacceptable technologies with valuable brands may result in loss of the brand rather than acceptance of the technology (Deliza et al. 2003a).

23.3.3 Characteristics of technologies

23.3.3.1 Risks

Consumers' attitudes toward a given technology largely depend on how they personally evaluate the benefits and risks associated with it. Since the development of the psychometric paradigm (Fischhoff et al. 1978; Slovic et al. 1985) it has been known that lay people evaluate the seriousness of risks based on stable criteria, albeit criteria that are different from the probabilistic ones used by the experts. In the above cited work, the perceived seriousness of various risks correlated with three factors: the seriousness of their consequences (dread factor), whether they were perceived to be known risks or not (unknown factor), and the extent of the exposed population (exposure factor). These labels cover a number of combined characteristics. The "dread" factor refers to the severity of consequences of a risk, its ability to cause death, its controllability, and its catastrophic potential. The "unknown" factor includes the fact that a risk is new, unknown to science and exposed persons, involuntary and has deferred, as opposed to immediate, effects in time. The "exposure" factor includes an assessment of the number of people exposed and of the threat personally felt by the respondent.

The fact that a consumer perceives that there is a risk correlates negatively with the intention to purchase a product (Siegrist et al. 2007), with interest in using a technology (Cardello et al. 2007) and with its overall acceptability (Henson et al. 2008). Risks to health top the list of concerns. According to Cardello et al. (2007) uncertainty over the existence of risks to health and the presence of toxic byproducts are the two risk statements that diminish interest in a given food technology the most, far more than the fact that it uses more energy or is hazardous to workers. Even though environmental hazards are cause for concern in the case of biotechnologies, risks to health are considered more worrisome, as shown by the fact that genetically modified crops are better accepted than genetically modified foods (Gaskell et al. 2000; Midden et al. 2002). Similarly, in the case of food irradiation, environmental concerns connected with irradiation facilities subsided between 1986 and 2002, and were overtaken by health-related concerns (Gauthier 2010). Interestingly, five decades after *Silent Spring,* synthetic pesticides and growth hormones are still the food risks most feared by consumers in the United States, followed by agricultural antibiotics, genetic engineering, and irradiation (Hwang et al. 2005).

A study of the acceptability of two food nanotechnologies in Germany (Marette et al. 2009) pointed to the same conclusions. Nanotechnologies used in a food[16] and thus intended to be ingested directly ("nano-inside") caused more fear than those used in packaging for a food ("nano-outside"), confirming earlier observations (Siegrist et al. 2007,2008). In both cases, 83.2% of respondents were concerned primarily with the health risks and benefits of these technologies, whereas the remainder were equally interested in social and environmental aspects.[17] Willingness to pay for these products was not significantly affected by the information disseminated about these latter two aspects.

---

[16] The two food applications of nanotechnologies used as examples in this study were orange juices, one of which was enriched with vitamin D (nano-inside), and the other whose bottle (nano-outside) was designed to protect the vitamin C from damage by UV rays.

[17] Environmental aspects presented (Marette et al. 2009): "It is worried that nanoparticles will be released in the environment during production, consumption and discharge of the products and may pose risks. It has, for example, been found that manufactured nanoparticles can have toxic effects on aquatic organisms. On the other hand, nanotechnology may help to reduce pollution from efficient use of materials and it is hoped that nanotechnology allows the development of new ways to transform and detoxify a wide variety of environmental contaminants."

The perceived degree of control over risk affects its acceptability, with involuntary risks not being as well accepted (Slovic 1987; Cardello 2003). Food technologies that do not alter the organoleptic properties of the food (e.g., irradiation and genetic modification) and are not subject to labeling or declarations by the manufacturer come under the category of risks that consumers do not choose to take knowingly, that is, involuntary risks. However, everything is relative: in the United Kingdom, genetically modified foods are less feared than cancer, terrorism, and climate change and are considered more controllable (Townsend et al. 2004).

Environmental risks therefore carry less weight than health risks when it comes to how consumers perceive food technologies. They are nonetheless significant in the rejection of certain technologies. Lyndhurst (2009) lists several of them cited by consumers in various studies as reasons for rejecting GMOs: contamination of non-GM plants and organisms; threat to native wildlife and the balance of nature; development of resistant "superbugs" and "superweeds" and increased use of pesticides; and unknown long-term effects on both human and environmental health. Generally, the fact that a technology is perceived as "unnatural" or upsetting the natural order (e.g., GMOs, animal cloning and synthetic biology) adds to concerns (Wagner et al. 2002; Siegrist 2008; Siegrist et al. 2008; Lyndhurst 2009).

23.3.3.2 Benefits

The acceptability of various food technologies to consumers was reported to be determined both by perceived risk and perceived benefit, with individual technologies lying along a continuum between the two (Henson et al. 2008). To be considered, the benefits must be direct and tangible (Costa-Font et al. 2008). Benefits to health and organoleptic improvements are most likely to improve acceptability (Hossain and Onyango 2004; Verbeke 2005), although environmental benefits are also considered, e.g., reduction of pollution in the case of some GMOs (Loureiro and Bugbee 2005). In a study among nutrition and health professionals (Delgado-Gutierrez and Bruhn 2008) on the acceptability of three emerging technologies (high pressure processing, ohmic heating and pulsed electric field), these professionals indicated that they would be willing to pay 10% more for foods produced using these technologies if they improved (in decreasing order of importance) their nutritional value as well as the taste and shelf life of the foods, and lastly, if these processes were more environmentally friendly. Reduction of environmental impact is therefore not the first benefit sought after by consumers when they think of food technologies, but it is among their expectations. Finally, the acceptability of a technology will vary depending on whether the benefits are for consumers or companies (Frewer et al. 1997b).

23.3.3.3 Novelty

Concerns raised by new food technologies is owing partly to their very novelty. Resistance was noted upon the introduction of formerly novel technologies, such as

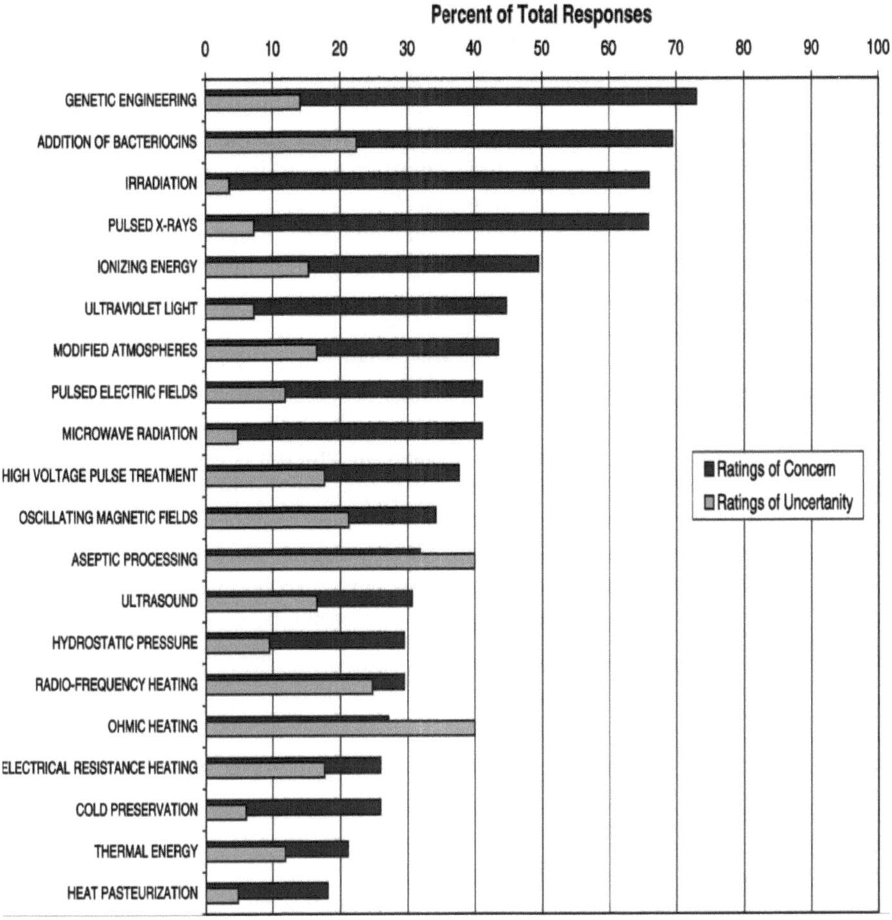

Fig. 23.1 Percentage of subjects reporting some degree of concern ("slight," "moderate" or "extreme") and the percentage of 'uncertain' responses for 20 different food processing technologies (Reproduced from Cardello 2003 with permission from Elsevier BV)

industrial canning (Young 2003) and pasteurization of milk (Jean 1999). Familiarity with a technology tends to increase its acceptability (Cardello et al. 2007; Chen and Li 2007). It also makes it possible for the consumer to personally assess its benefits, which is consistent with the previous point.

Since consumers are relatively unfamiliar with processing technologies, it is possible that technologies in use for decades will be considered novel, and thus disquieting, when consumers discover their existence. For example, 59% of respondents to a survey in the United Kingdom said that they were very or rather concerned about the use of modified atmospheres in bags of ready-to-eat salads (Clery and Bailey 2010). In a US study (Fig. 23.1), this technology caused concern among 45% of respondents (Cardello 2003). High pressure processing was well accepted by most respondents in a European study (Butz et al. 2003). In contrast, this process was very worrisome to 20% of respondents from a group of US military members (Cardello 2000, cited in Cardello 2003), whereas the term "high pressure" was poorly accepted by one in four respondents in a Brazilian study (Deliza et al. 2003b). In another survey, 31% of respondents said they were concerned about eating foods produced using a microwave, with that proportion increasing to 57% when a magnetron (a lesser-known word for a microwave) was involved. However, the description of the two processes given to the respondents was the same (Clery and Bailey 2010). This shows the influence of familiarity with the terminology. The concept of "minimally processed," used in the jargon of food science experts to refer to foods that are processed in such a way as to alter their quality and nutritional value as little as possible, was also seen negatively by participants in a US study (Cardello et al. 2007), as being synonymous with insufficiently processed and unsafe.

Numerous other factors may influence the acceptability of a technology in the eyes of consumers, such as the kind of product on which it is used or its impact on quality or taste. The results shown above can help to anticipate the reactions a food technology is likely to cause, but these reactions must be experimentally validated for each specific product and context. In the last section, we will take the issue of consumer understanding and acceptability of "green technologies" from another standpoint: when consumers look for green food products, what attributes do they think about?

23.3.4 Green food products: from the consumers' standpoint

According to a number of studies, consumers' desire to protect the environment is lumped in with other ethical issues (Howard and Allen 2010). The term "sustainable" food would thus appear more suitable in encompassing the desires of consumers than the concept of green food: for consumers, this term was shown to include locally grown products, production using practices that reduce the carbon footprint, organic production, minimal packaging, and general corporate responsibility, both in terms of the environment and social justice, particularly towards workers and communities (Demeritt 2008, cited in Bruhn 2009). In a global survey, respondents indicated that they were willing through the purchase of products to support proper treatment of workers first (respondents in the United States, the United Kingdom, and Germany) and then protection of the environment, followed in the United States by local buying. In Japan, respondents' desire to support corporate transparency came in first, followed by an interest in fair trade and environmental protection (Curry 2007, cited in Bruhn 2009).

The desire to protect the environment is also lumped in with health concerns, although the latter continue to dominate (Howard and Allen 2010). In a Strategy One survey in 2008 (cited in Bruhn 2009) 75% of respondents were in agreement with the statement that "eating green or sustainable foods will help me lead a life that is good for my body and the environment." This confluence may first have been initiated by environmental movements themselves (Lowe et al. 2008) and is still encouraged by certain health professional and social groups nowadays (Bruhn 2009). A nutritionist writes, "Eating healthier and eating greener are not two separate approaches to eating, but rather approaches that go hand-in-hand" (Geagan 2009). The Strategic Alliance for Healthy Food and Activity Environment (2009) calls on nutritionists to endorse a definition of healthy food that "comes from a food system where food is produced, processed, transported, and marketed in ways that are environmentally sound, sustainable, and just." This food must also be "minimally processed" and be free of "artificial colours, flavours, or unnecessary preservatives. Workers and natural resources should not be exploited, and animals should not be treated cruelly."

Packaging is one of the environmental attributes of products that consumers feel readily able to assess, even without knowing in detail the impact of the different kinds of materials used. Ninety percent of respondents in an Australian study considered it important for manufacturers to use less packaging, and 65% said they often or sometimes avoided purchasing products that were packaged in a non-environmentally friendly way (e.g., plastic, over packaging) (Lea and Worsley 2008). US consumers consider in particular hard packaging and overpackaging as harmful to the environment and respond favorably to companies that use recyclable packaging (Bruhn 2009) or biodegradable packaging, for example, recycled cardboard and corn-based biodegradable plastic (Porjes 2007).

Place of production is also among the environmental attributes weighed by consumers. Reference is made here to the "food miles" and "buy local" concepts, an action deemed important by 86% of Australian respondents in the study by Lea and Worsley (2008). The same percentage said that they bought local foods either occasionally or often. In the United States, 35% of consumers reported doing so, particularly because of the considerable environmental impact of transportation (Food Marketing Institute 2009). It is however very unlikely that consumers can compare the energy cost of local foods to that of their imported counterparts, for want of accurate information, for example, about the type of production (e.g., greenhouse or field), the method of transportation or the distance traveled.[18] The concept of "localness" can in fact, depending on the context, refer to the immediate neighborhood, a specific perimeter around the place of purchase[19] or a wider geographic area (e.g., province, state, country).

In the United States, 75% of consumers reported not knowing what products or companies were really sustainable (Hartman Group 2008, cited in Bruhn 2009). Ecolabels are one of the solutions proposed for this problem. Of the 115 ecolabels used on agricultural produce and food tracked thus far in the Ecolabel Index (2010), more than half belong to the category of agriculture and organic processing and its

[18] One mile by personal car is equivalent to 740 miles by highway truck, 2,400 miles by railroad, and 3,800 miles by ship (Thompson 2009, cited in Bruhn 2009).

[19] The Whole Foods chain of stores defines local foods as those produced less than 7 h away from the point of sale by truck (Porjes 2007).

close alternatives (e.g., Certified Naturally Grown [USA]; Wholesome Food Association [Europe]). Other ecolabels are specific to certain commodities (e.g., coffee, wine, avocados) and certify their environmental impact from production to sale. Another category concerns fisheries, for example, fight against overfishing or protection of specific species (e.g., "dolphin-safe" tuna). There are labels that certify fair trade, animal welfare, food safety, and catering. Lastly, some ecolabels are concerned with specific environmental impact indicators (e.g., greenhouse gas emissions and water use) or assess a series of these factors or the products' overall life cycle.

Organic production standards are primarily concerned with environmental protection. However, they are based on specifications that are not very explicit with regard to the environmental impact of processing technologies, save for certain specific processes (e.g., genetic modifications, cloning, irradiation, and fumigation) (e.g., USDA's National Organic Program, EU Organic products, Québec Vrai). Some agencies (e.g., Québec Vrai 2010) require, on the other hand, a written description of the efforts made to reduce the amount of solid or liquid waste and atmospheric emissions produced during handling and processing, and efforts devoted to recycling, such as the use of recycled materials and reduction of packaging. However, this requirement is not consistent from one standard to another. Moreover, it is not likely that consumers can easily access such information for a specific organic product they consider purchasing.

Organic standards are subject to interpretation by consumers, and these interpretations vary from country to country. When they buy organic products, consumers do so, among other things, to protect the environment. However, according to most studies, their primary motivations have to do with health, particularly the fear of ingesting pesticide residues and chemical inputs (Huang 1996; Grankvist and Biel 2001; Fotopoulos and Krystallis 2002; Lockie et al. 2002; Hartman Group 2008, cited in Bruhn 2009; Lea and Worsley 2008; Essoussi and Zahaf 2009). Consumers of organic products also justify their choice by the quality and taste of these foods and by their alignment with an agricultural model that matches their values: animal welfare, support for local producers, small-scale production, and trust in organic producers, who are not perceived as profit-driven (Essoussi and Zahaf 2009). Large companies that produce and market organically certified foods are therefore somewhat distrusted by consumers (Porjes 2007). In a UK study, organic foods were associated with "a *lack* of tampering and an *absence* of modernist, industrial technology" (Eden 2011). Participants also believed that organic certification rested on the specific testing of products themselves (e.g., to certify the absence of pesticides) rather than on-farm or plant audits to examine production processes.

Standards in the International Standards Organization (ISO) 14,000 series, which certify companies' obedience to their environmental management system, are not widely used in the food sector (Massoud et al. 2010). Consumers are also less familiar with them than organic standards. In a study conducted in Greece (Arvanitoyannis et al. 2003), 39.9% of respondents were able to identify these standards as an environmental management system, compared with 77% who were familiar with organic certification. Half of the respondents said they would buy products from companies that met these standards, and 55% would pay a premium for such products.

In summary, at the grocery store, consumers are left with relatively few indications on the actual environmental impact of food products to guide their choice. When evaluating whether a food product is green or not, they tend to bridge environmental, health, and social issues. Most of the current ecolabels used on food products do little to enlighten them on environmental issues since they do not necessarily carry precise information as to what exactly they are certifying. Consumers also construct their own representation of their meaning. Food companies thus face a challenge in communicating their efforts to reduce the environmental impact of their processes and practices to consumers. Bruhn (2009) has called on the food industry to define its standards in terms of durability, energy use (production, processing, and transportation), carbon emissions and natural resource use, and to develop a logo (e.g., a clover leaf) to communicate these good practices. It remains to be seen how consumers would interpret this new logo.

23.4 Conclusion

Concerns about the environment are definitely becoming universal. Some of these concerns regarding agriculture and food—first raised in the 1960s—are still very much alive today. It is not clear, however, that these environmental concerns are translating into spontaneous consumer support for just any environmental protection initiative taken on the part of industry, as we have seen since the "green marketing fever" of the 1990s. Nonetheless, the rise in ethical consumption is a major trend.

Companies introducing a new technology along their production lines to reduce their environmental impact or wishing to have their efforts in that regard recognized by consumers face certain challenges. Because consumers are largely unfamiliar with the various stages in the food production and processing chain, chances are that environmental efforts by companies will go unnoticed. When publishing such efforts, a company may meet unexpected scepticism or even cynicism among consumers. The social acceptability of a new, green technology, for example, cannot be taken for granted, no matter its potential to reduce waste production or energy costs. Just like for any other food technology, its acceptability will be a function of many variables such as individual factors, availability and interpretation of information, media coverage, and so on. Careful empirical validation of the impact of all relevant variables in the specific context where the technology is to be used is essential to enhancing acceptability. In addition, the positive or negative environmental impact of a product or a technology is often not the first characteristic that consumers take into consideration; the perceived impact on human health clearly comes in first. In that view, it may be useful for companies to try to link their environmental efforts in the minds of consumers with an ultimate goal of also protecting human health.

As a final comment, when communicating through advertising or public relations, food companies rarely unveil the technology and equipment they actually use in their daily operations. Images of food products that are prepared similar to their "home-made" equivalent using "traditional methods" are more frequently put forward, with an aim to increase consumer confidence. This choice is understandable. According to some of the studies cited above, consumers do not readily associate sophisticated technology with "natural foods" or sound environmental practices. However, the use of green food processing technologies holds the potential of a large scale, positive impact on the environment. It is ironical that the relative ignorance of consumers regarding actual operations involved in food production and processing, maintained by the industry itself, is now an obstacle to the acceptability of technologies that are aimed at protecting the environment and thus, at satisfying consumers' environmental concerns. This finding may be an opportunity for food companies to make their reality more visible to consumers. In doing so, they might open the way to a better understanding of the actual impact of food processes and practices on the environment and of the role of new green technologies in reducing these impacts.

Disclaimer: The opinions expressed are the author's own.

References

Anonymous 2009. Greenwash fears have equipment makers rolling ahead slowly. *Engineering News Record* 263(2): 68.

Ader, C.R. 1995. A longitudinal study of agenda setting for the issue of environmental pollution. *Journalism and Mass Communication* 72: 300–311.

Anderson, A. 1997. *Media, culture and the environment*. London: UCL Press.

Andrews, C., and D. DeVault. 2009. Green niche market development a model with heterogeneous agents. *Journal of Industrial Ecology* 13(2): 326–345.

Arvanitoyannis, I.S., A. Krystallis, and A. Kapirti. 2003. Health and environmental consciousness. Greek consumers' attitudes toward the organic, HACCP and ISO 14000 certifications on food. *International Journal of Food and Agribusiness Marketing* 15(1–2): 93–136.

Ausubel, J.H., and H.E. Sladovich. 1989. *Technology and environment*. Washington, DC: National Academy of Engineering.

Bauer, M.W., M. Kohring, A. Allansdottir, and J. Gutterling. 2001. The dramatisation of biotechnology in elite mass media. In *Biotechnology 1996–2000. The years of controversy*, ed. G. Gaskell and M.W. Bauer. London: Science Museum.

Bord, R.J., and R.E. O'Connor. 1990. Risk communication, knowledge, and attitudes. Explaining reactions to a technology perceived as risky. *Risk Analysis* 10(4): 499–506.

Borra, S. 2009. The convergence of health and wellness and the environment. Drivers behind consumer choice. *Paper presented at the IFT Annual Meeting & Food Expo*, June 7, 2009, Anaheim, CA.

Boykoff, M.T. 2008. Lost in translation? United States television news coverage of anthropogenic climate change, 1995–2004. *Climatic Change* 86(1–2): 1–11.

Boykoff, M.T., and J.M. Boykoff. 2004. Balance as bias. Global warming and the US prestige press. *Global Environmental Change* 14(2): 125–136.

Brooks, P. 1972. *The house of life. Rachel Carson at work*. Boston: Houghton Mifflin Company.

Bruhn, C.M. 2009. Understanding 'green' consumers. *Food Technology* 63(7): 28–35.

Bruhn, C.M., and J.W. Noell. 1987. Consumer in-store response to irradiated papayas. *Food Technology* 41: 83–87.

Butz, P., E.C. Needs, A. Baron, O. Bayer, B. Geisel, B. Gupta, U. Oltersdorf, and B. Tauscher. 2003. Consumer attitudes to high pressure food processing. *Journal of Food Agriculture and Environment* 1(1): 30–34.

Cardello, A.V. 2000. Consumer attitudes and expectations toward nonthermal and other novel food processing techniques. *Paper presented at the IFT Nonthermal processing division workshop on non-thermal processing of food*, January 26–27, Portland, OR.

Cardello, A.V. 2003. Consumer concerns and expectations about novel food processing technologies. Effects on product liking. *Appetite* 40(3): 217–233.

Cardello, A.V., H.G. Schutz, and L.L. Lesher. 2007. Consumer perceptions of foods processed by innovative and emerging technologies. A conjoint analytic study. *Innovative Food Science and Emerging Technologies* 8(1): 73–83.

Carrigan, M., and A. Attalla. 2001. The myth of the ethical consumer. Do ethics matter in purchase behaviour? *Journal of Consumer Market* 18(7): 560–577.

Champion, E., and C. Gendron. 2002. La "durabilité" selon Monsanto. Prémisses d'une privatisation des problèmes environnementaux pour un renforcement politique de l'entreprise privée. Cahiers de la Chaire de responsabilité sociale et de développement durable—Collection recherche (3-2002).

Chen, M.F., and H.L. Li. 2007. The consumer's attitude toward genetically modified foods in Taiwan. *Food Quality and Preference* 18(4): 662–674.

Clery, E., and R. Bailey. 2010. *Food technologies, findings from the 2008 British social attitudes survey*. London: Food Standards Agency.

Coddington, W. 1990. It's no fad. Environmentalism is now a fact of corporate life. *Marketing News* 24(21): 7.

Cormick, C. 2007. Public attitudes towards GM crops and food. *Agricultural Sciences* 21(2): 24–30.

Costa-Font, M., J.M. Gil, and W.B. Traill. 2008. Consumer acceptance, valuation of and attitudes towards genetically modified food. Review and implications for food policy. *Food Policy* 33(2): 99–111.

Crane, A. 2000. Facing the backlash. Green marketing and strategic reorientation in the 1990s. *Journal of Strategic Marketing* 8(3): 277–296.

Curry, A. 2007. *Consumer responses to environmental and ethical issues*. London: Durrell Wildlife Conservation Trust/Henley Centre HeadlightVision, p. 35.

De Boer, M., M. McCarthy, M. Brennan, A.L. Kelly, and C. Ritson. 2005. Public understanding of food risk issues and food risk messages on the island of Ireland. The views of food safety experts. *Journal of Food Safety* 25: 241–265.

de Jonge, J., H. van Trijp, E. Goddard, and L. Frewer. 2008. Consumer confidence in the safety of food in canada and the Netherlands. The validation of a generic framework. *Food Quality and Preference* 19(5): 439–451.

de Jonge, J., H. Van Trijp, R.J. Renes, and L.J. Frewer. 2010. Consumer confidence in the safety of food and newspaper coverage of food safety issues. A longitudinal perspective. *Risk Analysis* 30(1): 125–142.

Defra. 2008. A framework for pro-environmental behaviours. http://www.defra.gov.uk/evidence/social/behaviour/documents/behaviours-jan08-report.pdf. Accessed 18 May 2010.

Delgado-Gutierrez, C., and C.M. Bruhn. 2008. Health professionals' attitudes and educational needs regarding new food processing technologies. *Journal of Food Science Education* 7(4): 78–83.

Deliza, R., M.C. da Costa, A. Rosenthal, D. Hedderley, and L. Frewer. 2003a. Acceptance of novel food technologies. Comparison between Brazilian and British consumers. http://www.foodsciencecentral.com/fsc/ixid12294. Accessed 2 June 2010.

Deliza, R., A. Rosenthal, and A.L.S. Silva. 2003b. Consumer attitude towards information on nonconventional technology. *Trends in Food Science and Technology* 14(1–2): 43–49.

Demeritt, L. 2008. The new greening of the food industry. Sustainability from a consumer perspective. *Paper presented at the IFT Annual Meeting & Food Expo*, July 1st, New Orleans, LA.

Diamantopoulos, A., B.B. Schlegelmilch, R.R. Sinkovics, and G.M. Bohlen. 2003. Can sociodemographics still play a role in profiling green consumers? A review of the evidence and an empirical investigation. *Journal of Business Research* 56(6): 465–480.

Dunlap, R.E. 2002. En enduring concern. *Public Perspectives* 13(5): 10–14.

Dunlap, R., and K. van Liere. 1978. The new environmental paradigm. A proposed measuring instrument and preliminary results. *The Journal of Environmental Education* 9(10–19).

Dunlap, R.E., and R. York. 2008. The globalization of environmental concern and the limits of the postmaterialist values explanation. Evidence from four multinational surveys. *Sociological Quarterly* 49(3): 529–563.

Ecolabel index. 2010. http://www.ecolabelindex.com/. Accessed 8 Sept 2010.

Eden, S. 2011. Food labels as boundary objects. How consumers make sense of organic and functional foods. *Public Understanding of Science* 20(2):179–194.

Elkington, J., and J. Hailes. 1988. *The green consumer guide*. London: Victor Gollanz.

Entman, R.M. 1993. Framing: Toward clarification of a fractured paradigm. *Journal of Communication* 43(4): 51–58.

Essoussi, L.H., and M. Zahaf. 2009. Exploring the decision-making process of Canadian organic food consumers. Motivations and trust issues. *Qualitative Market Research* 12(4): 443–459.

European Commission. 2008. Special Eurobarometer 295. Attitudes of European citizens towards the environment. http://ec.europa.eu/public_opinion/archives/ebs/ebs_295_en.pdf. Accessed 9 Sept 2010.

Fierman, J. 1991. The big muddle in green marketing. *Fortune* 123(11): 91–101.

Fischhoff, B., P. Slovic, S. Lichtenstein, S. Read, and B. Combs. 1978. How safe is safe enough? A psychometric study of attitudes towards technological risks and benefits. *Policy Sciences* 9: 127–152.

Fishbein, M., and M.C. Yzer. 2003. Using theory to design effective health behavior interventions. *Communication Theory* 13(2): 164–183.

Food Marketing Institute. 1996–1999. *Trends in the United States. Consumer attitudes and the supermarket*. Washington, DC: Food Marketing Institute.

Food Marketing Institute. 2009. *Grocery shopper trends 2009*. Washington, DC: Food Marketing Institute.

Fotopoulos, C., and A. Krystallis. 2002. Purchasing motives and profile of Greek organic consumer. A countrywide survey. *British Food Journal* 104(9): 730–764.

Fox, J.A., C.M. Bruhn, and S.G. Sapp. 2001. Consumer acceptance of irradiated meats. In *Interdisciplinary food safety research*, ed. N.H. Hooker and E.A. Murano. Boca Raton: CRC Press.

Frentzen, P.D., A. Majchrowicz, J.C. Buzby, B. Imhoff, and Foodnet Working Group. 2000. Consumer acceptance of irradiated meat and poultry products. Issues in food safety economics. Agriculture Information Bulletin No. 757. United States Department of Agriculture, Economic Research Services, Washington, DC.

Frewer, L.J., M.M. Raats, and R. Shepherd. 1993. Modelling the media. The transmission of risk information in the British Quality Press. *IMA Journal of Mathematics Applied in Business and Industry* 5(1): 235–247.

Frewer, L.J., C. Howard, D. Hedderley, and R. Shepherd. 1996. What determines trust in information about food-related risks? Underlying psychological constructs. *Risk Analysis* 16(4): 473–485.

Frewer, L.J., D. Hedderley, C. Howard, and R. Shepherd. 1997a. "Objection" mapping in determining group and individual concerns regarding genetic engineering. *Agricuture and Human Values* 14: 67–79.

Frewer, L.J., C. Howard, and R. Shepherd. 1997b. Public concerns in the united kingdom about general and specific applications of genetic engineering. Risk, benefit and ethics. *Science Technology and Human Values* 22(1): 98–124.

Frewer, L.J., C. Howard, and J.I. Aaron. 1998. Consumer acceptance of transgenic crops. *Pesticide Science* 52(4): 388–393.

Frewer, L.J., S. Miles, and R. Marsh. 2002. The media and genetically modified foods. Evidence in support of social amplification of risk. *Risk Analysis* 22(4): 701–711.

Friedman, M. 1970. A Friedman doctrine. The social responsibility of business is to increase its profits. *New York Times Magazine*: (Sept 13) SM17.

Gaskell, G., N. Allum, M. Bauer, J. Durant, A. Allansdottir, H. Bonfadelli, D. Boy, S. de Cheveigne, B. Fjaestad, J.M. Gutteling, J. Hampel, E. Jelsoe, J.C. Jesuino, M. Kohring, N. Kronberger, C. Midden, T.H. Nielsen, A. Przestalski, T. Rusanen, G. Sakellaris, H. Torgersen, T. Twardowski, and W. Wagner. 2000. Biotechnology and the European public. *Nature Biotechnology* 18(9): 935–938.

Gaskell, G., A. Allansdottir, N. Allum, C. Corchero, C. Fischler, J. Hampel, J. Jackson, N. Kronberger, N. Mejlgaard, G. Revuelta, C. Schreiner, S. Stares, H. Thogersen, and W. Wagner. 2006. *Europeans and biotechnology in 2005. Patterns and trends*, Eurobarometer, vol. 64.3. Brussels: European Commission.

Gauthier, É. 2010. Social representations of risk in the food irradiation debate in Canada, 1986–2002. *Science Communication* 32(3): 295–329.

Gauthier, É. 2008. Les représentations sociales du risque dans le débat public sur l'irradiation des aliments au Canada. Thesis. Université du Québec à Montréal, Montréal.

Geagan, K. 2009. Why RDS are perfectly poised for going green. *American Dietetic Association Times* 6: 22.

Gitlin, T. 1980. *The whole world is watching. Mass media in the making and unmaking of the new left*. Berkeley: University of California Press.

GlobeScan, BBC World Service. 2009. Climate concerns continue to increase: Global poll. http://www.globescan.com/news_archives/bbc2009_climate_change/. Accessed 5 Jan 2010.

Grankvist, G., and A. Biel. 2001. The importance of beliefs and purchase criteria in the choice of eco-labeled food products. *Journal of Environmental Psychology* 21: 405–410.

Greenpeace. 2010. Stop greenwash. http://www.stopgreenwash.org/. Accessed 18 Feb 2010.

Grunert, K., H. Sondergaard, and J. Scholderer. 2004. *How can we know what we like when we don't understand it? Consumer attitude formation towards complex technical issues*, vol. 74. Aarhus V, Denmark: MAPP—Centre for Research on Customer Relations in the Food Sector, Aarhus School of Business.

Gunes, G., and M.D. Tekin. 2006. Consumer awareness and acceptance of irradiated foods. Results of a survey conducted on Turkish consumers. *LWT Food Science and Technology* 39: 443–444.

Hansen, A. 1993. Greenpeace and press coverage of environmental issues. In *The mass media and environmental issues*, ed. A. Hansen. Leicester: Leicester University Press.

Hartman Group. 2008. *Organic 2008. Topline insights*. Bellevue: The Hartman Group Inc.

Hayes, D.J., J.A. Fox, and J.F. Shogren. 2002. Experts and activists. How information affects the demand for food irradiation. *Food Policy* 27: 185–193.

Heller, R. 2003. *Gm nation? The findings of the public debate.* U.-K. Department for Trade and Industry, (DTI). London, p. 66.

Henson, S., M. Annou, J. Cranfield, and J. Ryks. 2008. Understanding consumer attitudes toward food technologies in Canada. *Risk Analysis* 28(6): 1601–1617.

Hirsh, J.B. 2010. Personality and environmental concern. *Journal of Environmental Psychology* 30: 245–248.

Hoffman, A. 2001. *From heresy to dogma. An institutional history of corporate environmentalism*. Stanford: Stanford University Press.

Holcomb, J., and Corporate greenwashers and environmental groups. 2008. Environmentalism and the internet. *Contemporary Justice Review* 11(3): 203–211.

Hossain, F., and B. Onyango. 2004. Product attributes and consumer acceptance of nutritionally enhanced genetically modified foods. *International Journal of Consumer Studies* 28(3): 255–267.

Howard, P.H., and P. Allen. 2010. Beyond organic and fair trade? An analysis of ecolabel preferences in the United States. *Rural Sociology* 75(2): 244–269.

Huang, C. 1996. Consumer preference and attitudes toward organically grown produce. *European Review of Agricultural Economics* 23: 331–342.

Hwang, Y.-J., B. Roe, and M.F. Teisl. 2005. An empirical analysis of United States consumers' concerns about eight food production and processing technologies. *AgBioforum* 8(1): 40–49.

Jackson, T. 2005. Motivating sustainable consumption. A review of evidence on consumer behaviour and behavioural change. Centre for Environmental Strategy, University of Surrey. http://www.c2p2online.com/documents/MotivatingSC.pdf. Accessed 5 Oct 2010.

Jean, F. 1999. L'empoisonnement par le lait. L'impact de la campagne du lait pur sur la lutte à la mortalité infantile au Québec, 1830–1930. Le cas de Montréal. Thesis, Sherbrooke University, Sherbrooke.
Johnston, J. 2008. The citizen-consumer hybrid. Ideological tensions and the case of whole foods market. *Theory and Society* 37(3): 229–270.
Jørgensen, S., G. Martín-Herrán, and G. Zaccour. 2010. Dynamic games in the economics and management of pollution. *Environment Modeling and Assessment* 15(6): 433–467.
Kangun, N., and M.J. Polonsky. 1995. Regulation of experimental marketing claims. A comparative perspective. *International Journal of Advertising* 14(1): 1–24.
Kasperson, R.E., O. Renn, P. Slovic, H.S. Brown, J. Emel, R. Goble, J.X. Kasperson, and S. Ratick. 1988. The social amplification of risk. A conceptual framework. *Risk Analysis* 8(2): 177–187.
Kilbourne, W., S. Beckmann, A. Lewis, and Y. van Dam. 2001. A multinational examination of the role of the dominant social paradigm in environmental attitudes of university students. *Environment and Behavior* 33(2): 209–228.
Knight, A. 2007. Do worldviews matter? Post-materialist, environmental, and scientific/technological worldviews and support for agricultural biotechnology applications. *Journal of Risk Research* 10(8): 1047–1063.
Lea, E., and A. Worsley. 2008. Australian consumers' food-related environmental beliefs and behaviours. *Appetite* 50(2–3): 207–214.
Lee, C.-J., D.A. Scheufele, and B.V. Lewenstein. 2005. Public attitudes toward emerging technologies. Examining the interactive effects of cognitions and affect on public attitudes toward nanotechnology. *Science Communication* 27(2): 240–267.
Lillford, P.J., and M.F. Edwards. 1997. Clean technology in food processing. *Philosophical Transactions of the Royal Society of London Series A Mathematical Physical and Engineering Sciences* 355(1728): 1363–1371.
Linné, O. 1993. Professional practice and organization. Environmental broadcasters and their sources. In *The mass media and environmental issues*, ed. A. Hansen. Leicester: Leicester University Press.
Lockie, S., K. Lyons, G. Lawrence, and K. Mummery. 2002. Eating 'green. Motivations behind organic food consumption in Australia. *Sociologia Ruralis* 42(1): 23–40.
Loureiro, M., and M. Bugbee. 2005. Enhanced GM foods. Are consumers ready to pay for the potential benefits of biotechnology? *Journal of Consumer Affairs* 39: 52–70.
Lowe, P. 1992. Industrial agriculture and environmental regulation. A new agenda for rural sociology. *Sociologia Ruralis* 32(1): 4–10.
Lowe, P., J. Phillipson, and R.P. Lee. 2008. Socio-technical innovation for sustainable food chains. Roles for social science. *Trends in Food Science and Technology* 19(5): 226–233.
Lyndhurst, B. 2009. *An evidence review of public attitudes to emerging food technologies*. London: Food Standard Agency/Social Science Research Unit.
Marette, S., J. Roosen, A. Bieberstein, S. Blanchemanche, and F. Vandermoere. 2009. Impact of environmental, societal and health information on consumers' choices for nanofood. *Journal of Agriculture and Food Industrial Organization* 7(2): Article 11. http://www.bepress.com/jafio/vol7/iss2/art11/.
Mason, C.F. 2006. An economic model of ecolabeling. *Environmental Modeling and Assessment* 11(2): 131–143.
Massoud, M.A., R. Fayad, M. El-Fadel, and R. Kamleh. 2010. Drivers, barriers and incentives to implementing environmental management systems in the food industry. A case of Lebanon. *Journal of Cleaner Production* 18(3): 200–209.
Mazur, A. 1987. Putting radon on the public's risk agenda. *Science Technology and Human Values* 12: 86–93.
Mazur, A., and J. Lee. 1993. Sounding the global alarm. Environmental issues in the US national news. *Social Studies of Science* 23(4): 681–720.
McCombs, M., and S.I. Ghanem. 2003. The convergence of agenda setting and framing. In *Framing public life. Perspectives on media and our understanding of the social world*, ed. S.D. Reese, O.H. Gandy, and A.E. Grant. Mahwah: Lawrence Erlbaum.

McGrath, A.J. 1992. Marketin' of the green. *Sales and Marketing Management*, 144(12): 31–32.
Mellman Group. 2006. *Public sentiment about genetically modified food. Pew Initiative on Food and Biotechnology*. Washington, DC: Mellman Group.
Mendleson, N., and M.J. Polonsky. 1995. Using strategic alliances to develop credible green marketing. *Journal of Consumer Market* 12(2): 4–18.
Menon, A., and A. Menon. 1997. Enviropreneurial marketing strategy. The emergence of corporate environmentalism as market strategy. *Journal of Marketing* 61(1): 51–67.
Midden, C.J.H., D. Boy, E.F. Einsiedel, B. Fjœstad, M. Liakopoulos, J.D. Miller, S. Öhman, and W. Wagner. 2002. The structure of public perceptions. In *Biotechnology. The making of a global controversy*, ed. M.W. Bauer and G. Gaskell. Cambridge, MA: Cambridge University Press.
Miles, S., and L.J. Frewer. 2001. Investigating specific concerns about different food hazards. *Food Quality and Preference* 12: 47–61.
Miles, S., D.S. Braxton, and L.J. Frewer. 1999. Public perceptions about microbiological hazards in food. *British Food Journal* 101(10): 744–762.
Mintel. 1995. *The second green consumer report*. London: Mintel.
Mintel. 2006. *Green living*. US Marketing Research Report, London.
Monsanto Chemical Company. 1962. The desolate year. *Monsanto Magazine* 40(10): 4–9.
National Consumer Council. 1996. *Green claims. A consumer investigation into marketing claims about the environment*. London: National Consumer Council.
Nelkin, D. 1987. *Selling science. How the press covers science and technology*. New York: Freeman.
Roper Organization. 1990. *The environment. Public attitudes and individual behaviour*. New York: Roper Organization/SC Johnson & Son.
Ottman, J.A. 1994. *Green marketing*. Chicago: NTC Business Books.
Page, G., and H. Fearn. 2005. Corporate reputation. What do consumers really care about? *Journal of Advertising Research* 45(3): 305–313.
Pearce, F. 1990. The consumers are not so green. *New Scientist* 1721: 13–14.
Peattie, K. 1995. *Environmental marketing management*. London: Pitman.
Peattie, K., and A. Crane. 2005. Green marketing. Legend, myth, farce or prophesy? *Qualitative Market Research* 8(4): 357–370.
Pfeifer, S., and R. Sullivan. 2008. Public policy, institutional investors and climate change. A UK case-study. *Climatic Change* 89(3–4): 245–262.
Plant, B. 2010. Comment se fier à une certification? http://ethiquette.info/fr/Comment-se-fier-%C3%A0-une-certification%3F. Accessed 25 May 2010.
Porjes, S. 2007. *Ethical consumers and corporate responsibility. The market and trends for ethical products in food and beverage, personal care and household items*, Packaged facts. Rockville: MarketResearch.com Inc.
Priest, S. 2006. The North American opinion climate for nanotechnology and its products. Opportunities and challenges. *Journal of Nanoparticle Research* 8: 563–568.
Protess, D.L., F.L. Cook, T.R. Curtin, M.T. Gordon, D.R. Leff, M.E. Mccombs, and P. Miller. 1987. The impact of investigative reporting on public opinion and policymaking targeting toxic waste. *Public Opinion Quarterly* 51(2): 166–185.
Prothero, A., P. McDonagh, and K. Peattie. 1994. Green marketing communications. Dressing windows or opening doors? In *Marketing: Unity in diversity. Proceedings of the 1994 marketing education group annual conference*, ed. J.E.A. Bell. University of Ulster.
Québec Vrai. 2010. Modèle de plan de préparation biologique. http://www.quebecvrai.org/getfile.php?file=pp--mod1003.doc. Accessed 11 June 2010.
Research Alert. 1990. *Environmentalism and Americans. A guide to the foremost trend of our time*. Long Island City: Research Alert.
Rimal, A.P., W. Moon, and S. Balasubramanian. 2005. Agro-biotechnology and organic food purchase in the United Kingdom. *British Food Journal* 107(2): 84–97.
Roberts, J.A. 1996. Green consumers in the 1990s. Profile and implications for advertising. *Journal of Business Research* 36(3): 217–231.

Roberts, J.A., and D.R. Bacon. 1997. Exploring the subtle relationships between environmental concern and ecologically conscious consumer behavior. *Journal of Business Research* 40(1): 79–89.

Robinson, R., and C. Smith. 2002. Psychosocial and demographic variables associated with consumer intention to purchase sustainably produced foods as defined by the Midwest food alliance. *Journal of Nutrition Education and Behavior* 34(6): 316–325.

Roth, B. 2009. Greenwashing's two-edged sword. Triple Pundit. People planet profit. http://www.triplepundit.com/2009/12/greenwashings-two-edged-sword/. Accessed 16 Sept 2010.

Sadat, T. 2004. Do we need x-rays? *Radiation Physics and Chemistry* 71(1–2): 545–549.

Sapp, S.G. 2003. A comparison of alternative theoretical explanations of consumer food safety assessments. *International Journal of Consumer Studies* 27(1): 34–39.

Sapp, S.G., and W.J. Harrod. 1990. Consumer acceptance of irradiated food. A study of symbolic adoption. *Journal of Consumer Studies and Home Economics* 14: 133–145.

Sapp, S.G., and P.F. Korsching. 2004. The social fabric and innovation diffusion. Symbolic adoption of food irradiation. *Rural Sociology* 69(3): 347–369.

Sapp, S.G., W.J. Harrod, and L.-J. Zhao. 1994. Social construction of consumer risk assessments. *Journal of Consumer Studies and Home Economics* 18: 97–106.

Sapp, S.G., W.J. Harrod, and L.-J. Zhao. 1995. Social demographic and attitudinal determinants of consumer acceptance of food irradiation. *Agribusiness* 11(2): 117–130.

Schot, J., and K. Fischer. 1993. The greening of the industrial firm. In *Environmental strategies for industry*, ed. K. Fischer and J. Schot. Washington, DC: Island Press.

Schutz, H.G., C.M. Bruhn, and K.V. Diaz-Knauf. 1989. Consumer attitudes towards irradiated foods. Effects of labeling and benefits information. *Food Technology* 43(10): 80–86.

Sen, S., and C.B. Bhattacharya. 2001. Does doing good always lead to doing better? Consumer reactions to corporate social responsibility. *Journal of Market Research* 38(2): 225–243.

Siegrist, M. 1998. Belief in gene technology. The influence of environmental attitudes and gender. *Personality and Individual Differences* 24: 861–866.

Siegrist, M. 2008. Factors influencing public acceptance of innovative food technologies and products. *Trends in Food Science and Technology* 19(11): 603–608.

Siegrist, M., M.-E. Cousin, H. Kastenholz, and A. Wiek. 2007. Public acceptance of nanotechnology foods and food packaging. The influence of affect and trust. *Appetite* 49: 459–466.

Siegrist, M., N. Stampfli, H. Kastenholz, and C. Keller. 2008. Perceived risks and perceived benefits of different nanotechnology foods and nanotechnology food packaging. *Appetite* 51(2): 283–290.

Simmons Market Research Bureau. 1991. *Study of media and markets*. New York: Syndicated division of Simmons.

Slovic, P. 1987. Perception of risk. *Science* 236: 280–285.

Slovic, P., B. Fischhoff, and S. Lichtenstein. 1985. Characterizing perceived risk. In *Perilous progress. Managing the hazards of technology*, ed. R.W. Kates, C. Hohenemser, and J.X. Kasperson. Boulder: Westview Press.

Soroka, S.N. 2002. Issue attributes and agenda-setting by media, the public, and policymakers in canada. *International Journal of Public Opinion Research* 14: 264–285.

Steger, M.A.E., J.C. Pierce, B.S. Steel, and N.P. Lovrich. 1989. Political culture, postmaterial values, and the new environmental paradigm. A comparative analysis of Canada and the United States. *Political Behaviour* 11(3): 233–254.

Stisser, P. 1994. A deeper shade of green. *American Demographics* 16: 24.

Strategic Alliance for Healthy Food and Activity Environment. 2009. Setting the record straight. Nutrition and health professionals define healthful food. http://eatbettermovemore.org/sa/documents/SA_SettingtheRecordStraight_090209.pdf. Accessed 13 June 2010.

Tankard, J.W. 2003. The empirical approach to the study of media framing. In *Framing public life. Perspectives on media and our understanding of the social world*, ed. S.D. Reese, O.H. Gandy, and A.E. Grant. Mahwah: Lawrence Erlbaum.

Teisl, M.F., B. Roe, and R.L. Hicks. 2002. Can eco-labels tune a market? Evidence from dolphin-safe labeling. *Journal of Environmental Economics and Management* 43(3): 339–359.

Ten Eyck, T.A. 1999. Shaping a food safety debate. Control efforts of newspaper reporters and sources in the food irradiation controversy. *Science Communication* 20(4): 426–447.
Ten Eyck, T.A. 2002. Food irradiation in the news. The cultural clash of a post-harvest technology. *Agricuture and Human Values* 19: 53–61.
Ten Eyck, T.A., and F.A. Deseran. 2001. In the words of the experts. The interpretive process of the food irradiation debate. *International Journal of Food Science and Technology* 36: 821–831.
Terrachoice Environmental Marketing. 2009. The seven sins of greenwashing. Environmental claims in consumer markets. Summary report North America. http://sinsofgreenwashing.org/findings/greenwashing-report-2009/. Accessed 9 Sept 2010.
Thøgersen, J. 2006. Media attention and the market for 'green' consumer products. *Business Strategy and Environment* 15: 145–156.
Thompson, J. 2009. *Energy use in transportation*. Davis: University of California.
Thomsen, M.R., M. Longstreth, and J.D. Miller. 2003. Media coverage of food irradiation. *Food Protection Trends* 23(3): 243–251.
Townsend, E., D.D. Clarke, and B. Travis. 2004. Effects of context and feelings on perceptions of genetically modified food. *Risk Analysis* 24: 1369–1384.
Traill, W.B., S.R. Jaeger, W.M.S. Yee, C. Valli, L.O. House, J.L. Lusk, M. Moore, and J.L. Morrow Jr. 2004. Categories of GM risk-benefit perceptions and their antecedents. *AgBioforum* 7(4): 176–186.
Van Kleef, E., L.J. Frewer, G.M. Chryssochoidis, J.R. Houghton, S. Korzen Bohr, T. Krystallis, J. Lassen, U. Pfenning, and G. Rowe. 2006. Perceptions of food risk management among key stakeholders. Results from a cross-European study. *Appetite* 47(1): 46–63.
Vandermerwe, S., and M. Oliff. 1990. Customers drive corporations green. *Long Range Planning* 23(6): 10–16.
Verbeke, W. 2005. Consumer acceptance of functional foods: Socio-demographic, cognitive and attitudinal determinants. *Food Quality and Preference* 16: 45–57.
Wagner, W. 2007. Vernacular science knowledge. Its role in everyday life communication. *Public Understanding of Science* 16(1): 7–22.
Wagner, W., N. Kronberger, N.C. Allum, S. de Cheveigné, C. Diego, G. Gaskell, M. Heinßen, C.J.H. Midden, M. Ødegaard, S. Öhman, B. Rizzo, T. Rusanen, and A. Stathopoulou. 2002. Padora's genes. Images of genes and nature. In *Biotechnology. The making of a global controversy*, ed. M.W. Bauer and G. Gaskell. Cambridge, MA: Cambridge University Press.
Waterviews, Globe Scan. 2009. Water issues research. http://www.circleofblue.org/waternews/waterviews/. Accessed 9 Sept 2010.
West, G.E., and B. Larue. 2005. Determinants of anti-GM food activism. *Journal of Public Affairs* 5(3/4): 236.
Wong, V., W. Turner, and P. Stoneman. 1996. Marketing strategies and market prospects for environmentally-friendly consumer products. *British Journal of Management* 7(3): 263–281.
Worcester, R. 1993. *Public and elite attitudes to environmental issues*. London: MORI.
Wynne, B. 1995. Public understanding of science. In *Handbook of science and technology studies*, ed. S. Jasanoff, G.E. Markle, J.C. Petersen, and T. Pinch. London: Sage.
Young, A.L. 2003. Food irradiation. After 35 years, have we made progress? A government perspective. *Environmental Science and Pollution Research* 10(2): 82–88.
Zucker, H.G. 1978. The variable nature of news media influence. In *Communication yearbook*, vol. 2. New Brunswick: Transaction.

Chapter 24
Food biodiversity and sustainable diets: implications of applications for food production and processing

Barbara Burlingame, U. Ruth Charrondiere, Sandro Dernini, Barbara Stadlmayr, and Stefano Mondovì

24.1 Introduction: biodiversity and nutrition: a common path

The Food and Agriculture Organization of the United Nations (FAO) actively promotes the conservation and sustainable use of biodiversity for food and nutrition. Biodiversity contributes to improving nutrition on all of the levels where it exists: genetic diversity, species diversity, and ecosystem diversity. Nutrition and biodiversity converge on a common path leading to food and nutrition security and sustainable development (Toledo and Burlingame 2006). They impact directly on two of the Millennium Development Goals:

- Goal 1—to eradicate extreme poverty and hunger
- Goal 7—to ensure environmental sustainability

Together, nutrition and biodiversity can provide the foundation for achieving these goals.

Agriculture began some 12,000 years ago. Since that time, approximately 7,000 plant species and several thousand animal species have been used for human food. Today, however, the worldwide trend is toward dietary simplification, with consequent negative impacts on food and nutrition security. That which is not used runs the risk of being lost, and the downturn in diverse diets has most certainly had an adverse effect on the world's biodiversity. Comparing the number of rice varieties cultivated today and in the past highlights this dramatic loss. In most Asian countries, the number of rice varieties has dropped from thousands to just a few dozen. In Thailand, for

B. Burlingame • U.R. Charrondiere • S. Dernini • B. Stadlmayr • S. Mondovì (✉)
Nutrition and Consumer Protection Division, Food and Agriculture Organization of the United Nations, Rome, Italy
e-mail: Barbara.Burlingame@fao.org; Ruth.Charrondiere@fao.org; Sandro.Dernini@fao.org; Barbara.Stadlmayr@fao.org; Stefano.Mondovi@fao.org

J.I. Boye and Y. Arcand (eds.), *Green Technologies in Food Production and Processing*, Food Engineering Series, DOI 10.1007/978-1-4614-1587-9_24,

example, the number of varieties under cultivation has fallen from more than 16,000 to just 37, and 50% of the rice-growing areas use only two varieties (Kennedy and Burlingame 2003; IRC 2006). Globalization, industrial development, population increases, and urbanization have changed patterns of food production and consumption in ways that profoundly affect both ecosystems and human diets. High-input industrial agriculture and long-distance transport have increased the availability and affordability of refined carbohydrates and fats, leading to a simplification of diets and a reliance on a limited number of energy-rich foods. Diets low in variety but high in energy contribute to the escalating problems of obesity and chronic disease, which are increasingly accompanied by micronutrient deficiencies and undernourishment. The causes and consequences of the dramatic reduction in food diversity and the simplification of diets are complex and not limited to any specific region or culture (FAO 2010a). The overall health of the population, agricultural practices, market conditions, and environmental stability in any given country all contribute to this complexity. However, the potential role that agricultural biodiversity can play in moderating nutritional problems is receiving increased attention.

The food systems of indigenous peoples show the important role that a diversified diet based on local plant and animal species and traditional foods can play to support health and well-being. By contrast, increasing the consumption of processed and commercial foods over time will have a negative impact on the quality of the diet. Countries, communities, and cultures that maintain traditional food systems are better able to conserve and access local food specialties based on a corresponding diversity of crops and animal breeds. As a result, they will be less likely to fall prey to diet-related diseases (FAO 2009a).

24.2 The relationship among food composition, food consumption, and biodiversity

Food consumption data on wild, underutilized, indigenous, and traditional plant and animal foods are limited and fragmented. Dietary surveys do not always collect intake information on species or varieties, partly because: (1) dietary assessment instruments have been developed to capture the habitual intake of foods as reported by subjects rather than detailed information on specific varieties of the foods consumed; (2) the corresponding compositional data are rarely available; and (3) it is widely believed that survey participants are generally not able to recognize the differences between foods at the taxonomic level below species. However, recent research suggests that this is not the case. A survey in Bangladesh (Kennedy et al. 2005) has shown that more than 80% of households were able to identify rice by cultivar and, in fact, were able to name 38 different cultivars.

If more data on the composition and consumption of traditional foods were available, the contribution from food biodiversity to nutrition could be evaluated better.

> Food biodiversity is defined as the diversity of plants, animals, and other organisms that are used for food. It includes the diversity within species, between species, and that which are found in ecosystems.

Table 24.1 Nutrient composition ranges among different varieties of the same species (per 100 g edible portion, raw) (Source: Burlingame et al. 2009)

	Protein (g)	Fibre (g)	Iron (mg)	Vitamin C (mg)	Beta-carotene (μg)
Rice	5.6–14.6	—	0.7–6.4	—	—
Cassava	0.7–6.4	0.9–1.5	0.9–2.5	25–34	<5–790
Potato	1.4–2.9	1–2.23	0.3–2.7	6.4–36.9	1–7.7
Sweet potato	1.3–2.1	0.7–3.9	0.6–14	2.4–35	100–23,100
Taro	1.1–3	2.1–3.8	0.6–3.6	0–15	5–2,040
Eggplant	—	9–19	—	50–129	—
Mango	0.3–1.0	1.3–3.8	0.4–2.8	22–110	20–4,320
Apricot	0.8–1.4	1.7–2.5	0.3–0.9	3.5–16.5	200–6,940 (beta carotene equivalent)
Banana	—	—	0.1–1.6	2.5–17.5	

Relevant, reliable, and up-to-date food composition data are of fundamental importance in nutrition, dietetics, and health, but also for other disciplines such as food science, biodiversity research, plant breeding, food industry, agricultural policy, trade, and food regulation (Greenfield and Southgate 2003; FAO 2008).

The nutrient contents of a single food can vary for many reasons, including influences such as feed and soil composition, climate, storage conditions, processing, fortification, market share, local food traditions, or differences in recipes. Differences in nutrient values may be introduced because of data quality, sampling, analytical methods, calculation methods, or data expression (Deharveng et al. 1999; Greenfield and Southgate 2003). Nevertheless, recent studies have shown varietal differences are responsible for the most dramatic variations in nutrient contents. Nutrient values may vary up to 1,000 times among different varieties of the same foods. Indeed, the nutrient content of foods can vary as much among different varieties of the same foods as among different foods. Sweet potato cultivars, for example, can differ in their carotenoid content by a factor of 200 or more. The protein content of rice varieties can range from 5% to 14% by weight. And the provitamin A carotenoid content of bananas can be less than 1 μg/100 g for some cultivars to as high as 8,500 μg/100 g for others. Intake of one variety rather than another can mean the difference between micronutrient deficiency and adequacy (Table 24.1).

For this reason, indicators of food biodiversity have recently been introduced into studies of food composition (Englberger et al. 2003a, b, c; Burlingame et al. 2009). Some might argue that foods will have a similar composition in different countries because of globalization. However, because the composition of foods is influenced by all of the factors mentioned in the preceding and these factors tend to vary from place to place, significant variations in nutrient values can normally be found in different countries. In addition, each country has its own traditional foods, varieties, and recipes. Commercial foods with the same brand name can have different compositions in different countries because of diverse tastes or fortification in some cases but not others. Consumption patterns also vary across countries and adapt in different ways over time.

In the past, generic food composition data were considered sufficient for most purposes. Today, there is greater awareness of the need for food composition studies that take biodiversity into account, although compositional data at the variety, cultivar, and breed level are not yet widely available. As a result, farmers and consumers are often not aware of the high nutrient values of certain plant cultivars and do not grow or consume these fruits or vegetables. Introducing more compositional data on biodiversity in food composition databases will improve the estimates of nutrient intake and assessments of dietary adequacy or inadequacy, especially with regard to micronutrients. Wrong decisions have been made in nutrition and health programs because the micronutrient values in a national food composition database did not reflect the nutrient content of the varieties actually consumed by that country's population. Data on nutrient content should be one of the criteria agriculture policy makers and practitioners use to ensure better and more nutritious crop varieties are available for consumption.

Achieving a worldwide system of compatible food composition databases lies at the heart of the INFOODS (International Network of Food Data Systems) program. Established in 1984, INFOODS operates under the auspices of FAO and United Nations University. Its goal is to stimulate and coordinate efforts to improve the quality and availability of compositional data globally. INFOODS has established standards and guidelines for generating and compiling food composition data (Greenfield and Southgate 2003). It has also produced a distance learning tool, the *Food Composition Study Guide* (FAO 2009b, c) of which an entire module is devoted to biodiversity and the growing need for knowledge about the composition of foods based on varieties, cultivars, and breeds. In addition, the first version of the INFOODS Food Composition Database for Biodiversity was launched in December 2010 (FAO/INFOODS 2010).

Diversity in aquatic animal species can also contribute to nutrition by securing high-quality protein and fatty acid intakes for populations that rely on wild or farmed fish or other animals that live in aquatic ecosystems (Halwart 2008). Calcium and vitamin A from small indigenous freshwater fish species can contribute from one third to one half of the total recommended intake of these nutrients.

Today, new market niches are being developed for a number of underutilized species, varieties and breeds of crops, livestock, and aquatic animals. Food biodiversity, with its offering of foods with unique, often superior nutritional properties, can supply products with added value that appeal to consumers and thus provide a source of income for rural people.

The Convention on Biological Diversity (CBD) Cross-cutting Initiative on Biodiversity for Food and Nutrition responds to an emerging global consensus that: (1) the simplification of diets, the growing incidence of chronic diseases related to nutrition-poor, energy-rich diets and the decline in the use of locally available foods are linked; and that (2) biodiversity is the source of many foods and dietary components that can reverse this unhealthy trend. As scientists and policymakers recognize that food biodiversity is essential for food and nutrition security and that it can make a major contribution to the achievement of the Millennium Development Goals, it is now becoming a feature in nutrition programs and interventions.

Identification and monitoring of nutrition indicators for biodiversity is critical for promoting sustainable diets. Together with Bioversity International and other partners, FAO has led an international process to develop two such indicators: one on food composition and another on food consumption.

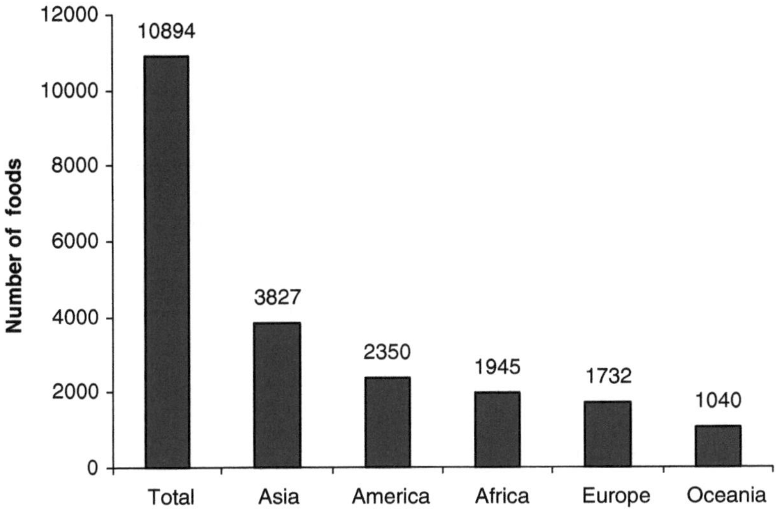

Fig. 24.1 Number of foods reported for the food composition indicator—status in 2010

The increase in the number of relevant foods reported by food composition databases and in food consumption surveys in recent years indicate that governments, researchers, farmers, the food industry, and consumers are becoming more aware of the role that biodiversity plays in nutrition. This increase can be expected to lead to greater conservation and sustainable use of biodiversity, to improve food and nutrition security (FAO 2008, 2010b).

24.2.1 Food composition indicator

The food composition indicator relates to the nutrients and bioactive non-nutrients that are provided by biodiversity. It is a count of the foods—with a sufficiently detailed description to identify genus, species, subspecies and variety, cultivar, or breed—with at least one value for a nutrient or bioactive component. There is an exception for wild and underutilized foods, for which information on the species level is satisfactory (FAO 2008). Since its development in 2007, more than 10,000 foods have been counted. Most of the available data have come from Asia (3,827), followed by the Americas (2,350), Africa (1,945), Europe (1,732), and Oceania (1,040) (Fig. 24.1). The amount of available data is steadily increasing (Stadlmayr et al. 2010). The documentation of more than 10,000 foods for the indicator to date implies that the importance of biodiversity for food composition is increasingly becoming recognized and reflected in publications. Nevertheless, a wider spectrum of local and traditional foods and their components should be analyzed and included in food composition databases.

24.2.2 Food consumption indicator

The food consumption indicator relates to the intake of food biodiversity. It is a count of the number of foods identified in dietary surveys as wild or underutilized and/or described as such at the taxonomic level below species. So far, 19 studies have contributed to the indicator, identifying 2,817 foods, most of them with origins in Asia, The studies were mainly undertaken through personal contacts because the scientific literature on food consumption surveys does not in general report these details, either because they were not recorded or were not felt to be important enough to be listed among the findings (FAO 2010b).

24.3 Sustainable diets

Sustainable diets are those diets with low environmental impacts that contribute to food and nutrition security and to healthy life for present and future generations. Sustainable diets are protective and respectful of biodiversity and ecosystems, culturally acceptable, accessible, economically fair and affordable, nutritionally adequate, safe, and healthy, while optimizing natural and human resources (FAO 2010c). Currently, FAO's activities on sustainable diets explore the links and synergies among food biodiversity, nutrition, agriculture, and sustainability as a means to improve food and nutrition security. They include studies to characterize different agro-ecological zones for the development of sustainable diets' models. These studies will serve to foster new ideas for building consensus on actions needed to improve nutrition, linking agriculture, health, and the environment sectors in a shared effort.

In the early 1980s, the notion of sustainable diets began to be explored as a way to investigate diets that would be healthier for the environment as well as for consumers (Gussow and Clancy 1986). With food globalization and the increased industrialization of agricultural systems, which tended to undermine the sustainability of agro-ecosystems, the concept of sustainable diets was neglected for many years.

There is growing institutional and academic recognition of the complexity of defining sustainability, and at the same time, a growing body of evidence of the unsustainable nature of current diets (Sustainable Development Commission 2009). Although good nutrition should be a goal of agriculture, it is imperative that concerns about sustainability are not lost in the process of meeting that goal. Many dietary patterns can be healthy, but they can vary substantially in terms of their cost to resources. The notion of sustainable diets promotes the use of traditional and local biodiversity as a source of foods with high nutritional value.

Improving food systems for sustainable diets requires an interdisciplinary effort to address the problems of malnutrition, the degradation of ecosystems, and the erosion of biodiversity caused, at least in part, by modern food systems and dietary patterns. As we face many challenges, ensuring sustainable diets requires the development of program activities and policies toward sustainable food production and consumption, with biodiversity and nutrition-sensitive agriculture at its core. The need to feed a growing population leads to continuous pressure on crop and livestock production systems, and fisheries and forestry resources, exacerbated by degraded natural resources, and difficulties arising from climate change. Agro-food systems as well as food consumption patterns need to change to permit communities to adapt to this scenario. FAO's multipartner initiative on sustainable diets addresses the sustainability of the whole food chain, while acknowledging the interdependencies of food production, food consumption, and nutrient requirements. The sustainable use, production, and consumption of food, along with education and policies, are fundamental to achieving the broader goals of sustainable development, connecting nutritional well-being of the individual and the community to the sustainability of feeding the planet.

Given the escalating population growth, land degradation, and increasing demands for food, achieving sustainable food systems and sustainable food supply chains is critical to food security and poverty alleviation. It is necessary to develop local capacity for assuring the equitable sharing of the benefits arising from the use of genetic resources along the agri-food value chain through new models for sustainable diets applied to the different agro-ecological zones.

The alarming pace of biodiversity loss and ecosystem degradation and its negative impact on poverty and health make a compelling case for re-examining food systems. Currently, close to 1 billion people suffer from hunger and another 2 billion suffer from micronutrient deficiencies. Diets that are low in variety but high in energy contribute to the escalating problems of obesity and chronic diseases that particularly threaten poor people in developing countries. Recent trends show an increase in these problems, highlighting the inadequacy of the present food supply and dietary patterns. Globalization, industrial agriculture, rural poverty, population pressures, and urbanization have all changed food production and consumption in ways that deeply affect ecosystems and have led to an overall simplification of human diets. This points to an urgent need to develop and promote strategies for the advancement of sustainable diets and biodiversity within an ecosystem approach.

The notion of sustainable diets posits nutrition and biodiversity as central to sustainable agricultural development. Sustainable diets use food biodiversity, including traditional foods of indigenous peoples, with their many sources of nutrient-rich species and varieties as readily accessible sources of quality nutrition. Sustainable diets enhance biodiversity, local food systems, household food security, and ecologically healthy and sustainable food supply chains that are crucial in building positive options for ecosystem services and human well-being (Millennium Ecosystem Assessment 2003).

The declaration of the 2009 World Summit on Food Security states that the FAO "will actively encourage the consumption of foods, particularly those available locally, that contribute to diversified and balanced diets, as the best means of addressing micronutrient deficiencies and other forms of malnutrition, especially among vulnerable groups" (FAO 2009d). At the 21st Session of the FAO Committee on Agriculture it was recommended "to accelerate the transition toward sustainability" (FAO 2009e). The sustainability of food and agriculture activities on which 2.4 billion people depend for their livelihoods—including the management and conservation of marine and forest resources—is a key driver in the transition toward a green economy, because of its positive impact on both ecosystem services and poverty alleviation.

Resilient, sustainable food systems are based on diversification of production and consumption patterns, adaptation to climate change, the security of terrestrial and marine tenure, and the empowerment of farmers, herders, fishers, forest dwellers, indigenous communities, and other primary producers.

Increasing the sustainability of food systems, from primary production to consumption, will improve food and nutrition security (in terms of availability, access, stability, and use), while minimizing the use of external inputs (e.g., fertilizers, water, and energy) and sustaining the health of ecosystems, biodiversity, and people.

Sustainable diets should be promoted by stakeholders (governments, UN organizations, civil societies, nongovernmental organizations, and the private sector) to assist populations in meeting nutritional requirements through the production and consumption of a biodiverse food supply with reliance on supplementations, fortification, and food aid only as short-term measures.

The achievement of sustainable diets linked to biodiversity requires a broad acknowledgment—in many sectors—that in an increasingly global, urban, and commercial environment, realizing the potential of food biodiversity requires integrating production, marketing, consumption, and the health concerns of both rural and urban people. The close involvement of civil society and the private sector will be needed to engage stakeholders in the fields of environment, agriculture, nutrition, health, education, culture, and trade.

Although the evidence base must be improved, existing knowledge warrants immediate action to promote the sustainable use of biodiversity in nutrition programs, as contributions to the achievement of food and nutrition security and the Millennium Development Goals (MDGs).

24.4 Case studies

In recent years, reduced access to land and natural resources, environmental degradation, climate change, globalization, and the westernization of diet and lifestyle have dramatically affected the role that traditional foods play in the lives of small farmers and indigenous societies. Studies have linked these changes to a wide range of negative consequences, including food insecurity, poor health, nutrition deficiencies, ecosystem deterioration, and cultural erosion. The consequences of the shift in production and consumption patterns are significant. Foods based on non-native resources are generally highly refined and processed, have higher concentrations of saturated fats, salt, and sugar, and are lower in micronutrients. These foods are appealing for a number of reasons. They often come ready to eat and are easy to prepare. They also tend to be inexpensive, a fact that is of particular relevance because many small farmers are from low-income households.

It is becoming increasingly important to reverse this trend by recovering and valorizing local food resources and identifying and promoting nutrition-rich varieties that meet both the needs of farmers and the preferences of consumers.

Vitamin A supplementation does not guarantee the health and nutritional benefits that have been claimed over the past couple of decades. On the other hand, there is increasing proof that health and nutrition can be improved or maintained through traditional foods and biodiversity (Englberger et al. 2003a, b, c; Kuhnlein 2010) or through biodiversity food-based interventions, especially when improved agricultural production is coupled with a nutrition component (FAO 2011a). The following case studies show examples of interventions using local foods with high nutritional value.

Example 1. Home garden project to address vitamin A deficiency in south africa

The Ndunakazi project (named for the province of Kwazulu-Natal in South Africa where it took place) sought to improve the vitamin A status of children through the cultivation and consumption of vegetables and fruits rich in provitamin A (FAO 2011b).

Dark green leafy vegetables (e.g., spinach and wild-growing leaves), carrot, orange-fleshed sweet potato, butternut squash, pumpkin, mango, and papaya were introduced into home gardens. These crops rich in vitamin A, bring substantial immediate health benefits, and longer-term benefits such as reduced risk of certain cancers, cardiovascular disease, cataracts, and macular degeneration. Some of these crops, for example, sweet potato, are ideal for sustainable agriculture because they grow on low-nitrogen soils, are drought-tolerant, crowd out weeds, and are susceptible to few pests.

The consumption of the vegetables introduced by the project has greatly improved the vitamin A status of the Ndunakazi people: About 85% of their intake of vitamin A comes from the new crops. A further positive impact has been the contribution of the vegetables to the intake of calcium and iron and, to a lesser extent, of magnesium, vitamin C, and riboflavin.

Example 2. Promotion of carotenoid-rich indigenous foods in Pohnpei, federated states of Micronesia

In spite of a wide diversity of local resources (55 banana, 133 breadfruit, and 171 yam varieties, and nearly 1,200 species of edible fish), the consumption of local traditional food by the population of Pohnpei has been steadily declining as their diet has shifted more and more toward imported foods (FAO 2009a). An increasingly large percentage of the people in Pohnpei now have serious nutrition-related problems, notably vitamin A deficiency, obesity, and diet-related chronic diseases.

A program aimed primarily at introducing green leafy vegetables into the Pohnpei diet over a period of 15 years was unsuccessful because these vegetables were neither indigenous nor well-liked. Indigenous foods, such as Karat and other local yellow-fleshed bananas and yellow-fleshed giant swamp taro varieties, were not promoted because there were no compositional data available for them.

Once proper analysis of the composition of the Karat variety bananas and giant swamp taro varieties revealed their high provitamin A carotenoid content, efforts got underway to promote local carotenoid-rich foods. A major awareness campaign based on the slogans "Go Yellow" and "Let's Go Local" has had great success in increasing the consumption of indigenous foods.

Example 3. Leafy vegetables in Kenya

In a peri-urban zone of Nairobi, a series of projects had the objective of promoting sustainable agriculture and improving the nutritional status of the population based on the reintroduction of highly nutritional local leafy vegetables (Bioversity International 2007). The neglect of these local products—considered "poor people's food" by young people and consumers in urban areas—had significantly reduced the intake of vitamins and micronutrients.

Leafy vegetables, such as Africa nightshade, vegetable amaranth, cowpea, spider plant, pumpkin leaves, slenderleaf, and jute mallow are rich in protein, calcium, and important micronutrients. The vegetables are also drought-tolerant: an essential added value, especially as the impact of climate change becomes more evident. Many of the vegetables need less water than hybrid varieties and some are resistant to diseases and pests as well.

The project undertook a campaign based on the development of agronomic and nutritional studies, distribution of seeds to farmers, promotion of local dishes, and importantly, the introduction of local leafy vegetables into Kenyan supermarkets, which bestowed the vegetables with a higher status. In the end, consumers grew proud of their local foods, which were linked to their culture and ethnicity. Traditional recipes were adapted to modern needs and lifestyles and to be suitable in an urban context.

The campaign had a positive impact on both livelihoods and health, enhancing the production and consumption of local biodiversity and increasing household incomes.

Example 4. Rice-based aquatic ecosystems

Studies of traditional rice farming in Asian communities found that the cultivation of rice in flooded systems is accompanied by a rich variety of fish and other aquatic organisms, and that wild and gathered foods from the aquatic habitat provide important diversity, and food and nutritional security (Halwart 2008). Such environments supply essential micronutrients that are not adequately found in rice, particularly calcium, iron, zinc, vitamin A, and some fatty acids and limiting amino acids. Many of the aquatic organisms found in rice ecosystems also serve as biological control agents for vectors and pests.

24.5 Policy issues and key recommendations

Organizations, institutions, NGOs, the private sector and other stakeholders are increasingly sensitive to the role that food biodiversity and sustainable diets can play in improving food and nutrition security. This sensitivity has resulted in a number of key policy initiatives and recommendations related to current FAO activities.

24.5.1 The convention on biological diversity cross-cutting initiative on biodiversity for food and nutrition

The framework for the Cross-cutting Initiative was adopted by the 8th Conference of the Parties of the Convention on Biological Diversity in Curitiba, Brazil (CBD 2006). The overall aim of the Cross-cutting Initiative is to promote and improve the sustainable use of biodiversity in programmes contributing to food security and human nutrition, as a contribution to the achievement of Millennium Development Goals 1 and 7 and related goals and targets and thereby to raise awareness of the importance of biodiversity, its conservation, and sustainable use.

The Initiative identifies the contribution of agricultural biodiversity as critical for improving the health and nutrition of the rural and urban poor. It addresses major global health issues and trends, such as micronutrient deficiencies, the decline of dietary diversity, and the concomitant rise of chronic diseases that are particularly affecting poor people in developing countries. The Initiative promotes the use of local biodiversity as readily accessible, empowering, and sustainable sources of nutrition. Furthermore, it recognizes that in an increasingly global, urban, and commercial environment, fulfilling the potential of local resources requires the successful integration of production, marketing, consumption, and the health concerns of rural and urban dwellers as components of sustainable food systems.

The Initiative's framework is built around four elements:

1. Developing and documenting knowledge
2. Integrating biodiversity, food, and nutrition issues into research and policy instruments
3. Conserving and promoting wider use of biodiversity for food and nutrition
4. Promoting public awareness and supporting activities

24.5.2 The international rice commission

During its 20th Session in 2002, the International Rice Commission (FAO 2002) recommended that:

- Existing biodiversity of rice varieties and their nutritional composition need to be explored *before* engaging in transgenics.
- Nutrient content needs to be among the criteria in cultivar promotion.
- Cultivar-specific nutrient analysis and data dissemination should be systematically undertaken.
- Member countries should promote the sustainable development of aquatic biodiversity in rice-based ecosystems and policy decisions and management measures should enhance the living aquatic resource base.
- In areas in which wild fish are depleted, rice-fish farming should be considered as a means of enhancing food security and securing sustainable rural development.
- Attention should be given to the nutritional contribution of aquatic organisms in the diet of rural people who produce or depend on rice.

During its 21st Session in 2006, the International Rice Commission (IRC 2006) recommended the following:

- The evaluation of the composition and consumption of rice cultivars should continue for the development of food biodiversity indicators to guide agrobiodiversity conservation and human nutrition. Increasing the availability and promoting the use of whole grain and moderately milled rice and rice products will provide human nutrition benefits, particularly related to micronutrients.

24.5.3 FAO regional conference for Europe

During the 26th FAO Regional Conference for Europe (June 2008) on "Promotion of traditional regional agricultural and food products: a further step towards sustainable rural development," two discussed points are worth noting:

- Many delegations highlighted the Mediterranean diet as rich in biodiversity and nutritionally healthy. The promotion of the Mediterranean diet could play a beneficial role in the sustainable development of agriculture in the Mediterranean region.
- The Conference remarked that the goal of increased global food production and biofuels, should be balanced against the need to protect biodiversity, ecosystems, traditional foods, and traditional agricultural practices.

24.5.4 International scientific symposium on biodiversity and sustainable diets

At the International Scientific Symposium on Biodiversity and Sustainable Diets organized by FAO and Bioversity International in Rome (3–5 November 2010), the following platform of action was issued (FAO 2010c):

> FAO, Bioversity International and the CBD Secretariat, in collaboration with other relevant organizations and institutions at international/regional/national/local level should establish a Task Force to promote and advance the concept of sustainable diets and the role of biodiversity within it, in the context of the CBD Cross-cutting Initiative on Biodiversity for Food and Nutrition, as contributions to the achievement of the MDGs and beyond.
>
> FAO and Bioversity International should encourage the UN System, Governments, International Organizations, International Food Security and Nutrition Initiatives, and other relevant bodies to finance and support research and development projects and programs on biodiversity and sustainable diets.
>
> Decision makers should give priority to and promote sustainable diet concepts in policies and programs in the agriculture, food, environment, trade, education, and health sectors. Nutrition should be given more emphasis by plant and animal breeders and research on nutrient content of food biodiversity should be encouraged. Food composition data should be compiled by FAO in the INFOODS databases and by regional and national institutions.

New projects and case studies should be encouraged to demonstrate the synergies among biodiversity, nutrition, and socioeconomic, cultural, and environment sustainability as well as to gather evidence about the potential of greater use of biodiversity for better nutrition and health and for poverty alleviation and improved livelihoods. The evidence gathered from these research efforts should be compiled by FAO and Bioversity International and made available on an open access web-based platform.

Food-based dietary guidelines and policies should give due consideration to sustainability when setting goals aimed at healthy nutrition. A guidance document on how to develop such guidelines and policies at the national level could be elaborated by FAO, in collaboration with Bioversity International and other partners.

Governments, UN agencies, civil society, research organizations, and the private sector should collaborate in the development of program activities and policies to promote sustainable diets in order to achieve sustainable food production, processing, and consumption, and to minimize environmental degradation and biodiversity loss.

The development of a Code of Conduct for Sustainable Diets was also strongly recommended.

24.6 Conclusion

Governments and consumers alike are increasingly aware of sustainability as well as nutrition issues. This awareness indicates a clear direction for the agricultural community, including smallholders, rural agriculturalists, and small and medium enterprises. Greener production and processing technologies, embracing biodiversity, and maximizing nutrient retentions should appeal to all policymakers and, at the same time, provide exciting research opportunities for scientists and technologists.

Improving nutrition through biodiversity could be the basis for the development of codes of practice, providing guidance to the health, agriculture, environmental, and food industry sectors and to consumers. Among the overarching principles are the recognition that the health of human beings cannot be isolated from the health of ecosystems and the affirmation of the need for all-inclusive cooperation in activities to improve human and environmental health through sustainable diets. Green technologies in food production and processing will provide a firm foundation for these efforts.

References

Bioversity International. 2007. Back by popular demand. Annual report 2007, Rome, Italy.

Burlingame, B., U.R. Charrondiere, and B. Mouille. 2009. Food composition is fundamental to the cross-cutting initiative on biodiversity for food and nutrition. *Journal of Food Composition and Analysis* 22(5): 361–365.

CBD. 2006. COP 8 decision VIII/23. Available at http://www.cbd.int/decision/cop/?id=11037.

CINE, FAO. 2009. Indigenous and tribal peoples: The importance of traditional foods for nutrition and health, Rome, Italy.

Deharveng, G., U.R. Charrondiere, N. Slimani, D.A.T. Southgate, and E. Riboli. 1999. Comparison of nutrients in the food composition tables available in the nine European countries participating in EPIC. *European Journal of Clinical Nutrition* 53: 60–79.

Englberger, L., W. Aalbersberg, M.H. Fitzgerald, G.C. Marks, and K. Chand. 2003a. Provitamin A carotenoid content of different cultivars of edible pandanus fruit. *Journal of Food Composition and Analysis* 16(2): 237–247.

Englberger, L., W. Aalbersberg, P. Ravi, E. Bonnin, F.M.H. MarksGC, and J. Elymore. 2003b. Further analyses on Micronesian banana, taro, breadfruit and other foods for provitamin A carotenoids and minerals. *Journal of Food Composition and Analysis* 16(2): 219–236.

Englberger, L., J. Schierle, G.C. Marks, and M.H. Fitzgerald. 2003c. Micronesian banana, taro, and other foods: newly recognized sources of provitamin A and other carotenoids. *Journal of Food Composition and Analysis* 16(1): 3–19.

FAO. 2002. Report of the International Rice Commission 20th Session (23–26 July 2002), Bangkok. FAO, Rome, Italy.

FAO. 2008. Expert consultation on nutrition indicators for biodiversity. 1. Food composition. Available at ftp://ftp.fao.org/docrep/fao/010/a1582e/a1582e00.pdf.

FAO. 2009a. Documentation of the traditional food system of Pohnpei. In *Indigenous peoples' food systems: The many dimensions of culture, diversity and environment for nutrition and health*, ed. L. Englberger, A. Lorens, A. Levendusky, P. Pedrus, K. Albert, W. Hagilmai, Y. Paul, D. Nelber, P. Moses, S. Shaeffer, M. Gallen, H.V. Kuhnlein, B. Erasmus, and D. Spigelski. Rome: FAO.

FAO. 2009b. Food Composition Study Guide. Questions and Exercises (vol. 1), by Charrondiere, U.R., B. Burlingame, S. Berman, and I. Elmadfa. Available at http://www.fao.org/infoods/publications_en.stm.

FAO. 2009c. Food composition study guide. Questions, exercises and answers (vol. 2), by Charrondiere, U.R., B. Burlingame, S. Berman, and I. Elmadfa. http://www.fao.org/infoods/publications_en.stm.

FAO. 2009d. Declaration of the world summit on food security. Available at http://www.fao.org/fileadmin/templates/wsfs/Summit/Docs/Final_Declaration/WSFS09_Declaration.pdf.

FAO. 2009e. Report of the twenty-first session of the Committee on Agriculture (COAG). Available at ftp://ftp.fao.org/docrep/fao/meeting/016/k4952e01.pdf.

FAO. 2010a. Report on technical workshop—Biodiversity in sustainable diets. Available at http://www.fao.org/ag/humannutrition/24994064a7cf9328fbe211363424ba7796919a.pdf.

FAO. 2010b. Expert consultation on nutrition indicators for biodiversity. 2. Food consumption. Available at http://www.fao.org/infoods/biodiversity/index_en.stm.

FAO. 2010c. International scientific symposium "biodiversity and sustainable diets," Final document. Available at http://www.fao.org/ag/humannutrition/25915-0e8d8dc364ee46865d5841c48976e9980.pdf.

FAO INFOODS. 2010. Food composition database for biodiversity. Available at http://www.fao.org/infoods/biodiversity/index_en.stm.

FAO. 2011a. Agricultural interventions and nutrition: Lessons from the past and new evidence, by Arimond M, Hawkes C, Ruel MT, Sifri Z, Berti PR, Leroy JL, Low JW, Brown LR, Frongillo EA. In *Combating micronutrient deficiencies: Food-based approaches*, ed. B. Thompson and L. Amoroso. Rome: FAO.

FAO. 2011b. Home gardening approach developed in South Africa to address vitamin A deficiency by Faber M, Laurie S. In *Combating micronutrient deficiencies: Food-based approaches*, ed. B. Thompson and L. Amoroso. Rome: FAO.

Greenfield, H., and D.A.T. Southgate. 2003. *Food composition data: Production, management and use*. Rome: FAO.

Gussow, J.D., and K.L. Clancy. 1986. Dietary guidelines for sustainability. *Journal of Nutrition Education* 18: 1–5.

Halwart, M. 2008. Biodiversity, nutrition and livelihoods in aquatic rice-based ecosystems. *Biodiversity* 9(1–2): 36–40. http://www.fao.org/infoods/publications_en.stm.

IRC. 2006. Report of the International Rice Commission twenty first session. Available at http://www.fao.org/ag/AGP/AGPC/doc/field/commrice/pages/finalreport.pdf.

Kennedy, G., and B. Burlingame. 2003. Analysis of food composition data on rice from a plant genetic resources perspective. *Food Chemistry* 80(4): 589–596.

Kennedy, G., O. Islam, P. Eyzaguirre, and S. Kennedy. 2005. Field testing of plant genetic diversity indicators for nutrition surveys: Rice-based diet of rural Bangladesh as a model. *Journal of Food Composition and Analysis* 18(4): 255–268.

Kuhnlein, H.V. 2010. Here is the good news, World Nutrition. *Journal of the World Public Health Nutrition Association* 1(2): 60–77. Available at http://www.wphna.org/wn_commentary HKjune2010.asp.

Millennium Ecosystem Assessment (MA). 2003. *Ecosystems and human well-being: A framework for assessment*. Washington, DC: Island Press.

Stadlmayr, B., E. Nilsson, B. Mouille, E. Medhammar, B. Burlingame, and U.R. Charrondiere. 2010. Nutrition indicator for biodiversity on food composition—A report on the progress of data availability. *Journal of Food Composition and Analysis*. doi:10.1016/j.jfca.2010.09.009.

Sustainable Development Commission. 2009. Setting the table: Advice to Government on priority elements of sustainable diets, London.

Toledo, Á., and B. Burlingame. 2006. Biodiversity and nutrition: A common path toward global food security and sustainable development. *Journal of Food Composition and Analysis* 19(6–7): 477–483.

ERRATUM

Green Technologies in Food Production and Processing

Joyce I. Boye and Yves Arcand, Editors

J.I. Boye and Y. Arcand (eds.), *Green Technologies in Food Production and Processing*,
Food Engineering Series, DOI 10.1007/978-1-4614-1587-9,

DOI 10.1007/978-1-4614-1587-9_25

The copyright holder of this book should be changed from Springer Science+Business Media, LLC 2012 to Her Majesty the Queen in Right of Canada, as represented by the minister of AAFC.

The online version of the original Book can be found at
http://dx.doi.org/10.1007/978-1-4614-1587-9

Author description

Joyce I. Boye, Ph.D., is a Research Scientist at the Food R & D Centre of Agriculture & Agri-Food Canada (AAFC) located in St Hyacinthe, Quebec. AAFC provides information, research and technology, and policies and programs to achieve security of the food system, health of the environment, and innovation for growth. Dr. Boye leads a comprehensive research study in value-added processing of plant-based foods and food allergies.

Yves Arcand, Ph.D., is a Research Scientist at the Food R & D Centre of Agriculture & Agri-Food Canada (AAFC) located in St Hyacinthe, Quebec. His areas of expertise include characterization and treatment of agri-food industrial effluents, biometric applications, system engineering, optimization of dehydration processes, and simulation and model building, including the design and building of an in vitro upper gastrointestinal tract model for nutritional studies.

J.I. Boye and Y. Arcand (eds.), *Green Technologies in Food Production and Processing*, Food Engineering Series, DOI 10.1007/978-1-4614-1587-9,

Book cover description

As we move into the future, decisions on food purchases will increasingly be influenced not only by price and quality but by social and environmental factors, such as the sustainability of technologies used for food production and processing and their environmental and health impacts. Growing consumer awareness about the impact of processing and production practices on the environment, the high energy consumption of certain processes, health impacts of some of the technologies used in processing, and a heightened social and industrial consciousness to reduce the carbon-footprint are examples of factors influencing food choice. These factors have been made quite evident in the recent "buy-local", "fair-trade", and "certified organic" trends. As a result, farmers and food manufacturers will increasingly be interested in identifying and using greener economically viable technologies for food production and processing. Some producers are already responding with the use of organic inputs in processing, use of recyclable and good-for-the-environment packaging, establishing just employer-employee relationships, and reducing animal testing in product development.

Green Technologies in Food Production and Processing provides a comprehensive review of the current status of the agriculture and agri-food sectors in regards to environmental sustainability and material and energy stewardship, and provides strategies that can be used by industries to enhance the use of environmentally-friendly technologies for food production and processing. The book further provides an in-depth look at some emerging analytical techniques for research and development which reduce solvent, chemical, and energy use. In addition, technologies to reduce the generation of process-induced toxins are also reviewed. In the last section of the book, a critical analysis of some of the challenges associated with the use of agricultural resources to grow biofuels and bio-based products are addressed. Furthermore, social factors that influence consumer perceptions about some of the current and emerging agri-food technologies, and the need and importance of biodiversity in maintaining sustainable diets of human populations are also discussed in detail.

Index

F

H

MIX
Papier aus verantwortungsvollen Quellen
Paper from responsible sources
FSC® C105338

If you have any concerns about our products,
you can contact us on
ProductSafety@springernature.com

In case Publisher is established outside the EU,
the EU authorized representative is:
Springer Nature Customer Service Center GmbH
Europaplatz 3, 69115 Heidelberg, Germany

Printed by Libri Plureos GmbH
in Hamburg, Germany